Integrated Models of Cognitive Systems

SERIES ON COGNITIVE MODELS AND ARCHITECTURES

Series Editor: Frank Ritter

Integrated Models of Cognitive Systems
Edited by Wayne D. Gray

Integrated Models of Cognitive Systems

Edited by Wayne D. Gray

OXFORD
UNIVERSITY PRESS
2007

OXFORD
UNIVERSITY PRESS

Oxford University Press, Inc., publishes works that further
Oxford University's objective of excellence
in research, scholarship, and education.

Oxford New York
Auckland   Cape Town   Dar es Salaam   Hong Kong   Karachi
Kuala Lumpur   Madrid   Melbourne   Mexico City   Nairobi
New Delhi   Shanghai   Taipei   Toronto

With offices in
Argentina   Austria   Brazil   Chile   Czech Republic   France   Greece
Guatemala   Hungary   Italy   Japan   Poland   Portugal   Singapore
South Korea   Switzerland   Thailand   Turkey   Ukraine   Vietnam

Published by Oxford University Press, Inc.
198 Madison Avenue, New York, New York 10016

www.oup.com

Oxford is a registered trademark of Oxford University Press

Library of Congress Cataloging-in-Publication Data
Integrated models of cognitive systems / edited by Wayne D. Gray.
      p. cm.—(Series on cognitive models and architectures)
    Includes bibliographical references and index.
    ISBN: 978-0-19-518919-3
  1. Cognition. 2. Cognitive science. I. Gray, Wayne D.
    BF311.I554 2007
    153.01'13—dc22   2006021297

9 8 7 6 5 4 3 2
Printed in the United States of America
on acid-free paper

The Rise of Cognitive Architectures

Frank E. Ritter

One way forward in studying human behavior is to create computer simulations that perform the tasks that humans perform by simulating the way humans process information. These simulations of behavior, called *cognitive models*, are theories of the knowledge and the mechanisms that give rise to behavior. The sets of mechanisms are assumed to be fixed across tasks, which allow them to be realized as a reusable computer program that corresponds to the architecture of cognition, or cognitive architecture. (More complete explanations are provided by Newell's [1990] *Unified Theories of Cognition*, by Anderson's ACT-R work [Anderson et al., 2004], and by ongoing work with connectionist and neural architectures.)

Because cognitive models increasingly allow us to predict behavior and explain the mechanisms behind behavior, they have many applications. They can support design activities, and they serve in many roles where intelligence is needed. As a result, interest in cognitive models and architectures can be found in several areas: Researchers in psychology and cognitive science are interested in them as theories. Researchers in human factors, in synthetic environments, and in intelligent systems are interested in them for applications and design. Researchers in applied domains such as video games and technical applications such as trainers are interested in them as simulated colleagues and opponents.

Although some earlier precursors can be found, the main work on cognitive models began in about 1960 (Newell, Shaw, & Simon, 1960). These models have now reached a new level of maturity. For example, a review commissioned by the National Research Council (Pew & Mavor, 1998) found that cognitive models had been developed to a level that made them useful in synthetic environments. A later review (Ritter et al., 2003) examined cognitive architectures created outside the United States and found similar results. Both reviews recommended a list of future projects, which are being undertaken by individual researchers. These and similar projects have been increasingly seen in requests for proposals put out by funding agencies around the world. Results and interest in cognitive models and architectures are rising.

A Series on Cognitive Models and Architectures

It would be useful to have access to larger sets of materials on cognitive models and cognitive architectures, including full explanations of the design, rationale, and use of a single architecture; comparisons of several architectures; and full explanations of a single model. A book series provides access to these larger sets of materials and allows readers to identify them more easily.

Topics for volumes in the series will be chosen to highlight the variety of advances in the field, to provide an outlet for advanced books (edited volumes and monographs), architecture descriptions, and reports on methodologies, as well as summaries of work in particular areas (e.g., memory) or of particular architectures (e.g., ACT-R, Soar). Each volume will be designed to for broad multidisciplinary appeal and will interest researchers and graduate students working with cognitive models and, as appropriate, related groups.

So, it is with great pleasure that we start this series with a book on control mechanisms in architectures edited by Wayne Gray. This book summarizes current work by leading researchers on how cognitive architectures control their information processing, the interaction between their mechanisms, and their interaction with the world. This book will be a valuable resource for those building and using architectures. It also serves as a repository of thinking on the mechanisms that control cognition.

References

Anderson, J. R., Bothell, D., Byrne, M. D., Douglass, S., Lebiere, C., & Qin, Y. (2004). An integrated theory of the mind. *Psychological Review, 111*(4), 1036–1060.

Newell, A. (1990). *Unified theories of cognition*. Cambridge, MA: Harvard University Press.

Newell, A., Shaw, J. C., & Simon, H. A. (1960). Report on a general problem-solving program for a computer. In *International Conference on Information Processing* (pp. 256–264). Paris: UNESCO.

Pew, R. W., & Mavor, A. S. (Eds.). (1998). *Modeling human and organizational behavior: Application to military simulations*. Washington, DC: National Academy Press.

Ritter, F. E., Shadbolt, N. R., Elliman, D., Young, R., Gobet, F., & Baxter, G. D. (2003). *Techniques for modeling human performance in synthetic environments: A supplementary review*. Wright-Patterson Air Force Base, OH: Human Systems Information Analysis.

Preface

It is with pleasure that I introduce researchers, teachers, and students to this volume on *Integrated Models of Cognitive Systems*. All such volumes present a snapshot of the time in which they are created; it is the intent of the contributors that this snapshot will grace a postcard to the future.

The history of cognitive studies is a history of trying to understand the mind by slicing and dicing it into functional components and trying to thoroughly understand each component. Throughout time, the size of the components has gotten smaller, and their shape has varied considerably, such that what was a whole, the human mind, has become a jigsaw puzzle of oddly shaped parts. The emphasis on *cognitive systems* shows how these pieces fit together to achieve "complete processing models" (Newell, 1973) or "activity producing subsystems" (Brooks, 1991). The emphasis on *integrated models* recognizes that the cognitive system is too large and complex for a single researcher or laboratory to model and that progress can only be made by developing our various parts so that they can fit together with the parts developed by other researchers in other laboratories.

As editor, it is my duty and pleasure to write a preface to this volume. I view my task as providing a succinct summary of how this volume came to be, an equally succinct overview of the volume, and thanks to the many people whose efforts contributed to its production and to the success of the workshop on which the volume is based. I will, however, avoid a more detailed discussion of integrated models of cognitive systems. That discussion is provided in chapter 1 and continues throughout this collection.

The Beginnings

This volume began with a conversation with Bob Sorkin during the spring 2004 workshop of the Air Force Office of Scientific Research's (AFOSR) Cognitive/Decision Making Program Review held in Mesa, Arizona. At that time, Bob was a program manager at AFOSR, on leave from the University of Florida. I approached him with an idea for a workshop and volume that would bring together elite members of the diverse cognitive modeling community. The basic notion was that those working on single-focus mathematical or computational models of cognitive functions and those interested in integrated models of cognitive systems should convene to discuss commonalities and differences in their approaches and the possibilities for synergy. Bob discussed this idea with his AFOSR colleagues John Tangney and Genevieve Haddad, and (after a formal written proposal, a formal review by AFOSR, and the usual amount of paperwork) AFOSR funded the workshop.

Deciding whom to invite to the workshop was one of the joys of my professional career. I solicited ideas from colleagues worldwide, searched for topics and citations on the Web of Science, and compiled a long list of people and seminal papers in key areas related to the control of integrated cognitive systems. This list included traditional areas such as visual attention and visual search, architectures of cognition very broadly defined, emerging areas such as the influence of emotion on cognition, and long-simmering areas such as the influence of the task environment on cognition. In searching through papers and deciding on topics, I received considerable assistance from doctoral students in my fall 2004 graduate seminar, COGS 6962, titled "Advanced Topics in Cognitive Science: Emotion, Control of Cognition, & Dynamic Decision Making." The students and I read through a list of 48 papers, of which about 30 were published in 2000 or later, and of those, 12 had been published in the preceding two years or in press at the time of the seminar.

At one time, the list of potential invitees approached 100. My goal was to narrow the list to about 20, at most, with a goal of 15. In meeting this goal, I wanted to mirror the intellectual diversity of the field, while sampling from the best and brightest of all current generations of cognitive scientists. Hence, my list began with brilliant postdoctorate students, who had published little but had shown great potential. It extended to assistant professors, junior researchers in military research laboratories, and midcareer and senior researchers whose ideas and intellectual vigor have shaped cognitive science over the past several decades.

Judging from my past experiences in organizing small workshops, large conferences, and recruiting contributors for special issues of journals, I thought that to end up with 15–20 researchers for a conference in March 2005 and a book the following year that I should start by inviting 35 participants. I contacted them by e-mail and telephone. To my delight and chagrin, 30 accepted. Hence, on March 3, 2005, a larger than expected group came together at the Sarotoga Inn in Saratoga Springs, New York, for an extended weekend of scientific discussions.

This Volume

An important feature of this volume is that each draft was reviewed and revised at least once. The core reviewers were a subset of the graduate students and faculty who had participated in my fall 2004 seminar as well as in workshop discussions (Prof. Michael Schoelles, Dr. Hansjörg Neth, Mr. Christopher Myers, Mr. Chris Sims, and Mr. Vladislav "Dan" Veksler). Hence, each reviewer had provided a substantial intellectual contribution to the volume even before they had received the chapters.

The workshop featured seven keynote speakers, and the talks were organized into five sessions. Each keynote speaker was given 75 minutes to speak (including questions) and asked write a chapter of no more than 12,000 words. All other speakers were given 20–30

minutes to speak and asked to write an 8,000-word chapter. Because the final versions of the chapters did not fit easily or naturally into the five-session organization of the workshop, my core group of reviewers helped reorganize the chapters into nine book sections and then assumed the additional responsibility of writing brief but cogent introductions to each section.

The result is a work in nine sections. Part I, "Beginnings," consists of three chapters each of which would be an excellent first chapter for this volume. As editor, I have given my chapter, "Composition and Control of Integrated Cognitive Systems" pride of place. However, the book could have started with the chapter by Kevin A. Gluck, Jerry T. Ball, and Michael A. Krusmark, which details their problems, pitfalls, and successes in modeling the integrated cognitive systems required by the pilot of an uninhabited air vehicle. Likewise, a good beginning would have been keynote speaker Richard W. "Dick" Pew's personal history of the field of human performance modeling from the mid-1950s onward.

Part II, "Systems for Modeling Integrated Cognitive Systems," focuses on four systems. The section begins with an introduction by Sims and Veksler and includes chapters by keynote speaker John R. Anderson, Ron Sun, Nicholas L. Cassimatis, and one by Randy J. Brou, Andrew D. Egerton, and Stephanie M. Doane. These chapters discuss ACT-R, CLARION, Polyscheme, and ADAPT, respectively. With the exception of the newest architecture, Polyscheme, these chapters do not attempt to provide an overview of the architecture but to provide a snapshot of research issues that the architecture is currently advancing.

Part III, "Visual Attention and Perception," begins with an introduction by Myers and Neth and includes chapters by Jeremy M. Wolfe, Marc Pomplun, and Ronald A. Rensink. Wolfe provides a review and update on his influential Guided Search 4.0 model of visual attention; Pomplun presents his area activation model that predicts eye movements during visual search; and Rensink sketches out a new and interesting *active vision* model of visual perception.

"Environmental Constraints on Integrated Cognitive Systems," part IV, begins with an introduction by Neth and Sims that is immediately followed by keynote speaker Peter M. Todd's chapter with Lael J. Schooler on how builders of "skyscraper cognitive models and of cottage decision heuristics" can work together once we understand the structure of the task environment in which cognition takes place. Wai-Tat Fu focuses on his recent rational-ecological approach to understanding the balance between exploration and exploitation as an organism adapts to a new environment. Michael C. Mozer, Sachiko Kinoshita, and Michael Shettel show how the sequential dependencies between repeated performance of the same task reflect the fine-tuning of cognitive control to the structure of the task environment. Keynote speaker Alex Kirlik's chapter concludes this section with a thoughtful discussion of the problems facing integrated models of cognitive systems as they begin to be applied to dynamic and interactive environments.

Part V, "Integrating Emotions, Motivation, Arousal Into Models of Cognitive Systems," was the hardest section to put together. Despite the recent surge of interest in the influence of emotion on cognition and of cognition on emotion, little of this work reflects both the state of the art in cognitive science research as well as the state of the art in emotion research. Even less of the current work reflects attempts by researchers to build more than box-diagram models in which an emotion box is connected to a cognitive box by a two-headed arrow; that is, very little of the work provides an integrated model of a cognitive-emotional system. But, where there is academic smoke, occasionally there may be intellectual fire! I am pleased to have assembled a group of papers that sheds light on the ways and means by which theories of emotion might be integrated with theories of cognition.

Part V is introduced by Veksler and Schoelles and consists of five chapters. Keynote speaker Jerome R. Busemeyer along with colleagues Eric Dimperio and Ryan K. Jessup present one of the most succinct and cogent discussions of emotions, motivation, and affect that I have read. They then proceed to show how affective state can be integrated into Busemeyer's influential decision field theory. Jonathan Gratch and Stacy Marsella provide an overview of a detailed implementation of appraisal theory in the Soar cognitive architecture. Glenn Gunzelman, Kevin A. Gluck, Scott Price, Hans P. A. Van Dongen, and David F. Dinges present human cognitive data from a sleep deprivation study. They model these data by first creating an ACT-R model of performance under normal conditions and then adjusting one parameter of their model to capture the effects of sleep deprivation on cognition. This work represents a major step along the path of integrated models of cognition and emotion. The chapter by Frank E. Ritter, Andrew L. Reifers, Laura Cousino Klein, and Michael L. Schoelles lays out a methodology for systematically exploring the ways in which emotion can be integrated into existing cognitive architectures.

Eva Hudlicka concludes this section with her chapter on the MAMID methodology for analysis and modeling of individual differences in cognition and affect.

Part VI, "Modeling Embodiment in Integrated Cognitive Systems," presents three chapters that reflect a moderate definition of embodiment as a cognitive system that integrates cognition with perception and action. Keynote speaker Dana Ballard and Nathan Sprague introduce Walter, a virtual agent in a virtual environment that navigates sidewalks and street crossings while avoiding obstacles and picking up trash (such a model virtual citizen!). As serious modelers, they recognize that there "is no free lunch in that the reason that embodied models can forego computation is that it is done implicitly by the body itself." Their goal is to simulate the body's "prodigious computational abilities." Laurence T. Maloney, Julia Trommershäuser, and Michael S. Landy show that speeded movement tasks can be described as formally equivalent to decision making under risk and, in doing so, provide one of the most cogent introductions to the topic of modeling decision making under risk that I have come across. Anthony Hornof's chapter ends this section. Hornof uses the EPIC architecture to build integrated models of visual search. His chapter provides a study of the issues faced by cognitive modelers when we attempt to account for the detailed control of an integrated cognitive system.

Part VII, "Coordinating Tasks Through Goals and Intentions," is all about control of integrated cognitive systems. The section begins with Schoelles's introduction and is followed by a trio of chapters by keynote speaker David Kieras, Dario D. Salvucci, and Niels Taatgen that frame the modern discussion of how control should be implemented in architectures of cognition. In the fourth chapter in this section, Erik M. Altmann shows that the construct of "goal" for the short-term control of behavior (as opposed to "goal" in a longer-term, motivational sense) can be reduced to more basic cognitive constructs. The section ends with Richard A. Carlson's interesting discussion of the costs and benefits of deictic specification in the real-time control of behavior.

The two chapters in "Tools for Advancing Integrated Models of Cognitive Systems," part VIII, share a common interest in advancing the cause of integrated modeling by providing tools to make the enterprise easier. A common problem in modeling human behavior is simply that there are many possible ways of performing the same task. Andrew Howes,

Richard L. Lewis, and Alonso Vera sketch an approach to predicting the optimal method or strategy, given a description of the task goal and task environment together with an explicit specification of the cognitive architecture and knowledge available for task performance. Where Howes et al. tackle a specialized issue of concern to expert modelers, Richard P. Cooper takes the opposite tack of creating a tool, COGENT, that provides a common graphical, object-oriented interface to a Swiss Army knife collection of modeling techniques. Rather than providing a software system that advances a particular cognitive theory, COGENT provides a means for students to become familiar with alternative modeling techniques and for researchers to quickly explore alternative theoretical approaches for modeling their phenomena of interest.

As there was a "Beginnings," so there must be an end. Part IX, "Afterword," consists of a short introduction followed by a single chapter. In this chapter, Michael D. Byrne provides a thoughtful and occasionally impassioned discussion of the issues discussed at the workshop and how best to advance cognitive science by the loose collaboration of those who would build integrated models of cognitive systems and those who see their task as building single-focus models of cognitive functions.

Thanks

Many people contributed to the different phases of this project. First there are the researchers whose chapters are published here. They believed that the goals of the workshop and volume were important enough to justify flying to Saratoga Springs in early March and writing a chapter after they got home. These are all people with research agendas whose individual goals may have been better served by spending more time in their laboratory or writing up research results to submit to good research journals. I managed to convince them that this project would be a better use of their time, and I thank them for letting me do so.

Hero stars and special commendations are due to Carol Rizzo and Cheryl Keefe of Rensselaer Polytechnic Institute. Carol worked tirelessly in dealing with the hotels and restaurants needed by the workshop. She dealt with each researcher to make travel arrangements. Together with Cheryl, she processed all of the paperwork so that each researcher's travel expenses were reimbursed. I could not have done this myself, and I sincerely thank them for doing it for me.

This brings me to my core reviewers and intellectual support staff. First and foremost, I thank Michael Schoelles, my close colleague and collaborator for many years. Next, I thank my junior colleagues Hansjörg Neth, Markus Guhe, Chris Myers, Chris Sims, and Dan Veksler. Their intellectual work on this project began before the fall 2004 seminar, was pushed hard during the seminar, extended to the workshop, into the reading and reviewing of each chapter (several twice), and continued to the writing of section introductions for this volume. Their work did not end there, as I called on several of them to critically review my chapter. (As now seasoned reviewers, they spared me no mercy and my chapter is better for their comments.) Beyond the intellectual work, this group also did the lifting and hauling—literally. On the busy days before the workshop, they greeted attendees as they arrived at the Albany Airport or Rensselaer train station and drove them to Saratoga Springs. They participated in workshop sessions, while attending to audiovisual needs and other errands as needed. Throughout all of this, they organized themselves and worked with energy and goodwill.

Outside this close circle, I called on several other colleagues to do some of the reviewing. These would include my former student, Wai-Tat Fu, and Glenn Gunzelmann.

I would be remiss if I did not thank Series Editor Frank Ritter or Oxford University Press Editor Catharine Carlin. As I was looking for an outlet for this volume, Frank's new Oxford University Press series, Series on Cognitive Models and Architectures, was looking for its first volume. I have known Frank for years and have known him to be that rare individual who spends as much time promoting good work that others are doing as he does promoting his own good work. I have also worked with him in my role as associate editor for several journals and knew that he could read critically and respond constructively—just the attributes needed for a series editor. I had also known Catharine for several years as the friendly face behind the Oxford University Press displays at Psychonomics, Cognitive Science, and other conferences. During the course of Psychonomics in Minneapolis (2004), I was pleased to realize that she possessed an in-depth knowledge of cognitive science as well as a nose for good wine.

Those of you who juggle careers and families know that thanks are due to my wife and companion of many years, Deborah Tong Gray, for putting up with yet another project. She knows that this is important for me and has accepted that, unlike football husbands who are obsessed for a season, academic ones are obsessed for 12 months of the year. For this project, at least, duty and family occasionally coincided as she directly assisted me with the not unpleasant task of sampling restaurants in Saratoga Springs in which to hold the various workshop dinners.

Above all I need to acknowledge my debt to the Air Force Office of Scientific Research that supported the workshop as well as my work on this volume through AFOSR grant F49620-03-1-0143. AFOSR is that rare federal agency that is staffed with people who know and understand research, researchers, and the research community. Beyond that, they are able to make the connections from basic cognitive science to its application as cognitive engineering in support of real-world problems that many of the researchers they fund are unable to make for themselves. In particular I need to thank Bob Sorkin, John Tangney, and Genevieve Haddad for their roles in making this happen.

This is an exciting time to be a cognitive scientist and an exciting time to be a modeler. From the psychology side, the paradigm of cognitive science is shifting from a focus on experiments-as-gold standard with box diagrams as models to quantitative and computational models that generate the behavior they model. From the artificial intelligence side, the paradigm of cognitive science is shifting from a computational philosophy with no connection to human behavior to one that emphasizes the rigorous comparison of model to empirical data. The researchers represented in this volume stand in the intersection where these two trends meet. Our postcard to the future is this volume dedicated to the proposition that our ultimate goal is to understand an integrated cognitive system.

References

Anderson, M. L. (2003). Embodied cognition: A field guide. *Artificial Intelligence, 149*(1), 91–130.

Brooks, R. A. (1991). Intelligence without representation. *Artificial Intelligence, 47*(1–3), 139–159.

Newell, A. (1973). You can't play 20 questions with nature and win: Projective comments on the papers of this symposium. In W. G. Chase (Ed.), *Visual information processing* (pp. 283–308). New York: Academic Press.

Wilson, M. (2002). Six views of embodied cognition. *Psychonomic Bulletin & Review, 9*(4), 625–636.

Contents

The Rise of Cognitive Architectures   v
  Frank E. Ritter

Contributors   xv

I BEGINNINGS   1
  Wayne D. Gray

1. Composition and Control of Integrated
   Cognitive Systems   3
   Wayne D. Gray

2. Cognitive Control in a Computational
   Model of the Predator Pilot   13
   *Kevin A. Gluck, Jerry T. Ball, & Michael A.
   Krusmark*

3. Some History of Human Performance
   Modeling   29
   Richard W. Pew

II SYSTEMS FOR MODELING
   INTEGRATED COGNITIVE SYSTEMS   45
   Chris R. Sims & Vladislav D. Veksler

4. Using Brain Imaging to Guide the Development
   of a Cognitive Architecture   49
   John R. Anderson

5. The Motivational and Metacognitive Control in
   CLARION   63
   Ron Sun

6. Reasoning as Cognitive Self-Regulation   76
   Nicholas L. Cassimatis

7. Construction/Integration Architecture: Dynamic
   Adaptation to Task Constraints   86
   *Randy J. Brou, Andrew D. Egerton, & Stephanie M.
   Doane*

III VISUAL ATTENTION AND
   PERCEPTION   97
   Christopher W. Myers & Hansjörg Neth

8. Guided Search 4.0: Current Progress With a
   Model of Visual Search   99
   Jeremy M. Wolfe

9. Advancing Area Activation Toward a General Model
   of Eye Movements in Visual Search   120
   Marc Pomplun

10. The Modeling and Control of Visual
    Perception   132
    Ronald A. Rensink

IV ENVIRONMENTAL CONSTRAINTS ON
   INTEGRATED COGNITIVE SYSTEMS   149
   Hansjörg Neth & Chris R. Sims

11. From Disintegrated Architectures of Cognition to
    an Integrated Heuristic Toolbox   151
    Peter M. Todd & Lael J. Schooler

12. A Rational–Ecological Approach to the
    Exploration/Exploitation Trade-Offs: Bounded
    Rationality and Suboptimal Performance   165
    Wai-Tat Fu

13. Sequential Dependencies in Human Behavior
Offer Insights Into Cognitive Control   180
Michael C. Mozer, Sachiko Kinoshita,
& Michael Shettel

14. Ecological Resources for Modeling Interactive
Behavior and Embedded Cognition   194
Alex Kirlik

V INTEGRATING EMOTIONS,
MOTIVATION, AROUSAL INTO MODELS
OF COGNITIVE SYSTEMS   211
Vladislav D. Veksler & Michael J. Schoelles

15. Integrating Emotional Processes Into Decision-
Making Models   213
Jerome R. Busemeyer, Eric Dimperio,
& Ryan K. Jessup

16. The Architectural Role of Emotion in Cognitive
Systems   230
Jonathan Gratch & Stacy Marsella

17. Decreased Arousal as a Result of Sleep Deprivation:
The Unraveling of Cognitive Control   243
Glenn Gunzelmann, Kevin A. Gluck, Scott Price,
Hans P. A. Van Dongen, & David F. Dinges

18. Lessons From Defining Theories of Stress for
Cognitive Architectures   254
Frank E. Ritter, Andrew L. Reifers, Laura Cousino
Klein, & Michael J. Schoelles

19. Reasons for Emotions: Modeling Emotions in
Integrated Cognitive Systems   263
Eva Hudlicka

VI MODELING EMBODIMENT IN
INTEGRATED COGNITIVE
SYSTEMS   279
Hansjörg Neth & Christopher W. Myers

20. On the Role of Embodiment in Modeling
Natural Behaviors   283
Dana Ballard & Nathan Sprague

21. Questions Without Words: A Comparison
Between Decision Making Under Risk and
Movement Planning Under Risk   297
Laurence T. Maloney, Julia Trommershäuser,
& Michael S. Landy

22. Toward an Integrated, Comprehensive Theory of
Visual Search   314
Anthony Hornof

VII COORDINATING TASKS THROUGH
GOALS AND INTENTIONS   325
Michael J. Schoelles

23. Control of Cognition   327
David Kieras

24. Integrated Models of Driver Behavior   356
Dario D. Salvucci

25. The Minimal Control Principle   368
Niels Taatgen

26. Control Signals and Goal-Directed
Behavior   380
Erik M. Altmann

27. Intentions, Errors, and Experience   388
Richard A. Carlson

VIII TOOLS FOR ADVANCING
INTEGRATED MODELS OF
COGNITIVE SYSTEMS   401
Wayne D. Gray

28. Bounding Rational Analysis: Constraints on
Asymptotic Performance   403
Andrew Howes, Richard L. Lewis, & Alonso Vera

29. Integrating Cognitive Systems: The COGENT
Approach   414
Richard P. Cooper

IX AFTERWORD   429
Wayne D. Gray

30. Local Theories Versus Comprehensive Architectures:
The Cognitive Science Jigsaw Puzzle   431
Michael D. Byrne

Author Index   445
Subject Index   457

Contributors

Erik M. Altmann
Department of Psychology
Michigan State University
East Lansing, Michigan

John R. Anderson
Psychology Department
Carnegie Mellon University
Pittsburgh, Pennsylvania

Jerry T. Ball
Air Force Research Laboratory
Mesa, Arizona

Dana Ballard
Department of Computer Science
University of Rochester
Rochester, New York

Randy J. Brou
Institute for Neurocognitive Science and Technology
Mississippi State University
Mississippi State, Mississippi

Jerome R. Busemeyer
Department of Psychological & Brain Sciences
Indiana University
Bloomington, Indiana

Michael D. Byrne
Department of Psychology
Rice University
Houston, Texas

Richard A. Carlson
Department of Psychology
Penn State University
University Park, Pennsylvania

Nicholas L. Cassimatis
Cognitive Science Department
Rensselaer Polytechnic Institute
Troy, New York

Richard P. Cooper
School of Psychology
University of London
London, United Kingdom

Eric Dimperio
Indiana University
Bloomington, Indiana

David F. Dinges
Division of Sleep and Chronobiology
University of Pennsylvania School of Medicine
Philadelphia, Pennsylvania

Stephanie M. Doane
Institute for Neurocognitive Science and Technology
Mississippi State University
Mississippi State, Mississippi

Andrew D. Egerton
Institute for Neurocognitive Science and Technology
Mississippi State University
Mississippi State, Mississippi

Wai-Tat Fu
University of Illinois at Urbana–Champaign
Human Factors Division and Beckman Institute
Savoy, Illinois

Kevin A. Gluck
Air Force Research Laboratory
Mesa, Arizona

Jonathan Gratch
Department of Computer Science
University of Southern California
Marina Del Rey, California

Wayne D. Gray
Cognitive Science Department
Rensselaer Polytechnic Institute
Troy, New York

Glenn Gunzelmann
Air Force Research Laboratory
Mesa, Arizona

Anthony Hornof
Department of Computer and Information Science
University of Oregon
Eugene, Oregon

Andrew Howes
School of Informatics
University of Manchester
Manchester, United Kingdom

Eva Hudlicka
Psychometrix Associates, Inc.
Blacksburg, Virginia

Ryan K. Jessup
Indiana University
Bloomington, Indiana

David Kieras
Artificial Intelligence Laboratory
Electrical Engineering and Computer Science
Department
University of Michigan
Ann Arbor, Michigan

Sachiko Kinoshita
MACCS and
Department of Psychology
Macquarie University
Sydney, New South Wales, Australia

Alex Kirlik
Human Factors Division and Beckman Institute
University of Illinois
Urbana, Illinois

Laura Cousino Klein
Biobehavioral Health
Penn State University
University Park, Pennsylvania

Michael A. Krusmark
Air Force Research Laboratory
Mesa, Arizona

Michael S. Landy
Department of Psychology and
Center for Neural Science
New York University
New York, New York

Richard L. Lewis
School of Psychology
University of Michigan
Ann Arbor, Michigan

Laurence T. Maloney
Department of Psychology and
Center for Neural Science
New York University
New York, New York

Stacy Marsella
Department of Computer Science
University of Southern California
Marina Del Rey, California

Michael C. Mozer
Department of Computer Science and
Institute of Cognitive Science
University of Colorado
Boulder, Colorado

Christopher W. Myers
Cognitive Science Department
Rensselaer Polytechnic Institute
Troy, New York

Hansjörg Neth
Cognitive Science Department
Rensselaer Polytechnic Institute
Troy, New York

Richard W. Pew
BBN Technologies
Cambridge, Massachusetts

Marc Pomplun
Department of Computer Science
University of Massachusetts at Boston
Boston, Massachusetts

Scott Price
Air Force Research Laboratory
Mesa, Arizona

Andrew L. Reifers
Applied Cognitive Science Lab
College of Information Sciences and Technology
Penn State University
University Park, Pennsylvania

Ronald A. Rensink
Departments of Computer Science and Psychology
University of British Columbia
Vancouver, BC, Canada

Frank E. Ritter
Applied Cognitive Science Lab
College of Information Sciences and Technology
Penn State University
University Park, Pennsylvania

Dario D. Salvucci
Department of Computer Science
Drexel University
Philadelphia, Pennsylvania

Michael J. Schoelles
Cognitive Science Department
Rensselaer Polytechnic Institute
Troy, New York

Lael J. Schooler
Center for Adaptive Behavior and Cognition
Max Planck Institute for Human Development
Berlin, Germany

Michael Shettel
Department of Computer Science
University of Colorado
Boulder, Colorado

Chris R. Sims
Cognitive Science Department
Rensselaer Polytechnic Institute
Troy, New York

Nathan Sprague
Department of Mathematics and Computer Science
Kalamazoo College
Kalamazoo, Michigan

Ron Sun
Computer Science Department
Rensselaer Polytechnic University
Troy, New York

Niels Taatgen
Carnegie Mellon University and
University of Groningen
Department of Psychology
Pittsburgh, Pennsylvania

Peter M. Todd
Department of Cognitive Science and Informatics
Indiana University
Bloomington, Indiana

Julia Trommershäuser
Department of Psychology
Giessen University
Giessen, Germany

Hans P. A. Van Dongen
Sleep and Performance Research Center
Washington State University
Spokane, Washington

Vladislav D. Veksler
Cognitive Science Department
Rensselaer Polytechnic Institute
Troy, New York

Alonso Vera
NASA Ames Research Center and
Carnegie Mellon University
Moffett Field, California

Jeremy M. Wolfe
Visual Attention Lab
Brigham and Womens Hospital
Professor of Ophthalmology
Harvard Medical School
Cambridge, Massachusetts

PART I

BEGINNINGS

Wayne D. Gray

The notion of building integrated models of cognitive systems did not originate with this book and did not spring up overnight. It is an issue with many beginnings, three of which are represented in this section.

My chapter (chapter 1) begins this part by proposing a taxonomy of control for integrated cognitive systems. As most contemporary cognitive scientists implicitly or explicitly accept the notion that functional subsystems exist, it is important to distinguish issues of control that are specific to the operation of a particular subsystem (Type 2) from control issues that entail either the coordination of various subsystems by a central system or more direct input and output relationships among subsystems (both of these alternatives are considered as Type 1 control). Additionally, in dealing with human behavior, it is always important to recognize that the methods and strategies (Type 3 control) brought to bear on task performance are influenced by prior knowledge, training, and the structure of the task environment.

Gluck, Ball, and Krusmark (chapter 2) provide another beginning to integrated models of cognitive systems. Their chapter details attempts to model the complex tasks performed by the uninhabited air vehicle (UAV) operator. A complete model that could take off, perform missions, and return safely would require the detailed integration of most, if not all, functional subsystems studied by cognitive scientists today as well as raise challenging issues regarding Type 1, 2, and 3 control. Although Gluck and colleagues do not solve this problem, what they have done is both interesting and important. Indeed, meeting the challenge presented by tasks such as modeling the UAV operator influenced this volume and was an important start to integrated models of cognitive systems.

The position of "the end of the beginning" is given to Pew's chapter, "Some History of Human Performance Modeling." It is all too easy for those working in this first decade of the 21st century to enmesh ourselves in the moment, ignoring those who have come before. Indeed, the mid-20th century was a time when giants strode the earth shaping the issues and assumptions of today. Pew offers a personal account of this time, its people, and their challenges and has written a chapter that is important not simply for the history but for our current understanding of integrated models of cognitive systems.

1

Composition and Control of Integrated Cognitive Systems

Wayne D. Gray

Integrated models of cognitive systems can be contrasted with the dominant variety of cognitive modeling that produces *single-focus models of cognitive functions* such as control of eye movements, visual attention, categorization, decision making, or memory. Such single-focus models are necessary but not sufficient for understanding human cognition. Although single-focused models are not usually created to be part of a larger, more integrated system, if cast in the right form, they can play strong roles in building integrated models of cognitive systems.

There are two key reasons why single-focus modelers might wish to see their work incorporated into an integrated model. The first is scientific impact. A model of eye movements, categorization, visual search, decision making, and so on that can be used as a component of an integrated model can extend the model's impact beyond the important, but narrow, debates with other models of the same phenomenon to the greater cognitive community.

The second reason is evaluation. Recent discussions have extended the focus of model evaluation from an obsession with goodness-of-fit (Roberts & Pashler, 2000; Rodgers & Rowe, 2002; Van Zandt, 2000) to an approach that combines considerations of goodness-of-fit with generality (Massaro, Cohen, Campbell, & Rodriguez, 2001; Pitt, Kim, & Myung, 2003; Pitt & Myung, 2002; Pitt, Myung, & Zhang, 2002). These new approaches are promising, but at present they are difficult (i.e., mathematically intractable) to apply to any but the simplest models. However, there is an older sense of generalization in terms of *external validity*

(Cook & Campbell, 1979; Gray & Salzman, 1998), that is, whether the findings can be generalized to different populations, different locations, and different settings. For example, it is one thing to have a model of visual attention that predicts human response times in detecting one target amid eight distracters when the items are displayed for 200 ms and immediately followed by a mask. It would be quite an important generalization of this model (and its underlying theory) to apply it successfully to predicting search times for finding an enemy fighter plane on a busy radar screen, where the distracters are commercial jetliners. Making the single-focus model available in a form that can be incorporated into integrated models could lead many other people to attempt to apply the model to conditions and tasks well beyond those envisioned by the original modeler. The best single-focus model will be the one that provides the best performance for an integrated model across a wide variety of tasks and task environments.

Any discussion has two sides, and builders of integrated models need to explain what they mean by integration. In this chapter, I propose a new vocabulary to do just that. I hope that, if adopted, this vocabulary will clarify for the modeling community some of the issues in building integrated models. It may also lead to ways of separating the evaluation of integrated models from their component single-focus models, thereby, helping us to highlight effectively the strengths and remediate the weaknesses in our current understanding of cognitive systems.

In the next section, I attempt to define *integrated cognitive system*. The heart of the chapter introduces and discusses three types of control and three types of components of integrated models of cognitive systems.

What Are Integrated Models of Cognitive Systems?

We never seem in the experimental literature to put the results of all the experiments together.... One picks and chooses among the qualitative summaries of a given experiment what to bring forward and juxtapose with the concerns of a present treatment. (Newell, 1973)

When researchers working on a particular module get to choose both the inputs and the outputs that specify the module requirements I believe there is little chance the work they do will fit into a complete intelligent system. (Brooks, 1991)

Much cognitive science research proceeds by encapsulating the topic of interest in a functional module that may be broadly grouped within larger functional modules such as perception, cognition, and action (see Figure 1.1). Researchers as diverse as Newell and Brooks have complained about this paradigm of single-focus research, with Newell enjoining us to build "complete processing models" (Newell, 1973) and Brooks instructing that "the fundamental slicing up of an intelligent system is in the orthogonal direction dividing it into activity producing subsystems" (Brooks, 1991) (Figure 1.1). (Ballard and Sprague, chapter 20, this volume, refer to the former as the "Marr paradigm" and the latter as the "Brooks paradigm.") These activity-producing subsystems provide simple behavioral programs that involve two or more functional modules. As such, they are similar in spirit to what I have called *basic activities* (Gray & Boehm-Davis, 2000) or, more recently, *interactive routines* (Gray, Sims, Fu, & Schoelles, 2006). Integrated models are those that incorporate interactive routines (a.k.a. activity-producing subsystems).

Any discussion of functional modules must be preceded by an important caveat. Cognitive science has an uncomfortable way of postulating novel mechanisms to account for functionality that is actually computed by more basic mechanisms or provided as an input by another module. As a result, the cognitive bestiary becomes populated by many real and

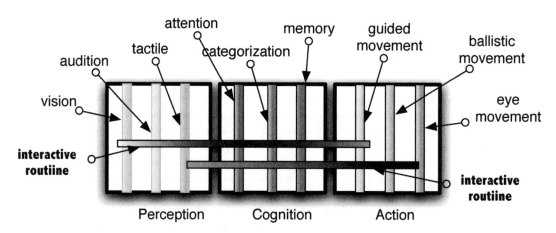

FIGURE 1.1 Notional organization of functional modules grouped within the higher-order functional modules of perception, cognition, and action. Two interactive routines are shown, each of which cuts across some of the functional modules.

imaginary creatures to the greater confusion of the field. Clearly, the proposal to talk about integrated models as integrating across a variety of functional modules (Figure 1.1) might be viewed as an invitation to name new lake-dwelling monsters rather than carefully observing and describing the play of light and waves on random configurations of wood and debris. This temptation should be resisted. Indeed, the call for building integrated models is a call to carefully observe and describe the functionality that emerges from the control of complex cognitive operations. An excellent example of trading off a module for a more nuanced understanding of control is provided by McElree (2006), who suggest that the decades-long debate over the tripartite architecture of memory (focal attention, working memory, and long-term memory) may be resolved as a bipartite architecture (focal attention vs. everything else), in which "the successful execution of complex cognitive operations" depends "more on our ability to shunt information between focal attention and memory than on the existence of a temporary store" (p. 194). In other words, McElree is asserting that getting the control issues right will allow us to reduce the cognitive bestiary by one imaginary beast. (See also Altmann's excellent chapter in this volume [chapter 26], which shows that special-purpose goal structures postulated to govern immediate behavior such as task switching can be reduced to more basic memory + control processes.)

An emphasis on integrated models is a call to refocus much of cognitive science on the careful observation and description of control processes that operate within and between valid functional modules. Although integrated models that get these control issues right promise much explanatory insight, it is undeniable that modeling control issues in even simple tasks is necessarily complex. The problem is that we have compounded an already hard problem. First are the hard problems of understanding single-focus cognitive functions such as categorization, memory, various types of perception, and movement. Second is the realization that the output of some of the functional modules depicted in Figure 1.1 serves as input to other functional modules. These output-input relationships imply that at some level of analysis the activity of the various functional modules must be coordinated. Third, we do not have a closed system; not only do the various functional modules interact with one other, but they also interact with the task environment (see part IV of this volume).

However, although the control problems posed by integrated models of cognitive systems are complex, they are no more complex than they need to be if our ultimate goal is to understand a cognitive system that is integrated.

Types of Control

This section introduces a vocabulary for understanding the necessary complexity of integrated cognitive systems. Specifically, three types of cognitive control are proposed, simply labeled Types 1, 2, and 3. Type 1 refers to the control exercised by one functional module on another, regardless of whether that influence is in the form of direct communication between modules or mediated by a central buffer or controller. Type 2 is control within a functional module such as memory, attention, or motor movement. Ideally, these functional modules constitute independently validated single-focus theories, but there is no compulsion that they be so. Type 3 is task-specific control—the strategies or methods that can be brought to bear on task performance given the task to achieve, a task environment in which to achieve it, a set of Type 2 functional modules, and Type 1 control among functional modules. Each type is elaborated on in the sections that follow.

Type 1 Control

At its simplest, Type 1 control is exerted by the output of one functional module being provided as an input to another functional module. This control may be direct as in a *society of mind* (Minsky, 1985) approach (see Figure 1.2A) or may be mediated by a central controller as in an *architectures of cognition* approach (Anderson, 1983, 1993; Newell, 1973, 1990, 1992) (see Figure 1.2B). For architectural approaches, the mechanisms imputed to the central controller vary greatly. Both ACT-R (adaptive control of thought–rational) (Anderson et al., 2004; chapter 4, this volume) and Soar postulate a number of special mechanisms to handle the selection and firing of production rules. In contrast, EPIC's (Kieras & Meyer, 1997; Kieras, chapter 23, this volume) central controller plays a very minor role in selecting which productions to fire and when to fire them. Instead, EPIC (executive process-interactive control) places the burden of control at the Type 3 level by using production rules to program explicitly different control strategies (e.g., Kieras, chapter 23, this volume) for different tasks. In production system

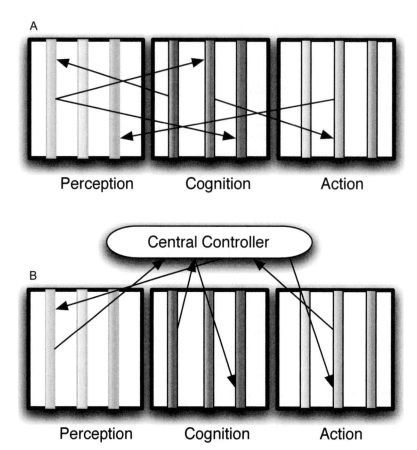

FIGURE 1.2 Two modes of control. (A) Inputs of one functional module go directly into one or more other functional modules. (B) Inputs and outputs of functional modules are mediated by a central controller.

architectures, inputs and outputs to many of the functional modules must pass through the central controller, but for other modules the communication may be direct.

This volume contains important discussions of Type 1 issues from the perspective of architectures of cognition by Kieras (chapter 23), Salvucci (chapter 24), and Taatgen (chapter 25). Although they do not focus on Type 1 control per se, Ballard and Sprague's (chapter 20) example of Walter seems to embody Type 1 control features associated with both approaches.

Functional Modules and Type 2 Input, Output, and Control

Type 2 control refers to the internal control of a given functional module. Inputs come from other functional modules or the task environment (for perception) and may be mediated by a central controller. Outputs go to other functional modules or act on the world (for action) and, likewise, may be mediated by a central controller.

In this section, I distinguish among three subtypes of functional modules based on their mechanism of internal control. Noncognitive modules predict, but do not explain, behavior; that is, they do not rest on cognitive theory. In addition, it seems convenient to distinguish between two types of theory-based functional modules based on whether internal control occurs during the run-time of the integrated model, or whether internal control occurs off-line with the results provided to the integrated model via some type of look up mechanism.

Noncognitive Modules: Empirical Regularities, Statistical Techniques, and Machine Learning Algorithms

Not all functional modules have a cognitive theory of their internal control. These modules must have inputs and outputs that connect them to other modules or a central controller but as for what occurs in-between, "then a miracle occurs."[1]

An example of a noncognitive, nontheoretic module is Fitts' law for predicting the time of guided motor movements (Fitts, 1954). Fitts' law takes two inputs: distance to the target and width of the target. Its internal control is provided by a simple formula that predicts movement time (*MT*) as,

$$MT = a + b * \log 2\left(\frac{D}{W} + 1\right),$$

where *D* is the distance to the target and *W* is the width of the target. The intercept parameter *a* and the slope parameter *b* are determined for different pointing devices such that the parameters for pointing with a finger differ from those for pointing with a head, and both differ from those for pointing with a mouse.

Fitts' law began as an information theoretic attempt to explain visually guided movement (Fitts, 1954). The theory that led to the law's original formulation has long since been discredited, and the reasons this equation usually works and an explanation of deviations when it does not work continue to be researched (Meyer, Smith, Kornblum, Abrams, & Wright, 1990). However, for many classes of movements, Fitts' law works very well. Indeed, at the dawn of the modern personal computer era, an evaluation of four types of computer pointing devices concluded that the mouse was superior because its "measured Fitts' law slope constant was close to that found in other eye-hand tasks" (Card, English, & Burr, 1978).

Fitts' law can be considered a functional module but one that does not constitute a cognitive theory. By itself, the empirical regularities captured by Fitts' law gives us no basis for reasoning about issues such as what happens if a second set of inputs comes into the *Fitts' module* after the first set is computed but before the movement duration calculated by Fitts' law has elapsed. These issues point out the limits of Fitts' law as a functional module and indicate the need for a complete Type 2 theory of motor control that includes issues such as interruptions, accuracy of motor movement, and preparation time for motor movement.

Noncognitive theoretic modules such as Fitts' law can be seen as representing a need for cognitive theory that research has not yet filled. Fitts' law exemplifies an empirical regularity that has stood the test of time and may confidently be used to predict, but not to explain, movement time. In other cases, it may be necessary to use a statistical or machine learning technique to provide the needed functionality as we may lack both cognitive theory and an ability to predict empirical regularities. For example, one could envision using a measure of semantic distance in a model of human information foraging (Landauer, Laham, & Derr, 2004; Lemaire & Denhiére, 2004; Turney, 2001) to compute the degree of association between the search target and the current label or link. Obviously, when a cognitive, perceptual, or motor function is computed by a nonpsychological module, the modeler needs to provide justification to convince the research community that the ersatz module has not undermined the particular theoretical points that the modeler is attempting to achieve.

Single-Focus Models of Cognitive Functions: Run-Time Control of Functional Modules

For a functional module to represent a cognitive theory, it must have a cognitive science based theory of internal control that may be expressed in either mechanistic or algorithmic (Anderson, 1987) terms. Examples might include single-focus models of cognitive functions such as the control of eye movements, visual saliency, or memory. The scope of the theory may not be settled, or its exact relationship to other cognitive theories may be disputed. For this chapter, however, the distinction captured by this category is that the theory has some cognitive validity and that its algorithms or mechanism compute its functionality at run-time to return an appropriate output to another module or to the central controller. In this section, two examples of how cognitive theories might work as functional modules in an integrated model are shown.

Rational Activation Theory of Declarative Memory
The rational activation theory[2] of Anderson and Schooler (1991) is based on a rational analysis

(Anderson, 1990, 1991) of the demands that the task environment makes on memory. The empirical regularities captured by the model successfully predict many phenomena that eluded other theories of memory such as the effect of massed versus distributed practice on retention.

The rational activation theory requires three types of inputs from a central controller: the number of encodings per item (i.e., frequency of encoding the same item), time between encodings (i.e., distributed vs. massed practice), and requests for retrieval. Its Type 2 control is provided at the algorithm level rather than the mechanism level (Anderson, 1987) by a series of mathematical equations that convert the frequency and recency of its inputs into measures of activation as well as into retrieval latencies and probabilities of retrieval success or failure. The estimate of retrieval latency and the item retrieved, or the failure to retrieve an item, are provided as outputs to the central controller.

Steering Angle The steering angle module of Salvucci and R. Gray (2004) is an interesting example of a single-focus model of a cognitive function that was developed to meet the needs of an integrated model of driving (Salvucci, 2006). Although the perception and action research community had identified control models that performed more complex tasks such as computing curvatures or more complex features, they had neglected to provide for the conceptually simpler task of relating road curvature to steering angle. The result of this work is a stand-alone, local theory that can also be used as a functional module within an integrated model.

The steering angle module takes three inputs: visual angle to the near point, visual angle to the far point, and time of the current samplings of near and far points. It calculates the differences between the current and immediately prior sample. In its current use in an ACT-R (adaptive control of thought–rational) model of driving, it then outputs the steering angle to the Type 1 central controller and this angle is then input, by the central controller, to the motor module. However, the steering angle module is independent of ACT-R's central controller, and if a theoretical basis for the change were justified, it would be possible to avoid the central controller by sending the output of this module directly to a motor module.

Single-Focus Models of Cognitive Functions: Off-Line Control of Functional Modules

The third category of functional module represents the task specific use of more basic processes to compute a particular output given a particular input. For example, Altmann (chapter 26, this volume) has developed a detailed, mechanistic theory of task switching in ACT-R. The theory makes extensive use of the rational activation theory of memory (Anderson & Schooler, 1991) as well as ACT-R's central controller. A modeler not working in ACT-R might, nonetheless, wish to use Altmann's theory to compute switch times. In this case, they would either develop an algorithmic form of Altmann's theory that approximated his mechanistic account or use the ACT-R form of the theory to precompute switch times for the desired range of inputs.

Another example is our use of ACT-R to precompute times to move visual attention and our use of Fitts' law to precompute movement times for use in a reinforcement learning model (Gray et al., 2006). In this case, we had developed a complete ACT-R model of a variation on Ballard's (Ballard, Hayhoe, & Pelz, 1995) blocks world task. However, to test the hypothesis that human resource allocation followed a strict cognitive cost accounting (the soft constraints hypothesis; see Gray et al., 2006), we wished to use a modeling paradigm that was formally guaranteed to find an optimal solution by minimizing time; namely, a variant on reinforcement learning called Q-learning (Sutton & Barto, 1998). Our estimates of the time to move visual attention included the times needed by the ACT-R model to acquire and send inputs to ACT-R's visual attention module as well as the processing time required by that module. Hence, we were able to claim that the time estimates for visual attention used in the reinforcement learning model inherited their cognitive validity from the ACT-R model. Furthermore, as the ACT-R model had been developed independently of the reinforcement learning model, we could also argue that these time estimates were not tweaked to obtain a good fit of the reinforcement learning model to the data.

These examples illustrate the off-line use of single-focus models in a functional module. In the preceding section, the module's Type 2 control computed its Type 1 output at run-time. In this section, that control is precomputed and the results supplied to the integrated model through a look-up table. In the two cases discussed here, the cognitive theory could not be easily run as part of the integrated model. However, in each case, the functional module would produce cognitively valid Type 1 output when given cognitively valid Type 1 inputs.

Issues for Type 2 Control

As the examples in this section illustrate, each of the functional modules takes some input and produces some output that can be used in an integrated model. Although highly desirable, it is not necessary that the functional module be based on cognitive theory. Noncognitive modules may be useful if they capture an empirical regularity (such as Fitts' law) or if they produce an outcome that is correlated with human performance. In these cases, the noncognitive module represents an issue that theory has not yet resolved. However, for integrated models of cognitive systems, it is obvious that theory-based modules are most desirable. Theory may be introduced in two different ways. First, the single-focus model of the cognitive function may itself be incorporated as a run-time component of the integrated model. Second, the functionality computed by the single-focus model may be computed offline so that theory-based outputs are provided in response to current inputs.

For both of these theory-based cases, cognitively valid results are returned to the integrated model from the functional module. In this way, the single-focus model gains external validity from its successful use as a component of an integrated model in new tasks and contexts. Likewise, the integrated model benefits by its ability to use theory developed independent of the current task to produce a reliable and valid result.

Beyond this discussion, it is important to note what is not implied by a functional module. All functional modules should not be viewed as representing different types of underlying cognitive processes. For example, it may well be that the mechanisms that underlie the computation of Fitts' law and task switching are composed of the same fundamental cognitive mechanisms and are most notably different because of their inputs and outputs. For example, one could imagine the same basic connectionist theory or production system architecture computing both outputs. Hence, the use of a given functional module is not necessarily a commitment to a distinctive or specialized type of cognitive processing.

Type 3

In complex adaptive behavior, the link between goals and environment is mediated by strategies and knowledge discovered or learned by the actor. Behavior cannot be predicted from optimality criteria alone without information about the strategies and knowledge agents possess and their capabilities for augmenting strategies and knowledge by discovery or instruction. (Simon, 1992, p. 157)

The most fundamental fact about behavior is that it is programmable. That is to say, behavior is under the control of the subject to shape in the service of his own ends. There is a sort of symbolic formula that we use in information processing psychology. To predict a subject you must know: (1) his goals; (2) the structure of the task environment; and (3) the invariant structure of his processing mechanisms. From this you can pretty well predict what methods are available to the subject; and from the method you can predict what the subject will do. Without these things, most importantly without the method, you cannot predict what he will do. (Newell, 1973, pp. 293–294)

Type 3 control combines task-specific knowledge with architectural universals inherited as constraints from Type 1 and Type 2 theories to develop strategies or methods (Type 3 theories) for accomplishing a given task in a given task environment (see part IV, this volume, "Task Environment"). In effect, Type 3 control entails a commitment to describing the play of light and waves on random configurations of wood and debris rather than naming new lake-dwelling monsters. As such, we might regard Type 3 theories as making normal use of the architecture.

In many laboratory tasks, the Type 3 methods or strategies are almost completely determined by hard constraints (Gray & Boehm-Davis, 2000) in the task environment. Consider, for example, the experimental straitjacket placed on human behavior by laboratory tasks such as the rapid serial visual presentation paradigm (Raymond, Shapiro, & Arnell, 1992) or the speed-accuracy tradeoff paradigm (McElree, 2006). For those familiar with these types of laboratory tasks, the variety of methods or strategies that subjects can bring to these tasks is constrained when compared with tasks such as walking down a sidewalk while avoiding obstacles (Ballard & Sprague, chapter 20, this volume), menu search (Hornof, chapter 22, this volume), working as a short-order cook (Kirlik, chapter 14, this volume), or piloting an uninhibited air vehicle (Gluck et al., chapter 2, this volume). Hence, to the extent that Type 3 control makes normal use of the human cognitive architecture, it follows that by design many laboratory tasks do not shed light on architectural control issues.

Those who study human acquisition and use of strategies and methods tend to study tasks more complex than the typical laboratory task. However, for integrated models of complex tasks, the problem is not simply that various methods or strategies can be used for performing the same task but, more complexly, that each of these methods or strategies will draw differentially on the set of available functional modules. These difficulties may seem to provide integrated modelers with a freedom to simply make up any strategy or method that fits the data.

This freedom may be more illusory than real; although often many patterns of interaction are possible, few patterns are typically employed (Gray, 2000). Rather, performance is constrained by the interactions among the task the user is attempting to accomplish, the design of the artifact (i.e., tool, device, or task environment) used to accomplish the task, and Type 1 and 2 controls. What emerges from this mixture is the interactive behavior needed to perform a given task on a given device. Indeed, proof of the productivity of taking architectural constraints seriously is provided by Hornof (chapter 22, this volume), who shows how the Type 1 and Type 2 theories embedded in EPIC can be used to constrain the generation of Type 3 strategies for a menu search task (see also Todd and Schooler, this volume).

Furthermore, as our knowledge of Type 1 and Type 2 controls grow it should be possible to provide constraints on the range of strategies or methods that plausibly can be developed. For example, our work on the soft constraints hypothesis (Gray et al., 2006) argues that at the 1/3- to 3-s time span that the allocation of cognitive, perceptual, and motor resources may be predictable from a least-cost analysis. Likewise, Howes, Lewis, and Vera (this volume) suggest that it may be possible to predict optimal strategies/methods based on optimality constraints on optimal performance.

Summary and Conclusions

As argued by thinkers as diverse as Newell and Brooks, cognitive science has had a regrettable tendency to postulate new mechanisms and modules to account for each new observation. This tendency has resulted in a cognitive bestiary populated by a various real and imaginary creatures. Integrated models of cognitive systems represent an attempt to step back and consider how much cognitive functionality might emerge from control mechanisms that work within and between what will turn out to be a large, but finite, set of functional modules. In this chapter, I have argued that as a working hypothesis it makes sense to distinguish between three broad types of control: Type 1 control between functional modules, Type 2 control within a given module, and Type 3 control of the strategies or methods brought to bear on task performance. To the extent that the cognitive system is more than a bushel of independent mechanisms, then we need to understand how these mechanisms are integrated to achieve cognitive functionality. Integrated models of cognitive systems are a necessary tool in understanding a cognitive system that is integrated.

Acknowledgments

The writing of this chapter benefited greatly from the many discussions and thoughtful reviews of Hansjörg Neth, Chris Sims, and Mike Schoelles. The writing was supported by Grant F49620-03-1-0143 from the Air Force Office of Scientific Research.

Notes

1. A reference to the famous cartoon by Sydney Harris in which one mathematician has filled the blackboard with an impressively complex looking formula except a bracketed middle area where this phrase occurs. The onlooker says, "I think you should be more explicit here in step 2."

2. Because of its long and close association with the ACT-R architecture of cognition (Anderson, 1993; Anderson et al., 2004; Anderson & Lebiere, 1998), rational activation theory is often regarded as "ACT-R's theory of memory." Although in my view this is a meritorious association, it has somehow diminished the status of rational activation as a stand-alone local theory.

References

Anderson, J. R. (1983). *The architecture of cognition.* Cambridge, MA: Harvard University Press.

———. (1987). Methodologies for studying human knowledge. *Behavioral and Brain Sciences, 10*(3), 467–477.

———. (1990). *The adaptive character of thought.* Hillsdale, NJ: Erlbaum.

———. (1991). Is human cognition adaptive? *Behavioral and Brain Sciences, 14*(3), 471–517.

———. (1993). *Rules of the mind.* Hillsdale, NJ: Erlbaum.

———, Bothell, D., Byrne, M. D., Douglas, S., Lebiere, C., & Quin, Y. (2004). An integrated theory of the mind. *Psychological Review, 111*(4), 1036–1060.

———, & Lebiere, C. (Eds.). (1998). *Atomic components of thought.* Hillsdale, NJ: Erlbaum.

———, & Schooler, L. J. (1991). Reflections of the environment in memory. *Psychological Science, 2,* 396–408.

Ballard, D. H., Hayhoe, M. M., & Pelz, J. B. (1995). Memory representations in natural tasks. *Journal of Cognitive Neuroscience, 7*(1), 66–80.

Brooks, R. A. (1991). Intelligence without representation. *Artificial Intelligence, 47*(1–3), 139–159.

Card, S. K., English, W. K., & Burr, B. J. (1978). Evaluation of mouse, rate-controlled isometric joystick, step keys and text keys for text selection on a CRT. *Ergonomics, 21*(8), 601–613.

Cook, T. D., & Campbell, D. T. (1979). *Quasi-experimentation: Design and analysis issues for field settings.* Chicago: Rand McNally.

Fitts, P. M. (1954). The information capacity of the human motor system in controlling the amplitude of movement. *Journal of Experimental Psychology, 47*(6), 381–391.

Gray, W. D. (2000). The nature and processing of errors in interactive behavior. *Cognitive Science, 24*(2), 205–248.

———, & Boehm-Davis, D. A. (2000). Milliseconds matter: An introduction to microstrategies and to their use in describing and predicting interactive behavior. *Journal of Experimental Psychology: Applied, 6*(4), 322–335.

———, & Salzman, M. C. (1998). Damaged merchandise? A review of experiments that compare usability evaluation methods. *Human-Computer Interaction, 13*(3), 203–261.

———, Sims, C. R., Fu, W.-T., & Schoelles, M. J. (2006). The soft constraints hypothesis: A rational analysis approach to resource allocation for interactive behavior. *Psychological Review, 113*(3).

Kieras, D. E., & Meyer, D. E. (1997). An overview of the EPIC architecture for cognition and performance with application to human-computer interaction. *Human-Computer Interaction, 12*(4), 391–438.

Landauer, T. K., Laham, D., & Derr, M. (2004). From paragraph to graph: Latent semantic analysis for information visualization. *Proceedings of the National Academy of Sciences of the United States of America, 101,* 5214–5219.

Lemaire, B., & Denhiére, G. (2004). Incremental construction of an associative network from a corpus. In K. D. Forbus, D. Gentner, & T. Regier (Eds.), *26th Annual Meeting of the Cognitive Science Society, CogSci2004.* Hillsdale, NJ: Erlbaum.

Massaro, D. W., Cohen, M. M., Campbell, C. S., & Rodriguez, T. (2001). Bayes factor of model selection validates FLMP. *Psychonomic Bulletin & Review, 8*(1), 1–17.

McElree, B. (2006). Accessing recent events. *The Psychology of Learning and Motivation, 46,* 155–200.

Meyer, D. E., Smith, J. E. K., Kornblum, S., Abrams, R. A., & Wright, C. E. (1990). Speed-accuracy trade-offs in aimed movements: Toward a theory of rapid voluntary action. In M. Jeannerod (Ed.), *Attention and performance, XIII: Motor representation and control* (pp. 173–225). Hillsdale, NJ: Erlbaum.

Minsky, M. (1985). *Society of mind.* New York: Touchstone.

Newell, A. (1973). You can't play 20 questions with nature and win: Projective comments on the papers of this symposium. In W. G. Chase (Ed.), *Visual information processing* (pp. 283–308). New York: Academic Press.

———. (1990). *Unified theories of cognition.* Cambridge, MA: Harvard University Press.

———. (1992). Precis of unified theories of cognition. *Behavioral and Brain Sciences, 15*(3), 425–437.

Pitt, M. A., Kim, W., & Myung, I. J. (2003). Flexibility versus generalizability in model selection. *Psychonomic Bulletin & Review, 10*(1), 29–44.

———, & Myung, I. J. (2002). When a good fit can be bad. *Trends in Cognitive Sciences, 6*(10), 421–425.

———, Myung, I. J., & Zhang, S. B. (2002). Toward a method of selecting among computational models of cognition. *Psychological Review, 109*(3), 472–491.

Raymond, J. E., Shapiro, K. L., & Arnell, K. M. (1992). Temporary suppression of visual processing in an RSVP task—an attentional blink. *Journal of Experimental Psychology—Human Perception and Performance, 18*(3), 849–860.

Roberts, S., & Pashler, H. (2000). How persuasive is a good fit? A comment on theory testing. *Psychological Review, 107*(2), 358–367.

Rodgers, J. L., & Rowe, D. C. (2002). Theory development should begin (but not end) with good empirical fits: A comment on Roberts and Pashler (2000). *Psychological Review, 109*(3), 599–604.

Salvucci, D. D. (2006). Modeling driver behavior in a cognitive architecture. *Human Factors*, 48(2), 362–380.

———, & Gray, R. (2004). A two-point visual control model of steering. *Perception*, 33(10), 1233–1248.

Simon, H. A. (1992). What is an "explanation" of behavior? *Psychological Science*, 3(3), 150–161.

Sutton, R. S., & Barto, A. G. (1998). *Reinforcement learning*. Cambridge, MA: MIT Press.

Turney, P. (2001). Mining the Web for Synonyms: PMI-IR versus LSA on TOEFL. In L. De Raedt & P. Flach (Eds.), *Proceedings of the Twelfth European Conference on Machine Learning (ECML-2001)* (pp. 491–502). Freiburg, Germany.

Van Zandt, T. (2000). How to fit a response time distribution. *Psychonomic Bulletin & Review*, 7(3), 424–465.

2

Cognitive Control in a Computational Model of the Predator Pilot

Kevin A. Gluck, Jerry T. Ball, & Michael A. Krusmark

This chapter describes four models of cognitive control in pilots of remotely piloted aircraft. The models vary in the knowledge available to them and in the aircraft maneuvering strategies that control the simulated pilot's interaction with the heads-up display. In the parlance of Gray's cognitive control taxonomy (chapter 1, this volume), these models are Type 3 (knowledge/strategy) variants. The first two models are successive approximations toward a valid model of expert-level pilot cognitive control. The first model failed because of a naïve flight control strategy, and the second succeeded because of an effective flight control strategy that is taught to Air Force pilots. The last two models are investigations of the relative contributions of different major components of our more successful model of pilot cognitive control. This investigation of knowledge and strategy variants produces an anomalous result in relative model performance, which we explore and explain through a sensitivity analysis across a portion of the Type 3 parameter space. The lesson learned is that seemingly innocuous assumptions at the Type 3 level can have large impacts in the performance of models that simulate human cognition in complex, dynamic environments.

The traditional scientific strategy in cognitive and experimental psychology has been to isolate specific perceptual, cognitive, and motor processes in simple, abstract laboratory tasks in order to measure and model certain effects and phenomena of interest. The strength of this approach is that it facilitates the development of detailed theories of the subprocesses and subcomponents of the total cognitive system. In Gray's "Composition and Control of Integrated Cognitive Systems" (chapter 1, this volume), these theories are of the Type 1 (central control) and Type 2 (functional processes) variety.

The major limitation of this "divide and conquer" approach is that it avoids, and therefore does not help us arrive at a better understanding of, the myriad interactions that exist among the subprocesses and subcomponents within the system. Wickens (2002) made the point that although a great deal of laboratory research has taken place to isolate and understand perceptual, cognitive, and psychomotor processes, "modeling the complex interactions among these phenomena

remains a critical challenge...to psychological researchers who are interested in 'scaling up' their theories to real-world problems" (p. 132).

An important role of cognitive architectures, such as those described elsewhere in this volume, is that they serve an integration function across these otherwise isolated models and theories of subprocesses and subcomponents. Architectures allow for computational exploration and quantitative prediction of the consequences of different assumptions about how the components interact. Developing a model with a cognitive architecture requires adding knowledge to it, which is the Type 3 portion of Gray's taxonomy. The complexity of the task largely determines the complexity of the Type 3 theory implemented (i.e., how much and what kind of knowledge is required) in a model. As we will see, Type 3 theories of cognitive control can involve complicated interactions that have unintended consequences for model performance.

This chapter describes four variations on models of cognitive control in pilots of remotely piloted aircraft.

The models vary in the knowledge available to them and in the aircraft maneuvering strategies that control the simulated pilot's interaction with the heads-up display (HUD). The first two models are successive approximations toward a valid model of expert-level pilot cognitive control. The last two models are investigations of the relative contributions of different major components of our more successful Type 3 theory of pilot cognitive control. This investigation of knowledge and strategy variants produced an anomalous result in relative model performance, which we investigate and explain through a sensitivity analysis across a portion of the Type 3 parameter space. The lesson learned is that seemingly innocuous assumptions at the Type 3 level can have large impacts in the performance of models that simulate human cognition in complex, dynamic environments. The chapter begins with a description of the chosen domain and task environment.

Predator Synthetic Task Environment

Piloting remote aircraft is the general domain context for the models described in this chapter. Empirical research and model development and validation have been possible through the use of a synthetic task environment (STE) that simulates the flight dynamics of the Predator RQ-1A, which is a reconnaissance aircraft. The simulation has a realistic core aerodynamic model; it has been used by the 11th Reconnaissance Squadron to train Predator pilots at Creech Air Force Base (previously known as Indian Springs Air Field). Built on top of the Predator simulation are three synthetic tasks: the basic maneuvering task, in which a pilot must make very precise, constant-rate changes to the aircraft's airspeed, altitude, and/or heading; the landing task, in which a pilot must fly a standard approach and landing; and the reconnaissance task, in which a pilot must maneuver the uninhabited air vehicle

(UAV) to obtain simulated video of a ground target through a small break in cloud cover.

A description of the philosophy and methodology used to design the Predator STE can be found in Martin, Lyon, and Schreiber (1998). Schreiber, Lyon, Martin, and Confer (2002) established the ecological validity of the STE. Their research shows that experienced Predator pilots perform better in the STE than highly experienced pilots who have no Predator experience, indicating that the STE taps Predator-specific pilot skill.

The focus of the empirical research and modeling described in this chapter is the basic maneuvering task, the implementation of which was inspired by an instrument flight task originally designed at the University of Illinois at Urbana–Champaign (Bellenkes, Wickens, & Kramer, 1997).[1] The task requires the operator to fly seven distinct instrument flight maneuvers. At the beginning of each maneuver is a 10-s lead-in, during which the operator is supposed to stabilize the aircraft in straight and level flight. Following the lead-in, a timed maneuver of 60 or 90 s begins, and the operator maneuvers the aircraft by making constant rate changes to altitude, airspeed, and/or heading, depending on the maneuver, as specified in Table 2.1. The goal of each maneuver is to minimize the deviation between actual and desired performance on airspeed, altitude, and heading.

During the basic maneuvering task, the operator sees only the HUD, which is presented on one of two computer monitors. Instruments displayed from left to right on the HUD monitor (see Figure 2.1) are angle of attack (AOA), airspeed, heading (bottom center), vertical speed, RPM (engine power setting), and altitude. The digital display of each instrument moves up and down in analog fashion as values change. Depicted at the center of the HUD are the reticle and horizon line, which together indicate the pitch and bank of the aircraft.

TABLE 2.1 Performance Goals for the Seven Basic Maneuvers

Maneuver	Airspeed	Heading	Altitude
1	Decrease 67–62 knots	Maintain 0°	Maintain 15,000 feet
2	Maintain 62 knots	Turn right 0–180°	Maintain 15,000 feet
3	Maintain 62 knots	Maintain 180°	Increase 15,000–15,200 feet
4	Increase 62–67 knots	Turn left 180–0°	Maintain 15,200 feet
5	Decrease 67–62 knots	Maintain 0°	Decrease 15,200–15,000 feet
6	Maintain 62 knots	Turn Right 0–270°	Increase 15,000–15,300 feet
7	Increase 62–67 knots	Turn left 270–0°	Decrease 15,300–15,000 feet

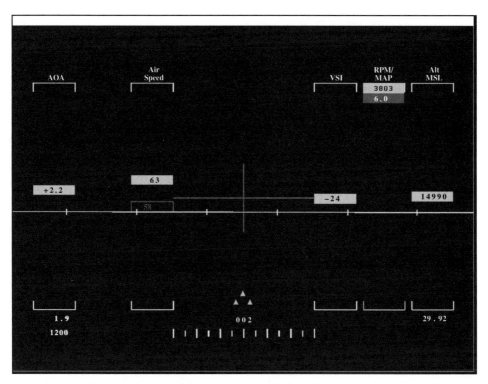

FIGURE 2.1 Predator synthetic task environment heads-up-display, with simulated video blacked out for instrument flight in the basic maneuvering task.

On a second monitor, there is a trial clock, a bank angle indicator, and a compass, which are presented from top to bottom on the far-right column of Figure 2.2. During a trial, the left side of the second monitor is blank. At the end of a trial, a feedback screen appears on the left side of the second monitor. The feedback depicts deviations between actual and ideal performance on altitude, airspeed, and heading plotted across time, as well as quantitative feedback in the form of root mean squared deviations (RMSDs).

The selection of Predator pilot modeling as a target domain for our cognitive modeling research was a decision made on the basis of its relevance to the needs of the U.S. Air Force. As an increasingly important military asset, Predator operations have been a strategic investment area within the Air Force Research Laboratory for several years. Development of the Predator STE, in fact, goes back to New World Vistas investments by the Air Force Office of Scientific Research (AFOSR) in the mid- to late 1990s. During this period, AFOSR invested in the development of a variety of different STEs related to command and control and Predator operations. These STEs were

intended to stimulate basic and applied research of value to the air force by government, academic, and industry researchers. The Predator STE provides an unclassified, yet highly relevant, simulation with built-in research tasks and data-collection capabilities that facilitate its use as a research tool.[2]

Another factor in our decision to use the Predator STE as a cognitive modeling research context was that it seemed to be an appropriately ambitious increment in the maturation of the cognitive modeling community, away from comparatively simple, static domains to more complex, dynamic ones. Some exemplary cognitive modeling efforts that had already been pushing the envelope in the direction of greater complexity and dynamics included Lee's models of cognitive processes in the Kanfer–Ackerman air traffic control task (Lee & Anderson, 2000, 2001), Schoelles's models of performance in the Argus task (Schoelles & Gray, 2000), and Salvucci's driver models (Salvucci, 2001; Salvucci, Boer, & Liu, 2001). We drew considerable inspiration because these scientists had previously used ACT-R (adaptive control of thought–rational) to develop computational explanations of human performance in

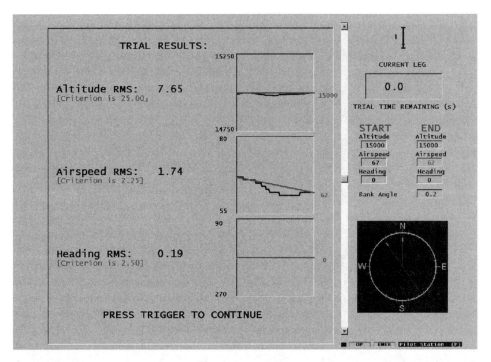

FIGURE 2.2 Predator synthetic task environment feedback screen for Maneuver 1 in the basic maneuvering task (decrease airspeed while holding altitude and heading constant).

those contexts. The next section describes our first attempt at developing an integrated cognitive model for Predator maneuvering.

Naïve Model

Our first model of pilot cognitive control was a failure. We refer to this as the naïve model because the model's cognitive control strategy was naïve, from the perspective of actual aircraft control. We faced three limitations in this first implementation, which we consider the reasons we ended up with a naïve implementation of pilot cognition. We review these briefly before describing the model implementation itself.

Limiting Factors

First, we were unable to find existing models in the modeling literature that could serve as guidance for the implementation of the model, or even as sources of code for model reuse. The first place we looked was the ACT-R Web site and workshop proceedings, but this produced no hits for existing ACT-R pilot models.

The search then broadened to other cognitive architectures. We had access to the code for the well-known Tac-Air Soar system (Jones et al., 1999), which we inspected in the hopes it would provide the basic maneuvering functionality we needed. After looking over the code, however, we determined that the implementation of Tac-Air Soar was not at the right grain size for our purposes. For instance, one of our primary concerns was how to implement an appropriate instrument scan strategy. It turns out Tac-Air Soar does not actually scan instrument representations in order to take in instrument values because that level of perceptual fidelity was not important in the applications for which it was intended. Even if Tac-Air Soar did model visual scan on instruments, the location and design of instruments available to it (in an F-16) would have been very different than the location and design of the HUD instruments in the Predator. Most of the knowledge actually required to pilot the aircraft through the tightly controlled, constant-rate-of-change maneuvers in the basic maneuvering task would be different as well because the flight dynamics of F-16s are very different than those of a Predator. Thus, unfortunately, Tac-Air Soar was not helpful in developing this model.

A second limiting factor was that the Predator STE was not originally developed to be used as a cognitive modeling test bed. Code that provided access to Predator STE state data had to be implemented and tested, which required a process model. We were concerned about whether we would be successful in interfacing ACT-R to the Predator STE because we needed to get some kind of model in place, and quickly. Solutions to the interface challenges are described briefly in Ball and Gluck (2003), and a review of them here is unnecessary. Overcoming the model-simulation interface issues required significant time and attention and we needed to develop a process model as quickly as possible to evaluate the interface implementation.

The third limiting factor was that we did not have the benefit of a subject matter expert (SME) on the modeling team. We clearly appreciated that this put us at a significant disadvantage, and we understood that the benefits of SME involvement in task analysis and model development are well documented (see Schraagen, Chipman, & Shalin, 2000, and especially Gray & Kirschenbaum, 2000). However, we reasoned that we could not let the absence of an SME on the team bring the cognitive modeling research completely to a stop. We resolved to rectify the situation as soon as an available and willing pilot could be found and, in the meantime, to press ahead.

With no leads on architecture-based process models of pilot cognition to reuse, time pressure mounting due to the distraction of model-simulation interface concerns, and no pilot SME on the modeling team, we set about implementing the model on the basis of our own analysis of the basic maneuvering task demands. A simple characterization of the task is that it requires visual attention to and encoding of flight instruments to maintain situation awareness regarding deviations from desired flight parameters, followed by adjustments to the aircraft flight controls to correct for any deviations. The goal of each maneuver and the desired aircraft parameters at timing checkpoints within the maneuvers are clearly defined for participants and could easily be represented in ACT-R's declarative knowledge. Appropriate rules for adjusting in response to deviations were naturally represented in productions. A challenging decision, however, was how to control the flow of visual attention across the instruments. What, precisely, should the model look at and when?

Unit Task Representation

The unit task construct was proposed by Card, Moran, and Newell (1983), and at the time we were developing our naïve model, the concept of a *unit task* had recently played a central role in the development of ACT-R models for simulated versions of radar operation (Schoelles & Gray, 2000), air traffic control (Lee & Anderson, 2001), and driving (Salvucci, Boer, & Liu, 2001). The success of this approach prompted us to consider how a flight maneuver might be decomposed into unit tasks, and we concluded that the three flight performance parameters (airspeed, altitude, and heading) that are explicitly called out in the task instructions and feedback mapped nicely into unit tasks. Thus, we had three unit tasks for any given flight maneuver, one each for airspeed, altitude, and heading. Figure 2.3 is a conceptual representation of these unit tasks, which support the superordinate goal of flying the aircraft.

This performance instrument–based unit task representation was the central cognitive control structure in our initial implementation of the basic maneuvering model. The model had a superordinate goal of flying the aircraft, from which it would select randomly among the three unit tasks. After completing the selected unit task, it would return to the goal of flying the aircraft, randomly select another unit task, and repeat. This cycle recurred continuously through the duration of a maneuver.

Within a unit task, the model completed a series of activities, always in the same sequence. First was the retrieval of the instrument location. An assumption was made in this model that experienced Predator pilots would have an existing declarative representation of the visual locations of the flight instruments. Selection of the unit task determined the target instrument, knowledge of the target instrument led to retrieval of the appropriate instrument location, and this resulted in firing the standard ACT-R productions for finding, attending, and encoding the value of the selected instrument.

The next step in the unit task was to retrieve the desired instrument value for that maneuver from declarative memory. The task instructions clearly defined the desired instrument values at timing checkpoints in each maneuver. The model then decided whether an action was required. This decision was made on the basis of the magnitude of deviations from desired performance instrument values. If the model

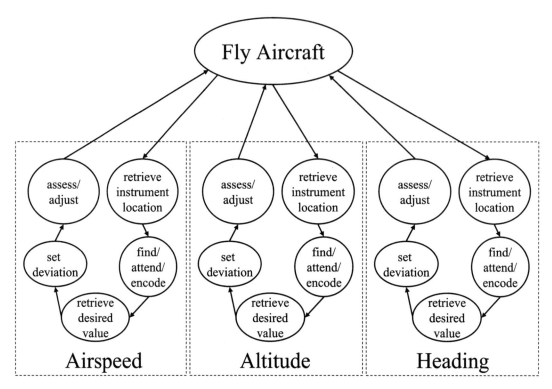

FIGURE 2.3 Conceptual flow of cognitive control in the naïve model, which attended to three performance instruments (airspeed, altitude, and heading) in random order and never attended to control instruments.

decided the performance parameters were acceptable (near desired values), it would change the goal back to flying the aircraft and select randomly again from the unit tasks. However, if the model detected a sufficiently significant deviation from the desired performance instrument values (e.g., attended to altitude and determined the plane was too high), it would change the goal to reduce the deviation, and then do so by adjusting the stick (pitch, bank) and/or the throttle (power). Following an adjustment, the model returned to the fly aircraft goal, and the unit task selection process began again.

Model Validity

To people with no aviation experience, such as the authors and most of the cognitive modeling community, the control strategy implemented in our naïve pilot model seems quite reasonable. At the very least, it provided a baseline model useful for testing and improving the implementation of the model-system interaction, and we were able to get performance data out of the model, which we could compare to human

Predator pilot data from an earlier study (Schreiber et al., 2002).

It was through careful comparison of this model with the Predator pilot data that we became aware of the problems created by the cognitive control processes implemented in the model. Outcome data showed model performance levels (RMSDs) that were consistently worse than those achieved by the human pilots. Inspection of the within-maneuver dynamics revealed that the model was generally too slow in responding to deviations from ideal performance, and when it did respond, the model tended to overcorrect. To get the model to respond more quickly, we tried decreasing the cognitive cycle time in ACT-R from 50 ms down to 10 ms. Although this change did have the desired effect of producing faster responses, it also had the negative effect of producing more overcorrections per maneuver, rather than eliminating them.

Just about the time that we were becoming concerned about the inability of this model to fit the human performance data and frustrated with our inability to understand why it would not, an expert pilot (an SME!) joined our model development team.

This SME had been a test pilot for the air force and had more than 3,000 hr of flight time in about 30 different aircraft. Amazingly, he also had a master's degree in computer science and a sincere interest in cognitive modeling. He was the perfect addition to the team at a critical juncture.

We learned that our initial pilot model was a terrific model of a terrible pilot. The critical flaw in the implementation was that the model was completely reactive. It would perceive an *existing* deviation from desired performance and react to that. In pilot vernacular, the model was always "behind the aircraft," and this produced model-induced oscillations, the computational model version of the phenomenon of pilot-induced oscillations (Wickens, 2003). Fortunately, our SME was not only able to diagnose the problems with the existing model, but he also prescribed a more pilot-like implementation of cognitive control that has proven considerably more successful at replicating expert human pilot processes and performance levels.

Control Focus and Performance Model

Control and Performance Concept

A critical piece of missing knowledge, for nonpilot cognitive modelers, was that air force pilots are taught to maneuver aircraft using a method known as the "control and performance concept." Before describing this aircraft control method, it is helpful to introduce the two key terms. Basic aircraft flight instrumentation falls into two categories: *performance* instruments and *control* instruments. Performance instruments reflect the behavior of the aircraft and include airspeed, heading, altitude, and vertical speed. Control instruments, such as pitch, bank, and engine speed (RPM) directly reflect the settings of the controls (i.e., stick and throttle), which in turn affect the behavior of the aircraft. Adjustments to the controls have a first-order (immediate) effect on the control instrument values and a second-order (delayed) effect on the performance instrument values.

The *Air Force Manual 11-217* on instrument flight (2000) provides the following high-level characterization of the recommended aircraft control method: "The control and performance concept of attitude instrument flying requires you to establish an aircraft attitude or power setting on the control instruments that should result in the desired aircraft performance" (p. 13).

Note the emphasis on the pilot's preexisting declarative knowledge of desired aircraft performance and appropriate attitude (bank, pitch) and power settings to establish desired control settings at the beginning of a maneuver. The benefit of using the control and performance concept to fly an aircraft is that it allows the pilot to stay "ahead" of the aircraft, rather than falling behind it. Knowledge of appropriate control settings allows the pilot to make control actions that will have predictable and desirable second-order effects on the aircraft's performance. For example, a pitch of 3 degrees and an engine RPM of 4,300 will maintain straight and level flight of the Predator at 67 knots over a range of altitudes and external conditions. The expert pilot need only set the appropriate pitch and engine RPM to obtain the desired performance, subject to monitoring and adjustment based on variable flight conditions like wind, air pressure, and other such perturbations.

In addition to having knowledge of desirable control instrument settings, expert pilots are vigilant in maintaining good aircraft situation awareness at all times. This is accomplished with a visual *cross-check* of instruments. The cross-check is a visual scan pattern across the instruments, intended to continuously update working memory regarding the state of the aircraft. Pilots typically employ either a hub-and-spoke pattern or a round-robin pattern, or some mixture of the two. During this cross-check, it is important not to focus too much attention on performance instruments and to keep control instruments in the cross-check.

The combination of this process description in *Air Force Manual 11-217* on instrument flight and ongoing guidance from our SME proved informative regarding the reimplementation of the basic maneuvering model. Figure 2.4 is a conceptual representation of the flow of cognitive control in the control focus and performance model (Model CFP).

Like expert pilots, Model CFP has knowledge of appropriate control instrument settings and uses that knowledge in its flight control strategy. At the beginning of a maneuver, the model focuses on establishing desired control instrument settings, and subsequently it performs a cross-check of both control and performance instruments. In the basic processing cycle, Model CFP selects an instrument to attend, finds it on the HUD, shifts attention to the instrument, encodes the value of the instrument, retrieves the desired instrument value from memory, assesses the encoded value against the desired value and sets a

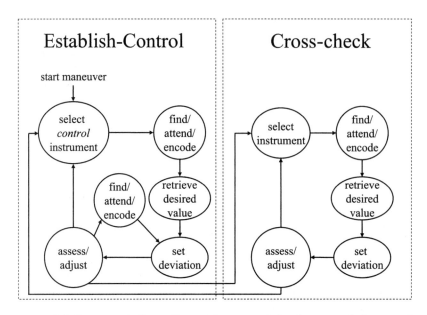

FIGURE 2.4 Conceptual flow of cognitive control in the control focus and performance model, which is based on the flight strategy taught in the *Air Force Manual 11-217*.

qualitative deviation, and makes an appropriate control adjustment.

In this more sophisticated implementation of cognitive control, there are actually two unit tasks: *establish-control* and *cross-check*. At the beginning of a maneuver, during the *establish-control* unit task (left side of Figure 2.4), the model focuses on each of the three control instruments (one at a time) and adjusts the stick and/or throttle until all three control instruments are at or very near their desired settings. Note the subloop in which Model CFP focuses on a single control instrument until the value of that instrument is qualitatively close enough to the desired value (e.g., a very small deviation). After the desired control settings are achieved, Model CFP switches to the *cross-check* unit task (right side of Figure 2.4), which includes all control and performance instruments. Once in the cross-check cycle, the model makes adjustments to the controls in response to perceived deviations from desired flight performance parameters and minor deviations in control settings. If the model detects a large deviation from desired control settings (which can result from changes to the controls caused by deviations from desired flight performance parameters), it will kick back to the establish-control unit task and reestablish control settings that are appropriate for the desired aircraft performance at that point in the maneuver.

Performance Comparison

To assess the validity of Model CFP, we collected performance data from seven aviation SMEs on each of the seven basic maneuvers (see Table 2.1). Subjects completed the seven maneuvers in order, completing all Maneuver 1 trials before moving on to Maneuver 2, and so on. The number of trials completed in each maneuver varied from 14 to 24, depending on difficulty, to minimize simultaneously the time requirements of the study (SMEs are in high demand) and to provide adequate opportunity for performance to improve and stabilize within each maneuver.

Recall from Figure 2.2 that the task has performance measures built into it, in the form of deviations from ideal performance on airspeed, altitude, and heading. To measure overall performance, a composite measure was computed from deviations between actual and desired airspeed, altitude, and heading. Results suggest that the model compares well with SMEs on overall mean performance and mean performance by maneuver (Gluck, Ball, Krusmark, Rodgers, & Purtee, 2003).

One result is an effect of maneuver complexity, wherein performance levels worsen for both the model and SMEs as maneuvers become more complex. The model predicts this effect even though it was not intentionally designed to do so. In another analysis, we computed goodness-of-fit estimates between model and SME performance and compared them to fit estimates of each SME's performance to the rest of the SMEs. As it turns out, the model is actually a better fit to the SME data than one of the SMEs is to the SME data. That fact, combined with the model's fits to overall expert performance and expert performance by maneuver, suggested that the model is a valid approximation to expert performance on this task (Gluck et al., 2003).

Process Comparison

Our interest is in getting not only the performance level right but also that the performance level is achieved using processes and procedures that reflect those used by pilots flying these maneuvers. Thus, we collected fine-grained process measures from SMEs while they were performing the basic maneuvering task. These measures include retrospective verbal reports, concurrent verbal reports, and eye movements. Retrospective verbal reports from the SMEs suggest they were using the control and performance strategy when performing the maneuvering task and that their application of the strategy was dependent on the demands of the task (Purtee, Krusmark, Gluck, Kotte, & Lefebvre, 2003).

More recently we have compared how the model and SMEs allocate their attention among instruments. We have done this by comparing the results from three measures: the frequency of Model CFP's shifts of visual attention to different instruments, the frequency of SME eye movements to different instruments, and the frequency of references to different instruments in the concurrent verbal protocols. Shifts of the model's visual attention are easy to record simply by keeping a running count of attention to each instrument and sending those data to an output file at the end of each trial. Analysis of human eye tracking and concurrent verbal protocol data requires a more complicated process, so we say a little about each of those next, before presenting the results.

El Mar's Vision 2000 Eye-Tracking System was used to collect eye movement data. The system is described in detail in Wetzel and Anderson (1998). The software estimates eye point of regard by recording horizontal and vertical eye position from the relative positions of corneal and pupil center reflections, and merging these data with a recording of the visual scene. El Mar's Fixation Analysis Software Technology was used to define eye fixations and generate data files that contain gaze sequences and times. Fixations were defined as periods of time during which the eye was stationary for a minimum of 167 ms, and the eye was considered stationary below a velocity of 30 deg/s. Fixations were associated with specific instruments on the UAV-STE interface when they occurred within the boundaries of a region of interest defined for each instrument.

Concurrent verbal protocols are another source of high-density data for studying human cognitive processes (Ericsson & Simon, 1993). The SMEs in our study provided concurrent verbal protocols on every other trial. Research associates transcribed, segmented, and coded the verbalizations. The coding system included five categories ("goals," "control instruments," "performance instruments," "actions, and "other"), which subdivided into 22 different codes. For instance, the control instruments category includes codes for "bank angle," "pitch," "RPM," "trim," and "general control verbalizations." To give the reader a sense for the grain size of the segmentation and the nature of the verbalizations, example statements coded as bank angle verbalizations include, "bank looks good," "ya, I've lost a little too much bank," and "fourteen degrees of bank." In all, there were 15,548 segments. One research assistant coded all of the segments and another coded a third of them as a reliability check. Agreement was high between coders ($\kappa = 0.875$).

Figure 2.5 compares SME and model attention to instruments relevant to the lateral axis of flight (changing the direction the plane is heading). These instruments include the heading indicator, the bank angle indicator, and the compass. Maneuvers 2, 4, 6, and 7 require a heading change (see Table 2.1); therefore, one would expect attention to instruments that display information about the plane's heading to be much greater on these maneuvers relative to nonheading change maneuvers. This is exactly what we find. Attention to lateral axis is greater on heading change relative to nonheading change maneuvers for model fixations, $F(1, 133) = 750.42$, $p < .001$; SME eye fixations $F(1, 246) = 1400.44$; $p < .001$; and SME verbalizations $F(1, 246) = 1084.07$, $p < .001$.

Previously we had shown that the level of *performance* we get out of Model CFP compares well with that of expert pilots, and now we also see that the *processes* that drive the model's behavior compare well to the processes used by expert pilots, as supported by verbal protocol and eye movement data.

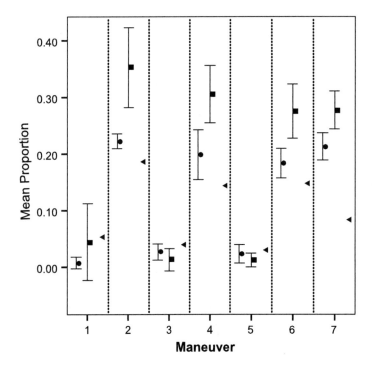

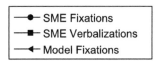

FIGURE 2.5 Comparison of proportions of human and model attention to the lateral axis across basic maneuvers in the Predator synthetic task environment. Instruments associated with the lateral axis include the heading indicator, the bank angle indicator, and the compass.

Variations in Knowledge and Strategy

Satisfied that we had a model (CFP) that is a good approximation of the cognitive control strategy that expert pilots use to maneuver an aircraft, we decided to experiment with some variations on Model CFP to better understand what the relative contributions are of different components of the Type 3 theory it represents. This was a test of the necessity of two components of the theory: (1) the control-focus strategy at the beginning of a maneuver and (2) the knowledge of control instruments and their target values. What happens to performance if we remove those components of the theory? Are they necessary in order to reproduce expert-level performance?

Three Type 3 Theories

We manipulated the model's knowledge of flying and instrument cross-check strategy, creating three distinct model variants (Ball, Gluck, Krusmark, & Rodgers, 2003):

1. The *control focus and performance model* (Model CFP) is the successful model described in the previous section. Model CFP knows what the desired control instrument settings are for specific maneuvers. It adheres to the control and performance concept by focusing on control instruments until they are properly set and then maintaining an effective cross-check of all control and performance instruments for the remainder of the maneuver.

2. The *control and performance model* (Model CP) is Model CFP, minus the F. It knows what the desired control instrument settings are for specific maneuvers, and it maintains an effective cross-check of all control and performance instruments throughout a maneuver, but it does not *focus* on getting control instruments set at the beginning of a maneuver.

3. The *performance only model* (Model P) is Model CFP, minus the F and minus the C. It does not know the desired control instrument settings. Of the three control instruments (bank,

pitch, and power), Model P attends to only bank angle, and this is only because the instructions explicitly mention bank angle as important for maneuvers that require a heading change. Model P's cross-check includes the performance instruments and bank angle, but not pitch or power.

With these Type 3 knowledge and strategy variants ready to run, we set out to compare their performance levels across the seven maneuvers to each other and to our sample of SME data. Here we are using SME data from the last 10 trials they completed in each maneuver. We use only the last 10 trials because there is a steep learning curve at the beginning of each maneuver, as people become familiar with the demands of that particular maneuver. Consequently, we wanted to compare these models with SME data that had stabilized to a point where little to no learning was observed. Presumably, the best and most stable performance we had from SMEs was on the last 10 trials for each maneuver. However, even on these trials we observed occasional outliers in performance. Extreme outliers on altitude, airspeed, and heading were identified by comparing studentized-deleted residuals to the critical value in Student's t-distribution with $\alpha = .05$ using a Bonferroni adjustment. Sixteen of 490 trials (last 10 trials $\times$ 7 maneuvers $\times$ 7 subjects) had extreme outliers on altitude, airspeed, and/or heading RMSDs.

Using the SME data from the last 10 trials screened for extreme outliers, we then computed a composite measure of performance. RMSDs for altitude, airspeed, and heading were converted to z-scores and then added together on each trial, resulting in a standardized sum RMSD. A similar composite measure was then computed for the model data. Using a simple linear transformation, model data for each performance measure on each trial were converted to z-scores using the same means and standard deviations that were used to compute z-scores for the SME data. This was done because we needed the human and model data on the same scale, so they were directly comparable. Standardized sum RMSDs were then computed for the models by adding z-scores for altitude, airspeed, and heading RMSDs.

Figure 2.6 shows results from all three of these model variants and the SME data, separately by

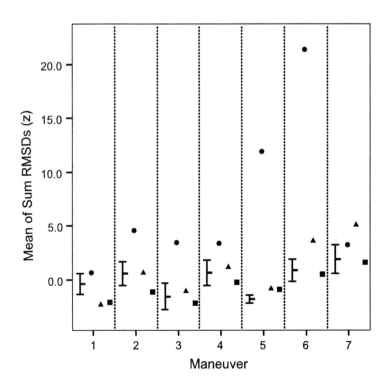

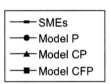

FIGURE 2.6 Comparison of composite performance levels among the three Type 3 theories (model knowledge variants) and the subject matter experts (SMEs).

maneuver. Better performance is toward the bottom of the graph (low on the ordinate) and worse performance is closer to the top of the graph (high on the ordinate). Results suggest that the performance of model CFP was most similar to expert pilots. This serves as additional validation of the theory that an appropriate model of expert pilot performance should include both knowledge of appropriate control settings and also the strategy of focusing on control settings until getting them set at the beginning of a maneuver.

Regarding model performances relative to each other, note that in general the pattern is clear and consistent, with Model CFP performing the best, Model CP a little worse, and Model P a little worse still. It is exactly the pattern of data one would expect as the cognitive control process implemented in the models deviates further and further from what is recommended in the *Air Force Manual 11-217* on instrument flight. Maneuver 7, however, shows a deviation from this pattern. The data in Maneuver 7 are noteworthy because Model P, which clearly had been the worst performer up until this point, actually performs slightly *better* than Model CP and nearly as well as Model CFP.

One might expect an interaction between maneuver difficulty and strategy, such that harder maneuvers, such as Maneuver 6, would exacerbate the negative impact of a poor cognitive control strategy like the one implemented in Model P. However, one also would expect such an effect to continue into Maneuver 7. According to our simulation results, it does not. What could explain this result?

Sensitivity Analysis

When running the model batches that produced the data for Figure 2.6, we had attempted to hold constant all Types 1, 2, and 3 cognitive control theory characteristics within a model type, across maneuvers, so that the only thing driving differences in performance (within a model type) would be changes in the maneuvers themselves. However, on reflection, we realized that a seemingly innocuous motor movement variable was, in fact, allowed to vary across maneuvers. This was due to an interesting lesson learned during the development and testing of the model variants: The models' initial manual motor movements (i.e., the direction and magnitude of the very first stick and throttle adjustments) at the onset of a maneuver could

have a significant effect on outcome. Through trial and error, and with guidance from our SME collaborator, we had settled on direction and magnitude values for initial stick and throttle movements that Model CFP would execute at the beginning of each maneuver. Naturally, these were different for each maneuver because the goals of the maneuvers differ. You do not want to push the stick to the left if one of the goals of the maneuver is to fly straight. When we implemented the other model variants (CP and P), these initial motor movements transferred over in the code base, but we did not reevaluate their appropriateness in the context of the new cognitive control strategies. We speculated that perhaps there was an interaction between the higher-level cognitive control strategies implemented in these models and the low-level initial motor movement knowledge that was provided to the model variants.

To test this idea, we conducted a sensitivity analysis that systematically manipulated the size of initial stick pitch and throttle movements made at the onset of Maneuver 7. Because it is assumed that these motor movements are learned, the initial motor movement directionality and magnitude are part of the Type 3 parameter space.

Figure 2.7 shows the results of the sensitivity analysis. The figure requires some explanation. The y-axis is the same composite performance measure described earlier. The x-axis values range from −2 (small movement of the throttle toward the pilot, which decreases power to the aircraft) to −38 (larger movement of the throttle toward the pilot, resulting in a more dramatic decrease in power). The z-axis values range from +12 (pushing the stick away from the pilot, which will pitch the nose of the aircraft down) to −8 (pulling the stick toward the pilot, which will pitch the nose of the aircraft up). The 10 throttle adjustment values and 11 stick pitch adjustment values produce a 110-cell grid. Because of the stochastic character of various Type 1 and Type 2 components of the ACT-R architecture, it is necessary to complete multiple model runs in each cell to achieve a valid mean estimate of performance. We ran each of the three models 30 times in each of the 110 cells, for a total of 9,900 model runs, or about 330 hr of computer processor time (the Predator STE cannot run faster than real time).

The primary result of interest in Figure 2.7 is the "sweet spot" in throttle adjustment values for Model P. Initial throttle adjustments in the range of −10 to −18 tended to result in very good performance. As we move

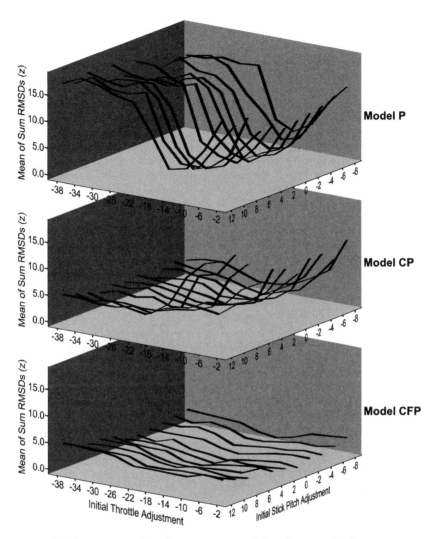

FIGURE 2.7 Sensitivity analysis showing mean model performance levels across an initial motor movement parameter space.

away from that range, in either direction, maneuvering performance gets dramatically worse. By contrast, Model CFP is robust across the entire range of initial stick and throttle adjustments explored here. This is yet more evidence of the effectiveness and adaptability of the control focus and performance strategy for instrument flight.

We do not plot the mean performance data by model variant, but it is clear from visual inspection of Figure 2.7 that Model CFP shows the best performance on average, with Model CP a little worse, and Model P worse still. This brings the relative model performance results for Maneuver 7 in line with the rest of the maneuvering results in Figure 2.6.

All of this begs the question, "Why did we get that anomalous result in the first place?" Figure 2.8 compares model performances when run using the initial motor input settings from Figure 2.6 to the initial motor input settings that produce the *best* performance for Model P in the sensitivity analysis results displayed in Figure 2.7. Note that the values actually used in the original model runs were similar to those that produce the best overall performance for Model P. This reinforces the point that what initially appeared to be superior performance of Model P in Maneuver 7 was actually a result of having stumbled on a sweet spot in the combination of initial motor movements that just happens to produce good model performance,

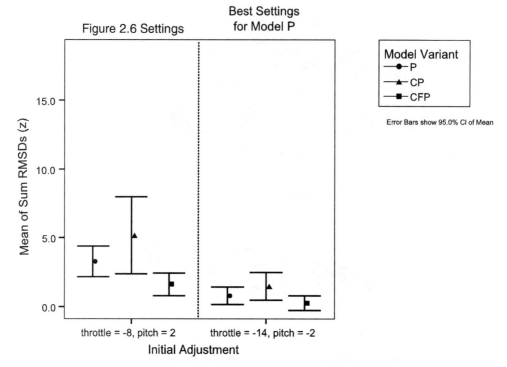

FIGURE 2.8 Initial throttle and stick pitch motor movement adjustments showing that the settings used to produce the anomalous performance of Model P in Maneuver 7 (see Figure 2.6) are very nearly the optimal settings for Model P.

regardless of which flight control strategy is guiding subsequent cognition.

Conclusion

Cognitive control resides at the intersection of architectural constraints and knowledge, at the intersection of Type 1, Type 2, and Type 3 theories. Although the Type 1 and Type 2 theories that form the persistent core of cognitive architectures do provide some constraints on the implementation of models, much of the control structure for cognition in complex, dynamic domains (such as aviation) actually comes from the knowledge, from the Type 3 theory, that modelers must add to the architecture to get situated performance. There are many degrees of freedom in the implementation of that knowledge.

In this chapter, we provided a retrospective on our explorations in the space of Type 3 theories of cognitive control in aircraft maneuvering. The failure of the

naïve model and subsequent success of the control focus and performance model demonstrate that an appropriate high-level cognitive control strategy is critical if the goal is to replicate human performance levels and processes in complex, dynamic tasks. We then took seriously Gray's notion of knowledge as theory and evaluated the necessity of components of that theory by subtracting out portions of the knowledge and strategy of the successful model, thereby creating two new model variants. Results from the comparison of these models to one another provided additional evidence that both knowledge of desired control instruments and a control-focus strategy early in a maneuver are necessary to achieve expert pilot-level performance in a cognitive modeling system with human limitations.

Finally, we conducted a sensitivity analysis across a low-level, motor control portion of the Type 3 parameter space, to explain an anomaly in the previous model variant performance comparison. The lesson from this analysis is that very different portions of a Type 3 theory can interact with each other in subtle ways and have the capacity to produce dramatic effects on behavior.

Acknowledgments

Cognitive model development was sponsored partly by the Air Force Research Laboratory's Warfighter Readiness Research Division and partly by Grant 02HE01COR from the Air Force Office of Scientific Research. Our sincere appreciation to Col. Stu "Wart" Rodgers for working with us to achieve a more sophisticated implementation of cognitive control for our model of expert human pilot performance. Thanks also to Wayne Gray and Mike Schoelles for candid and thorough feedback on earlier versions of this chapter, which is much improved because of their efforts.

Notes

1. The same set of maneuvers were used by Doane and Sohn (2000) in the development of construction-integration models of pilot performance, although we didn't learn of their research in this domain until after we had completed development of the models described in this chapter.

2. More information about the Predator STE, including distribution restrictions and requirements, is available at http://www.mesa.afmc.af.mil/UAVSTE.html

References

Air Force Manual 11–217. (2000). *Vol. 1. Instrument flight procedures.*

Anderson, J. R., Bothell, D., Byrne, M. D., Douglass, S., Lebiere, C., & Qin, Y. (2004). An integrated theory of the mind. *Psychological Review 111*, 1036–1060.

Ball, J. T., & Gluck, K. A. (2003). Interfacing ACT-R 5.0 to an uninhabited air vehicle (UAV) synthetic task environment (STE). In *Proceedings of the 2003 ACT-R workshop.* Retrieved from http://act-r.psy.cmu.edu/workshops/workshop-2003/proceedings/29.pdf

———, Gluck, K. A., Krusmark, M. A., & Rodgers, S. M. (2003). Comparing three variants of a computational process model of basic aircraft maneuvering. In *Proceedings of the 12th Conference on Behavior Representation in Modeling and Simulation* (pp. 87–98). Orlando, FL: Institute for Simulation and Training.

Bellenkes, A. H., Wickens, C. D., & Kramer, A. F. (1997). Visual scanning and pilot expertise: The role of attentional flexibility and mental model development. *Aviation, Space, and Environmental Medicine 68*(7), 569–579.

Card, S. K., Moran, T. P., & Newell, A. (1983). *The psychology of human-computer interaction.* Hillsdale, NJ: Erlbaum.

Doane, S. M., & Sohn, Y. W. (2000). ADAPT: A predictive cognitive model of user visual attention and action planning. *User Modeling and User-Adapted Interaction, 10*(1), 1–45.

Ericsson, K. A., & Simon, H. A. (1993). *Protocol analysis: Verbal reports as data* (Rev. ed.). Cambridge, MA: Bradford Books/MIT Press.

Gluck, K. A., Ball, J. T., Krusmark, M. A., Rodgers, S. M., & Purtee, M. D. (2003). A computational process model of basic aircraft maneuvering. In F. Detje, D. Doerner, & H. Schaub (Eds.), *Proceedings of the Fifth International Conference on Cognitive Modeling* (pp. 117–122). Bamberg: Universitaets-Verlag Bamberg.

Gray, W. D., & Kirschenbaum, S. S. (2000). Analyzing a novel expertise: An unmarked road. In J. M. C. Schraagen, S. F. Chipman, & V. L. Shalin (Eds.), *Cognitive task analysis* (pp. 275–290). Mahwah, NJ: Erlbaum.

Jones, R. M., Laird, J. E., Nielsen, P. E., Coulter, K. J., Kenny, P. G., & Koss, F. (1999). Automated intelligent pilots for combat flight simulation. *AI Magazine, 20*(1), 27–41.

Lee, F. J., & Anderson, J. R. (2000). Modeling eye-movements of skilled performance in a dynamic task. In N. Taatgen & J. Aasman (Eds.), *Proceedings of the 3rd International Conference on Cognitive Modelling.* Veenendaal, The Netherlands: Universal Press.

———. (2001). Does learning a complex task have to be complex? A study in learning decomposition. *Cognitive Psychology, 42*, 267–316.

Martin, E., Lyon, D. R., & Schreiber, B. T. (1998). Designing synthetic tasks for human factors research: An application to uninhabited air vehicles. *Proceedings of the Human Factors and Ergonomic Society 42nd Annual Meeting* (pp. 123–127). Santa Monica, CA: Human Factors and Ergonomics Society.

Purtee, M. D., Krusmark, M. A., Gluck, K. A., Kotte, S. A., & Lefebvre, A. T. (2003). Verbal protocol analysis for validation of UAV operator model. *Proceedings of the 25th I/ITSEC Conference*, 1741–1750. Orlando, FL: National Defense Industrial Association.

Salvucci, D. D. (2001). Predicting the effects of in-car interface use on driver performance: An integrated model approach. *International Journal of Human-Computer Studies, 55,* 85–107.

——, Boer, E. R., & Liu, A. (2001). Toward an integrated model of driver behavior in a cognitive architecture. *Transportation Research Record, 1779,* 9–16.

Schoelles, M., & Gray, W. D. (2000). Argus Prime: Modeling emergent microstrategies in a complex simulated task environment. In N. Taatgen & J. Aasman (Eds.), *Third International Conference on Cognitive Modeling* (pp. 260–270). Veenendal, The Netherlands: Universal Press.

Schraagen, J. M. C., Chipman, S. F., & Shalin, V. L. (Eds.). *Cognitive task analysis.* Mahwah, NJ: Erlbaum.

Schreiber, B. T., Lyon, D. R., Martin, E. L., & Confer, H. A. (2002). *Impact of prior flight experience on learning Predator UAV operator skills* (AFRL-HE-AZ-TR-2002–0026). Mesa, AZ: Air Force Research Laboratory, Warfighter Training Research Division.

Wetzel, P. A., & Anderson, G. (1998). *Portable eyetracking system used during F-16 simulator training missions at Luke AFB: Adjustment and calibration procedures.* Air Force Research Laboratory, Warfighter Training Research Division, Mesa, AZ. AFRL-HE-AZ-TP-1998-0111.

Wickens, C. D. (2002). Situation awareness and workload in aviation. *Current Directions in Psychological Science, 11*(4), 128–133.

——. (2003). Pilot actions and tasks: Selection, execution, and control. In P. Tsang & M. Vidulich (Eds.), *Principles and practice of aviation psychology* (pp. 239–263). Mahwah, NJ: Erlbaum.

3

Some History of Human Performance Modeling

Richard W. Pew

The history of modeling aspects of human behavior is as long as the history of experimental psychology. However, only since the 1940s have integrated models reflected human perceptual, cognitive, and motor behavior. This chapter describes three major threads to this history: (1) *manual control models* of human control in closed-loop systems; (2) *task networks models* that fundamentally predict probability of success and performance time in human–machine systems; and (3) *cognitive architectures* that typically capture theories of human performance capacities and limitations and the models derived from them tend to be more detailed in their representation of the substance of human information processing and cognition. In the past 15 years, interest in using these kinds of models to predict human–machine performance in applied settings has accelerated their development. Many of the concepts originated in the early models, such as "observation noise" and "moderator functions," live on in today's cognitive models.

Ideas have a way of being rediscovered. It seems that this is happening more frequently in recent years as the pace of research, development, and publication quickens. I am grateful to Wayne Gray for suggesting that a presentation (at the workshop) and a chapter of this book be devoted to the history of integrated modeling. I am also grateful that he asked me to write it. It is a longer history than some might suspect, is rooted in a practical need to support engineering design that was somewhat different from the current interests, and contains the seeds of many ideas and concepts that we take for granted today.

I have divided this sampling of history, of necessity it can be only a sampling, into three major threads: (1) *manual control*—models of human control in closed-loop systems; (2) *task networks*—models that fundamentally predict probability of success and performance time in human–machine systems; and (3) *cognitive architectures*—the architectures typically capture theories of human performance capacities and limitations. The models derived from them tend to be more detailed in their representation of the substance of human

information processing and cognition. My coverage of cognitive architectures will stop short of the detail of the other threads because it does not represent the core of my own expertise and because it is the "bread and butter" of the majority of the participants of the workshop and of the readers who are likely to be interested in this book.

Manual Control

Figure 3.1 provides an overview of the manual control thread. There are two main branches to this thread, one derived from the analysis of servomechanisms, or *classical control theory*, and one created with the advent of optimal control, usually referred to as *modern control theory*. I will describe both branches.

Early Studies of Human Movement Performance

During and immediately after the Second World War, a number of well-known psychologists in the United

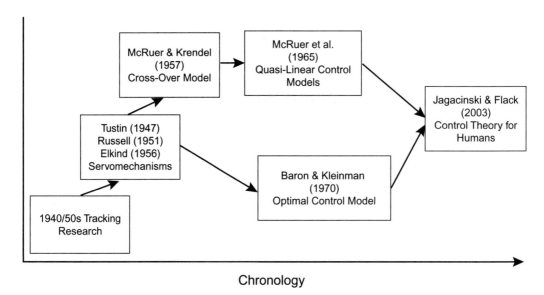

Chronology

FIGURE 3.1 The manual control thread.

States and Great Britain undertook studies of human movement performance. In the United States, names such as Robert Gagne, Lloyd Humphries, and Arthur Melton (see Fitts, 1947) come to mind. In Great Britain, there was J. K. W. Craik, E. R. F. W. Crossman, Christopher Poulton, and Margaret Vince (see Poulton, 1974). While their work took the form of basic research, they focused on the skill of flying. There was great interest in understanding what abilities contributed to success in flying, which was, of course, aimed at pilot selection tests and skill acquisition, to improve training effectiveness and efficiency. The early work was severely constrained by the available technology.

One prominent apparatus for studying skill acquisition, which dates back to the 1920s, was the pursuit rotor (Seashore, 1928). One version of this apparatus consisted of a phonograph turntable-like device with a small metal target embedded several inches from the center of the rotating disk The subject held a stylus with a wire "cat's whisker" on the end. As the disk rotated, the subject's task was to maintain contact between the cat's whisker and the metal

target. Standard electric time clocks recorded the "time-on-target" during a trial of fixed duration. The experimenter could vary the speed of the disk, the size of the target, the length of the trial, and schedules of practice. Before the war, this apparatus was used to contribute to the understanding of learning in general and psychomotor learning in particular. During and after the war, the applications to flying training and other needs for human motor control were studied.

Definition of Tracking

An important step forward was the development of the tracking paradigm. Initially, it was implemented by drawing a wavy line on a strip of paper, moving the paper past a narrow horizontal slit at a constant speed, and asking a subject to move a pointer (sometimes a pen or pencil) along the slit so that it followed the movement of the small segment of the line that was visible through the slit. As the technology advanced, the availability of strip chart recorders, cathode ray tube (CRT) displays, and control devices with electrical out-

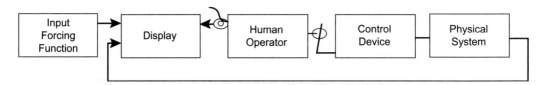

FIGURE 3.2 A block diagram illustrating the components of the tracking paradigm.

puts popularized and greatly improved the implementation and generality of the tracking paradigm (Fitts, 1952), which is illustrated in Figure 3.2. In the electronic version of the one-dimensional tracking task, the subject moves a control stick to superimpose a cursor over a target spot moving irregularly back and forth across the CRT. The instantaneous distance between the cursor and target is measured automatically and used to calculate an average error score for a trial, sometimes average absolute error, sometimes root-mean-square error.

The easiest way to understand the tracking task is to enlist a colleague and carry out the following simple demonstration: Hold out your finger and ask your colleague to follow the movement of your finger with her finger as closely as possible as you move it back and forth in an irregular pattern. Pick different irregular movement speeds, and see what happens. You are simulating a pursuit tracking task. In Figure 3.2, the input forcing function corresponds to the movement of your finger. The information is normally presented on a display, and your colleague takes the information in with her eyes and produces a motor response with her finger. Her response is imperfect, largely because the human reaction time is imposed between what she sees and what she does with her finger. In the figure, I show two additional blocks, the control device, which may be anything from a steering wheel to a stylus, and a physical system, the dynamical system being controlled. Moving vehicles have dynamical responses of their own. Flying an aircraft or controlling a submarine are challenging because their dynamics are complex. Controlling chemical processes or power plants also have physical dynamics, but the timescales of control are of the order of minutes and hours, not seconds and milliseconds. In the tracking paradigm, the subject is always comparing the input signal to the results of her movements and trying to minimize the error between the two. This is the essence of a feedback control system.

Early Applications of Classical Control Theory

At about this time, design engineers interested in the design of things like gun turrets used to track targets and aircraft flight control systems were developing the nascent field of servomechanisms. The guts of a servomechanism typically includes a powerful electric motor or a hydraulic pump that can develop high torque at low speeds and can be controlled by a low-power electrical signal. To improve their stability, they are almost always embedded in a feedback loop. Although one is interested in the relationship between sensed input and motor output as a function of time, engineers discovered that the properties of such devices are best characterized by understanding their frequency response, that is, the ratio of output to input amplitude and phase shift (the lag in response) for pure sine wave inputs over the range of frequencies to which they are sensitive. For systems that respond linearly to these inputs, the entire frequency response function can be characterized by a mathematical equation called the *transfer function*.

The first known publication devoted to understanding human control in the engineering language of servomechanisms was produced by Arnold Tustin, a well-known British electrical engineer (Tustin, 1947). During WWII, he was concerned with the design of massive gun turrets and wanted to make their servomechanism response compatible with human control. Through laboratory experiments and tedious paper and pencil analysis he demonstrated that the human response, which he acknowledged would be nonlinear in general, could be approximated with a "linear law" or transfer function, plus a "remnant," a random noise component. This representation has come to be called a quasi-linear *describing function*, which will be explained in more detail in a later section. If the linear part accounts for as much as 75% of the variance in the output, it is considered a useful representation. Think of it in terms of a linear correlation function: $R^2 = 0.75$. Tustin also explored various "aided gun-laying" feedback equalization schemes to improve aiming performance when a human operator was present in the control loop. It was truly pioneering work.

I was first introduced to these ideas while I was an electrical engineering student at Cornell University. I came upon an article by Franklin Taylor, a psychologist and Henry Birmingham, an electrical engineer, both at the U.S. Naval Research Lab (Birmingham & Taylor, 1954). They published the article in the *Journal of the Institute of Radio Engineers* (now the Institute of Electrical and Electronic Engineers, IEEE). It was entitled, "A Design Philosophy for Man-Machine Control Systems." Birmingham understood servomechanism theory and described conceptually how it would apply to human perceptual motor skills. The article discussed various control systems, particularly the manual control of submarines, which is a complex control problem because of the massiveness of the boat and the nature of the control surfaces. They also described some clever examples of how you could augment the display of information to improve the stability of control.

Collecting data to estimate the parameters of the human transfer function in the 1950s was a daunting task. If you asked subjects to track pure sine waves (i.e., follow a spot moving back and forth in one dimension in a sinusoidal pattern), they immediately detect that the waveform is regular and generate a waveform that approximates the desired pattern with no delay or phase shift, but in the real world, the patterns are typically irregular, not sinusoidal. When the patterns are unpredictable, the human reaction time severely limits the maximum frequencies that can be tracked with acceptable error to about 1 Hz (cycle per second), but with sine waves, that range can be extended to nearly 5 Hz because of their predictability (Pew, Duffendack, & Fensch, 1967). As a result, measuring human response requires using an input signal made up of either a randomly generated pattern having specified bandwidth or an irregular sum of multiple sine waves having different frequencies.

Furthermore, in the 1950s, fast Fourier transforms were not yet in use with digital computers (Cooley & Tukey, 1965), so the spectrum analysis had to be done tediously, with analog computers, one frequency component at a time.

In 1951, Lindsay Russell, a relatively obscure master's degree student at MIT (Massachusetts Institute of Technology), made one of the first systematic studies of the human transfer function (Russell, 1951). I say obscure because no one has heard of him in connection with manual control before or since. Nevertheless, he overcame the difficulties of estimating frequency response—power spectra and cross-power spectra—by building an ingenious real-time measurement system based on watt-hour meters just like the household electrical meters in use at the time. See appendix A for a description of how his measurement system worked.

Russell set up a human tracking task using a control stick, a CRT display and an analog computer to simulate the behavior of different physical systems. He measured human transfer functions used when controlling systems with these different simulated dynamics. Russell found that humans modified or adapted the parameters of their transfer functions when the system characteristics were changed to minimize the error they produced, an important insight that suggested there was no single human transfer function but a family of them, which adjusted in response to the type of system being controlled.

Jerry Elkind was a graduate student working with J. C. R. Licklider at MIT. He completed theses for both his master's and doctoral degrees in science on the topic of manual control (Elkind, 1953, 1956). For his Doctor of Science, he undertook the major challenge of mapping the characteristics of human response as a function of a wide range of input signal characteristics. To keep the control requirements as simple as possible, the subject tracked in one dimension by moving a handheld stylus with an embedded photocell on the surface of a CRT while viewing the dynamic moving spot appearing on another CRT, an early version of a light pen control. He created his input signals by adding together very large numbers of independent sine waves (40–144 in different conditions) in random phase relations. He set the amplitudes and frequencies of these individual sine waves to simulate input spectra having different overall amplitude and bandwidth characteristics. Some had square bandwidths for which the amplitude immediately dropped to zero at the cutoff frequency, and for others, the amplitude fell off gradually.

To process the mass of data he collected, he programmed an analog computer to behave as a very sophisticated spectrum analyzer that computed estimates of the required power spectra and cross-power spectra for each trial, This was a pioneering effort in itself. He then, with an assistant's help, analyzed each of a total of approximately three hundred-sixty 4-min trials among three different subjects. If I understand his method correctly, within each run, it was necessary to repeat the analysis at each of the sine wave input frequencies he wished to analyze serially, requiring as many as 40 passes for each run, although the analysis could be run faster than real time.

Elkind then derived analytic transfer function models for each condition he studied and proposed the adjustment rules necessary to characterize the different conditions. Taken together, this was a gigantic effort, worthy of at least three doctorates. Needless to say, his was, and still is, the definitive work on the effect of input signal characteristics on human manual control response. After finishing his degrees and spending a year working for RCA, Elkind rejoined Licklider, who had moved from MIT to Bolt, Beranek and Newman (BBN), and built a group that continued manual control research.

A distinguished aeronautical engineer became interested in manual control from a very practical point of view. Duane McRuer, known as Mac, was working for Northrup Aircraft as a key flight control engineer. He was one of the pioneers who promoted the idea of introducing the methods of control engineering into the analysis of flight control behavior (Bureau of Aeronautics, 1952). Before that, aircraft control behavior was described through a series of partial differential equations. McRuer was also interested in describing analytically the closed-loop behavior of aircraft incorporating the human pilot. To accomplish this required, in addition to the aircraft transfer function model, a representation, in transfer function terms, of the pilot (Bureau of Aeronautics, 1952). Early on, he established a collaboration with Dr. Ezra Krendel, a psychologist with the Franklin Institute in Philadelphia. Krendel was also studying human tracking behavior and had a laboratory where he could begin making frequency response measurements on people (Krendel & Barnes, 1954). McRuer teamed up with Krendel and won a contract from the U.S. Air Force[1] to undertake a comprehensive review of all the work, dating back to Tustin, that had sought quantitative control models of tracking and manual control. I was a newly minted second lieutenant in the air force, fresh out of Cornell University ROTC, assigned to the Psychology Branch of the Aeromedical Laboratory (now the Human Effectiveness Directorate of the Air Force Research Laboratory) at the time and was immediately enlisted to help Dr. John Hornseth monitor McRuer and Krendel's contract. It was in their best interest to educate me, a serendipitous opportunity that significantly shaped my career.

The resulting milestone report, entitled, "Dynamic Response of Human Operators" (McRuer & Krendel, 1957; see also McRuer, Graham, Krendel, & Reisener, 1965), solidified the interpretation of human response in terms of a "quasi-linear transfer function" and a "remnant" term. It presented the *crossover model* to explain human adaptation to changing physical dynamics, and it introduced a *precognitive model* to explain "programmed" behavior, such as response to signals that were predictable.

The Quasi-Linear Human Operator Model

On the basis of their review and interpretation, they defined the standard form for the quasi-linear human operator model. A transfer function is derived from a linear differential equation. No one believes that a human operator's response is truly linear, but it has been shown that a linear differential equation is a useful approximation, and when a random noise component is added to it, the equation can account for much of the variation in human output. Therefore, the model contains a transfer function and a noise term referred to as the "remnant"—that portion of the output that is not linearly correlated with the input—hence the name "quasi-linear model." The transfer function part, has four features—gain, time delay, smoothing, and anticipation—which translate into parameters of the model. They are adjusted by the human operator in response to the characteristics of the input signals and the dynamical system being controlled. The first feature is a gain or sensitivity constant associated with an individual's response strategy but constrained also by the need to maintain stability of control. The higher the gain, the more sensitive the operator is to signal variation. If the gain is too high, the operator is likely to overcorrect for errors. The time delay represents the operator's intrinsic reaction time. In the case of continuous control, values in the range from 0.13 to 0.20 s are typical. The smoothing term implies that the operator does not respond to every little detail in the input signal but rather filters or smoothes the output. Finally, the anticipation feature suggests that the operator introduces anticipation, or prediction, into the response based on the input signal time history. The trends in the signal as well as its actual position at each moment in time influence the response. These features are not immutable. While their form remains the same, experimental studies have shown that their values change from individual to individual and as a function of external system constraints. The human operator adapts his or her response to maintain as effective control as possible.

The Crossover Model

A particularly significant conclusion of these analyses was McRuer and Krendel's formulation of the crossover model that attempted to summarize in as simple a way as possible the nature of human adaptation to control of different physical, dynamical systems. The basic idea is to analyze the human controller and the physical system together as a single entity. Then the representation becomes much simpler because the human adapts performance to compensate for the dynamics of the physical system in such a way as to maintain the combination relatively fixed. The parameters of the crossover model are simply the gain, the time-delay parameter, and the smoothing parameter.

Understanding the details requires some knowledge of control engineering and system stability concepts (see, e.g., Jagacinski & Flach, 2003). It should be thought of as an approximation, or rule of thumb, because it is not accurate in every detail and only applies to a limited, but useful, range of physical system dynamical characteristics. For purposes of this chapter, it is enough to understand that the nature of human adaptation is to simplify and stabilize the overall system.

The work of McRuer, his collaborators, and many other colleagues who have shared his vision have had a significant effect on the world of aviation, especially with respect to aircraft handling qualities and flight control system design (see, e.g., National Research Council, 1997) and in the world of ground vehicles, their drivers, and steering control system design (e.g., Salvucci and Macuga, 2002).

The Optimal Control Model of Manual Control

In the 1960s, modern control theory was sufficiently developed that Shelly Baron and David Kleinman, at BBN, applied it to the manual control problem (Baron & Kleinman, 1969). The key developments in modern control theory were expressing the control problem in a matrix of state space variables and the idea that a closed-form solution to the equation representing optimality could be obtained by minimizing a very general evaluation function (Kalman & Bucy, 1961).

When applied to manual control the *evaluation function* is typically the mean square error or some weighted sum of error and control effort. The *ideal*, or optimal, controller is derived that minimizes this metric, subject to the constraints faced by human operators. It is then assumed that this optimal solution reflects the behavior of a well-trained operator faced with the same constraints and performance limitations. This is a normative model, just as the more familiar signal detection model is normative (Green &

Swets, 1966). Both describe the way the operator "ought to behave," and both happen to reflect quite veridically how they actually behave.

Figure 3.3 conceptualizes the optimal control model in block diagram form. The input signal is assumed to be corrupted by observation noise (similar to the perceptual noise introduced in the signal detection model) and delayed by the operator's reaction time. The core of the analysis is the Kalman estimator and predictor that operates on this noisy, delayed signal to create the best prediction of the control activity required to minimize the evaluation metric. Intrinsic to these two modules is a representation of the controlled system. This concept is not unlike the notion that operators have an internal or "mental" model of the system they are controlling. The remainder of the control loop includes some gain constants, further motor noise, and smoothing of the output. Although the motor noise term was inserted to get better data fits, it is also plausible that the human motor system introduces noise. Maloney, Trommershäuser, and Landy's (chapter 21, this volume) use of the term *motor variability* is similar in concept to this motor noise. The observation noise term and the motor noise term, taken together, are closely related to the remnant term in the classical control representations.

Current Status of Manual Control Models

Both classical quasi-linear models and optimal control models are still in use today. Table 3.1 provides a summary of the features and limitations of the two approaches. The bottom line is that classical models provide good intuition about conditions for stability and an understanding of the variables to which performance is sensitive. However, construction of the models requires extensive knowledge of transfer function behavior and involves extensive individual tailoring for specific cases. When applicable, the optimal control model produces a specific solution largely automatically but does not

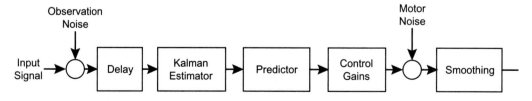

FIGURE 3.3 The block diagram form of the optimal control model of manual control performance.

TABLE 3.1 Comparison of Quasi-Linear and Optimal Control Models

Quasi-Linear Models	Optimal Control Models
Provide good intuition about behavior of system	Can be related to human information processing behavior
System analysis is a trial and error process based on many interactive criteria	Deal coherently with multivariable control
Much subjective adjustment required for multi-variable, multiloop systems	Derived automatically by computer program
	Require definition of quantitative performance metric

provide as much direct insight into the stability boundaries or the actual control behavior that will result.

There has not been much innovation in manual control models since about 1980. The quasi-linear model of McRuer et al., and the optimal control model still represent the state of the art. Automation has been introduced into many systems where manual control was critical, reducing the need for "inner loop" control. As a result, there are fewer demands for this class of model. However, there continue to be applications, mainly to aviation (fighter, civil aircraft) and vehicle driving. A recent book by Jagacinski and Flach (2003) captures the current state of theory and applications very well.

Network/Reliability Models

Figure 3.4 provides an overview of the genealogy of network-related models. There are two main thrusts, one derived from formal PERT networks, the kind used to monitor and control system development and manufacturing processes, and one focused on reliability assessment. I will discuss reliability models first.

Human Reliability Models

According to Miller and Swain (1987), as early as 1952 Alan Swain's group at Sandia National Laboratories attempted to analyze quantitatively the contribution of human reliability to overall system reliability in the context of a classified analysis of aircraft nuclear weapons systems, but the analysis suffered from a lack of reliable data. In 1962, the American Institutes for Research reported on an effort to create a database of reliability statistics, that is, the probability of error for elemental human actions, such as reading dials, turning valves, or operating controls. The document is referred to as the AIR Data Store (Payne & Altman, 1962). The author's goal was to support predictions about the probability of human error in routine operations. Performing typical human tasks involves the serial aggregation of collections of such elemental actions and, as task analysis reveals, the aggregation involves a contingent branching structure of possible paths through a network of such actions. Apply the

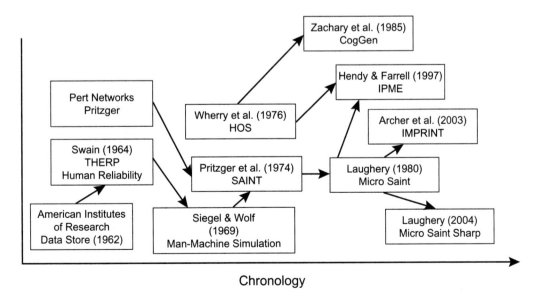

Chronology

FIGURE 3.4 The network model thread.

standard reliability equation (1) to this aggregation process and you have a simple model that could predict the probability of human error. In Equation 1, the $Q(e_k)$'s represent probabilities of error in each element in a particular path through the task. Then the probability of successfully completing each element, e_k, is one minus the error probability. Thus the aggregate probability of error is one minus the product of the individual probabilities of success:

$$P(\text{error}) = Q(e) = 1 - \left[\prod_{k=1}^{n} \{1 - Q(e_k)\} \right] \quad (1)$$

Swain has been the major innovator in this work, creating technique for human error rate prediction (THERP) (see Swain & Guttman, 1983). Subsequently, his group has pioneered the refinement and application of this technique in the nuclear power industry and elsewhere. It applies standard reliability equations to task analyses using databases like AIR Data Store and invokes performance shaping factors (PSFs) to account for human individual differences and environmental variables. The PSFs are adjustments to the database entries to take account of the specific contextual conditions that are postulated to exist in the task and application environment.

The Siegel and Wolf Network Model

It is only a small step from the early reliability analyses to the innovative modeling work of Siegel and Wolf (1969). Art Siegel was a psychologist at Applied Psychological Services, interested in predicting human performance in applied settings. He envisioned the possibility to create a Monte Carlo simulation version of a reliability network model that could incorporate performance times as well as reliabilities and that could predict a variety of measures derived from these, such as workload and productivity. He pursued this approach throughout the 1960s and early 1970s with support primarily from the Office of Naval Research. Jay Wolf was a computer specialist with a full-time job at the Burroughs Corporation. In the archives of the Charles Babbage Institute, he is credited with seminal contributions to several of Burrough's early computers. He was responsible for coding up Siegel's model. I assume he was paid for his work, but he did it in his spare time, more or less as a hobby.

Their approach was to create a task network, a branching series of network nodes, which captured the operations of a "man-machine" system. Each node, or *action unit*, had a probability of success and a statistical

distribution of completion times moderated by a series of PSFs or moderator functions. PSFs were implemented as scale factors applied to the action units and were implemented globally; that is, they were programmed to apply to all the relevant action units in a simulation. Aggregate probability of success and performance times were estimated by Monte Carlo simulation of the overall network. Siegel devoted significant effort to capture the effects of "psychosocial behavior," for example, "performance stress," "team cohesiveness," or "goal aspiration" in succinct PSF equations that could be coded into the model.

During the 1960s, he and Wolf created a series of models, starting with a single-operator single machine and working up to groups, or teams, operating in coordinated actions with larger-scale systems. The group model was validated using a realistic 21-day training mission of a nuclear submarine. Only data available to the system design personnel early in the system planning stage were employed as input data to the model, and the results were compared with actual mission results for typical 8- and 12-hr shifts. Quantitative data were available to compare with manning statistics, and actual submarine crew members' subjective assessments were used where appropriate performance measures were not available.

The model was originally programmed in SOAP (Symbolic Operating Assembly Program) on an IBM 650, later converted to FORTRAN to run on an IBM 7094. After some initialization, the program sequentially considered each action unit as an independent entity and simulated its completion through a series of sequential steps. Table 3.2 shows a sample of the computational steps undertaken and illustrates the extent to which the model attempted to capture more than just the raw task performance.

The nominal performance times and accuracies were drawn from prespecified distributions and averages were obtained by running repetitive trials, Monte Carlo style.

For each of these blocks of the program, detailed algorithms were specified. The algorithms were derived from Siegel's extensive review of the relevant social/psychological research. It is beyond the scope of this chapter to describe them in detail, but the several proficiency factors in Step 3 all involve linear equations—group proficiency is an average of the incoming proficiency of the individuals in the group; overtime adjustment decreases efficiency linearly with the average overtime hours worked; morale is calculated on a daily basis and downgrades efficiency by the

TABLE 3.2 Sample of the Siegel and Wolf Model's Computational Steps

1. Select crew members to form a group to accomplish next action unit (random selection within crew member specialties)
2. Calculate communications efficiency
3. Calculate action unit execution time based on performance times (sampled from specified distribution), group proficiency, overtime worked, morale and number of unavailable men
4. Adjust time worked by each group member
5. Calculate group orientation
6. Calculate psychological efficiency
7. Calculate psychosocial efficiency
8. Calculate efficiency of the environment
9. Calculate total action unit efficiency
10. Determine adequacy of group performance
11. Recalculate execution time and efficiency

amount it deviates from a reference value. One of the more complicated algorithms concerns the performance efficiency as a function of the percentage of work completed. The efficiency varies between 71% and 83%, starting low and rising to a peak of 83% at 20% complete, falling to 75% at 50% complete and showing an end spurt back to 83% for the last 20% of the task.

Siegel and Wolf (1969) describe the three modeling efforts in increasing levels of sophistication. The models were complex; validation was limited but not overlooked. Nevertheless, Siegel and his colleagues contributed a pivotal development in the history of integrated human performance modeling. A reader interested in more about the history of modeling during this time period might review Levy (1968).

SAINT and Micro Saint

The U.S. Air Force became interested in the modeling approach used by Siegel but realized that, while very promising, the expertise of Art Siegel and his proprietary code would be needed to accomplish it. To make the methodology more accessible, they funded the development of SAINT (systems analysis of integrated networks of tasks), a general purpose discrete simulation language, written in FORTRAN. It was designed specifically to capture the methods and innovations introduced by Siegel, particularly the capability to define global moderator functions that would affect multiple nodes (Wortman, Pritsker, Seum, Seifert, & Chubb, 1974). It was implemented by Pritsker and Associates, the organization that had a reputation for creating discrete simulation languages, especially the

very similar simulation language, GASP, used to quantify manufacturing and project development networks. SAINT was used in a number of air force studies and also by the Department of Transportation. An accessible example applied to a remotely piloted drone control facility (UAV in today's terminology) is Wortman, Duket, and Seifert (1975).

Then, in 1982 or so, Alan Pritsker hired Ronald Laughery, a recently minted PhD, to work on an army human factors application that was to use SAINT (R. Laughery, personal communication, 2005). It was not long before Ron found an outlet for his entrepreneurial genes. He wanted to start his own company, Micro Analysis and Design (MAAD), and sensed the value of rewriting SAINT in a simpler form that would run on a personal computer. Thus Micro Saint was born. The first commercial version was available in about 1986 and was written in C. It captured the functionality of SAINT and therefore traces its lineage to the Siegel and Wolf models. Micro Saint, like SAINT, is fundamentally a general purpose discrete simulation engine. Since 1986 it has been through several revisions, the most recent called Micro Saint Sharp, not surprisingly, since it is written in C-sharp. It has also spawned a family tree of special purpose applications of its own. These descendants, have varying degrees of commonality with Micro Saint and varying degrees of specificity. The most prominent thread is contained in the IMPRINT series of applications, mostly sponsored by the U.S. Army, which provide modeling templates specifically adapted to human performance modeling applications (Archer, Headley, & Allender, 2003).

HOS

HOS has a history of its own. In 1969, Robert Wherry Jr., then with the U.S. Navy at Point Magu, California, conceived of a human operator simulator (HOS) (Wherry, 1969). The ingenious idea was to have an easy-to-use, high-level procedure language (HOPROC) for programming task execution together with a collection of *micromodels* that could be "called" to represent individual human performance processes. The procedure language code, together with the micromodels would then be compiled to produce a runnable simulation of human–machine performance. Wherry moved to the navy's facility at Warminster, Pennsylvania, and in the early 1980s, funded the company, Analytics, to produce the early versions of HOS, which were used to model some naval air surveillance operations (Lane, Strieb, Glenn, & Wherry, 1981). Later, when Micro Analysis and Design took over further development, they produced a version, MS HOS, and HOS concepts began to appear in other MAAD products. IPME, a development primarily for Great Britian and Canada, employs the Micro Saint engine but used a HOS-like architecture (Hendy & Farrell, 1997).

Meanwhile, a number of the developers of HOS at Analytics migrated to the company, CHI Systems, newly formed by Wayne Zachary in 1985. Their first modeling product was heavily influenced by HOS. The current version of CHI Systems modeling software is COGNET/iGEN, which represents another major player in human performance modeling and simulation (Zachary, Santarelli, Ryder, Stokes, & Scolaro, 2000).

Cognitive Process Models

Of course the last of the historical threads, and the most contemporary, is that associated with cognitive architectures. An outline (maybe better described as a skeleton since it is clearly incomplete) is provided in Figure 3.5. A more substantive discussion is contained in Byrne (2003). Whereas both manual control and network models have their roots in applied needs, cognitive models have their roots in psychological theory. It is only relatively recently (in the past 15 years) that the potential usefulness of integrated human performance models has spurred considerable applied interest and the increase in funding support that often accompanies the promise of usefulness (Pew & Mavor, 1998). This discussion will focus on the historical precedents rather than the substance of these contributions.

Psychologists have been summarizing the results of their work in the form of models—verbal-analytical models (Broadbent, 1958), physical models (Bekesy, 1949), and mathematical models (Hull, 1943)—for most of psychology's relatively short history. They began capturing their theories in computer models almost as soon as digital computers became available (Feigenbaum, 1959; Feigenbaum, and Simon, 1962). The introduction of the *general problem solver* (Newell & Simon, 1963) placed the idea of computer models in the context of simulations of human information processing and cognition more generally, and these two individuals deserve much of the credit for kicking off the more general interest in the computer as a simulation tool for human performance. Of course, subsequent

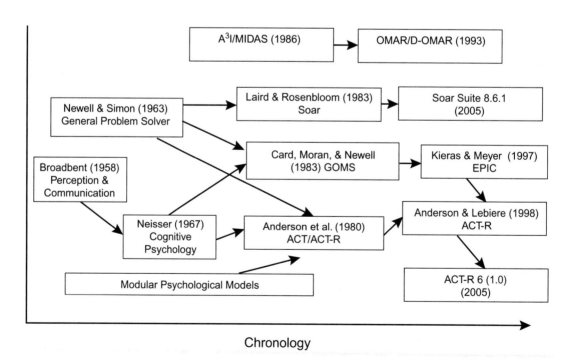

FIGURE 3.5 The cognitive architecture thread.

investigators have leveraged the empirical and conceptual contributions of Broadbent (1958, 1971), Fitts (1954, 1964), Neisser (1967), and many others.

Soar

The most direct spin-off of Alan Newell and Herbert Simon's work, especially Newell's, has been the cognitive architecture, Soar.[2] Newell students John Laird and Paul Rosenbloom completed theses in 1983 capturing the initial developments and stayed on at Carnegie Mellon to establish it as a cognitive architecture. The institution of annual Soar workshops and tutorials has been a significant driver in growing the community of interest since Laird and Rosenbloom left Pittsburgh (Rosenbloom, 2001).

GOMS

At about the same time, Stuart Card and Thomas Moran were also working with Alan Newell on the more applied implications of his perspectives. The seminal book, *The Psychology of Human-Computer Interaction* (Card, Moran, & Newell, 1983) introduces GOMS, which stands for goals, operators, methods, and selection rules. GOMS itself, is not a computer model, it is a systematic description of how to calculate the time to accomplish tasks by accounting for physical and mental actions required of the task. It catalogs "standard times," or simple algorithms, to compute them for various kinds of actions, and proposes how to assign and aggregate them to predict the overall performance time for a human–computer or human–system interaction task. The predicted times were derived from a comprehensive review of the behavioral literature and summarized in what the authors called the *model human processor*." A further valuable contribution was the idea of calculating "fast man," "slow man," and "middle man" for each action so that, in the aggregate, it would be possible to bracket the range of expected performance times.

Soon this recipe was converted into computer code that could be programmed to complete the calculations, and these versions were, indeed, computer models (Kieras, 2003). Kieras credits HOS as having influenced the structure he incorporated in NGOMSL, his language for generating GOMS models. In collaboration with David Meyer, this was soon followed by EPIC, executive-process/interactive control, a genuine cognitive architecture that elaborates and deepens the action descriptions, especially of perceptual-motor operations (Kieras & Meyer, 1997).

Human performance in a task is simulated by programming the cognitive processor with production rules organized as methods for accomplishing task goals. The EPIC model then is run in interaction with a simulation of the external system and performs the same task as the human operator would. The model generates events (e.g., eye movements, key strokes, vocal utterances) whose timing is accurately predictive of human performance. (quoted from Kieras, http://www.eecs.umich.edu/~kieras/epic.html)

The legacy that GOMS, and perhaps HOS as well, provide for EPIC should be clear from this description.

ACT-R

The provenance of ACT-R (adaptive control of thought–rational) began with John Anderson's work at Stanford with Gordon Bower on computer simulations of memory, most notably HAM (human associative memory; Anderson & Bower, 1973). Since then Anderson has dedicated his research to seeking a theory of cognition capable of being represented in a computer simulation. The most direct linkages to ACT-R reside in ACT theory (Anderson, 1976, 1983), but he also credits Newell and Simon for stimulating his interest in a production rule implementation (Anderson, 1995). The first published version using the ACT-R specifications was in Anderson (1993) and represented the culmination of 20 years of work. Together with his many distinguished students, he has broadened and deepened both the theory and application so that today ACT-R is perhaps the most widely used cognitive architecture, both for quantitative explorations of cognition and for applications in which human cognitive performance is paramount (Anderson & Lebiere, 1998). It is supported by annual workshops and tutorials. Several chapters of this book document much of this recent work and the challenges to be faced going forward.

MIDAS/D-OMAR

In the mid-1980s, NASA Ames Research Laboratory, together with the Army Aero-Flight Dynamics Directorate based at Ames, became interested in developing "a predictive methodology for use by designers of cockpits and training systems for advance technology rotocraft" (NASA Ames Human Factors Research Division, 1986). They referred to it as the Army-NASA

Aircraft Aircrew Integration Program, or A³I. Their initial interest was in modeling the human-systems integration of a future scout/attack helicopter with the idea of supporting the system acquisition process for the next army helicopter procurement. James Hartzell, representing the army, and Irving Statler, representing NASA, convened a three-day workshop, which I chaired, to review the state of the art, at the time, mostly concerned with manual control and human factors tools. On the basis of that unpublished study, a more serious review of the state of the art was funded in all the associated modeling areas of human performance and simulation, from modeling the visual system to motor control (Elkind, Card, Hochberg, & Huey, 1989). They were interested in something beyond what could be done with manual control, and, at the time, cognitive architectures were in their infancy (I suspect the term had not yet been invented even though the foundations existed) and were not adequate for a task of the scope of this requirement.

BBN, with Kevin Corker as the principal investigator, was awarded a small contract to produce the initial infrastructure for what eventually would become the MIDAS (Man-Machine Integrated Design and Analysis System) software. Hartzell and Statler should be credited with the initiative to create MIDAS. The infrastructure BBN produced was an early application of object-oriented programming paradigm using the Lisp Flavors system associated with Symbolics hardware and the IRUS 3-D graphics workstation. When Kevin Corker moved from BBN to NASA Ames, he initially took over the further development of MIDAS. Since that early work, it has undergone a number of transformations. Since one of the goals was to support designers with enhanced visualizations of the implications of cockpit design alternatives, it was decided to incorporate the JACK digital human anthropometric model into the system. Soon thereafter the entire system was rehosted on a more contemporary platform. In the mid-1990s, Corker left NASA for San Jose State University and took a version of MIDAS with him. Corker has continued to develop the software, emphasizing models of the air traffic control process, while Sandra Hart has continued to support further development and use at NASA.

Drawing on software components developed for the SIMNET semiautomated forces project at BBN and the BBN infrastructure that led to MIDAS, Stephen Deutsch began the development of a modeling and simulation environment, OMAR (operator model architecture) in 1993 (Deutsch, Adams, Abrett, Cramer, & Feehrer, 1993). Since then, OMAR has evolved into a distributed architecture version, D-OMAR. What distinguishes D-OMAR is that rather than being a particular cognitive architecture, it provides a suite of software tools from which to implement alternate architectures. D-OMAR provides a discrete event simulator and incorporates a frame language that forms the object-oriented substrate, a procedure language that implements goals and procedures as frame objects, and a rule language that operates on frame language and procedure language objects. The intent is to have a general infrastructure in which to implement and evolve a range of architectural alternatives. The core representation of human operator behaviors is in the procedure language. Because of the coverage provided by the procedural language, production rules, in the sense of most other cognitive architectures, have not found a place in the models developed to date. Following Glenberg (1997), who has identified memory for process, "the remembrance of what we know how to do"—as memory's principal function, the network of goals and procedures constitute procedural memory, while declarative memory is stored locally within procedures distributed across the network. D-OMAR has been used for a variety of modeling activities in the past five years, most notably in evaluating workplace design and understanding sources of human error in commercial aircrew and air traffic control environments (Deutsch & Pew, 2002; 2004; in press).

Hybrid Models

Hybrid models, that is, models that combine independent approaches to accomplish an applied goal have also played a role in this field. In the early 1990s, ONR sponsored a program that sought to bring together work in the computational field of machine learning with cognitive architectures and was focused on modeling human performance in two specific applied tasks, a simplified air traffic control task and a military command and control task (Gigley & Chipman, 1999). Models at multiple levels of granularity have been sought through combining network models and cognitive architectures (e.g., Imprint and ACT-R; Lebiere et al., 2002). I will describe two lesser-known hybrids that address the need for high-quality simulation of continuous control integrated with a representation of discrete task performance.

PROCRU

As early as the late 1970s, when more automation was being introduced into commercial cockpits, describing piloting performance involved less closed-loop control and more operation of discrete controls. Shelly Baron at BBN, with NASA support, generalized the optimal control model to include crew decision making and procedural activities. He produced a simulation of aircrew performance during approach and landing of a Boeing 727 type aircraft (before flight management computers and glass cockpits). When a procedural activity was required, the simulation switched attention from flying the aircraft to the new task. The control model continued to fly the aircraft, but without new sensory input from the pilot. Each procedural activity was represented as a submodel that could effect aircraft response and change the crew information state. Decisions among competing activities for what task to execute next were based on probabilistic assessments and an "expected gain" function derived from mission impact (Baron, Muralidharan, Lancraft, & Zacharias, 1980). There was never an opportunity to validate the model, but it represented a very significant innovation that captured many of the multitasking requirements addressed in later production rule models—and was integrated with continuous control.

The Integrated Driver Model

The second hybrid was prepared by Bill Levison and Nichael Cramer (1995) in connection with the intelligent transportation systems program of the Federal Highway Administration to assess impact of driver information systems on driving behavior. It combined a cognitive information processing model with the optimal control model of continuous driving control. Considerable effort was devoted to modeling task switching and attention sharing using then current theories, including Wickens' multiple resource theory (Wickens and Liu, 1988). The model compared favorably with simulator data on in-vehicle telephone use collected by Paul Green at the University of Michigan Transportation Research Institute (Serafin, Wen, Paelke, & Green, 1993) and predicted an interesting counterintuitive empirical result of Ian Noy (1990). Noy collected data on driver eye movements and found that the error in steering performance was smaller when the driver was not looking at the road, for example, looking at his in-vehicle

displays instead. While this seems a strange result, it can be rationalized on the grounds that the only time the driver is willing to look away from the road is when his error is small. The model exhibited this same behavior.

The only other control-cognitive architecture hybrid since these developments that I am aware of is the work of Dario Salvucci (2001), but they are very relevant, particularly to controller/vehicle/information system interactions where precision manual control is an integral part of the task. Such applications include safety of driver information systems, some tasks of unmanned aerial vehicles, and applications in civil aviation. I challenge the readers of this chapter to consider them in their own applications.

A Final Word

The success of modeling human performance depends on the constraints imposed by the environment. As more constraints are placed on an operator's performance, the more successful the models will be. The manual control models were, and are, successful because the required performance is very well defined and constrained, and this is the area where human performance modeling got its start. Similarly, network models are most successful when there is little discretionary time, that is, maximal constraint on what to do next at each moment in time. It is in the cognitive architectures and hybrid models that modelers have sought to extend the range of applicability to situations where there are potential choices of what to do next that are process constrained rather than time constrained, that elaborate alternative strategies, and that deepen the models to be more realistic with respect to internal perceptual and cognitive processes for which external environmental constraint is less useful. A great deal of progress has been made in the almost 60 years since Tustin first proposed an analysis of human performance in a closed-loop control system. I will leave it to the other chapter authors at this workshop, see particularly John Anderson's chapter, to forecast the most needed future developments.

Appendix: Lindsay Russell's Measurement System

For an input signal, Russell added together four sine waves of known frequency to produce a random appearing signal. When a commercial watt-hour meter is used for its intended purpose, its inputs are the three phases of a typical power distribution line, and its output is the integral of the "in phase" and "quadrature" or orthogonal component of the output power being used by the customer. He used a professional version of the same kind of device that had (at least in 1951) four dials and a rotating disk in it that was used in every residential electrical system. The way the watt-hour meters were used by Russell, each watt-hour meter was connected to an electrical signal, corresponding to one of the four sine waves that made up the input signal. The second input was the complex output wave form the subjects produced. The output of the watt-hour meters was then the integral over time (i.e., the duration of a trial) of that portion of the power in the output that was linearly correlated with the input signal at each particular frequency and when combined as a vector with the quadrature or orthogonal component, provided an estimate of the phase shift between the input and output at that frequency. Therefore, Russell's complete analyzer consisted of four watt-hour meters, one for each component sine wave in the input signal. The collection of all the required apparatus and wires to accomplish this analysis at the same time the subject was tracking the signal looked like one of Rube Goldberg's cartoons but produced interesting and reasonably reliable results.

Notes

1. It is interesting to note that this contract was entered into collaboratively between the Psychology Branch of the Aeromedical Laboratory and the Flight Control Laboratory of the U.S. Air Force at Wright Patterson AFB, OH.

2. According to the Web site http://acs.ist.psu.edu/soar-faq/soar-faq.html#G3, originally Soar stood for State, Operator, And Result. However, over time the community no longer considered it an acronym and eliminated the use of uppercase.

References

ACT-R 6 (1.0). (2005). Retrieved from http://act-r.psy .cmu.edu/actr6/

Anderson, J. R. (1976). *Language, memory, and thought.* Hillsdale, NJ: Erlbaum.

———. (1983). *The architecture of cognition.* Cambridge, MA: Harvard University Press.

———. (1993). *Rules of the mind.* Hillsdale, NJ: Erlbaum.

———. (1995). Biography of John R. Anderson. *American Psychologist, 50* (4), 213–215.

———, & Bower, G. H. (1973). *Human associative memory.* Washington, DC: Winston and Sons.

———, & Lebiere, C. (1998). *The atomic components of thought.* Mahwah, NJ: Erlbaum.

Archer, S., Headley, D., & Allender, L. (2003). Manpower, personnel, and training integration methods and tools. In H. Booher (Ed.), *Handbook of human systems integration* (pp. 379–431). New York: Wiley.

Baron, S., & Kleinman, D. L. (1969). The human as an optimal controller and information processor. *IEEE Transactions of Man-Machine Systems, 10,* 9–17.

———, Muralidharan, R., Lancraft, R., & Zacharias, G. (1980). *PROCRU: A model for analyzing crew procedures in approach to landing.* NASA CR-152397. Sunnyvale, CA: National Aeronautics and Space Administration.

Bekesy, G. von (1949). The vibration of the cochlear partition in anatomical preparations and in models of the inner ear. *Journal of the Acoustical Society of America, 21,* 233–245.

Birmingham, H. P., & Taylor, F. V. (1954). A design philosophy for man-machine control systems. *Proceedings of the Institute of Radio Engineers, 42,* 1748–1758.

Broadbent, D. (1958). *Perception and communications.* New York: Pergamon Press.

———. (1971). *Decision and stress.* New York: Academic Press.

Bureau of Aeronautics. (1952). *Methods of analysis of and synthesis of piloted aircraft flight control systems: Vol. 1. Fundamentals of design of piloted aircraft flight control systems.* (Report AE 61–4).

Byrne, M. D. (2003). Cognitive architecture. In J. A. Jacko & A. Sears (Eds.), *The human-computer interaction handbook* (pp. 98–117). Mahwah, NJ: Erlbaum.

Card, S., Moran, T., & Newell, A. (1983). *The psychology of human-computer interaction.* Hillsdale, NJ: Erlbaum.

Cooley, J. W., & Tukey, J. W. (1965). An algorithm for the machine calculation of complex Fourier series. *Mathematics of Computation, 19,* 297–301.

Deutsch, S. E., Adams, M. J., Abrett, G. A., Cramer, N. L., & Feehrer, C. E. (1993). Research, development, training and evaluation (RDT&E) support: Operator model architecture (OMAR) software functional specification. AL/HR-TR 1993–0027, Wright-Patterson AFB, OH: Human Resources Directorate, Air Force Research Laboratory.

———, & Pew, R. W. (2002). Modeling human error in a real-world teamwork environment. In *Proceedings of the Twentieth-fourth Annual Meeting of the Cognitive Science Society* (pp. 274–279). Fairfax, VA.

——, & Pew, R. W. (2004). Examining new flightdeck technology using human performance modeling. In *Proceedings of the 48th Meeting of the Human Factors and Ergonomic Society Meeting*. New Orleans, LA.

——, & Pew, R. W. (in press). Modeling the NASA scenarios in D-OMAR. In D. C. Foyle & B. Hooey (Eds.), *Human performance models in aviation*. Mahwah, NJ: Erlbaum.

Elkind, J. I. (1953). *Tracking response characteristics of the human operator*. Memorandum 40. Washington, DC: Human Factors Operations Research Laboratories, Air Research and Development Command, U.S. Air Force.

——. (1956). Characteristics of simple manual control systems. In *Technical Report III*. Lexington, MA: MIT Lincoln Laboratory.

——, Card, S. K., Hochberg, J., & Huey, B. M. (1989). *Human performance models for computer-aided engineering*. Washington, DC: National Academy Press.

Feigenbaum, E. A. (1959). An information processing theory of verbal learning (Report No. P-I 857). Santa Monica, CA: RAND Corporation.

——, & Simon, H. A. (1962). A theory of the serial position effect. *British Journal of Psychology, 53*, 307–320.

Fitts, P. M. (Ed.). (1947). *Psychological research on equipment design*. Washington, DC: Government Printing Office.

——. (1952). Engineering psychology and equipment design. In S. S. Stevens (Ed.), *Handbook of experimental psychology* (pp. 1287–1340). New York: Wiley.

——. (1954). The information capacity of the human motor system in controlling the amplitude of movement. *Journal of Experimental Psychology, 47*(6), 381–391.

——. (1964). Perceptual-motor skill learning. In A. W. Melton (Ed.), *Categories of human learning* (pp. 243–285), New York: Academic Press.

Gigley, H. M., & Chipman, S. F. (1999). Productive interdisciplinarity: The challenge that human learning poses to machine learning. In *Proceedings of the 21st Conference of the Cognitive Science Society*. Mahwah, NJ: Erlbaum.

Glenberg, A. M. (1997). What memory is for. *Behavioral and Brain Sciences, 20*, 1–55.

Green, D. M., & Swets, J. A. (1966). *Signal detection theory and psychophysics*. New York: Wiley.

Hendy, D. B., & Farrell, P. S. E. (1997). Implementing a model of human information processing in a task network simulation environment. In *DCIEM*

N0.97-R-71. Toronto, CA: Defense and Civil Institute of Environmental Medicine.

Hull, C. L. (1943). *Principles of behavior*. New York: Appleton Century Crofts.

Jagacinski, R. J., & Flach, J. M. (2003). *Control theory for humans: Quantitative approaches to modeling performance*. Mahwah, NJ: Erlbaum.

Kalman, R. E., & Bucy, R. S. (1961). New results in linear filtering and prediction theory. *ASME Journal of Basic Engineering. 80*, 193–196.

Kieras, D. (2003). Model-based evaluation. In J. A. Jacko & A. Sears (Eds.), *The human-computer interaction handbook* (pp. 1139–1151). Mahwah, NJ: Erlbaum.

——, & Meyer, D. E. (1997). An overview of the EPIC architecture for cognition and performance with application to human-computer interaction. *Human-Computer Interaction, 12*, 391–438.

Krendel, E. S., & Barnes, G. H. (1954). *Interim report on human frequency response studies* (Technical Report 54–370). Wright Air Development Center, OH: Air Materiel Command, USAF.

Lane, N., Strieb, M., Glenn, F., & Wherry, R. (1981). The human operator simulator: An overview. In J. Moraal & K.-F. Kraiss (Eds.), *Manned systems design: Methods, equipment, and applications* (pp. 121–152). New York: Plenum Press.

Lebiere, C., Biefeld, E., Archer, R., Archer, S., Allender, L., & Kelley, T. (2002). IMPRINT/ACT-R: Integration of a task network modeling architecture with a cognitive architecture and its application to human error modeling. In *Proceedings of the Advanced Technologies Simulation Conference*. San Diego, CA.

Levison, W. H., & Cramer, N. L. (1995). *Description of the integrated driver model*. FHWA-RD-94–092. Mclean, VA: Federal Highway Administration.

Levy, G. W. (Ed.). (1968). *Symposium on applied models of man-machine systems performance*. Columbus, OH: North American Aviation.

McRuer, D. T., Graham, D., Krendel, E., & Reisener, W., Jr. (1965). *Human pilot dynamics in compensatory systems*. Air Force Flight Dynamics Lab. AFFDL-65-15.

——, & Krendel, E. S. (1957). Dynamic response of human operators. In *WADC TR-56–523*. Wright Air Development Center, OH: Air Materiel Command, USAF.

Miller, D. P., & Swain, A. D. (1987). Human error and human reliability. In G. Salvendy (Ed.), *The handbook of human factors* (pp. 219–250). New York: Wiley.

National Research Council. (1997). *Aviation safety and pilot control*. Washington, DC: National Academy Press.

Neisser, U. (1967). *Cognitive psychology*. New York: Appleton Century Crofts.

Newell, A., & Simon, H. A. (1963). GPS, a program that simulates human thought. In E. A. Feigenbaum & J. Feldman (Eds.), *Computers and thought* (pp. 279–293). Cambridge, MA: MIT Press.

Noy, I. (1990). *Attention and performance while driving with auxiliary in-vehicle displays*. (Transport Canada Publication TP 10727 (E)). Ottawa, Ontario, Canada: Transport Canada, Traffic Safety Standards and Research, Ergonomics Division.

Payne, D., & Altman, J. W. (1962). *An index of electronic equipment operability: Report of development*. Pittsburgh, PA: American Institutes for Research.

Pew, R. W., Duffendack, J. C., & Fensch, L. K. (1967). Sine-wave tracking revisited. *IEEE Transactions on Human Factors in Electronics, HFE-8*(2), 130–134.

Pew, R., & Mavor, A. (1998). *Modeling human and organizational behavior*. Washington, DC: National Academy Press.

Poulton, E. C. (1974). *Tracking skill and manual control*. New York: Academic Press.

Rosenbloom, P. A. (2001). A brief history of Soar. Retrieved from http://www.cs.cmu.edu/afs/cs/project/soar/public/www/brief-history.html

Russell, L. (1951). *Characteristics of the human as a linear servo element*. Unpublished master's thesis, Massachusetts Institute of Technology, Cambridge, MA.

Salvucci, D. D. (2001). Predicting the effects of in-car interface use on driver performance: An integrated model approach. *International Journal of Human Computer Interaction, 55*, 85–107.

———, & Macuga, K. L. (2002). Predicting the effects of cellular-phone dialing on driving performance. *Cognitive Systems Research, 3*, 95–102.

Seashore, R. H. (1928). Stanford motor skills unit. *Psychology Monographs, 39*, 51–66.

Serafin, C., Wen, C., Paelke, G., & Green, P. (1993). *Development and human factors tests of car telephones* (Technical Report UMTRI-93-17). Ann Arbor: University of Michigan Transportation Research Institute.

Siegel, A. I., & Wolf, J. J. (1969). *Man-machine simulation models: Psychosocial and performance interaction*. New York: Wiley.

Soar Suite 8.6.1. (2005). Retrieved from http://sourceforge.net/projects/soar

Swain, A. D. (1964). "THERP." Albuquerque, NM: Sandia National Laboratories.

Swain, A. D., & Guttmann, H. E. (1983). *Handbook of human reliability analysis with emphasis on nuclear power plant applications*. (NUREG/CR 1278). Albuquerque, NM: Sandia National Laboratories.

Tustin, A. (1947). The nature of the human operators response in manual control and its implication for controller design. *Journal of the Institution of Electrical Engineers, 94*, 190–201.

Wherry, R. (1969). The development of sophisticated models of man-machine system performance. In *Symposium on Applied Models of Man-Machine Systems Performance* (Report No. NR-69H-591). Columbus, OH: North American Aviation.

Wickens, C. D., & Liu, Y. (1988). Codes and modalities in multiple resources: A success and a qualification. *Human Factors, 30*, 599–616.

Wortman, D. R., Duket, S., & Seifert, D. J. (1975). SAINT simulation of a remotely piloted vehicle/drone control facility. In *Proceedings of the 19th Annual Meeting of the Human Factors Society*. Santa Monica, CA: Human Factors Society.

———, Pritsker, A. A. B., Seum, C. S., Seifert, D. J., & Chubb, G. P. (1974). *SAINT: Vol. II User's Manual* (AMRL-TR-73-128). Wright Patterson AFB, OH: Aerospace Medical Research Laboratory.

Zachary, W., Santarelli, T., Ryder, J., Stokes, J., & Scolaro, D. (2000). *Developing a multi-tasking cognitive agent using the COGNET/iGEN integrative architecture* (Technical Report 001004.9915). Spring House, PA: CHI Systems.

PART II

SYSTEMS FOR MODELING INTEGRATED
COGNITIVE SYSTEMS

Chris R. Sims & Vladislav D. Veksler

Cognitive science attempts to understand the human mind through computational theories of information processing. This approach views the complexity of the human mind as an immediate consequence of the complexity of the information processing task that it must perform at every instant. The various sensory modalities (visual, auditory, tactile, etc.) are continuous sources of new information that must be integrated with prior knowledge to determine a course of action that is appropriate to a person's goals and motivations. Not only must this wealth of information be processed, but it also must be processed under environmentally relevant timescales—decisions must be made before the outcomes of those decisions are irrelevant. In many cases, such as avoiding obstacles while walking down a sidewalk (Ballard & Sprague, chapter 20, this volume) or deciding whether to slam on the brakes of a motorcycle (Busemeyer, chapter 15, this volume), the environmentally relevant timescale is on the order of tens or hundreds of milliseconds. Given these twin constraints of massive data and limited processing time, cognitive

systems must efficiently and effectively encode, route, and transmit information so that the information available to the central controller is pertinent to the immediate problem.

This section contains four chapters concerned primarily with Type 1 theories of cognition (Gray, chapter 1, this volume)—that is, theories of central cognitive control that address the question of how the flow of information is organized and coordinated to produce an integrated cognitive system capable of behaving intelligently and effectively in a complex and naturalistic environment. The concern here is not with developing specific theories of vision, motor control, or memory, but rather with theories that integrate and facilitate the flow of information between these components to achieve human-level intelligence.

A fundamental challenge to the task of developing Type 1 theories of human cognition is that behavioral measures in any one experiment—reaction time, performance, error rate, and so on—do not enable us to distinguish the contributions of the underlying cognitive architecture from the Type 3 contributions of strategies and methods adopted by our experimental participants. Consequently, in developing a Type 1 theory of cognition, researchers must actively seek constraints above and beyond the behavioral measures of traditional cognitive psychology.

In search of such constraints, Anderson (chapter 4) turns to predictive brain imaging. A fundamental hypothesis of the ACT-R cognitive architecture is the modular organization of human cognition. Type 2 systems, for example, vision, motor control, and memory, are viewed as consisting of independent modules that coordinate their activities through communication with a central procedural module. Each of these modules ensures the coherence of the overall system by communicating via a standardized system of buffers and symbolic encoding of information.

Recent work by Anderson and colleagues has addressed the challenge of mapping each of these modules and buffers onto specific brain areas. This mapping has the direct but profound consequence that cognitive models can be compared with human performance not only in terms of behavioral measures but also the precise temporal responses of the various brain areas predicted by the model. Specifically, Anderson and colleagues have begun the task of comparing the BOLD (blood oxygen level dependent) responses from participants performing an algebra equation solving in an fMRI (functional magnetic resonance imaging) experiment to the predictions made by an ACT-R model. The predictions of the model are surprisingly accurate and depend on remarkably few free parameters. This work represents the first example of such detailed predictive brain imaging and points the way toward a powerful new approach to understanding the human cognitive architecture.

Another source of constraints is to consider the function of cognition in meeting the basic needs and motivations of the human animal. This is the approach emphasized by Sun (chapter 5) in describing the interrelationships between cognitive control, the external task environment, and needs and drives such as hunger, thirst, curiosity, or social approval. Although few existing cognitive architectures consider the functional purpose for higher-level cognition, Sun points out that a primary function of cognition is to facilitate the attainment and satisfaction of these basic motivations and that, to do so effectively, cognition must take into account the regularities and structures of an environment. The idea that cognition serves as a bridge between the motivations of an agent and its environment

is a simple idea with many consequences for understanding the importance of incorporating both basic motivations and higher-level cognition within a unified cognitive architecture. The implementation of this idea, the CLARION cognitive architecture (chapter 5), also addresses fundamental questions concerning the relationship between explicit and implicit knowledge, top-down versus bottom-up learning, and the role of metacognition in cognitive control.

The recognition that humans do not approach each new challenge or learning task with a blank slate provides another approach to developing Type 1 theories. Cassimatis (chapter 6) argues that each human competence or expertise should not be viewed as an isolated and independent phenomenon, but rather, we should seek basic computational mechanisms that could underlie much of human intelligence. Cassimatis notes that reasoning pervades much of the seemingly disparate and independent aspects of human intelligence (e.g., theories of path planning, infants' models of physical causality, sentence processing, or even cognitive control). Studying these domains independently obscures the potential for uncovering a single cognitive mechanism underlying each domain and encourages a fractured view of the human cognitive architecture. Instead, Cassimatis argues that two basic computational principles, the common function and multiple implementation principles, can be used to motivate a Type 1 theory of human cognition that integrates multiple computational mechanisms and addresses the challenges for cognitive control posed by this integration. These ideas are implemented in the Polyscheme cognitive architecture, which can be used to test predictions for cognitive control across a broad range of tasks and experimental paradigms.

Beyond the architectural constraints, it is also important to consider the unconstrained manipulation of parameters and models (Type 3 control) that occurs within the cognitive system. Brou, Egerton, and Doane (chapter 7) hold that general and accurate Type 1 theories of human cognition will necessarily fall out given the following constraints: (1) the architecture must address a wide range of tasks with minimal parameter or model manipulation; (2) cognitive modeling must be done at the level of detailed individual performance, as opposed to the overall performance of groups of human subjects. They present the construction–integration (C-I) architecture as a sample system that integrates these constraints. Using generic plans to allow for dynamic adaptation, C-I claims to explain data from many tasks, from language comprehension to aviation piloting, without many assumptions or extensive parameter fitting.

Rather than treating each chapter in this section as competing explanations for human cognition, it is hoped that the reader will recognize the common goals and unique constraints on cognition emphasized by each approach. The task of developing systems-level theories of human cognition is a daunting one. The challenges include the ability to deal with massive amounts of information from the perceptual and memory systems, but to do so in a way that is efficient in terms of meeting the temporal demands imposed by the task environment, as well as the underlying goals and motivations of the cognitive agent. Although much progress has been made, much territory remains to be explored. The four chapters of this section represent the state of the art in this endeavor and, importantly, show a strong commitment to understanding how humans achieve this level of performance through computational information processing theories.

4

Using Brain Imaging to Guide the Development of a Cognitive Architecture

We have begun to use functional magnetic resonance imaging as a way to test and extend the ACT-R theory. In this chapter, we will briefly review where we are in these efforts, describe a new modeling effort that illustrates the potential of our approach, and then end with some general remarks about the potential of such data to guide modeling efforts and the development of a cognitive architecture generally. Brain imaging has grown hand in hand with the movement to a module-based representation of knowledge in the current ACT-R theory. In this chapter, we will first review the ACT-R architecture and its application to brain imaging. ACT-R is a general system, and it is possible to take a model developed for one domain and apply that same model to a second domain. We will describe an instance of this in the second section of the chapter. Then, in the third section, we will try to draw some lessons from this work about the connections between such a modeling framework and brain imaging.

We have begun to use functional magnetic resonance imaging (fMRI) brain imaging as a way to test and extend the adaptive control of thought–rational, or ACT-R theory (Anderson & Lebiere, 1998). In this chapter, I will briefly review where we are in these efforts, describe a new modeling effort that illustrates the potential of our approach, and then end with some general remarks about the potential of such data to guide modeling efforts and the development of a cognitive architecture generally. Brain imaging has grown hand in hand with the movement to a module-based representation of knowledge in the current ACT-R theory (Anderson et al., 2005). In this chapter, we will first review the ACT-R architecture and its application to brain imaging. ACT-R is a general system, and it is possible to take a model developed for one domain and apply that same model to a second domain. We will describe an instance of this in the second section of the chapter. Then, in the third section, we will try to draw some lessons from this work about the connections between such a modeling framework and brain imaging.

ACT-R and Brain Imaging

The ACT-R Architecture

According to the ACT-R theory, cognition emerges through the interaction of a number of independent modules. Figure 4.1 illustrates the modules relevant to solving algebraic equations:

1. A visual module that might hold the representation of an equation such as "$3x - 5 = 7$."
2. A problem state module (sometimes called an *imaginal module*) that holds a current mental representation of the problem. For instance, the student might have converted the original equation into "$3x = 12$."
3. A control module (sometimes called a *goal module*) that keeps track of one's current intentions in solving the problem — for instance, the model described in Anderson (2005) alternated between unwinding an equation and retrieving arithmetic facts.

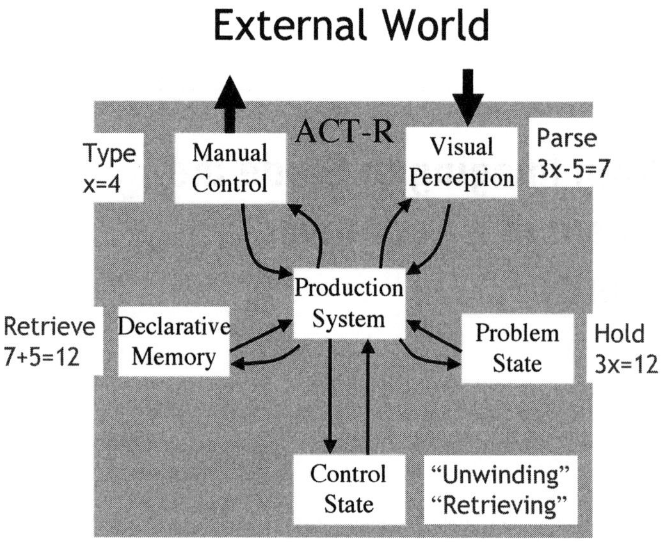

External World

FIGURE 4.1 The interconnections among modules in ACT-R 5.0.

4. A declarative module that retrieves critical information from declarative memory such as that "7 + 5 = 12."

5. A manual module that programs manual responses such as the key presses to give the response "x = 4."

Each of these modules is capable of massively parallel computation to achieve its objectives. For instance, the visual module is processing the entire visual field and the declarative module searches through large databases. However, each of these modules suffers a serial bottleneck such that only a small amount of information can be put into a buffer associated with the module—a single object is perceived, a single problem state represented, a single control state maintained, a single fact retrieved, or a single program for hand movement executed. Formally, each buffer can only hold what is called a *chunk* in ACT-R, which is a structured unit bundling a small amount of information. ACT-R does not have a formal concept of a working memory, but the current state of the buffers constitutes an effective working memory. Indeed, there is considerable similarity between these buffers and Baddeley's (1986) working memory "slave" systems.

Communication among these modules is achieved via a procedural module (production system in Figure 4.1). The procedural module can respond to information in the buffers of other modules and put information into these buffers. The response tendencies

of the central procedural module are represented in ACT-R by production rules. For instance, the following might be a production rule for transforming an equation:

IF the goal is to solve the equation

and the equation is of the form Expression − number1 = number2

and number1 + number2 = number3 has been retrieved,

THEN transform the equation to Expression = number3

This production responds when the control chunk encodes the goal to solve an equation (first line), when the problem state chunk represents an equation of the appropriate form (second line, for example, $3[x-2] - 4 = 5$), when a chunk encoding an arithmetic fact has been retrieved from memory (third line—in this case, $4 + 5 = 9$), and appropriately changes the problem representation chunk (fourth line—in this case to $3[x - 2] = 9$).

The procedural module is also capable of massive parallelism in sorting out which of its many competing rules to fire, but as with the other modules, it has a serial bottleneck in that it can only fire a single rule at a time. Since it is responsible for communication among the other modules, the production system comprises the central bottleneck (Pashler, 1994) in the ACT-R theory. Therefore, cognition can be slowed when there are simultaneous demands to process information in distinct modules. As already noted, the other modules themselves also have bottlenecks. All of the bottlenecks are in the communication among modules; within modules things are massively parallel. (Figure 4.4, later in the chapter, illustrates in some considerable detail how this parallelism and seriality mix.) Documenting the accuracy of this characterization of human cognition has been one of the preoccupations of research on ACT-R (e.g., Anderson, Taatgen, & Byrne, 2005).

Until recently, the problem state and the control state were merged into a single goal system. There have been a number of developments to improve ACT-R's goal system (Altmann & Trafton, 2002; Anderson & Douglass, 2001), and the splitting of the goal system into a control module and a problem state module is another development. There were two reasons for choosing to separate control state (goal module) and problem state knowledge (imaginal module). First (and this was the source of the idea to separate the two aspects), our imaging data indicated that the parietal region of the brain reflected changes to problem state information, while the anterior cingulate reflected control state changes. Later, the chapter will elaborate on the neural basis for this distinction. Second, the distinction offered a solution to a number of nagging problems we had with the existing system that merged the two types of knowledge. One problem was that our goal chunks often seemed too large, violating the spirit of the claim that chunks were supposed to only contain a little information. This is because they contained both problem-state information and control-state information, which both could involve a number of elements. Also, the control information was getting in the way of storing useful information about the problem solution in declarative memory. For instance, arithmetic facts such as $3 + 4 = 7$ might represent the outcome of a counting process or of an effort to comprehend a sentence. Because the control information would be different for these two sources for the same arithmetic fact, we effectively were creating parallel memories storing the same essential information. Now, with control and problem state separated, the differences between the counting and comprehension can be represented in different control chunks, while the common result would be represented identically in single problem solution chunk. By factoring control information away (in what we are now calling the goal module), one can accumulate abstract memories of the information achieved in the problem state.

Use of Brain Imaging to Provide Converging Data

We have associated these modules with specific brain regions, and fMRI allows us to track these modules individually and provide converging evidence for assumptions of the ACT-R theory. We have now completed a large number of fMRI studies of many aspects of higher-level cognition (Anderson, Qin, Sohn, Stenger, & Carter, 2003; Anderson, Qin, Stenger, & Carter, 2004; Qin et al., 2003; Sohn, Goode, Stenger, Carter, & Anderson, 2003; Sohn et al., 2005) and based on the patterns over these experiments we have made the following associations between a number of brain regions and modules in ACT-R. In this chapter, we will be concerned with five brain regions and their ACT-R associations:

1. Caudate (procedural): Centered at Talairach coordinates $x = -15, y = 9, z = 2$. This is a subcortical structure.
2. Prefrontal (retrieval): Centered at $x = -40$, $y = 21, z = 21$. This includes parts of Brodmann Areas 45 and 46 around the inferior frontal sulcus.
3. Anterior cingulate (goal): Centered at $x = -5$, $y = 10, z = 38$. This includes parts of Brodmann Areas 24 and 32.
4. Parietal (problem state or imaginal): Centered at $x = -23, y = -64, z = 34$. This includes parts of Brodmann Areas 7, 39, and 40 at the border of the intraparietal sulcus.
5. Motor (manual): Centered at $x = -37, y = -25$, $z = 47$. This includes parts of Brodmann Areas 2 and 4 at the central sulcus.

We have defined these regions once and for all and use them over and over again in predicting different experiments. This has many advantages over the typical practice in imaging research of using exploratory analyses to find out what regions are significant in particular experiments. The exploratory approach has substantial problems in avoiding false positives because there are so many experimental tests being done looking for significance in each brain voxel. To the extent that the exploratory approach can cope with this, it winds up setting very conservative criteria and fails to find many effects that occur in experiments. This had lead to the impression (e.g., Uttal, 2001) that results do not replicate over experiments.

Beyond these issues, determining regions by exploratory means is not suitable for model testing.

Being selected to pass a very conservative threshold of significance, these regions give biased estimates of the actual effect size. Also the exploratory analyses typically look for effects that are significant and not whether they are the same. This can lead to merging brain regions that actually display two (or more) different effects that are both significant. For instance, if one region shows a positive effect of a factor and an adjacent region shows a negative effect, they will be merged, and the resulting aggregate region may show no effect.

Predicting the BOLD Response

We have developed a methodology for relating the profile of activity in ACT-R modules to the blood oxygen level dependent (BOLD) responses from the brain regions that correspond to these modules. Figure 4.2 illustrates the general idea about how we map from events in an information-processing model onto the predictions of the BOLD function. Each time an information-processing component is active it will generate a demand on associated brain regions. In this hypothetical case, we assume that an ACT-R module is active for 150 ms from 0.5 to 0.65 s, for 600 ms from 1.5 to 2.1 s, and for 300 ms from 2.5 to 2.8 s. The bars at the bottom of the graph indicate when the module is active.

A number of researchers (e.g., Boyton, Engel, Glover, & Heeger, 1996; Cohen, 1997; Dale & Buckner, 1997) have proposed that the hemodynamic response to an event varies according to the following function of time, t, since the event:

$$h(t) = t^a e^{-t}, \qquad (1)$$

where estimates of the exponent have varied between 2 and 10. This is essentially a gamma function that will reach maximum at a time units after the event. As illustrated in Figure 4.2, this function is slow to rise, reflecting the lag in the hemodynamic response to neural activity.

We propose that while a module is active it is constantly producing a change that will result in a BOLD response according the above function. The observed fMRI response is integrated over the time that the module is active. Therefore, the observed BOLD response will vary with time as

$$B(t) = M \int_0^t d(x)\, h\!\left(\frac{t-x}{s}\right) dx, \qquad (2)$$

where M is the magnitude scale for response, s is the latency scale, and $d(x)$ is a "demand function" that reflects the probability that the module will be in use at time t. Note because of the scaling factor, the prediction is that the BOLD function will reach maximum at roughly $t = a \times s$ seconds.

As Figure 4.2 illustrates, one can think of the observed BOLD function in a region as reflecting the sum of separate BOLD functions for each period of time the module is active. Each period of activity is going to generate a BOLD function according to a gamma function as illustrated. The peak of the BOLD functions reflects roughly when the module was active but is offset because of the lag in the hemodynamic response. The height of the BOLD function reflects the duration of the event since the integration makes the height of the function proportional to duration over short intervals.

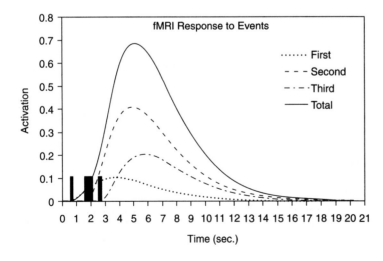

FIGURE 4.2 An illustration of how three BOLD functions from three different events result in an overall BOLD function.

Note that this model does not reflect a frequent assumption in the literature (e.g., Just, Carpenter, & Varma, 1999) that a stronger BOLD signal reflects a higher rate of metabolic expenditure. Rather, our assumption is that it reflects a longer duration of metabolic expenditure. The two assumptions are relatively indistinguishable in the BOLD functions they produce, but the time assumption more naturally maps onto an information-processing model that assumes stages taking different durations of activity. Since these processes are going to take longer, they will generate higher BOLD functions without making any extra assumptions about different rates of metabolic expenditure. The total area under the curve in Figure 4.2 will be directly proportional to the period of time that the module is active. If a module is active for a total period of time T, the area under the BOLD function will be $M \times \Gamma(a + 1) \times T$, where Γ is the gamma function (in the case of integer a, note that $\Gamma[a + 1] = a!$).

Application of an Existing Model to a New Data Set

The Anderson (2005) Algebra Model

Anderson (2005) described an ACT-R model of how children learned to solve algebra equations in an experiment reported by Qin, Anderson, Silk, Stenger, and Carter (2004). That model successfully predicted how children would speed up in their equation solving over a five-day period. The model used the general instruction-following approach described in Anderson et al. (2004) to model how children learned. Thus, it did not require handcrafting production rules specifically for the task. Rather the model used the same general instruction-following procedures described in Anderson et al. (2004) for learning of anti-air warfare coordinator (AAWC) system. That model was just given a declarative representation of the instructions that children received rather than a declarative representation of the AAWC instructions. The model initially interpreted these declarative instructions, but with practice, it built its own productions to perform the task directly. Only two parameters were estimated in Anderson (2005) to fit the model the model to latency data. One parameter, for the visual module, concerned the time to encode a fragment of instruction from the screen into an internal representation. The other parameter scaled the amount of time it took to perform

retrievals in declarative memory as a function of level of activation. All the remaining parameters were default parameters of the ACT-R model as described in Anderson et al. (2004).

Given these time estimates, that model predicted when the various modules of the ACT-R theory would be active and for how long. Moreover, it predicted how these module activities would change over the five-day course of the experiment. Thus, it generated the demand functions we needed to predict the BOLD responses in these brain regions and how these BOLD functions varied with equation complexity and practice. In general, these predictions were confirmed.

Adult Learning of Artificial Algebra

This chapter proposes to go one step further than Anderson (2005). It proposes to take the model in Anderson (2005), including the time estimates and make predictions for another experiment (Qin et al., 2003). This can be seen as a further test of the underlying model of instruction and as a further demonstration of how brain imaging can provide converging data for a theory. Participants in this experiment were adults performing an artificial algebra task (based on Blessing & Anderson, 1996) in which they had to solve "equations."[1] To illustrate, suppose the equation to be solved was

$$②P③4↔5, \tag{3}$$

where the solution means isolating the P before the "↔." In this case, the first step is to move the "③4" over to the right, inverting the "③" operator to a "②"; the equation now looks like

$$②P↔②5②4. \tag{4}$$

Then the ② in front of the P is eliminated by converting ②s on the right side into ③s so that the "solved" equation looks like:

$$P↔③5③4. \tag{5}$$

Participants were asked to perform these transformations in their heads and then key out the final answer—this involved pressing the thumb key to indicate that they had solved the problem and then keying 3, 5, 3, and 4 in this example (2 was mapped to the index finger, 3 to middle finger, 4 to ring finger, and 5 to little finger). The problems required 0, 1, or 2 (as in this example) transformations to solve. The experiment looked at how participants speed up over five days of practice. Figure 4.3 shows time to hit the first

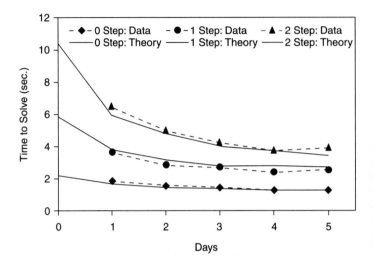

FIGURE 4.3 Mean solution times (and predictions of the ACT-R model) for the three types of equations as a function of delay. Although the data were not collected, the predicted times are presented for the practice session of the experiment (Day 0).

key (thumb press) in various conditions as a function of days.[2] The figure shows a large effect of number of transformations but also a substantial speed up over days. It also presents the predictions from the ACT-R model, which will now be described.

The ACT-R Model

Table 4.1 gives an English rendition of the instructions that were presented to the model. The general strategy of the model was to form an image of the items to the right of the "↔" and then transform that image according to the information to the left of the "↔." In addition to the instructions, we provided the model with the knowledge

TABLE 4.1 English Rendition of Task Instructions Given to ACT-R

1. To solve an equation, first find the "↔," then encode the first pair that follows, then shift attention to the next pair if there is one, then encode the second pair.
2. If this is a simple equation, output it; otherwise process the left side.
3. To process the left side, first find the P.
4. If "↔" immediately follows, then work on the operator that precedes the P; otherwise, first encode the pair that follows, then invert the operator, and then work on the operator that precedes the P.
5. To process the operator that preceded the P, first retrieve the transformation associated with that operator, then apply the transformation, and then output.
6. To output press the thumb, output the first item, output the next, output the next, and then output the next.

1. that ② and ③ were inverses of each other as were the operators ④ and ⑤.
2. the specific rules for getting rid of the ②, ③, ④, and ⑤ operators when they occurred in front of a P

These instructions and other information are encoded as declarative structures and ACT-R has general interpretative productions for converting these instructions to behavior. For instance, there is a production rule that retrieves the next step of an instruction:

IF one has retrieved an instruction for achieving a goal,

THEN retrieve the first step of that instruction

There are also productions for performing reordering operations such as

IF one's goal is to apply a transformation to an image

and that transformation involves inverting the order of the second and fourth terms

and the image is of the form "a b c d,"

THEN change the image to "a d c b"

Using such general instruction-following productions is laborious and accounts for the slow initial performance of the task.

Production compilation (see Anderson et al., 2004; Taatgen & Anderson, 2002) is one reason the model is speeding up. This is a process by which new production rules are learned that collapse what was originally done by multiple production rules. In this situation, the initial instruction-following productions are compiled over time to produce productions to embody procedures that efficiently solve equations. For instance, the following production rule is acquired:

IF the goal is to transform an image

and the prefix is ③

and the image is of the form "a b c d"

THEN change the image to "a d c b"

The model was given the same number of trials of practice as the participants received over the course of the experiment. Thus, we can look at changes in the model's performance on successive days. Figure 4.4a compares the encoding portion of a typical trial at the beginning of the Day 1 and with a typical trial at the end of the Day 5. In both cases, the model is solving the two-step equation:

$$②P③④↔②5$$

The figure illustrates when the various modules were active during the solution of the equation and what they were doing. Some general features of the activity in the figure include:

1. Multiple modules can be active simultaneously. For instance, on Day 5 there is a point where the visual module detects nothing beyond the ②5 (encode null right), while an instruction is being retrieved, while the goal module notes that it is in the encoding phase and while an image of the response "2 5" is being built up.
2. Much of the speed up in processing is driven by collapsing multiple steps into single steps. A particularly dramatic instance of this is noted in Figure 4.4 where five production firings and five retrievals on Day 1 (between "encode null right" and "encode equation ②P③④") are collapsed into one each. Production compilation can compress these internal operations without limit.

Figure 4.4b compares the transforming portion of a typical trial at the beginning of the Day 1 and with a typical trial at the end of the Day 5. The reduction in time is even more dramatic here because this portion of the trial involves the retrieval of inverse and transformation rules for getting rid of prefixes. These retrieval times show considerable speed up because of the

growth in base-level activation in the declarative representation of these basic facts. Figure 4.4c shows the output portion of a typical trial, which is identical on Days 1 and 5 since production compilation cannot collapse productions that would skip over external actions. Note, however, that the times reported in Figure 4.3 correspond to the time of the thumb press, which is the first key press. Nonetheless, the rest of Figure 4.4c will affect the BOLD response that we will see.

Brain Imaging Data

Participants were scanned on Days 1 and 5. Participants had 18 s for each trial. Figure 4.5 shows how the BOLD signal in different brain regions varies over the 18-s period beginning 3 s before the onset of the stimulus and continuing for 15 s afterward. Activity was measured every 1.5 s. The first two scans provide an estimate of baseline before the stimulus comes on. These figures also display the ACT-R predictions. The BOLD functions displayed are typical in that there is some inertia in the rise of the signal after the critical event and then decay. The BOLD response is delayed so that it reaches a maximum about 4–5 s after the brain activity. In each part of Figure 4.5 we provide a representation of the effect of problem complexity averaging over number of days and a representation of the effect of practice, averaging over problem complexity. None of the regions showed a significant interaction between practice and number of steps or between practice, number of steps, and scan.

Figure 4.5a shows the activity around the left central sulcus in the region that controls the right hand. The effect of complexity is to delay the BOLD function (because the first finger press is delayed in the more complex condition), but there is no effect on the basic shape of the BOLD response because the same response sequence is being generated in all cases. The effect of practice is also just to move the motor BOLD response forward in time.

Figure 4.5b shows the activity around the left inferior frontal sulcus, which we take as reflecting the activity of the retrieval module. It shows very little rise in the zero transformation condition because there are few retrievals (only of a few instructions) in this condition. The lack of response in this condition distinguishes this region from most others. The magnitude of the response decreases after five days, reflecting that the declarative structures have been greatly strengthened and the retrievals are much quicker.

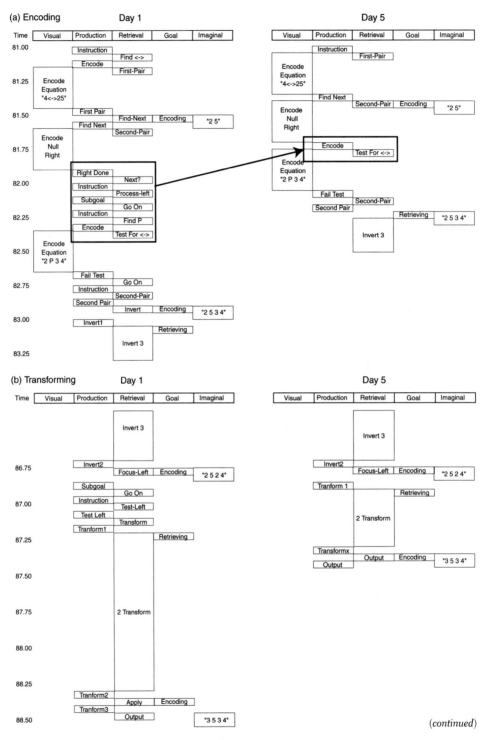

FIGURE 4.4 Module activity during the three phases of a trial: (a) encoding, (b) transforming, and (c) outputting. In the first two phases, the module activity changes from Day 1 to Day 5.

(continued)

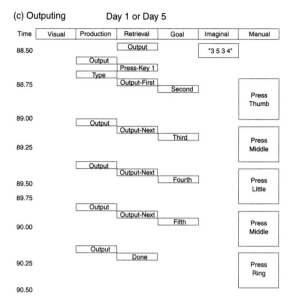

FIGURE 4.4 Continued.

Figure 4.5c shows activity in the left anterior cingulate, which we take as reflecting control activity, and Figure 4.5d shows activity around the left intraparietal sulcus, which we take as reflecting changes to the problem representation. Both of these regions show large effects of problem complexity and little effect of number of days of practice. Unlike the prefrontal region, they show a large response in the condition of zero transformations. There is virtually no effect of practice on the anterior cingulate. According to the ACT-R theory, this is because the model still goes through the same control states, only more rapidly on Day 5. In the case of the parietal region and its association with problem representation, there is a considerable drop out of intermediate problem representations, but most of this happens early in the learning and therefore not much further learning occurs from Day 1 to Day 5.

Figure 4.5e shows the activity in the caudate, which is taken to reflect production firing. The signal is rather weak, here but there appears to be little effect of complexity and a substantial effect of practice. The effect of complexity is predicted to be weak by the model because most of the time associated with transformation is taken up in long retrievals and not many additional productions are required. The model underpredicts the effect of learning for much the same reason it predicts a weak effect of practice in the parietal. The effects of practice on number of productions

tends to happen early in this experiment and there is not that much reduction after Day 1.

Comments on Model Fitting

The model that yields the fits displayed in these figures was run without estimating any time parameters. This makes the fit to the latency data in Figure 4.3 truly parameter free, and it is remarkable how well that data does fit, given that we estimated parameters with children and now are fitting them to adults. At some level, this indicates that the children were finding learning real algebra as much of a novel experience as these adults were finding learning the artificial algebra and were taking about as long to do the task.

In the case of fitting the BOLD functions, however, we had to allow ourselves to estimate some parameters that describe the underlying BOLD response. To review, there were three parameters—an exponent a that governs the shape of the BOLD response; a timescale parameter s that, along with a, determines the time to peak ($a \times s = $ peak); and a magnitude parameter m that determines just how much increase there is in a region. Table 4.2 summarizes the values of these parameters for this experiment with adults and artificial algebra and the previous experiment with children and real algebra.

We used the same value of a for both experiments and all regions. This value is 3 and it seems to give us

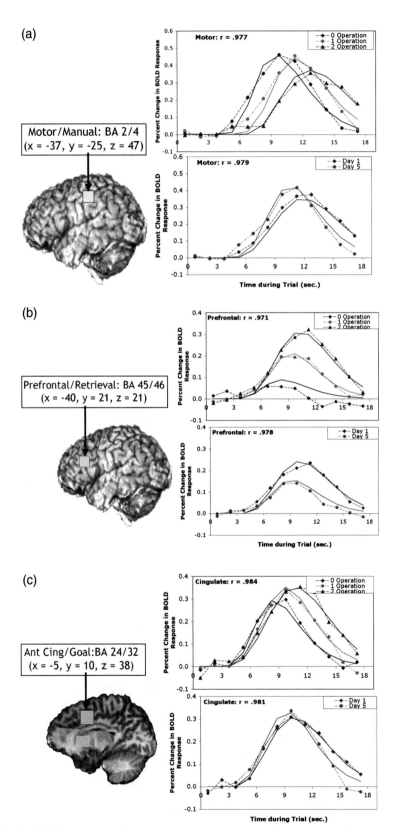

(continued)

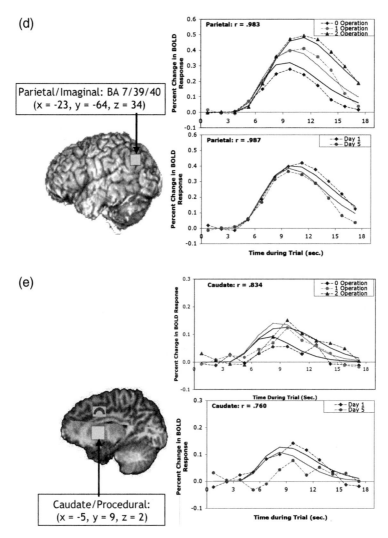

FIGURE 4.5 Use of module behavior to predict BOLD response in various regions: (a) manual module predicts motor region; (b) retrieval module predicts prefrontal region; (c) control/goal module predicts anterior cingulate region; (d) imaginal/problem state module predicts parietal region; (e) procedural module predicts caudate region.

a pretty good fit over a wide range of situations. The value of the latency scale parameter was estimated separately for each region in both experiments. It shows only modest variability and has a value of approximately 1.5 s, which would be consistent with the general observation that it is about 4.5 s for the BOLD response to peak. There is some variability in the BOLD response across subjects and regions (e.g., Huettel & McCarthy, 2000; Kastrup, Krüger, Glover, Neumann-Haefelin, & Moseley, 1999).

The situation with the magnitude parameter, however, does reveal some discrepancies that go beyond

naturally expected variation. In particular, our experiment has estimated a motor magnitude that is less than 40% of the magnitude estimated for the children and a parietal magnitude that is almost four times as large. It is possible that these reflect differences in population, perhaps related to age, but such an explanation does not seem very plausible.

In the case of the parietal region, we think that the difference in magnitude may be related to the difficulty in manipulating the expressions. While this is the first time the children were exposed to equations, these expressions had a lot of similarity to other sorts of

TABLE 4.2 Parameters Estimated and Fits to the Bold Response $B(t) = m \left(\dfrac{t}{s} \right)^a e^{-(t/s)}$

		Motor/ Manual	Prefrontal/ Retrieval	Parietal/ Imaginal	Cingulate/ Goal	Caudate/ Procedural
Magn(m)	Children	0.531	0.073	0.231	0.258	0.207
	Adults	0.197	0.078	0.906	0.321	0.120
Exponent(a)		3	3	3	3	3
Scale(s)	Children	1.241	1.545	1.645	1.590	1.230
	Adults	1.360	1.299	1.825	1.269	1.153

arithmetic expressions children had seen before in their lives. In contrast, the expressions in the artificial algebra that the adults saw were quite unlike anything experienced before. One might have expected that this would be reflected in different times to parse them but we used the same estimates as with the children—0.1s for each box in the imaginal columns of Figure 4.4. If we increased this estimate, however, we would have had to decrease some other time estimate to fit the latency data.

In the case of the motor region, we think that the difference in magnitude may be related to the different number of key presses. The adults in this experiment had to press five keys to indicate their answer, while the children had only to press one key. There is some indication (e.g., Glover, 1999) that the BOLD response may be subadditive.

Both discrepancies reflect on fundamental assumptions underlying our modeling effort. In the case of the parietal region, it may be that the same region working for the same time may produce a different magnitude response, depending on how "difficult" the task is. In the case of the motor region, it may be the case that our additivity assumption is flawed.

While acknowledging that there might be some flies in the ointment with respect to parameter estimates, it is still worth asking how well the model does fit the data. We have presented in these figures measures of correlation between data and theory. While these are useful qualitative indicants, they really do not tell us whether the deviations from data are "significant." Addressing this question is both a difficult and questionable enterprise, but I thought it would be useful to report our approach. We obtained from an analysis of variance how much the data varied from subject

to subject. This is measured as the subject-by-condition interaction term, where the conditions are the 72 observations obtained by crossing difficulty (3 values) with days (2 values) with scans (12 values). This gives us an error of estimate of the mean numbers going into the figures as data (although in these figures we have averaged over one of the factors). We divided the sum of the squared deviations by this error term and obtained a chi-square quantity:

$$\chi^2 = \frac{\sum_i (\hat{X}_i - \bar{X}_i)^2}{S_{\bar{X}}^2}, \quad (6)$$

which has degrees of freedom equal to the number of observations being summed (72) minus the number of parameters estimated (2—latency scale and magnitude). With 70 degrees of freedom, this statistic is significant if greater than 90.53. The chi-square values for four of the five regions are not significant (motor, 70.42; prefrontal, 46.91; cingulate, 48.25; parietal, 88.86), but the estimate for the caudate is with a chi-square measure of 99.56. It turns out that a major discrepancy for the caudate is that the BOLD function rises too fast. If we allow an exponent of 5 (and so change the shape of the BOLD response), we get a chi-square deviation of only 79.23 for the caudate.

It is wise not to make too much of these chi-square tests as we are just failing to reject the null hypothesis. There may be real discrepancies in the model's fit that are hidden by noise in the data. The chi-square test is just one other tool available to a modeler and sometimes (as in the case of the caudate) it can alert one to a discrepancy between theory and data.

Conclusions

The use of fMRI brain imaging has both influenced the development of the current ACT-R theory and provided support for the state of that theory. For instance, it was one of the reasons for the separation of the previous goal structure into a structure that just held control information (currently called the *goal*) and a structure that contained information about the problem state (now called an *imaginal module*). Besides giving us a basis for testing a model fit, the data provided some converging evidence for major qualitative claims of the model—such as that there was little retrieval in the zero transformation condition and that there was little effect of learning in this experiment on control information.

While things are encouraging at a general level, our discussion of the details of the model fitting suggested that there are some things that remain to be worked out. We saw uncertainty about a key assumption that magnitude of the BOLD response only reflects time a module is active. Differences in the magnitude of response in the two experiments in the parietal region suggested that there be different magnitude of effort in a fixed time. Again differences in magnitude of response in the motor region suggested that BOLD effects might be subadditive. On another front, problems in fitting the caudate raised the question of whether all the regions are best fit by the same shape parameter. While use of brain imaging data is a promising tool, it is apparent we are still working out how to use that tool.

We should note that there is no reason such data and methodology should be limited to testing the ACT-R theory. Many other information-processing theories could be tested. The basic idea is that the BOLD response reflects the duration for which various cognitive modules are active. The typical additive-factors information-processing methodology has studied how manipulations of various cognitive components affect a single aggregate behavioral measure like total time. If we can assign these different components to different brain regions, we have essentially a separate dependent measure to track each component. Therefore, this methodology promises to offer strong guidance in the development of any information-processing theory. Finally, we want to comment on the surprising match of fMRI methodology to the study of complex tasks. A problem with fMRI is its poor temporal resolution. However, as is particularly apparent in the behavior of our manual module, the typical effect size in a complex mental task is such that one can still make temporal discriminations in fMRI data. One might have thought the outcome of such a complex task would be purely uninterpretable. However, with the guidance of a strong information-processing model and well-trained participants one not only can interpret but also predict the BOLD response in various regions of the brain.

Acknowledgments

This research was supported by the National Science Foundation Grant ROLE: REC-0087396 and ONR Grant N00014–96–1–0491. I would like to thank Jennifer Ferris, Wayne Gray, and Hansjörg Neth for their comments on this chapter. Correspondence concerning this chapter should be addressed to John R. Anderson, Department of Psychology, Carnegie Mellon University, Pittsburgh, PA 15213. Electronic mail may be sent to ja+@cmu.edu.

Notes

1. The reason for using an artificial algebra is that these participants already knew high school algebra, and we wanted to observe learning.

2. Note that there is a Day 0 when subjects practiced the different aspects of the task but were not metered in a regular task set; see Qin et al. (2003) for details.

References

Altmann, E. M., & Trafton, J. G. (2002). Memory for goals: An activation-based model. *Cognitive Science, 26,* 39–83.

Anderson, J. R. (2005). Human symbol manipulation within an integrated cognitive architecture, *Cognitive Science, 29,* 313–342.

——, Bothell, D., Byrne, M. D., Douglass, S., Lebiere, C., & Qin, Y. (2004). An integrated theory of mind. *Psychological Review, 111,* 1036–1060.

——, & Douglass, S. (2001). Tower of Hanoi: Evidence for the cost of goal retrieval. *Journal of Experimental Psychology: Learning, Memory, & Cognition, 27,* 1331–1346.

——, & Lebiere, C. (1998). *The atomic components of thought.* Mahwah, NJ: Erlbaum.

——, Qin, Y., Sohn, M.-H., Stenger, V. A., & Carter, C. S. (2003). An information-processing model of the BOLD response in symbol manipulation tasks. *Psychonomic Bulletin & Review, 10,* 241–261.

——, Qin, Y., Stenger, V. A., & Carter, C. S. (2004). The relationship of three cortical regions to an information-processing model. *Journal of Cognitive Neuroscience, 16,* 637–653.

——, Taatgen, N. A., & Byrne, M. D. (2005). Learning to achieve perfect time sharing: architectural implications of Hazeltine, Teague, & Ivry (2002). *Journal of Experimental Psychology: Human Perception and Performance, 31,* 749–761.

Baddeley, A. D. (1986). *Working memory.* Oxford: Oxford University Press.

Blessing, S., & Anderson, J. R. (1996). How people learn to skip steps. *Journal of Experimental Psychology: Learning, Memory and Cognition, 22,* 576–598.

Boyton, G. M., Engel, S. A., Glover, G. H., & Heeger, D. J. (1996). Linear systems analysis of functional magnetic resonance imaging in human V1. *Journal of Neuroscience, 16,* 4207–4221.

Cohen, M. S. (1997). Parametric analysis of fMRI data using linear systems methods. *NeuroImage, 6,* 93–103.

Dale, A. M., & Buckner, R. L. (1997). Selective averaging of rapidly presented individual trials using fMRI. *Human Brain Mapping, 5,* 329–340.

Glover, G. H. (1999). Deconvolution of impulse response in event-related BOLD fMRI. *NeuroImage, 9,* 416–429.

Huettel, S., & McCarthy, G. (2000). Evidence for refractory period in the hemodynamic response to visual stimuli as measured by MRI. *NeuroImage, 11,* 547–553.

Just, M. A., Carpenter, P. A., & Varma, S. (1999). Computational modeling of high-level cognition and brain function. *Human Brain Mapping, 8,* 128–136.

Kastrup, A., Krüger, G., Glover, G. H., Neumann-Haefelin, T., & Moseley, M. E. (1999). Regional variability of cerebral blood oxygenation response to hypercapnia. *NeuroImage, 10,* 675–681.

Pashler, H. (1994). Dual-task interference in simple tasks: Data and theory. *Psychological Bulletin, 116,* 220–244.

Qin, Y., Sohn, M.-H., Anderson, J. R., Stenger, V. A., Fissell, K., Goode, A., et al. (2003). Predicting the practice effects on the blood oxygenation level-dependent (BOLD) function of fMRI in a symbolic manipulation task. *Proceedings of the National Academy of Sciences of the United States of America, 100,* 4951–4956.

——, Anderson, J. R., Silk, E., Stenger, V. A., & Carter, C. S. (2004). The change of the brain activation patterns along with the children's practice in algebra equation solving. *Proceedings of National Academy of Sciences, 101,* 5686–5691.

Sohn, M.-H., Goode, A., Stenger, V. A., Carter, C. S., & Anderson, J. R. (2003). Competition and representation during memory retrieval: Roles of the prefrontal cortex and the posterior parietal cortex. *Proceedings of National Academy of Sciences, 100,* 7412–7417.

——, Goode, A., Stenger, V. A., Jung, K.-J., Carter, C. S., & Anderson, J. R. (2005). An information-processing model of three cortical regions: Evidence in episodic memory retrieval. *NeuroImage, 25,* 21–33.

Taatgen, N. A., & Anderson, J. R. (2002). Why do children learn to say "broke"? A model of learning the past tense without feedback. *Cognition, 86,* 123–155.

Uttal, W. R. (2001). *The new phrenology: The limits of localizing cognitive processes in the brain.* Cambridge, MA: MIT Press.

5

The Motivational and Metacognitive Control in CLARION

Ron Sun

This chapter presents an overview of a relatively recent cognitive architecture and its internal control structures, that is, its motivational and metacognitive mechanisms. The chapter starts with a look at some general ideas underlying this cognitive architecture and the relevance of these ideas to cognitive modeling of agents. It then presents a sketch of some details of the architecture and their uses in cognitive modeling of specific tasks.

This chapter presents an overview of a relatively recent cognitive architecture and its internal control structures (i.e., motivational and metacognitive mechanisms) in particular. We will start with a look at some general ideas underlying this cognitive architecture and the relevance of these ideas to cognitive modeling.

In the attempt to tackle a host of issues arising from computational cognitive modeling that are not adequately addressed by many other existent cognitive architectures, CLARION, a modularly structured cognitive architecture, has been developed (Sun, 2002; Sun, Merrill, & Peterson, 2001). Overall, CLARION consists of a number of functional subsystems (e.g., the action-centered subsystem, the metacognitive subsystem, and the motivational subsystem). It also has a dual representational structure—implicit and explicit representations in two separate components in each subsystem. Thus far, CLARION has been successful in capturing a variety of cognitive processes in a variety of task domains based on this division of modules (Sun, 2002; Sun, Slusarz, & Terry, 2005).

A key assumption of CLARION, which has been argued for amply before (see Sun, 2002; Sun et al., 2001; Sun et al., 2005), is the dichotomy of implicit and explicit cognition. In general, implicit processes are less accessible and more "holistic," while explicit processes are more accessible and crisper (Reber, 1989; Sun, 2002). This dichotomy is closely related to some other well-known dichotomies in cognitive science: the dichotomy of symbolic versus subsymbolic processing, the dichotomy of conceptual versus subconceptual processing, and so on (Sun, 1994). The dichotomy can be justified psychologically, by the voluminous empirical studies of implicit and explicit learning, implicit and explicit memory, implicit and explicit perception, and so on (Cleeremans, Destrebecqz, & Boyer, 1998; Reber 1989; Seger, 1994; Sun, 2002). In social psychology, there are similar dual-process models, for describing socially relevant cognitive processes (Chaiken & Trope, 1999). Denoting more or less the same distinction, these dichotomies serve as justifications for the more general notions of implicit versus explicit cognition, which is the focus of CLARION. See Sun (2002) for an extensive treatment of this distinction.

Besides the above oft-reiterated point about CLARION, there are also a number of other characteristics that are especially important. For instance, one particularly pertinent characteristic of this cognitive architecture is its focus on the cognition–motivation–environment interaction. Essential motivations of an agent, its biological needs in particular, arise naturally, before cognition (but interact with cognition of course). Such motivations are the foundation of action and cognition. In a way, cognition is evolved to serve the essential needs of an agent. Cognition, in the process of helping to satisfy needs and following motivational forces, has to take into account environments, their regularities, and structures. Thus, cognition bridges the needs and motivations of an agent and its environments (be it physical or social), thereby linking all three in a "triad" (Sun, 2004, 2005).

Another important characteristic of this architecture is that multiple subsystems interact with each other constantly. In this architecture, these subsystems have to work closely with each other to accomplish cognitive processing. The interaction among these subsystems may include metacognitive monitoring and regulation. The architecture also includes motivational structures and therefore the interaction also includes that between motivational structures and other subsystems. These characteristics are significantly different from other cognitive architectures such as ACT-R (adaptive control of thought–rational) and Soar.

Yet another important characteristic of this cognitive architecture is that an agent may learn on its own, regardless of whether there is a priori or externally provided domain knowledge. Learning may proceed on a trial-and-error basis. Furthermore, through a bootstrapping process, or *bottom-up learning*, as it has been termed (Sun et al., 2001), explicit and abstract domain knowledge maybe developed, in a gradual and incremental fashion (Karmiloff-Smith 1986). This is significantly different from other cognitive architectures (e.g., Anderson & Lebiere, 1998).

Although it addresses trial-and-error and bottom-up learning, the architecture does not exclude innate biases and innate behavioral propensities from being represented within the architecture. Innate biases and propensities may be represented, implicitly or even explicitly, and they interact with trial-and-error and bottom-up learning by way of constraining, guiding,

and facilitating learning. In addition to bottom-up learning, top-down learning, that is, assimilation of explicit/abstract knowledge from external sources into implicit forms, is also possible in CLARION (Sun, 2003).

In the remainder of this chapter, first, justifications for CLARION are presented in the next section. Then, the overall structure of CLARION is presented. Each subsystem is presented in subsequent sections. Together, these sections substantiate all of the characteristics of CLARION previously discussed. Various prior simulations using CLARION are summarized in the section following them. Some concluding remarks then complete this chapter.

Why Model Motivational and Metacognitive Control

It is not too far-fetched to posit that cognitive agents must meet the following criteria in their activities (among many others):

- Sustainability: An agent must attend to its basic needs, such as hunger and thirst. The agent must also know to avoid danger and so on (Toates, 1986).
- Purposefulness: The action of an agent must be chosen in accordance with some criteria, instead of completely randomly (Anderson & Lebiere, 1998; Hull, 1951). Those criteria are related to enhancing sustainability of an agent (Toates, 1986).
- Focus: An agent must be able to focus its activities in some ways, with respect to particular purposes. Its actions need to be consistent, persistent, and contiguous, to fulfill its purposes (Toates, 1986). However, an agent needs to be able to give up some of its activities, temporally or permanently, when necessary (Simon, 1967; Sloman, 2000).
- Adaptivity: An agent must be able to adapt its behavior (i.e., to learn) to improve its purposefulness, sustainability, and focus.

Within an agent, two types of control are present: the primary control of actions affecting the external environment and the secondary (internal) control by motivational and metacognitive mechanisms. To meet these criteria above, motivational and metacognitive processes are necessary, especially to deal with issues of purpose and focus. Furthermore, to foster integrative work to counteract the tendency of fragmentation in cognitive science into narrow and isolated subdisciplines, it is necessary to consider seriously the overall architecture of the mind that incorporates, rather than excludes, important elements such as motivations and metacognition. Furthermore, it is beneficial to translate into architectural terms the understanding that has been achieved of the interrelations among cognitive, metacognitive, motivational, and emotional aspects of the mind (Maslow, 1962, 1987; Simon, 1967; Toates, 1986; Weiner, 1992). In doing so, we may create a more complete picture of the structuring of the mind, and an overall understanding of the interaction among cognition, motivation, metacognition, and so on.

Compared with other existent cognitive architectures, CLARION is unique in that it contains (1) built-in motivational constructs and (2) built-in metacognitive constructs. These features are not commonly found in other existing cognitive architectures. Nevertheless, we believe that these features are crucial to the enterprise of cognitive architectures, as they capture important elements in the interaction between an agent and its physical and social world.

For instance, without motivational constructs, a model agent would be literally aimless. It would wander around the world aimlessly accomplishing hardly anything. Or it would have to rely on knowledge hand coded into it, for example, regarding goals and procedures (Anderson & Lebiere, 1998), to accomplish some relatively minor things, usually only in a controlled environment. Or it would have to rely on external "feedback" (reinforcement, reward, punishment, etc.) to learn. But the requirement of external feedback begs the question of how such a signal is obtained in the natural world. In contrast, with a motivational subsystem as an integral part of CLARION, it is able to generate such feedback internally and learn on that basis, without requiring a "special" external feedback signal or externally provided and hand coded a priori knowledge (Edelman, 1992).

This mechanism is also important for social interaction. Each agent in a social situation carries with it its own needs, desires, and motivations. Social interaction is possible in part because agents can understand and appreciate each other's (innate or acquired) motivational structures (Bates, Loyall, & Reilly, 1992; Tomasello, 1999). On that basis, agents may find ways to cooperate.

Similarly, without metacognitive control, a model agent may be blindly single-minded: It will not be able to flexibly and promptly adjust its own behavior. The ability of agents to reflect on, and to modify dynamically, their own behaviors is important to achieve effective behaviors in complex environments. Note also that social interaction is made possible by the (at least partially) innate ability of agents to reflect on, and to modify dynamically, their own behaviors (Tomasello, 1999). The metacognitive self-monitoring and control enables agents to interact with each other and with their environments more effectively, for example, by avoiding social impasse, which are created because of the radically incompatible behaviors of multiple agents (see, e.g., Sun, 2001). Such cognitive–metacognitive interaction has not yet been fully addressed by other cognitive architectures such as ACT-R or Soar (but see, e.g., Sloman, 2000).

Note that the duality of representation, and the concomitant processes and mechanisms, are present in, and affect thereby, both the primary control of actions and also the secondary control, that is, the motivational and metacognitive processes. Computational modeling may capture details of the duality of representation, in both the primary and the secondary control processes.

Furthermore, to understand computational details of motivational and metacognitive processes, many questions specific to the computational understanding of motivation and metacognition need to be asked. For example, how can the internal drives, needs, and desires of an agent be represented? Are they explicitly represented (as symbolist/logicist AI would suggest), or are they implicitly represented (in some ways)? Are they transient, or are they relatively invariant temporally? How do contexts affect their status? How do their variations affect performance? How can an agent exert control over its own cognitive processes? What factors determine such control? How is the control carried out? Is the control explicit or implicit? In the remainder of this chapter, details of motivational and metacognitive processes will be developed. Computational modeling provides concrete and tangible answers to many of these questions. That is why computational modeling of motivational and metacognitive control is useful.

The Overall Architecture

CLARION is intended for capturing essential cognitive processes within an individual cognitive agent.

CLARION is an integrative architecture, consisting of a number of distinct subsystems, with a dual representational structure in each subsystem (implicit vs. explicit representations). Its subsystems include the action-centered subsystem (the ACS), the non-action-centered subsystem (the NACS), the motivational subsystem (the MS), and the metacognitive subsystem (the MCS). See Figure 5.1 for a sketch of the architecture. The role of the ACS is to control actions, regardless of whether the actions are for external physical movements or internal mental operations. The role of the NACS is to maintain general knowledge, either implicit or explicit. The role of the MS is to provide underlying motivations for perception, action, and cognition, in terms of providing impetus and feedback (e.g., indicating whether outcomes are satisfactory). The role of the MCS is to monitor, direct, and modify the operations of the ACS dynamically as well as the operations of all the other subsystems.

Each of these interacting subsystems consists of two levels of representation (i.e., a dual representational structure): Generally, in each subsystem, the top level encodes explicit knowledge and the bottom level encodes implicit knowledge; this distinction has been argued for earlier (see also Reber, 1989; Seger, 1994; Cleeremans et al., 1998). Let us consider the representational forms that need to be present for encoding these two types of knowledge. Notice that the relatively inaccessible nature of implicit knowledge may be captured by subsymbolic, distributed representation provided, for example, by a back-propagation network (Rumelhart, McClelland, & PDP Research Group, 1986). This is because distributed representational units in the hidden layer(s) of a back-propagation network are capable of accomplishing computations but are subsymbolic and generally not individually meaningful (Rumelhart et al., 1986; Sun, 1994). This characteristic of distributed representation, which renders

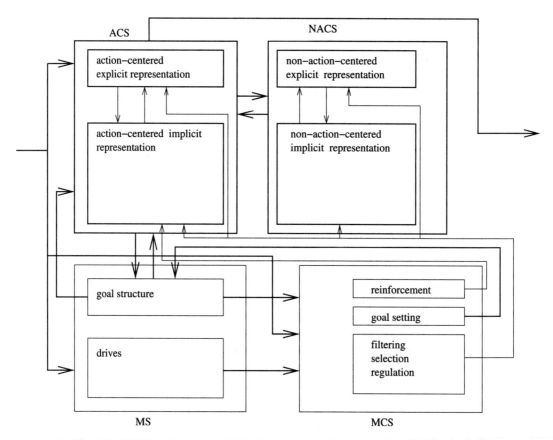

FIGURE 5.1 The CLARION architecture. ACS denotes the action-centered subsystem, NACS the non-action-centered subsystem, MS the motivational subsystem, and MCS the metacognitive subsystem.

the representational form less accessible, accords well with the relative inaccessibility of implicit knowledge (Cleeremans et al., 1998; Reber, 1989; Seger, 1994). In contrast, explicit knowledge may be captured in computational modeling by symbolic or localist representation (Clark and Karmiloff-Smith, 1993), in which each unit is more easily interpretable and has a clearer conceptual meaning. This characteristic of symbolic or localist representation captures the characteristic of explicit knowledge being more accessible and more manipulable (Sun, 1994).

Accessibility here refers to the direct and immediate availability of mental content for the major operations that are responsible for, or concomitant with, consciousness, such as introspection, forming higher-order thoughts, and verbal reporting. The dichotomous difference in the representations of the two different types of knowledge leads naturally to a two-level architecture, whereby each level uses one kind of representation and captures one corresponding type of process (implicit or explicit).

Let us now turn to learning. First, there is the learning of implicit knowledge at the bottom level. One way of implementing a mapping function to capture implicit knowledge is to use a multilayer neural network (e.g., a three-layer back-propagation network). Adjusting parameters of this mapping function to change input/output mappings (i.e., learning implicit knowledge) may be carried out in ways consistent with the nature of distributed representation (e.g., as in back-propagation networks), through trial-and-error interaction with the world. Often, reinforcement learning can be used (Sun et al., 2001), especially Q-learning (Watkins, 1989), implemented using back-propagation networks. In this learning setting, there is no need for a priori knowledge or external teachers providing desired input/output mappings. Such (implicit) learning may be justified cognitively. For instance, Cleeremans (1997) argued at length that implicit learning could not be captured by symbolic models but neural networks. Sun (1999) made similar arguments.

Explicit knowledge at the top level can also be learned in a variety of ways (in accordance with localist/symbolic representation used there). Because of its representational characteristics, one-shot learning (e.g., based on hypothesis testing) is preferred during interaction with the world (Bruner, Goodnow, & Austin, 1956; Busemeyer & Myung, 1992; Sun et al., 2001). With such learning, an agent explores the world

and dynamically acquires representations and modifies them as needed.

The implicit knowledge already acquired in the bottom level may be used in learning explicit knowledge at the top level, through bottom-up learning (Sun et al., 2001). That is, information accumulated in the bottom level through interacting with the world is used for extracting and then refining explicit knowledge. This is a kind of "rational reconstruction" of implicit knowledge at the explicit level. Conceivably, other types of learning of explicit knowledge are also possible, such as explicit hypothesis testing without the help of the bottom level. Conversely, once explicit knowledge is established at the top level, it may be assimilated into the bottom level. This often occurs during the novice-to-expert transition in instructed learning settings (Anderson and Lebiere, 1998). The assimilation process, known as top-down learning (as opposed to bottom-up learning), may be carried out in a variety of ways (Anderson and Lebiere, 1998; Sun, 2003).

Figure 5.1 presents a sketch of this basic architecture of a cognitive agent, which includes the four major subsystems interacting with each other. The following four sections will describe, one by one and in more detail, these four subsystems of CLARION. We will first look into the ACS, which is mostly concerned with the control of the interaction of an agent with its environment, as well as the NACS, which is under the control of the ACS. On the basis of these two subsystems, we will then focus on the MS and the MCS, which provide another layer of (secondary) control on top of the ACS and the NACS.

The Action-Centered Subsystem

The action-centered subsystem (ACS) of CLARION is meant to capture the action decision making of an individual cognitive agent in its interaction with the world, that is, the primary control of actions of an agent. The ACS is the most important part of CLARION. In the ACS, the process for action decision making is essentially the following: Observing the current state of the world, the two levels of processes within the ACS (implicit or explicit) make their separate decisions in accordance with their own knowledge, and their outcomes are somehow "combined." Thus, a final selection of an action is made and the action is then performed. The action changes the world in some way. Comparing the changed state of the world with the previous state, the agent learns (e.g., in accordance with Q-learning of Watkins, 1989). The cycle then repeats itself.

In this subsystem, the bottom level is termed the *implicit decision networks* (IDNs), implemented with neural networks involving distributed representations, and the top level is termed the *action rule store* (ARS), implemented using symbolic/localist representations.

The overall algorithm for action decision making during the interaction of an agent with the world is as follows:

1. Observe the current state x.
2. Compute in the bottom level (the IDNs) the "value" of each of the possible actions (a_i's) associated with the state x: $Q(x, a_1)$, $Q(x, a_2)$, . . . , $Q(x, a_n)$. Stochastically choose one action according to these values.
3. Find out all the possible actions ($b_1, b_2, . . . , b_m$) at the top level (the ARS), based on the current state x (which goes up from the bottom level) and the existing rules in place at the top level. Stochastically choose one action.
4. Choose an appropriate action by stochastically selecting the outcome of either the top level or the bottom level.
5. Perform the action, and observe the next state y and (possibly) the reinforcement r.
6. Update the bottom level in accordance with an appropriate algorithm (to be detailed later), based on the feedback information.
7. Update the top level using an appropriate algorithm (for extracting, refining, and deleting rules, to be detailed later).
8. Go back to Step 1.

The input (x) to the bottom level consists of three sets of information: (1) sensory input, (2) working memory items, (3) the selected item of the goal structure. The sensory input is divided into a number of input dimensions, each of which has a number of possible values. The goal input is also divided into a number of dimensions. The working memory is divided into dimensions as well. Thus, input state x is represented as a set of dimension–value pairs: (d_1, v_1) (d_2, v_2) . . . (d_n, v_n).

The output of the bottom level is the action choice. It consists of three groups of actions: working memory actions, goal actions, and external actions.[1]

In each network (encoding implicit knowledge), actions are selected based on their values. A Q value is an evaluation of the "quality" of an action in a given state: $Q(x, a)$ indicates how desirable action a is in state x. At each step, given state x, the Q values of all the actions (i.e., $Q[x, a]$ for all a's) are computed. Then the Q values are used to decide probabilistically on an action to be performed, through a Boltzmann distribution of Q values:

$$p(a \mid x) = \frac{e^{Q(x,a)/\alpha}}{\sum_i e^{Q(x,a_i)/\alpha}} \qquad (1)$$

where α controls the degree of randomness (temperature) of the decision-making process. (This method is also known as Luce's choice axiom; Watkins, 1989.)

The Q-learning algorithm (Watkins, 1989), a reinforcement learning algorithm, is used for learning implicit knowledge at the bottom level. In the algorithm, $Q(x, a)$ estimates the maximum (discounted) total reinforcement that can be received from the current state x on. Q values are gradually tuned, on-line, through successive updating, which enables reactive sequential behavior to emerge through trial-and-error interaction with the world. Q-learning is implemented in backpropagation networks (see Sun, 2003, for details).

Next, explicit knowledge at the top level (the ARS) is captured by *rules* and *chunks*. The condition of a rule, similar to the input to the bottom level, consists of three groups of information: sensory input, working memory items, and the current goal. The output of a rule, similar to the output from the bottom level, is an action choice. It may be one of the three types: working memory actions, goal actions, and external actions. The condition of a rule constitutes a distinct entity known as a chunk; so does the conclusion of a rule.

Specifically, rules are in the following form: *state-specification action*. The left-hand side (the condition) of a rule is a conjunction (i.e., logic AND) of individual elements. Each element refers to a dimension x_i of state x, specifying a value range, for example, in the form of $x_i \in (v_{i1}, v_{i2}, . . . , v_{in})$. The right-hand side (the conclusion) of a rule is an action recommendation.

The structure of a set of rules may be translated into that of a network at the top level. Each value of each state dimension (i.e., each feature) is represented by an individual node at the bottom level (all of which together constitute a distributed representation). Those bottom-level feature nodes relevant to the condition of a rule are connected to the single node at the top level, representing that condition, known as a chunk node (a localist representation). When given a set of rules, a rule network can be wired up at the top level, in which conditions and conclusions of rules are represented by respective chunk nodes and links representing rules are established that connect corresponding pairs of chunk nodes.

To capture the bottom-up learning process (Karmiloff-Smith, 1996; Stanley, Mathews, Buss, & Kotler-Cope, 1989), the rule–extraction–refinement (RER) algorithm learns rules at the top level using information in the bottom level. The basic idea of bottom-up learning of action-centered knowledge is as follows: If an action chosen (by the bottom level) is successful (i.e., it satisfies a certain criterion), then an explicit rule is extracted at the top level. Then, in subsequent interactions with the world, the rule is refined by considering the outcome of applying the rule: If the outcome is successful, the condition of the rule may be generalized to make it more universal; if the outcome is not successful, then the condition of the rule should be made more specific and exclusive of the current case.

An agent needs a rational basis for making these decisions. Numerical criteria have been devised for measuring whether a result is successful, used in deciding whether to apply these operations. The details of the numerical criteria measuring whether a result is successful can be found in Sun et al. (2001). Essentially, at each step, an information gain measure is computed, which compares different rules. The aforementioned rule learning operations (extraction, generalization, and specialization) are determined and performed based on the information gain measure (see Sun, 2003, for details).

However, in the opposite direction, the dual representation (implicit and explicit) in the ACS also enables top-down learning. With explicit knowledge (in the form of rules), in place at the top level, the bottom level learns under the guidance of the rules. That is, initially, the agent relies mostly on the rules at the top level for its action decision making. But gradually, when more and more knowledge is acquired by the bottom level through "observing" actions directed by the rules (based on the same Q-learning mechanism as described before), the agent becomes more and more reliant on the bottom level (given that the interlevel stochastic selection mechanism is adaptable). Hence, top-down learning takes place.

For the stochastic selection of the outcomes of the two levels, at each step, with probability P_{BL}, the outcome of the bottom level is used. Likewise, with probability P_{RER}, if there is at least one RER rule indicating a proper action in the current state, the outcome from that rule set (through competition based on rule utility)

is used; otherwise, the outcome of the bottom level is used (which is always available). Other components may be included in a like manner. The selection probabilities may be variable, determined through a process known as *probability matching*; that is, the probability of selecting a component is determined based on the relative success ratio of that component. There exists some psychological evidence for such intermittent use of rules; see, for example, Sun et al. (2001).

This subsystem has been used for simulating a variety of psychological tasks, including process control tasks in particular (Sun, Zhang, Slusarz, & Mathews, in press). In process control tasks, participants were supposed to control a (simulated) sugar factory. The output of the sugar factory was determined by the current and past inputs from participants into the factory, often through a complex and nonsalient relationship. In the ACS of CLARION, the bottom level acquired implicit knowledge (embodied by the neural network) for controlling the sugar factory, through interacting with the (simulated) sugar factory in a trial-and-error fashion. However, the top level acquired explicit action rules for controlling the sugar factory, mostly through bottom-up learning (as explained before). Different groups of participants were tested, including verbalization groups, explicit instruction groups, and explicit search groups (Sun et al., in press). Our simulation succeeded in capturing the learning results of different groups of participants, mainly through adjusting one parameter that was hypothesized to correspond to the difference among these different groups (i.e., the probability of relying on the bottom level; Sun et al., in press).

Besides simulating process control tasks, this subsystem has been employed in simulating a variety of other important psychological tasks, including artificial grammar learning tasks, serial reaction time tasks, Tower of Hanoi, minefield navigation, and so on, as well as social simulation tasks such as organizational decision making.

The Non-Action-Centered Subsystem

The non-action-centered subsystem (the NACS) is used for representing general knowledge about the world that is not action-centered, for the purpose of making inferences about the world. It stores such knowledge in a dual representational form (the same as in the ACS): that is, in the form of explicit "associative rules" (at the top level), as well as in the form of implicit "associative memory" (at the bottom level). Its operation is under the control of the ACS.

First, at the bottom level of the NACS, associative memory networks (AMNs) encode non-action-centered implicit knowledge. Associations are formed by mapping an input to an output. The regular back-propagation learning algorithm, for example, can be used to establish such associations between pairs of input and output (Rumelhart et al., 1986).

At the top level of the NACS, however, a general knowledge store (the GKS) encodes explicit non-action-centered knowledge (see Sun, 1994). As in the ACS, chunks are specified through dimensional values. The basic form of a chunk consists of a chunk id and a set of dimension-value pairs. A node is set up in the GKS to represent a chunk (which is a localist representation). The chunk node connects to its constituting features (i.e., dimension-value pairs) represented as individual nodes in the bottom level (a distributed representation in the AMNs). Additionally, in the GKS, links between chunks encode explicit associations between pairs of chunk nodes, which are known as associative rules. Such explicit associative rules may be formed (i.e., learned) in a variety of ways in the GKS of CLARION (Sun, 2003).

In addition, similarity-based reasoning may be employed in the NACS. A known (given or inferred) chunk may be compared automatically with another chunk. If the similarity between them is sufficiently high, then the latter chunk is inferred.

Similarity- and rule-based reasoning can be intermixed. As a result of mixing similarity- and rule-based reasoning, complex patterns of reasoning may emerge. As shown by Sun (1994), different sequences of mixed similarity-based and rule-based reasoning capture essential patterns of human everyday (mundane, commonsense) reasoning.

As in the ACS, top-down or bottom-up learning may take place in the NACS, either to extract explicit knowledge in the top level from the implicit knowledge in the bottom level or to assimilate the explicit knowledge of the top level into the implicit knowledge in the bottom level.

The NACS of CLARION has been used to simulate a variety of psychological tasks. For example, in artificial grammar learning tasks, participants were presented with a set of letter strings. After memorizing these strings, they were asked to judge the grammaticality of new strings. Despite their lack of complete explicit knowledge

about the grammar underlying the strings, they nevertheless performed well in judging new strings. Moreover, they were also able to complete partial strings in accordance with their implicit knowledge. The result showed that participants acquired fairly complete implicit knowledge although their explicit knowledge was fragmentary at best (Domangue, Mathews, Sun, Roussel, & Guidry, 2004). In simulating this task, while the ACS was responsible for controlling the overall operation, the NACS was used for representing most of the relevant knowledge. The bottom level of the NACS acquired implicit associative knowledge that enabled it to complete partial strings. The top level of the NACS recorded explicit knowledge concerning sequences of letters in strings. When given partial strings, the bottom level or the top level might be used, or the two levels might work together, depending on circumstances. On the basis of this setup, our simulation succeeded in capturing fairly accurately human data in this task across a set of different circumstances (Domangue et al., 2004). In addition, many other tasks have been simulated involving the NACS, including alphabetic arithmetic tasks, categorical inference tasks, and discovery tasks.

The Motivational Subsystem

Now that we dealt with the primary control of actions within CLARION (through the ACS and the NACS), we are ready to explore details of motivational and metacognitive control within CLARION. In CLARION, secondary internal control processes over the operations of the ACS and the NACS are made up of two subsystems: the motivational subsystem and the metacognitive subsystem.

The motivational subsystem (MS) is concerned with drives and their interactions (Toates, 1986). That is, it is concerned with why an agent does what it does. Simply saying that an agent chooses actions to maximize gains, rewards, or payoffs leaves open the question of what determines gains, rewards, or payoffs. The relevance of the motivational subsystem to the main part of the architecture, the ACS, lies primarily in the fact that it provides the context in which the goal and the reinforcement of the ACS are determined. It thereby influences the working of the ACS, and by extension, the working of the NACS.

As an aside, for several decades, criticisms of commonly accepted models of human motivations, for example in economics, have focused on their overly narrow views regarding motivations, for example, solely in terms of simple economic reward and punishment (economic incentives and disincentives). Many critics opposed the application of this overly narrow approach to social, behavioral, cognitive, and political sciences. Complex social motivations, such as desire for reciprocation, seeking of social approval, and interest in exploration, also shape human behavior. By neglecting these motivations, the understanding of some key social and behavioral issues (such as the effect of economic incentives on individual behaviors) may be hampered. Similar criticisms may apply to work on reinforcement learning in AI (e.g., Sutton & Barto, 1998).

A set of major considerations that the motivational subsystem of an agent must take into account may be identified. Here is a set of considerations concerning drives as the main constructs (cf. Simon, 1967; Tyrell, 1993):

- Proportional activation. The activation of a drive should be proportional to corresponding offsets, or deficits, in related aspects (such as food or water).
- Opportunism. An agent needs to incorporate considerations concerning opportunities. For example, the availability of water may lead to preferring drinking water over gathering food (provided that food deficits are not too great).
- Contiguity of actions. There should be a tendency to continue the current action sequence, rather than switching to a different sequence, in order to avoid the overhead of switching.
- Persistence. Similarly, actions to satisfy a drive should persist beyond minimum satisfaction, that

is, beyond a level of satisfaction barely enough to reduce the most urgent drive to be slightly below some other drives.[2]
- Interruption when necessary. However, when a more urgent drive arises (such as "avoid danger"), actions for a lower-priority drive (such as "get sleep") may be interrupted.
- Combination of preferences. The preferences resulting from different drives should be combined to generate a somewhat higher overall preference. Thus, a compromise candidate may be generated that is not the best for any single drive but the best in terms of the combined preference.

A bipartite system of motivational representation is as follows (cf. Nerb, Spada, & Ernst, 1997; Simon, 1967). The explicit goals (such as "finding food") of an agent (which is tied to the working of the ACS, as explained before) may be generated based on internal drive states (e.g., "being hungry") of the agent. This explicit representation of goals derives from, and hinges on, (implicit) drive states. See Figure 5.2.[3]

Specifically, we refer to as *primary drives* those drives that are essential to an agent and are most likely built-in (hardwired) to begin with. Some sample low-level primary drives include (see Tyrell, 1993):

Get food. The strength of this drive is determined by two factors: food deficit felt by the agent, and the food stimulus perceived by it.

Get water. The strength of this drive is determined by water deficit and water stimulus.

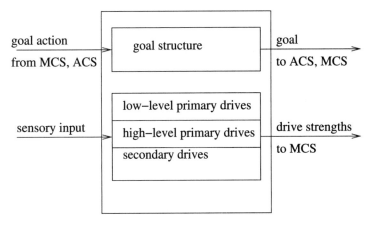

FIGURE 5.2 Structure of the motivational subsystem.

Avoid danger. The strength of this drive is proportional to the danger signal: its distance, intensity, severity (disincentive value), and certainty.

In addition, other drives include "get sleep," "reproduce," and a set of "avoid saturation" drives, for example, "avoid water saturation" or "avoid food saturation." There are also drives for "curiosity" and "avoid boredom." See Sun (2003) for further details.

Beyond such low-level drives (concerning physiological needs), there are also higher-level drives. Some of them are primary, in the sense of being hardwired. The "need hierarchy" of Maslow (1987) identifies some of these drives. A few particularly relevant high-level drives include belongingness, esteem, and self-actualization (Sun, 2003).

These drives may be implemented in a (pretrained) back-propagation neural network, representing evolutionarily prewired tendencies.

While primary drives are built-in and relatively unalterable, there are also "derived" drives. They are secondary, changeable, and acquired mostly in the process of satisfying primary drives. Derived drives may include (1) gradually acquired drives, through "conditioning" (Hull, 1951) and (2) externally set drives, through externally given instructions. For example, because of the transfer of the desire to please superiors into a specific desire to conform to his/her instructions, following the instructions becomes a (derived) drive.

Explicit goals may be set based on these (primary or derived) drives, as will be explored in the next section (Nerb et al., 1997; Simon, 1967).

The Metacognitive Subsystem

Metacognition refers to one's knowledge concerning one's own cognitive processes and their outcomes. Metacognition also includes the active monitoring and consequent regulation and orchestration of these processes, usually in the service of some concrete goal (Flavell, 1976; Mazzoni & Nelson, 1998). This notion of metacognition is operationalized within CLARION.

In CLARION, the metacognitive subsystem (MCS) is closely tied to the motivational subsystem. The MCS monitors, controls, and regulates cognitive processes for the sake of improving cognitive performance (Simon, 1967; Sloman, 2000). Control and regulation may be in the forms of setting goals for the ACS, interrupting and changing ongoing processes in the

ACS and the NACS, and setting essential parameters of the ACS and the NACS. Control and regulation are also carried out through setting reinforcement functions for the ACS on the basis of drive states.

In this subsystem, many types of metacognitive processes are available for different metacognitive control purposes. Among them are the following types (Mazzoni & Nelson, 1998; Sun, 2003):

1. Behavioral aiming:
   setting of reinforcement functions
   setting of goals
2. Information filtering:
   focusing of input dimensions in the ACS
   focusing of input dimensions in the NACS
3. Information acquisition:
   selection of learning methods in the ACS
   selection of learning methods in the NACS
4. Information utilization:
   selection of reasoning methods in the ACS
   selection of reasoning methods in the NACS
5. Outcome selection:
   selection of output dimensions in the ACS
   selection of output dimensions in the NACS
6. Cognitive mode selection:
   selection of explicit processing, implicit processing, or a combination thereof (with proper integration parameters), in the ACS
7. Setting parameters of the ACS and the NACS:
   setting of parameters for the IDNs
   setting of parameters for the ARS
   setting of parameters for the AMNs
   setting of parameters for the GKS

Structurally, the MCS may be subdivided into a number of modules. The bottom level consists of the following (separate) networks: the goal setting network, the reinforcement function network, the input selection network, the output selection network, the parameter setting network (for setting learning rates, temperatures, etc.), and so on. In a similar fashion, the rules at the top level (if they exist) can be correspondingly subdivided. See Figure 5.3 for a diagram of the MCS. Further details, such as monitoring buffer, reinforcement functions (from drives), goal setting (from drives), and information selection can be found in Sun (2003).

This subsystem may be pretrained before the simulation of any particular task (to capture evolutionary prewired instincts, or knowledge/skills acquired from prior experience).

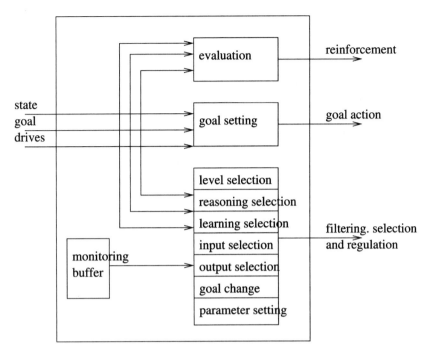

FIGURE 5.3 The structure of the metacognitive subsystem.

Simulations Conducted with CLARION

CLARION has been successful in simulating a variety of psychological tasks. These tasks include serial reaction time tasks, artificial grammar learning tasks, process control tasks, categorical inference tasks, alphabetical arithmetic tasks, and the Tower of Hanoi task (see Sun, 2002). Some tasks have been explained earlier. In addition, extensive work has been done on a complex minefield navigation task (Sun et al., 2001). We have also tackled human reasoning processes through simulating reasoning data.

Therefore, we are now in a good position to extend the effort on CLARION to capturing various motivational and metacognitive control phenomena. Simulations involving motivational structures and metacognitive processes are under way. For instance, in the task of Metcalfe (1986), subjects were given a story and asked to solve the puzzle in the story. They were told to write down every 10 s a number between 0 and 10, whereby 0 meant that they were "cold" about the problem and 10 meant that they were certain that they had the right solution. The general finding was that subjects who came up with the correct solution gave lower warmth ratings than subjects with incorrect solutions. In our simulation involving the MCS, those

variants of the models that generated the correct solution gave lower warmth ratings than those that generated incorrect solutions because of the more diverse range of potential solutions they generated. Thus, the simulation model accounted for the counterintuitive findings in the experimental data of Metcalfe (1986).

For another instance, in Gentner and Collins (1991), inferences were shown to be made based on (1) the lack of knowledge about something and (2) the importance/significance of that knowledge. To make such inferences, metacognitive monitoring of one's own reasoning process is necessary. However, beyond metacognitive monitoring (as in the previous task), active metacognitive intervention is also necessary. Our model was shown to be able to capture such inferences.

Let us also take a brief look at some rather preliminary applications of CLARION to social simulation, which also involve motivational and metacognitive control to some extent. In one instance, tribal societies were simulated on the basis of CLARION modeling individual cognitive processes. In the simulation, different forms of social institutions (such as food distribution, law, political system, and law enforcement) were investigated and related to factors of individual cognition. The interaction of social institutions and cognition is important both theoretically and practically

(Sun, 2005). Social institutions affect agents' actions and behaviors, which in turn affect social institutions. In this interaction, individual motivational factors are considered, which include social norms, ethical values, social acceptance, empathy, and imitation. The role of metacognitive control is also being investigated in this process. It has been suggested that such simulations are the best way to understand or to validate the significance of contributing cognitive, motivational, and metacognitive factors (Sun, 2005).

Concluding Remarks

In summary, this chapter covered the essentials of the CLARION cognitive architecture and focused, in particular, on the motivational and metacognitive control within CLARION. CLARION is distinguished by its inclusion of multiple, interacting subsystems: the action-centered subsystem, the non-action-centered subsystem, the motivational subsystem, and the metacognitive subsystem. It is also distinguished by its focus on the separation and the interaction of implicit and explicit knowledge (in these different subsystems, respectively). With these mechanisms, especially the motivational and metacognitive mechanisms, CLARION has something unique to contribute to cognitive modeling— it attempts to capture the motivational and metacognitive aspects of cognition, and to explain their functioning in concrete computational terms.

For the full technical details of CLARION, see Sun (2003), which is available at http://www.cogsci.rpi .edu/~rsun/clarion-pub.html. CLARION has been implemented as a set of Java packages, available at http://www.cogsci.rpi.edu/~rsun/clarion.html.

Acknowledgments

The work on CLARION has been supported in part by Army Research Institute contracts DASW01-00-K-0012 and W74V8H-04-K-0002 (to Ron Sun and Robert Mathews). The writing of this chapter was supported by AFOSR Contract F49620-03-1-0143 (to Wayne Gray). Thanks are due to Xi Zhang, Isaac Naveh, Paul Slusarz, Robert Mathews, and many other collaborators, current or past. Thanks are also due to Jonathan Gratch, Frank Ritter, Chris Sims, and Bill Clancey for their comments on an early draft, and to Wayne Gray for the invitation to write this article.

Notes

1. Note that aforementioned working memory is for storing information temporarily for the purpose of facilitating subsequent decision making (Baddeley, 1986). Working memory actions are used either for storing an item in the working memory or for removing an item from the working memory. Goal structures, a special case of working memory, are for storing goal information specifically.

2. For example, an agent should not run toward a water source and drink only a minimum amount, then run toward a food source and eat a minimum amount, and then go back to the water source to repeat the cycle.

3. Note that it is not necessarily the case that the two types of representations directly correspond to each other (e.g., one being extracted from the other), as in the case of the ACS or the NACS.

References

Anderson, J., & Lebiere, C. (1998). *The atomic components of thought*. Mahwah, NJ: Erlbaum.

Baddeley, A. (1986). *Working memory*. New York: Oxford University Press.

Bates, J., Loyall, A., & Reilly, W. (1992). Integrating reactivity, goals, and emotion in a broad agent. *Proceedings of the 14th Meeting of the Cognitive Science Society*. Mahwah, NJ: Erlbaum.

Bruner, J., Goodnow, J., & Austin, J. (1956). *A study of thinking*. New York: Wiley.

Busemeyer, J., & Myung, I. (1992). An adaptive approach to human decision making: Learning theory, decision theory, and human performance. *Journal of Experimental Psychology: General, 121*(2), 177–194.

Chaiken, S., & Trope, Y. (Eds.), (1999). *Dual process theories in social psychology*. New York: Guilford Press.

Clark, A., & Karmiloff-Smith, A. (1993). The cognizer's innards: A psychological and philosophical perspective on the development of thought. *Mind and Language. 8*(4), 487–519.

Cleeremans, A. (1997). Principles for implicit learning. In D. Berry (Ed.), *How implicit is implicit learning?* (pp. 195–234). Oxford: Oxford University Press.

———, Destrebecqz, A., & Boyer, M. (1998). Implicit learning: News from the front. *Trends in Cognitive Sciences, 2*(10), 406–416.

Domangue, T., Mathews, R., Sun, R., Roussel, L., & Guidry, C. (2004). The effects of model-based and memory-based processing on speed and accuracy of grammar string generation. *Journal of Experimental Psychology: Learning, Memory, and Cognition, 30*(5), 1002–1011.

Edelman, G. (1992). *Bright air, brilliant fire*. New York: Basic Books.

Flavell, J. (1976). Metacognitive aspects of problem solving. In B. Resnick (Ed.), *Nature of intelligence*. Hillsdale, NJ: Erlbaum.

Gentner, D., & Collins, A. (1981). Studies of inference from lack of knowledge. *Memory and Cognition, 9*, 434–443.

Hull, C. (1951). *Essentials of behavior*. New Haven, CT: Yale University Press.

Karmiloff-Smith, A. (1986). From metaprocesses to conscious access: Evidence from children's metalinguistic and repair data. *Cognition, 23*, 95–147.

Maslow, A. (1987). *Motivation and personality* (3rd ed.). New York: Harper & Row.

Mazzoni, G., & Nelson, T. (Eds.). (1998). *Metacognition and cognitive neuropsychology*. Mahwah, NJ: Erlbaum.

Metcalfe, J. (1986). Dynamic metacognitive monitoring during problem solving. *Journal of Experimental Psychology: Learning, Memory and Cognition, 12*, 623–634.

Nerb, J., Spada, H., & Ernst, A. (1997). A cognitive model of agents in a common dilemma. In *Proceedings of the 19th Cognitive Science Conference* (pp. 560–565). Mahwah, NJ: Erlbaum.

Reber, A. (1989). Implicit learning and tacit knowledge. *Journal of Experimental Psychology: General, 118*(3), 219–235.

Rumelhart, D., McClelland, J., & PDP Research Group. (1986). Parallel distributed processing: Explorations in the microstructures of Cognitive. Cambridge, MA: MIT Press.

Seger, C. (1994). Implicit learning. *Psychological Bulletin, 115*(2), 163–196.

Simon, H. (1967). Motivational and emotional controls of cognition. *Psychological Review, 74*, 29–39.

Sloman, A. (2000). Architectural requirements for human-like agents both natural and artificial. In K. Dautenhahn (Ed.), *Human cognition and social agent technology*. Amsterdam: John Benjamins.

Stanley, W., Mathews, R., Buss, R., & Kotler-Cope, S. (1989). Insight without awareness: On the interaction of verbalization, instruction and practice in a simulated process control task. *Quarterly Journal of Experimental Psychology, 41A*(3), 553–577.

Sun, R. (1994). Integrating rules and connectionism for robust common-sense reasoning. New York: Wiley.

———. (2001). Meta-learning in multi-agent systems. In N. Zhong, J. Liu, S. Ohsuga, & J. Bradshaw (Eds.), *Intelligent agent technology: Systems, methodologies, and tools* (pp. 210–219). Singapore: World Scientific.

———. (2002). *Duality of the mind*. Mahwah, NJ: Erlbaum.

———. (2003). *A tutorial on CLARION 5.0*. http://www.cogsci.rpi.edu/~rsun/sun.tutorial.pdf.

———. (2004). Desiderata for cognitive architectures. *Philosophical Psychology, 17*(3), 341–373.

———. (Ed.). (2005). *Cognition and multi-agent interaction: From cognitive modeling to social simulation*. New York: Cambridge University Press.

———, Merrill, E., & Peterson, T. (2001). From implicit skills to explicit knowledge: A bottom-up model of skill learning. *Cognitive Science, 25*(2), 203–244.

———, Slusarz, P., & Terry, C. (2005). The interaction of the explicit and the implicit in skill learning: A dual-process approach. *Psychological Review, 112*(1), 159–192.

———, Zhang, X., Slusarz, P., & Mathews, R. (in press). The interaction of implicit learning, explicit hypothesis testing, and implicit-to-explicit knowledge extraction. *Neural Networks*.

Toates, F. (1986). *Motivational systems*. Cambridge: Cambridge University Press.

Tomasello, M. (1999). *The cultural origins of human cognition*. Cambridge, MA: Harvard University Press.

Tyrell, T. (1993). *Computational mechanisms for action selection*. Unpublished doctoral dissertation, Oxford University, Oxford, United Kingdom.

Watkins, C. (1989). *Learning with delayed rewards*. Unpublished doctoral dissertation, Cambridge University, Cambridge, United Kingdom.

Weiner, B. (1992). *Human motivation: Metaphors, theories, and research*. Newbury Park, CA: Sage.

6

Reasoning as Cognitive Self-Regulation

Nicholas L. Cassimatis

Comprehensive models of reasoning require models of cognitive control and vice versa. This raises several questions regarding how reasoning is integrated with other cognitive processes, how diverse reasoning strategies are integrated with each other, and how the mind chooses which reasoning strategies to use in any given situation. A major difficulty in answering these questions is that the cognitive architectures used to model the cognitive processes involved in reasoning and control are based on many different computational formalisms that are difficult to integrate with one another. This chapter describes two computational principles and five hypotheses about human cognitive architecture that (together with empirical studies of cognition) motivate a solution to this problem. These hypotheses posit an integrative focus of cognitive attention and conceive of reasoning strategies as the generalization of attention control strategies from visual perception (e.g., habituation and negative priming). Further, a unifying theme among these cognitive attention control strategies is that they can each be seen as the mind's way of regulating its own activity and addressing cognitive problems (such as contradiction or uncertainty) that arise during normal cognition and perception. These principles and hypotheses enable an integrated view of cognitive architecture that explains how cognitive and perceptual processes that were previously difficult to model within one computational framework can exist and interact within the human mind.

Reasoning as a Control Problem and Solution

This chapter outlines a theory of human cognitive architecture based on the hypothesis that much human reasoning is the manifestation of very general cognitive and perceptual attention control mechanisms. The work reported here is motivated by the hypotheses that (1) reasoning is an important part of control and (2) a better understanding of how the mind integrates diverse cognitive and perceptual mechanisms is required to fully understand its role. The theory presented in this chapter aims to explain how multiple reasoning strategies, currently modeled using difficult-to-integrate computational formalisms, are integrated with each other and how they integrate with other cognitive processes. It is based on two computational principles that enable many reasoning strategies to be modeled within the same computational framework and five hypotheses about human cognitive architecture that are motivated by these principles and by many empirical

studies of human cognition. This theory of cognitive architecture enables cognitive models of reasoning and control that explain human behavior in situations where the mind integrates cognitive and perceptual mechanisms that were formally difficult to study within a single modeling framework.

Reasoning

Before proceeding, it will be helpful to describe how the word *reasoning* is used in this chapter. Reasoning will refer to a set of more "open-ended" cognitive processes that make inferences and solve problems that are not specifically addressed by specific cognitive processes. Some examples illustrate this distinction.

There is a sense in which retrieving the location of an object in a cognitive map is a much more definite, less open-ended problem than, say, planning a path from the office to one's home. The retrieval process involves a map and a method of cueing the map for an object's location. It is not very open-ended in that factors such as the weather or one's personal social relationships do not affect the course of the retrieval. The retrieval process does not vary much from case to case. In a task such as planning a set of movements to go from one location to another, the process is much more open-ended. There is not a fixed path plan one can retrieve with a few parameters to be filled in to determine the course of action for a given situation. Factors such as your relationship with your spouse (should you drive by the grocery store to pick up milk, or the florist for your anniversary, etc.) can have a significant effect on the path that is chosen.

In this chapter, the term "reasoning" will be used to distinguish these forms of more open-ended inference and problem solving from other cognitive processes. Many psychologists, consistent with common practice in philosophy, have implicitly or explicitly adopted the assumption that all reasoning processes are in some sense conscious or deliberate. However, reasoning as conceived in this chapter often occurs unconsciously and automatically. For example, many explanations of behavior in the infant physics and language acquisition literature presuppose some form of reasoning, for example, search (finding a continuous path traversed by an object (Spelke, Kestenbaum, Simons, & Wein, 1995) falsification (i.e., where an incorrect belief leads to the rejection of an assumption that generated it [Baillargeon, 1998]), and mutual exclusivity (where objects are assumed to have unique verbal labels [Markman, Wasow, & Hansen, 2003]). Problems in understanding garden path sentences (Frazier & Rayner, 1982), for example, where an initial interpretation must be retracted and a sentence reanalyzed, correspond to problems in the reasoning literature such as belief revision and truth maintenance (Doyle, 1992). Thus, this chapter rejects the distinction between conscious and deliberate reasoning and other cognitive processes and deals with cognition in many situations that are not normally referred to by most psychologists under with the term *reasoning*.

Thus conceiving of reasoning reveals how much more common and ubiquitous it is than often supposed. For example, the cases of infant physical reasoning, path planning, and sentence processing demonstrate that reasoning is involved in what are normally thought of as "only" perceptual, motor, or linguistic processes and rarely explicitly studied by reasoning researchers.

Thus, both because it is part of so many important and ubiquitous domains and because it is often integrated with other cognitive and perceptual processes, good models of reasoning can have a broad impact throughout cognitive science.

Reasoning and Control Are Related

Issues of reasoning and control are related for at least three reasons. First, humans often engage in reasoning to resolve control issues that arise during cognition. Second, humans in any given situation often have available to them more than one reasoning strategy. Choosing an appropriate strategy is in part a control problem. Third, the execution of reasoning strategies often involves multiple control issues.

First, one way to solve a control problem is to reason it through. The need for reasoning is well illustrated in the case of control strategy adaptation. In novel situations, people cannot rely on existing strategies or use learning methods that involve incrementally adjusting behavior over multiple instances of trial and error. They must instead use some form of reasoning to formulate a strategy that is likely to achieve their goal in that situations.

Many problems involve different possible reasoning strategies. For example, in chess, people choose moves in part by performing a search and by using pattern recognition (Chase & Simon, 1973). Many problem domains present situations where more than one reasoning strategy applies. Choosing which strategy, or combination thereof, to apply is a control problem.

Finally, many reasoning strategies have their own control issues. For example, in backtracking search, there are in general, at any given moment, multiple actions or operators to explore. Search strategies differ by choice. In models of reasoning in the Soar (Laird, Newell, & Rosenbloom, 1987) and adaptive control of thought–rational, or ACT-R (Anderson & Lebiere, 1998) production systems, which operator to choose (Soar) or which production rule to fire and chunk to retrieve (ACT-R) involves a conflict resolution strategy that is one of the distinguishing features of each of those architectures. That conflict resolution in ACT-R and Soar is used in models well beyond reasoning suggests a close relation between the control of reasoning and the control of cognition generally.

The Problem of Integration

If control and reasoning are thus related, several questions about integration in cognitive architecture arise. Underlying each of these questions is the fundamental puzzle of how to create models of complex human cognition that involves cognitive and perceptual processes currently best modeled using algorithms and data structures that are very difficult to integrate.

Integrating the Mechanisms of Higher- and Lower-Order Cognition

First, if the mind engages in reasoning to resolve control issues that arise among attention, memory, perception, and motor mechanisms, how does reasoning interact with these processes? Does not the reasoning itself require some measure of memory and attention? Further, computational formalisms used to model reasoning (e.g., logic, Bayesian networks, and search) are very different from computational formalisms often used to model memory and attention (e.g., production rules and spreading of activation).[1] How do such seemingly different mechanisms integrate? For the purposes of this chapter, the former category of processes will often be called *higher-order cognition* and the latter *lower-order cognition*, not to prejudice their importance or complexity, but the level of abstraction at which these mechanisms normally (are thought to) operate.

Integrating the Mechanisms of Higher-Order Cognition

The problem of integrating algorithms and data structures used to model higher-order and lower-order cognitive processes also arises among higher-order cognitive processes themselves. For example, infant physical reasoning requires reasoning about spatial and temporal relations, which are currently modeled using techniques such as cognitive maps and constraint graphs. However, physical reasoning often involves uncertain states and outcomes. Such uncertain reasoning is often best modeled with methods such as Bayesian networks (Pearl, 1998). Yet, it is not at all obvious how to integrate these two classes of computational methods into one integrated model of physical reasoning. This difficulty of integrating qualitatively different models of reasoning makes it difficult to create models of control involving more than one form of reasoning.

Strategy Choice

Once we understand how the mind integrates different reasoning strategies, the question arises: Which one does it choose in any particular situation? In situations where more than one reasoning strategy is needed, what is the process that decides which strategy is used for which part of the situation? Furthermore, different reasoning strategies tend to require that knowledge about a situation be represented using different representational formalisms. For example, backtracking search-based problem-solving strategies often require actions in a domain to be specified in a propositional format that describes the preconditions and consequences of taking an action. Strategies based on Bayesian networks require conditional probabilities between elements of a domain to be represented in graphical networks. In advance of choosing which strategy to apply for (part of) a situation, which representational formalism is used to represent the problem? If more than one strategy is used per problem, how does information from one knowledge representation scheme become shared with that from another?

For all these reasons, to understand the control of cognition, especially when reasoning is involved, some significant integration puzzles need to be solved. At bottom, the puzzle is how to integrate the data structures and algorithms used to model different cognitive processes into one model. For example, how do chunks and production rules in ACT-R integrate with nodes, edges, and conditional probabilities in Bayesian networks? A comprehensive model of the control of cognition would be difficult to develop without a solution to this problem.

This chapter presents a theory of human cognitive architecture that explains how the mind integrates multiple mechanisms and cognitive processes and how it deals with control issues that arise from this integration. Since one of the main goals of this theory is to explain how reasoning is integrated with the rest of cognition, the chapter begins with an overview of two computational principles (Cassimatis, 2005), which enable reasoning strategies currently modeled using very different computational formalisms to be conceived of within the same framework. These two computational principles, together with empirical data on the human cognition, motivate several hypotheses about human cognitive architecture. These principles explain how the mind integrates multiple reasoning strategies based on different computational methods and how they integrate with "lower-order" cognition and perception, but they do not explain which strategies are chosen in any particular situation. Additional architectural hypotheses are introduced to explain this.

An Integrated Theory of Cognitive Architecture

Computational Principles

Cassimatis (2005) described two insights: the common function and multiple implementation principles, which motivate a theory of cognitive architecture that explains how the mind integrates cognitive processes currently best modeled using heretofore difficult-to-integrate computational methods.

The common function principle states that many algorithms, especially those used to model reasoning, from many different subfields of computational cognitive modeling can be conceived of as different ways of selecting sequences of the same basic set of *common functions*. The following is a preliminary list of these functions:

- Forward inference: Given a set of beliefs, infer other beliefs that follow from them.
- Subgoaling: Given the goal of establishing the truth of a proposition, P, make a subgoal of determining the truth values of propositions that would imply or falsify P.
- Simulate alternate worlds: Represent and make inferences about alternate, possible, hypothetical, or counterfactual states of the world.
- Identity matching: Given a set of propositions about an object, find other objects that might be identical to it.

The common function principle can be justified (e.g., Cassimatis, 2005; Cassimatis, Trafton, Bugajska, & Schultz) by showing how these common functions can implement a variety of algorithms. The following rough characterizations of two widely used algorithms in cognitive modeling illustrates how methods from different branches of formal cognitive science can be implemented using the same set of common functions:

- Search: "When uncertain about whether A is true, *represent the world* where A is true, perform *forward inference, represent the world* where A is not true, perform *forward inference*. If forward inference leads to further uncertainty, repeat."
- Stochastic simulation (used widely in Bayes network propagation): "When A is more likely than not-A, *represent the world* where A is true and *perform forward inference* in it more often than you do so for the world where not-A is true."

Explaining the integration of cognitive processes best modeled using different cognitive modeling frameworks can be difficult because these frameworks are often based on algorithms that are difficult to reconcile with each other. If the mind executes each strategy using a sequence of common functions, the integration of different reasoning strategies becomes a matter of explaining how the mind combines and interleaves sequences of common functions. The multiple implementation principle states that multiple computational and representational mechanisms can implement each common function. For example, forward inference can be implemented using at least these four mechanisms:

- Production rule firing: Can involve matching a set of rules against a set of known facts to infer new facts.
- Feed-forward neural networks: Take the facts represented by the activation of the input units, propagate these activations forward, and output new facts represented by the values of the output units.
- Memory: The value of a slow-changing attribute (e.g., a mountain's location) at time T2 can be inferred by recalling its value at an earlier time, T1, so long as the interval between T1 and T2 is sufficiently brief.
- Perception: The value of a slow-changing attribute at some point in the near future can be inferred after perceiving the value of that attribute in the present.

The multiple implementation principle, together with the common function principle, suggests a way to explain how reasoning strategies integrate with lower-order perceptual, motor, and mnemonic processes. If the mind implements reasoning strategies using sequences of common functions and if (according to the multiple implementation principle) each common function can be implemented by multiple lower-order mechanisms, then those mechanisms can influence every step of reasoning.

Architectural Hypotheses

Cassimatis (2005) explains how these computational principles motivate several hypotheses about human cognitive architecture that can explain the integration this chapter set out to address. The fundamental hypothesis guiding this line of thought, the *higher-order cognition through common functions hypothesis*, states that the mind implements higher-order reasoning strategies by executing sequences of common functions. This hypothesis, together with a large body of empirical evidence, motivates that rest of the architectural hypotheses in this chapter.

First, several lines of evidence (reviewed by Baars, 1988) converge to suggest that the mind has specialized processors, which are here called *specialists*, for perceiving, representing, and making inferences about various aspects of the world. The *specialist-common function implementation hypothesis* states that the mind is made up of specialized processors that implement the common functions using computational mechanisms that are different from specialist to specialist. The ability of multiple specialized mechanisms to implement the same common functions has been established by the multiple implementation principle. Example specialists can include a place memory specialist that keeps track of object locations with a cognitive map and a temporal specialist that maintains representations of relations among temporal intervals using a constraint graph.

If the mind implements common functions using specialists and if (as follows from the multiple implementation principle), any specialist can potentially be relevant to the execution of a particular common function, which subset of the specialists are involved in any particular common function execution? Evidence points to the hypothesis that all specialists are involved in the execution of each common function execution.

Integrative Cognitive Focus of Attention Hypothesis

The mind uses all specialists simultaneously to execute each common function, and the mind has an integrative cognitive focus of attention that at once forces the specialists to execute a particular common function on the current focus, integrates the results of this computation, and distributes these results to each of the specialists.

Interference in the Stroop effect (Stroop, 1935) between processing of multiple attributes of stimuli (e.g., word and color recognition) suggests that multiple cognitive processes (i.e., specialists) engage perceptual input at the same time. Such interference can be found among emotional, semantic, and many other nonperceptual (MacLeod, 1991) aspects of stimuli.

This suggests that, most or all, not just those involving perception, specialists process the same information at the same time. In the present context, this means that each common function is executed by each specialist. If interference in Stroop-like tasks is a result of the mind's attempt to integrate information from multiple cognitive processes, then it is possible that the mechanism for achieving this integration is a focus of attention. Also, Treisman and Gelade (1980) support the hypothesis that visual attention is the main medium for integrating information from multiple perceptual modalities. This hypothesis can be generalized to posit a cognitive focus of attention. The *integrative cognitive focus of attention hypothesis* is based on the notion that just as the perceptual Stroop effect generalizes to nonperceptual cognition the notion of an integrative perceptual focus of attention generalizes to a hypothesis about the existence of a not-just-perceptual cognitive focus of attention that integrates multiple forms of information. Whether the mind's perceptual and cognitive focus of attention are the same mechanism remains, for now, an open question.

The hypothesis that the mind implements higher-order reasoning and problem-solving strategies by sequences of the individual functions specified in the common function principle and the hypothesis that these are executed by a cognitive focus of attention imply the following hypothesis:

Higher-Order Cognition as Attention Selection Hypothesis

The mind's mechanisms for choosing the cognitive focus of attention determine which higher-order reasoning strategies it executes.

The following formulation of well-known reasoning strategies as attention strategies show that attention selection can indeed be used to characterize a broad array of human reasoning.

- Alternate world simulation: Implemented by focusing on imagined worlds.
  When uncertain whether A is true, *avoid focusing on the world where not-A* and focus on the world where A is true until you reach a contradiction or confirming evidence that clarifies A's truth value or *until no new inferences are made*.
  If you have not reached any conclusion, repeat for the world where not-A is true.

- Backtracking search: Implemented by repeated alternate world simulation.
- Stochastic simulation (for Bayesian inference): Implemented by focusing on more probable outcome.

    When A is more likely than not-A, focus on the world where A is true more often to not-A *in proportion to how much more likely it is.*

    $P(A) \approx$ (number of world where A is true)/ (number of worlds where A is false)

- Truth maintenance: Focusing on change:

    When your belief about A *changes*, focus on all the events that involved A.

This formulation of important reasoning strategies as methods of selecting attention suggest a correspondence between attention selection in reasoning and more familiar attention selection strategies people employ in controlling their visual attention.

Each of the bolded phrases in this formulation of reasoning strategies refers to an element of attention control in reasoning that has a direct analogue to an aspect of visual attention control. For example, "avoiding focus on the world where not-A," while focusing on the world where A is analogous to negative priming, where attention to a distracting element in the visual field is inhibited to maintain focus on a target. To continue focus on a possible world "until no more new inferences are made" is analogous to habituation, where an unchanging object is less likely to be continued to be focused on. To focus on something "in proportion to how much more likely it is" characterizes visual attention as well (Nakayama, Takahashi, & Shimizu, 2002) in some cases. Finally, to focus on something when it changes is analogous to the visual system's proclivity to focus on change in the visual field.

All these considerations motivate the *domain-general attention selection hypothesis*, which states that the mechanisms used to control visual attention are the same as those used to guide cognitive attention and thus that much human reasoning is the manifestation of attention control mechanisms studied in perception. Another way to formulate this point, which will be elaborated on below, is that a model of human cognition that contains a focus of attention controlled by the mechanisms of visual cognition and that can concentrate on mental images as well as visual scenes is ipso facto a model of reasoning.

Cognitive Self-Regulation

The computational principles and architectural hypotheses described so far help explain how the mind integrates multiple reasoning strategies with each other and with lower-order cognitive processes. What they do not explain is which strategy (or combination thereof) is chosen in any given situation. The formulation of the attention control (and hence, reasoning control) strategies of the last section motivate an explanation.

Notice that each of the attention selection strategies discussed so far can be seen as a reaction to a problem or a less-than-desirable cognitive state, which we shall call a *cognitive problem*. For example, habituation is a strategy for dealing with the problem of a currently holding state or repeated action not leading to any new information. Focusing on changes will give the mind information about an attribute of an object whose current representation is now inaccurate and needs to be changed. Inhibiting distractors in negative priming deals with the problem of having more than one item demanding attention. Focusing on a more likely outcome is a way of dealing with the problem of having more than one alternative for the next focus of attention in a situation with uncertainty.

Thus, the attention/reasoning strategies of the last section can be seen as strategies for dealing with cognitive problems. This is one explanation of which reasoning strategies the mind chooses to execute and in what order: at any given moment, the mind chooses an attention-control/reasoning strategy applicable to the current cognitive problem. This recalls Soar's modeling of "weak" reasoning methods as strategies for dealing with impasses in operator selection. Because Soar is entirely rule based and does not represent probabilistic relationships, the reasoning methods it integrates are confined mostly to variations on search (and, e.g., do not include Bayesian inference) and do not enable models that integrate with cognitive processes (especially perceptual and spatial) that are not best modeled with production rules. One consequence of this, then, is that the Soar community has not made the connection between addressing impasses and attention control strategies one finds in visual cognition.

Explaining Integration, Control, and Reasoning

This chapter set out to understand three aspects of how the mind integrates multiple cognitive processes in a way that sheds light on reasoning and control. As discussed throughout, the computational principles and architectural hypotheses provide an explanation of how the mind achieves each form of integration.

Integrating Reasoning With Other Cognitive Processes

If the mind implements reasoning and problem-solving strategies using sequences of fixations, all the computation is performed by the computation of the specialists during each attention fixation. In other words, according to this theory, much higher-order cognition is nothing more then the guided focus of lower-level cognitive and perceptual processes. This helps explain how symbolic and serial cognitive processes are grounded (in the sense of Harnad, 1990) in lower-level processes and to the extent these lower-level mechanism are sensorimotor, constitutes an embodied theory of higher-order reasoning. Also, since every focus of attention can be influenced by memory and perceptual and sensorimotor mechanisms, the architectural principles explain how reasoning can be interrupted or guided by these at any moment. This view helps reconcile embodied and symbolic theories of cognitive architecture. Reasoning processes such as search and Bayesian inference, which generally are not thought to be "embodied," are in this view being executed by the mind insofar as it uses attention control strategies such as habituation, frequency, and negative priming to guide a focus of attention that concentrates on mental images as well as the visual field.

Integrating Diverse Reasoning Strategies

If the mind executes each algorithm using a sequence of attention fixations, then we can explain how the mind integrates more than one reasoning strategy in a single situation. Integrating reasoning strategies would then simply be a matter of integrating the attention fixations that execute them. For example, suppose the mind executes a probabilistic reasoning strategy P in a situation using the sequence of common functions P1 . . . Pm and executes a searched-based reasoning strategy in the same situation using S1 . . . Sn. By hypothesizing that the mind executes these two sequences together, for example, S1, S2, P1, S3 . . . SnPm-1Pm, it becomes much easier to explain how the mind integrates search and probabilistic inference. The key to this form of explanation is to take algorithms that were

formally understood in cognitive modeling using very different computational formalisms (e.g., probabilistic reasoning and search) and recognize that (as the common function principle states) their execution can each be understood as sequences of the same small basic set of common functions.

Deciding Which Strategies to Deploy in a Given Situation

Conceiving of reasoning as cognitive self-regulation provides an explanation of how the mind chooses reasoning strategies. Since reasoning strategies are a reaction to cognitive problems, the mind chooses the reasoning strategy to a cognitive problem that arises in any given moment. For example, if in reasoning through a situation, the mind is unsure about whether A is true or false and has no reason to presume one alternative is more likely than the other, then it will engage in counterfactual reasoning. If in the process of considering a hypothetical world, the mind is unsure about whether B is true or false and has no reason to believe either alternative more likely than the other, it will perform counterfactual reasoning on B and thus be engaging in backtracking search. Alternatively, if it does believe, say, B is more likely than not-B, then it will simulate the world where B is more likely than not-B, in effect integrating search and stochastic simulation. This example illustrates how the mind arrives at a mix of reasoning strategies by applying an attention/reasoning control strategy most appropriate for the cognitive problem it faces at any given moment.

An example will illustrate how the architectural hypotheses proposed in this chapter help explain how reasoning is a result of cognitive self-regulation. Suppose a person, Bob, is attempting to find a route to drive from point A to his home at point H and that Bob also has the goal of acquiring milk. Suppose further that his mind includes the following specialists:

- V uses the visual system to detect objects in the environment.
- P uses production rules to propose actions that are potentially the next step in achieving a goal.
- M uses a cognitive map to keep track of objects in the environment.
- G includes a memory of Bob's intentions and goals.
- U detects conflicts among the specialists and asserts subgoals to help resolve these.

The following is rough characterization of Bob's attention fixations in planning a path home. More precise and specific accounts of cognition based on this framework are available in (Cassimatis, 2002). The description of each step begins with the proposition Polyscheme focuses on during that step. (Location [x, l, t, w] means that x is at Location 1 during time t in world w. Move [a, b, t, w] means that the person being modeled moves from location a to b during time t in world w. Get [x, o, t, w] means that x gets object o at time t in world w. Each execution of a common function, e.g., *forward inference*, is italicized.)

1. Location (Bob, A, now, R). V asserts (by making it the focus of attention and assenting to its truth) the proposition that Bob is presently at A.
2. Location (Bob, H, t, g). G asserts that in the desired state of the world, g, Bob is at H at some time t.
3. Move (A, A1, t + 1, R). P, through *forward inference*, infers that moving to A1 will bring Bob closer to his goal.
4. Move (A, A2, t + 1, R). P, through *forward inference*, infers that moving to A2 will also bring Bob closer to his goal.
5. Move (A, A1, t + 1, w1). U notices the contradiction between Fixations 3 and 4 and suggests (to the focus manager) *simulating the alternate world, w1*, in which Bob moves to A2.
6. A number *forward inferences* simulate a moving toward B.
7. Location (Bob, H, t + n, w1). *Forward inference* ultimately determines that in w1 Bob reaches his goal.
8. Move (A, A2, t + 1, w2). Since there was uncertainty about whether to move toward A1 or A2, U suggests *simulating the alternate world, w2*, where Bob's next act is to move to A2.
9. Move (A, A3, t + 2, w2). P's *forward inference* states that the next step toward B in w2 is A3.
10. Location (milk, A3, t + 2, w2). M states that there is milk at t + 2.
11. Get (Bob, milk, t, g). G states that in the goal world, g, Bob gets milk.
12. Get (Bob, milk, t + 3, w2). P, through *forward inference*, determines that Bob can get milk in w2.
13. . . .
14. Location (Bob, H, t + m, w2). P, through *forward inference*, states that Bob can reach his desired goal B in w2.
15. Move (A, A2, now, R). P chooses to move to A2 since this will lead to both goals being satisfied.

This example illustrates several points.

Executing Reasoning and Problem-Solving Strategies as Sequences of Fixations

This example includes two instances of problem solving: (1) Bob finding a way home and (2) Bob acquiring some milk. Each instance of problem solving is executed as a sequence of attention fixations. Planning a path home is executed by Fixations 1–9, 13–15. Determining how to acquire milk is executed by Fixations 9–12.

Reasoning as Attention Selection and Cognitive Self-Regulation: No Need for Reasoning and Problem-Solving Modules

The reasoning in this example is a simple case of backtracking search. Notice that search was not modeled as a problem solving strategy encapsulated inside a specialist, but as the result of a focus control strategy for dealing with a cognitive problem (in this case, contradiction). This is an example of how reasoning and problem solving can be implemented as methods of attention selection whose aim it is to deal with cognitive problems.

Integrating Different Reasoning and Problem-Solving Strategies

This example shows how modeling the integration of two instances of problems solving is as easy as interleaving the sequence of fixations that model them.

Integrating Reasoning and Problem Solving With Lower-Order Cognition

The integration of problem solving with lower-order cognitive processes in this example is explained by the attention fixations that make up the execution of a problem-solving strategy each involve lower-order processes, inside of specialists, such as perception, memory, and cognitive map maintenance.

Conclusions

The computational principles and architectural hypotheses presented in this paper help explain how the mind integrates multiple reasoning, problem solving, memory, and perceptual mechanisms in a way that sheds light on both the control of reasoning and of cognition generally. The main theses of this chapter is that much human reasoning is the manifestation of mechanisms for dealing with cognitive problems and that these mechanisms are either identical to or behave in the same way as mechanisms for the control of visual attention.

The best current evidence that this view of cognitive architecture explains cognitive integration is the success so far achieved in constructing integrated models of human reasoning. This work has been embodied in research surrounding the Polyscheme cognitive architectures. Polyscheme was initially developed to build a model of infant physical reasoning (Cassimatis, 2002) that combined neural networks (for object recognition and classification), production rules (for causal inference), constraint propagation (for keeping track of temporal and spatial constraints), cognitive maps (for object location memory), and search (for finding plausible models of unseen events and for finding continuous paths). This model demonstrates how even apparently simple physical cognition can require sophisticated reasoning, which could be modeled as the guided focus of attention of cognitive and perceptual processes empirically known to exist in infants. This model was adapted to construct a model of syntactic understanding (Cassimatis, 2004) and models of human-robot interaction (Cassimatis et al., 2004). The model of human–robot interaction demonstrated that implementing every step of human reasoning as a focus explains how, for example, perceptual information, spatial cognition and social reasoning could be continually integrated during language use. For example, one model enabled human nominal references to be instantly resolved using information about the speaker's spatial perspective by implementing this linguistic process as a focus of attention on the location of an object that the models' spatial perspective specialist could refine. Thus, by implementing a language understanding algorithm, not as a process encapsulated in a module, but as a sequence of fixations, every step of that algorithm could be refined by perception. This thus enables an explanation of how integration is achieved in human dialogue.

This view of cognitive architecture implies that a separate "reasoning module" need not be added to theories or models of cognition to explain much reasoning. Instead, reasoning is the result of the application of control strategies designed to correct cognitive problems. Reasoning is a form of cognitive self-regulation.

Note

1. It is often mistakenly thought that because ACT-R contains a subsymbolic system for conflict resolution based on Bayes theorem that it performs Bayesian reasoning somehow related to belief propagation in Bayesian networks. Although both frameworks use Bayes theorem, they differ vastly in many other ways and solve different computational problems. Bayesian networks propagate probabilities given what is known about the world and given a set of prior and conditional probabilities. ACT-R resolves conflicts in chunk retrieval or rule firing using a process that involves no notion of prior or conditional probability.

References

Anderson, J. R., & Lebiere, C. (1998). *The atomic components of thought.* Hillsdale, NJ: Erlbaum.

Baars, B. J. (1988). A *cognitive theory of consciousness.* Cambridge: Cambridge University Press.

Baillargeon, R. (1998). Infants' understanding of the physical world. In M. Sabourin, F. Craik, & M. Robert (Eds.), *Advances in psychological science* (Vol. 2, pp. 503–509). London: Psychology Press.

Cassimatis, N. L. (2002). *Polyscheme: A cognitive architecture for integrating multiple representation and inference schemes.* Unpublished doctoral dissertation, Media Laboratory, Massachusetts Institute of Technology, Cambridge.

———. (2004). Grammatical processing using the mechanisms of physical inferences. In K. Forbus, D. Gentner, & T. Reiger (Eds.), *Proceedings of the 26th Annual Cognitive Science Society.* Mahwah, NJ: Erlbaum.

———. (2005). Modeling the integration of multiple cognitive processes in dynamic decision making. In B. Bara, B. Barsalou, & M. Bucciarelli (Eds.), *Proceedings of the 27th Annual Cognitive Science Conference.* Mahwah, NJ: Erlbaum.

———, Trafton, J., Bugajska, M., & Schultz, A. (2004). Integrating cognition, perception and action through mental simulation in robots. *Journal of Robotics and Autonomous Systems, 49*(1–2), 13–23.

Chase, W. G., & Simon, H. A. (1973). Perception in chess. *Cognitive Psychology, 4,* 55–81.

Doyle, J. (1992). Reason maintenance and belief revision: Foundations vs. coherence theories. In P. Gardenfors

(Ed.), *Belief revision* (pp. 29–51). Cambridge: Cambridge University Press.

Frazier, L., & Rayner, K. (1982). Making and correcting errors during sentence comprehension: Eye movements in the analysis of structurally ambiguous sentences. *Cognitive Psychology, 14*(2), 178–210.

Harnad, S. (1990). The symbol grounding problem. *Physica D 42,* 335–346.

Laird, J. E., Newell, A., & Rosenbloom, P. S. (1987) Soar: An architecture for general intelligence. *Artificial Intelligence, 33,* 1–64.

MacLeod, C. M. (1991). Half a century of research on the Stroop effect: An integrative review. *Psychological Bulletin, 109,* 163–203.

Markman, E. M., Wasow, J. L., & Hansen, M. B. (2003). Use of the mutual exclusivity assumption by young word learners. *Cognitive Psychology, 47,* 241–275.

Nakayama, M., Takahashi, K., & Shimizu, Y. (2002). The act of task difficulty and eye-movement frequency for the "oculo-motor indices." In *Eye Tracking Research & Applications (ETRA) Symposium,* ACM, 43–51.

Pearl, J. (1988). *Probabilistic reasoning in intelligent systems: Networks of plausible inference.* San Mateo, CA: Morgan Kaufmann.

Spelke, E. S. (1990). Principles of object perception. *Cognitive Science, 14,* 29–56.

Spelke, E. S., Kestenbaum, R., Simons, D., & Wein, D. (1995). Spatiotemporal continuity, smoothness of motion and object identity in infancy. *British Journal of Developmental Psychology, 13,* 113–142.

Stroop, J. R. (1935). Studies of interference in serial verbal reactions. *Journal of Experimental Psychology, 18,* 622–643.

Treisman, A. M., & Gelade, G. (1980). A feature integration theory of attention. *Cognitive Psychology, 12,* 97–136.

Tversky, A., & Kahneman, D. (1982). Evidential impact of base rates. In D. Kahneman, P. Slovic, & A. Tversky (Eds.), *Judgment under uncertainty: Heuristics and biases* (pp. 153–160). New York: Cambridge University Press.

7

Construction/Integration Architecture

Dynamic Adaptation to Task Constraints

Randy J. Brou, Andrew D. Egerton, & Stephanie M. Doane

Since the late 1980s, much effort has been put into extending the construction–integration (C-I) architecture to account for learning and performance in complex tasks. The C-I architecture was originally developed to explain certain aspects of discourse comprehension, but it has proved to be applicable to a broader range of cognitive phenomena, including complex task performance. One prominent model based on the C-I architecture is ADAPT. ADAPT models individual aviation pilot performance in a dynamically changing simulated flight environment. The model was validated experimentally. Individual novice, intermediate, and expert pilots were asked to execute a series of flight maneuvers using a flight simulator, and their eye fixations, control movements, and flight performance were recorded. Computational models of each of the individual pilots were constructed, and the individual models simulated execution of the same flight maneuvers performed by the human pilots. Rigorous tests of ADAPT's predictive validity demonstrate that the C-I architecture is capable of accounting for a significant portion of individual pilot eye movements, control movements, and flight performance in a dynamically changing environment.

There are numerous theories of how cognitive processes constrain performance in problem-solving tasks, and several have been implemented as computational models. In Soar, problem solving is constrained by the organization, or *chunking*, of results of searches through memory (e.g., Rosenbloom, Laird, Newell, & McCarl, 2002). Anderson's adaptive control of thought–rational, or ACT-R, theory assumes that goal-directed retrievals from memory and production utilities constrain problem-solving performance (e.g., Anderson et al., 2004). Alternatively, our theoretical premise is that comprehension-based mechanisms identical to those used to understand a list of words, narrative prose, and algebraic word problems constrain problem-solving episodes as well.

The construction–integration (C-I) theory (Kintsch, 1988) was initially developed to explain certain phenomena of text comprehension, such as word-sense disambiguation. The theory describes how contextual information is used to decide on an appropriate meaning for words that have multiple meanings. For example,

the appropriate assignment of meaning for the word *bank* is different in the context of conversations about paychecks (money *bank*) and about swimming (river *bank*). Kintsch's theory shows how this can be explained by representing memory as an associative network with nodes containing propositional representations of knowledge about the current state of the world (context-dependent), general (context-independent) declarative facts, and if/then rules that represent possible plans of action (Mannes & Kintsch, 1991). The declarative and plan knowledge are similar to declarative and procedural knowledge contained in ACT-R (e.g., Anderson, 1993).

When a C-I model simulates comprehension in the context of a specific task (e.g., reading a paragraph for a later memory test), a set of weak symbolic production rules *construct* an associative network of knowledge interrelated on the basis of superficial similarities between propositional representations of knowledge without regard to task context. For example, after reading the sentence, "I went to the bank to deposit my check," a C-I model with propositions related to all possible meanings of *bank* would spread activation to each of the superficially related propositions. That is, propositions related to riverbanks, the bank of an aircraft, or any other type of bank would initially receive activation based on the fact that a superficial match exists.

After the initial spread of activation, the associated network of knowledge is then *integrated* via a constraint–satisfaction algorithm. This algorithm propagates activation throughout the network, strengthening connections between items relevant to the current task context and inhibiting or nullifying connections between irrelevant items. Thus, because a proposition related to *check* would have received activation during the construction phase, the connection between the propositions related to *check* and the appropriate meaning of *bank* would be strengthened. At the same time, propositions containing inappropriate meanings such as *riverbank* would be inhibited because they lack links to other activated propositions. Once completed, the integration phase results in context-sensitive knowledge activation constrained by commonalities among activated propositions and current task relevance.

Kintsch's C-I theory has been used to explain a variety of behavioral phenomena, including narrative story comprehension (Kintsch, 1988), algebra story problem comprehension (Kintsch, 1988), the solution of simple computing tasks (Mannes & Kintsch, 1991), and completing the Tower of Hanoi task (Schmalhofer & Tschaitschian, 1993). The C-I theory has also proved fruitful for understanding human–computer interaction skills (e.g., Doane, Mannes, Kintsch, & Polson, 1992; Kitajima & Polson, 1995; Mannes & Doane, 1991) and predicting the impact of instructions on computer user performance (Doane, Sohn, McNamara, & Adams, 2000; Sohn & Doane, 1997, 2000, 2002). The breadth of application suggests that the comprehension processes described in Kintsch's theory play a central role in many tasks and, as such, may be considered a general architecture of cognition (Kintsch, 1998; Newell, 1987).

At a high level, C-I is similar to other architectures such as ACT-R or Soar in several ways. For example, C-I uses something like declarative and procedural memory as do ACT-R and Soar, although these are represented differently across the architectures. Further, C-I is goal driven and employs the use of subgoals to reach higher-level goals, as do the other architectures. Despite the high-level similarities, C-I departs from ACT-R and Soar in a number of important ways. Like ACT-R, but

unlike Soar, C-I has a serial bottleneck for executing actions. Each C-I cycle results in the firing of the most activated plan element (analogous to a production in ACT-R) whose preconditions are met. For example, a plan element to "feed Nick" may have as a precondition that Nick's food is in hand. However, even if the food is in hand, another plan element such as "call Nick" may fire if it is more highly activated due to the current task conditions.

Another distinguishing feature of the C-I architecture is the relatively unstructured way that declarative facts are represented. C-I uses individual propositions such as "Nick exists," "Nick is gray," and "Nick lives in water" to represent declarative facts about an animal named Nick, whereas ACT-R might represent information about Nick with a structured declarative chunk containing slots for color and habitat. The unstructured representation of information in the C-I architecture allows for activation to be spread "promiscuously" during the construction phase of the C-I cycle (Kintsch, 1988). Nonsensical or conflicting interpretations of a situation may be formed during construction (e.g., contemplating that a building is being chewed on after reading, "Nick likes to chew mints"), but the integration phase will take context into account, leaving only the relevant propositions activated (e.g., "mints are candy," and "candy can be chewed").

Because the C-I architecture can simulate context-sensitive knowledge activation, it is ideal for modeling dynamic adaptations to task constraints (Holyoak & Thagard, 1989; Mannes & Doane, 1991; Thagard, 1989). Recent C-I modeling efforts have included the construction of adaptive, novel plans of action in dynamic situations rather than retrieval of known routine procedures (e.g., Holyoak, 1991). This chapter will detail some of the recent modeling efforts using the C-I architecture and demonstrate how the architecture supports the development of individual performance models in real-time dynamic situations involving complex tasks.

ADAPT

The C-I architecture has been applied to modeling cognition in the complex and dynamically changing environment of airplane piloting (Doane & Sohn, 2000). ADAPT is a C-I model of piloting skill that can predict a significant amount of the variance in individual pilot visual fixations, control manipulations, and flight performance during simulated flight maneuvers. ADAPT has been used to simulate the performance of 25 human pilots on seven segments of flight. Human and modeled performance data have been compared to determine the predictive validity of ADAPT. In the following sections, the knowledge base, execution, and validation of ADAPT will be described.

ADAPT Knowledge Representation

ADAPT represents the three classes of knowledge as proposed by Kintsch (1988, 1998): world knowledge, general knowledge, and plan element knowledge.

TABLE 7.1   Examples of Knowledge in ADAPT

Type of Knowledge
World Knowledge
Desired altitude is 3500 ft
Current altitude is 3000 ft
Desired altitude is greater than current altitude
General Knowledge
Control-performance relationship:
Power controls altitude
Flight dynamics:
Pitch up causes airspeed decrease
Primary-supporting display:
VSI supports altimeter
Display instrument:
Altimeter indicates altitude
Control movement:
Pushing forward throttle increases power
Plan Element Knowledge
Cognitive plan
Name:
Increase altitude
Preconditions:
Desired altitude is greater than current altitude
Altimeter indicates altitude
Power controls altitude
Pushing forward throttle increases power
Outcome(s):
Need to look at altimeter
Need to push forward throttle
Action plan
Name:
Look at altimeter
Preconditions:
Need to look at altimeter
Altimeter indicates altitude
Outcome(s):
Looked at altimeter
Know current altitude

In ADAPT, each class of knowledge is limited to aviation-specific information, but the knowledge included in another C-I model could represent any other domain. Table 7.1 lists examples of each class of knowledge in a plain English format. In the ADAPT model, the knowledge displayed in Table 7.1 would be written in a propositional format such as "Know altimeter indicate altitude."

World Knowledge

In ADAPT, world knowledge represents the modeled pilot's current state of the world. Examples of world knowledge in ADAPT include the pilot's knowledge of the current and desired states of the airplane, determined relationships between the current and desired states (e.g., altitude is higher than desired value), and flight segment goals. World knowledge is contextually sensitive and fluid. That is, it changes with the state of the world throughout the simulated flight performance.

General Knowledge

This knowledge refers to factual information about flying an aircraft. In ADAPT, general knowledge represents facts about the relationships between control inputs and plane performance, as well as knowledge of flight dynamics, display instruments, and control movements.

Plan Element Knowledge

The final class of knowledge, plan elements, represents "executable" (procedural) knowledge. Plan elements are analogous to productions in ACT-R (Anderson, 1993). Plan elements describe actions and specify the conditions under which actions can be taken. Thus, individuals have condition–action rules that can be executed if conditions "in the world" (i.e., in the current task context) match those specified in the plan element. Plan elements consist of three parts: name, preconditions, and outcome fields (see Table 7.1). The name field is self-explanatory. The preconditions refer to world knowledge or general knowledge that must exist before a plan element can be executed. For example, a plan element that has a pilot look at his or her altimeter requires that the pilot has a need to look at the altimeter existing "in the world" before it can be fired. Plan element outcome fields contain propositions that are added to or update the model's world knowledge when the plan element is fired. For example, once the pilot has looked at the altimeter, the world knowledge will change to reflect that the pilot knows the plane's current altitude.

The preconditions of plan elements are not "bound" to specific in-the-world situations until the construction phase is completed. For example, a plan element representing the procedural knowledge for increasing altitude includes the precondition that a need to increase altitude exists in the world, but the specific magnitude of the control movement is determined by relating the plan element to currently existing world knowledge regarding flight conditions, task instructions, and so on. The general condition of a plan element does not become specific except in the context of a particular task and the process of specification is automated.

Plan elements in ADAPT can be categorized as either cognitive or action plan elements (see Table 7.1 for examples). Cognitive plan elements represent mental operations or thought processes hypothesized to motivate explicit behaviors. An example of a cognitive plan element would be "increase altitude." This plan element represents the mental operation of recognizing the need to increase altitude (i.e., understanding that current altitude is lower than desired) and setting a goal to pull back on the elevator control and/or push forward on the throttle. Action plan elements, however, represent the explicit pilot behaviors (i.e., eye fixations and control movements). Thus, the action plan "pull back elevator" directly results in an increase in altitude, but the cognitive plan "increase altitude" was necessary to set the goal to accomplish the action. Cognitive and action plans were separated in ADAPT to model the failure of action even when a goal has been activated to accomplish that action. For example, pilots know that successful flight requires monitoring the status of various attributes of plane performance (e.g., airspeed, altitude, and heading). However, sometimes pilots fail to look at relevant displays during critical periods of flight.

Constructing Individual Knowledge Bases

The goal of developing ADAPT was to model individual pilots, and to do so, a knowledge base for each individual had to be generated. Understanding the full range of knowledge necessary for the simulated flight tasks was essential to constructing these knowledge bases, so a "prototypical" expert knowledge base was created using various sources such as flight instructors and flight manuals. ADAPT models were built for 25 human pilots (8 novices, 11 intermediates, and 6 experts) that completed seven flight maneuvers in simulated flight conditions. Pilots' time-synched eye movements, control manipulations, and flight performance were recorded during an empirical study (Doane & Sohn, 2000).

Knowledge bases for each pilot were constructed after observing a small portion of the pilot's eye scan, control movement, and airplane performance data. Data were sampled from six different 7- to 15-s time blocks ("windows") for each individual pilot. Thus, we sampled 56s (i.e., 7s for three blocks, 10s for two blocks, and 15s for one block) of empirical performance data to score missing knowledge using an overlay method (see VanLehn, 1988) and build individual knowledge bases that were then used to "predict" approximately 11 min of recorded individual pilot behavior during simulated flight maneuvers. Explicit scoring rules were used during the data sampling. For example, if a pilot manipulated the elevator and looked at the altimeter in parallel, then the knowledge base for that pilot's model would include knowledge that the elevator control is used to change altitude and that the altimeter indicates altitude. In addition, the knowledge base would include a plan element for changing aircraft altitude that included manipulation of the elevator and fixation on the altimeter.

Model Execution

Plan Selection

ADAPT simulated pilot performance in simulated flight tasks. Figure 7.1 depicts the procedures used for accomplishing the simulations. First, a given pilot's knowledge base was accessed by ADAPT, and the flight goals for the first flight segment were added to the world knowledge. The model executed a C-I cycle, found the most activated plan element, and determined whether its preconditions existed in world or general knowledge. If they existed, then the plan was selected to fire, and its outcome propositions were added to the world knowledge. If one or more preconditions did not exist, then the process was repeated using the next-most-activated plan element until a plan could be fired. In the event that an action plan was fired, a separate hardware simulator interpreted the impact of pilot control manipulations on the status of the aircraft following each C-I cycle. The simulator sent the updated plane status to the

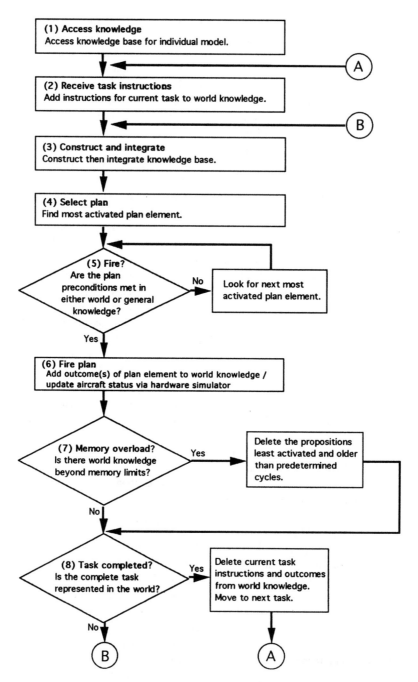

FIGURE 7.1 Schematic representation of ADAPT simulation procedures (adapted from Doane & Sohn, 2000).

ADAPT model, and this was added to the world knowledge. Then the C-I cycles began again with the modified knowledge base until the model represented a plan of action (made up of a sequence of fired plan elements) that would accomplish the specified task.

The context-sensitive nature of the C-I architecture played a vital role in the selection of plan elements in that the spread of activation was constrained by the specific situation in which the model was acting. For example, if the model became aware that the current airspeed in the world was below the desired value, the model's plans for adjusting airspeed would begin to receive more activation than they would if the current and desired values matched. Plan selection was not, however, deterministic in the sense that the model always responded in the same way to a given situation. In fact, analyses of ADAPT have shown that during any given C-I cycle, the models' degrees of freedom ranged from 16 to 46 plan elements whose preconditions were met and could be fired (Sohn & Doane, 2002).

ADAPT Memory Constraints

Working memory serves an important function for complex task performance in a dynamically changing environment (e.g., Durso & Gronlund, 1999; Sohn & Doane, 2003). To account for the impact of working memory limitations on piloting performance, two memory components were incorporated in the ADAPT model to represent capacity and decay constraints (see Step 7 in Figure 7.1).

Capacity Function

When the number of in-the-world propositions exceeded working memory capacity limitations (represented as a parameter value), ADAPT began to delete propositions, starting with those having the lowest activation. For example, if the capacity limits are set to 4, the in-the-world propositions that were not among the fourth most activated were deleted. This procedure simulated context-sensitive working memory limitations because proposition activation was constrained by relevance to the current task context. This capacity function was applied only to the in-the-world propositions.

Decay Function

ADAPT also incorporated a decay component. In the dynamic context of flight, world information must be updated in a timely manner. Decay was represented by tracking the age of each proposition in the world, where propositional age increased by one after each C-I cycle. A decay threshold was used to delete old propositions automatically following each C-I cycle. For example, if the decay threshold was set at 7, in-the-world propositions older than seven C-I cycles were deleted. Note that decay only applied to needs and traces existing in the world. Other in-the-world knowledge such as the

current values for airspeed, heading, and altitude did not decay but could be replaced as the situation changed.

Training and Testing Individual Models

Using procedures more frequently encountered in connectionist and mathematical modeling, models of individual pilots were "trained" by using the initial knowledge base to simulate the first 10 s of performance. At the beginning of the training period, values for the decay and capacity parameters were set to their minimum values. The first several seconds of flight of the first segment were then simulated. If a mismatch between the state of the model's flight and that of the human pilot occurred, the mismatch was noted, and the model's flight state was reset to match that of the human pilot. The cause of the mismatch was then examined. If the mismatch was due to a lack of general or plan knowledge, then the knowledge scoring was reexamined for any errors (e.g., a rater missed a control manipulation or an eye fixation in the pilot's performance data and, as a result, a piece of knowledge was missing from the knowledge base). If an error was found, then the knowledge base was corrected to include the appropriate knowledge. If the mismatch was due to in-the-world knowledge falling out of working memory too soon, then the decay and/or capacity parameters were increased. If the in-the-world knowledge remained in working memory longer than it should, then the decay and/or capacity parameters were decreased.

Following training, the model was "tested" by simulating pilot performance for the remaining 11 min of simulated flight completed by the human pilots. After accessing the knowledge base for an individual pilot, C-I cycles were executed and the state of the world was updated after each cycle. Note that following each cycle, ADAPT determined whether working memory capacity and decay thresholds had been exceeded (see Step 7 in Figure 7.1). If so, the model retained the most activated (capacity) and recent (decay) propositions that fell within the limits set for the individual model during the training phase. This procedure was repeated until the model obtained the desired flight goals or exceeded arbitrary time (cycle) limits (approximately 150 cycles). The entire procedure was automated, and no experimenter intervention took place during the testing period.

Model Validation

Fit Between Human and Modeled Pilot Performance

To quantify the fit between human and modeled pilot behavior, the match between the sequence of actions observed for human pilots and the sequence of corresponding plan elements fired by their ADAPT models was calculated. For clarity, actions observed for human pilots will be referred to as *human pilot plan elements*. The match between human and modeled pilot plan elements was calculated only for comparable flight situations. Although the status of the actual and modeled flight situation matched at the beginning of each segment, no intervention took place to maintain this match. As a result, if the model executed a plan element that the human pilot did not, then the modeled and actual flight situations could diverge. The mismatch of behavior in dissimilar flight situations is not of interest, and as a result, the data analyzed represent those of the human and modeled pilot in identical flight situations.

To synchronize the human and modeled pilot data, a goal-based unit of processing time called *coding time* was introduced. This was important because the human pilot's behavior was measured as a function of time, whereas the model's behavior was measured in cycles. Coding time refers to all activities taking place while a particular cognitive goal is active. This coding time increases by one when a cognitive goal is accomplished. For example, if the cognitive goal to change airspeed was active, all behaviors that took place until the change was accomplished were considered to take place within the same coding time. Once the change in airspeed was accomplished, the cognitive change plan was removed from world knowledge, and coding time was incremented by one. A pilot could interleave goals, creating a single coding time that contained two or more cognitive goals. In this event, coding time was incremented by one when all the established goals were completed. Matching coding times between human and model pilots accounted for between 33% and 44% of each segment of flight. Note that the models were never "reset" after the initial training period, so the matching coding time percentages were not inflated by experimenter intervention.

As an example of how the match between human and modeled pilot plan elements was calculated, consider a situation in which the human pilot executes action A, then B, C, and D, while the model executes action A, then D, B, and C. If these actions were executed during the same coding time, the match between human and model pilot plan elements would be 100%. Although the serial order of actions was not identical, there is no functional difference in performance. If a pilot needs to look at both the altimeter and the airspeed indicator, there is nothing to say which should be viewed first. If the model, however, had executed action A, then D, B, and A, the match would be 75% because it failed to execute action C that the human pilot executed.

The mean percentage of matches for the sequence of cognitive plan elements fired by the model and hypothesized to be active for the human were 89%, 88%, and 88% for novices, intermediates, and experts, respectively. Keep in mind that there is no singular observable behavior to indicate that a cognitive goal is active for a human, but if a pilot is looking at the altimeter and manipulating the elevator when the aircraft is not at the desired altitude, then one can hypothesize that changing altitude is an active cognitive goal. The same matches for the action plans were 80%, 79%, and 78% for the three expertise groups, respectively. For action plans, there is observable human pilot behavior that can be used to calculate the match for each plan element (e.g., a model execution of the plan element "look at altimeter" can be matched to a human pilot's eye fixation).

Given that 25 pilots have been modeled using a very small window of human pilot data to build individual knowledge bases, the match obtained between the human and modeled pilots is impressive. The average fit was greater for the cognitive plans than for the action plans. This is essentially an artifact of the superordinate nature of cognitive plans compared with action plans. For example, many different combinations of the monitor-display action plans (e.g., the altimeter, the attitude indicator, and the vertical speed indicator) could be executed in the process of accomplishing one monitor-status cognitive plan (e.g., monitor altitude).

Fit Between Human and Modeled Pilot Correct Performance

In addition to predicting the individual actions taken by the pilots during flight, the ADAPT models also predicted correct performance of the human pilots as a function of expertise and task complexity. Pilot performance

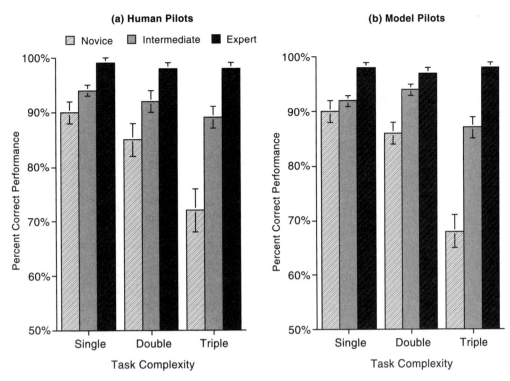

FIGURE 7.2 Mean percent correct performance for novice, intermediate, and expert human and corresponding model pilot groups as a function of task complexity.

was scored as correct if the status of the airplane was within the predetermined error limits by the end of a flight segment. The status of the airplane is characterized by three variables (current value, direction of change, and rate of change) for three flight axes (altitude, heading, and airspeed). For example, the error limits for current values were ±50 feet of the desired altitude, ±5 degrees of the desired heading, and ±5 knots of the desired airspeed.

Percentage of correct performance was measured as a function of task complexity for the human and modeled pilots in each expertise group. Task complexity refers to the number of flight axes requiring change to

obtain desired values. "Single," "double," and "triple" tasks require changes in one, two, and three axes, respectively. Figure 7.2 depicts the mean percentage of correct performance for novice, intermediate, and expert human and corresponding modeled pilots as a function of task complexity. Notice that the percentage of correct performance decreases as the task complexity increases, but this effect attenuates as expertise increases. ANOVAs (analysis of variances) were used to examine the effect of expertise and task complexity on the percentage of correct performance for the human and modeled pilots separately. Table 7.2 shows the human and modeled pilot performance analyses result

TABLE 7.2   ANOVA Results for Human Pilot and Model Pilot Performance

Source	Human Pilots			Model Pilots		
	df	F	MSE	df	F	MSE
Expertise (E)	(2, 22)	29.7*	0.004	(2, 22)	37.5*	0.003
Task complexity (T)	(2, 44)	14.7*	0.002	(2, 44)	12.7*	0.004
E × T	(4, 44)	6.4*	0.002	(4, 44)	6.9*	0.004

*$p < .01$.

in analogous main and interaction effects of expertise level and task complexity.

Further analyses of the match between human and modeled pilots are ongoing. One interesting set of findings relates to the use of *chains* across expertise levels (Doane, Sohn, & Jodlowski, 2004). Chains are sequence-dependent plans fired in response to a particular goal. Modeling results indicate that a positive relationship exists between expertise and the use of chains. That is, for expert pilots, the selection of a given plan is strongly tied to the sequence of plans that have previously been executed. For novices, the selection of a given plan has less to do with the sequence of plans previously executed and more to do with the immediate situation. At the intermediate level, pilots are more heterogeneous in their use of chains. This may be one reason why intermediate human pilots appear to be harder to accurately model than novices and experts.

Summary

The focus of this chapter has been on the use of Kintsch's C-I theory as a cognitive architecture for modeling complex task performance. C-I, ACT-R, and Soar architectures share many attributes, including the use of declarative and procedural knowledge. What distinguishes the three architectures is how problem-solving context is represented and influences knowledge activation. In Soar, episodic knowledge is used to represent actions, objects, and events that are present in the modeled agent's memory (e.g., Rosenbloom, Laird, & Newell, 1991). This knowledge influences the use of procedural and declarative knowledge by affecting the activation of knowledge based on the context of historical use. In ACT-R, the utility of productions based on previous experience guides the selection of steps taken in a problem-solving context.

In the C-I architecture, context acts to constrain the spread of knowledge activation based on the configural properties of the current task situation using low-level associations. The C-I architecture may cover a more modest range of cognitive behaviors than those examined by Soar and ACT-R researchers (e.g., VanLehn, 1991). However, the C-I architecture is parsimonious, and rigorous tests of predictive validity suggest that this simplistic approach to understanding adaptive planning has significant promise.

An important strength of the C-I architecture is that it has been applied to many cognitive phenomena

using very few assumptions and very little parameter fitting. One weakness is that greater parsimony has lead to less than perfect model fits to the human data. Thus, C-I is a relatively parsimonious architecture that has provided reasonable fits to highly complex human performance in a dynamically changing environment. In addition, the ADAPT model highlights the importance of turning toward the building of predictive individual models of individual human performance, rather than simply describing more aggregate performance.

References

Anderson, J. R. (1993). *Rules of the mind.* Hillsdale, NJ: Erlbaum.

———, Bothell, D., Byrne, M., Douglass, S., Lebiere, C., & Qin, Y. (2004). An integrated theory of the mind. *Psychological Review, 111,* 1036–1060.

Doane, S. M., Mannes, S. M., Kintsch, W., & Polson, P. G. (1992). Modeling user command production: A comprehension-based approach. *User Modeling and User Adapted Interaction, 2,* 249–285.

———, & Sohn, Y.W. (2000). ADAPT: A predictive cognitive model of user visual attention and action planning. *User Modeling and User Adapted Interaction, 10,* 1–45.

———, Sohn, Y. W., & Jodlowski, M. (2004). Pilot ability to anticipate the consequences of flight actions as a function of expertise. *Human Factors, 46,* 92–103.

———, Sohn, Y. W., McNamara, D. S., & Adams, D. (2000). Comprehension-based skill acquisition. *Cognitive Science, 24,* 1–52.

Durso, F. T., & Gronlund, S. D. (1999). Situation awareness. In F. T. Durso, R. Nickerson, R. Schvaneveldt, S. Dumais, S. Lindsay, & M. Chi (Eds.), *The handbook of applied cognition* (pp. 283–314). New York: Wiley.

Holyoak, K. J. (1991). Symbolic connectionism: Toward third-generation theories of expertise. In K. A. Ericsson & J. Smith (Eds.), *Toward a general theory of expertise* (pp. 301–336). Cambridge: Cambridge University Press.

———, & Thagard, P. (1989). Analogical mapping by constraint satisfaction. *Cognitive Science, 13,* 295–355.

Kintsch, W. (1988). The use of knowledge in discourse processing: A construction-integration model. *Psychological Review, 95,* 163–182.

———. (1998). *Comprehension: A paradigm for cognition.* New York: Cambridge University Press.

Kitajima, M., & Polson, P. G. (1995). A comprehension-based model of correct performance and errors in skilled, display-based, human-computer interaction.

International Journal of Human-Computer Studies,
43, 65–99.

Mannes, S. M., & Doane, S. M. (1991). A hybrid model of script generation: Or getting the best of both worlds. *Connection Science, 3*(1), 61–87.

———, & Kintsch, W. (1991). Routine computing tasks: Planning as understanding. *Cognitive Science, 15,* 305–342.

Newell, A. (1987). *Unified theories of cognition* (The 1987 William James Lectures). Cambridge, MA: Harvard University Press.

Rosenbloom, P. S., Laird, J. E., & Newell, A. (1991). Toward the knowledge level in SOAR: The role of architecture in the use of knowledge. In K. VanLehn (Ed.), *Architectures for intelligence* (pp. 75–112). Hillsdale, NJ: Erlbaum.

———, Laird, J. E., Newell, A., & McCarl, R. (2002). A preliminary analysis of the SOAR architecture as a basis for general intelligence. *Artificial Intelligence, 47,* 289–325.

Schmalhofer, F., & Tschaitschian, B. (1993). The acquisition of a procedure schema from text and experiences. *Proceedings of the 15th Annual Conference of the Cognitive Science Society* (pp. 883–888). Hillsdale, NJ: Erlbaum.

Sohn, Y. W., & Doane, S. M. (1997). Cognitive constraints on computer problem-solving skills. *Journal of Experimental Psychology: Applied, 3*(4), 288–312.

———, & Doane, S. M. (2000). Predicting individual differences in situation awareness: The role of working memory capacity and memory skill. *Proceedings of the Human Performance, Situation Awareness and Automation Conference* (pp. 293–298), Savannah, GA.

———, & Doane, S. M. (2002). Evaluating comprehension-based user models: Predicting individual user planning and action. *User Modeling and User Adapted Interaction, 12*(2–3), 171–205.

———, & Doane, S. M. (2003). Roles of working memory capacity and long-term working memory skill in complex task performance. *Memory and Cognition, 31*(3), 458–466.

Thagard, P. (1989). Explanatory coherence. *Brain and Behavioral Sciences, 12,* 435–467.

VanLehn, K. (1988). Student modeling. In M. C. Polson & J. J. Richardson (Eds.), *Foundations of intelligent tutoring systems* (pp. 55–76). Hillsdale, NJ: Erlbaum.

———. (Ed.). (1991). *Architectures for intelligence* (pp. 75–112). Hillsdale, NJ: Erlbaum.

PART III

VISUAL ATTENTION AND PERCEPTION

Christopher W. Myers & Hansjörg Neth

Visual attention serves to direct limited cognitive resources to a subset of available visual information, allowing the individual to quickly search through salient environmental stimuli, detect changes in their visual environment, and single out and acquire information that is relevant to the current task. Visual attention shifts very rapidly and is influenced from the bottom, up, through features of the visual stimulus, such as salience and clutter (Franconeri & Simons, 2003). Visual attention is also influenced from the top, down, through intentional goals of the individual, such as batting a ball or making a sandwich (Land & McLeod, 2000). Determining how visual attention is modulated through the interaction of bottom-up and top-down processes is key to understanding when shifts of attention occur and to where attention is shifted. The current section presents three high-fidelity, state-of-the-art formal (mathematical or computational) models of visual attention.

These three models focus on relatively different levels of visual attention and behavior. The section begins with Jeremy M. Wolfe's chapter, which addresses the interaction of bottom-up and top-down influences on visual attention and provides a progress report on his *guided search* model of visual search. In chapter 9, Marc Pomplun presents a computational approach to predicting eye movements and scanning patterns based on the *areas of activation* within a presented scene. In chapter 10, Ronald A. Rensink proposes an organizing principle for several levels of visual perception through the intimate connection between movements of the eye and movements of attention, providing a plausible explanation for the exciting and puzzling phenomenon of change blindness.

In "Guided Search 4.0: Current Progress With a Model of Visual Search" (chapter 8), Wolfe provides an update on his guided search model of visual search. Guided Search 4.0 is a model of the tight bottleneck between early massively parallel input stages and later massively parallel object recognition processes. Specifically, Guided Search is a model of the workings of that bottleneck. The bottleneck and recognition processes are modeled using an asynchronous diffusion process, capturing a wide range of empirical findings.

"Advancing Area Activation Toward a General Model of Eye Movements in Visual Search" (chapter 9) provides an overview of Pomplun's area activation model of eye movements. Pomplun argues that understanding the mechanisms behind movements of attention during visual search is crucial to understanding elementary functions of the visual system. In turn, this understanding will enable the development of sophisticated computer vision algorithms. The area activation model is presented as a promising start to developing such a model. The basic assumption of the model is that eye movements in visual search tasks tend to target areas of the display that provide a maximum amount of task-relevant information.

In "The Modeling and Control of Visual Perception" (chapter 10), Ronald A. Rensink surveys some recent developments in vision science and sketches the potential implications for the way to which vision is modeled and controlled. Rensink emphasizes the emerging view that visual perception involves the sophisticated coordination of several quasi-independent systems, each with its own kind of intelligence, and provides a basis for his coherence theory of attention. Several consequences of this view are discussed, including new and exciting possibilities for human-machine interaction, such as the notion of *coercive graphics*.

Developing integrated models of attention and perception requires theories powerful enough to incorporate the multitude of phenomena reported by the vision research community. To be ultimately successful, these emerging models of visual attention and perception must be integrated with other cognitive science domains such as memory, categorization, and cognitive control to yield complete architectures of cognition. The three chapters composing this section on "Models of Visual Attention and Perception" are steps toward the objective of within-domain incorporation, and all are viable candidate modules of visual attention for integrated models of cognitive systems.

References

Franconeri, S. L., & Simons, D. J. (2003). Moving and looming stimuli capture attention. *Perception & Psychophysics, 65*(7), 999–1010.

Land, M. F., & McLeod, P. (2000). From eye movements to actions: How batsmen hit the ball. *Nature Neuroscience, 3,* 1340–1345.

8

Guided Search 4.0

Current Progress With a Model of Visual Search

Jeremy M. Wolfe

Visual input is processed in parallel in the early stages of the visual system. Later, object recognition processes are also massively parallel, matching a visual object with a vast array of stored representation. A tight bottleneck in processing lies between these stages. It permits only one or a few visual objects at any one time to be submitted for recognition. That bottleneck limits performance on visual search tasks when an observer looks for one object in a field containing distracting objects. Guided Search is a model of the workings of that bottleneck. It proposes that a limited set of attributes, derived from early vision, can be used to guide the selection of visual objects. The bottleneck and recognition processes are modeled using an asynchronous version of a diffusion process. The current version (Guided Search 4.0) captures a range of empirical findings.

Guided Search (GS) is a model of human visual search performance, specifically of search tasks in which an observer looks for a target object among some number of distracting items. Classically, models have described two mechanisms of search: *serial* and *parallel* (Egeth, 1966). In serial search, attention is directed to one item at a time, allowing each item to be classified as a target or a distractor in turn (Sternberg, 1966). Parallel models propose that all (or many) items are processed at the same time. A decision about target presence is based on the output of this processing (Neisser, 1963). GS evolved out of the two-stage architecture of models like Treisman's feature integration theory (FIT; Treisman & Gelade, 1980). FIT proposed a parallel, preattentive first stage and a serial second stage controlled by visual selective attention. Search tasks could be divided into those performed by the first stage in parallel and those requiring serial processing. Much of the data comes from experiments measuring reaction time (RT) as a function of set size. The RT is the time required to respond that a target is present or absent. Treisman

proposed that there was a limited set of attributes (e.g., color, size, motion) that could be processed in parallel, across the whole visual field (Treisman, 1985, 1986; Treisman & Gormican, 1988). These produced RTs that were essentially independent of the set size. Thus, slopes of RT × set size functions were near zero.

In FIT, targets defined by two or more attributes required the serial deployment of attention. The critical difference between preattentive search tasks and serial tasks was that the serial tasks required a serial "binding" step (Treisman, 1996; von der Malsburg, 1981). One piece of brain might analyze the color of an object. Another might analyze its orientation. Binding is the act of linking those bits of information into a single representation of an object—an object file (Kahneman, Treisman, & Gibbs, 1992). Tasks requiring serial deployment of attention from one item to the next produce RT × set size functions with slopes markedly greater than zero (typically, about 20–30 ms/item for target-present trials and a bit more than twice that for target-absent).

The original GS model had a preattentive stage and an attentive stage, much like FIT. The core of GS was the claim that information from the first stage could be used to guide deployments of selective attention in the second (Cave & Wolfe, 1990; Wolfe, Cave, & Franzel, 1989). Thus, if observers searched for a red letter *T* among distracting red and black letters, preattentive color processes could guide the deployment of attention to red letters, even if no front-end process could distinguish a *T* from an *L* (Egeth, Virzi, & Garbart, 1984). This first version of GS (GS1) argued that *all* search tasks required that attention be directed to the target item. The differences in task performance depended on the differences in the quality of guidance. In a simple feature search (e.g., a search for red among green), attention would be directed toward the red target before it was deployed to any distractors, regardless of the set size. This would produce RTs that were independent of set size. In contrast, there are other tasks where no preattentive information, beyond information about the presence of items in the field, is useful in guiding attention. In these tasks, as noted, search is inefficient. RTs increase with set size at a rate of 20–30 ms/item on target-present trials and a bit more than twice that on the target-absent trials (Wolfe, 1998). Examples include searching for a 2 among mirror-reversed 2s (5s) or searching for rotated *T*s among rotated *L*s. GS1 argued that the target is found when it is sampled, at random, from the set of all items.

Tasks where guidance is possible (e.g., search for conjunctions of basic features) tend to have intermediate slopes (Nakayama & Silverman, 1986; Quinlan & Humphreys, 1987; Treisman & Sato, 1990; Zohary, Hochstein, & Hillman, 1988). In GS1, this was modeled as a bias in the sampling of items. Because it had the correct features, the target was likely to be picked earlier than if it had been picked by random sampling but later than if it had been the only item with those features.

GS has gone through major revisions yielding GS2 (Wolfe, 1994) and GS3 (Wolfe & Gancarz, 1996). GS2 was an elaboration on GS1 seeking to explain new phenomena and to provide an account for the termination of search on target-absent trials. GS3 was an attempt to integrate the covert deployments of visual attention with overt deployments of the eyes. This paper describes the current state of the next revision, uncreatively dubbed Guided Search 4.0 (GS4). The model is not in its final state because several problems remain to be resolved.

What Does Guided Search 4.0 Seek to Explain?

GS4 is a model of simple search tasks done in the laboratory with the hope that the same principles will scale up to the natural and artificial search tasks that are performed continuously by people outside of the laboratory. A set of phenomena is described here. Each pair of figures illustrates an aspect of the data that any comprehensive model of visual search should strive to account for (see Figure 8.1). The left-hand member of the pair is the easier search in each case.

In addition, there are other aspects of the data, not illustrated here, that GS4 seeks to explain. For example, a good model of search should account for the distributions and not merely the means of reaction times and it should explain the patterns of errors (see, e.g., Wolfe, Horowitz, & Kenner, 2005).

The Structure of GS4

Figure 8.2 shows the current large-scale architecture of the model. Referring to the numbers on the figure, parallel processes in early vision (1) provide input to object recognition processes (2) via a mandatory selective bottleneck (3). One object or, perhaps, a group of objects can be selected to pass through the bottleneck at one time. Access to the bottleneck is governed by visual selective attention. *Attention* covers a very wide range of processes in the nervous system (Chun & Wolfe, 2001; Egeth & Yantis, 1997; Luck & Vecera, 2002; Pashler, 1998a, 1998b; Styles, 1997). In this chapter, we will use the term *attention* to refer to the control of selection at this particular bottleneck in visual processing. This act of selection is mediated by a "guiding representation," abstracted from early vision outputs (4). A limited number of attributes (perhaps 1 or 2 dozen) can guide the deployment of attention. Some work better than others. Guiding attention on the basis of a salient color works very well. Search for a red car among blue and gray ones will not be hard (Green & Anderson, 1956; Smith, 1962). Other attributes, such as *opacity* have a weaker ability to guide attention (Mitsudo, 2002; Wolfe, Birnkrant, Horowitz, & Kunar, 2005). Still others, like the presence of an intersection, fail to guide altogether (Wolfe & DiMase, 2003). In earlier versions of GS, the output of the first, preattentive stage guided the second attentive stage. However, GS4 recognizes that guidance is a control

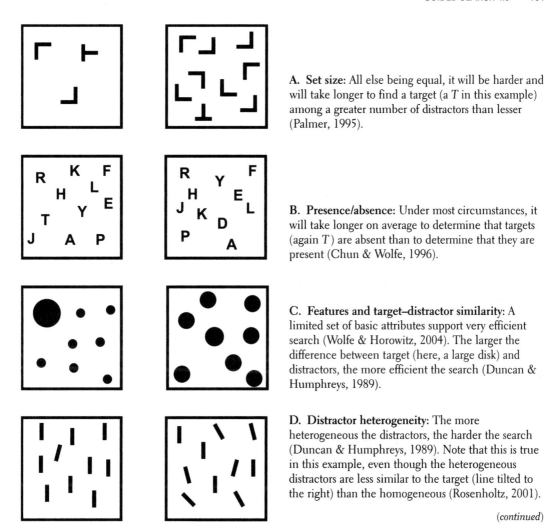

A. **Set size:** All else being equal, it will be harder and will take longer to find a target (a *T* in this example) among a greater number of distractors than lesser (Palmer, 1995).

B. **Presence/absence:** Under most circumstances, it will take longer on average to determine that targets (again *T*) are absent than to determine that they are present (Chun & Wolfe, 1996).

C. **Features and target–distractor similarity:** A limited set of basic attributes support very efficient search (Wolfe & Horowitz, 2004). The larger the difference between target (here, a large disk) and distractors, the more efficient the search (Duncan & Humphreys, 1989).

D. **Distractor heterogeneity:** The more heterogeneous the distractors, the harder the search (Duncan & Humphreys, 1989). Note that this is true in this example, even though the heterogeneous distractors are less similar to the target (line tilted to the right) than the homogeneous (Rosenholtz, 2001).

(continued)

FIGURE 8.1 Eight phenomena that should be accounted for by a good model of visual search.

signal, derived from early visual processes. The guiding control signal is not the same as the output of early vision and, thus, is shown as a separate guiding representation in Figure 8.2 (Wolfe & Horowitz, 2004).

Some visual tasks are not limited by this selective bottleneck. These include analysis of image statistics (Ariely, 2001; Chong & Treisman, 2003) and some aspects of scene analysis (Oliva & Torralba, 2001). In Figure 8.2, this is shown as a second pathway, bypassing the selective bottleneck (5). It seems likely that selection can be guided by scene properties extracted in this second pathway (e.g., where are people likely to be in this image?) (Oliva, Torralba, Castelhano, & Henderson, 2003) (6). The notion that scene statistics

can guide deployments of attention is a new feature of GS4. It is clearly related to the sorts of top-down or *reentrant* processing found in models like the Ahissar and Hochstein reverse hierarchy model (Ahissar & Hochstein, 1997; Hochstein & Ahissar, 2002) and the DiLollo et al. reentrant model (DiLollo, Enns, & Rensink, 2000). These higher-level properties are acknowledged but not explicitly modeled in GS4.

Outputs of both selective (2) and nonselective (5) pathways are subject to a second bottleneck (7). This is the bottleneck that limits performance in attentional blink (AB) tasks (Chun & Potter, 1995; Shapiro, 1994). This is a good moment to reiterate the idea that attention refers to several different processes, even in

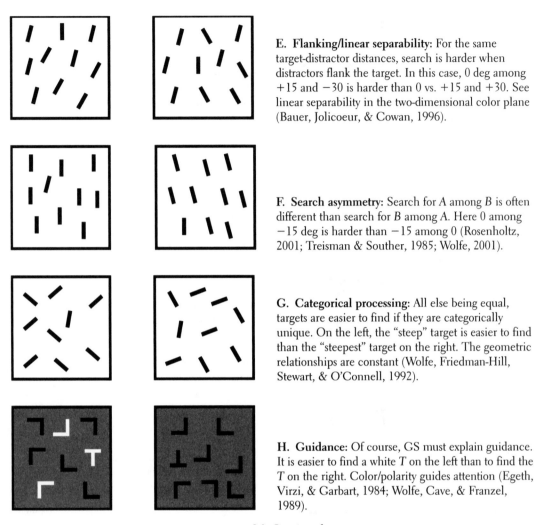

E. Flanking/linear separability: For the same target-distractor distances, search is harder when distractors flank the target. In this case, 0 deg among +15 and −30 is harder than 0 vs. +15 and +30. See linear separability in the two-dimensional color plane (Bauer, Jolicoeur, & Cowan, 1996).

F. Search asymmetry: Search for A among B is often different than search for B among A. Here 0 among −15 deg is harder than −15 among 0 (Rosenholtz, 2001; Treisman & Souther, 1985; Wolfe, 2001).

G. Categorical processing: All else being equal, targets are easier to find if they are categorically unique. On the left, the "steep" target is easier to find than the "steepest" target on the right. The geometric relationships are constant (Wolfe, Friedman-Hill, Stewart, & O'Connell, 1992).

H. Guidance: Of course, GS must explain guidance. It is easier to find a white T on the left than to find the T on the right. Color/polarity guides attention (Egeth, Virzi, & Garbart, 1984; Wolfe, Cave, & Franzel, 1989).

FIGURE 8.1 Continued.

the context of visual search. In AB experiments, directing attention to one item in a rapidly presented visual sequence can make it difficult or impossible to report on a second item occurring within 200–500 ms of the first. Evidence that AB is a late bottleneck comes from experiments that show substantial processing of "blinked" items. For example, words that are not reported because of AB can, nevertheless, produce semantic priming (Luck, Vogel, & Shapiro, 1996).

Object meaning does not appear to be available before the selective bottleneck (3) in visual search (Wolfe & Bennett, 1997), suggesting that the search bottleneck lies earlier in processing than the AB bottleneck (7). Moreover, depending on how one uses the term, *attention*, a third variety occurs even earlier in visual search. If an observer is looking for something

red, all red items will get a boost that can be measured psychophysically (Melcher, Papathomas, & Vidnyánszky, 2005) and physiologically (Bichot, Rossi, & Desimone, 2005). Melcher et al. (2005) call this *implicit attentional selection*. We call it *guidance*. In either case, it is a global process, influencing many items at the same time—less a bottleneck than a filter. The selective bottleneck (3) is more local, being restricted to one object or location at a time (or, perhaps, more than one; McMains & Somers, 2004). Thus, even in the limited realm illustrated in Figure 8.2, attentional processes can be acting on early parallel stages (1) to select features, during search to select objects (3), and late, as part of decision or response mechanisms (7).

Returning to the selective pathway, in GS, object recognition (2) is modeled as a diffusion process where

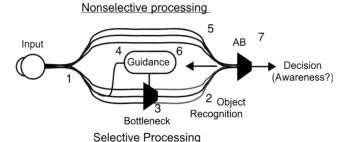

FIGURE 8.2 The large-scale structure of GS4. Numbers refer to details in text. Multiple lines illustrate parallel processing.

information accumulates over time (Ratcliff, 1978). A target is identified when information reaches a target threshold. Distractors are rejected when information reaches a distractor threshold. Important parameters include the rate and variability of information accrual and the relative values of the thresholds. Many parallel models of search show similarities to diffusion models (Dosher, Han, & Lu, 2004). Effects of set size on reaction time are assumed to occur either because accrual rate varies inversely with set size (limited-capacity models; Thornton, 2002; Figure 8.3) or because, to avoid errors, target and distractor thresholds increase with set size (e.g., Palmer, 1994; Palmer & McLean, 1995).

In a typical parallel model, accumulation of information begins for all items at the same time. GS differs from these models because it assumes that information accumulation begins for each item only when it is selected (Figure 8.3). That is, GS has an *asynchronous* diffusion model at its heart. If each item needed to wait for the previous item to finish, this

becomes a strict serial process. If N items can start at the same time, then this is a parallel model for set sizes of N or less. In its general form, this is a hybrid model with both serial and parallel properties. As can be seen in Figure 8.3, items are selected, one at a time, but multiple items can be accumulating information at the same time. A carwash is a useful metaphor. Cars enter one at a time, but several cars can be in the carwash at one time (Moore & Wolfe, 2001; Wolfe, 2003). (Though note that Figure 8.3 illustrates an unusual carwash where a car entering second could, in principle, finish first.)

As noted at the outset, search tasks have been modeled as either serial or parallel (or, in our hands, *guided*). It has proved difficult to use RT data to distinguish serial from parallel processes (Townsend, 1971, 1990; Townsend & Wenger, 2004). Purely theoretical considerations aside, it may be difficult to distinguish parallel from serial in visual search tasks because those tasks are, in fact, a combination of both sorts of process. That, in any case, is the claim of GS4, a model that could be described as a parallel–serial hybrid. It has a parallel front end, followed by an attentional bottleneck with a serial selection rule that then feeds into parallel object recognition processes.

Modeling Guidance

In GS4, objects can be recognized only after they have been passed through the selective bottleneck between early visual processes and object recognition processes. Selection is controlled by a guiding representation. That final guiding representation is created from bottom-up and top-down information. Guidance is not based directly on the contents of early visual processes but on a coarse and categorical representation derived from those processes. Why argue that guidance is a control process, sitting, as it were, to the side of the main selective pathway? The core argument is that information

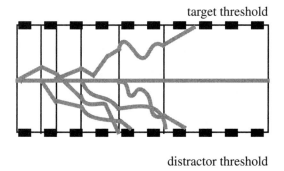

target threshold

distractor threshold

time ⟶

FIGURE 8.3 In GS4, the time course of selection and object recognition is modeled as an asynchronous diffusion process. Information about an item begins to accumulate only after that item has been selected into the diffuser.

that is available in early vision (Figure 8.2, no. 1) and later (2) is not available to guidance (4). If guidance were a filter in the pathway, we would need to explain how information was lost and then regained (Wolfe & Horowitz, 2004).

Consider three examples that point toward this conclusion:

1. Even in simple feature search, efficient guidance requires fairly large differences between targets and distractors. For example, while we can resolve orientation differences on the order of a degree (Olzak & Thomas, 1986), it takes about a 15-deg difference to reliably attract attention (Foster & Ward, 1991b; Moraglia, 1989). Fine-grain orientation information is available before attentional selection and after but not available to the guidance mechanism.
2. Search is more efficient if a target is categorically unique. For example, it is easier to find a line that is the only "steep" item as illustrated in Figure 8.1. There is no categorical limitation on processing outside of the guidance mechanism.
3. Intersection type (*t*-junction vs. *x*-junction) does not appear to guide attention (Wolfe & DiMase, 2003). It can be used before selection to parse the field into preattentive objects (Rensink & Enns, 1995). Intersection type is certainly recognized in attentive vision, but it is not recognized by guidance.

Thus, we suggest that the guiding representation should be seen as a control module sitting to one side of the main selective pathway rather than as a stage within that pathway. In the current GS4 simulation, guidance is based on the output of a small number of broadly tuned channels. These can be considered to be channels for *steep, shallow, left,* and *right* (for steep and shallow, at least, see Foster & Ward, 1991a). Only orientation and color are implemented, but other attributes are presumed to be similar. In orientation, the four channels are modeled as the positive portion of sinusoidal functions, centered at 0 (vertical), 90, 45, and −45 deg and raised to a power less than 1.0 to make the tuning less sharp. Thus, the steep channel is defined as $\max[\cos(2 \times \deg), 0]^{0.3}$. The precise shape is not critical for the qualitative performance of the model. In color, a similar set of channels covers a red-green axis with three categorical channels for *red, yellow,* and *green.* Color, of course, is a three-dimensional feature space.

Restricting modeling to one red-green axis is merely a matter of convenience.

Another major simplification needs to be acknowledged. Selection is presumed to select objects (Wolfe & Bennett, 1997). As a consequence, the "receptive field" for the channels described above is an object, conveniently handed to the model. The model does not have a way to parse a continuous image into "preattentive object files" (our term) or "proto-objects" (Rensink & Enns, 1995, 1998).

Bottom-Up Guidance

The more an item differs from its neighbors, the more attention it will attract, all else being equal. This can be seen in Figure 8.4. The vertical line "pops out" even though you were not instructed to look for vertical. That this pop-out is the result of local contrast can be intuited by noticing that the other four vertical lines in this image do not pop-out. They are not locally distinct (Nothdurft, 1991, 1992, 1993).

In GS4, bottom-up salience for a specific attribute such as orientation is based on the differences between the channel response for an item and the other items in the field. Specifically, for a given item, in orientation, we calculate the difference between the response to the item and the response to each other item for each of the four categorical channels. For each pairwise comparison, it is the maximum difference that contributes to bottom-up salience. The contribution of each pair is divided by the distance between the items. Thus, closer neighbors make a larger contribution to bottom-up

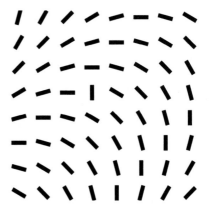

FIGURE 8.4 Local contrast produces bottom-up guidance. Note that there are five vertical lines in this display. Only one is salient.

activation of an item than do more distant items (Julesz, 1981, 1984). The distance function can be something other than linear distance. In the current simulation, we actually use the square root of the linear distance. Further data would be needed to strongly constrain this variable.

$$\sum_{b=1}^{setsize} \left\{ \max[(Ch_1(a) - Ch_1(b)) \ldots \right.$$

$$\left. (Ch_n(a) - Ch_n(b))]/d_{ab} \right\} \quad \text{Bottom-up activation}$$

Thus, this bottom-up calculation will create a bottom-up salience map where the signal at each item's location will be a function of that item's difference from all other items scaled by the distance between items.

Local differences are the basis for many models of stimulus salience (e.g., Itti & Koch, 2000; Koch & Ullman, 1985; Li, 2002). Many of these use models of cells in early stages of visual processing to generate signals. In principle, one of these salience models could replace or modify the less physiologically driven bottom-up guidance modules in GS4.

Top-Down Guidance

If you were asked to find the targets in Figure 8.5, it would be reasonable to ask, "What targets?" However, if told to find the horizontal items, you can rapidly locate them. Thus, in Figure 8.5, bottom-up salience does not define targets, but efficient search is still

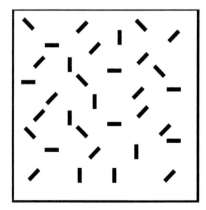

FIGURE 8.5 Bottom-up information does not define a target here, but top-down guidance can easily direct attention to a specified orientation (e.g., horizontal).

possible, guided by top-down information. In GS4, top-down guidance is based on the match between a stimulus and the desired properties of the target. For each item, the channel responses are the signals out of which top-down guidance is created. The steep channel would respond strongly to the vertical lines, the "right" channel to 45-deg lines and so on. Top-down guidance results when higher weight is placed on the output of one channel than on others. In the current formulation of GS4, the model picks one channel for each attribute by asking which channel contains the largest signal favoring the target over the mean of the distractors. For example, consider a search for an orange line, tilted 22 deg off vertical. If the distractors were yellow and vertical, GS4 would place its weights on the red channel (targets and distractors both activate the yellow but only orange activates red) and the right-tilted channel (for similar reasons). If the same target were placed among red 45-deg lines, then it would be the yellow and steep channels that would contain the best signal.

The Activation Map

In GS, the activation map is the signal that will guide the deployment of attention. For each item in a display, the guiding activation is simply a weighted sum of the bottom-up activation and the activity in each channel (composed of the top-down activation) plus some noise. In the current version of GS, the weights are constrained so that one weight for a particular dimension (color or orientation) is set to 1.0 and the others are set to 0. This is the formal version of the claim that you can only select one feature in a dimension at a time (Wolfe et al., 1990). If you set the bottom-up weight to 0, you are making the claim that a salient but irrelevant distractor can be ignored. If you declare that it cannot go to 0, you are holding out the possibility of true *attentional capture* against the desires of the searcher. There is an extensive and inconclusive literature on this point (e.g., Bacon & Egeth, 1994; Folk et al., 1992; Lamy & Egeth, 2003; Theeuwes, 1994; Todd & Kramer, 1994; Yantis, 1998) that has been usefully reviewed by Rauschenberger (2003). GS4 does not allow the bottom-up weight to go to 0.

$$[wt_{bu}(BU)] + \sum^{Channels} [wt_{ch}(CH)]] + \text{noise}$$

Activation Map

In earlier versions of GS, the activation map was fixed for a trial. Attention was deployed in order of activation strength from highest down until the target was found or until the search was abandoned. This assumes perfect memory for which items have been attended. Subsequent work has shown this to be incorrect (Horowitz & Wolfe, 1998, 2005). More will be said on this topic later. For now, the relevant change in GS4 is that the added noise is dynamic and each deployment of attention is directed to the item with the highest current activation.

Guidance and Signal Detection Theory

Note that GS4, to this point, is very similar to a signal detection theory (SDT) model (Cameron, Tai, Eckstein, & Carrasco, 2004; Palmer & McLean, 1995; Palmer, Verghese, & Pavel, 2000; Verghese, 2001). Consider the standard SDT-style experiment. A search display is presented for 100 ms or so and masked. The distractors can be thought of as noise stimuli. The target, if present, is signal plus noise. In a standard SDT account, the question is how successfully the observer can distinguish the consequence of N(noise) from $[(N-1)$ (noise) + signal], where N is the set size. As N gets larger, this discrimination gets harder and that produces set size effects in brief exposure experiments. SDT models generally stop here, basing a decision directly on the output of this parallel stage. In GS, the output of this first stage guides access to the second stage. However, for brief stimulus presentation, GS4, like SDT models, would show a decrease in accuracy, albeit via a somewhat different mechanism. With a brief exposure, success in GS depends on getting attention to the target on the first deployment (or in the first few deployments). If there is no guiding signal, the chance of deploying to the target first is $1/N$. Performance drops as N increases. As the guiding signal improves, the chance of deploying to the target improves. If the signal is very large, the effect of increasing N becomes negligible and attention is deployed to the target first time, every time. There is more divergence between the models when stimulus durations are long. The rest of the GS model deals with deployments of attention over a more extended period. SDT models have not typically addressed this realm (but see Palmer, 1998). GS rules make different quantitative predictions than SDT "max" or "sum" rules but these have not been tested as yet.

Why Propose a Bottleneck?

GS is a two-stage model with the activation map existing for the purpose of guiding access to the second stage where object recognition occurs. Why have two stages? Why not base response on a signal derived, like the activation map, in parallel from early visual processes? Single-stage models of this sort account for much search performance, especially for briefly presented stimuli (Baldassi & Burr, 2000; Baldassi & Verghese, 2002; McElree & Carrasco, 1999; Palmer & McLean, 1995). Is there a reason to propose a bottleneck in processing with access controlled by guidance? Here are four lines of argument, which, taken together, point to a two-stage architecture.

1. *Targets may be easy to identify but hard to find.* Consider the search for a *T* among *L*s in Figure 8.1A and the search for tilted among vertical in Figure 8.1D. In isolation, a *T* is trivially discriminable from an *L* and tilted is trivially discriminable from vertical. However, search for the *T* is inefficient while search for tilted is efficient. The GS, two-stage account is fairly straightforward. The first stage registers the same vertical and horizontal elements for *T*s and *L*s. However, the intersection type is not available to guide attention (Wolfe & DiMase, 2003). The best that guidance can do is to deliver one object after another to the second stage. The relationship between the vertical and horizontal elements that identifies an object as *T* or *L* requires second-stage binding. The lack of guidance makes the search inefficient. The orientation search in Figure 8.1D, in contrast, is easy because the first stage can guide the second stage. This argument would be more convincing if the single T and the tilted line were equated for discriminability. Even so, a single stage model must explain why one easy discrimination supports efficient search and another does not.

2. *Eye movements.* Saccadic eye movements impose an obvious seriality on visual processing (Sanders & Houtmans, 1985). Attention is deployed to the locus of the next saccade before it is made (Kowler, Anderson, Dosher, & Blaser, 1995), and guidance mechanisms influence the selection of eye movement targets (Bichot & Schall, 1999; Motter & Belky, 1998; Shen, Reingold, & Pomplun, 2003; Thompson & Bichot, 2004).

Invoking the control of saccades as an argument for a model of covert deployments of attention is a double-edged sword. Numerous researchers have argued that overt deployment of the eyes is what needs to be explained and that there is no need for a separate notion of covert deployments (Deubel & Schneider, 1996; Findlay & Gilchrist, 1998; Maioli, Benaglio, Siri, Sosta, & Cappa, 2001; Zelinsky & Sheinberg, 1995, 1996). If true, the link between attention and eye movements is not trivially simple. Take the rate of processing for example. The eyes can fixate on 4–5 items per second. Estimates of the rate of processing in visual search are in the range of 10 to 30 or 40 per second (based, e.g., on search slopes). The discrepancy can be explained by assuming that multiple items are processed, in parallel, on each fixation. Indeed, it can be argued that eye movements are a way for a parallel processor to optimize its input, given an inhomogeneous retina (Najemnik & Geisler, 2005).

Eye movements are not required for visual search. With acuity factors controlled, RTs are comparable with and without eye movements (Klein & Farrell, 1989; Zelinsky & Sheinberg, 1997), and there is endless evidence from cueing paradigms that spatial attention can be deployed away from the point of fixation (for useful reviews, see Driver, 2001; Luck & Vecera, 2002). Nevertheless, the neural circuitry for eye movements and for deployment of attention are closely linked (Moore, Armstrong, & Fallah, 2003; Schall & Thompson, 1999), so the essential seriality of eye movements can point toward the need for a serial selection stage in guided search.

3. *Binding.* The starting point for Treisman's feature integration theory was the idea that attention was needed to *bind* features together (Treisman & Gelade, 1980). Failure to bind correctly could lead to *illusory conjunctions*, in which, for example, the color of one object might be perceived with the shape of another (Treisman & Schmidt, 1982). While the need for correct binding can be seen as a reason for restricting some processing to one item at a time, it is possible that multiple objects could be bound at the same time. Wang, for example, proposes an account where correlated oscillations of activity are the mechanism for binding and where several oscillations can coexist (Wang, 1999), and Hummel and Stankiewicz (1998) showed that a single parameter that varies the amount of overlap between oscillatory firings acts a lot like attention. The oscillation approach requires that when several oscillations coexist, they must be out of synchrony with each other to prevent errors like illusory conjunctions. Given some required temporal separation between oscillating representations, this places limit on the number of items that can be processed at once, consistent with an attentional bottleneck.

4. *Change blindness.* In change blindness experiments, two versions of a scene or search display alternate. If low-level transients are hidden, observers are poor at detecting substantial changes as long as those changes do not alter the gist, or meaning, of the display (Rensink, O'Regan, & Clark, 1997; Simons & Levin, 1997; Simons & Rensink, 2005). One way to understand this is to propose that observers only recognize changes in objects that are attended over the change and that the number of objects that can be attended at one time is very small, perhaps only one. In a very simple version of such an experiment, we asked observers to examine a display of red and green dots. On each trial, one dot would change luminance. The Os' task was to determine whether it also changed color at that instant. With 20 dots on the screen, performance was 55% correct. This is significantly above the 50% chance level but not much. It is consistent with an ability to monitor the color of just 1–3 items (Wolfe, Reinecke, & Brawn, 2006).

Early vision is a massively parallel process. So is object recognition. A stimulus (e.g., a face) needs to be compared with a large set of stored representations in the hopes of a match. The claim of two-stage models is that there are profound limitations on the transfer of information from one massively parallel stage to the next. Those limitations can be seen in phenomena such as change blindness. At most, it appears that a small number of objects can pass through this bottleneck at one time. It is possible that the limit is one. Guidance exists to mitigate the effects of this limitation. Under most real-world conditions, guidance allows the selection of an intelligently chosen subset of all possible objects in the scene.

Modeling the Bottleneck

In earlier versions of GS, object recognition was regarded as something that happened essentially instantaneously when an item was selected. That was never intended to be realistic. Data accumulating from many labs since that time has made it clear that the time required to identify and respond to a target is an important constraint on models of the bottleneck in the selective pathway. If it is not instantaneous, how long is selective attention occupied with an item after that item is selected? Measures of the *attentional dwell time* (Moray, 1969) have led to apparently contradictory results. One set of measures comes from attentional blink (Raymond, Shapiro, & Arnell, 1992; Shapiro,

1994) and related studies (Duncan, Ward, & Shapiro, 1994; Ward, Duncan, & Shapiro, 1996, 1997). These experiments suggest that, once attention is committed to an object, it is tied up for 200–500 ms (see also Theeuwes, Godijn, & Pratt, 2004). This dwell time is roughly consistent with the time required to make voluntary eye movements and volitional deployments of attention (Wolfe, Alvarez, & Horowitz, 2000). It would seem to be incompatible with estimates derived from visual search. In a classic, serial self-terminating model of search, the time per item is given by the slope of target-absent trials or twice the slope of the target-present trials. Typical estimates are in the range of 30–60 ms/item. Efforts have been made to find a compromise position (Moore, Egeth, Berglan, & Luck, 1996), but the real solution is to realize that slopes of RT × set size functions are measures of the rate of processing, not of the time per item. We have made this point using a carwash metaphor (Moore & Wolfe, 2001; Wolfe, 2002; cf. Murdock, Hockley, & Muter, 1977). The core observation is that, while cars might enter (or emerge from) a carwash at a rate of 50 ms/item, they might be in this very fast carwash for 200–500 ms. Of course, a necessary corollary of this observation is that more than one car can be in the carwash at one time.

In GS4, as noted earlier, the carwash is formally modeled with an asynchronous diffusion model. Asynchronous diffusion is really a class of models with a large number of parameters, as illustrated in Figure 8.6. Having many parameters is not usually seen as a strength of a model (Eckstein, Beutter, Bartroff, & Stone, 1999). However, complex behaviors are likely to have complex underpinnings. The goal of this modeling effort is to constrain the values of the parameters so that variation in a small subset can account for a large body of data.

The assumption of diffusion models is that information begins to accumulate when an item is selected into the diffuser. The time between successive selections is labeled SSA for *stimulus selection asynchrony*. It could be fixed or variable. In either case, the average SSA is inversely related to the rate of processing that, in turn, is reflected in the slope of RT × set size functions. Because search RT distributions are well described as gamma distributions, we have used exponentially distributed interselection intervals. However, it is unclear that this produces a better fit to the data than a simple, fixed interval of 20–40 ms/item.

In the case of visual search, the goal is to determine if the item is a target or a distractor and the answer is established when the accumulating information crosses a *target threshold* or *distractor threshold*. Both of those thresholds need to be set. It would be possible to have either or both thresholds change over time (e.g., one might require less evidence to reject a distractor as time progresses within a search trial). In the present version of GS4, the target threshold, for reasons described later, is about 10 times the distractor threshold. The *start* point for accumulation might be fixed or variable to reflect a priori assumptions about a specific item. For example, contextual cueing effects might be modeled by assuming that items in the cued location start at a point closer to the target threshold (Chun, 2000; Chun & Jiang, 1998). In the current GS4, the start point is fixed.

Items diffuse toward a boundary at some average *rate*. In principle, that rate could differ for different items in a display (e.g., as a function of eccentricity Carrasco, Evert, Chang, & Katz, 1995; Carrasco & Yeshurun, 1998; Wolfe, O'Neill, & Bennett, 1998). The rate divided into the distance to the threshold gives the average time in the diffuser for a target or distractor. The diffusion process is a continuous version of a random walk model with each step equal to the rate plus some *noise*. In the current GS4, the rate parameter is used to account for differences between Os, but is set so that the time for a target to diffuse to the target boundary is on the order of 150–300 ms. Ratcliff has pointed out that noise that is normally distributed around the average path will produce a positively skewed *distribution* of finishing times (Ratcliff, 1978; Ratcliff, Gomez, & McKoon, 2004). This is a useful property since search RTs are positively skewed. An asynchronous diffusion model assumes that infor-

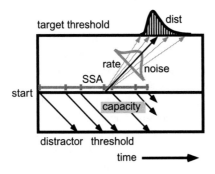

FIGURE 8.6 The parameters of an asynchronous diffusion model. SSA = stimulus selection asynchrony.

mation about items can start accumulating at different times.

The diffuser is assumed to have some *capacity*. This brings with it a set of other choices that need to be made. If the capacity is K, then the $K + 1$th item cannot be selected until the one of the K items is dismissed. At the start of a search, can K items be selected simultaneously into an empty diffuser? If items are selected one at a time, then there will be periods when the number of items in the diffuser is less than K. This will also occur, of course, if the set size is less than K. When the diffuser contains fewer than K items, is the rate of information accumulation fixed or is it proportional to the number of items in the diffuser? That is, if $K = 4$ and the set size is 2, does the processing rate double? In GS4, we typically use a capacity of 4 items (inspired, in part, by the ubiquity of the number 4 in such capacity estimates; Cowan, 2001). Small changes in N do not produce large changes in the behavior of the model. At present, in GS4, if there are fewer than the maximum number of items in the diffuser or if the same item is selected more than once (hard for cars in a car wash but plausible here), then the rate of information accrual increases.

Memory in Search

If capacity, K, is less than the set size, then the question of memory in search arises. If an item has been dismissed from the diffuser, can it be reselected in the same search? The classic serial, self-terminating model (FIT and earlier versions of GS) had a capacity of one (i.e., items are processed in series) and an assumption that items were not reselected. That is, visual search was assumed to be sampling without replacement. In 1998, we came to the conclusion that visual search was actually sampling with replacement—that there was no restriction on reselection of items (Horowitz & Wolfe, 1998). Others have argued that our claim that "visual search has no memory" was too strong and that selection of some number of recently attended items is inhibited (Kristjansson, 2000; Peterson, Kramer, Wang, Irwin, & McCarley, 2001; Shore & Klein, 2000). In our work, we have been unable to find evidence for memory in search. Nevertheless, we have adopted a middle position in our modeling. Following Arani, Karwan, and Drury (1984), the current version of GS inhibits each distractor as it is rejected. At every cycle of the model thereafter, there is some probability that the inhibition

will be lifted. Varying that probability changes the average number of items that are inhibited. If that parameter is 1, then visual search has no memory. If it is 0, search has perfect memory. We typically use a value of 0.75. This yields an average of about three inhibited items at a time during a search trial. These are not necessarily the last three rejected distractors. Rigid N-back models of memory in search tend to make strong predictions that are easily falsified (e.g., that search through set sizes smaller than N will show perfect memory). Modest variation in this parameter does not appear to make a large difference in model output.

Constraining Parameter Values

At this point, the reader would be forgiven for declaring that a model with this many parameters will fit all possible data and that some other model with fewer parameters must be preferable. If all of the parameters could vary at will, that would be a fair complaint. However, GS assumes that most of these are fixed in nature; we just do not know the values. Moreover, other apparently simple models are simple either by virtue of making simplifying assumptions about these (or equivalent) parameters or by restriction of the stimulus conditions. For example, if stimuli are presented briefly, then many of the issues (and parameters) raised by an asynchronous diffusion process become moot.

The data provide many constraints on models of search. At present, it must be said that these constraints are better at ruling out possibilities than they are at firmly setting parameters, but modeling by exclusion is still progress. We have obtained several large data sets in an effort to understand normal search behavior. Figure 8.7 shows average RTs for 10 Os, tested for 4,000 trials on each of three search tasks: A simple feature search for a red item among green distractors, a color X orientation conjunction search, and a "spatial configuration" search for a 2 among 5s, the mirror reverse of the 2. The 2 versus 5 search might have been called a *serial* search in the past but that implies a theoretical position. Calling it an inefficient spatial configuration search is neutrally descriptive. This data set, confirming other work (Wolfe, 1998), shows that the ratio of target-absent to target-present slopes is greater than 2:1 for spatial configuration searches. This violates the assumptions of a simple serial, self-terminating search model with complete memory for rejected distractors. The variance of the RTs increases with set size and is greater for target-absent

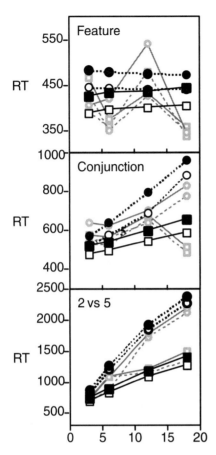

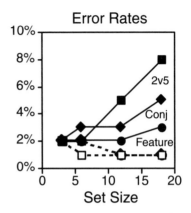

FIGURE 8.8 Average error rates for data shown in Figure 8.7. Closed symbols are miss errors as a percentage of all target-present trials. Open symbols are false alarms (all false alarm rates are low and similar).

FIGURE 8.7 Average reaction times for 10 observers tested for 1,000 trials per set size in three tasks: Feature (red among green), conjunction (red vertical among red horizontal and green vertical), and a search for a 2 among 5 (the mirror reversed item). The black, bold lines represent correct responses. Light-gray lines are the corresponding error trials. Squares are hits, circles are correct absent responses; closed symbols are means, open are medians (always slightly faster than the means). In the gray error trials, squares are false alarms (very rare), circles are misses. Note the very different y-axes.

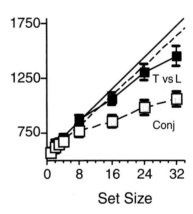

FIGURE 8.9 RT (reaction time) × set size functions with linear regression lines fitted to just Set Sizes 1–4 to illustrate the nonlinearity of these functions.

than for target-present trials. Error rates increase with set size in all conditions (Figure 8.8). The great bulk of errors are miss errors: False alarms are rare in RT search studies. Miss-error RTs tend to be somewhat faster than correct absent RTs (Figure 8.7). Thus, if a model predicts a large number of false alarms or predicts that errors are slow, it is failing to capture the

shape of the data. GS4 produces qualitatively correct patterns of RTs as described later.

In a separate set of experiments, we tested the same three tasks on a wider and denser range of set sizes than is typical. As shown in Figure 8.9, the salient finding is that RT × set size functions are not linear (Michod, Wolfe, & Horowitz, 2004). They appear to be compressive with small set sizes (1–4) producing very steep slopes. The cause of the nonlinearity is not clear but the result means that models (like earlier versions of GS) that produce linear RT × set size functions are missing something. In GS4, a nonlinearity is produced by allowing the rate of information accrual to be proportional to the number of items in the diffuser (up to the capacity limit). Small set sizes will benefit more from this feature than large, causing small set size RTs to be somewhat faster than they would otherwise be.

The large number of trials that we ran to collect the data in Figures 8.7 and 8.8 allows us to look at RT distributions. Search RT distributions, like so many other RT distributions, are positively skewed (Luce, 1986; Van Zandt, 2002). This general shape falls out of diffusion models (Ratcliff et al., 2004). In an effort to compare distributions across Os, set sizes, and search tasks, we normalized the distributions using a nonparametric equivalent of a z-transform. Specifically, the 25th and 75th percentiles of the data were transformed to −1 and +1, respectively, and the data were scaled relative to the interquartile distance. As shown in Figure 8.10, the first striking result of this analysis is how similar the distribution shapes are. To a first approximation, distributions for feature and conjunction searches are scaled copies of each other with no qualitative change

in the shape of the distribution with set size. Models that predict that the shape of the normalized RT distribution changes with set size would, therefore, be incorrect. Moreover, after this normalization, there is little or no difference between target-present (thick lines) and target-absent (thin lines)—also a surprise for many models (e.g., FIT and earlier version of GS).

RT distributions from the 2 versus 5 task are somewhat different. They are a bit more rounded than the feature and conjunction distributions. They change a little with set size and absent distributions are somewhat different from present distributions. A number of theoretical distributions (gamma, Weibull, lognormal, etc.) fit the distributions well, and there does not seem to be a data-driven reason to choose between these at the present time. GS4 produces RT distributions that are qualitatively consistent with the pattern of Figure 8.10.

Hazard functions appear to magnify these differences. Hazard functions give the probability of finding the target at one time given that it has not been found up until that time. In Figure 8.11, we see that the hazard functions are clearly nonmonotonic. All tasks at all set sizes, target present or absent, seem to have the same initial rise. (The dashed line is the same in all three panels.) The tasks differ in the later portions of the curve, but note that data beyond an x-value of 3 come from the few trials in the long tail of this RT distribution. Gamma and ex-Gaussian distributions have monotonic hazard functions and, thus, are imperfect models of these RT distributions. Van Zandt and Ratcliff (2005) note that "the increasing then decreasing hazard is ubiquitous" and are an indication that the RT distribution is a mixture of two or more underlying distributions. This seems entirely plausible in the case of a complex behavior like search.

From the point of view of constraining models of search, a model should not predict qualitatively different shapes of RT distributions, after normalization, as a function of set size, task, or target presence or absence for reasonably efficient searches. Some differences between more efficient (feature and conjunction) and less efficient (2 vs. 5) search are justified. Moreover, inefficient search may produce some differences in distributions as a function of set size and target presence/absence.

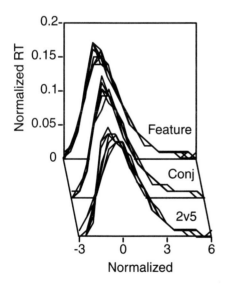

FIGURE 8.10 Probability density functions for normalized RT distributions for four set sizes in three search tasks. Thicker lines are target-present; thinner are target-absent. Note the similarity of the probability density functions, especially for the feature and conjunction tasks.

Target-Absent Trials and Errors

In some ways, modeling the process that observers use to find a target is the easy part of creating a model of

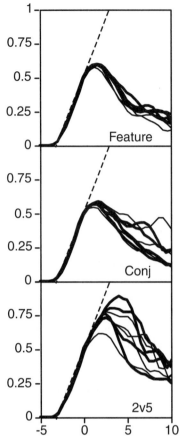

FIGURE 8.11 Hazard functions for the probability density functions in Figure 8.10.

visual search. After all, once attention has been guided to the target, the model's work is done. What happens when no target is present? When do you terminate an unsuccessful search? Simple serial models have a clear account. When you have searched all items, you quit. Such models predict lower variance on target-absent trials than on target-present trials because target-absent trials should always require observers to attend to N items where N is the set size. On target-present trials, observers might find a target on the first deployment of attention or on the Nth. That prediction is not correct. Moreover, we had observers search through displays in which items were continuously being replotted in random locations and found that observers can terminate search under these conditions even though dynamic search displays would make it impossible to know when everything had been examined (Horowitz & Wolfe, 1998). (Note, compared with standard search tasks,

dynamic search conditions do lead to longer target-absent RTs and more errors, suggesting some disruption of target-absent search termination.)

Instead of having a method of exhaustively searching displays, observers appear to establish a quitting rule in an adaptive manner based on their experience with a search task. Observers speed subsequent responses after correct responses and slow subsequent responses after errors (Chun & Wolfe, 1996). An adaptive rule of this sort can be implemented in many ways. Observers could adjust the time spent searching per trial. They could adjust the number of items selected or the number of items rejected. Whatever is adjusted, the resulting quitting threshold must be scaled by set size. That is, the threshold might specify quitting if no target has been found after some percentage of the total set size has been selected, not after some fixed number of items had been selected regardless of set size.

In GS4, miss errors occur when the quitting threshold is reached before the target is found. As shown in Figure 8.7, miss RTs are slightly faster than RTs for correct absent trials. Misses occur when observers quit too soon. As shown in Figure 8.8, false alarms are rare and must be produced by another mechanism. If observers produced false alarms by occasionally guessing "yes" when the quitting threshold was reached, then false alarms and miss RTs should be similar, which they are not. False alarms could be produced when information about distractor items incorrectly accumulates to the target boundary. There may also be some sporadic fast guesses that produce false alarms. At present, GS4 does not produce false alarms at even the infrequent rate that they are seen in the data.

The data impose a number of constraints on models of search termination. Errors increase with set size, at least for harder search tasks. One might imagine that this is a context effect. The quitting threshold gets set to the average set size and is, therefore, conservative for smaller set sizes and liberal for larger. This cannot be the correct answer because the patterns of RTs and errors do not change in any qualitative way when set sizes are run in blocks rather than intermixed (Wolfe, Palmer, Horowitz, & Michod, 2004). Slopes for target-absent are reliably more than twice as steep as slopes for target-present trials (Wolfe, 1998).

One of the most interesting constraints on search termination is that observers appear to successfully terminate target-absent trials too fast. Suppose that observers terminated trials at time T, when they were convinced

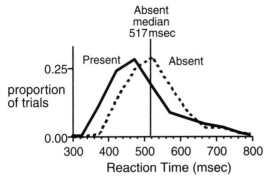

FIGURE 8.12 Reaction time (RT) distributions for one observer, Set Size 3, conjunction task. Note the degree of overlap between target-present and target-absent RTs. Twenty-five percent of correct present trials lie above the median for the correct absent trials. Miss error rate in this condition is 1.9%. How can observers answer "no" so quickly?

that only X% of targets would require more than T ms to find, where X% is the error rate (for the condition illustrated in Figure 8.12, the miss error rate is approximately 1.9%). While the details depend on the particulars of the model (e.g., assumptions about guessing rules and RT distributions), the median of the target-absent RTs should cut off about X% of the target-present distribution. A glance at Figure 8.12 shows that this is not true for one O's conjunction data for Set Size 3. More than 25% of the correct target-present RTs lie above absent median. This is merely an illustrative example of a general feature of the data. The mean/median of the absent RTs falls far too early. This is especially true for the

smaller set sizes where 30% of target-present RTs can fall above the target-absent mean. There are a variety of ways to handle this. Returning to Figure 8.6, it is reasonable to assume that the target threshold will be much higher than the distractor threshold. A Bayesian way to think about this is that an item is much more likely to be a distractor than a target in a visual search experiment. It is therefore reasonable to dismiss it as a distractor more readily than to accept it as a target. If observers can successfully quit after N distractors have been rejected, it is possible that a fast target-absent search could end in less time than a slow target-present search. The present version of GS uses this difference in thresholds to capture this aspect of the data. The ratio of target to distractor threshold is generally set to 10:1. Nevertheless, while we can identify these constraints in the data, we are still missing something in our understanding of blank trial search termination. Modeling the pattern of errors is the least successful aspect of GS4 at the present time. Parameters that work in one condition tend to fail in others.

State of the Model

To what extent does GS4 capture the diverse empirical phenomena of visual search? Figure 8.13 shows data for the 2 versus 5 task for a real O (solid symbols) and for the model using parameters as described above (same diffusion parameters, same error rules, etc.). The free parameter is a rate parameter that is used to equate target-present slopes so the excellent match between model

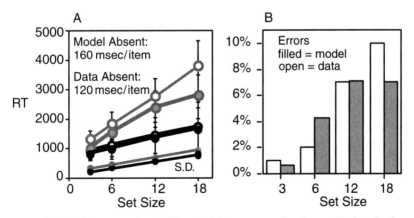

FIGURE 8.13 (A) An example of GS4 model data compared with one O's data for the 2 versus 5 task. Solid symbols indicate the O, open symbols the model. Target-present trials are in black, target-absent in gray. Small symbols denote standard deviations. (B) Miss error rates: open bars are data; filled are model results.

and data for the target-present data is uninteresting. Target-absent RTs produced by the model are a reasonable approximation of the data, though the slopes are too steep. Standard deviations of the RTs (shown at the bottom of the figure) are very close for data and model. The model and the observer had very similar errors rates, rising from about 2% to about 8% as a function of set size. Model RT distributions are positively skewed and qualitatively similar to the real data.

If we now use exactly the same parameters for the conjunction tasks, the model produces slopes of 12 ms/item on target-present and 24 ms/item for target-absent trials. This compares well to 9 and 26, respectively, for this O's data. However, the model's target-absent RTs are significantly too fast. Moving the distractor threshold is one way to compensate, but this disrupts slopes and errors. The model does not quite capture the O's rules for search termination. The heart of the problem seems to relate to the point illustrated by Figure 8.12. Real observers are somehow able to abandon unsuccessful searches quickly without increasing their error rates unacceptably. We have not developed a mechanism that allows GS4 to avoid this speed-accuracy tradeoff.

GS4 does capture other qualitative aspects of the search data, however. Returning to the checklist in Figure 8.1, the model certainly produces appropriate set size effects (Figure 8.1A) and differences between target-present and target-absent trials (Figure 8.1B). The structure of the first, guiding stage produces most of the other properties listed here. Search becomes less efficient as target-distractor similarity increases (Figure 8.1C) and as distractor heterogeneity increases (Figure 8.1D). If the target is flanked by distractors, the setting of top-down weights is less successful and efficiency declines (Figure 8.1E). If the target is defined by the presence of a categorical attribute, search is more efficient than if it is defined by the absence of that attribute (Figure 8.1G). Thus, for example, in search for 15 deg among 0 deg, GS4 can place its weight on the right-tilted channel and find a signal that is present in the target and absent in the distractors. If the target is 0 deg and the distractors are 15 deg, the best that can be done is to put weight on the "steep" channel. The 0-deg signal is bigger than the 15-deg signal in that channel but not dramatically. As a result, search is less efficient—a search asymmetry (Figure 8.1F). And, of course, guidance (Figure 8.1H) is the model's starting point. If the target is red, search will be guided toward red items.

Summary

The current implementation of GS4 captures a wide range of search behaviors. It could be scaled up to capture more. The front end is currently limited to orientation and color (and only the red-green axis of color, at that). Other attributes could be added. This would allow us to capture findings about triple conjunctions, for example (Wolfe et al., 1989). Ideally, one of the more realistic models of early vision could be adapted to provide the front end for GS. At present, the guiding activation map is a weighted sum of the various sources of top-down and bottom-up guidance. The weights are set at the start of a block of trials. This is a bit simple-minded. A more complete GS model would learn its weights and would change them in response to changes in the search task. A more adaptive rule for setting weights could capture many of the priming effects in search. Observers would be faster to find a target if the target was repeated because the weights would have been set more effectively for that target (Hillstrom, 2000; Maljkovic & Nakayama, 1994; Wolfe, Butcher, Lee, & Hyle, 2003; though others might differ with this account; Huang, Holcombe, & Pashler, 2004). A more substantive challenge is presented by the evidence that attention is directed toward objects. While it would not be hard to imagine a GS front-end that expanded the list of guiding attributes beyond a cartoon of color and orientation processing, it is hard to envision a front end that would successfully parse a continuous image into its constituent *objects of attention*. The output of such front-end processing could be fed through a GS-style bottleneck to an object recognition algorithm. Such a model might be able to find what it was looking for but awaits significant progress in other areas of vision research. In the meantime, we believe that the GS architecture continues to serve as a useful model of the bottleneck between visual input and object recognition.

References

Ahissar, M., & Hochstein, S. (1997). Task difficulty and visual hierarchy: Counter-streams in sensory processing and perceptual learning. *Nature, 387*(22), 401–406.

Arani, T., Karwan, M. H., & Drury, C. G. (1984). A variable-memory model of visual search. *Human Factors, 26*(6), 631–639.

Ariely, D. (2001). Seeing sets: Representation by statistical properties. *Psychological Science, 12*(2), 157–162.

Bacon, W. F., & Egeth, H. E. (1994). Overriding stimulus-driven attentional capture. *Perception and Psychophysics, 55*(5), 485–496.

Baldassi, S., & Burr, D. C. (2000). Feature-based integration of orientation signals in visual search. *Vision Research, 40*(10–12), 1293–1300.

———, & Verghese, P. (2002). Comparing integration rules in visual search. *Journal of Vision, 2*(8), 559–570.

Bauer, B., Jolicœur, P., & Cowan, W. B. (1996). Visual search for colour targets that are or are not linearly-separable from distractors. *Vision Research, 36*(10), 1439–1466.

Bichot, N. P., Rossi, A. F., & Desimone, R. (2005). Parallel and serial neural mechanisms for visual search in macaque area V4. *Science, 308*(5721), 529–534.

———, & Schall, J. D. (1999). Saccade target selection in macaque during feature and conjunction. *Visual Neuroscience, 16*, 81–89.

Cameron, E. L., Tai, J. C., Eckstein, M. P., & Carrasco, M. (2004). Signal detection theory applied to three visual search tasks—identification, yes/no detection and localization. *Spatial Vision, 17*(4–5), 295–325.

Carrasco, M., Evert, D. L., Chang, I., & Katz, S. M. (1995). The eccentricity effect: Target eccentricity affects performance on conjunction searches. *Perception and Psychophysics, 57*(8), 1241–1261.

———, & Yeshurun, Y. (1998). The contribution of covert attention to the set size and eccentricity effects in visual search. *Journal Experimental Psychology: Human Perception and Performance, 24*(2), 673–692.

Cave, K. R., & Wolfe, J. M. (1990). Modeling the role of parallel processing in visual search. *Cognitive Psychology, 22*, 225–271.

Chong, S. C., & Treisman, A. (2003). Representation of statistical properties. *Vision Research, 43*(4), 393–404.

Chun, M. M. (2000). Contextual cueing of visual attention. *Trends in Cognitive Sciences, 4*, 170–178.

———, & Jiang, Y. (1998). Contextual cuing: Implicit learning and memory of visual context guides spatial attention. *Cognitive Psychology, 36*, 28–71.

———, & Potter, M. C. (1995). A two-stage model for multiple target detection in RSVP. *Journal of Experimental Psychology: Human Perception & Performance, 21*(1), 109–127.

———, & Wolfe, J. M. (1996). Just say no: How are visual searches terminated when there is no target present? *Cognitive Psychology, 30*, 39–78.

———. (2001). Visual Attention. In E. B. Goldstein (Ed.), *Blackwell's handbook of perception* (pp. 272–310). Oxford: Blackwell.

Cowan, N. (2001). The magical number 4 in short-term memory: A reconsideration of mental storage capacity. *Behavioral and Brain Sciences, 24*(1), 87–114; discussion, 114–185.

Deubel, H., & Schneider, W. X. (1996). Saccade target selection and object recognition: Evidence for a common attentional mechanism. *Vision Research, 36*(12), 1827–1837.

DiLollo, V., Enns, J. T., & Rensink, R. A. (2000). Competition for consciousness among visual events: the psychophysics of reentrant visual pathways. *Journal of Experimental Psychology: General, 129*(3), 481–507.

Dosher, B. A., Han, S., & Lu, Z. L. (2004). Parallel processing in visual search asymmetry. *Journal of Experimental Psychology: Human Perception and Performance, 30*(1), 3–27.

Driver, J. (2001). A selective review of selective attention research from the past century. *British Journal of Psychology, 92*, 53–78.

Duncan, J., & Humphreys, G. W. (1989). Visual search and stimulus similarity. *Psychological Review, 96*, 433–458.

———, Ward, R., & Shapiro, K. (1994). Direct measurement of attention dwell time in human vision. *Nature, 369*(26), 313–314.

Eckstein, M., Beutter, B., Bartroff, L., & Stone, L. (1999). Guided search vs. signal detection theory in target localization tasks. *Investigative Ophthalmology & Visual Science, 40*(4), S346.

Egeth, H. E., Virzi, R. A., & Garbart, H. (1984). Searching for conjunctively defined targets. *Journal of Experimental Psychology: Human Perception and Performance, 10*, 32–39.

———, & Yantis, S. (1997). Visual attention: Control, representation, and time course. *Annual Review of Psychology, 48*, 269–297.

Egeth, H. W. (1966). Parallel versus serial processes in multidimensional stimulus discrimination. *Perception and Psychophysics, 1*, 245–252.

Findlay, J. M., & Gilchrist, I. D. (1998). Eye guidance and visual search. In G. Underwood (Ed.), *Eye guidance in reading and scene perception* (pp. 295–312). Amsterdam: Elsevier.

Folk, C. L., Remington, R. W., & Johnston, J. C. (1992). Involuntary covert orienting is contingent on attentional control settings. *Journal of Experimental Psychology: Human Perception and Performance, 18*(4), 1030–1044.

Foster, D. H., & Ward, P. A. (1991a). Asymmetries in oriented-line detection indicate two orthogonal filters in early vision. *Proceedings of the Royal Society (London B), 243,* 75–81.

——, & Ward, P. A. (1991b). Horizontal-vertical filters in early vision predict anomalous line-orientation frequencies. *Proceedings of the Royal Society (London B), 243,* 83–86.

Green, B. F., & Anderson, L. K. (1956). Color coding in a visual search task. *Journal of Experimental Psychology, 51*(1), 19–24.

Hillstrom, A. P. (2000). Repetition effects in visual search. *Perception and Psychophysics, 62*(4), 800–817.

Hochstein, S., & Ahissar, M. (2002). View from the top: Hierarchies and reverse hierarchies in the visual system. *Neuron, 36,* 791–804.

Horowitz, T. S., & Wolfe, J. M. (1998). Visual search has no memory. *Nature, 394*(August 6), 575–577.

——, & Wolfe, J. M. (2005). Visual search: The role of memory for rejected distractors. In L. Itti, G. Rees, & J. Tsotsos (Eds.), *Neurobiology of attention* (pp. 264–268). San Diego, CA: Academic Press/Elsevier.

Huang, L., Holcombe, A. O., & Pashler, H. (2004). Repetition priming in visual search: episodic retrieval, not feature priming. *Memory & Cognition, 32*(1), 12–20.

Hummel, J. E., & Stanikiewicz, B. J. V. C. (1998). Two roles for attention in shape perception: A structural description model of visual scrutiny. *Visual Cognition,* (1–2), 49–79.

Itti, L., & Koch, C. (2000). A saliency-based search mechanism for overt and covert shifts of visual attention. *Vision Research, 40*(10–12), 1489–1506.

Julesz, B. (1981). A theory of preattentive texture discrimination based on first order statistics of textons. *Biological Cybernetics, 41,* 131–138.

——. (1984). A brief outline of the texton theory of human vision. *Trends in Neuroscience, 7*(February), 41–45.

Kahneman, D., Treisman, A., & Gibbs, B. (1992). The reviewing of object files: Object-specific integration of information. *Cognitive Psychology, 24,* 179–219.

Klein, R., & Farrell, M. (1989). Search performance without eye movements. *Perception and Psychophysics, 46,* 476–482.

Koch, C., & Ullman, S. (1985). Shifts in selective visual attention: Towards the underlying neural circuitry. *Human Neurobiology, 4,* 219–227.

Kowler, E., Anderson, E., Dosher, B., & Blaser, E. (1995). The role of attention in the programming of saccades. *Vision Research, 35*(13), 1897–1916.

Kristjansson, A. (2000). In search of rememberance: Evidence for memory in visual search. *Psychological Science, 11*(4), 328–332.

Lamy, D., & Egeth, H. E. (2003). Attentional capture in singleton-detection and feature-search modes. *Journal of Experimental Psychology: Human Perception and Performance, 29*(5), 1003–1020.

Li, Z. (2002). A salience map in primary visual cortex. *Trends in Cognitive Sciences, 6*(1), 9–16.

Luce, R. D. (1986). *Response times.* New York: Oxford University Press.

Luck, S. J., & Vecera, S. P. (2002). Attention. In H. Pashler & S. Yantis (Eds.), *Stevens' handbook of experimental psychology: Vol. 1. Sensation and perception* (3rd ed., pp. 235–286). New York: Wiley.

——, Vogel, E. K., & Shapiro, K. L. (1996). Word meanings can be accessed but not reported during the attentional blink. *Nature, 382,* 616–618.

Maioli, C., Benaglio, I., Siri, S., Sosta, K., & Cappa, S. (2001). The integration of parallel and serial processing mechanisms in visual search: evidence from eye movement recording. *European Journal of Neuroscience, 13*(2), 364–372.

Maljkovic, V., & Nakayama, K. (1994). Priming of popout: I. Role of features. *Memory and Cognition, 22*(6), 657–672.

McElree, B., & Carrasco, M. (1999). The temporal dynamics of visual search: Evidence for parallel processing in feature and conjunction searches. *Journal of Experimental Psychology: Human Perception and Performance, 25*(6), 1517–1539.

McMains, S. A., & Somers, D. C. (2004). Multiple spotlights of attentional selection in human visual cortex. *Neuron, 42*(4), 677–686.

Melcher, D., Papathomas, T. V., & Vidnyánszky, Z. (2005). Implicit attentional selection of bound visual features. *Neuron, 46,* 723–729.

Michod, K. O., Wolfe, J. M., & Horowitz, T. S. (2004). *Does guidance take time to develop during a visual search trial?* Paper presented at the Visual Sciences Society, Sarasota, FL, April 29–May 4.

Mitsudo, H. (2002). Information regarding structure and lightness based on phenomenal transparency influences the efficiency of visual search. *Perception, 31*(1), 53–66.

Moore, C. M., Egeth, H., Berglan, L. R., & Luck, S. J. (1996). Are attentional dwell times inconsistent with serial visual search? *Psychonomic Bulletin & Review, 3*(3), 360–365.

——, & Wolfe, J. M. (2001). Getting beyond the serial/parallel debate in visual search: A hybrid approach. In K. Shapiro (Ed.), *The limits of attention: Temporal constraints on human information processing* (pp. 178–198). Oxford: Oxford University Press.

Moore, T., Armstrong, K. M., & Fallah, M. (2003). Visuomotor origins of covert spatial attention. *Neuron, 40*(4), 671–683.

Moraglia, G. (1989). Display organization and the detection of horizontal lines segments. *Perception and Psychophysics, 45,* 265–272.

Moray, N. (1969). *Attention: Selective processing in vision and hearing.* London: Hutchinson.

Motter, B. C., & Belky, E. J. (1998). The guidance of eye movements during active visual search. *Vision Research*, 38(12), 1805–1815.

Murdock, B. B., Jr., Hockley, W. E., & Muter, P. (1977). Two tests of the conveyor-belt model for item recognition. *Canadian Journal of Psychology*, 31, 71–89.

Najemnik, J., & Geisler, W. S. (2005). Optimal eye movement strategies in visual search. *Nature*, 434(7031), 387–391.

Nakayama, K., & Silverman, G. H. (1986). Serial and parallel processing of visual feature conjunctions. *Nature*, 320, 264–265.

Neisser, U. (1963). Decision time without reaction time: Experiments in visual scanning. *American Journal of Psychology*, 76, 376–385.

Nothdurft, H. C. (1991). Texture segmentation and pop-out from orientation contrast. *Vision Research*, 31(6), 1073–1078.

——. (1992). Feature analysis and the role of similarity in pre-attentive vision. *Perception and Psychophysics*, 52(4), 355–375.

——. (1993). The role of features in preattentive vision: Comparison of orientation, motion and color cues. *Vision Research*, 33(14), 1937–1958.

Oliva, A., & Torralba, A. (2001). Modeling the shape of the scene: A holistic representation of the spatial envelope. *International Journal of Computer Vision*, 42(3), 145–175.

——, Torralba, A., Castelhano, M. S., & Henderson, J. M. (2003). *Top-down control of visual attention in object detection*. Paper presented at the Proceedings of the IEEE International Conference on Image Processing, September 14–17, Barcelona, Spain.

Olzak, L. A., & Thomas, J. P. (1986). Seeing spatial patterns. In K. R. Boff, L. Kaufmann, & J. P. Thomas (Eds.), *Handbook of perception and human performance* (Chap. 7). New York: Wiley.

Palmer, J. (1994). Set-size effects in visual search: the effect of attention is independent of the stimulus for simple tasks. *Vision Research*, 34(13), 1703–1721.

——. (1995). Attention in visual search: Distinguishing four causes of a set size effect. *Current Directions in Psychological Science*, 4(4), 118–123.

——. (1998). Attentional effects in visual search: relating accuracy and search time. In R. D. Wright (Ed.), *Visual attention* (pp. 295–306). New York: Oxford University Press.

——, & McLean, J. (1995). *Imperfect, unlimited-capacity, parallel search yields large set-size effects*. Paper presented at the Society for Mathematical Psychology, Irvine, CA.

——, Verghese, P., & Pavel, M. (2000). The psychophysics of visual search. *Vision Research*, 40(10–12), 1227–1268.

Pashler, H. E. (1998a). *Attention*. Hove, East Sussex, UK: Psychology Press.

Pashler, H. (1998b). *The psychology of attention*. Cambridge, MA: MIT Press.

Peterson, M. S., Kramer, A. F., Wang, R. F., Irwin, D. E., & McCarley, J. S. (2001). Visual search has memory. *Psychological Science*, 12(4), 287–292.

Quinlan, P. T., & Humphreys, G. W. (1987). Visual search for targets defined by combinations of color, shape, and size: An examination of the task constraints on feature and conjunction searches. *Perception and Psychophysics*, 41, 455–472.

Ratcliff, R. (1978). A theory of memory retrieval. *Psychological Review*, 85(2), 59–108.

——, Gomez, P., & McKoon, G. (2004). A diffusion model account of the lexical decision task. *Psychological Review*, 111(1), 159–182.

Rauschenberger, R. (2003). Attentional capture by auto-and allo-cues. *Psychonomic Bulletin & Review*, 10(4), 814–842.

Raymond, J. E., Shapiro, K. L., & Arnell, K. M. (1992). Temporary suppression of visual processing in an RSVP task: An attentional blink? *Journal of Experimental Psychology: Human Perception and Performance*, 18(3), 849–860.

Rensink, R. A., & Enns, J. T. (1995). Pre-emption effects in visual search: evidence for low-level grouping. *Psychological Review*, 102(1), 101–130.

——, & Enns, J. T. (1998). Early completion of occluded objects. *Vision Research*, 38, 2489–2505.

——, O'Regan, J. K., & Clark, J. J. (1997). To see or not to see: The need for attention to perceive changes in scenes. *Psychological Science*, 8, 368–373.

Rosenholtz, R. (2001a). Search asymmetries? What search asymmetries? *Perception and Psychophysics*, 63(3), 476–489.

——. (2001b). Visual search for orientation among heterogeneous distractors: experimental results and implications for signal-detection theory models of search. *Journal of Experimental Psychology: Human Perception and Performance*, 27(4), 985–999.

Sanders, A. F., & Houtmans, M. J. M. (1985). Perceptual modes in the functional visual field. *Acta Psychologica*, 58, 251–261.

Schall, J., & Thompson, K. (1999). Neural selection and control of visually guided eye movements. *Annual Review of Neuroscience*, 22, 241–259.

Shapiro, K. L. (1994). The attentional blink: The brain's eyeblink. *Current Directions in Psychological Science*, 3(3), 86–89.

Shen, J., Reingold, E. M., & Pomplun, M. (2003). Guidance of eye movements during conjunctive visual search: the distractor-ratio effect. *Canadian Journal of Experimental Psychology*, 57(2), 76–96.

Shore, D. I., & Klein, R. M. (2000). On the manifestations of memory in visual search. *Spatial Vision*, 14(1), 59–75.

Simons, D. J., & Levin, D. T. (1997). Change blindness. *Trends in Cognitive Sciences*, 1(7), 261–267.

———, & Rensink, R. A. (2005). Change blindness: past, present, and future. *Trends in Cognitive Sciences*, 9(1), 16–20.

Smith, S. L. (1962). Color coding and visual search. *Journal of Experimental Psychology*, 64, 434–440.

Sternberg, S. (1966). High-speed scanning in human memory. *Science*, 153, 652–654.

Styles, E. A. (1997). *The psychology of attention*. Hove, East Sussex, UK: Psychology Press.

Theeuwes, J. (1994). Stimulus-driven capture and attentional set: Selective search for color and visual abrupt onsets. *Journal of Experimental Psychology: Human Perception and Performance*, 20(4), 799–806.

———, Godijn, R., & Pratt, J. (2004). A new estimation of the duration of attentional dwell time. *Psychonomic Bulletin & Review*, 11(1), 60–64.

Thompson, K. G., & Bichot, N. P. (2004). A visual salience map in the primate frontal eye field. *Progress in Brain Research*, 147, 249–262.

Thornton, T. (2002). *Attentional limitation and multiple-target visual search*. Unpublished doctoral dissertation, University of Texas at Austin.

Todd, S., & Kramer, A. F. (1994). Attentional misguidance in visual search. *Perception and Psychophysics*, 56(2), 198–210.

Townsend, J. T. (1971). A note on the identification of parallel and serial processes. *Perception and Psychophysics*, 10, 161–163.

———. (1990). Serial and parallel processing: Sometimes they look like Tweedledum and Tweedledee but they can (and should) be distinguished. *Psychological Science*, 1, 46–54.

Townsend, J. T., & Wenger, M. J. (2004). The serial-parallel dilemma: A case study in a linkage of theory and method. *Psychonomic Bulletin & Review*, 11(3), 391–418.

Treisman, A. (1985). Preattentive processing in vision. *Computer Vision, Graphics, and Image Processing*, 31, 156–177.

———. (1986). Features and objects in visual processing. *Scientific American*, 255, 114B–125.

———. (1996). The binding problem. *Current Opinion in Neurobiology*, 6, 171–178.

———, & Gelade, G. (1980). A feature-integration theory of attention. *Cognitive Psychology*, 12, 97–136.

———, & Gormican, S. (1988). Feature analysis in early vision: Evidence from search asymmetries. *Psychological Review*, 95, 15–48.

———, & Sato, S. (1990). Conjunction search revisited. *Journal of Experimental Psychology: Human Perception and Performance*, 16(3), 459–478.

———, & Souther, J. (1985). Search asymmetry: A diagnostic for preattentive processing of seperable features. *Journal of Experimental Psychology: General*, 114, 285–310.

Treisman, A. M., & Schmidt, H. (1982). Illusory conjunctions in the perception of objects. *Cognitive Psychology*, 14, 107–141.

Van Zandt, T. (2002). Analysis of response time distributions. In H. Pashler & J. Wixted (Eds.), *Stevens' handbook of experimental psychology: Vol. 4. Methodology in experimental psychology* (3rd ed., pp. 461–516). New York: Wiley.

———, & Ratcliff, R. (2005). Statistical mimicking of reaction time data: Single-process models, parameter variability, and mixtures. *Psychonomic Bulletin & Review*, 2(1), 20–54.

Verghese, P. (2001). Visual search and attention: A signal detection approach. *Neuron*, 31, 523–535.

von der Malsburg, C. (1981). *The correlation theory of brain function*. Göttingen, Germany: Max-Planck-Institute for Biophysical Chemistry.

Wang, D. L. (1999). Object selection based on oscillatory correlation. *Neural Networks*, 12(4–5), 579–592.

Ward, R., Duncan, J., & Shapiro, K. (1996). The slow time-course of visual attention. *Cognitive Psychology*, 30(1), 79–109.

———. (1997). Effects of similarity, difficulty, and nontarget presentation on the time course of visual attention. *Perception and Psychophysics*, 59(4), 593–600.

Wolfe, J. M. (1994). Guided Search 2.0: A revised model of visual search. *Psychonomic Bulletin & Review*, 1(2), 202–238.

———. (1998). What do 1,000,000 trials tell us about visual search? *Psychological Science*, 9(1), 33–39.

———. (2001). Asymmetries in visual search: An Introduction. *Perception and Psychophysics*, 63(3), 381–389.

———. (2003). Moving towards solutions to some enduring controversies in visual search. *Trends in Cognitive Sciences*, 7(2), 70–76.

Wolfe, J., Alvarez, G., & Horowitz, T. (2000). Attention is fast but volition is slow. *Nature*, 406, 691.

Wolfe, J. M., & Bennett, S. C. (1997). Preattentive object files: Shapeless bundles of basic features. *Vision Research*, 37(1), 25–43.

———, Birnkrant, R. S., Horowitz, T. S., & Kunar, M. A. (2005). Visual search for transparency and opacity: Attentional guidance by cue combination? *Journal of Vision*, 5(3), 257–274.

———, Butcher, S. J., Lee, C., & Hyle, M. (2003). Changing your mind: On the contributions of top-down and bottom-up guidance in visual search for feature singletons. *Journal of Experimental Psychology: Human Perception and Performance*, 29(2), 483–502.

——, Cave, K. R., & Franzel, S. L. (1989). Guided Search: An alternative to the Feature Integration model for visual search. *Journal of Experimental Psychology: Human Perception and Performance*, 15, 419–433.

——, & DiMase, J. S. (2003). Do intersections serve as basic features in visual search? *Perception*, 32(6), 645–656.

——, Friedman-Hill, S. R., Stewart, M. I., & O'Connell, K. M. (1992). The role of categorization in visual search for orientation. *Journal of Experimental Psychology: Human Perception and Performance*, 18(1), 34–49.

——, & Gancarz, G. (1996). Guided Search 3.0: A model of visual search catches up with Jay Enoch 40 years later. In V. Lakshminarayanan (Ed.), *Basic and clinical applications of vision science* (pp. 189–192). Dordrecht, Netherlands: Kluwer Academic.

——, & Horowitz, T. S. (2004). What attributes guide the deployment of visual attention and how do they do it? *Nature Reviews Neuroscience*, 5(6), 495–501.

——, Horowitz, T. S., & Kenner, N. M. (2005). Rare items often missed in visual searches. *Nature*, 435, 439–440.

——, O'Neill, P. E., & Bennett, S. C. (1998). Why are there eccentricity effects in visual search? *Perception and Psychophysics*, 60(1), 140–156.

——, Palmer, E. M., Horowitz, T. S., & Michod, K. O. (2004). Visual search throws us a curve. *Abstracts of the Psychonomic Society*, 9. (Paper presented at the meeting of the Psychonomic Society, Minneapolis, MN.)

——, Reinecke, A., & Brawn, P. (2006). Why don't we see changes? The role of attentional bottlenecks and limited visual memory. *Visual Cognition*, 19(4–8), 749–780.

——, Yu, K. P., Stewart, M. I., Shorter, A. D., Friedman-Hill, S. R., & Cave, K. R. (1990). Limitations on the parallel guidance of visual search: Color X color and orientation X orientation conjunctions. *Journal of Experimental Psychology: Human Perception and Performance*, 16(4), 879–892.

Yantis, S. (1998). Control of visual attention. In H. Pashler (Ed.), *Attention* (pp. 223–256). Hove, East Sussex, UK: Psychology Press.

Zelinsky, G., & Sheinberg, D. (1995). Why some search tasks take longer than others: Using eye movements to redefine reaction times. In J. M. Findlay, R. Walker, & R. W. Kentridge (Eds.), *Eye movement research: Mechanisms, processes and applications: Vol. 6. Studies in visual information processing* (pp. 325–336). Amsterdam, Netherlands: Elsevier.

Zelinsky, G. J., & Sheinberg, D. L. (1996). Using eye saccades to assess the selectivity of search movements. *Vision Research*, 36(14), 2177–2187.

——. (1997). Eye movements during parallel/serial visual search. *Journal of Experimental Psychology: Human Perception and Performance*, 23(1), 244–262.

Zohary, E., Hochstein, S., & Hillman, P. (1988). Parallel and serial processing in detecting conjunctions. *Perception*, 17, 416.

9

Advancing Area Activation Toward a General Model of Eye Movements in Visual Search

Marc Pomplun

Many everyday tasks require us to perform visual search. Therefore, an adequate model of visual search is an indispensable part of any plausible approach to modeling integrated cognitive systems that process visual input. Because of its quantitative nature, absence of freely adjustable parameters, and support from empirical research results, the area activation model is presented as a promising starting point for developing such a model. Its basic assumption is that eye movements in visual search tasks tend to target display areas that provide a maximum amount of task-relevant information for processing. To tackle the shortcomings of the current model, an empirical study is briefly reported that provides a variety of quantitative data on saccadic selectivity in visual search. How these and related data will be used to develop the area activation model toward a general model of eye movements in visual search is discussed.

During every single day of our lives we perform thousands of elementary tasks, which often are so small and require so little effort that we hardly even recognize them as tasks. Think of a car driver looking for a parking spot, a gardener determining the next branch to be pruned, a painter searching for a certain color on her palette, or an Internet surfer scrutinizing a Web page for a specific link. These and other routine tasks require us to perform visual search. Given this ubiquity of visual search, it is not surprising that we excel at it. Without great effort, human observers clearly outperform every current artificial vision system in tasks such as finding a particular face in a crowd or determining the location of a designated item on a desk. Understanding the mechanisms underlying visual search behavior will thus not only shed light on crucial elementary functions of the visual system, but it may also enable us to devise more efficient and more sophisticated computer vision algorithms. Moreover, an adequate model of visual search is an indispensable part of any plausible approach to modeling integrated cognitive systems that consider

visual input. As a consequence, for several decades visual search has been one of the most thoroughly studied paradigms in vision research.

In a visual search task, subjects usually have to decide as quickly and as accurately as possible whether a visual display, composed of multiple search items, contains a prespecified target item. Many of these studies analyzed the dependence of response times and error rates on the number of search items in the display. Although rather sparse, such data led to the development of numerous theories of visual search. These theories differ most significantly in the function they ascribe to visual attention and its control in the search process. For an introduction to the questions and approaches in the field of visual search, see chapter 8 in this volume by Jeremy M. Wolfe. The same author also wrote a comprehensive review on visual search (Wolfe, 1998).

Furthermore, it was Jeremy Wolfe and his colleagues who proposed one of the most influential theories of visual search, *guided search theory* (e.g., Cave & Wolfe, 1990; Wolfe 1994, 1996; Wolfe, Cave, & Franzel, 1989;

see also chapter 8). The basic idea underlying this theory is that visual search consists of two consecutive stages: an initial stage of preattentive processing that guides a subsequent stage of serial search. After the onset of a search display, a parallel analysis is carried out across all search items, and preattentive information is derived to generate an *activation map* that indicates likely target locations. The overall activation at each stimulus location consists of a top-down and a bottom-up component. A search item's top-down (goal-driven) activation increases with greater similarity of that item to the target, whereas its bottom-up (data-driven) activation increases with decreasing similarity to other items in its neighborhood. This activation map is used to guide shifts of attention during the subsequent serial search process. First, the subject's focus of attention is drawn to the stimulus location with the highest activity. If the target actually is at this location, the subject manually reports target detection, and the search trial terminates. Otherwise, the subject's attention moves on to the second-highest peak in the activation map, and so on, until the subject either detects the target or decides that the display does not contain a target.

The guided search theory has been shown to be consistent with a wide variety of psychophysical visual search data (e.g., Brogan, Gale, & Carr, 1993). Besides the standard measures of response time and error rate, these data also encompass more fine-grained measures, most importantly eye-movement patterns. In static scenes such as standard search displays, eye movements are performed as alternating sequences of saccades (quick *jumps*, about 30–70 ms) and fixations (almost motionless phases, about 150–800 ms). Interestingly, information from the display is extracted almost entirely during fixations (for a review of eye-movement research, see Rayner, 1998). Therefore, the positions of fixations—or saccadic endpoints—can tell us which display items subjects looked at during a visual search trial before they determined the presence or absence of the target. Analyzing the features of the inspected items and relating them to the features of the target item can provide valuable insight into the search process. On the basis of this idea, several visual search studies have examined saccadic selectivity, which is defined as the proportion of saccades directed to each type of nontarget item (distractor), by assigning each saccadic endpoint to the nearest item in the search display. The guided search theory received support from several of these studies, which revealed that those distractors sharing a certain feature such as

color or shape with the target item received a disproportionately large number of saccadic endpoints (e.g., Findlay, 1997; Hooge & Erkelens, 1999; Motter & Belky, 1998; Pomplun, Reingold, & Shen, 2001b; Scialfa & Joffe, 1998; Shen, Reingold, & Pomplun, 2000; Williams & Reingold, 2001; but see Zelinsky, 1996).

At this point, however, a closer look at the relationship between eye movements and visual attention is advisable. We implicitly assumed that the items subjects look at are also the ones that receive their attention. From our everyday experience, we know that this does not always have to be correct: First, we can direct our gaze to an object in our visual field without paying attention to the object or inspecting it—we could simply think of something completely unrelated to the visual scene, maybe an old friend we have not seen in years. Second, even if we inspect the visual scene, we are able to process both the item that we are currently fixating and its neighboring items. For example, when fixating on any of the bar items in Figure 9.1a, we can sequentially shift our attention to each of its neighboring items, without moving our eyes, and thereby determine their brightness and orientation. These covert shifts of attention work efficiently for items near the fixation but become less feasible with increasing retinal eccentricity; for instance, while fixating on the center of Figure 9.1a, we cannot examine any specific item in Figure 9.1b (see Rayner, 1998). It has been shown by several studies that subjects typically process multiple items within a single fixation during visual search tasks (e.g., Bertera & Rayner, 2000; Pomplun, Reingold, & Shen, 2001a).

The first phenomenon—inattention—can be accounted for reasonably well by asking subjects to perform the visual search task as quickly and as accurately as possible and only analyzing those trials in which subjects gave a correct manual response within three standard deviations from the mean response time. The second phenomenon—covert shifts of attention—however, is inherent to eye-movement research: It is impossible to infer from a subject's gaze trajectory the exact sequence of items that were processed. We can only estimate this sequence, as saccades roughly follow the focus of attention to provide high visual acuity for efficient task performance. So it is important to notice that the nearest-item definition of saccadic selectivity, while it can identify features that guide search, does not measure attentional selectivity, that is, the proportion of attention directed to each distractor type.

Undoubtedly, modeling visual attention through guided search has already advanced the field of visual search quite substantially—so what would be the additional benefit of quantitatively predicting the position and selectivity of saccadic endpoints? First, quantitative modeling is important for the evaluation and comparison of different models. Since eye movements—unlike shifts of attention—can be directly measured, the empirical testing of models predicting eye movement data is especially fruitful. Moreover, if we want to integrate our model into an embodied computational cognitive architecture, the model needs to accept quantitative input and produce quantitative output in order to interact with the other components in the architecture. The most important argument for the quantitative modeling of eye movements, however, is the fact that for executing elementary tasks we use our gaze as a pointer into visual space. We thereby create an external coordinate system for that particular task that greatly reduces working memory demands (see Ballard, 1991; Ballard, Hayhoe, Pook, & Rao, 1997). For instance, the task of grasping an object is typically performed in two successive steps: First, we fixate on the object, and second, we move our hand toward the origin of the gaze-centered coordinate system (Milner & Goodale, 1995). Since visual search is a component of so many elementary tasks, having a quantitative model of eye-movement control in visual search is crucial for simulating and fully understanding the interaction of small-scale processes that enable us to perform many natural tasks efficiently. Such a model could serve as a valuable visual search module for the more ambitious endeavor of accurately modeling integrated cognitive systems.

Rao, Zelinsky, Hayhoe, and Ballard (2002) propose a computational eye-movement model for visual search tasks that uses iconic scene representations derived from oriented spatiochromatic filters at multiple scales. In this model, visual search for a target item proceeds in a coarse-to-fine fashion. The first saccadic endpoint is determined based on the target's largest scale filter responses, and subsequent endpoints are based on filters of decreasing size. This coarse-to-fine processing, which is supported by psychophysical data (e.g., Schyns & Oliva, 1994), makes this model a biologically plausible and intuitive approach. However the psychophysical data also show the duration of the coarse-to-fine transition not to exceed a few hundred milliseconds after stimulus onset, which makes it especially relevant to short-search processes. Accordingly,

the model was tested on a visual search task that was easier and, therefore shorter, than typical tasks in the literature. It will be interesting to see the further development of this model toward a greater variety of search tasks.

A very simple, quantitative approach to modeling the spatial distribution and feature selectivity of eye movements in longer visual search tasks is the area activation model (Pomplun, Reingold, Shen, & Williams, 2000; Pomplun, Shen, & Reingold, 2003). Its original version only applies to artificial search displays with discrete search items, and it requires substantial a priori information about the guiding features and the task difficulty to work accurately. Nevertheless, the model made novel predictions that were successfully tested in an empirical study. Because of its straightforward nature and ability of precise prediction, the area activation model—once its restrictions have been tackled—can be considered a promising candidate for a general visual search component for integrated models of cognitive systems. Section 1 will briefly introduce the original area activation model, while section 2 will describe current efforts and preliminary results regarding the improvement of the model toward a general visual search module.

The Original Area Activation Model

The original area activation model (Pomplun et al., 2000, 2003) is related to the guided search theory because it also assumes a preattentively generated activation map that determines feature guidance in the subsequent search process. The most important difference between the two models is the functional role of the activation map. Guided search assumes an activation map containing a single peak for each relevant search item; this map is used to guide visual attention. Area activation proposes an activation map containing for every position in the search display the amount of relevant information that could be processed during a fixation at that position; this map determines the position of fixations. To be precise, the area activation model is based on assumptions concerning three aspects of visual search performance: (1) the extent of available resources for processing, (2) the choice of fixation positions, and (3) the scan-path structure. These assumptions will be briefly described.

Regarding the resources available for visual processing, it is assumed that during a fixation the distribution of these resources on the display can be approximated by a two-dimensional Gaussian function centered at the fixation point (see Pomplun, Ritter, & Velichkovsky, 1996). The region in the display covered by this distribution is also called the *fixation field*. We define it as the area from which task-relevant information is extracted during a fixation. The size of this area—technically speaking, the standard deviation of the Gaussian function—depends on numerous stimulus dimensions such as task difficulty, item density, and item heterogeneity. For example, in displays with widely dispersed items, where the distractor items are clearly different from the target, the fixation field is expected to be larger than in displays with densely packed items and distractors that are very similar to the target (see Bertera & Rayner, 2000). The model further assumes that a smaller fixation field requires more fixations to process a search display. This is quite intuitive because if a smaller area can be processed with each fixation, more fixations are needed to process the entire display. Such a functional relationship makes it possible to estimate the fixation field size based on the empirically observed number of fixations in a particular experimental condition. Consequently, the model must be given an estimate of the number of fixations that subjects will generate, ideally obtained through a pilot study. An iterative gradient-descent algorithm is used on a trial-by-trial basis to determine the fixation field size in such a way that the number of simulated fixations matches the number of empirical ones.

Concerning fixation positioning, the model assumes that fixations are distributed in such a way that each fixation processes a local maximum of task-relevant information. To determine these positions, the model first computes the informativeness of positions across the display, that is, the amount of task-relevant information that can be processed during a fixation at that position. The model calculates informativeness of a fixation position as the sum of visual processing resources—according to the Gaussian function centered at that position—that is applied to each display item having one or more of the guiding features (these items will be referred to as guiding items). In other words, the value of the Gaussian function for the position of each of guiding items is determined, and the sum of these values equals the informativeness. This means that positions in the display within dense groups of guiding items are most informative because if observers fixate on those positions, they can process many guiding items and are more likely to find the target than during a fixation on a less informative position. Computing the informativeness

for every pixel in a search display results in a smooth activation function that represents the saccade-guiding activation map of the area activation model (see the examples in Figure 9.1c and 9.1d). If the fixation field is sufficiently large, local groups of guiding search items will induce a single activation peak in their center. Consequently, the model will predict a fixation in the center of this group to be more likely than a fixation on any individual guiding item. This center-of-gravity effect has been empirically observed in numerous studies (e.g., Findlay, 1997; Viviani & Swensson, 1982). Once the activation map has been generated, the choice of fixation positions by the model is a statistical process: More highly activated—that is, more informative—positions are more likely to be fixation targets than less activated positions. Notice that to perform all of these computations the model must be *told* which feature or features of search items guide the search process, which can be determined through a pilot study.

Finally, to compute a visual scan path, the model needs to determine not only the fixation positions but also the order in which they are visited. As indicated by empirical studies (e.g., Zelinsky, 1996), determination of such an order is geared toward minimizing the length of the scan path. The following simple rule in the area activation model reflects this principle: The next fixation target is always the novel activation peak closest to the current gaze position. This method of local minimization of scan path length (*greedy heuristic*) was shown to adequately model empirical scanning sequences (e.g., Pomplun, 1998).

A complete mathematical description of the area activation model is provided in Pomplun et al. (2003), and its application to another visual search task involving simultaneous guidance by multiple features is described in Pomplun et al. (2000). In the present context, however, the operation of the area activation model will only be demonstrated by outlining the empirical testing of one of its major predictions. This prediction is rather counterintuitive and clearly distinct from those made by other models such as guided search—it states that saccadic selectivity depends on the spatial arrangement of search items in the display. In other words, according to the area activation model, two displays that show identical, but differently arranged sets of items can induce substantially different patterns of saccadic selectivity as measured by the nearest-item method. This is because the peaks in the activation map do not necessarily coincide with the positions of the guiding distractors but may have a distractor of another type as their nearest neighbor. As a consequence, a spatial arrangement of search items that leads to a closer match between the positions of activation peaks and distractors of a particular type will produce higher saccadic selectivity toward this distractor type. In contrast, models assuming a single activation peak for each guiding distractor predict a pattern of saccadic selectivity that is independent of the arrangement of search items.

To test this prediction, two groups of visual search displays were created. Although all of these displays contained the same set of search items, they were—according to the area activation model—hypothesized to induce relatively strong saccadic selectivity ("high guidance displays") or relatively weak saccadic selectivity

("low guidance displays") toward the guiding distractor type. Subjects showing a significant selectivity difference in the predicted direction between these two groups of displays would yield strong support for the model.

To demonstrate the model's performance, Figure 9.1 shows a simplified variant of the stimuli used in the original study. The search items are bars of different brightness (bright vs. dark) and orientation (vertical vs. horizontal), with a bright vertical bar serving as the target. In a preliminary study, it was shown that only brightness but not orientation guided the search process. This means that the subjects' attention was attracted by the bright horizontal bars but not by the dark vertical or dark horizontal bars. In the high-guidance display

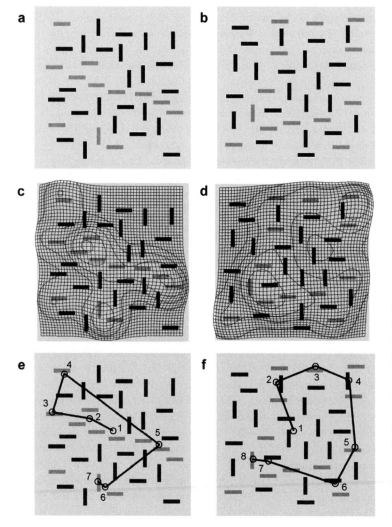

FIGURE 9.1 Demonstration of the original area activation model for displays created to induce high brightness guidance (left column) or low brightness guidance (right column). (a), (b) Sample displays with the target being a bright vertical bar; (c), (d) activation function computed for each of the two displays; (e), (f) predicted scan path for each display with circles marking fixation positions and numbers indicating the fixation sequence.

shown in Figure 9.1a, the search items are arranged in such a way that all the peaks in the activation function (see Figure 9.1c) computed by the model coincide with guiding items, that is, bright horizontal bars. The low-guidance display in Figure 9.1b, however, includes several activation peaks whose nearest neighbors are non-guiding items, that is, dark vertical bars (see Figure 9.1d). Figures 9.1e and 9.1f show predicted scan paths for the high- and low-guidance display, respectively, with circles indicating fixation positions and numbers marking their temporal order. Notice that the model assumes a first fixation at the center of the display and a final fixation on the detected target, neither of which is included in the computation of saccadic selectivity.

In the actual study, each of eight subjects performed the visual search task on a set of 480 displays, which was composed of 240 high-guidance and 240 low-guidance displays. The analysis of the subjects' eye-movement data revealed that their saccadic selectivity toward the guiding items in the high-guidance displays was about 33% greater than in the low-guidance displays. Moreover, these values closely matched the ones predicted by the area activation model (see Pomplun et al., 2003).

These and other empirical tests (Pomplun et al., 2000) provided supporting evidence for the area activation model. Its other strong points are its straightforward nature, its consistency with environmental principles, and the absence of any freely adjustable model parameters, at least in the common case of single-feature guidance. The more such parameters a model includes, the more difficult it is to assess its performance, because these parameters can be adjusted to match simulated with empirical data, even if the underlying model is inadequate.

However, the shortcomings of the model are just as obvious as its advantages. First, it needs a priori information about empirical data—the number of fixations per trial and the features guiding search—before it can generate any eye-movement predictions. Second, the model assumes that only target features guide visual attention, although it is well known that conspicuous features in the display attract attention through bottom-up activation, even if these features are not shared with the target (e.g., Thompson, Bichot, & Sato, 2005). Finally, like most visual search models, the original Area Activation model can only be applied to artificial search images with discrete search items, each of which has a well-defined set of features. These shortcomings need to be dealt with in order to further develop the area activation model.

Toward a General Model of Eye Movements in Visual Search

The first shortcoming of the original area activation model, namely, its need for a priori information, is due to its inability to estimate the task difficulty or the pattern of saccadic selectivity based on the features of the target and distractor items. How could such estimates be derived? Based on visual search literature (e.g., Wolfe, 1998), we can safely assume two things: (1) Saccadic selectivity for a certain distractor type increases with greater similarity of that distractor type with the search target, and (2) task difficulty increases both with greater similarity between target and distractor items and with decreasing similarity between different distractor types in the display. However, the crucial question is: What does the word *similarity* mean in this context? Consequently, the first aim must be to derive an operational definition of similarity and quantify its influence on saccadic selectivity.

With regard to our second aim, the introduction of bottom-up effects to the model, we should quantify how different features attract attention even without being target features. Moreover, there is another effect that may involve both bottom-up and top-down influences and should be considered, namely the distractor-ratio effect. In the present context, this effect is best explained by the eye-movement study by Shen et al. (2000), who had subjects detect a target item among two types of distractors, each of which shared a different feature with the target. While the total number of search items was held constant, the ratio between the two distractor types was varied across trials (see Figure 9.2 for sample stimuli and gaze trajectories taken from a related study). Saccadic selectivity toward a particular feature (brightness or orientation) was found to increase with fewer display items sharing this feature with the target, indicating that participants tended to search along the stimulus dimension shared by fewer distractors (e.g., brightness with few same-brightness distractors and orientation with few same-orientation distractors). This indicated that subjects were able to change their pattern of visual guidance to take advantage of more informative dimensions, demonstrating the flexibility of the preattentive processing. An adequate model of visual search should account for this important effect.

Finally, the model's restriction to artificial search displays can be tackled at the same time as the term *similarity* is quantified. Since a generalized definition of similarity is most useful, searches in both complex

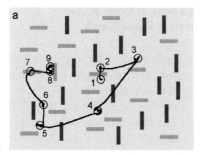

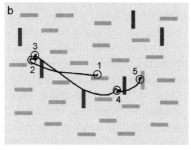

FIGURE 9.2 Illustration of the distractor-ratio effect on saccadic selectivity with circles marking fixation positions and numbers indicating the temporal order of fixations. The target is a bright vertical bar. (a) If the same proportion of the two distractor types is given, subjects are usually guided by brightness, that is, they search through the bright horizontal distractors. (b) If there is a much larger proportion of bright horizontal distractors, subjects typically switch their dimension of guidance.

and real-world images must be studied. The features to be explored should not include specific shapes or patterns but rather should be general features. More specifically, features should be selected that are known to be relevant to the early stages of the human visual processing hierarchy such as intensity, contrast, spatial frequency, and orientation. The area activation concept can easily be applied to these features and their continuous distribution instead of categorical features associated with discrete items. To compute bottom-up activation, we do not need discrete search items either but can base this computation solely on the features in the image. For instance, on average we might expect a display region of high contrast to receive more attention than a no-contrast (*empty*) region, just because of its greater information content. The proportion of features in the search display can also be used to account for a

continuous-image equivalent of the distractor-ratio effect, if it exists.

To address these issues, a visual search study on complex images was conducted, whose details are described in Pomplun (2006). In this study, each of 16 subjects performed 200 visual search trials. Of the 200 search displays, 120 contained real-world images that were randomly rotated by 90, 180, or 270 deg to prevent subjects from applying context-based search and trained scanning patterns (see Figure 9.3a for a sample display). The other 80 displays showed complex artificial images such as fractals or abstract mosaics. All images were in grayscale format using 256 gray levels. For this exploratory study of visual guidance in complex images, it was prudent to eliminate color information in order to avoid the strong attentional capture by color features. Obviously, color is an important feature that guides visual search in everyday tasks, but in order to get a better assessment of other—possibly less guiding—features, color was not included in the study described.

Each of the 200 trials started with a 4-s presentation of a small 64×64 pixel image at the center of the screen. The subjects' task was to memorize this image. Subsequently, the small image was replaced by a large 800×800 pixel search image subtending a visual angle of about 25 deg horizontally and vertically. The subjects knew that the previously shown small image was contained somewhere in this large search display. Their task was to find the position of the small image within the large one as quickly as possible. As soon as they were sure to have found this position, subjects were to fixate on that position and press a designated button to terminate the trial. If they did not press the button within 5 s after the onset of the large display, the trial was ended automatically. In either case, subjects received feedback about the actual position immediately after the end of the trial.

During the search phase, the subjects' eye movements were recorded with the head-mounted SR Research EyeLink-II eye tracking system. The obtained eye-movement data made it possible to assess both the accuracy of the subjects' visual search performance and—most importantly—their saccadic selectivity. For the saccadic selectivity analysis, the fixation positions generated by all 16 subjects were accumulated for each search display. Subsequently, the distribution of visual processing across each display was calculated as follows: Every fixation in the display was associated with a Gaussian function centered at the fixation position. The maximum value of the Gaussian function was

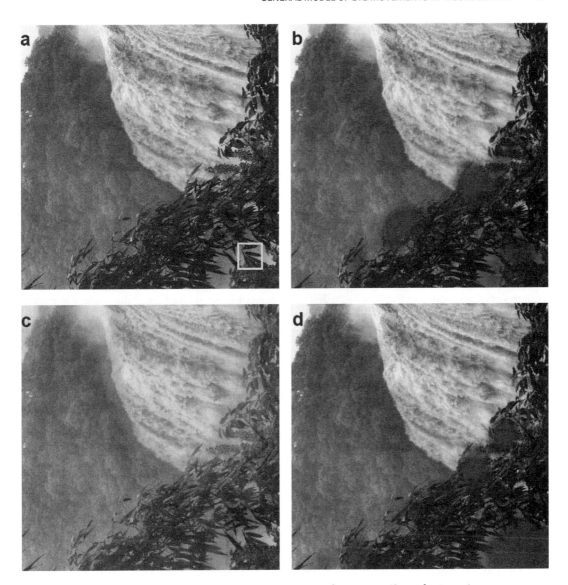

FIGURE 9.3 Study on saccadic selectivity in complex images. (See color insert.)

proportional to the duration of its associated fixation, and its standard deviation was one degree of visual angle. This value was chosen to match the approximate angle subtended by the human fovea. Although the area activation model assumes a variable fixation field size (see the section "The Original Area Activation Model"), at this point a constant size had to be used because no estimate for the actual size could yet be computed. Finally, all Gaussian functions for the same display were summed, resulting in a smooth function that indicated the amount of visual processing across positions in the display. Figure 9.3b illustrates this function for the sample stimulus in Figure 9.3a. The more processing a local region in the image received, the more strongly

it is overlaid with the color purple. As can clearly be seen, those regions in the image that attract most saccadic endpoints are very similar to the target area in that they contain leaves of comparable size.

The next step was to define appropriate basic image features for the analysis of saccadic selectivity. At the current state of this research, features along four dimensions have been used: intensity (average brightness of local pixels), contrast (standard deviation of local brightness), dominant spatial frequency (most elevated frequency interval in a local area as compared to baseline data), and preferred orientation (angle of dominant orientation of local edges). For a mathematical definition of these variables, see Pomplun (under review).

The values along each dimension were scaled to range from 0 to 1, and divided into 20 same-size intervals. Henceforth, the word *feature* will refer to one such interval within a given dimension, for example, brightness-3 or contrast-17. The visual search targets were chosen in such a way that their features varied along the full range of all dimensions.

Because of space limitations, only the analysis of contrast will be described below. The other dimensions showed similar functional behavior. Figure 9.3c visualizes the contrast values computed across the sample stimulus shown in Figure 9.3a. The more pronounced the color green is at a point in the image, the greater is the local contrast value. Notice the square at the target position containing no contrast information. To avoid artifacts in the analysis of saccadic selectivity, no feature information was computed or analyzed near the target positions because whenever subjects detect a target, they look at it for a certain duration before terminating the trial. So if their fixations on the target area were included in our selectivity analysis, we would find an elevated number of saccadic endpoints aimed at the target features, indicating visual guidance toward those features, regardless of whether such guidance actually exists during the search process.

The first question we can now ask is whether there is a contrast-related bottom-up effect: Do certain contrast features attract more saccadic endpoints than others, independent of the contrast in the target area? To find out, we can analyze the average amount of processing as a function of the local contrast across all displays and subjects. The result is shown in Figure 9.4a. Clearly, regions of high contrast (greater than 0.6) receive more processing than areas of low contrast. Given that the contrast of the target regions was distributed in the range from 0 to 1 approximately evenly, this finding indicates a general bias in favor of high-contrast regions. This is not surprising as there usually is less—or less easily available—information in low-contrast areas. So we can state that there are preferred features, that is, feature-based bottom-up effects, in searching complex images, and they can be quantified with the method previously described.

The next, and of course crucial, question is whether there is also feature guidance. Increased processing of those areas that share certain features with the target would indicate this type of guidance. To investigate this, the search displays were separated into three groups, namely, those with low-contrast targets (0 to 0.33), medium-contrast targets (0.34 to 0.66), and high-contrast

targets (0.67 to 1). Figure 9.4b presents the amount of processing as a function of the local contrast relative to the values shown in Figure 9.4a for the three groups of displays. Clearly, for the displays with low-contrast targets, there is a bias toward processing low-contrast regions, and we find similar patterns for medium- and high-contrast displays. This observation is strong evidence for visual feature guidance in complex images, which we can now quantify for any given dimension. One possible definition of the amount of guidance exerted by a particular feature dimension is the average bias in processing across all 20 feature values of that dimension, given a target of the same feature value. For example, to compute contrast guidance, we first determine the average amount of processing for the feature contrast-1 for all trials in which the target was of contrast-1 as well. To obtain the processing bias for contrast-1, we subtract from this value the average amount of processing that contrast-1 received across all 200 trials (and thus across targets of all contrast features). Then contrast guidance is calculated as the arithmetic mean of the 20 bias values derived for contrast-1 to contrast-20.

Having obtained this operational measure of guidance, one can ask whether in complex, continuous images there is a counterpart to the distractor-ratio effect in item-based search images. This can be studied by comparing the average visual guidance for trials in which the search display contains a large proportion of the target feature in a particular dimension with those trials in which this proportion is small. In analogy to the distractor-ratio effect, the small-proportion features should receive more processing than the large-proportion ones. Figure 9.4c shows such an analysis for the contrast dimension. The left bar shows the guidance exerted by target features with above-average presence in the displays, whereas the right bar shows the corresponding value for target features with below-average presence. The guidance for features with above-average presence is substantially smaller than for features with below-average presence, providing evidence for a continuous counterpart to the distractor-ratio effect (*feature-ratio effect*).

The incorporation of these functional relationships into the area activation model is currently in progress. Carefully select the most appropriate set of feature dimensions to be used in the model. By simply having a linear combination of the four dimensions intensity, contrast, spatial frequency, and orientation determine the activation function, the current version of the area activation

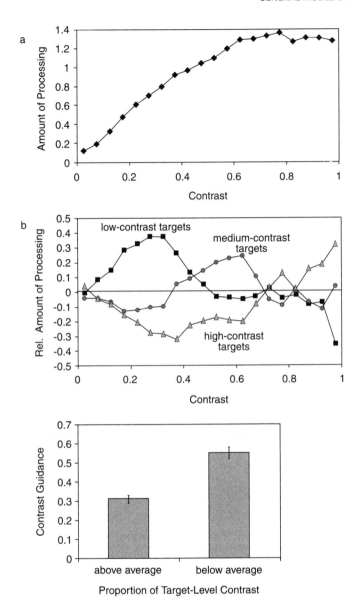

FIGURE 9.4 Selected results of the saccadic selectivity study. (a) Amount of processing in a display region as a function of the local contrast, indicating a bottom-up effect; (b) relative amount of processing (as compared to average values) as a function of local contrast and depending on contrast in the target area, indicating feature guidance; (c) contrast guidance (see text for definition) for above- and below-average proportion of the target-level contrast in the display, indicating a feature-ratio effect.

model can often—but not reliably—approximate the distribution of actual saccadic endpoints. Figure 9.3d illustrates the current model's prediction of this distribution in the same way as the actual distribution is shown in Figure 9.3b.

Conclusions

This chapter has presented current work on the area activation model aimed at developing a quantitative model of eye-movement control in visual search tasks.

Such a model would be of great scientific utility because it could be employed as a general visual search module in integrated models of cognitive systems. While its approach originated from the guided search theory, the area activation model and its current development presented here clearly exceed the scope of guided search. Most significantly, guided search focuses on explaining shifts of attention during search processes, and although Guided Search 4.0 (chapter 8, this volume) supports the simulation of eye movements, guided search largely understands eye movements as by-products of attentional shifts. In contrast, the area activation model considers an observer's gaze behavior as a central part of visual task performance. The model makes quantitative predictions about the locations of saccadic endpoints in a given search display, which have been shown to closely match empirical data. Despite the current work on the area activation model, it still does not nearly reach the breadth and complexity of guided search. However, its quantitative prediction of overt visual behavior makes area activation more suitable as a visual search module for integration into cognitive architectures.

Another promising candidate for development into a general visual search component is the model by Rao et al. (2002). Just like area activation, this model also predicts the location of saccadic endpoints based on the similarity of local image features with target features. It differs from area activation in its use of spatiochromatic filters at multiple scales that are applied to the visual input in a temporal coarse-to-fine sequence. This is a plausible approach for rapid search processes, and its simulation generates saccades that closely resemble empirical ones. However, during longer search processes in complex real-world scenes, it seems that the coarse-to-fine mechanism is less relevant for the quantitative prediction of eye movements. At the same time, factors such as the spectrum of local display variables, the proportion of features in the display, and the difficulty of the search task become more important. All of these factors are currently not considered by the Rao et al. model but are being incorporated into the evolving area activation model.

This ongoing work on the area activation model addresses significant shortcomings of its original version (Pomplun et al., 2003). As a first step, an empirical visual search study on complex images was conducted and briefly reported here. On the basis of the data obtained, various aspects of the influence of display and target features on eye-movement patterns have been quantified. This information has been used to tackle questions unanswered by the original area activation model and to develop it further toward a more generalizable visual search model. The crucial improvements so far include the elimination of required empirical a priori information, the consideration of bottom-up activation, and the applicability of the model to search displays beyond artificial images with discrete items and features.

Future research on the area activation model will focus on determining the feature dimensions to be represented. The goal of this endeavor will be the selection of a small set of dimensions that appropriately characterizes visual guidance and allows precise predictions while being simple enough to make the model transparent. Further steps in the development of the model will include the investigation of color guidance in complex displays, which was omitted in the study presented here, and the prediction of fixation field size. Ideally, the resulting model will be straightforward, consistent with natural principles, and carefully avoid any freely adjustable model parameters to qualify it as a streamlined and general approach to eye-movement control in visual search. It will certainly be a long and challenging road toward a satisfactory model for the integration into a unified cognitive architecture, but the journey is undoubtedly worthwhile.

Acknowledgments

I would like to thank Eyal M. Reingold, Jiye Shen, and Diane E. Williams for their valuable help in devising and testing the original version of the area activation model. Furthermore, I am grateful to May Wong, Zhihuan Weng, and Chengjing Hu for their contribution to the eye-movement study on complex images, and to Michelle Umali for her editorial assistance.

References

Ballard, D. H. (1991). Animate vision. *Artificial Intelligence Journal, 48,* 57–86.

——, Hayhoe, M., Pook, P., & Rao, R. (1997). Deictic codes for the embodiment of cognition. *Behavioral and Brain Sciences, 20,* 723–767.

Bertera, J. H., & Rayner, K. (2000). Eye movements and the span of the effective visual stimulus in visual search. *Perception and Psychophysics, 62,* 576–585.

Brogan, D., Gale, A., & Carr, K. (1993). *Visual search 2.* London: Taylor & Francis.

Cave, K. R., & Wolfe, J. M. (1990). Modeling the role of parallel processing in visual search. *Cognitive Psychology, 22,* 225–271.

Findlay, J. M. (1997). Saccade target selection during visual search. *Vision Research, 37,* 617–631.

Hooge, I. T., & Erkelens, C. J. (1999). Peripheral vision and oculomotor control during visual search. *Vision Research, 39,* 1567–1575.

Milner, A. D., & Goodale, M. A. (1995). *The visual brain in action.* Oxford: Oxford University Press.

Motter, B. C., & Belky, E. J. (1998). The guidance of eye movements during active visual search. *Vision Research, 38,* 1805–1815.

Pomplun, M. (1998). *Analysis and models of eye movements in comparative visual search.* Göttingen: Cuvillier.

———. (2006). Saccadic selectivity in complex visual search displays. *Vision Research, 46,* 1886–1900.

Pomplun, M., Reingold, E. M., & Shen, J. (2001a). Investigating the visual span in comparative search: The effects of task difficulty and divided attention. *Cognition, 81,* B57–B67.

———, Reingold, E. M., & Shen, J. (2001b). Peripheral and parafoveal cueing and masking effects on saccadic selectivity in a gaze-contingent window paradigm. *Vision Research, 41,* 2757–2769.

———, Reingold, E. M., Shen, J., & Williams, D. E. (2000). *The area activation model of saccadic selectivity in visual search.* In L. R. Gleitman & A. K. Joshi (Eds.), *Proceedings of the Twenty Second Annual Conference of the Cognitive Science Society* (pp. 375–380). Mahwah, NJ: Erlbaum.

———, Ritter, H., & Velichkovsky, B. M. (1996). Disambiguating complex visual information: Towards communication of personal views of a scene. *Perception, 25,* 931–948.

———, Shen, J., & Reingold, E. M. (2003). Area activation: A computational model of saccadic selectivity in visual search. *Cognitive Science, 27,* 299–312.

Rao, R. P. N., Zelinsky, G. J., Hayhoe, M. M., & Ballard, D. H. (2002). Eye movements in iconic visual search. *Vision Research, 42,* 1447–1463.

Rayner, K. (1998). Eye movements in reading and information processing: 20 years of research. *Psychological Bulletin, 124,* 372–422.

Schyns, P. G., & Oliva, A. (1994). From blobs to edges: evidence for time and spatial scale dependent scene recognition. *Psychological Science, 5,* 195–200.

Scialfa, C. T., & Joffe, K. (1998). Response times and eye movements in feature and conjunction search as a function of eccentricity. *Perception & Psychophysics, 60,* 1067–1082.

Shen, J., Reingold, E. M., & Pomplun, M. (2000). Distractor ratio influences patterns of eye movements during visual search. *Perception, 29,* 241–250.

Thompson, K. G., Bichot, N. P., & Sato, T. R. (2005). Frontal eye field activity before visual search errors reveals the integration of bottom-up and top-down salience. *Journal of Neurophysiology, 93,* 337–351.

Viviani, P., & Swensson, R. G. (1982). Saccadic eye movements to peripherally discriminated visual targets. *Journal of Experimental Psychology: Human Perception and Performance, 8,* 113–126.

Williams, D. E., & Reingold, E. M. (2001). Preattentive guidance of eye movements during triple conjunction search tasks. *Psychonomic Bulletin and Review, 8,* 476–488.

Wolfe, J. M. (1994). Guided search 2.0: A revised model of visual search. *Psychonomic Bulletin & Review, 1,* 202–238.

———. (1996). Extending guided search: Why guided search needs a preattentive "item map." In A. F. Kramer, M. G. H. Coles, & G. D. Logan (Eds.), *Converging operations in the study of visual attention* (pp. 247–270). Washington, DC: American Psychological Association.

———. (1998). Visual search. In H. Pashler (Ed.), *Attention* (pp. 13–71). Hove, UK: Psychology Press.

———, Cave, K. R., & Franzel, S. L. (1989). Guided search: An alternative to the feature integration model for visual search. *Journal of Experimental Psychology: Human Perception and Performance, 15,* 419–433.

Zelinsky, G. J. (1996). Using eye saccades to assess the selectivity of search movements. *Vision Research, 36,* 2177–2187.

10

The Modeling and Control of Visual Perception

Ronald A. Rensink

Recent developments in vision science have resulted in several major changes in our understanding of human visual perception. For example, attention no longer appears necessary for "visual intelligence"—a large amount of sophisticated processing can be done without it. Scene perception no longer appears to involve static, general-purpose descriptions but instead may involve dynamic representations whose content depends on the individual and the task. And vision itself no longer appears to be limited to the production of a conscious "picture"—it may also guide processes outside the conscious awareness of the observer. This chapter surveys some of these new developments and sketches their potential implications for how vision is modeled and controlled. Emphasis is placed on the emerging view that visual perception involves the sophisticated coordination of several quasi-independent systems, each with its own intelligence. Several consequences of this view will be discussed, including new possibilities for human–machine interaction.

When we view our surroundings, we invariably have the impression of experiencing it via a "picture" formed immediately and containing a great amount of detail. This impression is the basis of three strong intuitions about how visual perception works: (1) Because visual experience is immediate, it must result from a relatively simple system. (2) Because the picture we experience is unitary, perception must involve a single system whose only goal is to generate this picture. (3) Because this picture contains enough information to let us react almost immediately to any sudden event in front of us, it must contain a complete description of almost everything in sight.

But recent research has shown that each of these intuitions is wrong: (1) The immediacy of an output is no guarantee that the underlying processes are simple. In fact, recent work has shown visual perception to be a highly complex activity, with a considerable amount of sophisticated processing done extremely rapidly. (2) If we experience a unitary percept, this does not necessarily mean that a single integrated system created it.

Indeed, visual perception increasingly appears to involve several quasi-independent subsystems, only some of which are responsible for the picture we experience. (3) If we can quickly access information about something whenever needed, this does not imply that all of it is represented at all times. Indeed, recent work shows that the picture we experience contains far less information at any given moment than our conscious impressions indicate, particularly in regard to dynamic events.

This new view of vision, then, suggests that the processes involved are more sophisticated than previously believed, and may do more than just provide a picture to our minds. This view provides considerable support for dynamic models that emphasize interaction and coordination of component processes and that have a strong sensitivity to what the operator knows and the task they are engaged in. It also points toward the possibility that vision itself might be controlled in interesting ways, allowing its component processes to be seamlessly incorporated into systems that extend beyond the physical body of the operator.

This chapter provides an overview of this new view of vision, focusing on several of the results and theories that have emerged. The first section, "Component Systems," discusses individual processes, characterizing them in terms of how they relate to attention. The second proposes how these internal systems might be integrated to produce the picture we consciously experience. The section "Integration of External Systems" then discusses some possible ways that these integration mechanisms might interact with external systems, creating more effective forms of human–machine interaction.

Component Systems

Visual perception results from the operation of a highly complex and heterogeneous set of processes, some of which remain poorly understood to this day. This section briefly describes several of these processes, along with some of the theories and models put forward to account for their operation. (For a more complete discussion, see Palmer, 1999.)

Since many of the new findings about visual perception involve attention—either in terms of what it is or how various operations relate to it—processes are grouped here into three largely disjoint sets: those that act before visual attention operates, those involved with attention itself, and those in which attention—and perhaps consciousness—may never be involved at all.

Preattentive Processes

When light enters the eye, it strikes the retina and is transformed into an array of neural signals that travels along the optic nerve, maintaining a retinotopic organization. What happens next is less well understood, but the prevailing view is that this marks the beginning of *early vision* (Marr, 1982), a stage of vision characterized by processes that are *low level* (i.e., operating locally on each point of the retinotopic input) and *rapid* (i.e., completed within about 200 ms). These are believed to operate automatically, without any need for attention (Figure 10.1).

Simple Properties

Early visual processing is thought to create a set of "primitives" on which all subsequent processing is based. Information concerning the nature of these primitives has largely been obtained via two kinds of study. The first is texture perception (e.g., Julesz, 1984), where

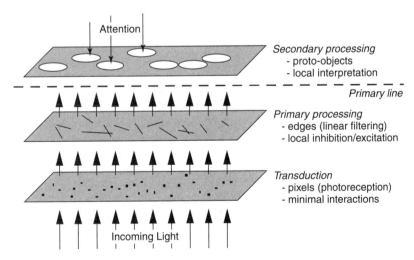

FIGURE 10.1 Schematic of early visual processing. The first stage is *transduction*, where photoreception occurs (i.e., the retina). The next is *primary processing*, where linear or quasi-linear filters measure image properties. This is followed by *secondary processing*, which applies "intelligent" nonlinear operations. Processing in all three stages is carried out rapidly and in parallel across the visual field. The outputs of the secondary stage constitute the 2½D sketch. The contents of this sketch are also the operands for subsequent attentional processes—the limit to immediate attentional access is given by the *primary line* (see Rensink, 2000).

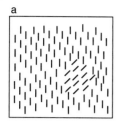 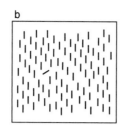

FIGURE 10.2 Tests for visual primitives. (a) Texture perception. Elements differing in orientation lead to the effortless segmentation of patch from the rest of the display. (b) Visual search. An item with a unique orientation immediately "pops out" from among the other items.

textons are defined as elements whose properties support effortless texture segmentation (Figure 10.2a). The second is visual search (e.g., Treisman, 1988; Wolfe, Cave, & Franzel, 1989; Wolfe, chapter 8, this volume), where basic *features* are properties that "pop-out"; that is, they can be quickly detected if they have a value unique to the display (Figure 10.2b).

In both cases, the set of primitives includes color, motion, contrast, and orientation. These properties have a common computational nature in that they can be determined on the basis of the limited local information around each point; this allows them to be computed rapidly and in parallel across the image (see Wolfe, chapter 8 and Pomplun, chapter 9, this volume). The explicit redescription of the image in terms of such elements is sometimes referred to as a *primal sketch* (Marr, 1982). Most models of the underlying processes involve an initial stage of linear filtering, followed by various nonlinear operations (see Palmer, 1999).

Complex Properties

Although many early visual properties are simple, the structures they measure may be relatively complex. For example, an isolated line fragment will pop-out if it has a distinctive length. But what structure is measured to get this value? If a distinctive line segment becomes part of a group (e.g., a drawing), pop-out of this segment is no longer guaranteed—this will now depend on the overall length of the group (Rensink & Enns, 1995). This indicates two things. First, some forms of grouping occur preattentively, with "length" being that of the overall group. As such, this indicates a fair degree of *visual intelligence* at this level. Second, the components of a group are inaccessible to higher-level processes, at least over the periods of time characteristic of this stage.

As such, the elements of early vision may be better characterized as *proto-objects* (i.e., precursors of objects) rather than the outputs of the very first stages of processing (Rensink, 2000). Features would then be those properties of proto-objects capable of affecting performance.

Another example of such visual intelligence is the ability to compensate for occlusion. For example, if a bar is occluded by a cube midway along its length, the visible portions will project to two line segments separated by the image of the occluder. But these segments can be linked preattentively, the resultant proto-object reflecting that they correspond to the same object in the scene (Rensink & Enns, 1998). Scene-based properties themselves can also be encoded. For example, search can be affected by three-dimensional orientation, direction of lighting, and shadow formation, with these estimates apparently formed on the basis of "quick and dirty" assumptions true only most of the time (Rensink & Cavanagh, 2004). Such behavior is in accord with theories that postulate the goal of early vision to be a 2½D *sketch*, a viewer-centered description of the world in which scene properties are represented in a fragmented way (Marr, 1982).

Control

Most models of early vision assume a unidirectional flow of information from the retina to the 2½D sketch, without any influence from higher-level factors, such as the nature of the task or knowledge about the scene (Marr, 1982). But anatomical studies show that there are a huge number of return connections from higher levels to lower ones, and psychological studies have shown these connections to have perceptual consequences, such as the image of an item being "knocked out" of iconic memory under the right conditions (DiLollo, Enns, & Rensink, 2000). Thus, the possibility arises that the set of elements at this stage may not be invariant for all tasks, but might be at least partly subject to higher-level control.

This possibility receives some support on computational grounds, in that it would be needlessly complex to have dedicated early-level processes for every possible aspect of scene structure. Rather, it might be more efficient to simply invoke instructions to calculate these whenever necessary.

Other Open Issues

Commonality of Visual Elements The properties that govern texture segmentation are neither a subset nor a superset of the properties that govern pop-out in visual search (Wolfe, 1992). This is difficult to reconcile with a single set of basic elements. It may be that different systems are involved; each with its own set of elements.

Reference Frame   Most models of early vision assume it to be based on a retinotopic frame of reference (see Palmer, 1999). However, visual search appears unaffected by sudden changes in position or size of the display, suggesting that it may be based upon a more abstract spatiotopic frame that is invariant to changes in size, and perhaps to other transformations as well (Rensink, 2004a).

Influence of Top-Down Control   Few experiments to date have investigated how the operation of early vision might be affected by the knowledge of the observer or the task they are carrying out. It has been proposed that early vision is not susceptible to these factors (Pylyshyn, 2003). But it is not clear how this can be reconciled with results showing the effects of connections from higher levels (DiLollo et al., 2000). Perhaps only particular types of control are possible.

Attentional Processes

Although observers can easily understand a request to "pay attention," it has proved extraordinarily difficult to determine what is happening when they do so. Earlier models considered attention to be a unitary faculty, or homogeneous "stuff." However, an emerging view is that *visual attention* is better characterized as the selective control of information in the visual system, which can be carried out in various ways by various processes (Rensink, 2003). As such, there exist—at least functionally—several kinds of attention, which may or may not be directly related to each other. (For an extensive set of current perspectives on attention, see Itti, Rees, & Tsotsos, 2005).

Selective Access

The simplest form of attention is *selective access*—the selective routing of some aspect of the input (usually involving a simple property) to later processes. For example, observers can detect a target more quickly and more accurately if cued to its location (Posner, Snyder, & Davidson, 1980). This has been explained by a *spotlight of attention* that amplifies inputs from the selected area and/or suppresses inputs from others.

It was once thought that selective access simply protected processors at higher levels from being overwhelmed by the sheer amount of information at early levels. More recently, selective access has been thought to serve a number of additional purposes. For example, it can improve the signal-to-noise ratio of the incoming signal and so improve performance (Treisman & Gormican, 1988). It can also delimit control of various actions—for example, by focusing on the particular part of an item to be grasped (Neumann, 1990). Appropriate use of selective access can also enable low-complexity approximations of high-complexity visual tasks (Tsotsos, 1990).

Selective Integration

Another form of attention is *selective integration*—the binding of selected parts or properties into more complex structures. For example, experiments have shown that it is difficult to detect a single L-shaped item among a set of T-shaped items. Similar difficulties are experienced for unique combinations of orientation and color, or of most other features. An influential account of this is *feature integration theory* (Treisman, 1988), which posits that attention acts via a spotlight that integrates the features at each location into an *object file*. If an item contains a unique feature, this spotlight is drawn to it automatically and the items is seen; otherwise, the spotlight must travel from item to item, integrating the corresponding feature clusters at a rate of about 50 ms/item.

The earlier belief was that to make good use of limited processing "resources" such integration needed to be selective. However, more recent work tends to view integration in terms of the *coordination* of the processes involved with each feature; selection can greatly simplify the management of this (Rensink, 2003).

Selective Hold/Change Detection

Recent work shows that observers can have difficulty noticing a sudden change made simultaneously with an eye movement, a brief flash in the image, or a sudden occlusion of the changed item (e.g., Figure 10.3).[1] Such *change blindness* (Rensink, O'Regan, & Clark, 1997) occurs under a variety of conditions and can happen even when the changes are large, repeatedly made, and in full knowledge that they will occur. It can be accounted for by the hypothesis that attention is necessary to see change. A change will then be difficult to see

FIGURE 10.3 Flicker paradigm. One way to induce change blindness is by alternating an original and a modified version of an image, with a brief blank or mask between each presentation. Performance is measured by time required to see the change. Even at a rate of two to three alternations per second, observers typically need several seconds to see the change. (In this example, the appearance/disappearance of the aircraft engine.)

whenever the motion transients that accompany it cannot draw attention to its location (e.g., if they are swamped by other motion signals in the image).

In this view, seeing a change requires a selective process that involves several steps: (1) the item from the original image is entered into a short-term store — presumably visual short-term memory; (2) it is held there briefly; and (3) it is then compared with the corresponding item in the new image. Selection could potentially occur at any or all of these stages. It is currently unclear which — if any — is the critical one.

One proposed account is *coherence theory* (Rensink, 2000), in which attention corresponds to the establishment of coherent feedback between lower-level proto-objects and a higher-level collection point, or *nexus* (see "Integration of Component Systems" section). Here, attention is characterized by the *selective hold* of items in a short-term store. In addition to change detection (essentially the tracking of an item across time), this form of attention may also be involved in the tracking of items across space (Pylyshyn, 2003). The reason for its selectivity may arise from the difficulty of establishing or maintaining two or more distinct feedback circuits

that connect arbitrary (and possibly disparate) parts of the brain.

Control

Visual attention is subject to two different kinds of control. The first is *exogenous* (or low level), which automatically draws attention to a particular item or location. This is governed by *salience*, a scalar quantity that reflects the priority of attentional allocation. Salience is usually modeled as a function of the spatial gradient of early-level features (e.g., changes in the average orientation in a certain part of the field), with large changes in density (such as those at the borders of objects or regions) having the highest salience (Itti, 2005). Exogenous control is believed to be largely independent of high-level factors. Some aspects may be affected by task and instruction set, although this has not yet been firmly established (Egeth & Yantis, 1997; Theeuwes & Godijn, 2002).

The second kind of control is *endogenous* (or high level). This is a slower, more effortful form of control that is engaged voluntarily on the basis of more abstract, context-sensitive factors such as task instruction. The relation of both exogenous and endogenous control to the various types of attention has not been worked out completely, nor has the way that exogeneous and endogenous control interact (see Egeth & Yantis, 1997).

Other Open Issues

Relation Between Attention and Eye Movements Experiments show that attention does not have to coincide with eye fixation: People can shift their attention without moving their eyes. Some form of attention is needed to select the target of an eye movement, but this need not be accompanied by a withdrawal of attention from other items. Thus, attention (of any type) and eye fixation are distinct processes and should not be conflated. However, a detailed model of the interaction between attention shifts and eye movements has not yet been developed (see Henderson, 1996).

The Basis of Selection   Visual search can be guided via the selection of features such as color or motion (Wolfe, Cave, & Franzel, 1989). For location, the situation is less clear: selection could be of a particular point in space (at which some object happens to be), an object (at some point in space), or perhaps both. Both spatial and object factors appear to play a role (see Egeth & Yantis, 1997). The relation of space- and object-based selection is not yet clear; these may be related to different types of attention.

Capacity   For space-based selection, a natural model is the spotlight of attention. Some models allow the intensity and size of the spotlight to be varied continuously; others do not (see Cave & Bichot, 1999). Most models posit one spotlight, although recent experiments suggest that it can be divided for some tasks (McMains & Somers, 2004). For object-based processes, capacity is usually about four items (Pylyshyn, 2003), although for some operations it is only one (Rensink, 2001, 2002a). There is currently no general consensus as to how all these accounts can be reconciled.

Nonattentional Processes

In the past, it was generally believed that the sole purpose of vision was to produce a sensory experience of some kind (i.e., a picture) and that attention was the "central gateway" to this. However, evidence is increasing that a good deal of sophisticated processing can be done without attention even beyond the early stage and that some of these processes can result in outputs having nothing to do with visual experience.

Rapid Vision

Although recent work shows a considerable amount of visual intelligence at early levels (see "Preattentive Processes" section), this is not limited to processes based on local information. Instead, such intelligence is found throughout *rapid vision*—that aspect of perception carried out during the first few hundred milliseconds throughout the entire visual system, as the initial sweep of information travels from the eyes to the highest cortical levels, and perhaps back down again (Rensink & Enns, 1998).[2]

One quantity determined this way is the abstract meaning of a scene, or *gist* (e.g., whether it is a city, port, or farm). Gist can be ascertained within 100 ms of presentation, a time insufficient for attending to more than a few items. It can be extracted from blurred images and without attention; indeed, two different gists can be determined simultaneously. Gist is likely determined on the basis of simple measures such as the distribution of line orientations or colors in the image; other properties of early-level proto-objects may also be used. Some aspects of scene composition, such as how open or crowded it is, can also be obtained this way, without the involvement of coherent object representations. (For a more comprehensive review, see Oliva, 2005.)

A possibly related development is the finding that observers are extremely good at obtaining *statistical summaries* of a group of briefly presented items. For example, observers can match the average size of a group of disks to an individual disk about as accurately as they can match the sizes of two individual disks (Ariely, 2001).

"Medium-Term" Memory

Although attention (in the form of selective hold) appears to be involved with visual short-term memory, there also appears to exist a form of memory that does not require attention, at least for its maintenance. This is not the same as long-term memory, since it can dissipate after a few minutes, once there is no further need for it. As such, it might be considered to be a separate, *medium-term* memory.

One possibility in this regard is memory for *layout*—the spatial arrangement of objects in the scene, without regard to their visual properties or semantic identities

(Hochberg, 1968). Some layout information may be extracted within several seconds of viewing—likely via eye movements—and can be held over intervals of several seconds without a constant application of attention. (Tatler, 2002). Interestingly, memory for repeated layouts can be formed in the complete absence of awareness that such patterns are being repeated (Chun & Jiang, 1998); such memory appears to help guide attention to important locations in that layout.

Visuomotor Guidance

It has been proposed (Milner & Goodale, 1995) that vision involves two largely separate systems: a fast *on-line* stream, concerned with the guidance of visually guided actions such as reaching and eye movement, and a slower *off-line* stream, concerned with the conscious perception and recognition of objects. Evidence for this *two-systems theory* is largely based on patients with brain damage: some can see objects but have great difficulty grasping them, while others cannot see objects, but (when asked to) can nevertheless grasp them easily and accurately.

Effects also show up in normal observers. For example, if a dot is shifted the moment an observer moves his eye to it, his eye always makes a corrective jump to the new location, even if the observer has no awareness of the shift (Bridgeman, Hendry, & Stark, 1975). And if a target is displaced during an eye movement, the hand of an observer reaching toward it will correct its trajectory, even if the observer does not consciously notice the displacement (Goodale, Pelisson, & Prablanc, 1986).

Implicit Perception

Although somewhat controversial, consensus is increasing that *implicit* processes exist; that is, performance is affected in some way, even though no conscious picture of the stimuli is involved. An example of this is *subliminal* perception, where a stimulus is embedded among irrelevant items and presented very briefly (typically less than 50 ms), making it difficult to see. Presentation of an unseen stimulus under these conditions nevertheless has several effects, such as speeding the conscious recognition of it in a subsequent display. (For a discussion of this, see Norretranders, 1999.) It is thought that subliminal perception may be a form of perception without attention (Merikle & Joordens, 1997).

Results from other approaches are consistent with this position. In *inattentional blindness,* observers can fail to see an unexpected stimulus if their attention is focused elsewhere (Mack & Rock, 1998). Stimuli with strong emotional impact are exceptions, indicating that some degree of semantic processing can occur in the absence of attention.

Implicit perception can be explored by comparing performance for stimuli reported as "seen" against performance for stimuli reported as "unseen." If these are not the same (e.g., have different sensitivities to color), the implicit processes must differ from those that gave rise to the conscious picture (Merikle & Daneman, 1998). This approach has uncovered several distinctive characteristics of implicit processing, such as a strong response to emotionally charged stimuli, a sensitivity to semantic meaning (but not to geometric structure), and an inability to exclude stimuli, for example, an inability to choose any word other than the one subliminally presented (Merikle & Daneman, 1998).

Control

In principle, it might be possible to control various aspects of a nonattentional process. For example, if it is possible to control the kind of outputs at the preattentive level (see "Preattentive Processes"), it might also be possible to control the outputs of other nonattentional processes, such as different kinds of statistical summaries (e.g., means or standard deviations). Inputs to nonattentional processes might also be selectable, for example, visuomotor operations might be able to act only on stimuli of a particular color or shape. If attention is defined as a selective process, then it might be that various forms of nonconscious attention exist. But separating out these from the known forms of *conscious* attention would be difficult.

Other Open Issues

Commonality of Processes   Although the perceptual processes in this subsection have been grouped together on the basis of not involving attention, such a negative definition says little about how—or even whether—they are related to one another. Issues such as the extent to which these processes are based on similar elements or involve common reference frames are still to be investigated.

Mindsight   Some observers can have a "feeling" that a change is occurring, even though they do not have a visual picture of the change itself (Rensink, 2004b). Although this phenomenon—*mindsight*—has been replicated, disagreement exists about how to best interpret it

(Simons, Nevarez, & Boot, 2005). The mechanisms involved are poorly understood. One possibility is that mindsight is a form of alert, relying on nonattentional processes such as layout perception (Rensink, 2004b).

Summary

Several major advances have recently occurred in our understanding of individual visual processes. One of these is the finding that considerable visual intelligence exists at early levels, with sophisticated processing carried out rapidly in the absence of attention. This opens up the possibility of interfaces that allow tasks traditionally done by high-level thinking to be off-loaded onto these faster, less effortful, and possibly more capable systems (Card, Mackinlay, & Shneiderman, 1999; Ware, 2004). The findings that intelligence also exists in rapid vision and visuomotor guidance opens up even more possibilities (see "Integration of External Systems" section).

Another set of advances concerns the nature of attention itself. Attention seems neither as pervasive, powerful, or unitary as originally believed. However, it still appears to be critical for particular operations, such as integrating information from selected items. Importantly, a delineation of the various processes (or at least functions) grouped under this label is now emerging, with some understanding of the characteristics of each. Among other things, this provides a much better grounding for the design of interface systems in which attention must be used in an appropriate way (see "Integration of External Systems" section).

Interestingly, recent work also shows that vision may involve more than just attention and consciousness— systems may exist that operate entirely without the involvement of either. The existence of such systems has major implications for the modeling and control of visual perception, in that they indicate that the conscious picture experienced by an observer is only a part of a wider-ranging system. To appreciate what this might mean, we must consider how these component systems might be integrated.

Integration of Component Systems

How can the component systems of vision be integrated such that an observer can experience a unitary picture of their surroundings? It was originally believed that these processes—acting via attention—constructed a complete, detailed description of the scene, for example,

accumulating representations in a visual buffer of high-information density (see Rensink, 2002a). Models of this kind, however, have great difficulty explaining induced failures of perception, such as inattentional blindness and change blindness (see "Component Systems" section).

Recent work tends to view the integration of systems in dynamic rather than static considerations—as *coordination* rather than *construction*. Among other things, this has the consequence that different people can literally see the same scene in different ways, depending on their expectations and the task they are engaged in.

Coherence Theory

To illustrate the idea of coordination, consider how it might apply to visual attention (or at least, to selective hold). According to some models, attention *welds* visual features into relatively long-lasting representations. But if so, why aren't all visible items welded within the first few seconds of viewing, allowing detection of all objects and events under all conditions?

Rather than assuming that focused attention acts by forming new structures that last indefinitely, it may be that it simply endows existing structures with a degree of coherence, with this lasting only as long as attention is directed at them. Developing this line of thought leads to a *coherence theory* of attention (Rensink, 2000).

Basics

Coherence theory is based on three related hypotheses (Figure 10.4):

1. Before attention, early-level proto-objects are continually formed rapidly and in parallel across the visual field. Although these can contain detailed descriptions and be quite complex (see "Component Systems" section), they are *volatile*, lasting only a few hundred milliseconds. As such, they are constantly in flux, with any proto-object simply being *replaced* when a new stimulus appears at its location.
2. Attention selects a small number of proto-objects from this flux and stabilizes them into an object representation. This is done via reciprocal links between the selected items and a higher-level *nexus*. The resulting circuit (or *coherence field*) forms a representation coherent across space and time. A new stimulus at the attended location is

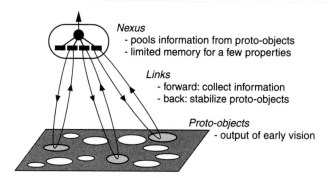

FIGURE 10.4 Coherence theory. Early vision continually creates proto-objects rapidly and in parallel across the visual field. Attention selects a subset of these, incorporating them into a circuit called the *coherence field*. As long as the proto-objects are "held" in this field, they provide the visual content of an individuated object that has both temporal and spatial coherence.

then perceived as the *change* of an existing structure rather than the appearance of a new one.

3. After attention is released (i.e., after the circuit is broken), the field loses its coherence and the object representation dissolves back into its constituent proto-objects. There is little or no *after-effect* of having attended to an item, at least in terms of the structures that underlie "here-and-now" perception.[3] (Also see Wolfe, 1999.)

According this view, then, attention is not stuff that helps create a static representation. Rather, it is the establishment (and maintenance) of a coordinated information flow that can span several levels of processing. The components that enter into the coherence field remain at the processing level where they were formed; what is added are the links that allow these components to be treated as part of the same object.

Implications

No Buildup of Attended Items   An important part of coherence theory is its assertion that attention corresponds to the establishment of a circuit of information flow. Thus, no buildup results from attention having been allocated to an item—once attention is withdrawn, the components of the coherence field revert to their original status as volatile proto-objects. Since attention is limited to just a few items (see Rensink, 2002a), most parts of a scene will therefore not have a coherent, detailed representation at any given moment. Of course, more durable, longer-term representations can exist, but these do not have such coherence.

No Complete Coherent Representation   According to coherence theory, some representations (those of early vision) are *complete*, in that they cover the entire visual field, and some representations (those resulting from

attention) are *coherent* over space and time. However since there is no buildup of coherence fields, no representation can be both complete *and* coherent.[4]

No Dense Coherent Representation   Although a proto-object might have a high density of information, only a small amount of this is held in the nexus (Rensink, 2001, 2002a). Thus, a coherence field cannot hold much detail about the attended item. If one of the properties represented in the nexus is one of the properties changing in the world, the change will be seen. Otherwise, it will not, even if the object is attended.

Virtual Representation

If only a few objects in a scene can have a coherent representation at any time, and if only a few properties of these objects are encoded in each representation, why do observers have the impression of seeing all events in the scene in great detail?

One way to account for this is the idea of a *virtual representation*: instead of a coherent representation of all the objects in an observer's surroundings, a coherent representation is created only of the object—and its properties—needed for the task at hand (Rensink, 2000).

Basics

If a coherent representation of an object can be created whenever needed, and if this representation contains those aspects required for the task at hand, the representation of the scene will appear to higher levels as if *real*, as if all objects are represented in complete detail simultaneously. Such a representation will have all the power of a real one, while using much less in the way of resources.

This strategy has been successfully applied to information systems. For example, it is the basis of virtual memory in computers, where—if coordination of memory access is successful—more memory appears available than is physically present at any given time. Browsing Web sites on a computer network can also be characterized this way (Rensink, 2000).

Requirements for Successful Operation

Virtual representation reduces complexity in space by trading it off for increased complexity in time. Only certain types of task can take advantage of this trade-off. For visual perception, what is required is

1. only a few objects need to be represented at any moment, and
2. only a few properties of these objects need to be represented at that moment, and
3. the appropriate object(s) can always be selected, and
4. the appropriate information about that object is always available when requested.

The first requirement is easily met for most tasks. Most operators need to control only one object (e.g., a steering wheel) or monitor one information source (e.g., a computer display) at a time. Tasks involving several independent objects or events can usually be handled by time-sharing, that is, rapidly switching between the objects or events. The second requirement is likewise easily met, in that most tasks only involve a few properties of an object at any given time (e.g., its overall size or color). Time-sharing can again be used if several properties are needed.

The third requirement can also be met, provided that three conditions hold. The first is having the ability to respond to any sudden event, and create the appropriate representation. As discussed in the section "Attentional Processes," this ability—in the form of the exogenous control of attention—does exist in humans. The second is having the ability to anticipate events so that nothing important is missed, even if other events are occurring. This can be done if the observer has a good understanding of the scene (i.e., knows what to expect) to direct endogenous attentional control appropriately. Third, the average time between important events must be at least as great as the average switching time of the control mechanisms. This is generally true for the world in which we live (or, at least, our ancestral environment), where important events almost never occur several times a second.

The fourth requirement is also met under most conditions of normal viewing. Provided that eye fixation and attention can be directed to the location of a selected object and that sudden occlusions are not common, it will usually be possible to obtain visual detail from the stream of incoming light, with the relevant properties then extracted from this stream. Thus, a high-capacity internal memory for objects is not needed: detailed information is almost always available from the world itself, which acts as an *external repository* (or *external memory*).[5]

Implications

Dependence on Knowledge and Task   In this view, the visual perception of a scene—including all events taking place in it—rests on a dynamic "just in time" system that represents only what is needed for the task at hand. The degree to which this is successful will depend on how well the creation of appropriate object representations is managed. Since this in turn strongly depends on the knowledge of the observer and the task being carried out, different people will literally see the same scene in different ways.

Change Blindness Blindness   If the creation of object representations is managed well, virtual representation will capture most of the relevant events in an environment. Meanwhile, any failure to attend to an appropriate object will not be noticed, and will usually have few consequences. As such, observers will generally become susceptible to *change blindness blindness*—they greatly overestimate their ability to notice any large changes that might occur (Levin, 2002).

Distributed Perception   Virtual representation implies a partnership between observer and environment: rather than an internal re-presentation containing all details of the scene, the observer uses an external repository (i.e., the world), trusting that it can provide detailed information whenever needed. In an important sense, then, observer and surroundings form a single system, with perception *distributed* over the components involved.

Triadic Architecture

The successful use of virtual representation requires that eye movements and attentional shifts can be made to the appropriate object at the appropriate time. But how might this be done? One possibility consistent with what is known of human vision is the *triadic architecture* (Rensink, 2000).

Basics

This architecture involves three separate systems, each of which operates somewhat independently of the others (Figure 10.5):

1. An *early vision* system that rapidly creates detailed, volatile proto-objects in parallel across the visual field (see "Preattentive Processes").
2. A limited-capacity *attentional* system that links these structures into coherent object representations (see "Coherence Theory" section).
3. A nonattentional *setting system* that provides a context to guide attention to the appropriate objects (see "Nonattentional Processes" section).

These largely correspond to the groups of systems in the "Component Systems" section, except that the setting system contains only those nonattentional processes that control visual attention. In addition to these three systems, there is a connection to long-term knowledge—such as schemas and particular skills—that helps direct high-level (endogenous) control, and

so influences perception. But since most of long-term memory is effectively off-line at any instant it is not considered part of here-and-now visual perception (Rensink, 2000).

Of the three systems involved in this architecture, the one concerned with setting is perhaps the least articulated. It likely involves at least two aspects of scene structure useful for the effective endogenous control of attention:

1. The abstract meaning (or *gist*) of the scene, for example, whether it is a forest or barnyard (see "Nonattentional Processes" section). This quantity is invariant over different eye positions and viewpoints, and, to some degree, over changes in the composition of objects. Consequently, it could provide a stable constraint on the kinds of objects expected and perhaps even indicate their importance for the task at hand.
2. The spatial arrangement (or *layout*) of objects in the scene. This quantity is, at least from an allocentric point of view, invariant to changes in eye position, and as such could help direct eye movements and attentional shifts. If held in a medium-term memory, the location of many objects could be represented. Some additional information

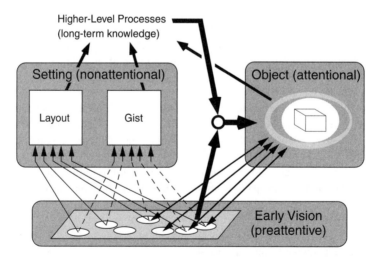

FIGURE 10.5 Triadic architecture. Visual perception is carried out by three interacting systems: (1) Early vision creates volatile proto-objects. (2) Attention "grabs" these structures and forms an object with temporal and spatial coherence. (3) Setting information—together with long-term knowledge and salience estimates obtained from early visual processing—guides attentional management.

concerning each item may also be possible; such information need not be extensive to be useful.

Interaction of Systems

Although there is currently little empirical evidence concerning the way that these systems interact, one possibility is as follows:

1. Early vision provides a constantly regenerating sketch of the scene visible to the observer.
2. Gist, layout, and perhaps some object semantics are determined without attention; these invoke a scene schema in long-term memory which provides constraints on the types of objects that might be present, possible actions, and so on.
3. The invoked schema is verified, beginning with a simple checking of expected features. Items consistent with the schema at this level need not be examined (and therefore need not be encoded) in detail.
4. If an unexpected structure in the image is encountered or an (unknown) salient item is suddenly detected at early levels, attentional processes form a coherent representation of it, attempt to determine its identity, and possibly reevaluate the gist. Layout can be used both to check the current interpretation as well as help guide attention to a requested object.

Such interaction involves a complex combination of exogenous and endogenous control, as well as of immediate, relatively changeable information about the scene and longer-term, more stable knowledge.

Implications

Construction Versus Coordination   Representations beyond early vision are no longer dense structures *constructed* via eye movements and attentional shifts; instead, they may be better viewed as sparse structures that *coordinate* the use of detailed information from the world. Thus, early-level representations are not *replaced* by more complex representations, but are *incorporated* into circuits spanning several levels of processing.

Role of Attentional Processing   Rather than being the *central gateway* of visual perception, attention (in the form of selective hold) may only be one of several concurrent streams—the one concerned with the conscious perception of coherent objects. Other streams may operate in complete independence of it. Indeed, the role of attention (or at least of consciousness) itself may even be somewhat restricted in regards to the control of action, being mostly involved with initiation of actions in unfamiliar situations, and perhaps learning (see Norretranders, 1999).

Role of Nonattentional Processes   In this view, nonattentional processes do not rely on attention for their "intelligence"—they in fact help guide it. Nonattentional processes beyond the setting system may enable aspects of perception having nothing to do with the production of a conscious picture, such affecting emotional state, or guiding visuomotor actions (see the "Integration of External Systems" section).

Summary

The integration of the component systems underlying visual perception appears to be achieved on the basis of *coordination* rather than *construction*. The nature of this more dynamic view can be seen, for example, in the coherence theory of visual attention. Here, attention is treated as a linkage between selected components, without the need for a separate representation of the attended object. The components are simply incorporated into a circuit along which information circulates, with a relatively small nexus serving to stabilize this linkage.

This style may also apply to our experience of a scene. Rather than being based on a static, dense representation, our experience may be based on a *virtual representation* that encodes only what is needed at each moment. As such, what is perceived will depend strongly on the particular observer and on the task they are carrying out. This account also leads to a view in which observer and environment form a single system, with perception distributed over the components involved.

One possible implementation of this in human vision is the *triadic architecture*, where a critical role is played by the mechanisms that control attention. Here, perception is distributed over several component systems, with the nonattentional (and presumable nonconscious) mechanisms that control attention playing a critical role. If these mechanisms can be properly controlled, it might be possible to integrate the component systems of human vision not only with each other but also with external systems. The next section explores some of these possibilities.

Integration of External Systems

It increasingly appears that visual perception may be based on several quasi-independent systems, each with its own kind of intelligence (see "Component Systems" and "Integration of Component Systems" sections). Given that these systems are integrated via dynamic coordination rather than static fusion, and that this coordination can be influenced by external factors, the possibility arises that these systems can be integrated with external systems as well.

If so, consideration of the mechanisms that carry out this coordination would provide a basis for the design of more effective visual display systems. It also opens up some genuinely new prospects for human–machine interaction.

Reduced Change Blindness

Given that attention is needed to see change (see "Attentional Processes" section), an observer will be blind to most unattended transitions in a display, resulting in informative transitions being missed. This would be especially important in displays in which information is conveyed dynamically, for example, if an operator is tracking the location of an item or following the orientation of an indicator needle. During such times, a transition could easily occur elsewhere in the display—for example, an alert appearing—without the operator noticing it. High-level (endogenous) control could lower the likelihood of such change blindness, but even if the observer could maintain a full state of alertness, the likelihood of missing something will still be considerable.

Change blindness can be induced in a number of ways (see Rensink, 2002a); a system should reduce each of these contributions as much as possible. For example, change blindness can be induced by eye movements, which make up about 10% of total viewing time on average[6]; any transition will therefore have about a 10% chance of being missed because of this factor alone. One way of lowering this likelihood is by minimizing the need for (or the size of) eye movements, for example, by keeping important sources of information close together. In addition, displays could minimize the number of dynamic events occurring

elsewhere, since these could draw attention to themselves, diverting attention away from the main information source. Moreover, although a single event can be attended without problems, two cannot—their contents will be pooled (Rensink, 2002a). Consequently, only a single source of dynamic information should ever be used at any time.

Coercive Graphics

A more speculative possibility involves the use of nonconscious processing to control what the observer consciously experiences. Given that the visual experience of an observer depends on the coordination of attention, and given that this coordination is strongly affected by what is shown to the eyes, the possibility arises of *coercive graphics*—displays that can control attention to make the observer see (or not see) a particular part of the image (Rensink, 2002b).

Coercion has long been used by magicians and filmmakers to achieve a variety of striking effects. Three means of control are commonly used:

1. High-level interest. Semantic factors that influence the semi-voluntary control of attention, for example, stories that interest the observer in a particular object or event.
2. Mid-level directives. Cues that require some intelligence, but then cause attention to rapidly move to a given location. Examples are the direction of eye gaze of another person (or image), and the direction of finger pointing.
3. Low-level salience. Simple scalar quantity that is the basis of exogenous control (see "Attentional Processes" section). Attention is automatically drawn—often involuntarily—to items such as those with a unique color, motion, orientation, or contrast.

All of these can be highly effective when done by humans (see Sharpe, 1988). If a system could make effective use of these, it could lead to *magical displays* capable of effects even more powerful than those produced by professional magicians.

A coercive display could ensure that important events would not be missed. It might also speed up operation by directing attention to required locations or items. Coercion would also be useful for older observers, acting as a form of "glasses" to compensate for the reduction in attentional abilities that generally happens with increasing age. Again, the user would notice nothing unusual—they would simply never miss anything important that occurred.

Emotional Control/Vigilance

In the past, visual displays were concerned only with the visual experience of the observer. But according to the triadic architecture (see "Triadic Architecture" section), this experience involves just one perceptual stream— the attentional system. However, other systems may also operate in tandem with this, and carry out a significant (albeit nonconscious) part of perception. As such, the potential arises for displays expressly designed to work with such processes, and influence aspects of an observer other than the visual percept they experience.

One such example is the control of emotional state. Nonattentional (and nonconscious) processes have a pronounced sensitivity to emotionally laden words and pictures (see "Nonattentional Processes" section). Moreover, some of these processes can affect the physiological mechanisms underlying the associated emotions, even though the stimuli involved are unseen (e.g., Liddell et al., 2005; Whalen et al., 1998). As such, it may be possible to develop displays that could, e.g., calm an operator down or increase their level of vigilance, all on the basis of stimuli that are not consciously experienced.

Soft Alerts

For many tasks, the system must allow the operator to respond quickly to unexpected events. This is typically done via an alert, which attempts to draw attention to its location, thereby ensuring it will be seen. Although such alerts can be successful, they can also be dangerous (especially for time-critical tasks), since they have the potential to divert attention away from important objects or operations.

A somewhat speculative alternative to these involves the phenomenon of mindsight—the feeling of something happening without an accompanying picture (see "Nonattentional Processes" section). This phenomenon is poorly understood; it may be related to feelings generated by emotional states, although this is far from certain (Rensink, 2004b). In any event, if visual displays could be designed to invoke this feeling whenever desired, it would make an extremely useful form of alert, a *soft alert*, that would not disturb existing attentional

control (Rensink, 2002b). Such an alert would be useful for situations where the arrival of a new event does not require immediate attention, for example, the arrival of email while the operator is monitoring a changing situation.

Direct Support of Action

As discussed in the "Nonattentional Processes" section, considerable evidence exists that actions such as reaching and grasping are guided by nonattentional systems having nothing to do with conscious visual experience (see Milner & Goodale, 1995). It is likely that activities such as moving a mouse or pointing are guided similarly. More generally, the set of visuomotor systems (along with other motor systems and perhaps some rapid perceptual processes) may be coordinated to result in an *inner zombie* capable of carrying out operations in a highly sophisticated way, even though consciousness is not involved (see Norretranders, 1999).

If this view is correct, it suggests the need for displays designed expressly for the *direct support* of action, that is, displays to act directly on the nonconscious visuomotor systems rather than only on the systems that produce conscious visual experience. For example, pointing without visual feedback may help a user aim a laser pointer at a given location, even though this is counterintuitive from the viewpoint of conscious perception (Po, Fisher, & Booth, 2003). In such situations, there may be no awareness that the display is providing such guidance; the user simply does the right thing.

Cognitive Extension

The type of dynamic representation discussed here is a special case of the more general notion of *deictic* (or *indexical*) representation. Here, the goal is not to *construct a copy* of the world, but rather to *coordinate* component systems so as to carry out actions in it (Ballard, Hayhoe, Pook, & Rao, 1997; Clancey, 1997; Ballard & Sprague, chapter 20, this volume). It does not matter whether the components involved are internal or external—all that matters is that they are part of a circuit of information flow under the control of the user (see Clark, 2003).

If the coordination with an operator's visual system is done properly, external processors (e.g., a calculator or an information visualization system) could become part of such a circuit, allowing sophisticated processing to be incorporated in a seamless manner. External effectors (e.g., a car or an airplane) could likewise become part of such a circuit, with each system treated as a visuomotor system of the operator. Indeed, when such an interface functions well, the operator can experience a literal extension of themselves into the task domain (e.g., becoming part of the car or airplane), resulting in highly effective control of all component systems (see Clark, 2003).

Summary

An emerging view is that the operation of the human visual system is based on the coordination of several quasi-independent systems. If the coordination mechanisms within an operator can be applied to external systems as well, highly effective forms of human–machine interaction could result. For example, systems might be designed to reduce the likelihood of change blindness, to ensure that the operator will always see what they need to see (assuming it is in the display), or even to bring other internal systems (e.g., emotions) into play. In addition, the possibility also exists of incorporating external systems—not only information sources but also processing elements and effectors—in a similar fashion, allowing human perceptual and cognitive abilities to be extended in a highly natural way.

Conclusions

This chapter has surveyed some of the main developments that have recently occurred in our understanding of human perception and discussed some of their implications for the modeling and control of visual perception. Among these developments is the increasing recognition that visual attention may not be the central gateway to visual perception but may instead be simply one of several quasi-independent systems, each capable of sophisticated processing. It also appears that these systems are not integrated via dense, static representations that accumulate results but rather via dynamic coordination that depends on such factors as the knowledge of the observer and the nature of the task they are engaged in. This kind of coordination— if done properly—can result in a virtual representation that provides the observer with a unitary picture of the scene. Such coordination may also enable purely nonconscious systems to act coherently, resulting in an *inner zombie* capable of intelligent on-line control of actions without any involvement of consciousness.

From this viewpoint, then, human visual perception appears to be based on the coordination of several quasi-independent systems, each with its own form of intelligence. It may be useful to view human–machine interaction in a similar way, with the operation of a human–machine system based on the coordination of several quasi-independent systems (some internal to the operator, some external), each with *its* own form of intelligence. Such a perspective not only suggests ways of improving existing display and control systems, but also points to new possibilities for increasing the effectiveness and scope of human–machine interaction.

Acknowledgments

I would like to thank Nissan Motor Co. (Japan) and Natural Sciences and Engineering Research Council (Canada) for supporting the work described in this chapter. Thanks also to Wayne Gray, Chris Myers, and Hans Neth for their helpful comments on an earlier version. I would also like to thank Wayne Gray and Air Force Office of Scientific Research (USA) for giving me the opportunity to present these ideas in this context.

Much of the work described here—both the particular results and the general approach—had its origins during the years 1994–2000 at Cambridge Basic Research (CBR), a laboratory of Nissan Motor Co. in Cambridge Massachusetts. Thanks to my colleagues at CBR for their encouragement and support during that time: Jack Beusmans, Erwin Boer, Jim Clark, Rob Gray, Andy Liu, Simon Rushton, and Ian Thornton. Also thanks to Takao Noda and Akio Kinoshita for maintaining a wonderful environment that greatly fostered research. This chapter is dedicated to them.

Notes

1. This and other examples can be downloaded from www.cs.ubc.ca/~rensink/flicker/download or from www.psych.ubc.ca/~rensink/flicker/download.

2. *Rapid vision* can be defined as that occurring during the first 200 ms or so of visual processing; it can involve mechanisms throughout the visual system. *Low-level vision* occurs via low-levels mechanisms, which generally operate in parallel in a spatiotopic array and without any influence of stimulus-specific knowledge. *Early vision* can be defined as the intersection of these two, that is, processing that is both rapid and low level (Rensink & Enns, 1998). As such, rapid

vision comprises several different—and coordinated—processing systems, of which early vision is one. *Preattentive processes* are the set of processes at early levels; these operate without attention, before any attentional application. Although all preattentive processes are nonattentional, not all nonattentional processes are preattentive.

3. There may be effects such as entry into long-term memory. But long-term memory is not considered to be among the mechanisms that directly underlie "here-and-now" (or "working") perception (Rensink, 2000).

4. It may be that a relatively complete representation of the *static* aspects of a scene is built up—experimental evidence to date is not sufficient to rule out this possibility (Simons & Rensink, 2005). However, the existence of change blindness clearly shows the existence of severe limits on how much of its *dynamic* aspects are represented at any time.

5. The more usual term is *external memory* (see, e.g., Clark, 2003). But *memory* does not entirely capture the situation because what is available from the world is not a remnant of any information that disappeared from the environment. Even if information might have disappeared from the observer, it is still problematic how *remnant* would apply before or during the first time the observer accessed this information.

6. The duration of each ballistic movement (or *saccade*) of the eye depends on the angle A traversed, according to $D = 21 + 2.2A$, where D is in milliseconds, and A is in degrees (Carpenter, 1988). Such movements can sometimes take more than 100 ms. However, on average these take about 30 ms, at an average rate of about three to four per second (see Palmer, 1999). Thus, the amount of time spent in ballistic movement—where blur induced by the eye movement destroys the automatic drawing of attention to the location of a change—is typically 90–120 ms per second, with greater durations for movements through greater angles.

References

Ariely, D. (2001). Seeing sets: Representation by statistical properties. *Psychological Science, 12,* 157–162.

Ballard, D. H., Hayhoe, M. M., Pook, P. K., & Rao, R. P. (1997). Deictic codes for the embodiment of cognition. *Behavioral and Brain Sciences, 20,* 723–767.

Bridgeman, B., Hendry, D., & Stark, L. (1975). Failure to detect displacement of the visual world during saccadic eye movements. *Vision Research, 15,* 719–722.

Card, S. K., Mackinlay, J. D., & Shneiderman, B. (1999). Information visualization. In S. K. Card, J. D. Mackinlay, & B. Shneiderman B. (Eds.), *Readings in information visualization: Using vision to think* (pp. 1–34). San Francisco: Morgan Kaufman.

Carpenter, R. H. S. (1988). *Movements of the eyes* (2nd ed., p. 72). London: Pion.

Cave, K. R., & Bichot, N. P. (1999). Visuo-spatial attention: Beyond a spotlight model. *Psychonomic Bulletin and Review, 6,* 204–223.

Chun, M. M., & Jiang, Y. (1998). Contextual cueing: Implicit learning and memory of visual context guides spatial attention. *Cognitive Psychology, 36,* 28–71.

Clancey, W. J. (1997). *Situated cognition: On human knowledge and computer representations.* Cambridge: Cambridge University Press.

Clark, A. J. (2003). *Natural-born cyborgs: Minds, technologies, and the future of human intelligence.* Cambridge, MA: MIT Press.

DiLollo, V., Enns, J. T., & Rensink, R. A. (2000). Competition for consciousness among visual Events: The psychophysics of reentrant visual processes. *Journal of Experimental Psychology: General, 129,* 481–507.

Egeth, H. E., & Yantis, S. (1997). Visual attention: Control, representation, and time course. *Annual Review of Psychology, 48,* 269–297.

Goodale, M. A., Pelisson, D., & Prablanc, C. (1986). Large adjustments in visually guided reaching do not depend on vision of the hand or perception of target displacement. *Nature, 320,* 748–750.

Henderson, J. M. (1996). Visual attention and the attention-action interface. In K. Akins (Ed.), *Perception* (pp. 290–316). Oxford: Oxford University Press.

Hochberg, J. E. (1968). In the mind's eye. In R.N. Haber (Ed.), *Contemporary theory and research in visual perception* (pp. 309–331). New York: Holt, Rinehart & Winston.

Itti, L. (2005). Models of bottom-up attention and saliency. In L. Itti, G. Rees, & J. K. Tsotsos (Eds.), *Neurobiology of attention* (pp. 576–582). San Diego, CA: Elsevier.

———, Rees, G., & Tsotsos, J. K. (Eds.). (2005). *Neurobiology of attention.* San Diego, CA: Elsevier.

Julesz, B. (1984). A brief outline of the texton theory of human vision. *Trends in Neuroscience, 7,* 41–45.

Levin, D. T. (2002). Change blindness blindness as visual metacognition. *Journal of Consciousness Studies, 9,* 111–130.

Liddell, B. J., Brown, K. J., Kemp, A. H., Barton, B. J., Das, P., & Peduto, A., et al. (2005). A direct brainstem–amygdala–cortical "alarm" system for subliminal signals of fear. *NeuroImage, 24,* 235–243.

Mack, A., & Rock, I. (1998). *Inattentional blindness.* Cambridge, MA: MIT Press.

Marr, D. (1982). *Vision: A computational investigation into the human representation and processing of visual information.* San Francisco: Freeman.

McMains, S., & Somers, D. C. (2004). Multiple spotlights of attentional selection in human visual cortex. *Neuron, 42,* 677–686.

Merkle, P. M., & Daneman. M. (1998). Psychological investigations of unconscious perception. *Journal of Consciousness Studies, 5,* 5–18.

————, & Joordens, S. (1997). Parallels between perception without attention and perception without awareness. *Consciousness and Cognition, 6,* 219–236.

Merkle, P., & Reingold, E. (1992). Measuring unconscious perceptual processes. In R. Bornstein & T. Pittman (Eds.), *Perception without awareness: Cognitive, clinical, and social perspectives* (pp. 55–80). New York: Guilford.

Milner, A. D., & Goodale, M. A. (1995). *The visual brain in action.* Oxford: Oxford University Press.

Neumann, O. (1990). Visual attention and action. In O. Neumann & W. Prinz (Eds.), *Relationships between perception and action: Current approaches* (pp. 227–267). Berlin: Springer.

Norretranders, T. (1999). *The user illusion: Cutting consciousness down to size* (Chaps. 6, 7, 10). New York: Penguin Books.

Oliva, A. (2005). Gist of a scene. In L. Itti, G. Rees, & J. K. Tsotsos (Eds.), *Neurobiology of attention* (pp. 251–256). San Diego, CA: Elsevier.

Palmer, S. E. (1999). *Vision science: Photons to phenomenology.* Cambridge, MA: MIT Press.

Po, B. A., Fisher, B. D., & Booth, K. S. (2003). Pointing and visual feedback for spatial interaction in large-screen display environments. In *Proceedings of the 3rd International Symposium on Smart Graphics* (pp. 22–38). Heidelberg: Springer.

Posner, M. I., Snyder, C. R., & Davidson, B. J. (1980). Attention and the detection of signals. *Journal of Experimental Psychology: General, 109,* 160–174.

Pylyshyn, Z. (2003). *Seeing and visualizing: It's not what you think.* Cambridge, MA: MIT Press.

Rensink, R. A. (2000). The dynamic representation of scenes. *Visual Cognition, 7,* 17–42.

————. (2001). Change blindness: Implications for the nature of attention. In M. R. Jenkin & L. R. Harris (Eds.), *Vision and attention* (pp. 169–188). New York: Springer.

————. (2002a). Change detection. *Annual Review of Psychology, 53,* 245–277.

————. (2002b). Internal vs. external information in visual perception. *Proceedings of the Second International Symposium on Smart Graphics* (pp. 63–70). New York: ACM Press.

————. (2003). Visual attention. In L. Nadel (Ed.), *Encyclopedia of cognitive science* London: Nature Publishing Group.

————. (2004a). The invariance of visual search to geometric transformation. *Journal of Vision, 4,* 178a. http://journalofvision.org/4/8/178.

————. (2004b). Visual sensing without seeing. *Psychological Science, 15,* 27–32.

————, & Cavanagh, P. (2004). The influence of cast shadows on visual search. *Perception, 33,* 1339–1358.

————, & Enns, J. T. (1995). Preemption effects in visual search: Evidence for low-level grouping. *Psychological Review, 102,* 101–130.

————, & Enns, J. T. (1998). Early completion of occluded objects. *Vision Research, 38,* 2489–2505.

————, O'Regan, J. K., & Clark, J. J. (1997). To see or not to see: The need for attention to perceive changes in scenes. *Psychological Science, 8,* 368–373.

Sharpe, S. H. (1998). *Conjurer's psychological secrets.* Calgary, Alberta, Canada: Hades.

Simons, D. J., Nevarez, G., & Boot, W. R. (2005). Visual sensing is seeing: Why "mindsight," in hindsight, is blind. *Psychological Science, 16,* 520–524.

————, & Rensink, R. A. (2005). Change blindness: Past, present, and future. *Trends in Cognitive Sciences, 9,* 16–20.

Tatler, B. W. (2002). What information survives saccades in the real world? In J. Hyönä, D. P. Munoz, W. Heide, & R. Radach (Eds.), *The brain's eye: Neurobiological and clinical aspects of occulomotor research* (pp. 149–163). Amsterdam: Elsevier.

Theeuwes, J., & Godijn, R. (2002). Irrelevant singletons capture attention: Evidence from inhibition of return. *Perception & Psychophysics, 64,* 764–770.

Treisman, A. (1988). Features and objects: The fourteenth Bartlett memorial lecture. *Quarterly Journal of Experimental Psychology, 40A,* 201–237.

————, & Gormican, S. (1988). Feature analysis in early vision: Evidence from search asymmetries. *Psychological Review, 95,* 15–48.

Tsotsos, J. K. (1990). Analyzing vision at the complexity level. *Behavioral and Brain Sciences, 13,* 423–445.

Ware, C. (2004). *Information visualization: Perception for design* (2nd ed.). San Francisco: Morgan Kaufman.

Whalen, P. J., Rauch, S. L., Etcoff, N. L., McInerney, S. C., Lee, M. B., & Jenike, M. A. (1998). Masked presentations of emotional facial expressions modulate amygdala activity without explicit knowledge. *Journal of Neuroscience, 18,* 411–418.

Wolfe, J. M. (1992). "Effortless" texture segmentation and "parallel" visual search are not the same thing. *Vision Research, 32,* 757–63.

————, Cave, K. R., & Franzel, S. L. (1989). Guided search: An alternative to the feature integration model for visual search. *Journal of Experimental Psychology: Human Perception and Performance, 15,* 419–433.

————. 1999. Inattentional amnesia. In V. Coltheart (Ed.), *Fleeting memories* (pp. 71–94). Cambridge, MA: MIT Press.

PART IV

ENVIRONMENTAL CONSTRAINTS ON INTEGRATED COGNITIVE SYSTEMS

Hansjörg Neth & Chris R. Sims

The four chapters of this section stand in different theoretical traditions, study different phenomena, and subscribe to different methodological frameworks. However, they share the conviction that cognition routinely exploits environmental regularities. This common focus on the role of the environment allows their authors to transcend the boundaries that are traditionally drawn between the fields of decision making, interactive behavior, sequential learning, and expertise. Finally, their shared commitment to precise formalisms shows that an emphasis on environmental contributions to the control of cognition does not imply a purely narrative account but is most powerful when combined with computational and mathematical models.

Todd and Schooler (chapter 11) demonstrate precisely this idea—that an adaptive agent would do well to lean on the environment as an aid to cognitive control and decision making. At first glance, the view of cognition presented by the authors appears alarmingly fragmented. Rather than identifying general computational principles, their *adaptive toolbox* contains a loose collection of computationally simple decision heuristics,

each designed to solve a different type of task or decision problem. However, the chapter contains an interesting double twist. The first twist is that performance that relies on the most famous heuristic from the adaptive toolbox, the recognition heuristic, can be predicted from the way in which the human memory system works. The second twist is that the theory of memory used to make these predictions was inspired by the insight that no storage system could store an unlimited number of memories for an unlimited period of time. This insight led the theorists (Anderson & Schooler, 1991) to perform a detailed analysis of the pattern of demands that the environment makes on human memory and to build a theory of memory that would match those demands. Hence, the success of this theory in predicting human performance using the recognition heuristic is, first, an interesting and important validation of the theory. Second, it takes us beyond the simple fact that the recognition heuristic is easy to apply and correct a large percentage of the time, to the insight that the success of the heuristic, like the success of the memory theory, stems from an adaptation of human cognition to the characteristics of its task environment.

Fu (chapter 12) uses the statistical properties of the environment to predict performance on information foraging tasks. In doing so, the author reconceptualizes search through a problem space as sequential decision making. For example, when searching for information on the Web, deciding which link, if any, to click requires us to evaluate the probability that the link will lead to the desired information. Following a link incurs costs in terms of the time required to load and search the next page. A central problem in this class of tasks is deciding when to quit. When is an answer good enough? When should you continue searching for a better answer? As in the Todd and Schooler chapter, the answer to seemingly insurmountable computational complexity lies in subtle statistical properties of the task environment. In the case of Web navigation, Fu assumes a law of diminishing returns, such that increasing search efforts are met with ever-decreasing gains in the quality of the final solution. Although this effect all but eliminates the chance of achieving optimal performance, it greatly facilitates the chance of obtaining near-optimal performance. An adaptive agent can use a simple decision rule for determining when to stop searching: stop when the incremental gain for evaluating an additional alternative is likely to fall below the cost of its evaluation. Fu implements this approach using a Bayesian satisficing model and compares its performance to humans across several task domains.

The *environment* explored by Mozer, Kinoshita, and Shettel (chapter 13) is constituted by the sequential context in which a particular task is performed. Sequential dependencies occur whenever current experience affects subsequent behavior and, from an adaptive perspective, reflect the fact that the *drift* in most task environments is slow enough that sequential decisions are unlikely to be independent. Apart from highlighting the temporal plasticity of the human mind, sequential dependencies provide a new look on experimental paradigms in which blocked designs are seen as factors of control, rather than as the source of potential effects. The remarkable scope and explanatory power of deceptively simple models underline the authors' suggestion that sequential dependencies reflect a continuous adaptation of the brain to the ongoing stream of experience and allow a glimpse into the fine-tuning mechanisms of cognitive control on a timescale of seconds.

In chapter 14, Kirlik explores the potential of ecological resources for modeling interactive cognition and behavior. Although Kirlik recognizes that modeling techniques can be applied to complex real-world tasks, he points to obstacles in modeling human expertise with dynamic systems using tools developed to model simple laboratory tasks. For example, in his functional analysis of short-order cooks, Kirlik shows that humans do not passively adapt to their task environments but actively adapt these task environments to the service of their current tasks. Capturing this dynamic interplay of users, tasks, and task environments requires a comprehensive analysis of the entire interactive system, not just high-fidelity models of individuals.

All chapters in part IV attempt to solve the issue of cognitive control by partially dissolving it: If the external environment exerts control on cognitive processes by triggering context-specific heuristics and mechanisms, the controlling homunculus is partially put out of business. But this perspective does not imply that cognitive science is outsourced to ecological analysts. Instead, the focus on environmental contributions to cognition raises many new questions. For instance, how did our arsenal of context- and task-specific tools or intricate fine-tuning mechanisms develop? How is it organized? How do we select what we need when we need it? More fundamentally, how can we predict which features of the environment are relevant for any given task? Throughout this section, *environment* appears in many guises—candidates include the background conditions under which particular strategies were acquired, the physical or temporal context of their execution, and externally imposed constraints (e.g., of speed or accuracy) on task performance.

It may seem ironic that cognitive science discovered the merits of looking outside the human head for constraints on cognition at a time when neuroscience strives toward the localization of cognitive functions within the brain. While the notion of *integrated* cognitive systems may be at full stretch when the system includes the environment, it is still not overinflated. Cognition routinely recruits external resources to achieve its goals, thereby making the constraints of adaptive organisms and their environments complementary. A viable theory of cognition must encompass more than just the mind or brain. The following chapters aptly demonstrate that we can learn a lot about cognition by studying how it adapts to its environments as well as how it adapts its environments to itself.

Reference

Anderson, J. R., & Schooler, L. J. (1991). Reflections of the environment in memory. *Psychological Science*, 2, 396–408.

11

From Disintegrated Architectures of Cognition to an Integrated Heuristic Toolbox

Peter M. Todd & Lael J. Schooler

How can grand unified theories of cognition be combined with the idea that the mind is a collection of disparate simple mechanisms? In this chapter, we present an initial example of the possible integration of these two perspectives. We first describe the "adaptive toolbox" model of the mind put forth by Gigerenzer and colleagues: a collection of simple heuristic mechanisms that can be used to good effect on particular tasks and in particular environments. This model is aimed at describing how humans (and other animals) can make good decisions despite the limitations we face in terms of information, time, and cognitive processing ability—namely, by employing *ecological rationality*, that is, using heuristics that are fit to the structure of information in different task environments, and letting the environment itself exert significant control over what components of cognition are employed. Yet such a disintegrated and externally driven view of cognition can still ultimately come together within an integrated model of a cognitive system, as we demonstrate via an implementation within the ACT-R cognitive architecture of two simple decision heuristics that exploit patterns of recognition and familiarity information. We conclude by pointing the challenges remaining in developing an integrated simple heuristics conception of cognition, such as determining which decision mechanism to use in a particular situation.

In the center of Berlin, the Sony Corporation has recently erected an architectural marvel: a complex of buildings comprising 130,000 square meters of glass and steel, in parts over twenty-five stories tall, topped by a volcano-like sail structure that is lit in different colors throughout the night. The design of architect Helmut Jahn calls for the Sony Center to capture all aspects of modern human life in a single continuous edifice. One plainly sees the architect's vision for how we should live, work, play, shop, and eat in the carefully orchestrated spaces all under the same hue-shifting umbrella.

A few kilometers south and 800 years older, the reconstructed medieval village of Düppel stands in stark contrast. Modest huts seem haphazardly strewn about a central green, each built independently and dedicated to a different task essential to the villagers' survival: textile weaving and sewing in one; pottery in another; and shoe making, tool forging, and tar making in still others. Animals were tended and crops raised in adjoining areas, making the village a self-sufficient working whole that adapted to the challenges of its surroundings as they came and changed with the seasons.

Ever since Allan Newell warned that the only way to achieve progress in understanding human behavior is to produce unified theories of cognition (Newell, 1973), a number of cognitive scientists have taken up the challenge to build models of the architecture of cognition that call to mind the Sony Center: grand visions aiming to account for the wide range of human (mental) life under one overarching framework, from perception and memory to planning and decision making. This approach, by bringing together the constraints of many separate systems to bear on a central architectural design, has advanced our understanding of how the system as a whole and its individual parts can work together (Anderson & Lebiere, 1998).

At the same time that the Sony Center was being constructed in Berlin, a group of researchers down the road were building a seemingly different model of human cognition, one that harkened to the structure of the medieval village. Gerd Gigerenzer and colleagues developed a view of the mind composed of a set of simple heuristics, each dedicated to a single task and working in tune with the structure of the environment to achieve its goals in a quick and efficient manner, much as the medieval specialists tackled different tasks in their separate huts (Gigerenzer, Todd, & the ABC Research Group, 1999). Is such a vision doomed to the dark ages of disconnected psychological modeling that Newell decried? Or can it be brought together with the unified cognitive architecture models in a useful way?

In this chapter, we show that the two approaches are indeed compatible, and we present an initial example of their successful integration. We start by describing the model put forth by Gigerenzer and colleagues of a collection of simple heuristic mechanisms that can be used to good effect on particular tasks and in particular environments. This approach allows considerable progress to be made in understanding how humans (and other animals) can make good decisions in spite of the limitations we face in terms of information, time, and cognitive processing ability. The main message is that different structures of information in different task environments call for different heuristics with appropriately matched processing structures, producing *ecological rationality*; in this way, the environment itself exerts significant control over what components of cognition are employed. Yet such a disintegrated and externally driven view of cognition can still ultimately come together within an integrated model of a cognitive system, as we demonstrate via an implementation of two simple decision heuristics within the adaptive control of thought–rational (ACT-R) framework (Anderson & Lebiere, 1998). We conclude by pointing out some of the challenges remaining in developing an integrated simple heuristics conception of cognition, such as determining which decision mechanism to use in a particular situation.

A Dis-integrated View?

The traditional view of human decision making is one of *unbounded* rationality, dictating that decisions be made by gathering and processing all available information, without concern for the human mind's computational speed or power. This view is found surprisingly often in perspectives ranging from *homo economicus* in economics to the GOFAI (good old-fashioned AI)

school of artificial intelligence (for a multidisciplinary review, see Goodie, Ortmann, Davis, Bullock, & Werner, 1999). It leads to the idea that people use one big processing tool—an integrated inference system based on the laws of logic and probability—applied to all the data we can muster to make decisions in all the domains that we encounter.

But this unbridled approach to information processing certainly fails to capture how most people make most decisions most of the time. Herbert Simon, noting that people must usually make their choices and inferences despite limited time, limited information, and limited computational abilities, championed the view of *bounded* rationality: studying how people (and other animals) can make reasonable decisions given the constraints that they face. Simon argued that because of the mind's limitations, humans "must use approximate methods to handle most tasks" (Simon, 1990, p. 6). These methods include recognition processes that largely obviate the need for further information; mechanisms that guide the search for information or options and determine when that search should end; simple rules that make use of the information found; and combinations of these components into decision heuristics. Under this view, the mind does not use one all-powerful tool, but rather a number of less sophisticated, though more specialized, gadgets.

Simon's notion of bounded rationality, originally developed in the 1950s, was enormously influential on the psychologists and economists who followed. However, it was interpreted in two distinct and conflicting ways. First, a number of researchers accepted Simon's assertion that the mind relies on simple decision heuristics and shortcuts, but they assumed, at the same time, that it is often flawed in doing so: rather, we should all be unboundedly rational, if only we could. Under this view, the simple heuristics that we so commonly use frequently lead us astray, making us reach biased decisions, commit fallacies of reason, and suffer from cognitive illusions (Piattelli-Palmarini, 1996). The very successful "heuristics-and-biases" research program of Tversky and Kahneman (1974) and Kahneman, Slovic, and Tversky (1982) embodied this interpretation of bounded rationality and led to much work on how to *de-bias* people so they could overcome their erroneous heuristic decision making.

In stark contrast, a growing number of researchers are finding that people can and do often make *good* decisions with simple rules or heuristics that use little information and process it in quick ways (Gigerenzer &

Selten, 2001; Gigerenzer et al., 1999; Payne, Bettmann, & Johnson, 1993). This second view of bounded rationality argues that our cognitive limits do not stand in the way of adaptive decision making; in fact, not only are these bounds not always hindrances, they can even be beneficial in various ways (Hertwig & Todd, 2003). To see how these cognitive limits impact on the kinds of decision mechanisms we use, we must consider the source of our bounded rationality (Todd, 2001). The usual assumption is that the constraints that bound our rationality are internal ones, such as limited memory and computational power. But this view leaves out most of the picture—namely, the external world and the constraints that *it* imposes on decision makers.

There are two particularly important classes of constraints that stem from the nature of the world. First, because the external world is uncertain—we never face exactly the same situation twice—our mental mechanisms must be robust, that is, they must generalize well from old instances to new ones. This robustness can be achieved by being simple, as in a mechanism that uses few variable parameters. As a consequence, external uncertainty can impose a bound of simplicity on our mental mechanisms. Second, because the world is competitive, our decision mechanisms must generally be fast. The more time we spend on a given decision, the less time we have available for other activities, and the less likely we are to outcompete our rivals. To be fast, we must minimize the information or alternatives we search for in making our decisions. That is, the external world also constrains us to be frugal in what we search for.

However, the external world does not just impose the bounds of simplicity, speed, and frugality on us—it also provides means for staying within these bounds. A decision mechanism can stay simple and robust by relying on some of its work being done by the external world—that is, by counting on the presence of certain useful patterns of information in the environment. Some observable cues are useful indicators of particular aspects of the world, such as the color red usually indicating ripe fruit. Our minds are built to seek and exploit such useful cues and thereby reduce the need for gathering and processing extra information. What the research in the heuristics-and-biases program demonstrated is that such reliance on particular expected information patterns can lead us astray if we are presented with environments that violate our expectations. When we operate in the natural environments we are most likely to encounter outside of psychology labs,

our decision heuristics are typically well matched to the situations and information structures we encounter.

Emphasizing the role of the environment for bounding, constraining, and enabling human cognition leads to the conception of *ecological* rationality (Todd, Fiddick, & Krauss, 2000). The goal in studying ecological rationality is to explore how simple mental mechanisms can yield good decisions by exploiting the structure inherent in the particular decision environments where they are used. Because different environment structures are best fit to different information-processing strategies, the ecological rationality perspective implies that the mind draws upon an *adaptive toolbox* of specific simple heuristics designed to solve different inference and choice tasks (Gigerenzer et al., 1999; Todd & Gigerenzer, 2000). There is no claim that the optimal or best decision mechanism will be chosen for any given task, but rather that an appropriate mechanism will be selected that will tend to make good choices within the time, information, and computation constraints impinging on the decision maker.

So what could be the contents of this dis-integrated image of successful decision making? Two main types of simple heuristics in the adaptive toolbox have been explored so far: those that make decisions among currently available options or alternatives by limiting the amount of information they seek about the alternatives and those that search for options themselves in a fast and frugal way. Both types rely on even simpler building blocks that guide the search for information or options, stop that search in a quick and easily computed manner, and then decide on the basis of the search's results. Next, we will consider three examples of the first sort of information-searching decision heuristics along with the types of information structures to which they are matched, before covering heuristics for sequential search among alternatives.

Pulling Apart the Adaptive Toolbox

The Recognition Heuristic

We start pulling out what's in the adaptive toolbox by considering mechanisms for handling one of the simplest decisions that can be made: selecting one option from two possibilities, according to some criterion on which the two can be compared. How this decision is made depends on the available information. If the environment constrains the decision maker so that the only information at hand is whether she recognizes one of the alternatives, and if it is structured so that recognition is positively correlated with the criterion, then she can do little better than rely on her own partial ignorance, choosing recognized options over unrecognized ones. This kind of "ignorance-based reasoning" is embodied in the *recognition heuristic* (Goldstein & Gigerenzer, 2002), which for two-alternative choice tasks can be stated as follows:

> IF one of two objects is recognized and the other is not,

> THEN infer that the recognized object has the higher value with respect to the criterion.

This minimal strategy may not sound like much for a decision maker to go on, but often information is implicit in the failure to recognize something, and this failure can be exploited by the heuristic. Goldstein and Gigerenzer (1999, 2002) conducted studies to find out whether people actually use the recognition heuristic. For instance, they presented U.S. students with pairs of U.S. cities and with pairs of German cities. The task was to infer which city in each pair had the most inhabitants. The students performed about equally well on city-pairs from both countries. This result is counterintuitive because the students had accumulated a lifetime of facts about U.S. cities that could be useful for inferring population, but they knew little or nothing about the German cities beyond merely recognizing about half of them—so how could their inferences about both sets of cities be equally accurate? According to Goldstein and Gigerenzer (2002), the students' lack of knowledge of German geography is just what allowed them to employ the recognition heuristic to infer that the German cities that they recognized were larger than those they did not. The students could not use this heuristic when comparing U.S. cities, because they recognized all of them and thus had to rely on other, often fallible, methods for making their decisions. In short, the recognition heuristic worked in this situation, and works in other environments as well, because our lack of recognition knowledge is often not random, but systematic and useable. (How we determine if a particular heuristic will work in a given environment is discussed later in this chapter.)

Thus, following the recognition heuristic will yield correct responses more often than would random choice only in environments with a particular type of structure:

namely, those decision environments in which exposure to different possibilities is positively correlated with their ranking along the decision criterion being used. Such useable correlations are likely to be present in environments where important objects are communicated about and unimportant ones are more often ignored, such as when discussing cities—big ones tend to dominate discussions because more interesting things are found and occur there. In such environments, a lack of knowledge about some objects allows individuals to make accurate decisions about the relative importance or size of many of the objects. In fact, adding more knowledge for the recognition heuristic to use, by increasing the proportion of recognized objects in an environment, can even decrease decision accuracy. This *less-is-more effect*, in which an intermediate amount of (recognition) knowledge about a set of objects can yield the highest proportion of correct answers, is straightforward from a mathematical perspective (the chance of picking a pair of objects in which one is recognized and one is unrecognized, so that the recognition heuristic can be applied, is minimized when all objects are recognized or all are unrecognized, and must rise for intermediate rates of recognition), but surprising from a cognitive one. Knowing more is not usually thought to *decrease* decision-making performance, but when using simple heuristics that rely on little knowledge, this is exactly what is theoretically predicted and found experimentally, as well (Goldstein & Gigerenzer, 1999, 2002).

There is growing evidence of the use and usefulness of the recognition heuristic in a variety of domains, from predicting sports matches (Andersson, Ekman, & Edman, 2003; Pachur & Biele, in press) to navigating the complex dynamic environment of the stock market. When deciding which companies to invest in from among those trading in a particular exchange, the recognition heuristic would lead us to choose just those that we have heard of before. Such a choice can be profitable assuming that more-often-recognized companies will typically have better-performing stocks. This assumption has been tested (Borges, Goldstein, Ortmann, & Gigerenzer, 1999) by asking several sets of people what companies they recognized and forming investment portfolios based on the most familiar firms. Nearly 500 people in the United States and Germany were asked which of 500 American and 298 German publicly owned companies they recognized. To form portfolios based on very highly recognized companies, we used the American participants' responses to select their top 10 most-recognized German companies, and the German responses to choose the top 10 most-recognized American firms. In this trial performed during 1996–1997, the simple ignorance-driven recognition heuristic beat highly trained fund managers using all the information available to them, as well as randomly chosen portfolios (which fund managers themselves do not always outperform). When a related study was performed during the bear market of 2000 (Boyd, 2001), the recognition heuristic did not perform as well as other strategies, showing the importance of the match between a particular decision heuristic and the structure of the environment in which it is used. In this case, the talked-about and hence recognized companies may have been those that were failing spectacularly, rather than those that were on the rise as in the earlier trial.

One-Reason Decision Mechanisms

When a decision maker encounters an environment in which she recognizes all of the objects or options, she cannot use the recognition heuristic to make comparisons between those objects. Instead, she is constrained to use some decision mechanism that relies on other pieces of information or cues. There are many possible mechanisms that could be employed, and again the particular structure of the environment will enable some heuristics to work well while others founder. The fastest and simplest heuristics would rely on just a single cue to make a decision but can this possibly work well in any environments? The answer turns out to be yes, so long as the cue to use is itself chosen properly. Imagine that we again have two objects to compare on some criterion and several cues that could be used to assess each object on the criterion. A one-reason heuristic that makes decisions on the basis of a single cue could then work as follows: (1) Select a cue dimension and look for the corresponding cue values of each option; (2) compare the two options on their values for that cue dimension; (3) if they differ, then (a) stop and (b) choose the option with the cue value indicating a greater value on the choice criterion; (4) if the options do not differ, then return to the beginning of this loop (Step 1) to look for another cue dimension. Such a heuristic will often have to look up more than one cue before making a decision, but the simple stopping rule (in Step 3) ensures that as few cues as possible will be sought, minimizing the time taken to search for information. Furthermore, ultimately

only a single cue will be used to determine the choice, minimizing the amount of computation that must be done.

This four-step loop incorporates three important building blocks of simple heuristics (Todd & Gigerenzer, 2000): a search rule that indicates in what order we should look for information (Step 1, selecting each successive cue); a stopping rule that says when we should stop looking for more cues (Step 3a, stopping when a cue is found with different values for the two options); and a decision rule that directs how the information found determines the choice to be made (Step 3b, deciding in favor of the option to which the single discriminating cue *points*). Two intuitive search rules can be incorporated in a pair of simple decision heuristics that have been tested in a variety of task environments (Gigerenzer & Goldstein, 1996, 1999): The *Take The Best* heuristic searches for cues in the order of their ecological validity, that is, their correlation with the decision criterion, whereas the *Minimalist* heuristic selects cues in a random order. Again, both stop their information search as soon as a cue is found that allows a decision to be made between the two options.

Despite (or often because of) their simplicity and disregard for most of the available information, these two fast and frugal heuristics can make very accurate choices in appropriate environments. A set of 20 environments was collected to test the performance of these heuristics, varying in number of objects and number of available cues and ranging in content from the German cities data set mentioned earlier to fish fertility to high-school dropout rates (Czerlinski, Gigerenzer, & Goldstein, 1999). The decision accuracies of Take The Best and Minimalist were compared against those of two more traditional decision mechanisms that use all available information and combine it in more or less sophisticated ways: multiple regression, which weights and sums all cues in an optimal linear fashion, and Dawes's rule, which tallies the positive and negative cues and subtracts the latter from the former. All of these mechanisms were first trained on some part of a given data set to set their parameters (i.e., to find the cue directions, whether a cue's presence indicates a greater or smaller criterion value for Minimalist and Dawes's rule, cue directions, and cue validity order for Take The Best and regression weights for multiple regression) and then were tested on some part of the same dataset. The two fast and frugal heuristics always came close to, and often exceeded, the performance of the traditional algorithms when all were tested on just the data they were trained on—the overall average performance across all 20 data sets is shown in Table 11.1 (under "Fitting"). This surprising performance on the part of Take The Best and Minimalist was achieved even though they only looked through a third of the cues on average (and only decided using one of them), while multiple regression and Dawes's rule used them all (see Table 11.1, "Frugality"). The advantages of simplicity grew in the more important test of generalization performance, where the decision mechanisms were tested on a portion of each dataset that they had not seen during training. Here, Take The Best outperformed the other algorithms by at least two percentage points (see Table 11.1, "Generalization").

What kinds of environments then are particularly well matched to these simple one-reason decision heuristics? Take The Best works well in environments where the most highly valid cue is considerably better than the cue with the second highest validity, which is also considerably better than the third cue, and so on (and, in fact, in such *noncompensatory* environments, Take The Best cannot be beaten by multiple linear regression, the usual gold standard for multiple-cue decision making; see Martignon & Hoffrage, 1999). Minimalist works well in environments where the cues to be selected from at random all have high validity. These and other lexicographic (one-reason) heuristics are also particularly suited to environments in which there is severe time pressure or other costs that make it expensive to search for a lot of information (Payne et al., 1993; Rieskamp & Hoffrage, 1999). In contrast, they will not

have much advantage in environments where information is cheap and cues can be seen simultaneously or where multiple cues must be combined in nonlinear ways (such as in the exclusive or logical relationship).

Related simple decision strategies were extensively tested in environments of risky choices by Payne et al. (1993). These comparisons included two one-reason decision heuristics, LEX (like Take The Best, with cue order determined by some measure of importance) and LEXSEMI (the same as LEX except for the requirement that cue values must differ by more than a specified amount before one alternative can be chosen over another—LEX merely requires any inequality in cue values). The heuristics were tested in a variety of environments consisting of risky choices between gambles with varying probabilities over a set of payoffs—there were two, five, or eight alternatives to choose between, each with two, five, or eight possible payoffs. Environments varied further in how the payoff probabilities were generated, and in whether time pressure prevented heuristics from completing their calculations. Different heuristics performed best (in terms of accuracy, which meant how well they reproduced the choices made by an expected utility maximizing weighted additive model) in different environments, but overall LEX and LEXSEMI performed very well in comparison with the weighted additive model, usually having the highest accuracy relative to their required effort (number of operations needed to reach a decision). When time pressure was severe, LEX performed particularly well, usually winning the accuracy competition because it could make a choice with fewer operations—as a consequence of looking for fewer cues—than the other heuristics. These results on the importance of the match between environment and decision strategy supported Payne et al.'s contention that decision makers will choose among a set of alternative simple strategies, depending on the particular task environment they face, again highlighting the role of the environment in controlling cognition.

TABLE 11.1   Performance of Different Decision Strategies Across 20 Data Sets

| | | Accuracy (% correct) | |
Strategy	Frugality	Fitting	Generalization
Minimalist	2.2	69	65
Take The Best	2.4	75	71
Dawes's rule	7.7	73	69
Multiple regression	7.7	77	68

Note. The performance of two fast and frugal heuristics (Minimalist, Take The Best) and two linear strategies (Dawes's rule, multiple regression) is shown for a range of data sets. The mean number of predictors available in the 20 data sets was 7.7. "Frugality" indicates the mean number of cues actually used by each strategy. "Fitting" indicates the percentage of correct answers achieved by the strategy when fitting data (test set = training set). "Generalization" indicates the percentage of correct answers achieved by the strategy when generalizing to new data (cross validation, i.e., test set ≠ training set).

Simple Categorizing Mechanisms That Use More Than One Cue

The efficacy of simple heuristics in structured environments is not restricted to tasks where decisions must be made between two objects. More generally, fast and frugal heuristics can also be found that use as few cues as possible to categorize objects. *Categorization by elimination* (Berretty, Todd, & Blythe, 1997; Berretty,

Todd, & Martignon, 1999), similar to Tversky's (1972) *elimination by aspects* model of preference-based choices on which it is based, uses one cue after another in a particular order to narrow down the set of remaining possible categories until only a single one remains. When cues are ordered in terms of their usefulness in predicting the environment, accurate categorization can be achieved using only the first few of the available cues. Payne et al. included a version of elimination by aspects in the comparisons described earlier and found it to be "the most robust procedure as task conditions grow more difficult" (1993, p. 140), maintaining reasonable accuracy even in large problems with severe time constraints. Even more accurate performance can arise when categorization is based on only a single cue through the use of a fine-grained cue-value-to-category map (Holte, 1993), though with a trade-off with memory usage for storing this map.

Estimation can also be performed accurately by a simple algorithm that exploits environments with a particular structure. The *QuickEst* heuristic (Hertwig, Hoffrage, & Martignon, 1999) is designed to estimate the values of objects along some criterion while using as little information as possible. To estimate the criterion value of a particular object, the heuristic looks through the available cues or features in a criterion-determined order, until it comes to the first one that the object does not possess. At this point, QuickEst stops searching for any further information and produces an estimate based on criterion values associated with the absence of the last cue. QuickEst proves to be fast and frugal, as well as accurate, in environments characterized by a distribution of criterion values in which small values are common and big values are rare (a so-called J-shaped distribution). Such distributions characterize a variety of naturally occurring phenomena, including many formed by accretionary growth. This growth pattern applies to cities and indeed big cities are much less common than small ones. As a consequence, when applied to the data set of German cities, for example, QuickEst is able to estimate rapidly and accurately the small sizes that most of them have.

Sequential Search Heuristics

All of the choice heuristics discussed so far are applicable when choosing a single option from two or more alternatives, assuming that all alternatives are presently available. But different heuristics are needed when alternatives (as opposed to cue values) appear sequentially

over an extended period or spatial region. In this type of choice environment, the stopping rule must specify which object stops search, just as in the previously described heuristics the stopping rule specified which cue stops search. An instance of this type of problem is found in the context of individuals who are searching for a mate, a house, or a job from a stream of potential possibilities that are seen one at a time—when should the searcher stop and stay with the current option?

To address such sequential search problems, agents can *satisfice* (Simon, 1955, 1990). Satisficing works by setting an aspiration level and searching until a candidate is found that exceeds that aspiration. Satisficing eliminates the need to compare a large number of possible outcomes with one another, thus saving time and the need to acquire large amounts of information. But how is the aspiration level to be set? One way is to examine a certain number of alternatives and use the best criterion value seen in that sample as the aspiration level for further search. Consider the problem of finding the single best alternative from a sequence of fixed length drawn from an unknown distribution—an extreme form of sequential search. In this scenario, using an initial sample of 37% of all available alternatives for setting the aspiration level provides the highest likelihood of picking the best (the optimal solution to the so-called *secretary*, or *dowry problem*; see Ferguson, 1989). However, much less search (e.g., setting an aspiration level using 10% of the available alternatives) is required for attaining other more realistic goals such as maximizing the mean criterion value found across multiple searches (Dudey & Todd, 2002; Todd & Miller, 1999). Other search rules have also been explored, such as stopping search after encountering a long gap between attractive candidates (Seale & Rapoport, 1997). In a mutual search setting, for instance, where both males and females are searching for a suitable mate or employers and employees are searching for a suitable job match, heuristics that learn an aspiration level based on the rejections and offers one receives can lead to successful matching of the two populations of searchers, again with relatively little information (Todd, Billari, & Simão, 2005; Todd & Miller, 1999).

Putting the Toolbox Together Again

Of course, all of the cognitive tools in the adaptive toolbox are ultimately implemented within a single brain using common sorts of underlying neural components, so at a sufficiently abstract level, the toolbox appears integrated again. In fact, as Newell (1973) proposed, it is useful to study the different decision heuristics within a common unified modeling framework, so that their underlying similarities and their connections to other components of behavior such as memory and perception can be made clear and explored further. Just this step has been taken by Schooler and Hertwig (2005) in their work on modeling the recognition heuristic with the framework of ACT-R (Anderson & Lebiere, 1998). Their results show how implementing a simple heuristic within a broader psychological system elucidates the influence of other cognitive components (here, memory) on decision making and points toward other heuristics that people may be using, making predictions which can then be tested experimentally.

Modeling the Recognition Heuristic Within ACT-R

According to Goldstein and Gigerenzer (2002), the recognition heuristic works because of the chain of correlations linking the criterion (e.g., city population), via environmental frequencies (e.g., how often a city is mentioned), to recognition. ACT-R's activation tracks just such environmental regularities, so that activation differences reflect, in part, these frequency differences. Thus, it appears that inferences—such as deciding which of two cities is larger—could be based directly on the activation of associated chunks (e.g., city representations). However, this is prohibited in the ACT-R modeling framework for reasons of psychological plausibility: subsymbolic quantities, such as activation, are held not to be directly accessible, just as people presumably cannot make decisions on the basis of differences in the long-term potentiation of neurons in their hippocampus. Instead, though, the system could still capitalize on activation differences associated with various objects by gauging how it responds to them. The simplest measure of the system's response is whether a chunk associated to a specific object can be retrieved at all, and this is what Schooler and Hertwig used to implement the recognition heuristic in ACT-R.

To create their model, Schooler and Hertwig first determined the activations of the chunks associated with various German cities. Following Goldstein and Gigerenzer's (2002) original assumption that the frequency with which a city is mentioned in newspapers mirrors its overall environmental frequency, they constructed environments consisting of German cities such that the probability of encountering a city name on any given simulated day was proportional to the overall frequency with which the city was mentioned in the *Chicago Tribune*. The model learned about these simulated environments by strengthening memory chunks associated with each city according to ACT-R's activation equation. In ACT-R, the activation of a chunk increases with each encounter of the item and decays as a function of time.

Second, the model's recognition rates for the German cities were determined. Following Anderson, Bothell, Lebiere, and Matessa (1998), recognizing a city was considered to be equivalent to retrieving the chunk associated with it. The model's recognition rate for a particular city was obtained by fitting the ACT-R equation that yields the probability that a chunk will be retrieved (given its activation learned in Step 1) to the empirical recognition rates that Goldstein and Gigerenzer (2002) observed. These empirical recognition rates were the proportion of University of Chicago participants who recognized the city.

Third, the model was tested on pairs of German cities. To do this, the model's recognition rates were used to determine the probability that it would successfully retrieve a memory chunk associated with a city when it was presented with the city name as a retrieval cue. The successful retrieval of the chunk was taken to be equivalent to recognizing the associated city. Finally, the production rules for the recognition heuristic dictated that whenever one city was recognized and the other was not, the recognized one was selected as being larger, and in all other cases (both cities recognized or unrecognized) a guess was made. These decisions closely matched the observed human responses.

This implementation showed that the recognition heuristic could easily be modeled within the broader ACT-R framework with the appropriate assumptions about how recognition could be determined in the system. But once this model was in place, Schooler and Hertwig (2005) proceeded to ask a much more interesting question: Can forgetting help memory-based inferences, such as those made by the recognition heuristic, to be more accurate? The notion that forgetting serves an adaptive function has repeatedly been put forth in the history of the analysis of human memory (part of the broader idea that cognitive limits may carry benefits; see Hertwig & Todd, 2003; Todd, Hertwig, & Hoffrage, 2005). Bjork and Bjork (1988), for instance, have argued that forgetting prevents obsolete information from interfering with the recall of more current information. Altmann and Gray (2002; see also Altmann, chapter 26, this volume) make a similar point for the short-term goals that govern our behavior. From this perspective, forgetting prevents the retrieval of information that is likely obsolete.

Schooler and Hertwig were interested in whether forgetting could enhance decision making by strengthening the usefulness of recognition. To find out, they varied forgetting rates in terms of how quickly chunk activation decays in memory (i.e., ACT-R's parameter d) and looked at how this affects the accuracy of the recognition heuristic's inferences. The results are plotted in Figure 11.1, showing that the performance of the recognition heuristic peaks at intermediate decay rates. In other words, the recognition heuristic does best when the individual forgets some of what she knows—with too little forgetting, performance actually declines (as it does with too much forgetting as well, though this is what one would normally expect). This happens because intermediate levels of forgetting maintain a distribution of recognition rates that are highly correlated with the criterion, and as stated earlier, it is just these correlations on which the recognition heuristic relies.

Using Continuous Recognition Values: The Fluency Heuristic

The recognition heuristic (and accordingly its ACT-R implementation) relies on a binary representation of recognition: an object is either simply recognized (and retrieved by ACT-R) or it is unrecognized (and not retrieved). But this heuristic essentially throws away information when two objects are both recognized but one is recognized more strongly than the other—a difference that could be used by some other mechanism to decide between the two objects, but which the recognition heuristic ignores. Considering this situation, Schooler and Hertwig (2005) noted that recognition could also be assessed within ACT-R in a continuous fashion in terms of how quickly an object's chunk can be retrieved. This information can then be used to make inferences with a related simple mechanism, the *fluency heuristic*. Such a heuristic for using the fluency of reprocessing as a cue in inferential judgment has been suggested earlier (e.g., Jacoby & Dallas, 1981), but Schooler and Hertwig define it more precisely for the same context as the recognition heuristic (i.e., selecting one of two alternatives based on some criterion on which the two can be compared). Following this version of

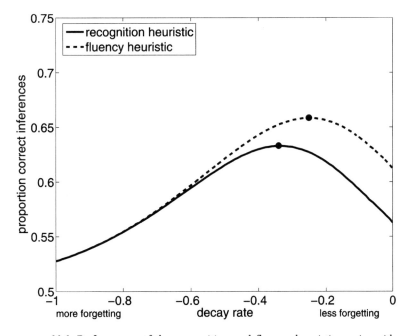

FIGURE 11.1 Performance of the recognition and fluency heuristics varies with decay rate. (Reprinted with permission from Schooler & Hertwig, 2005.)

the fluency heuristic, if one of two objects is more fluently reprocessed, then infer that this object has the higher value with respect to the criterion.

For such a heuristic to be psychologically plausible, individual decision makers must be sensitive to differences in recognition times, able, for instance, to tell the difference between recognizing "Berlin" instantaneously and taking a moment to recognize "Stuttgart." Schooler and Hertwig (2005) then proposed that these differences in recognition time partly reflect retrieval time differences, which, in turn, reflect the base-level activations of the corresponding memory chunks, which correlate with environmental frequency, and finally with city size. Further, rather than assuming that the system can discriminate between minute differences in any two retrieval times, they allowed for limits on the system's ability to do this: If the retrieval times of the two alternatives are within a just noticeable difference of 100 ms, then the system cannot distinguish its fluency for the alternatives and must guess between them.

The performance of the fluency heuristic turns out to be influenced by forgetting in much the same way as the recognition heuristic, as shown by the upper line in Figure 11.1. In the case of the fluency heuristic, intermediate amounts of forgetting increase the chances that differences in the retrieval times of two chunks will be detected. The explanation for this is illustrated in Figure 11.2, which shows the exponential function that relates a chunk's activation to its retrieval time. Forgetting lowers the range of activations to levels that correspond to retrieval times that can be more easily discriminated. In other words, a given difference in activation at a lower range results in a larger, more easily detected difference in retrieval time than the same difference at a higher range.

Both the recognition and fluency heuristics can be understood as means to indirectly tap the environmental frequency information locked in the activations of chunks in ACT-R. These heuristics will be effective to the extent that the chain of correlations—linking the criterion values, environmental frequencies, activations and responses—is strong. By modifying the rate of memory decay within ACT-R, Schooler and Hertwig (2005) demonstrated the surprising finding that forgetting actually serves to improve the performance of these heuristics by strengthening the chain of correlations on which they rely.

Keeping the Toolbox Under Control

What we have ended up with, then, is a proposal for how the architects of skyscraper cognitive models and of cottage decision heuristics can work together. Building up and empirically validating a collection of individual heuristics can show how our minds go

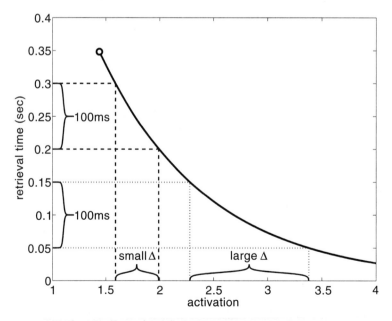

FIGURE 11.2 A chunk's activation determines its retrieval time. (Reprinted with permission from Schooler & Hertwig, 2005.)

about using specific pieces of information from appropriately structured environments to solve particular problems. Then implementing these heuristics within a broader cognitive modeling framework can show how they all fit together, and fit with other components of the mind. From a disintegrated beginning, an integrated view of the adaptive toolbox can emerge.

More specifically, we have shown the benefits of this combined approach through a particular example: By implementing one of the tools from the adaptive toolbox, the recognition heuristic, within the broader cognitive modeling framework of ACT-R, a second heuristic, the fluency heuristic, was suggested for further study. In addition, the connections from these two heuristics to other cognitive systems, particularly memory, were illuminated in a way that would not have happened without such modeling.

What we still have not specified, though, is how the multiple heuristic components, within either architectural perspective, are controlled. How are heuristics selected from the adaptive toolbox in the first place? If the wrong heuristic is chosen for use, then all of the arguments earlier about their power and robustness will fall by the wayside. Somehow, the mind must solve this metaproblem as well (Todd, Gigerenzer, & the ABC Research Group, 2000).

But the mind need not solve this metaproblem optimally, making extensive cost–benefit calculations to determine which tool will best solve a given problem. Just as the heuristic tools in the toolbox provide a satisficing, good-enough decision with little computation and information, the heuristic-selection method should also arrive at a choice of which tool to use in a fast and frugal manner. Furthermore, the mind need not solve this metaproblem of tool selection alone. The environmentally oriented perspective of ecological rationality points out that the world can also help with the solution. This can happen in a number of ways.

First, heuristic selection may not always be a problem at all, for example, when the use of a particular heuristic has been hardwired by evolution whenever a particular environment structure is encountered, allowing little or no choice between strategies. (While lower animals may have many such hardwired heuristics, for humans this may be restricted to perceptual judgments, e.g., depth perception.) When there *is* more than one available heuristic, the set of possibilities may still be small. One reason is that the heuristics in the adaptive toolbox are designed for specific tasks rather than general-purpose strategies; like screwdrivers

and wrenches, they only fit certain of the tasks set by the environment. This specificity goes a long way to reduce the selection problem: For instance, when a task calls for estimating a quantitative variable, QuickEst is a candidate, but Take The Best, meant for a different type of task, will not be. A second factor that reduces the set of possible heuristics from which to choose is the availability of particular knowledge for the decision maker. For instance, Minimalist and Take The Best are both designed for the same type of choice task, but if a person cannot assess the environment to determine a rank ordering of cues based on their validity as necessary for Take The Best, then that heuristic will not be selected and the random ordering of Minimalist may be used instead.

Still, even after these task- and knowledge-specific reductions of the choice set, there may remain a number of heuristics that could be used in a given situation, varying in their performance according to how well their information-processing structure fits the environment's information structure. How then can people choose between these candidates? Morton (2000) suggested an answer consistent with the notion of an adaptive toolbox: a metaheuristic that chooses between heuristics using the same principles as the fast and frugal heuristics themselves. Just as Take The Best looks up cues in a particular order, a metaheuristic can select heuristics (perhaps probabilistically) according to a particular order. Furthermore, just as the cue order that Take The Best follows is not arrived at through complex computations, the ordering of heuristics is not achieved via extensive calculations either, but by using simple and robust criteria, such as past success. The process of ordering itself can be modeled by a straightforward reinforcement learning mechanism such as that described by Erev and Roth (1998). Rieskamp and Otto (2006) have shown that related reinforcement learning mechanisms can capture how participants select between heuristics in ways that are adaptive for particular environments. Similarly, Nellen (2003) found that associative learning mechanisms used in ACT-R can achieve an adaptive match between heuristics and environment structure as well.

Empirical evidence for the use of simple heuristics in appropriate environmental circumstances supports the idea of such an effective tool selection mechanism (see Payne et al., 1993, for related results). For instance, Bröder (2003) showed that Take The Best predicted participants' inferences more frequently when it produced the highest payoff compared with other strategies (including situations with relatively high information acquisition costs; see Bröder, 2000). More direct evidence that people can learn to select strategies based on their fit to the structure of the current task environment comes from a study by Rieskamp and Otto (in press). In their experiment, participants repeatedly had to decide which of two companies was more creditworthy, using up to six cues. Participants received immediate feedback on whether each inference was correct. One group of participants saw an environment in which a lexicographic heuristic such as Take The Best could achieve the highest accuracy, while another group experienced an environment where a strategy that integrates all available information reached the highest accuracy. The crucial outcome was that participants were able to adapt intuitively their selection of a decision strategy to the particular environment they faced: After some learning, the strategy that best predicted participants' choices in a given environment was also the strategy that performed best in that environment.

When a new environment is encountered where people have not had the benefit of prior learning, heuristics that worked well in similarly structured environments can be called upon. But to do this, the mind needs to have a way of judging the similarity of environments and knowing what aspects of their structure matter when choosing a tool to employ. We still do not know how this is accomplished, nor what the relevant dimensions of environment structure are—are they the statistical measures of information usefulness such as cue validity and discrimination rate? Other aspects such as information cost and time pressure? More domain-specific measures such as the spatial dispersion of physical resources or the connection patterns of social networks? We need to develop a theory of the environment structures that matter behaviorally before we can fully answer just how the utilization of different mechanisms from the adaptive toolbox is controlled.

In the meantime, until we have such an environment theory providing a global view of the lay of the land, we should continue to build both models of small heuristic structures designed to work in sync with the local environment's features and grander cognitive architectures that can reintegrate the smaller components in illuminating ways.

References

Altmann, E. M., & Gray, W. D. (2002). Forgetting to remember: The functional relationship of decay and interference. *Psychological Science, 13*(1), 27–33.

Anderson, J. R., Bothell, D., Lebiere, C., & Matessa, M. (1998). An integrated theory of list memory. *Journal of Memory and Language, 38,* 341–380.

———, & Lebiere, C. (1998). *The atomic components of thought.* Mahwah, NJ: Erlbaum.

Andersson, P., Ekman, M., & Edman, J. (2003). *Forecasting the fast and frugal way: A study of performance and information-processing strategies of experts and non-experts when predicting the World Cup 2002 in soccer.* SSE/EFI Working Paper Series in Business Administration No. 2003:9. Stockholm School of Economics, Stockholm.

Berretty, P. M., Todd, P. M., & Blythe, P. W. (1997). Categorization by elimination: A fast and frugal approach to categorization. In M. G. Shafto & P. Langley (Eds.), *Proceedings of the Nineteenth Annual Conference of the Cognitive Science Society* (pp. 43–48). Mahwah, NJ: Erlbaum.

———, Todd, P. M., & Martignon, L. (1999). Categorization by elimination: Using few cues to choose. In G. Gigerenzer, P. M. Todd, & the ABC Research Group (Eds.), *Simple heuristics that make us smart* (pp. 235–254). New York: Oxford University Press.

Bjork, E. L., & Bjork, R. A. (1988) On the adaptive aspects of retrieval failure in autobiographical memory. In M. M. Gruneberg, P. E. Morris, & R. N. Sykes (Eds.), *Practical aspects of memory II* (pp. 283–288). London: Wiley.

Borges, B., Goldstein, D. G., Ortmann, A., & Gigerenzer, G. (1999).Can ignorance beat the stock market? Name recognition as a heuristic for investing. In G. Gigerenzer, P. M. Todd, & the ABC Research Group, *Simple heuristics that make us smart* (pp. 59–72). New York: Oxford University Press.

Boyd, M. (2001). On ignorance, intuition, and investing: A bear market test of the recognition heuristic. *Journal of Psychology and Financial Markets, 2*(3), 150–156.

Bröder, A. (2000). Assessing the empirical validity of the "take-the-best" heuristic as a model of human probabilistic inference. *Journal of Experimental Psychology: Learning, Memory, and Cognition, 26,* 1332–1346.

———. (2003). Decision making with the "adaptive toolbox": Influence of environmental structure, intelligence, and working memory load. *Journal of Experimental Psychology: Learning, Memory, and Cognition, 29,* 611–625.

Czerlinski, J., Gigerenzer, G., & Goldstein, D. G. (1999). Accuracy and frugality in a tour of environments. In G. Gigerenzer, P. M. Todd, & the ABC Research Group, *Simple heuristics that make us smart* (pp. 97–118). New York: Oxford University Press.

Dudey, T., & Todd, P. M. (2002). Making good decisions with minimal information: Simultaneous and sequential choice. *Journal of Bioeconomics*, 3, 195–215.

Erev, I., & Roth, A. (1998). Prediction how people play games: Reinforcement learning in games with unique strategy equilibrium. *American Economic Review*, 88, 848–881.

Ferguson, T. S. (1989). Who solved the secretary problem? *Statistical Science*, 4, 282–296.

Gigerenzer, G., & Goldstein, D. G. (1996). Reasoning the fast and frugal way: Models of bounded rationality. *Psychological Review*, 103, 650–669.

——, & Goldstein, D. G. (1999). Betting on one good reason: Take the best and its relatives. In G. Gigerenzer, P. M. Todd, & the ABC Research Group, *Simple heuristics that make us smart* (pp. 75–95). New York: Oxford University Press.

——, & Selten, R. (Eds.). (2001). *Bounded rationality: The adaptive toolbox.* (Dahlem Workshop Report). Cambridge, MA: MIT Press.

——, Todd, P. M., & the ABC Research Group. (1999). *Simple heuristics that make us smart.* New York: Oxford University Press.

Goldstein, D. G., & Gigerenzer, G. (1999). The recognition heuristic: How ignorance makes us smart. In G. Gigerenzer, P. M. Todd, & the ABC Research Group, *Simple heuristics that make us smart* (pp. 37–58). New York: Oxford University Press.

——, & Gigerenzer, G. (2002). Models of ecological rationality: The recognition heuristic. *Psychological Review*, 109, 75–90.

Goodie, A. S., Ortmann, A., Davis, J. N., Bullock, S., & Werner, G. M. (1999). Demons versus heuristics in artificial intelligence, behavioral ecology and economics. In G. Gigerenzer, P. M. Todd, & the ABC Research Group, *Simple heuristics that make us smart* (pp. 327–355). New York: Oxford University Press.

Hertwig, R., Hoffrage, U., & Martignon, L. (1999). Quick estimation: Letting the environment do some of the work. In G. Gigerenzer, P. M. Todd, & the ABC Research Group, *Simple heuristics that make us smart* (pp. 209–234). New York: Oxford University Press.

——, & Todd, P. M. (2003). More is not always better: The benefits of cognitive limits. In D. Hardman & L. Macchi, *Thinking: Psychological perspectives ·on reasoning, judgment and decision making* (pp. 213–231). Chichester, UK: Wiley.

Holte, R. C. (1993). Very simple classification rules perform well on most commonly used datasets. *Machine Learning*, 3(11), 63–91.

Jacoby, L. L., & Dallas, M. (1981). On the relationship between autobiographical memory and perceptual learning. *Journal of Experimental Psychology: General*, 110, 306–340.

Kahneman, D., Slovic, P., & Tversky, A. (1982). *Judgement under uncertainty: Heuristics and biases.* Cambridge: Cambridge University Press.

Martignon, L., & Hoffrage, U. (1999). Why does one-reason decision making work? A case study in ecological rationality. In G. Gigerenzer, P. M. Todd, & the ABC Research Group, *Simple heuristics that make us smart* (pp. 119–140). New York: Oxford University Press.

Morton, A. (2000). Heuristics all the way up? Commentary on Todd and Gigerenzer. *Behavioral and Brain Sciences*, 23(5), 758–759.

Nellen, S. (2003). The use of the "take-the-best" heuristic under different conditions, modeled with ACT-R. In F. Detje, D. Dörner, & H. Schaub (Eds.), *Proceedings of the Fifth International Conference on Cognitive Modeling* (pp. 171–176). Bamberg, Germany: Universitätsverlag Bamberg.

Newell, A. (1973). You can't play 20 questions with nature and win: Projective comments on the papers of this symposium. In W. G. Chase (Ed.), *Visual information processing* (pp. 283–308). New York: Academic Press.

Pachur, T., & Biele, G. (in press). Forecasting from ignorance: The use and usefulness of recognition in lay predictions of sports events. *Acta Psychologica*.

Payne, J. W., Bettman, J. R., & Johnson, E. J. (1993). *The adaptive decision maker.* New York: Cambridge University Press.

Piattelli-Palmarini, M. (1994). *Inevitable illusions: How mistakes of reason rule our minds.* New York: Wiley.

Rieskamp, J., & Hoffrage, U. (1999). When do people use simple heuristics and how can we tell? In G. Gigerenzer, P. M. Todd, & the ABC Research Group, *Simple heuristics that make us smart* (pp. 141–167). New York: Oxford University Press.

——, & Otto, P. E. (2006). SSL: A theory of how people learn to select strategies. *Journal of Experimental Psychology: General*, 135(2), 207–236.

Schooler, L. J., & Hertwig, R. (2005). How forgetting aids heuristic inference. *Psychological Review*, 112, 610–628.

Seale, D. A., & Rapoport, A. (1997). Sequential decision making with relative ranks: An experimental investigation of the "secretary problem." *Organizational Behavior and Human Decision Processes*, 69(3), 221–236.

Simon, H. A. (1955). A behavioral model of rational choice. *Quarterly Journal of Economics*, 69, 99–118.

——. (1990). Invariants of human behavior. *Annual Review of Psychology*, 41, 1–19.

Todd, P. M. (2001). Fast and frugal heuristics for environmentally bounded minds. In G. Gigerenzer & R. Selten (Eds.), *Bounded rationality: The adaptive toolbox* (Dahlem Workshop Report, pp. 51–70). Cambridge, MA: MIT Press.

———, Billari, F. C., & Simão, J. (2005). Aggregate age-at-marriage patterns from individual mate-search heuristics. *Demography*, 42(3), 559–574.

———, Fiddick, L., & Krauss, S. (2000). Ecological rationality and its contents. *Thinking and Reasoning*, 6(4), 375–384.

———, & Gigerenzer, G. (2000). Simple heuristics that make us smart. *Behavioral and Brain Sciences*, 23(5), 727–741.

———, Gigerenzer, G., & the ABC Research Group. (2000). How can we open up the adaptive toolbox? (Reply to commentaries). *Behavioral and Brain Sciences*, 23(5), 767–780.

———, Hertwig, R., & Hoffrage, U. (2005). The evolutionary psychology of cognition. In D. M. Buss (Ed.), *The handbook of evolutionary psychology* (pp. 776–802). Hoboken, NJ: Wiley.

———, & Miller, G. (1999). From pride and prejudice to persuasion: Satisficing in mate search. In G. Gigerenzer, P. M. Todd, & the ABC Research Group, *Simple heuristics that make us smart* (pp. 287–308). New York: Oxford University Press.

Tversky, A. (1972). Elimination by aspects: A theory of choice. *Psychological Review*, 79, 281–299.

———, & Kahneman, D. (1974). Judgment under uncertainty: Heuristics and biases. *Science*, 185, 1124–1131.

12

A Rational–Ecological Approach to the Exploration/Exploitation Trade-Offs

Bounded Rationality and Suboptimal Performance

Wai-Tat Fu

This chapter describes a rational–ecological approach to derive the processes underlying the balance between exploration and exploitation of actions as an organism adapts to a new environment. The approach uses a two-step procedure: First, an analysis of the general environment is conducted to identify its invariant properties; second, a set of adaptive mechanisms are proposed that exploit these invariant properties. The underlying assumption of the approach is that cognitive algorithms are adapted to the invariant properties of the general environment. When faced with a new environment, these cognitive algorithms will collect information samples to update the internal representation of the new environment. The current proposal is that suboptimal performance can be often explained by the interaction of the cognitive algorithms, information samples, and the specific properties of the new environment so that the obtained samples of the environment may provide a biased representational input to the cognitive algorithms. The current approach is applied to analyze behavior in two information-seeking tasks. A Bayesian satisficing model was derived, which combines a global Bayesian learning mechanism and a local decision rule to decide when to stop. The model correctly predicted that subjects adaptively traded off exploration against exploitation in response to different costs and utilities of information in most information environments. However, when presented with a local-minimum environment, the model correctly predicted that subjects underexplored the problem space, and as a consequence, performance was suboptimal. Suboptimal performance is often an emergent property of the dynamic interactions between cognition, information samples, and the characteristics of the environment.

How do humans or animals adapt to a new environment? After years of research, it is embarrassing how little we understand the underlying processes of adaptation: just look at how difficult it is to build a robot that learns to navigate in a new environment or to teach someone to master a second language. It is amazing how seagulls and vultures have learned to be landfill scavengers in the last century and be able to sort through human garbage to dig out edible morsels. At the time this chapter is written, an alligator is found in the city park of Los Angeles, outwitting licensed hunters who tried to trap the alligator for over 2 months. The ability to adapt to new environments goes beyond hardwired processes and relies on the ability to acquire new knowledge of the environment. An important step in the

adaptation process is to sample the effects of possible actions and world states so that the right set of actions can be chosen to attain important goals in the new environment.

The acquisition of new knowledge of the environment is often achieved through the dynamic interactions between an organism and the environment, in which actions are performed and their effects evaluated based on the outcomes of actions. In most cases, the organism has to deal with a *probabilistically textured* environment (Brunswik, 1952), in which the outcomes of actions are uncertain. The evaluation of different actions is therefore similar to the process of sampling from probability distributions of possible effects of the actions (e.g., see Fiedler & Juslin, 2006). The sampling process can therefore be considered an interface between the organism's cognitive representation of the environment and the probabilistically textured environment (see Figure 12.1).

A central problem in the adaptation process is how to balance *exploration* of new actions against *exploitation* of actions that are known to be good. The benefit of exploration is often measured as the *utility of information*—the expected improvement in performance that might arise from the information obtained from exploration. Exploring the environment allows the agent to observe the results of different actions, from which the agent can learn to estimate the utility of information by some forms of reinforcement-learning algorithms (see Fu & Anderson, 2006; Sutton & Barto, 1998; and Ballard & Sprague, chapter 20, this volume). The estimates allow "good" actions to be differentiated from "bad" actions, and that exploitation of good actions will improve performance in the future. On the one hand, the agent should keep on exploring, as exploiting the good actions too early may settle on suboptimal performance; on the other hand, the cost of exploring all possible actions may be too large to be justified. The balance between the expected cost and benefit of exploration and exploitation is therefore critical to performance in the adaptation process.

Reinforcement learning is one of the important techniques in artificial intelligence (AI) and control theory that allows an agent to adapt to complex environments. However, most reinforcement-learning techniques either require perfect knowledge of the environment or extensive exploration of the environment to reach the optimal solution. Because of these requirements, these computationally extensive techniques often fail to provide a good descriptive account of human adaptation. Instead, theories have been proposed that humans often adopt simple heuristics or *cognitive shortcuts* given the cognitive and knowledge

constraints they face (see Todd & Schooler, chapter 11, this volume, and Kirlik, chapter 14, this volume). These heuristics seem to work reasonably well, presumably because they were well adapted to the *invariants* of the environment (e.g., Anderson, 1990; Simon, 1996). The major assumption is that these invariants arise from the statistical structure of the environment that cognition has adapted to through the lengthy process of evolution. By exploiting these invariant properties, simple heuristics may perform reasonably well in most situations within the limits of knowledge and cognition.

The study of the constraints imposed by the environment to behavior is often referred to as the *ecological approach* that emphasizes the importance of the interactions between cognition and the environment and has shown considerable success in the past (e.g., chapters 11 and 14, this volume). A similar, but different, approach called the rational approach further assumes that cognition is adapted to the constraints imposed by the environment, thus allowing the construction of adaptive mechanisms that describe behavior (e.g., Anderson, 1990; Oaksford & Chater, 1998). The *rational approach* has been applied to explain a diverse set of cognitive functions such as memory (Anderson & Milson, 1989; Anderson & Schooler, 1991), categorization (Anderson, 1991), and problem solving (Anderson, 1990). The key assumption is that these cognitive functions optimize the adaptation of the behavior of the organism to the environment. In this chapter, I combine the ecological and rational approaches to perform a two-step procedure to construct a set of adaptive mechanisms that explain behavior. First, I perform an analysis to identify invariant properties of the environment; second, I construct adaptive mechanisms that exploit these invariant properties and show how they attain performance at a level comparable to that of computationally heavy AI algorithms. The major

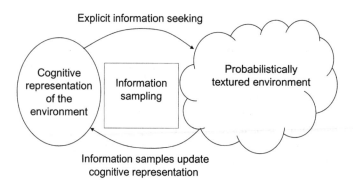

FIGURE 12.1 The information sampling process as an interface between the cognitive representation of the environment and the external environment.

Explicit information seeking

Cognitive representation of the environment

Information sampling

Probabilistically textured environment

Information samples update cognitive representation

advantage of this rational–ecological approach is that, instead of constructing mechanisms based on complex mathematical tricks, one is able to provide answers to why these mechanisms exist in the first place, and how the mechanisms may interact with different environments.

To explain how cognition adapts to new environments, the rational–ecological approach assumes that, if cognition is well adapted to the invariant properties of the general environment, cognition should have a high tendency to use the same set of mechanisms that work well in the general environment when adapting to a new environment, assuming (implicitly) that the new environment is likely to have the same invariant properties. The implication is that, when the new environment has specific properties that are different from those in the general environment, the mechanisms that work well in the general environment may lead to suboptimal performance. Traditionally, the information samples collected from the environment are often considered unbiased and suboptimal performance or judgment biases are often explained by cognitive heuristics that fail to process the information samples according to some normative standards. In the current proposal, suboptimal performance can be explained by dynamic interactions between the cognitive processes that collect information samples and the cognitive representation of the environment that is updated by the information samples obtained. Indeed, as I will show later in two different tasks, suboptimal performance often emerges as a natural consequence of this kind of dynamic interaction among cognition, information samples, and the characteristics of the environment.

In this chapter, I will present a model of how humans adapt to complex environments based on the rational–ecological approach. In the next section, I will first cast the exploration/exploitation trade-off as a general sequential decision problem. I will then focus on the special case where alternatives are evaluated sequentially and each evaluation incurs a cost. I will then present a Bayesian satisficing model (BSM) that decides when exploration should stop. I will then show that the BSM provided good match to human performance in two different tasks. The first task is a simple map-navigation task, in which subjects had to figure out the best route between two cities. In the second task, subjects were asked to search for a wide range of information using the World Wide Web (WWW). In both tasks, the BSM matched human data well and provided good explanations of human performance,

suggesting that the simple mechanisms in the BSM provide a good descriptive account of human adaptation.

The Exploration/Exploitation Trade-off

A useful concept to study human activities in unfamiliar domains is the construct of a "problem space" (Newell & Simon, 1972). A problem space consists of four major components: (1) a set of states of knowledge, (2) operators for changing one state into another, (3) constraints on applying operators, and (4) control knowledge for deciding which operator to apply next. The concept of a problem space is useful in characterizing how a problem solver searches for (exploration) different operators in the connected states of knowledge and how the accumulation of experiences in the problem space allows the problem solver to accumulate search control knowledge for deciding which operator to apply next in the future (exploitation). The concept of the problem space is similar to a Markov decision process (MDP), which has been studied extensively in the domain of machine learning and AI in the last 20 years (e.g., see Puterman, 2005, for a review). A MDP is defined as a discrete time stochastic control process characterized by a set of states, actions, and transition probability matrices that depend on the actions chosen within a given state. Extensive sets of algorithms, usually in some forms of dynamic programming and reinforcement learning, have been derived by defining a wide range of optimization problems as MDPs. Although these algorithms are efficient, they often require extensive computations that make them psychologically implausible. However, the ideas of a MDP and the associated algorithms have provided a useful set of terminologies and methods for constructing a descriptive theory of human performance. Indeed, applying ideas from machine learning to psychological theories (or vice versa) has a long history in cognitive science. By relating the concepts of operator search in a problem space to that of MDPs, another goal of the current analyses is to bridge the gap between research in cognitive psychology and machine learning.

In this section, I will borrow the terminologies from MDPs to characterize the problem of balancing exploration and exploitation and apply the rational–ecological approach to replace the complex algorithms by a BSM. I will show that the BSM uses simple, psychologically plausible mechanisms that successfully describe human behavior as they adapt to new environments.

Sequential Decision Making

Finding the optimal exploration/exploitation trade-off in a complex environment can be cast as a sequential decision-making (SDM) problem in a MDP. In general, a SDM problem is characterized by an agent choosing among various actions at different points in time to fulfill a particular goal, usually at the same time trying to maximize some form of total reinforcement (or minimize the total costs) after executing the sequence of actions. The actions are often interdependent, so that later choice of actions depends on what actions have been executed. In complex environments, the agent has to choose from a vast number of combinations of actions that eventually may lead to the goal. In situations where the agent does not have complete knowledge of the environment, finding the optimal sequence of actions requires exploring the possible sequences of actions while learning which of them are better than the others. A good balance of exploration and exploitation is necessary when the utility of exploring does not justify the cost of exploring all possible sequences, as in the case of a complex environment such as the WWW.

Many cognitive activities, such as skill learning, problem solving, reasoning, or language processing, can be cast as SDM problems, and the exploration/exploitation trade-off is a central problem to these activities.[1] The optimal solution to the SDM problem is to find the sequence of actions so that the total reinforcement obtained is maximized. This can often be done by some forms of reinforcement learning, which allows learning of the values of the actions in each problem state so that the total reinforcement received is maximized after executing the sequence of actions that lead to the goal (see, e.g., Watkins, 1989). These algorithms, however, often require perfect knowledge of the environment; even if the knowledge is available, complex computations are required to derive the optimal solution. The goal of the rational–ecological approach is to show that the requirement of perfect knowledge of the environment and complex computations can be replaced by simple mechanisms with certain assumption of the properties of the environment.

When Search Costs Matters: A Rational Analysis

Algorithms for many SDM problems use the *softmax* method to select actions in each state. Interestingly, the softmax method by itself offers a simple way to tackle the problem of balancing exploration and exploitation. Specifically, the softmax equation is based on the Gibbs, or Boltzmann, distribution:

$$P(a_k|s) = \frac{\exp(v(a_k,s)/t)}{\sum_i \exp(v(a_i,s)/t)}, \tag{1}$$

in which $P(a_k|s)$ is the probability that the action a_k will be selected in state s, $v(a_k,s)$ is the value of action a in state s, t is positive parameter called the temperature, and the summation is over all possible actions in state s. The equation has the property that when the temperature is high, actions will be (almost) equally likely to be chosen (full exploration). As the temperature decreases, actions with high values will be more likely to be chosen (a mix of exploration and exploitation), and in the limiting case, where $t \rightarrow 0$, the action with the highest value will always be chosen (pure exploitation). The balance between exploration and exploitation can therefore be controlled by the temperature parameter in the softmax equation. In fact, the softmax method has been widely used in different architectures to handle the exploration/exploitation tradeoffs, including adaptive control of thought–rational (ACT-R; see Anderson et al., 2004). Recently, it has also been shown that the use of the softmax equation in reinforcement learning is able produce a wide range of human and animal choice behavior (Fu & Anderson, 2006).

Although the softmax method can lead to reasonable exploration/exploitation trade-offs, it assumes that the values of all possible actions are immediately available without cost. The method is therefore only useful in simple or laboratory situations where the alternatives are presented at the same time to the decision maker; in that case, the search cost is negligible. In realistic situations, the evaluation process itself may often be costly. For example, in a chess game, the number of possible moves is enormous, and it is unlikely that a person will exhaust the exploration of all possible moves in every step. A more plausible model is to assume that alternatives are considered sequentially, in that case a stopping rule is required to determine when evaluation should stop (e.g., Searle & Rapoport, 1997). The problem of deciding when to stop searching can be considered a special case of the exploration/ exploitation trade-offs discussed earlier: At the point where the agent decides to stop searching, the best item encountered so far will be selected (exploitation) and the search for potential better options (exploration) will stop.

Finding the optimal stopping rule (thus the optimal exploration/exploitation trade-off) is a computationally expensive procedure in SDM problems. The goal of the following analyses is to replace these complex computations by simple mechanisms that exploit certain characteristics of the environment. The basic idea is to use a local stopping rule based on some estimates of the environment so that when an alternative is believed to be good enough no further search may be necessary. This is the essence of *bounded rationality* (Simon, 1955), a concept that assumes that the agent does not exhaust all possible options to find the optimal solution. Instead, the agent makes choices based on the mechanism of *satisficing*, that is, the goodness of an option is compared to an aspiration level and the evaluation of options will stop once an option that reaches the aspiration level is found. There are a number of ways the aspiration level can be estimated. Here, I will show how the aspiration level can be estimated by an adaptation process to an environment based on the rational analysis framework.

Optimal Exploration in a Diminishing-Return Environment

One major assumption in the current analysis is that when the agent is searching sequentially for the right actions, the potential benefits of obtaining a better action tend to diminish as the search cost increases. This kind of diminishing-return environment is commonly found in the natural world, as well as in many artificial environments. For example, in his seminal article, Stigler (1961) shows that most economic information in the market place has this diminishing-return property. Research on animal foraging has found that food patches in the wild seem to have this characteristic of diminishing returns, as the more of the patch the animal consumes, the lower the rate of return will be for the remainder of the patch because the food supply is running out (Stephens & Krebs, 1986). Recently, Pirolli and Card (1999) also found that large information structures tend to have this diminishing-return characteristic. To further illustrate the generality of this diminishing-return property, I will give a real-world example of information-seeking task below.

Consider a person looking for a plane ticket from Pittsburgh to Albany on the internet. Assume that the P value of each link is calculated by the following simple preference function:

$$P = \text{Time} + \text{Stopover} + \text{Layover}, \qquad (2)$$

in which Time, Stopover, and Layover are variables that take values from 1 (least preferable) to 5 (most preferable). For example, a flight that leaves at 11 a.m.,

makes one stopover and has a layover of 5 hours has Time = 5, Stopover = 3, and Layover = 1 (thus P = 9). By using a simple set of rules that transform each flight encounter on the Web to a P value, Figure 12.2 shows the P values of the flights in their order of encounter from a popular Web site that sells plane tickets. We can see that a few desirable flights are found in the first few encounters, but the likelihood of finding a better flight is getting smaller and smaller, as shown by the line in the figure. It can be shown that this property of diminishing-return is robust with different preference functions or Web sites.[2]

If we assume the simplistic view that the evaluation of each action incurs a constant cost,[3] and the information obtained from each evaluation (i.e., exploration of a new action) reduces the expected execution cost required to finish a task, we can calculate the relationship among the number of evaluations (n), the evaluation costs ($n \times C$), the expected execution costs $f(n)$, and the total costs $f(n) + n \times C$. As shown in Figure 12.3, the positively sloped straight line represents the increase of evaluation costs with the number of evaluations. The curve $f(n)$ represents the expected execution costs as a function of the number of actions evaluated. The function $f(n)$ has the characteristic of diminishing return, so that more evaluations will lead to smaller savings in execution costs. The U-shape curve is the total costs, which equals the sum of evaluation costs and execution costs. The U-shape curve implies that optimal performance is associated with a moderate number of evaluations. In other words, too much or too little evaluations may lead to suboptimal performance (as measured by the total costs).

The Bayesian Satisficing Model (BSM)

With the assumption of the invariant property of diminishing return in the general information environment, the next step is to propose a set of adaptive mechanisms that exploit this property. I will show that the Bayesian satisficing model, which combines a Bayesian learning mechanism and a simple, local decision rule, produces good match to how humans adaptively balance exploration and exploitation in a general environment with this diminishing-return property (Fu & Gray, 2006). The Bayesian learning mechanism in the BSM calculates the expected utility of information in terms of the expected improvement in performance resulting from the new information. The local decision rule then compares the evaluation cost with the utility of information and will stop evaluating actions

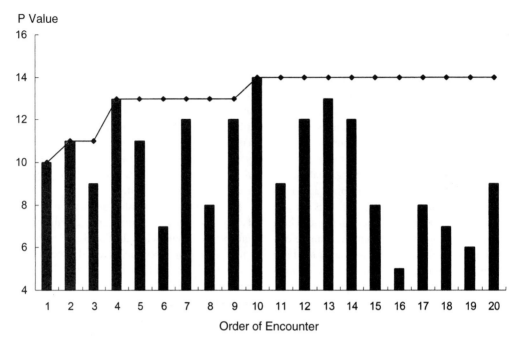

FIGURE 12.2 The P(reference) value of the links encountered on a Web page. The line represents the P value of the best link encountered so far.

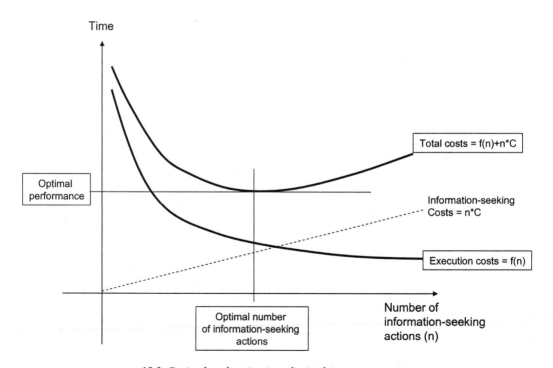

FIGURE 12.3 Optimal exploration in a diminishing-return environment.

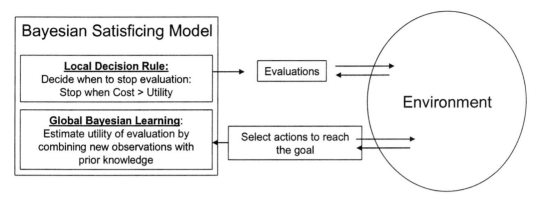

FIGURE 12.4 The structure of the Bayesian satisficing model.

when the cost exceeds the utility. The logic of the model is that if cognition is well adapted to the characteristic of diminishing returns in which a local decision rule performs well, then cognition should have a high tendency to use the same rule when adapting to a new environment, assuming that the new environment is likely to have the same diminishing-return characteristic.

The details of the BSM are illustrated in Figure 12.4, which shows the two major processes that allow the model to adapt to the optimal level of evaluations in a diminishing-return environment that maps to the variables in Figure 12.3: (1) the estimation of the function $f(n)$ and (2) the decision on when to stop evaluating further options. The first process requires the understanding of how people estimate the utility of additional evaluations based on experience. The second process requires the understanding of how the decision to stop

evaluating further options is sensitive to the cost structure of the environment. In the global learning process, the model assumes that execution costs can be described by a diminishing-return function of the number of evaluations (i.e., $f[n]$). A local decision rule is used to decide when to stop evaluating the next option (see Figure 12.5) based on the existing estimation of $f(n)$. Specifically, when the estimated utility of the next evaluation (i.e., $f[N] - f[N + 1]$) is lower than its cost, the model will stop evaluating the next option. This local decision rule decides how many evaluations are performed. The time spent to finish the task given the particular number of evaluations is then used to update the existing knowledge of $f(n)$ based on Bayes's theorem.

Fu and Gray (2006) ran a number of simulations of the BSM using a variety of diminish-return environments, and showed that the BSM made a number of

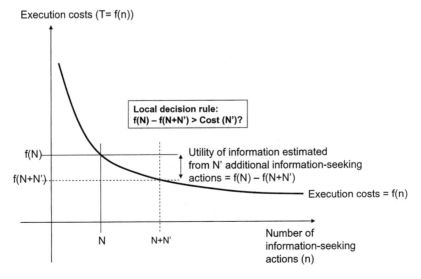

FIGURE 12.5 The local decision rule in the Bayesian satisficing model.

interesting predictions on behavior. In summary, the simulation results show that (1) with sufficient experience, people make good trade-offs between exploration and exploitation and converge to a reasonably good level of performance in a number of diminishing-return environments, (2) people respond to changes in costs faster than changes in utility of evaluation, and (3) in a local-minimum environment, high cost may lead to premature termination of exploration of the problem space, thus suboptimal performance.

Figure 12.6 illustrates the third prediction of the BSM. The flat portion of $f(n)$ (i.e., region B) represents what we refer to as a *local-minimum* environment, in which the marginal utility of exploration (i.e., the slope of $f[n]$) varies with the number of evaluations. The marginal utility is high during initial exploration, becomes flat with intermediate number of evaluations, but then becomes high again with greater number of evaluations. Using the local decision rule, exploration is likely to stop at the flat region (i.e., when the marginal utility of evaluation is lower than the cost), especially when the cost is high. We therefore predict that in a local-minimum environment the use of a local decision rule will predict poor exploration of the task space, especially when the cost is high.

Testing the BSM Against Human Data

In this section, I will summarize how the BSM matched human performance in two tasks. In the first task, subjects were given a simple map-navigation task in which they were asked to find the best route between two points on the map (Fu & Gray, 2006). Subjects were given the option to obtain information on the speeds of different routes before they started to navigate on the map. The cost and utility of information was manipulated to study how these factors influenced the decision on when subjects would stop seeking information. To directly test whether subjects were using the local decision rule, a local-minimum environment was constructed. The local-minimum environment had an uneven diminish-return characteristic, so that the use of a local decision rule would be more likely to prematurely stop seeking information, leaving the problem space underexplored and, as a result, performance would be suboptimal. Indeed, the human data confirmed the prediction, providing strong support for the use of a local decision rule. The second task was a real-world task in which subjects were asked to search for information using the WWW. We combined the BSM with the measure of information scent (Pirolli & Card,

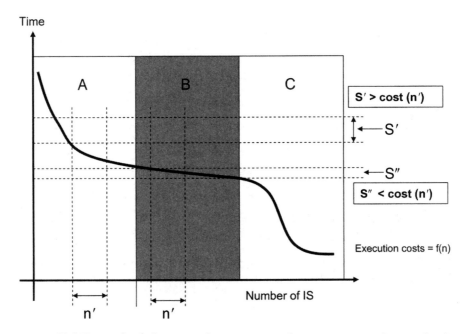

FIGURE 12.6 How a local decision rule may stop exploration prematurely in a local-minimum environment. In the figure, when the saving in execution costs is smaller than the cost of the exploration costs ($S'' < \text{cost } [n']$), exploration will stop, leaving a large portion of the task space unexplored (i.e., task space C).

1999) to predict link selections and the decision on when to leave a particular Web page in two real-world Web sites. In both task, we found that the model fit the data well, suggesting that the adaptive exploration/ exploitation trade-offs produced by the simple mechanisms of the BSM matched human performance well in different types of information environment. To preview our conclusions, the results from both tasks provided strong support for the BSM. The success of the BSM in explaining performance in the local-minimum environment also suggests that stable suboptimal performance is likely a result of the dynamic interactions between bounded rationality and specific properties of the environment.

The Map-Navigation Task

In the map-navigation task, subjects were presented with different maps on a computer screen and were asked to go from a start point to an endpoint on the map. A simple hill-climbing strategy (usually the shortest route) was always applicable and sufficient to accomplish the task (and any path can eventually lead to the goal), but the hill-climbing strategy was not guaranteed to lead to the best (i.e., fastest) path. With sufficient experience, subjects learned the speeds of different routes and turns and improved performance by a better choice of solution paths. The problem of finding the best path in a map could therefore be considered a SDM problem, in which each junction in the map was a discrete state, each of the possible routes passing through the junction defined a possible action in the state, and finding the fastest path defined a standard optimization problem.

The speed of the path chosen was experienced in real time (a red line went from one point to another on the map, at an average rate of approximately 1 cm/ s), but the speed of a path could also be checked beforehand by a simple mouse click (i.e., an information-seeking action). The number of information-seeking actions therefore served as a direct measure of how much exploration subjects were willing to do in the task, and the corresponding execution costs could be measured by the actual time spent to go from the start to the endpoint. We manipulated the exploration cost by introducing a lock-out time to the information-seeking action. Specifically, in the high-cost condition, after subjects clicked on the map to obtain the speed information of a path, they had to wait 1 s before they could see the information. The utility of information was

manipulated by varying the difference between the fastest and slowest paths in the map. For example, when the difference was large, the potential saving in execution costs (i.e., the utility) per information-seeking action would be higher (i.e., the curve $f[n]$ in Figure 12.3 or Figure 12.5 is steeper).

To match behavior of the model to human data, the model was implemented in the ACT-R architecture (Anderson & Lebiere, 1998). The decision rule is implemented by having two competing productions, one abstractly representing exploration, and the other representing exploitation.[4] In ACT-R, each production has a utility value, which determines how likely that it will fire in a given cycle by the softmax equation stated above. The utility value of each production is updated after each cycle according to a Bayesian learning mechanism (see Anderson & Lebiere, 1998, for details) as in the BSM (see Figure 12.4). In general, when the utility of the exploration production is higher than that of the exploitation production, the model is likely to continue to explore. However, when the utility of the exploration production falls below that of the exploitation production, the model will likely stop exploring. The competition between the two productions through the softmax equation therefore serves as a stochastic version of the local decision rule in the BSM.

Because of the space limitation, only a briefly summary of the major findings of the three experiments was presented here (for details, see Fu & Gray, 2006). First, in diminishing-return environments with different costs and utilities of information, subjects were able to adapt to the optimal levels of exploration. The BSM model provided good fits to the data, suggesting that the local decision rule in the BSM was sufficient to lead to optimal performance. Second, in environments where the costs or utilities of exploration were changed, subjects responded to changes in costs faster than changes in utilities of information. Finally, when the cost was high in a local-minimum environment, subjects prematurely stopped seeking information and stabilized at suboptimal performance.

The empirical and simulation results suggest that subjects used a local decision rule to decide when to stop seeking information. Perhaps the strongest evidence for the use of a local decision rule was the finding that in the local-minimum environment, high cost of exploration led to "premature" stopping of information seeking, and as a result, performance stabilized at a suboptimal level. Although the BSM was effective in finding the right level of information seeking in most situations, the nature of local processing inherently limits the exploration of the environment. Indeed, we found that the same model, when interacting with environments with different properties, exhibited very different behavior. In particular, in a local-minimum environment, the local decision rule often results in "insufficient" information seeking when high information-seeking costs discourage exploration of the environment. On the basis of this result, it is concluded that suboptimal performance may emerge as a natural consequence of the dynamic interactions between bounded rationality and the specific properties of the environment.

A Real-World Information-Seeking Task: Searching on the WWW

To further test the behavior of the BSM, a real-world task was chosen and human performance on this task was compared with that of the BSM. Similar to the map-navigation task, searching on the World Wide Web is a good example of a SDM problem: Each Web page defines a state in the problem space, and clicking on any of the links on the Web page defines a subset of all possible actions in that state (other major actions include going back to the previous pages or going to a different Web site). The activities on the WWW can therefore be analyzed as a standard MDP. Because the number of Web pages on the Internet is enormous, exhaustive search of Web pages is impossible. Before I present how to model the exploration/exploitation trade-off in this task, I need to digress to discuss a measure that captures the user's estimation of how likely a link will lead to the target information. One such measure is called *information scent*, which will be described next.

Information Scent

Pirolli and Card (1999) developed the information foraging theory (IFT) to explain information-seeking behavior in different user interfaces and WWW navigation (Fu & Pirolli, in press; Pirolli & Fu, 2003). The IFT assumes that information-seeking behavior is adaptive within the task environment and that the goal of the information-seeker is to maximize information gain per unit cost. The concept of information scent measures the mutual relevance of text snippets (such as the link text on a Web page) and the information goal. The measure of information scent is based on a Bayesian estimate of the relevance of a distal source of information

conditional on the proximal cues. Specifically, the degree to which proximal cues predict the occurrence of some unobserved target information is reflected by the strength of association between cues and the target information. For each word i involved in the user's information goal, the accumulated activation received from all associated information scents for word j is calculated by

$$IS(Link_k) = \sum_i \sum_j W_j \log\left(\frac{\Pr(i \mid j)}{\Pr(i)}\right), \qquad (3)$$

where $\Pr(i \mid j)$ is the probability (based on past experience) that word i has occurred when word j has occurred in the environment; W_j represents the amount of attention devoted to word j; and $\Pr(i)$ is the base rate probability of word i occurring in the environment. Equation 3 is also known as *pointwise mutual information* (Manning & Schuetze, 1999) or PMI.[5] The actual probabilities are often estimated by calculating the co-occurrence of word i and j and the base frequencies of word i from some large text corpora (see Pirolli & Card, 1999). The measure of information scent therefore provides a way to measure how subjects evaluate the utility of information contained in a link on a Web page.

The SNIF-ACT Model

On the basis of the IFT, Fu and Pirolli (in press) developed a computational model called SNIF-ACT (scent-based navigation and information foraging in the ACT architecture) that models user–WWW interactions. The newest version of the model, SNIF-ACT 2.0, is based on a rational analysis of the information environment. I will focus on the part where the model is facing a single Web page and has to decide when to stop evaluating links on the Web page. In fact, the basic idea of this part of the SNIF-ACT model was identical to that of the BSM, which was composed of a Bayesian learning mechanism and a local decision rule (Figure 12.4). Specifically, the model assumed that when users evaluated each link on a Web page, they incrementally updated their perceived relevance of the Web page to the target information according to a Bayesian learning process. A local decision rule then decided when to stop evaluating link: the evaluation of the next link continued until the perceived relevance was lower than the cost of evaluating for the next link. At that point, the best link encountered so far will be selected. Details of the model will be presented below.

When the model is facing a single Web page, it had the same exploration/exploitation trade-off problem: to balance between the utility of evaluating the next link and the cost of doing so. However, in contrast to the map-navigation task, the utility of information was not measured by time. Instead, the utility of information (from evaluating the next link) is measured by the likelihood that the next link will lead to the target information. Details of the analysis can be found in Fu and Pirolli (in press). The probability that the current Web page will eventually lead to the target information after the evaluation of a set of links L_n is

$$P(\text{Target}|L_n) = K \sum_{j=0}^{n} \frac{\Gamma(\alpha + J)}{\Gamma(\alpha + n)} x(O_j), \qquad (4)$$

where X is a variable that measures the closeness to the target; O_j is the observation of link j on the current Web page; K, α, and n are parameters to be estimated. The link likelihood equation is derived from the Bayes's theorem and thus is identical to the Bayesian learning mechanism in BSM (see Figure 12.4). As explained, $X(O_j)$ can be substituted by the measure of information scent (i.e., the information scent equation) of each link j. The link likelihood equation provides a way to incrementally update the probability that a given Web page will eventually lead to the target information after each link is evaluated (i.e., $f[n]$ in Figures 12.3 and 12.5).

The model is again implemented in the ACT-R architecture. To illustrate the behavior of the model, we will focus on the case where the model is facing a single Web page with multiple links. There are three possible actions, each represented by a separate production: *attend-to-link*, *click-link*, and *backup-a-page*. Similar to the BSM model in the map-navigation task, these productions compete against each other according to the softmax equation (which implements the local decision rule in the BSM; see Figure 12.3). In other words, at any time, the model will attend to the next link on the page (exploration), click on a link on a page (exploitation), or decide to leave the current page and return to the previous page. The utilities of the three productions are derived from the link likelihood equation, and they can be shown as:

$$\text{Attend - to - Link} : U(n+1) = \frac{U(n) + IS(link)}{1 + N(n)}$$

$$\text{Click - Link} : \qquad U(n+1) = \frac{U(n) + IS(bestlink)}{1 + k + N(n)} \qquad (5)$$

$$\text{Backup - a - Page} : U(n + 1) = MIS(\textit{Previous Pages}) \\ - MIS(\textit{links 1 to n}) \cdot \text{GoBackCost}.$$

In the equations above, $U(n)$ represents the utility of the production at cycle n, $IS(link)$ represents the information scent of the currently attended link, $N(n)$ represents the number of links already attended on the Web page after cycle n (one link is attended per cycle), $IS(bestlink)$ is the link with the highest information scent on the Web page, k is a scaling parameter, $MIS(page)$ is the mean information scent of the links on the Web page, and $GoBackCost$ is the cost of going back to the previous page. The values of k and GoBackCost are estimated to fit the data. The equation for backup-a-page assumes that the model is keeping a moving average of the information scent encountered in previous pages. It can be easily shown that the utility of backup-a-page will increase as the information scent of the links encountered on the current Web page declines.

Figure 12.7 shows a hypothetical situation when the model is processing a Web page in which the information scent decreases from 10 to 2 as the model attends and evaluates Links 1 to 5. The information scent of the links from 6 onward stays at 2. The mean information scent of the previous page was 10, and the noise parameter t in the softmax equation was set to 1.0. The value of k and GoBackCost were both set to 5. The initial utilities of all productions were set to 0. We can see that initially, the probability of choosing attend-to-link is high. This is based on the assumption that when a Web page is first processed, there is a bias in learning the utility of links on the page before a decision is made. However, as more links are evaluated, the utilities of the productions decreases (as the denominator gets larger as $N[n]$ increases). Since the utility of attend-to-link decreases faster than that of click-link (since $IS[Best] = 10$, but $IS[link]$ decreases from 10 to 2), the probability of choosing attend-to-link decreases but that of click-link increases. The implicit assumption of the model is that since evaluation of links takes time, the more links that are evaluated, the more likely that the best link evaluated so far will be selected (otherwise the time cost may outweigh the benefits of finding a better link). As shown in Figure 12.7, after four links on the hypothetical Web page have been evaluated, the probability of choosing click-link is larger than that of attend-to-link. At this point, if click-link is selected, the model will choose the best (in this case the first) link and the model will continue to process the next page. However, as the selection process is stochastic

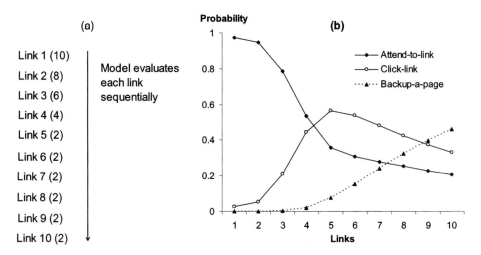

FIGURE 12.7 (a) A hypothetical Web page in which the information scent of links decreases linearly from 10 to 2 as the model evaluated links 1 to 5. The information scent of the links from 6 onward stays at 2. The number in parenthesis represents the value of information scent. (b) The probability of choosing each of the competing productions when the model processes each of the link in (a) sequentially. The mean information scent of the previous pages was 10. The noise parameter t was set to 1.0. The initial utilities of all productions were set to 0. k and GoBackCost were both set to 5.

(because of the softmax equation), attend-to-link may still be selected. If this is the case, as more links are evaluated (i.e., as $N[n]$ increases), the probability of choosing attend-to-link and click-link decreases. However, the probability of choosing backup-a-page is low initially because of the high GoBackCost. However, as the mean information scent of the links evaluated (i.e., $MIS[links\ 1\ to\ n]$) on the page decreases, the probability of choosing backup-a-page increases. This happens because the mean information scent of the current page is *perceived* to be dropping relative to the mean information scent of the previous page. In fact, after eight links are evaluated, the probability of choosing backup-a-page becomes higher than that of attend-to-link and click-link, and the probability of choosing backup-a-page keeps on increasing as more links are evaluated (as the mean information scent of the current page decreases). We can see how the competition between the productions can serve as a local decision rule that decides when to stop exploration.

The Tasks

Data from tasks performed at two Web sites in the Chi et al. (2003) data set were selected: (1) help.yahoo.com (the help system section of Yahoo!) and (2) parcweb .parc .com (an intranet of company internal information).

We will refer to these sites as *Yahoo* and *ParcWeb*, respectively, for the rest of the article. Each of these Web sites (Yahoo and ParcWeb) had been tested with a set of eight tasks, for a total of $8 \times 2 = 16$ tasks. For each site, the eight tasks were grouped into four categories of similar types. For each task, the user was given a specific information goal in the form of a question (e.g., "Find the 2002 Holiday Schedule"). The Yahoo and ParcWeb data sets come from a total of $N = 74$ adult users (30 users in the Yahoo data set and 44 users in the ParcWeb data set). Of all the user sessions collected, the data were cleaned to throw out any sessions that employed the site's search engine as well as any sessions that did not go beyond the starting home page.

In general, we found that in both sites, there were only a few (<10) "attractor" pages visited by most of the users, but there were also many pages visited by fewer than 10 users. In fact, many Web pages in both sites were visited only once. To set our priorities, we decided that it was more important to test whether the model was able to identify these attractor pages. In fact, Web pages that were visited fewer than five times among all users seemed more random than systematic, and thus were excluded from our analyses. These Web pages amounted to approximately 30% of all the Web pages visited by the users.

To test the predictions of the model on its selection of links, we first started the model on the same pages as the participants in each task. The model was then run the same number of times as the number of participants in each task and the selection of links were recorded. After the recordings, in case the model did not pick the same Web page as participants did, we forced the model to follow the same paths as participants. This process repeated until the model had made selections on all Web pages visited by the participants. The selections of links by the model were then aligned with those made by participants. The model provided good fits to the data ($R^2 = 0.90$ for Yahoo and $R^2 = 0.72$ for ParcWeb).

One unique feature of the WWW in the exploration of actions was the ability to go back to a previous state. Indeed, the decision to go back to the previous state indicated that the user believed further search along the same path might not be justified. It was therefore important that the model was able to match when users decided to go back to the previous state. In the model, when the information scent of a page dropped below the mean information scent of previous pages, the probability of going back increased. Indeed, the model's decisions to go back a page were highly correlated with human decisions to go back for both the Yahoo ($R^2 = 0.80$) and ParcWeb ($R^2 = 0.73$) sites. These results provided further support for the adaptive trade-offs between exploration and exploitation implemented by the model.

When searching for information on the WWW, the large number of Web pages makes exhaustive search impossible. When faced with a Web page with a list of links, the decision on which link to follow can be considered a balance between exploration and exploitation. I showed that the BSM matched the behavior of the users well. Since the study was not a controlled experiment, it was hard to manipulate the information environment to test directly whether suboptimal performance would result from the use of a local decision rule. However, it was promising that the BSM, combining a Bayesian learning mechanism and the use of a local decision rule, was able to match the human data well when users interact with a large information structure such as the WWW.

Summary and Conclusions

When an organism adapts to a new environment, the central problem is how to balance exploration and exploitation of actions. The idea of exploration is similar to the traditional concept of *search in a problem space* (Newell & Simon, 1972), in which the problem solver needs to know when to stop searching and choose actions based on limited search control knowledge. Recently, the idea has also been studied extensively in the area of machine learning in the form of an SDM problem, and complex algorithms have been derived for finding the optimal trade-offs between exploration and exploitation in different environments.

A rational–ecological approach to the problem of balancing exploration and exploitation was described. The approach adopts a two-step procedure: (1) identify invariant properties of the general environment and (2) construct adaptive mechanisms that exploit these properties. The underlying assumption is that cognition is well adapted to the invariant properties of the general environment; when faced with a new environment, cognition tends to apply the same set of mechanisms that work well in the general environment to perform in the new environment.

It is assumed that the general information environment has an invariant property of diminishing-return. A BSM was then derived to exploit this property. The BSM dynamically obtains information samples from the new environment to update its internal representation of the new environment according to the Bayesian learning mechanism. A local decision rule is then applied to decide when to stop exploration of actions. The model matched human data well in two very different tasks that involved different information environments, showing that the simple mechanisms in the BSM can account for the adaptive trade-offs between exploration and exploitation when adapting to a new environment, a problem that usually requires complex algorithms and computations.

One major advantage of the current approach is that one is able to provide an explanation for why certain mechanisms compute the way they do. In the BSM, the local decision is effective based on the assumption of the assumed invariant property of diminishing return. Another major advantage is that complex computations can be replaced by simple heuristics that exploit the statistical properties of the environment. Indeed, finding the optimal solution in each new environment has been a tough problem for research in the area of AI and machine learning that focuses on various kinds of optimization problems in SDM. It is promising that the single set of simple mechanisms in BSM seems to be sufficient to replace complex computational algorithms by providing good match to human performance in two diversely different task environments.

Insufficient exploration often leads to suboptimal performance, as better actions are unexplored and thus not used. The model demonstrates nicely how a simple mechanism that exploits the invariant properties of the general environment may fail to provide an unbiased representation of the new environment. In fact, elsewhere we argued that this is the major reason for why inefficient procedures persist even after years of experience with the various artificial tools in the modern world, such as the many computer applications that people use everyday (Fu & Gray, 2004). We found that many of these artificial tools have the characteristics of a local-minimum environment as shown in Figure 12.6. Since the cost of exploring new (and often more efficient) procedures is often high in these computer applications, users tend to stop exploring more efficient procedures and stabilize at suboptimal procedures even after years of experience.

Notes

1. In the machine learning literature, the SDM problem is often solved as a Markov decision problem over the set of information states S, and the agent has to choose one of the possible actions in the set A. After taking action $a \in A$ from state $s \in S$, the agent's state becomes some state s' with the probability given by the transition probability $P(s'|s,a)$. However, the agent is often not aware of the current state (because of lack of complete knowledge of the environment). Instead, the agent only knows the information state i, which is a probability distribution over possible states. We can then define $i(s)$ as the probability that the person is in state s. After each transition, the agent makes an observation o of its current state from the set of possible observations O. We can define $P(o|s',a)$ as the probability that observation o is made after action a is taken and state s' is reached. We can then calculate the next information state as:

$$i(s'|o,a) = \frac{P(o|s',a)\sum_{s \in S} P(s'|s,a)x(s)}{\sum_{s' \in S} P(o|s',a)\sum_{s \in S} P(s'|s,a)x(s)}$$

2. In fact, if one considers the value of P as a normally distributed variable, then the likelihood of finding a better alternative will naturally decrease as the sampling process continues, as one gets more to the tail of the distribution.

3. One may argue that the cost of exploration is likely to be an increasing function, which is probably true. However, the actual function does not play a crucial role in the current analyses (one still gets a U-shaped curve for the total costs in Figure 12.3). For the sake of simplicity, a linear relationship is assumed in this analysis.

4. The productions were called hill-climbing (exploitation) and information-seeking (exploration) in Fu and Gray (2006).

5. The PMI calculations can also be found at http://glsa.parc.com.

References

Anderson, J. R. (1990). *The adaptive character of thought.* Hillsdale, NJ: Erlbaum.

———. (1991). The adaptive nature of human categorization. *Psychological Review, 98,* 409–429.

———, Bothell, D., Byrne, M. D., Douglass, S., Lebiere, C., & Qin, Y. (2004). An integrated theory of mind. *Psychological Review, 11*(4), 1036–1060.

———, & Lebiere, C. (1998). *The atomic components of thought.* Mahwah, NJ: Erlbaum.

———, & Milson, R. (1989). Human memory: An adaptive perspective. *Psychological Review, 96,* 703–719.

———, & Schooler, L. J. (1991). Reflections of the environment in memory. *Psychological Science, 2,* 396–408.

Barto, A., Sutton, R., & Watkins, C. (1990). Learning and sequential decision making. In M. Gabriel & J. Moore (Eds.), *Learning and computational neuroscience: Foundations of adaptive networks* (pp. 539–602). Cambridge, MA: MIT Press.

Brunswik, E. (1952). *The conceptual framework of psychology.* Chicago: University of Chicago Press.

Chi, E. H., Rosien, A., Suppattanasiri, G., Williams, A., Royer, C., Chow, C., et al. (2003). The Bloodhound Project: Automating discovery of Web usability issues using the InfoScent simulator. *CHI 2003, ACM Conference on Human Factors in Computing Systems, CHI Letters, 5*(1), 505–512.

Fiedler, K., & Juslin, P. (2006). *Information sampling and adaptive cognition.* Cambridge: Cambridge University Press.

Fu, W. T., & Anderson, J. R. (2006). From recurrent choice to skill learning: A reinforcement-learning model. *Journal of Experimental Psychology: General, 135*(2), 184–206.

———, & Gray, W. D. (2004). Resolving the paradox of the active user: Stable suboptimal performance in interactive tasks. *Cognitive Science, 28*(6).

———, & Gray, W. D. (2006). Suboptimal tradeoffs in information-seeking. *Cognitive Psychology, 52,* 195–242.

———, & Pirolli, P. (in press). SNIF-ACT: A model of information-seeking on the World Wide Web. *Human-Computer Interaction*. Accepted for publication.

Manning, C. D., & Schuetze, H. (1999). *Foundations of statistical natural language processing*. Cambridge, MA: MIT Press.

Newell, A., & Simon, H. A. (1972). *Human problem solving*. Englewood Cliffs, NJ: Prentice-Hall.

Oaksford, M., & Chater, N. (Eds.). (1998). *Rational models of cognition*. Oxford: Oxford University Press.

Pirolli, P., & Card, S. K. (1999). Information foraging. *Psychological Review 106*(4), 643–675.

Pirolli, P. L., & Fu, W.-T. (2003). *SNIF-ACT: A model of information foraging on the World Wide Web*. Ninth International Conference on User Modeling, Johnstown, Pennsylvania.

Puterman, M. L. (2005). *Markov decision processes*. Hoboken, NJ: Wiley.

Simon, H. A. (1955). A behavioral model of rational choice. *Quarterly Journal of Economics, 69*, 99–118.

———. (1996). *The sciences of the artificial* (3rd ed.). Cambridge, MA: MIT Press.

Stephens, D. W., & Krebs, J. R. (1986). *Foraging theory*. Princeton, NJ: Princeton University Press.

Stigler, G. J. (1961). The economics of information. *Journal of Political Economy, 69*, 213–225.

Sutton, R., & Barto, A. (1998). *Reinforcement learning: An introduction*. Cambridge, MA: MIT Press.

Watkins, C. (1989). *Learning from delayed rewards*. Unpublished doctoral dissertation, King's College, Oxford.

13

Sequential Dependencies in Human Behavior Offer Insights Into Cognitive Control

Michael C. Mozer, Sachiko Kinoshita, & Michael Shettel

We present a perspective on cognitive control that is motivated by an examination of *sequential dependencies* in human behavior. A sequential dependency is an influence of one incidental experience on subsequent experience. Sequential dependencies arise in psychological experiments when individuals perform a task repeatedly or perform a series of tasks, and one task trial influences behavior on subsequent trials. For example, in a naming task, individuals are faster to name a word after having just named easy (e.g., orthographically regular) words than after having just named difficult words. And in a choice task, individuals are faster to press a response key if the same response was made on recent trials than if a different response had been made. We view sequential dependencies as reflecting the fine tuning of cognitive control to the structure of the environment. We discuss the two sequential phenomena just mentioned, and present accounts of the phenomena in terms of the adaptation of cognitive control. For each phenomenon, we characterize cognitive control in terms of constructing a predictive model of the environment and using this model to optimize future performance. This same perspective offers insight not only into adaptation of control but also into how task instructions can be translated into an initial configuration of the cognitive architecture.

In this chapter, we present a particular perspective on cognitive control that is motivated by an examination of *sequential dependencies* in human behavior. At its essence, a sequential dependency is an influence of one incidental experience on subsequent experience. Sequential dependencies arise both in naturalistic settings and in psychological experiments when individuals perform a task repeatedly or perform a series of tasks and performing one task trial influences behavior on subsequent trials. Measures of behavior are diverse, including response latency, accuracy, type of errors produced, and interpretation of ambiguous stimuli.

To illustrate, consider the three columns of addition problems in Table 13.1. The first column is a series of easy problems; individuals are quick and accurate in naming the sum. The second column is a series of hard problems; individuals are slower and less accurate in responding. The third column contains a mixture of easy and hard problems. If sequential dependencies arise in repeatedly naming the sums, then the response time or accuracy to an easy problem will depend on

the preceding context, that is, whether it appears in an easy or mixed block; similarly, performance on a hard problem will depend on whether it appears in a hard or mixed block. Exactly this sort of dependency has been observed (Lupker, Kinoshita, Coltheart, & Taylor, 2003): Responses to a hard problem are faster but less accurate in a mixed block than in a pure block; similarly, responses to an easy problem are slower and more accurate in a mixed block than in a pure block of easy trials. Essentially, the presence of recent easy problems causes response-initiation processes to treat a hard problem as if it were easier, speeding up responses but causing them to be more error prone; the reverse effect occurs for easy problems in the presence of recent hard problems.

Sequential dependencies reflect cortical adaptation operating on the timescale of seconds, not—as one usually imagines when discussing learning—days or weeks. Sequential dependencies are robust and nearly ubiquitous across a wide range of experimental tasks. Table 13.2 presents a catalog of sequential dependency

TABLE 13.1 Three Blocks of Addition Problems

Easy Block	Hard Block	Mixed Block
3 + 2	9 + 4	3 + 2
1 + 4	7 + 6	7 + 6
10 + 7	8 + 6	10 + 7
5 + 5	6 + 13	6 + 13

effects, spanning a variety of components of the cognitive architecture, including perception, attention, language, stimulus-response mapping, and response initiation. Sequential dependencies arise in a variety of experimental paradigms. The aspect of the stimulus that produces the dependency—which we term the *dimension of dependency*—ranges from the concrete, such as color or identity, to the abstract, such as cue validity and item difficulty. Most sequential dependencies are fairly short lived, lasting roughly five intervening trials, but some varieties span hundreds of trials and weeks of passing time (e.g., global display configuration, Chun & Jiang, 1998; syntactic structure, Bock & Griffin, 2000).

Sequential dependencies may be even more widespread than Table 13.2 suggests, because they are ignored in the traditional psychological experimental paradigm.

In a typical experiment, participants perform dozens of practice trials during which data are not collected, followed by experimental trials that are randomized such that when aggregation is performed over trials in a particular experimental condition, sequential effects are cancelled. When sequential effects are studied, they are often larger than other experimental effects explored in the same paradigm; for example, in visual search, sequential effects can modulate response latency by 100 ms given latencies in the 700 ms range (e.g., Wolfe et al., 2003).

Sequential dependencies are often described as a sort of *priming*, facilitation of performance due to having processed similar stimuli or made similar responses in the past. We prefer not to characterize sequential dependencies using the term *priming* for two reasons. First, priming is often viewed as an experimental curiosity used to diagnose the nature of cognitive representations, one which has little bearing on naturalistic tasks and experience. Second, many sequential dependencies are not due to repetitions of specific stimulus identities or features, but rather to a more abstract type of similarity. For example, in the arithmetic problem difficulty manipulation described earlier, problem *difficulty*, not having experience on a specific problem,

TABLE 13.2 A Catalog of Sequential Dependency Effects

Component of Architecture	Experimental Paradigm	Dimension of Dependency	Example Citations
Perception	Figure-ground Identification	Stimulus color Stimulus shape and identity	Vecera (2005) Bar & Biederman (1998); Ratcliff & McKoon (1997)
	Intensity judgement Categorization	Stimulus magnitude Stimulus features	Lockhead (1984, 1995) Johns & Mewhort (2003); Stewart et al. (2002)
	Ambiguous motion	Previous judgments	Maloney et al. (2005)
Stimulus-response mapping	Task switching	Task set	Rogers & Monsell (1995)
Language	Semantic judgment	Syntactic structure	Bock & Griffin (2000)
Response initiation	Word naming	Task difficulty	Taylor & Lupker (2001)
	Choice		Kiger & Glass (1981); Strayer & Kramer (1994a)
		Response	Jentszch & Sommer (2002); Jones et al. (2003)
Attention	Cued detection and identification	Cue validity	Bodner & Masson (2001); Posner (1980)
	Visual search	Stimulus features	Maljkovic & Nakayama (1996); Wolfe et al. (2003)
		Scene configuration and statistics	Chun & Jiang (1998, 1999)

induces sequential dependencies; and in language, syntactic structure induces sequential dependencies, not particular words or semantic content.

Cognitive Control

We view sequential dependencies as a strong constraint on the operation of *cognitive control*. Cognitive control allows individuals to flexibly adapt behavior to current goals and task demands. Aspects of cognitive control include the deployment of visual attention, the selection of responses, forming arbitrary associations between stimuli and responses, and using working memory to subserve ongoing processing. At its essence, cognitive control involves translating a task specification into a configuration of the cognitive architecture appropriate for performing that task. But cognitive control involves a secondary, more subtle, ability—that of fine-tuning the operation of the cognitive architecture to the environment. For example, consider searching for a key in a bowl of coins versus searching for a key on a black leather couch. In the former case, the environment dictates that the most relevant feature is the size of the key, whereas in the latter case, the most relevant feature is the metallic luster of the key.

We adopt the perspective that sequential dependencies reflect this fine-tuning of cognitive control to the structure of the environment. We discuss two distinct sequential phenomena and present accounts of the phenomena in terms of the adaptation of cognitive control. For each phenomenon, we assume that cognitive control involves constructing a predictive model of the environment and using this model to optimize future performance.

Sequential Effects Involving Response Repetition

In this section, we model a speeded discrimination paradigm in which individuals are asked to classify a sequence of stimuli (Jones, Cho, Nystrom, Cohen, & Braver, 2002). The stimuli are letters of the alphabet, A–Z, presented in rapid succession, and individuals are asked to press one response key if the letter is an X or another response key for any letter other than X (as a shorthand, we will refer to the alternative responses as R_1 and R_2). Jones et al. (2003) manipulated the relative frequency of R_1 and R_2; the ratio of presentation frequency was either 1:5, 1:1, or 5:1. Response conflict

arises when the two stimulus classes are unbalanced in frequency, resulting in more errors and slower reaction times. For example, when R_1s are frequent but R_2 is presented, individuals are predisposed toward producing the R_1 response, and this predisposition must be overcome by the perceptual evidence from the R_2. Cognitive control is presumed to be required in situations involving response conflict. In this task, response repetition is key, rather than stimulus repetition, because effects are symmetric for R_1 and R_2, even though one of the responses corresponds to many distinct stimuli, and those stimuli are *not* repeated.

A Probabilistic Information Transmission Model

The heart of our account is an existing model of probabilistic information transmission (PIT) that explains a variety of facilitation effects that arise from long-term repetition priming (Colagrosso, 2004; Colagrosso & Mozer, 2005; Mozer, Colagrosso, & Huber, 2003), and more broadly, that addresses changes in the nature of information transmission in neocortex due to experience. We give a brief overview of the aspects of this model essential for the present work.

The model posits that the cognitive architecture can be characterized by a collection of information-processing *pathways*, and any act of cognition involves coordination among pathways. To model a simple discrimination task, we might suppose a *perceptual pathway* to map the visual input to a semantic representation, and a *response pathway* to map the semantic representation to a response. The model is framed in terms of probability theory: pathway inputs and outputs are random variables and inference in a pathway is carried out by Bayesian belief revision.

To elaborate, consider a pathway whose input at time t is a discrete random variable, denoted $X(t)$, which can assume values 1, 2, 3, . . . , n_x corresponding to alternative input states. Similarly, the output of the pathway at time t is a discrete random variable, denoted $Y(t)$, which can assume values 1, 2, 3, . . . , n_y. For example, the input to the perceptual pathway in the discrimination task is one of $n_x = 26$ visual patterns corresponding to the letters of the alphabet, and the output is one of $n_y = 26$ letter identities.[1] To present a particular input alternative, i, to the model for T time steps, we clamp $X(t) = i$ for $t = 1 . . . T$. The model computes a probability distribution over Y given X, that is, $P(Y(t) \mid X(1) . . . X(t))$, the probability over the output states given the input sequence.[2]

A pathway is modeled as a dynamic Bayes network; the minimal version of the model used in the present simulations is simply a hidden Markov model, where the $X(t)$ are observations and the $Y(t)$ are inferred state (see Figure 13.1, left panel).[3] To understand the diagram, ignore the directionality of the arrows, and note simply that $Y(t)$ is linked to both $Y(t-1)$ and $X(t)$, meaning that $Y(t)$ is constrained by these other two variables. To compute $P(Y(t) \mid X(1) \ldots X(t))$, it is necessary to specify three probability distributions. The particular values hypothesized for these three distributions embody the knowledge of the model and give rise to predictions from the model. The three distributions are

1. $P(Y(t)|Y(t-1))$, which characterizes how the pathway output evolves over time, that is, how the output at time t, $Y(t)$, depends on the output at time $t-1$, $Y(t-1)$;
2. $P(X(t) \mid Y(t))$, which characterizes the *strength of association* between inputs and outputs, that is, how likely it is to observe a given state of the input at some point in time, $X(t)$, if the correct output at that time, $Y(t)$, is a known state; and
3. $P(Y(0))$, the *prior* distribution over outputs, that is, in the absence of any information about the relative likelihood of the various output states.

To give a sense of how PIT operates, the right panel of Figure 13.1 depicts the time course of inference in a single pathway, which has 26 input and output alternatives, with one-to-one associations. The solid line in the Figure 13.1 (right panel) shows, as a function of time t, $P(Y(t) = 1 \mid X(1) = 1 \ldots X(t) = 1)$, that is, the probability that a given input will produce its target output. Because of the limited association strengths,

perceptual evidence must accumulate over many iterations for the target to be produced with high probability. The densely dashed line shows the same target probability when the target prior is increased, and the sparsely dashed line shows the target probability when the association strength to the target is increased. Increasing either the prior or the association strength causes the speed-accuracy curve to shift to the left. In our previous work, we proposed a mechanism by which priors and association strengths are altered following each experience. This mechanism gives rise to sequential effects; we will show that it explains the response-repetition data described earlier.

PIT is a generalization of random walk models and has several advantages. It provides a mathematically principled means of handling multiple alternative responses (necessary for naming) and similarity structure among elements of representation, and characterizes perceptual processing, not just decision making. The counter model (Ratcliff & McKoon, 1997) or connectionist integrator models (e.g., Usher & McClelland, 2001) could also serve us, although PIT has an advantage in that it operates using a currency of probabilities—versus more arbitrary units of *counts* or *activation*—which has two benefits. First, fewer additional assumptions are required to translate model output to predictions of experimental outcomes: If the tendency to make responses is expressed as a probability distribution over alternatives, stochastic sampling can be used to obtain a response, whereas if response tendency is expressed as activation, an arbitrary transformation must be invoked to transform activation into a response (e.g., a normalized exponential transform is often used in connectionist models). Second, operating in a currency of probability leads to explicit, interpretable

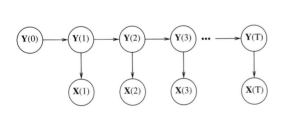

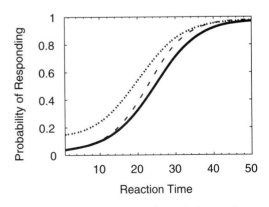

FIGURE 13.1 Basic pathway architecture (left panel); time course of inference in a pathway (right panel).

decision criteria and learning mechanisms; for example, Bayes's rule can be used to determine an optimal decision criterion or update of beliefs after obtaining evidence, whereas the currency of activation in connectionist models allows for arbitrary threshold and learning rules.

Model Details

The simulations we report in this chapter use a cascade of two pathways. A perceptual pathway maps visual patterns (26 alternatives) to a letter-identity representation (26 alternatives), and a response pathway maps the letter identity to a response. For the choice task, the response pathway has two outputs, corresponding to the two response keys. The interconnection between the pathways is achieved by copying the output of the perceptual pathway, $Y^p(t)$, to the input of the response pathway, $X^r(t)$, at each time. The free parameters of the model are mostly task and experience related. Nonetheless, in the current simulations, we used the same parameter values as Mozer et al. (2003), with one exception: Because the speeded perceptual discrimination task studied here is quite unlike the tasks studied by Mozer et al., we allowed ourselves to vary the association-strength parameter in the response pathway. This parameter has only a quantitative, not qualitative, influence on predictions of the model.

In our simulations, we also use the priming mechanism proposed by Mozer et al. (2003). Essentially, this mechanism constructs a *model of the environment*, which consists of the prior probabilities of the various stimuli and responses. To elaborate, the priors for a pathway are internally represented in a nonnormalized form: the nonnormalized prior for alternative i is p_i, and the normalized prior is

$$P(Y(0) = i) = p_i \Big/ \sum_i p_i .$$

The priming mechanism maintains a running average of recent experience. On each trial, the priming mechanism increases the nonnormalized prior of alternative i in proportion to its asymptotic activity at final time T, and all priors undergo exponential decay:

$$\Delta p_i = \gamma P(Y(T) = i \,|\, X(1) \ldots X(T)) - \varepsilon p_i,$$

where γ is the strength of priming, and ε is the decay rate. (The Mozer et al. model also performs priming in the association strengths by a similar rule, which is included in the present simulation although it has a negligible effect on the results here.)

This priming mechanism yields priors on average that match the presentation probabilities in the task, for example, .17 and .83 for the two responses in the 1:5 condition of the Jones et al. experiment. Consequently, when we report results for overall error rate and reaction time in a condition, we make the assumption of rationality that the model's priors correspond to the true priors of the environment. Although the model yields the same result when the priming mechanism is used on a trial-by-trial basis to adjust the priors, the explicit assumption of rationality avoids any confusion about the factors responsible for the model's performance. We use the priming mechanism on a trial-by-trial basis to account for performance conditional on recent trial history, as explained later.

Control Processes and the Speed-Accuracy Trade-Off

The response pathway of the model produces a speed-accuracy performance function much like that in the right panel of Figure 13.1. This function characterizes the operation of the pathway, but it does not address the control issue of when in time to initiate a response. A control mechanism might simply choose a threshold in accuracy or in reaction time, but we hypothesize a more general, rational approach in which a *response utility* is computed, and control mechanisms initiate a response at the point in time when a maximum in utility is attained.

When stimulus $\mathbf{S}$ is presented and the correct response is $\mathbf{R}$, we posit a utility of responding at time T following stimulus onset:

$$U(T \,|\, \mathbf{S}, \mathbf{R}) = \int_{t=0}^{T} [P(Y^r(t) = \mathbf{R} \,|\, \mathbf{S}) - \kappa t] \, dt \qquad (1)$$

This utility involves two terms, the accuracy of response and the reaction time. Utility increases with increasing accuracy and decreases with response time. The relative importance of the two terms is determined by κ. This form of utility function leads to an extremely simple stopping rule, which we'll explain shortly.

We assume that κ depends on task instructions: if individuals are told to make no errors, κ should be small to emphasize the error rate; if individuals are told to respond quickly and not concern themselves with occasional errors, κ should be large to emphasize the reaction time. We picked a value of κ to obtain the best fit to the human data.

The utility cannot be computed without knowing the correct response $\mathbf{R}$. Nonetheless, the control mechanism could still compute an *expected* cost over the n_y^r alternative responses based on the model's current estimate of the likelihood of each:

$$\bar{U}(T|\mathbf{S}) = \sum_r P(Y'(T) = r|\mathbf{S})U(T\,|\,\mathbf{S}, r) \qquad (2)$$

The optimal point in time at which to respond is the value of T that yields the maximum utility. This point in time can be characterized in a simple, intuitive manner by rearranging Equations 3 and 4. Based on the response probability distribution $P(Y'(\mathbf{T}) = \|\mathbf{S})$, an *estimate of response accuracy* for the current stimulus $\mathbf{S}$ at time T can be computed, even without knowing the correct response:

$$\bar{A}(T\,|\,\mathbf{S}) = \sum_r P(Y'(T){=}r\,|\,\mathbf{S})^2. \qquad (3)$$

This equation is the expectation, under the current response distribution, of a correct response assuming that the actual probability of a response being correct matches the model's internal estimate. In terms of $\bar{A}$, the optimal stopping time according to Equation 2 occurs at the earliest time T when

$$\bar{A}(T|\mathbf{S}) = 1 - \kappa T. \qquad (4)$$

The optimal stopping time can be identified by examination of $\bar{A}\,(T|\mathbf{S})$ and T at two consecutive time steps, satisfying the essential requirement for real-time performance.

Results

Figure 13.2 illustrates the model's performance on the choice task when presented with a stimulus, $\mathbf{S}$, associated with a response, R_1, and the relative frequency of R_1 and the alternative response, R_2, is 1:5, 1:1, or 5:1 (left, center, and right columns, respectively). The top row plots the probability of R_1 and R_2 against time. Although R_1 wins out asymptotically in all three conditions, it must overcome the effect of its low prior in

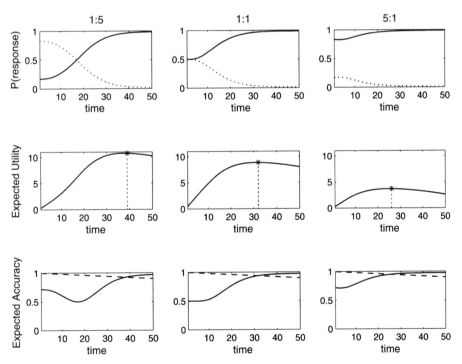

FIGURE 13.2 (top row) Output of probabilistic information transmission (PIT) response pathway as a function of time when stimulus $\mathbf{S}$, associated with response R_1, is presented, and relative frequency of R_1 (solid line) and the alternative response, R_2 (dotted line), is 1:5, 1:1, and 5:1. (middle row) Expected cost of responding; the asterisk shows the optimal point in time. (bottom row) PIT's internal estimate of accuracy over time (solid line) and time-decreasing criterion, $1 - \kappa T$ (dashed line).

the 1:5 condition. The middle row plots the expected utility over time. Early on, the high error rate leads to low utility; later on, reaction time leads to decreasing utility. Our rational analysis suggests that a response should be initiated at the global maximum—indicated by asterisks in the figure—implying that both the reaction time and error rate will decrease as the response prior is increased. The bottom row plots the model's estimate of its accuracy, $\bar{A}\,(T|S)$, as a function of time. Also shown is the $1 - \kappa T$ line (dashed), and it can be seen that the utility maximum is obtained when Equation 4 is satisfied.

Figure 13.3 presents human and simulation data for the choice task. The data consist of mean reaction time and accuracy for the two target responses, R_1 and R_2, for the three conditions corresponding to different R_1:R_2 presentation ratios. The qualities of the model giving rise to the fit can be inferred by inspection of Figure 13.3; namely, accuracy is higher and reaction times are faster when a response is expected.

The model provides an extremely good fit not only to the overall pattern of results but also sequential effects. Figure 13.3 reveals how the recent history of experimental trials influences reaction time and

error rate. The trial *context* along the *x*-axis is coded as $v_4 v_3 v_2 v_1$, where v_i specifies that trial $n - i$ required the same (S) or different (D) response as trial $n - i + 1$. For example, if the five trials leading up to and including the current trial are—in forward temporal order—R_2, R_2, R_2, R_1, and R_1, the current trial's context would be coded as "SSDS." The correlation coefficient between human and simulation data is .960 for reaction time and .953 for error rate.

The simple priming mechanism proposed previously by Mozer et al. (2003), which aims to adapt the model's priors rapidly to the statistics of the environment, is responsible for the model's performance: On a coarse timescale, the mechanism produces priors in the model that match priors in the environment. On a fine timescale, changes to and decay of the priors results in a strong effect of recent trial history, consistent with the human data: The graphs in Figure 13.4 show that the fastest and most accurate trials are clearly those in which the previous two trials required the same response as the current trial (the leftmost four contexts in each graph). The fit to the data is all the more impressive given that Mozer et al. priming mech-

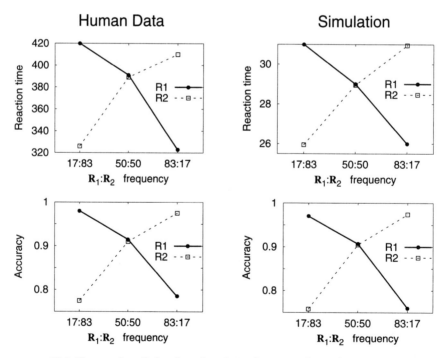

FIGURE 13.3 Human data (left column) and simulation results (right column) for the choice task. Human data from Jones et al. (2003). The upper and lower rows show mean reaction time and accuracy, respectively, for the two responses R_1 and R_2 in the three conditions corresponding to different R_1:R_2 frequencies.

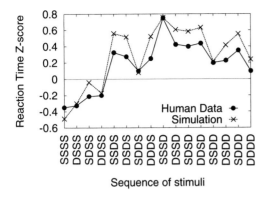
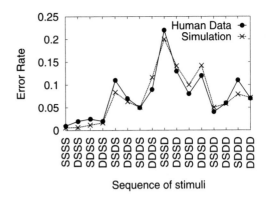

FIGURE 13.4 Reaction time (left curve) and accuracy (right curve) data for humans (solid line) and model (dotted line), contingent on the recent history of experimental trials.

anism was used to model perceptual priming, and here the same mechanism is used to model response priming.

Discussion

We introduced a model that accounts for sequential effects of response repetition in a simple choice task. The model was based on the principle that control processes incrementally estimate response prior probabilities. The PIT model, which performs Bayesian inference, uses these response priors to determine the optimal point in time at which to initiate a response. The probabilistic framework imposes strong constraints on the model and removes arbitrary choices and degrees of freedom that are often present in psychological models.

Jones et al. (2003) proposed a neural network model to address response conflict in a speeded discrimination task. Their model produces an excellent fit to the data too but involves significantly more machinery, free parameters, and ad hoc assumptions. In brief, their model is an associative net mapping activity from stimulus units to response units. When response units R_1 and R_2 both receive significant activation, noise in the system can push the inappropriate response unit over threshold. When this conflict situation is detected, a control mechanism acts to lower the baseline activity of response units, requiring them to build up more evidence before responding and thereby reducing the likelihood of noise determining the response. Their model includes a priming mechanism to facilitate repetition of responses, much as we have in our model. However, their model also includes a secondary priming mechanism to facilitate *alternation* of responses, which our model does not require. Both models address additional

data; for example, a variant of their model predicts a neurophysiological marker of conflict called error-related negativity (Yeung, Botvinick, & Cohen, 2004).

Jones et al. (2003) also performed an functional magnetic resonance imaging (fMRI) study of this task and found that anterior cingulate cortex (ACC) becomes activated in situations involving response conflict. Specifically, when one stimulus occurs infrequently relative to the other, event-related fMRI response in the ACC is greater for the low frequency stimulus. According to the Jones et al. model, the role of ACC is to conflict detection. Our model allows for an alternative interpretation of the fMRI data: ACC activity may reflect the expected utility of decision making on a fine time grain. Specifically, the ACC may provide the information needed to determine the optimal point in time at which to initiate a response, computing curves such as those in the bottom row of Figure 13.2. If ACC activity is related to the height of the utility curves, then fMRI activation—which reflects a time integral of the instantaneous response—should be greater when the response prior is lower, that is, when conflict is present. Recent neuropsychological data have shown a deficit in performance with a simple RT task following ACC damage (Fellows & Farah, 2005). These data are consistent with our interpretation of the role of ACC but not with the conflict-detection interpretation.

Sequential Effects Involving Task Difficulty

In this section, we return to the sequential dependency on item difficulty described in the introduction to the chapter. To remind the reader, Table 13.1 shows three columns of addition problems. Some problems are intrinsically easier than others, for example, $10 + 3$ is easier than $5 + 8$, whether because of practice or the number of cognitive operations required to determine the sum. By definition, individuals have faster RTs *and* lower error rates to easy problems. However, when items are presented in a sequence or *block*, reaction time *(RT)* and error rate to an item depend on the composition of the block. When presented in a mixed block (column 3 of Table 13.1), easy items slow down relative to a pure block (column 1 of Table 13.1) and hard items speed up relative to a pure block (column 2 of Table 13.1). However, the convergence of RTs for easy and hard items in a mixed block is not complete. Thus, RT depends both on the stimulus type and the composition of the block.

This phenomenon, sometimes called a *blocking effect*, occurs across diverse paradigms, including naming, arithmetic verification and calculation, target search, and lexical decision (e.g., Lupker, Brown, & Columbo, 1997; Lupker et al., 2003; Taylor & Lupker, 2001). It is obtained when stimulus or response characteristics alternate from trial to trial (Lupker et al., 2003). Thus, the blocking effect is not associated with a specific stimulus or response pathway. Because blocking effects influence the speed-accuracy trade-off, they appear to reflect the operation of a fundamental form of cognitive control—the mechanism that governs the initiation of a behavioral response. The blocking effect shows that control of response initiation depends not only on information from the current stimulus, but also on recent stimuli in the trial history.

Explaining the Blocking Effect

Any explanation of the blocking effect must specify how response-initiation processes are sensitive to the composition of a block. Various mechanisms of control adaptation have been proposed, including domain specific mechanisms (Meyer, Roelofs, & Levelt, 2003; Rastle & Coltheart, 1999), adjustment of the rate of processing (Kello & Plaut, 2003), and adjustment of an evidence criterion in a random walk model (e.g., Strayer & Kramer, 1994b). In Mozer and Kinoshita (in preparation), we present a detailed critique of these accounts.

We propose an alternative account. By this account, response-initiation mechanisms are sensitive to the statistical structure of the environment for the following reason. An accurate response can be produced only

when the evidence reaching the response stage from earlier stages of processing is reliable. Because the point in time at which this occurs will be earlier or later depending on item difficulty, some estimate of the difficulty is required. This estimate can be explicit or implicit; an implicit estimate might indicate the likelihood of a correct response at any point in time given the available evidence. If only noisy information is available to response systems concerning the difficulty of the current trial, a rational strategy is to increase reliability by incorporating estimates of difficulty from recent—and presumably similar—trials.

We elaborate this idea in a mathematical model of response initiation. The model uses the PIT framework described previously to characterize the temporal dynamics of information processing, and the optimal decision criterion used for response initiation. As described earlier, PIT proposes that the transmission of stimulus information to response systems is gradual and accumulates over time, and that control mechanisms respond at the point in time that maximizes a utility measure that depends on both expected accuracy and time. In the previous model we described, we assumed that the response distribution is available for control processes to estimate the expected accuracy, $\overline{A}$. (Equation 3, and depicted in the bottom row of Figure 13.2, solid lines). However, if the response distribution obtained is noisy, $\overline{A}$ will be a high variance estimate of accuracy. Rather than relying solely on $\overline{A}$, the variance can be lowered by making the ecological assumption that the environment is relatively constant from one trial to the next, and therefore, the estimates over successive trials can be averaged.[4]

We use the phrase *current accuracy trace* (CAT) to denote the complete time-varying trace of $\overline{A}$, that is,

$$\text{CAT} \equiv \{\overline{A}(t \mid S), t = 1 \dots T\}.$$

To implement averaging over trials, the model maintains a *historical accuracy trace* (HAT), and the trace used for estimating utility—the *mean accuracy trace* (MAT)— is a weighted average of CAT and HAT, that is,

$$\text{HAT}(n) = \lambda \text{CAT}(n - 1) + (1 - \lambda)\text{HAT}(n - 1),$$

where n is an index over trials, and

$$\text{MAT}(n) = \theta \text{CAT}(n) + (1 - \theta)\text{HAT}(n);$$

λ and θ are averaging weights. Figure 13.5a depicts the CAT, HAT, and MAT. The thin and thick solid curves represent CATs for easy and hard trials, respectively; these same curves also represent the MATs for

(a)

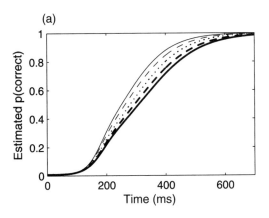

(b)

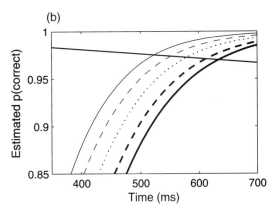

FIGURE 13.5 (a) Easy current accuracy trace (CAT; thin solid line, which is also the mean accuracy trace [MAT] in a pure block of easy items) and hard CAT (thick solid, which is also the MAT in a pure block of hard items); mixed block historical accuracy trace (HAT; dotted) and easy and hard MAT in a mixed block (thin and thick dashed); (b) close-up of the traces, along with the time threshold (gray solid).

pure blocks. The dotted curve represents the expected HAT in a mixed block—an average of easy and hard CATs. The thin and thick dashed curves represent the MATs for easy and hard trials in a mixed block, respectively, formed by averaging the HAT and corresponding CAT. Because the CAT and HAT are time-varying functions, the notion of averaging is ambiguous; possibilities include averaging the accuracy of points with the same time value and times of points with the same accuracy value. It turns out that the choice has no qualitative impact on the simulation results we present.

Results

Figure 13.5b provides an intuition concerning the model's ability to replicate the basic blocking effect. The mean RT for easy and hard items in a pure block is indicated by the point of intersection of the CAT with the time threshold (Equation 4). The mean RT for easy and hard items in a mixed block is indicated by the point of intersection of the MAT with the time threshold. The easy item slows down, the hard item speeds up. Because the rate of processing is not affected by the blocking manipulation, the error rate will necessarily drop for easy items and rise for hard items. Although the RTs for easy and hard items come together, the convergence is not complete as long as $\theta > 0$.

A signature of the blocking effect concerns the relative magnitudes of easy-item slow down and hard-item speed up. Significantly more speed up than slow down is never observed in experimental studies. The trend is that speed up is less than slow down; indeed,

some studies show no reliable speed up, although equal magnitude effects are observed. Empirically, the model we propose never yields more speed up than slow down. The slow down is represented by the shift of the easy MAT in mixed versus pure blocks (the thin dashed and thin solid lines in Figure 13.5b, respectively), and the speed up is represented by the shift of the hard MAT in mixed versus pure blocks (the thick dashed and thick solid lines). Comparing these two sets of curves, one observes that the hard MATs hug one another more closely than the easy MATs, at the point in time of response initiation. The asymmetry is due to the fact that the easy CAT reaches asymptote before the hard CAT. Blocking effects are more symmetric in the model when responses are initiated at a point when both easy and hard CATs are ascending at the same rate. The invalid pattern of more speed up than slow down will be obtained only if the hard CAT is more negatively accelerated than the easy CAT at the point of response initiation; but by the definition of the easy and hard items, the hard CAT should reach asymptote after the easy CAT, and therefore should never be more negatively accelerated than the easy CAT.

The theory thus explains the key phenomena of the blocking effect. The theory is also consistent with three additional observations: (1) Blocking effects occur across a wide range of tasks and even when tasks are switched trial to trial; and (2) blocking effects occur even in the absence of overt errors; and (3) blocking effects occur only if overt responses are produced; if responses are not produced, the response-accuracy curves need not

TABLE 13.3 Experiment 1 of Taylor and Lupker (2001): Human Data and Simulation

| | Human Data | | | Simulation | | |
	Pure	Mixed	Difference	Pure	Mixed	Difference
Easy	519 ms (0.6%)	548 ms (0.7%)	29 ms (0.1 %)	524 ms (2.4%)	555 ms (1.7%)	31 ms (−0.7%)
Hard	631 ms (2.9%)	610 ms (2.9%)	−21 ms (0.0%)	634 ms (3.0%)	613 ms (3.7%)	−21 ms (0.7%)

be generated, and the averaging process that underlies the effect cannot occur.

Beyond providing qualitative explanations for key phenomena, the model fits specific experimental data. Taylor and Lupker (2001, Experiment 1) instructed participants to name high frequency words (easy items) and nonwords (hard items). Table 13.3 compares mean RTs and error rates for human participants and the simulation. One should not be concerned with the error-rate fit, because measuring errors in a naming task is difficult and subjective. (Over many experiments, error rates show a speed-accuracy trade-off.) Taylor and Lupker further analyzed RTs in the mixed block conditional on the context—the 0, 1, and 2 preceding items. Figure 13.6 shows the RTs conditional on context. The model's fit is excellent. Trial n is most influenced by trial $n-1$, but trial $n-2$ modulates behavior as well; this is well modeled by the exponentially decaying HAT.

Simulation Details

Parameters of the PIT model were chosen to obtain pure-block mean RTs comparable to those obtained in the experiment and asymptotic accuracy of 100% for both easy and hard items. We added noise to the transmission rates to model item-to-item and trial-to-trial variability but found that this did not affect the

expected RTs and error rates. We fixed the HAT and MAT averaging terms, λ and θ, at 0.5, and picked κ to obtain error rates in the pure block of the right order. Thus, the degrees of freedom at our disposal were used for fitting pure block performance; the mixed block performance (Figure 13.6) emerged from the model.

Testing Model Predictions

In the standard blocking paradigm, the target item is preceded by a context in which roughly half the items are of a different difficulty level. We conducted a behavioral study in which the context was maximally different from the target. Each target was preceded by a context of 10 items of homogeneous difficulty, either the *same* or *different* difficulty as the target. This study allows us to examine the asymptotic effect of context switching. We performed this study for two reasons. First, Taylor and Lupker (2001) obtained results suggesting that a trial was influenced by only the previous two trials; our model predicts a cumulative effect of all context, but diminishing exponentially with lag. Second, several candidate models we explored predict that with a strong context, speed up of hard is significantly larger than slow down of easy; the model we've described does not.

The results are presented in Table 13.4. The model provides an excellent fit to the data. Significantly larger context effects are obtained than in the previous

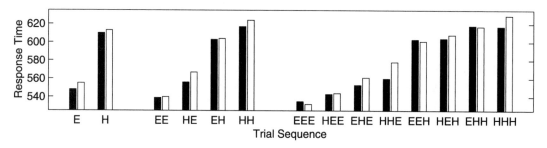

FIGURE 13.6 RTs from human subjects (black) and simulation (white) for easy and hard items in mixed block, conditional on 0, 1, and 2 previous item types. Last letter in a string indicates the current trial and first letters indicate context. Thus, "EHH" means a hard item preceded by another hard item preceded by an easy item.

TABLE 13.4 Context Experiment: Human Data and Simulation

	Human Data			Simulation		
	Same Context	Different Context	Switch Effect	Same Context	Different Context	Switch Effect
Easy	432 ms	488 ms	56 ms	437 ms	493 ms	56 ms
Hard	514 ms	467 ms	−47 ms	514 ms	470 ms	−44 ms

simulation (~50 ms in contrast to ~25 ms), and, given the strong context, the easy items become slower than the hard (although this effect is not statistically reliable in the experimental data). Further, both data and model show more slow down than speed up, a result that allowed us to eliminate several competing models.[5]

We have conducted a variety of other behavioral experiments testing predictions of the model. For example, in Kinoshita and Mozer (2006), we explore the conditions giving rise to symmetric versus asymmetric blocking effects. We have also shown that various other phenomena involving blocked performance comparisons can be interpreted as blocking effects (Mozer & Kinoshita, in preparation).

Conclusions

Theories in cognitive science often hand the problem of cognitive control to an unspecified homunculus. Other theories consider cognitive control in terms of a central, unitary component of the cognitive architecture. In contrast, we view cognitive control as a collection of simple, specialized mechanisms. We described two such mechanisms in this chapter, one that determines the predisposition to produce specific responses, and another that determines how long to wait following stimulus onset before initiating a response. We characterized the nature and adaptation of these bottom-up control mechanisms by accounting for two types of sequential dependencies. The central claim of our accounts is that bottom-up cognitive control constructs a predictive model of the environment—response priors in one case, item difficulty in the other case—and then uses this model to optimize performance on subsequent trials. Although we focused on mechanisms of response initiation, predictive models of the environment can be useful for determining where in the visual field to look, what features to focus attention on, and how to interpret and categorize objects the visual field (e.g., Mozer, Shettel, & Vecera, 2005).

Our accounts are based on the premise that the goal of cognition is optimal and flexible performance across a variety of tasks and environments. In service of this goal, cognition must be sensitive to the statistical structure of the environment, and must be responsive to changes in the structure of the environment. We view sequential dependencies as reflecting continual adaptation to the ongoing stream of experience, wherein each sensory and motor experience can affect subsequent behavior. Sequential dependencies suggest that learning should be understood not only in terms of changes that occur on the timescale of hours or days, but also in terms of changes that occur from individual incidental experiences that occur on the scale of seconds.

Acknowledgments

This research was supported by National Science Foundation IBN Award 9873492, NSF BCS Award 0339103, and NIH/IFOPAL R01 MH61549–01A1. This chapter greatly benefited from the thorough reviews and critiques by Hansjörg Neth and Wayne Gray. We also thank Andrew Jones for generously providing the raw data from the Jones et al. (2003) study.

Notes

1. This model is highly abstract: The visual patterns are enumerated, but the actual pixel patterns are not explicitly represented in the model. Nonetheless, the similarity structure among inputs can be captured, but we skip a discussion of this issue because it is irrelevant for the current work.

2. A brief explanation of probability notation: If V is a random variable, then $P(V)$ denotes a distribution over values that the variable can take on. In the case of a discrete random variable, $P(V)$ denotes a vector of values. For example, if V can take on the values v_1, v_2, and v_3, then $P(V)$ might represent the probability vector [.3, .6, .1], meaning that V has value v_1 with probability .3, and so forth. To denote the probability of V taking on a certain

value, we use the standard notation $P(V = v_1)$, and in this example, $P(V = v_1) = .3$. The notation $P(V|W)$ denotes the probability vector for V given a specific, yet unspecified value of W.

3. In typical usage, a hidden Markov model (HMM) is presented with a sequence of distinct inputs, whereas we maintain the same input for many successive time steps. Further, in typical usage, an HMM transitions through a sequence of distinct hidden states, whereas we attempt to converge with increasing confidence on a single state. Thus, our model captures the time course of information processing for a single event.

4. The claim of noise in the response system's estimation of evidence favoring a decision is also made in what is perhaps the most successful model of decision processes, Ratcliff's (1978) diffusion model. This assumption is reflected both in the diffusion process itself, and the assumption of trial to trial variability in drift rates.

5. For this simulation, we fit parameters of the PIT model to the same-context results. We also treated the MAT averaging constant, θ, as a free parameter on the rational argument that this parameter can be tuned to optimize performance: if there is not much variability among items in a block, there should be more benefit to suppressing noise in the CAT using the HAT, and hence θ should be smaller. We used 0.35 for this simulation, in contrast to 0.5 for the first simulation.

References

Bar, M., & Biederman, I. (1998). Subliminal visual priming. *Psychological Science, 9,* 464–469.

Bock, K., & Griffin, Z. M. (2000). The persistence of structural priming: Transient activation or implicit learning? *Journal of Experimental Psychology: General, 129,* 177–192.

Bodner, G. E., & Masson, M. E. (2001). Prime validity affects masked repetition priming: Evidence for an episodic resource account of priming. *Journal of Memory & Language, 45,* 616–647.

Chun, M., & Jiang, Y. (1998). Contextual cueing: Implicit learning and memory of visual context guides spatial attention. *Cognitive Psychology, 36,* 28–71.

———. (1999). Top-down attentional guidance based on implicit learning of visual covariation. *Psychological Science, 10,* 360–365.

Colagrosso, M. (2004). *A Bayesian cognitive architecture for analyzing information transmission in neocortex.* Unpublished doctoral dissertation, University of Colorado, Boulder.

Colagrosso, M. D., & Mozer, M. C. (2005). Theories of access consciousness. In L. K. Saul, Y. Weiss, & L. Bottou (Eds.), *Advances in neural information processing systems 17* (pp. 289–296). Cambridge, MA: MIT Press.

Fellows, L., & Farah, M. (2005). Is anterior cingulate cortex necessary for cognitive control? *Brain, 128,* 788–796.

Jentzsch, I., & Sommer, W. (2002). Functional localization and mechanisms of sequential effects in serial reaction time tasks. *Perception & Psychophysics, 64,* 1169–1188.

Johns, E. E., & Mewhort, D. J. K. (2003). The effect of feature frequency on short-term recognition memory. *Memory & Cognition, 31,* 285–296.

Jones, A. D., Cho, R. Y., Nystrom, L. E., Cohen, J. D., & Braver, T. S. (2002). A computational model of anterior cingulate function in speeded response tasks: Effects of frequency, sequence, and conflict. *Cognitive, Affective, & Behavioral Neuroscience, 2,* 300–317.

Kello, C. T., & Plaut, D. C. (2003). Strategic control over rate of processing in word reading: A computational investigation. *Journal of Memory and Language, 48,* 207–232.

Kiger, J. I., & Glass, A. L. (1981). Context effects in sentence verification. *Journal of Experimental Psychology: Human Perception and Performance, 7,* 688–700.

Kinoshita, S., & Mozer, M. C. (2006). How lexical decision is affected by recent experience: Symmetric versus asymmetric frequency blocking effects. *Memory and Cognition, 34,* 726–742.

Lockhead, G. R. (1984) Sequential predictors of choice in psychophysical tasks. In S. Kornblum and J. Requin (Eds.), *Preparatory states and processes,* Hillsdale, NJ: Erlbaum.

———. (1995) Context Determines Perception. In F. Kessel (Ed.), *Psychology, science, and human affairs: Essays in honor of William Bevan* (pp. 125–137). New York: Westview Press.

Lupker, S. J., Brown, P., & Colombo, L. (1997). Strategic control in a naming task: Changing routes or changing deadlines? *Journal of Experimental Psychology: Learning, Memory, and Cognition, 23,* 570–590.

———, Kinoshita, S., Coltheart, M., & Taylor, T. E. (2003). Mixing costs and mixing benefits in naming words, pictures, and sums. *Journal of Memory and Language, 49,* 556–575.

Maloney, L. T., Dal Martello, M. F., Sahm, C., & Spillmann, L. (2005). Past trials influence perception of ambiguous motion quartets through pattern completion. *Proceedings of the National Academy of Sciences, 102,* 3164–3169.

Maljkovic, V., & Nakayama, K. (1996). Priming of popout: II. Role of position. *Perception & Psychophysics, 58,* 977–991.

Meyer, A. S., Roelofs, A., & Levelt, W. J. M. (2003). Word length effects in object naming: The role of a response criterion. *Journal of Memory and Language, 48,* 131–147.

Mozer, M. C., Colagrosso, M. D., & Huber, D. E. (2003). Mechanisms of long-term repetition priming and skill refinement: A probabilistic pathway model. In *Proceedings of the Twenty Fifth Annual Conference of the Cognitive Science Society.* Hillsdale, NJ: Erlbaum.

——, & Kinoshita, S. (in preparation). *Control of the speed-accuracy trade off in sequential speeded-response tasks: Mechanisms of adaptation to the stimulus environment.*

——, Shettel, M., & Vecera, S. P. (2006). Control of visual attention: A rational account. In Y. Weiss, B. Schoelkopf, & J. Platt (Eds.), *Neural information processing systems 18* (pp. 923–930). Cambridge, MA: MIT Press.

Posner, M. I. (1980). Orienting of attention. *Quarterly Journal of Experimental Psychology, 32,* 3–25.

Rastle, K., & Coltheart, M. (2000). Lexical and nonlexical print-to-sound translation of disyllabic words and nonwords. *Journal of Memory and Language, 42,* 342–364.

Ratcliff, R. (1978). A theory of memory retrieval. *Psychological Review, 85,* 59–108.

——, & McKoon, G. (1997). A counter model for implicit priming in perceptual word identification. *Psychological Review, 104,* 319–343.

Rogers, R. D., & Monsell, S. (1995). Costs of a predictable switch between simple cognitive tasks. *Journal of Experimental Psychology: General, 124,* 207–231.

Stewart, N. Brown, G. D. A., & Chater, N. (2002). Sequence effects in categorization of simple perceptual stimuli. *Journal of Experimental Psychology: Learning, Memory, and Cognition, 28,* 3–11.

Strayer, D. L., & Kramer, A. F. (1994a). Strategies and automaticity. I: Basic findings and conceptual framework. *Journal of Experimental Psychology: Learning, Memory, and Cognition, 20,* 318–341.

——, & Kramer, A. F. (1994b). Strategies and automaticity. II: Dynamic aspects of strategy adjustment. *Journal of Experimental Psychology: Learning, Memory, and Cognition, 20,* 342–365.

Taylor, T. E., & Lupker, S. J. (2001). Sequential effects in naming: A time-criterion account. *Journal of Experimental Psychology: Learning, Memory, and Cognition, 27,* 117–138.

Usher, M., & McClelland, J. L. (2001). On the time course of perceptual choice: The leaky competing accumulator model. *Psychological Review, 108,* 550–592.

Vecera, S. P. (2005). Sequential effects in figure-ground assignment. Manuscript in preparation.

Wolfe, J. M., Butcher, S. J., Lee, C., & Hyle, M. (2003). Changing your mind: On the contributions of top-down and bottom-up guidance in visual search for feature singletons. *Journal of Experimental Psychology: Human Perception and Performance, 29,* 483–502.

Yeung, N., Botvinick, M. M., & Cohen, J. D. (2004). The neural basis of error detection: Conflict monitoring and the error-related negativity. *Psychological Review, 111,* 931–959.

14

Ecological Resources for Modeling Interactive Behavior and Embedded Cognition

Alex Kirlik

A recent trend in cognitive modeling is to couple cognitive architectures with computer models or simulations of dynamic environments to study interactive behavior and embedded cognition. Progress in this area is made difficult because cognitive architectures traditionally have been motivated by data from discrete experimental trials using static, noninteractive tasks. As a result, additional theoretical problems must be addressed to bring cognitive architectures to bear on the study of cognition in dynamic and interactive environments. I identify and discuss three such problems dealing with the need to model the sensitivity of behavior to environmental constraints, the need to model highly context-specific adaptations underlying expertise, and the need for environmental modeling at a functional level. I illustrate these problems and describe how we have addressed them in our research on modeling interactive behavior and embedded cognition.

An emerging trend in the study of interactive behavior and embedded cognition is to couple a cognitive model implemented in a cognitive architecture with a computational model or simulation of a dynamic and interactive environment such as a flight simulator, military system, or video game (Byrne & Kirlik, 2005; Foyle & Hooey, in press; Gluck, Ball, & Krusmark, chapter 2, this volume; Gluck & Pew, 2005; Gray, Schoelles, & Fu, 2000; Shah, Rajyagura, St. Amant, & Ritter, 2003; Salvucci, chapter 24, this volume). Some of the impetus for this research is a growing interest in prospects for using computational cognitive modeling as a technique for engineering analysis and design. These attempts follow by about a generation a set of related attempts to model closed-loop cognition and behavior in the field of human–machine systems engineering (Rouse, 1984, 1985; Sheridan & Johannsen, 1976; also see Pew, chapter 3, this volume). As noted by Sheridan (2002), these systems engineering models represented a desire "to look at information, control, and decision making as a continuous process within a closed loop

that also included physical subsystems—more than just sets of independent stimulus-response relations" (Sheridan, 2002, p. 4).

The cognitive architectures available to today's modeling community such as ACT-R (Anderson, chapter 4, this volume), COGENT (Cooper, chapter 29, this volume), ADAPT (Brou, Egerton, & Doane, chapter 7, this volume), EPIC (Hornoff, chapter 22, this volume), Soar (Ritter, Reifers, Klein, & Schoelles, chapter 18, this volume), or Clarion (Sun, chapter 5, this volume) are better suited than were their engineering-based predecessors for describing the internal processes underlying behavior beyond merely "sets of independent of stimulus-response relations" (Sheridan, 2002, p. 4). So why is it still so difficult to model a (typically experienced) pilot, driver, or video game player with a cognitive architecture? My aim in this chapter is to address this question by providing some distinctions and modeling techniques that will hopefully accelerate progress in modeling interactive behavior and embedded cognition.

Theoretical Issues in Modeling Embedded Cognition

Difficulties in what is sometimes called "scaling up" cognitive modeling to the complexities of dynamic and interactive contexts such as aviation and driving largely have their origins in tasks and data. In particular, there are qualitative differences between the types of tasks and data sets that gave rise to many of the better-known cognitive architectures and the types of tasks and data sets characteristic of many dynamic and interactive contexts. A central goal of this chapter is to bring some clarity to the description of these qualitative differences and their implications. My hope is that clarifying these distinctions will be useful in moving beyond vague and not particularly informative discussions on the need to *scale up* models, to bridge theory and application, or to model more *real-world* behavior.

As I will try to show in the following, what is at issue here is not so much a scaling up as a scaling over. Modeling interactive behavior and embedded cognition raises interesting and challenging theoretical questions that are distinct from the types of theoretical questions that provided the traditional empirical foundation for cognitive architectures. By "distinct" I mean that many of the theoretical questions that arise when modeling dynamic and interactive tasks are not reducible in any interesting sense to the questions that motivated the design of many current cognitive architectures. New and different questions arise, along with their attendant modeling challenges and opportunities.

In the following sections, I discuss three types of theoretical issues that emerge when examining mismatches between the types of empirical data that have typically motivated the design of cognitive architectures and the types of data confronting modelers of interactive and embedded cognition in operational contexts. The first issue deals with the fact that cognitive architectures have chiefly been designed to model cognition in discrete and static tasks (i.e., laboratory trials), whereas data on embedded cognition often reflects performance in continuous and dynamic tasks. I suggest that modeling cognition and behavior in the latter type of tasks creates a need to model the manner in which behavior is dynamically sensitive to environmental constraints and opportunities. Doing so may require expanding one's view of the functional contribution of perception to intelligent behavior. Rather than viewing perception to be devoted solely to reporting the existence of objects and their properties to cognition in objective or task-neutral terms, it may be increasingly important to also view perception as capable of detecting information that specifies opportunities for behavior itself.

The second issue concerns the fact that the design of cognitive architectures has mainly been motivated by data from largely task-naive subjects or often with subjects with no more than a few hours of task-relevant experience. In contrast, modeling cognition in operational contexts such as aviation and driving often involves data from highly experienced performers. It is impossible to create a good model of a performer who knows more about the task environment than does the modeler. As a result, modeling experienced cognition requires not only expertise in cognitive modeling but also an ability to obtain expert knowledge of the relevant task and environment. While modeling students acquiring Lisp programming or arithmetic skills allows one to obtain this expert knowledge from books, modeling performers in interactive and dynamic domains typically requires detailed empirical study (e.g., Gray & Kirschenbaum, 2000). This knowledge is required not only to guide the development of a cognitive model but also to develop a detailed model of the task environment[1] with which the cognitive model can interact. Such environmental models play a key role in modeling the highly context-specific cognitive and behavioral adaptations underlying expert performance in dynamic, interactive tasks.

Finally, I discuss theoretical questions that arise out of the profoundly interactive nature of much behavior and embedded cognition in operational contexts. In particular, I suggest that interactive tasks create a need to view and model the environment much more functionally than may be required when modeling noninteractive contexts. This suggests that a largely physicalistic approach to environmental modeling, for example, in terms of the types, locations, and features of perceptible objects on a display is likely to be insufficient for understanding cognition and behavior as a functional interaction with the world. Richer techniques for functional-level environmental modeling are needed to marry the functional accounts of cognition provided by cognitive modeling with functional accounts of the environment. When modeling interactive behavior and embedded cognition, one can get only so far by trying to couple functional models of cognition with physical models of the environment. A functional perspective must be adopted for both.

After each of these issues is discussed in greater detail, I then present a set of modeling projects from our previous research touching, in one way or another, on these issues. Each project represents an explicit attempt to model computationally interactive behavior and embedded cognition in a dynamic and interactive environment.

Modeling Sensitivity to Environmental Constraints and Opportunities

One axiom within the engineering-oriented modeling tradition discussed previously concerned the necessity of modeling the environment as a prerequisite to modeling cognition and behavior. As Baron (1984) put it:

> Human behavior, either cognitive or psychomotor, is too diverse to model unless it is sufficiently constrained by the situation or environment; however, when these environmental constraints exist, to model behavior adequately, one must include a model for that environment. (Baron, 1984, p. 6)

Baron's comment places a spotlight on the *constraining* (note: not *controlling*) nature of the environment as an important source of variance that must be known when modeling behavior. Understanding how environmental constraints and opportunities determine the playing field of behavior is such a mundane exercise in everyday life that we often forget or overlook the important role that it plays. You will obviously not be swimming in the next minute unless you are already sitting near a pool or on a beach. In experimental research, a modeler typically would not get any credit for explaining all the variance associated with things that our subjects do and not do because a task does and does not provide the opportunity to do those things. Instead, the focus is on explaining variance above and beyond what could be *trivially* predicted by examining the carefully equated opportunities for behavior an experiment affords.

All of the cognitive architectures of which I am aware, because of their origins in describing data from experimental psychology, have built into them this focus on explaining variance in behavior above and beyond environmental constraints on that behavior. This can be seen from what these models predict: reaction times that, if the experiment is well designed, represent solely internal constraints but not external task constraints (a potential confound); the selection of an action from a set of actions all of which are carefully designed to be equally available to the subject (another

potential confound). Cognitive experimentalists typically take great pains to equate the availability of the various actions (e.g., keypresses) presented to participants. It is easy to overlook how this tenet of experimental design limits generalization to contexts in which the detection of action opportunities themselves and variance associated with the possibly differing levels of the availability of various actions contribute to variance in behavior.

I am hardly the first to note the many differences between the largely static, noninteractive environment of the discrete laboratory trial and environments such as video games, aviation, and driving. But note the implications regarding the necessity of environmental modeling in the two cases. To explain variance in the static laboratory experiment, since credit is given only for explaining or predicting variance above and beyond what is environmentally constrained, no attention need be given to modeling how behavioral variance is environmentally constrained. As such, cognitive architectures typically provide no resources explicitly dedicated to this ubiquitous aspect of cognition and behavior in everyday situations. In modeling experimental data, determining which actions are appropriate given the environmental context is a task performed *by the modeler* and encoded once and for all in the model: it is rarely if ever a modeled inference. This only works because the environment of the laboratory trial is presumed to be static in the sense that all (relevant) actions are always equally available.

So the modeler who would like to apply cognitive architectures motivated almost solely by data from such experiments to dynamic, interactive situations is largely on his or her own when determining how to make the model sensitive to environmental constraints and opportunities in a dynamic and interactive fashion. Modeling this type of sensitivity will be necessary any time a performer is interacting with a dynamic and especially uncertain environment. Both dynamism and uncertainty place a premium on perception to aid in determining the state of the environment in terms of which behaviors are and are not appropriate at a given time. As such, the modeler will be faced with questions concerning the design of perceptual mechanisms to aid in performing this task (e.g., Fajen & Turvey, 2003). If *primitive* perceptual mechanisms are provided by the architecture, the modeler will be faced with questions about which environmental information these mechanisms should be attuned to, and additional primitive mechanisms may need to be invented (e.g.,

Runeson, 1977). This may well require reference to an environmental model that represents perceptually available information at a high level of fidelity and the task of defining perceptual units or objects may present nontrivial problems. All of these issues speak to the question of why it has proved to be difficult to use computational cognitive architectures to model performers in dynamic, interactive environments.

Knowing as Much or More Than the Performer

I have already discussed perhaps the most primitive aspect of adaptation to an environment: ensuring that behavior is consistent with environmental constraints on behavior. Assume for a moment that this problem is solved and we are interested solely in examinations of cognition and behavior above and beyond what is so constrained. One finding from the human–machine systems tradition discussed previously is that a good step toward predicting the behavior of experienced performers in dynamic, interactive contexts is to analyze a task in terms of what behavior would be optimal or most adaptive (see Pew, chapter 3, this volume). At first blush, this approach would seem to dovetail quite nicely with modeling approaches with origins in either rational analysis (Anderson, chapter 4, this volume) or ecological rationality (Todd & Schooler, chapter 11, this volume).

Appeals are made to different quarters, however, when one assumes the rationality or optimality of basic cognitive mechanisms and when one assumes the rationality or optimality of experienced behavior. The rationality underlying the design of ACT-R's memory, categorization, and inference mechanisms and Gigerenzer, Todd, and the ABC Research Group's (1999) toolbox of fast and frugal heuristics appeals to evolutionary arguments rather than to learning or experience per se. The subjects in experiments performed from the perspective of both these adaptive approaches to cognition are not typically presumed to have any firsthand experience with the tasks studied. The hypothesis that memory exhibits a Bayesian design or that some decisions are made by a recognition heuristic are intended as claims about the human cognitive architecture independent of any *task-specific* experience. In fact, one can look at learning to be accumulating the additional adaptations necessary to perform a given task like an experienced performer instead of like a task-naive novice.

Much, if not most, modeling research done in dynamic, interactive environments is oriented toward understanding and supporting skilled performance. Much, if not most, experimental research done to inform the design of cognitive architectures uses largely task-naive subjects or, at best, subjects with only a few hours of instruction or training. It is hardly surprising, then, that researchers interested in modeling the behavior of automobile drivers, video game players, and pilots have to invent their own methods for identifying and codifying the experiential adaptations underlying skilled behavior. This is true even if they select and use a cognitive architecture informed by rationality or optimality considerations and even if the behavior to be modeled is highly rational or even optimal.

Modeling task-naive behavior can be done by similarly task-naive scientists. The main requirement is expertise in cognitive modeling. But modeling expert performance also requires expert knowledge of the task environment to which the expert is adapted. Neisser (1976) put the matter of modeling expert performance as follows:

> What would we have to know to predict how a chess master will move his pieces, or his eyes? His moves are based on information he has picked up from the board, so they can only be predicted by someone who has access to the same information.
>
> In other words, an aspiring predictor would have to understand the position at least as well as the master does; he would have to be a chessmaster himself! If I play chess against the master he will always win, because he can predict and control my behavior while I cannot do the reverse. To change this situation I must improve my knowledge of chess, not of psychology. (Neisser, 1976, p. 183)

Our own experiences in modeling expert performers, detailed in the examples to follow, have taught us that one must often spend as much, if not more, time studying and explicitly modeling the external task environment as is spent modeling inner cognition. As Neisser suggested, one cannot successfully model a performer who has access to more information or knowledge about a task environment than does the modeler. As such, we have found that a deep analysis of environmental structure and the use of abstract formalisms to represent this structure is a fundamental prerequisite to modeling experienced performers in dynamic, interactive tasks. Only then can the often highly context-specific cognitive adaptations the environment characteristic of expert interaction be discovered and modeled. The modeling examples presented in the following provide many detailed examples of these context-specific adaptations.

Mind and World Function in Concert

In his wonderfully researched and written biography of the late Nobel Prize–winning physicist Richard Feynman, James Gleick relates an episode in which MIT historian Charles Weiner was conducting interviews with Feynman at a time when Feynman had considered working with Weiner on a biography. Gleick writes that Feynman, after winning the Nobel Prize, had begun dating his scientific notes, "something he had never done before" (Gleick, 1992, p. 409). In one discussion with Feynman, "Weiner remarked casually that his new parton notes represented 'a record of the day-to-day work,' and Feynman reacted sharply" (p. 409). What was it about Weiner's comment that drew a "sharp" reaction from this great scientist? Did he not like his highly theoretical research described merely as "day-to-day work"?

No, and the answer to this question reflects, to me at least, something of Feynman's ability to have deep insights, not only into physics but into other systems as well. Feynman's reaction to Weiner describing his notes as "a record" was to say: "I actually did the work on the paper." (p. 409). To which an apparently uncomprehending Weiner responded, "Well, the work was done in your head, but the record of it is still here" (p. 409). One cannot fail to sense frustration in Feynman's retort: "No, it's not a *record*, not really. It's *working*. You have to work on paper, and this is the paper. Okay?" (p. 409, italics in the original).

My take on this interchange is that Feynman had a deep understanding of how his work was composed of a functional transaction (Dewey, 1896) between his huge accumulation of internal cognitive tools as well as his external, cognitive tools of pencil and paper, enabling him to perform functions such as writing, reflecting upon, and amending equations, diagrams, and so on (cf. Donald, 1991; Vygotsky, 1981). Most importantly, note Feynman's translation from Weiner's description of the world in terms of physical form ("No, it's not a *record*, not really.") into a description in terms of function ("It's *working*.").

Why did Weiner have such a difficult time understanding Feynman? External objects, such as Feynman's notes, do of course exist as things, typically described by nouns. Yet, in our functional transactions with these objects, the manner in which they contribute to

cognition and behavior requires that these things also be understood in functional terms, that is, in terms of their participation in the operation of the closed-loop, human-environment system (cf. Monk, 1998, on "cyclic interaction"). Weiner, like so many engineering students through the ages, apparently had difficulty in viewing the external world not only in terms of form (nouns) but also in terms of function (verbs).

I share this anecdote here because I believe it to be an exceptional illustration of the fact that studying expert behavior not only presents challenges for understanding what the expert knows but also challenges for understanding how the expert's environment contributes to cognition and how that contribution should be described (Hutchins, 1995). As the examples presented below will demonstrate, we have found in our own modeling of interactive behavior and embedded cognition a need to understand a performer's environment in functional terms, as a dynamic system in operation. Human-environment interaction is then understood in terms of a functional coupling between cognition and the environment functionally described. When modeling experienced performers engaged in interactive behavior and embedded cognition, I suggest that one has a much greater chance of identifying regularities in behavior by analysis at the functional level than by searching for these regularities in patterns of responses to stimuli described in physical terms. Modeling the environment in functional terms is also critically important when trying to model how a person might use tools in the performance of cognitive tasks, as the following examples will hopefully demonstrate.

I highlight the importance of adopting a functional perspective on environmental modeling for a number of reasons. As mentioned in the opening of this chapter, a trend currently exists to couple models with simulations of dynamic and interactive environments such as flight simulators, video games, and the like. While this is an important technical step in the evolution of cognitive modeling, having such an external simulation does not of course obviate the need for addressing the theoretical problem of modeling the environment in functional terms relevant to psychology. A bitmap model of the visual environment, for example, could be helpful in identifying the information technically available to a model's perceptual (input) mechanisms. This environmental model, however, is insufficient for determining the dimensions of information a model should "perceive" to mimic human cognition and performance.

I have little else to say about the importance of functional modeling of the environment at a general level other than to alert the reader to attend to its prevalence in the modeling examples that follow. These examples hopefully demonstrate how functional analysis allowed us to gain at least some insight into issues such as:

- Timing issues associated with the dynamic coupling of cognition and environment.
- How skilled performers might come to perceive an environment in functional terms, that is, as opportunities for action.
- How complex behavior can arise from the coupling of simple heuristics with a complex environment.
- How people might functionally structure their environment to reduce cognitive demands;
- How making cost–benefit analyses of decision making may require extremely task-specific adaptations to environmental contingencies;
- How human error might arise from generally adaptive heuristics operating in ecologically atypical situations.

Models illustrating these points and others are described in the following section.

Modeling Interactive Behavior and Embedded Cognition

In this section, I describe a set of cognitive models sharing a few common themes. Each represents an attempt to computationally model human cognition and behavior in a dynamic and interactive environment. None of the models were created in an attempt to develop a unified cognitive architecture. Instead, the central reason modeling was performed was to try to shed light on how experienced performers could have possibly managed to meet the demands of what we believed to be extremely complex dynamic and interactive tasks. In other words, in none of these cases were we in the possession of knowledge of how the task could even possibly be performed in a manner consistent with known cognitive limitations prior to analysis and modeling.

Our focus on modeling *experienced* performers in *dynamic* and *interactive* tasks placed a premium on addressing the three theoretical questions discussed earlier in this chapter concerning sensitivity of behavior to environmental constraints, the need to identify and describe highly context-specific adaptations and the need for detailed functional analysis and modeling.

The Scout World: Modeling the Environment with Dynamic Affordance Distributions

The first modeling example illustrates the use of a finely grained, functional description of an environment in terms of Gibson's (1979) theory of affordances, that is, a functional description of the environment in terms of opportunities for action. This study shed light into understanding the fluency of behavior in a highly complex, dynamic task, plausible explanations of the differences between high and low performers, and insights into why we believe that some knowledge underlying skill or expertise may appear to take on a tacit (Polanyi, 1966), or otherwise unverbalizable, form.

Consider Figure 14.1, which depicts an experimental participant performing a dynamic, interactive simulation of a supervisory control task described here as the *Scout World*. This laboratory simulation required the participant to control not only his or her own craft, called the Scout, but also four additional craft over which the participant exercised supervisory control (Sheridan, 2002), by entering action plans at a keyboard (e.g., fly to a specified waypoint, conduct patrol, load cargo, return to a home base). The left monitor in Figure 14.1 depicts a top-down situation display of the partially forested, 100-square-mile world to which activity was confined. The display on the right shows an out-the-window scene (lower half) and a set of resource and plan information for all vehicles under control (upper half). The participant's task was to control the activities of both the Scout and the four other craft to score points in each 30-min session by processing valued objects that appeared on the display once sighted by Scout radar. See Kirlik, Miller, and Jagacinski (1993) for details.

Our goal was to create a computer simulation capable of performing this challenging task and one that would allow us to reproduce, and thus possibly explain, differences between the performance of both one- and two-person crews, and novice and expert crews. At the time, the predominant cognitive modeling architectures, such as Soar (Newell, 1990; Ritter et al., chapter 18, this volume), ACT-R (Anderson, chapter 4, this volume; Anderson & Lebiere, 1998) and the like did not have mature perception and action resources

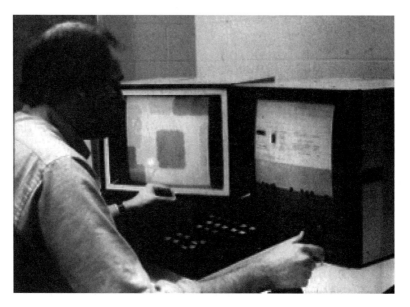

FIGURE 14.1 Experimental participant performing the Scout World task. (See color insert.)

allowing them to be coupled with external environments, nor had they been demonstrated to be capable of performing dynamic, uncertain, and interactive tasks (a limitation Newell agreed to be a legitimate weakness of these approaches; see Newell, 1992). In addition, modeling techniques drawn from the decision sciences would have provided an untenably enumerative account of participants' decision processes and were rejected because of bounded rationality considerations (Simon, 1956).

Instead, and what was a relatively novel idea at the time, we observed that our participants seemed to be relying heavily on the external world (the interface) as "its own best model" (Brooks, 1991). This was suggested not only by intimate perceptual engagement with the displays but also by self-reports (by participants) of a challenging, yet deeply engaged and often enjoyable sense of "flow" (Csikszentmihalyi, 1993) during each 30-min session (not unlike any other "addictive" video game or sport). We thus began to entertain the idea that if we were going to model the function of our human performers, we would have to model their world in functional terms as well, if we were to demonstrate how the two functioned collectively and in concert. This turned us to the work of Gibson (1979/1986), whose theory of affordances provided an account of how people might be attuned to perceiving the world functionally; in this case, in terms of actions that

could be performed in particular situations in the Scout World.

Following through on this idea entailed creating descriptions of the environment using the experimental participant's capacities for action as a frame of reference to achieve a functional description of the Scout World environment. That is, instead of creating solely perceptually oriented descriptions in terms of, say, object locations and colors, we described spatiotemporal regions or *slices* of the environment as *fly-throughable, land-onable, load-able,* and so on. A now classic example of this technique was presented by Warren (1984), who measured the riser heights of various stairs in relation to the leg lengths of various stair climbers and found, in this ratio, a functional invariance in people's ability to detect perceptually whether a set of stairs would be climbable (for them). Warren interpreted this finding to mean that people could literally perceive the *climbability* of the stairs; that is, people can perceive the world not only in terms of form but also in terms of function.

Like Warren, we created detailed, quantitative models of the Scout World environment in terms of the degree to which various environmental regions and objects afforded locomotion, searching (discovering valued objects by radar), processing those objects (loading cargo, engaging enemy craft), and returning home to unload cargo and reprovision. Because participants' actions influenced the course of events experienced, they shaped or partially determined the affordances of their own worlds. Flying the Scout through virgin forest to sight and discover cargo, for example, created new action opportunities (cargo loading), and once cargo was loaded these opportunities in turn ceased to exist. In such situations, the state of the task environment is, in experimental psychology terminology, both a *dependent* and *independent variable.*[2] This observation is useful for understanding the need for functional-level modeling of the environment to describe closed-loop dynamics. Not only must "S-R" relations be described (with some theory of cognition), but so must "R-S" relations be described, the latter requiring a model of environmental dynamics to depict how the environment changes as a function of human activity.

Figure 14.2 contains a set of four maps of the same Scout World layout, including a representation purely in terms of visual form, and as shown to participants (a), and functional representations in terms of affordances for actions of various types (b, c, d). For the Scout, for example, locomotion (flying) was most readily afforded in open, unforested areas (the white areas in Figure 14.2a) and less readily afforded as forest density grew. As such, Figure 14.2b shows higher locomotion affordances as dark and lower affordances as lighter. (Here we are of course simply using grayscale coding to represent these affordance values to the reader; in the actual model, the dark regions had high quantitative affordance values, and the light regions had relatively low quantitative affordance values.) Since the Scout radar for sighting objects (another action) had a 1.5-mile radius and valued objects were more densely scattered in forests, the interaction between the Scout's capacity for sighting and the forest structure was more graded and complex, as shown in Figure 14.2c (darker areas again indicating higher sighting affordance values). Considering that the overall affordance for searching for objects was composed of both locomotion and sighting affordances (searching was most readily afforded where one can most efficiently locomote

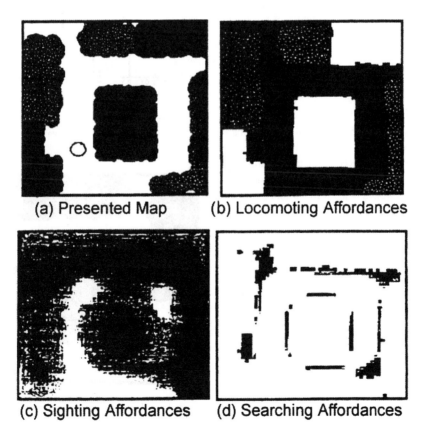

(a) Presented Map   (b) Locomoting Affordances

(c) Sighting Affordances   (d) Searching Affordances

FIGURE 14.2 The presented world map (a), a map of affordances for locomotion (b), a map of affordances for sighting objects (c), and a final searching affordance map (d).

and sight objects), the final searching affordance map in Figure 14.2d was created by superimposing Figures 14.2b and 14.2c. Figure 14.2d thus depicts ridges and peaks that maximally afforded the action of searching.

As explained in Kirlik et al. (1993), this functional, affordance-based differentiation of the environment provided an extremely efficient method for mimicking the search paths created by participants. We treated the highest peaks and ridges in this map as successive way-points that the Scout should attempt to visit at some point during the mission, thus possessing an attractive "force." Detailed Scout motion was then determined by a combination of these waypoint forces and the entire, finely graded, search affordance structure, or field. As one might expect, placing a heavy weight on the attractive forces provided by the waypoint peaks (as opposed to the entire field of affordances) resulted in Scout motion that looked very goal oriented in its ignorance of the immediately local search affordance field. However, reversing these weights resulted in relatively meandering, highly opportunistic Scout motion that was strongly shaped by the local details of the finely grained search affordance field.

In an everyday situation such as cleaning one's house, the first case would correspond to rigidly following a plan to clean rooms in a particular order, ignoring items that could be opportunistically straightened up or cleaned along the way. The second case would correspond to having a general plan, but being strongly influenced by local opportunities for cleaning or straightening up as one moved through one's house. In the actual, computational Scout World model, this biasing parameter was set in a way that resulted in scout search paths that best mimicked the degree of goal-directedness versus opportunism in the search paths observed.

For object-directed rather than region-directed actions, such as loading cargo or visiting home base, the Scout World's affordances were centered on those objects rather than distributed continuously in space. As shown in Figure 14.3, we created a set of dynamic affordance distributions for these discrete, object-directed actions for both the Scout and the four craft under supervisory control (F1–F4 in Figure 14.3a). Each of the 15 distributions shown in Figure 14.3a indicates the degree to which actions directed toward each of the environmental objects that can be seen in Figure 14.3b were afforded at a given point in an action-based (rather than time-based) planning horizon.

Space precludes a detailed explanation of how these distributions were determined (see Kirlik et al., 1993, for more detail). To take one example, consider the craft F1 over which the participant had supervisory control by entering action plans via a keyboard. F1 appears in the northwest region of the world as shown in Figure 14.3b, nearby a piece of cargo labeled C1. The *first action* affordance distribution for F1 indicates that loading C1 is the action most highly afforded for this craft, and a look down the column for all of the other craft, including the Scout, indicates that the affordance for loading this cargo is no higher for any craft other than F1. Thus, the model would in this case *decide* to assign the action of loading this piece of cargo to F1.

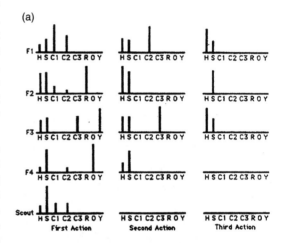

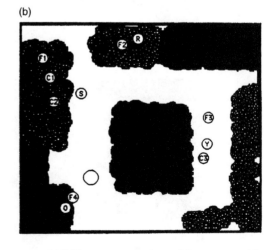

FIGURE 14.3 Two representations of the same world state: (a) functional representation in terms of dynamic affordance distributions; (b) representation in terms of visual form.

Given that F1 had been committed in this fashion, the model was then able to determine what the affordances for F1 would be at the time it had completed loading this cargo. This affordance distribution for F1 is shown in the *second action* column of distributions. Notice there is no longer any affordance for loading C1 (as this action will have been completed), and now the action of loading the cargo labeled C2 is most highly afforded. In this case, a plan to load this cargo allowed the model to generate a *third action* affordance distribution for F1, in this case indicating that the action of visiting home base H would be most highly afforded at that time, due to the opportunity to then score points by unloading two pieces of cargo.

What is absolutely crucial to emphasize, however, is that Figure 14.3 provides a mere snapshot of what was actually a dynamic system. Just moments after the situation represented by this snapshot, an event could have occurred that would have resulted in a radical change in the affordance distributions shown (such as the detection of an enemy craft by radar). Although I have spoken as if the model had committed to plans, these plans actually functioned solely as a resource for prediction, anticipation, and scheduling, rather than as prescriptions for action (cf. Suchman, 1987). The *perceptual* mechanisms in the model, tuned to measure the value of the environmental affordances shown in Figures 14.2 and 14.3, could be updated 10 times per second, and the actual process of selecting actions was always determined by the affordances in the first action distribution for all craft. Thus, even though the model would plan when enough environmental and participant-provided constraint on the behavior of the controlled system allowed it to do so, it abandoned many plans as well. A central reason for including a planning horizon in the model was to avoid conflicts among the four craft and the Scout: For example, *knowing* that another craft had a plan to act on some environmental object removed that object from any other craft's agenda, and *knowing* that no other craft's plans did not include acting on some other object increased the affordance for acting on that object for the remaining craft.

The components of the model intended to represent functions performed by internal cognition consisted of the previously mentioned perceptual mechanisms for affordance detection, a conflict resolution mechanism, and a simple mechanism for combining the affordance measures with priority values keyed to the task payoff structure (e.g., points awarded per type of object processed). Notably, as described in Kirlik et al. (1993),

these priority values turned out to be largely unnecessary since an experimental manipulation varying the task payoff structure (emphasizing either loading cargo or engaging enemy craft) by a ratio of 16:1 had *no* measurable effect on the behavior of participants (evidence suggested that they tried to process all of the discovered objects regardless of payoff). This finding lent credence to the view that participants' behavior was intimately tailored to the dynamic affordance structure of the Scout World, a set of opportunities for action that performers' actions themselves played a role in determining. Because that behavior involved a continual shaping of the environment, any causal arrow between the two would have to point in both directions (Dewey, 1896; Jagacinski & Flach, 2003). The general disregard of payoff information in favor of exploiting affordances is also consistent with the (or at least my) everyday observation that scattering water bottles around one's home is much more likely to prompt an increase of one's water consumption than any urging by a physician.

Additionally, we manipulated the planning horizon of the model and found that the variance that resulted was not characteristic of expert–novice differences in human performance. This task apparently demanded less *thinking-ahead* than it did *keeping-in-touch*. In support of this view, what *did* turn out to be the most important factor in determining the model's performance and a plausible explanation for expert–novice differences in this task, was the time required for each perceptual update of the world's affordance structure. As this time grew (from 0.5 s to 2 s), the model (and participants, our validation suggested) got further and further behind in their ability to exploit opportunistically the dynamic set of action opportunities provided by the environment, in a cascading, positive-feedback fashion. This result highlights that many, if not most, dynamic environments, or at least those we have studied, favor fast but fallible, rather than accurate but slow, methods for profitably conducting one's transactions with the world.

A final observation concerning our affordance-based modeling concerns the oft-stated finding that experts or skilled performers are frequently unable to verbalize rules or strategies that presumably *underlie* their behavior. When shown a concrete situation or problem, in contrast, these same experts are typically able to report a solution with little effort. This phenomenon is often interpreted using constructs such as *tacit knowledge* (Polanyi, 1966) or automaticity (e.g., Shiffrin & Dumais, 1981). If one does assume, for the sake of discussion, that much procedural knowledge exists in the form of if p, then q conditionals or rules, then our *Scout World* modeling provides a different explanation of why experts may often be unable to verbalize knowledge. Rather than placing such if p, then q rules in the *head* of our model, we instead created perceptual mechanisms that functioned to *see* the world functionally, as affordances, which we interpret as playing the roles of the p terms in the if p, then q construction. The q, however, is the internal response to assessing the world in functional terms, and as such, the if p, then q construct is distributed across the boundary of the human-environment system. Or at least this was the case in our computational model.

As such, even if the capability existed to allow our model to introspect and report on its "knowledge," like human experts it could not have verbalized any if p, then q rules either, since it contained only the "then q" parts of these rules. But if we instead showed the model any particular, concrete Scout World situation, it would have been able to readily select an intelligent course of action. Perhaps human experts and skilled performers have difficulty reporting such rules for the same reason: At high levels of skill, these conditionals, considered as knowledge, become distributed across the person-context system, and are thus not fully internal entities (cf. Greeno, 1987, on situated knowledge). Simon (1992) discussed the need to consider not only production rules triggered by symbol structures in working memory but also productions triggered by conditions in the external world to model situated action. By using both types of rules, Simon noted that, "Productions can implement either situated action or internally planned action, or a mixture of these" (Simon, 1992, p. 125). Our Scout World modeling demonstrates that it is certainly possible to computationally model situated action using conditionals in which the p elements of if p, then q rules exist in the (modeled) external environment rather than in the (modeled) head. The important point is that computationally modeling the external environment is necessary to give a modeler choice over whether the condition sides of condition–action rules should be located in the model of the head or in the model of the world. Making choices of this type is a hallmark of modeling truly distributed cognition.

Using Tools and Action to Shape One's Own Work Environment

In Kirlik (1998a, 1998b), I presented a field study of short-order cooking showing how more skilled cooks used strategies for placing and moving meats to create novel and functionally reliable information sources unavailable to cooks of lesser skill. We observed a variety of different cooks using three different strategies to ensure that each piece of meat (hamburgers) placed on the grill were cooked to the specified degree of doneness (rare, medium, or well). The simplest ("brute force") strategy observed involved the cook randomly placing the meats on the grill and using no consistent policy for moving them. As a result, this cook's external environment contained relatively little functionally relevant information. The second ("position control") strategy we observed was one where the cook placed meats to be cooked to specified levels at specified locations on the grill. As such, this strategy created functionally relevant perceptual information useful for knowing how well each piece of meat should be cooked, thus eliminating the demand for the cook to keep this information in internal memory. Under the most sophisticated (position + velocity control) strategy observed, the cook used both an initial placement strategy as well as a dynamic strategy for moving the meats over time.

Specifically, the cook placed meats to be cooked well done at the rear and right-most section of the grill. Meats to be cooked medium were placed toward the center of the grill (back to front) and not as far to the right as the meats to be cooked well done. Meats to be cooked rare were placed at the front and center of the grill. Interspersed with his other duties (cooking fries, garnishing plates, etc.), this cook then intermittently "slid" each piece of meat at a relatively fixed rate toward the left border of the grill, flipping them about halfway in their journey across the grill surface. Using this strategy, everything that the cook needed to know about the task was perceptually available from the grill itself, and thus, the meats signaled their own completion when they arrived at the grill's left boundary.

To abstract insights from this particular field study that could potentially be applied in other contexts (such as improving the design of frustratingly impenetrable information technology), we decided to model this behavioral situation formally, "to abstract away many of the surface attributes of work context and then define the deep structure of a setting" (Kirsh, 2001, p. 305). To do so, we initially noted that the function of the more sophisticated strategies could perhaps best be understood, and articulated, as creating constraints or correlations to exist between the value of environmental variables that could be directly observed and thus considered *proximal*, and otherwise unobservable, covert, or *distal* variables. As such, we were drawn to

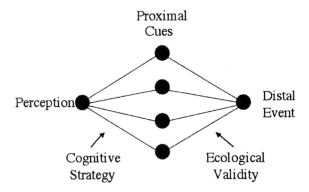

FIGURE 14.4 Brunswik's lens model of perception.

consider Brunswik's theory of probabilistic functionalism, which represents the environment in terms of exactly these functional, proximal–distal relations (Brunswik, 1956; Hammond & Stewart, 2001; Kirlik, 2006). These ideas are articulated within Brunswik's lens model, shown in Figure 14.4.

Brunswik advanced the lens model as a way of portraying perceptual adaptation as a "coming to terms" with the environment, functionally described as probabilistic relations between proximal cues and a distal stimulus. As illustrated in Hammond and Stewart (2001), this model has been quite influential in the study of judgment, where the cues may be the results of medical observations and tests, and the judgment (labeled "Perception" in Figure 14.4) is the physician's diagnosis about the covert, distal state of a patient (e.g., whether a tumor is malignant or benign). In our judgment research, we have extended this model to dynamic situations (Bisantz et al., 2000) and also to tasks in which cognitive strategies are better described by rules or heuristics rather than by statistical (linear regression based) strategies (Rothrock & Kirlik, 2003). Note that the lens model represents a distributed cognitive system, where half the model represents the external proximal–

distal relations to which an agent must adapt to function effectively, and the other half represents the internal strategies or knowledge by which adaptation is achieved.

Considering the cooking case, one deficiency of the lens model should become immediately apparent: In its traditional form, it lacks resources for representing the proximal–distal structure of the environment for action, that is, the relation between proximal means and distal ends or goals. The conceptual precursor to the lens model, originally developed by Tolman and Brunswik (1935), actually did place equal emphasis on proximal–distal functional relations in both the cue–judgment and means–ends realms. As such, we sought to extend the formalization of at least the *environmental* components of the lens model to include both the proximal–distal structure of the world of action, as well as the world of perception and judgment. The structure of the resulting model is shown in Figure 14.5.

This extended model represents the functional structure of the environment, or what Brunswik termed its *causal texture*, in terms of four different classes of variables, as well as any lawful or statistical relationships among them, representing any structure in the manner in which they may co-vary. The first [PP,PA] variables

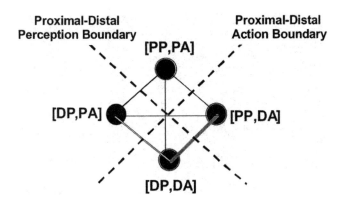

FIGURE 14.5 A functional model of the environment for perception and action.

are proximal with respect to both perception and action: Given an agent's perceptual and action capacities, their values can be both directly measured and manipulated (in Gibson's terms, they are directly perceptible affordances). [PP,DA] variables can be directly perceived by the agent but cannot be directly manipulated. [DP,PA] variables, however, can be directly manipulated but cannot be directly perceived. Finally, [DP,DA] variables can be neither directly perceived nor manipulated. Distal inference or manipulation occurs through causal links with proximal variables.

Note the highlighted link between the [PP,DA] variables and the [DP,DA] variables. These two variable types, and the single link between them, are the only elements of environmental structure that appear in the traditional lens model depicted in Figure 14.4. All of the additional model components and relations represented in Figure 14.5 have been added to be able to represent both the functional, perceptual, and action structure of the environment in a unified system. See Kirlik (1998b, 2006) for a more complete presentation.

To analyze the cooking case formally, we used this model to describe whether each functionally relevant environmental variable (e.g., the doneness of the underside of a piece of meat) is either proximal (directly perceivable; directly manipulable) or distal (must be inferred; must be manipulated by manipulating intermediary variables), under each of the three cooking strategies observed. Entropy-based measurement (multidimensional information theory; see McGill, 1954, for

the theory, see Kirlik, 1998b, 2006, for the application to the cooking study), revealed that the most sophisticated cooking strategy rendered the dynamically controlled grill surface not its "own best model" (Brooks, 1991) but rather a fully informative external model of the covert meat cooking process. This perceptible model allowed cooks to offload memory demands to the external world.

Quantitative modeling revealed that the most sophisticated (position + velocity) strategy resulted in by far the greatest amount of variability or entropy in the proximal, perceptual variables in the cook's ecology. This variability, however, was tightly coupled with the values of variables that were covert, or distal, to other cooks and thus this strategy had the function of reducing the uncertainty associated with this cook's distal environment nearly to zero. More generally, we found that knowledge of the demands this workplace task placed on internal cognition would be *underdetermined* without a precise, functional analysis of the proximal and distal status of both perceptual information and affordances, along with a functional analysis of how workers used tools to adaptively shape their own cognitive ecologies.

Modeling the Origins of Taxi Errors at Chicago O'Hare

Figure 14.6 depicts an out-the-window view of the airport taxi surface in a high-fidelity NASA Ames Research

FIGURE 14.6 Simulated view of the Chicago O'Hare taxi surface in foggy conditions. Courtesy of NASA Ames Research Center. (See color insert.)

Center simulation of a fogbound Chicago O'Hare airport. The pilot is currently in a position where only one of these yellow lines constitutes the correct route of travel. Taxi navigation errors and especially errors known as runway incursions are a serious threat to aviation safety. As such, NASA has pursued both psychological research and technology development to reduce these errors and mitigate their consequences. In my recent collaborative research with Mike Byrne, we completed a computational modeling effort using ACT-R (Anderson, chapter 4, this volume; Anderson & Lebiere, 1998) aimed at understanding why experienced airline flight crews may have committed particular navigation errors in the NASA simulation of taxiing under these foggy conditions (for more detail on the NASA simulation and experiments, see Hooey & Foyle, 2001; Foyle & Hooey, in press; for more detail on the computational modeling, see Byrne & Kirlik, 2005).

Notably, our resulting model was composed of a dynamic, interactive simulation, not only of pilot cognition but also of the external, dynamic visual scene, the dynamic taxiway surface, and a model of aircraft (B-767) dynamics. In our task analyses with subject-matter experts (working airline captains), we discovered five strategies pilots could have used to make turn-related decisions in the NASA simulation: (1) accurately remember the set of clearances (directions) provided by air traffic control (ATC) and use signage to follow these

directions; (2) derive the route from a paper map, signage, and what one can remember from the clearance; (3) turn in the direction of the destination gate; (4) turn in the direction that reduces the maximum of the X or Y (cockpit-oriented) distance between the aircraft and destination gate; (5) guess.

We were particularly intrigued by the problem of estimating the functional validity of the two "smart heuristics" (Raab & Gigerenzer, 2004; Todd & Schooler, chapter 11, this volume), involving simply turning in the direction of the destination gate. As such, we provided one of our expert pilots with taxiway charts from all major U.S. airports, and he selected those with which he was most familiar. He then used a highlighter to draw the taxi clearance routes he would likely expect to receive at each of these airports (258 routes were collected). We then analyzed these routes in terms of their consistency with the two "fast and frugal" heuristic strategies and found levels of effectiveness as presented in Figure 14.7.

We were quite surprised at the effectiveness, or functional validity, of these simple heuristic strategies over such a variety of airports. For example, at Sea-Tac (Seattle–Tacoma), these results suggest that a pilot could largely forget the clearance provided by ATC and simply make a turn toward the destination gate at every decision option and have ended up fully complying with the clearance that he or she would have most

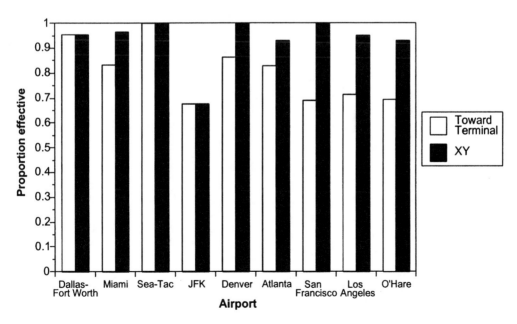

FIGURE 14.7 Accuracy of the two "fast and frugal" heuristics at nine major U.S. airports.

likely been given by ATC. After assembling similar information for all five decision strategies, the ACT-R Monte Carlo analysis of our integrated, functional model resulted in information indicating the frequency with which each of the five strategies would be selected as a function of the decision horizon for each turn in the NASA simulation (see Byrne & Kirlik, 2005).

Specifically, we found that for decision horizons between 2 and 8 s, our model predicted that pilots in the NASA experiments would have selected either the "toward terminal" or "minimize XY distance" heuristics, since within this time interval these heuristics had the highest relative accuracy. Furthermore, an examination of the NASA error data showed that a total of 12 taxi navigation errors were committed. Verbal transcripts indicated that eight of these errors involved decision making, while the other four errors involved flight crews losing track of their location on the airport surface (these "situation awareness" errors were beyond the purview of our model of turn-related decision making).

In support of our functional modeling, every one of the eight decision errors in the NASA data set involved either an incorrect or premature turn toward the destination gate. Finally, we found that at *every* simulated intersection in which the instructed clearance violated *both* heuristics, at least one decision error was made. In these cases, the otherwise functionally adaptive strategies used by pilots for navigating under low visibility conditions steered them astray because of atypical structure that defeated their typically rewarded experiential knowledge. Errors did not then result from a general lack of adaptation to the environment but rather from an overgeneralization of adaptive rules. Generally, adaptive decision rules, as measured by their mesh with environmental structure (Todd & Schooler, chapter 11, this volume) were defeated by ecologically atypical situations.

Discussion

Earlier in this chapter, I suggested that modeling interactive behavior and embedded cognition raises theoretical questions that are distinct from the types of theoretical questions that provided the traditional empirical foundation for many cognitive architectures. By "distinct" I meant that some of the theoretical questions that arise when modeling dynamic and interactive tasks are not necessarily reducible in any interesting

sense to the questions that motivated the design of these cognitive architectures. I hope that the three modeling examples presented in the previous section are at least somewhat convincing on this point. Each project required us to grapple with problems in cognitive and environmental modeling that I believe to be distinct from the types of questions normally addressed by many cognitive architectures. While one might make the observation that our first two modeling examples, the Scout World and short-order cooking could have benefited by our use of a cognitive architecture, I would not necessarily disagree. The important point to note is that some sort of detailed functional analyses of those tasks, either those presented or some alternative would have been required whether or not cognitive architectures were used as the partial repository for the information gained.

These examples illustrate the three general points presented in early sections of this chapter on the need to deal head-on with theoretical questions arising from the dynamic and interactive nature of embedded cognition. These include the need to model environmental sensitivity to environmental constraints and opportunities, the need to model highly context-specific cognitive and behavioral adaptations, and the need to analyze and model the environment of cognition and behavior in functional terms. I hope that our work demonstrates that even if one had available the scientifically ideal architecture for modeling internal cognition, many important and challenging theoretical questions about the nature of interactive behavior and embedded cognition would still remain. Addressing these questions will play a key role in advancing cognitive modeling in a very broad range of practically relevant contexts. While I certainly do not believe that the approach we have taken to addressing these questions is the final word by any means, I do hope that our ecological–cognitive modeling research has highlighted the need to address them.

Notes

1. At the outset, I wish to stress that my many comments in this chapter on the need for environmental modeling refer to models of the external world of the performer and not to models of the performer's internal representations of that world, although the latter may also be needed.

2. Actually, I believe this raises the question of whether the logic underlying the notion of *independent* and *dependent* variables is even appropriate in such situations (Dewey, 1896), but this issue is beyond the current scope.

References

Anderson, J. R., & Lebiere, C. (1998). *The atomic components of thought.* Mahwah, NJ: Erlbaum.

Baron, S. (1984). A control theoretic approach to modeling human supervisory control of dynamic systems. In W. B. Rouse (Ed.), *Advances in man-machine systems research* (Vol. 1, pp. 1–48). Greenwich, CT: JAI Press.

Bisantz, A., Kirlik, A., Gay, P., Phipps, D., Walker, N., & Fisk, A.D. (2000). Modeling and analysis of a dynamic judgment task using a lens model approach. *IEEE Transactions on Systems, Man, and Cybernetics,* 30(6), 605–616.

Brooks, R. (1991). Intelligence without representation. *Artificial Intelligence, 47,* 139–159.

Brunswick, E. (1956). *Perception and the representative design of psychological experiments.* Berkeley: University of California Press.

Byrne, M., & Kirlik, A. (2005). Using computational cognitive modeling to diagnose possible sources of aviation error. *International Journal of Aviation Psychology,* 15(2), 135–155.

Csikszentmihalyi, M. (1993). *Flow: The psychology of optimal experience.* New York: HarperCollins.

Dewey, J. (1896). The reflex arc in psychology. *Psychological Review, 3,* 357–370.

Donald, M. (1991). *Origins of the modern mind: Three stages in the evolution of culture and cognition.* Cambridge, MA: Harvard University Press.

Fajen, B. R., & Turvey, M. T. (2003). Perception, categories, and possibilities for action. *Adaptive Behavior, 11*(4), 279–281.

Foyle, D., & Hooey, R. (in press). *Human performance models in aviation: Surface operations and synthetic vision systems.* Mahwah, NJ: Erlbaum.

Gibson, J. J. (1979/1986). *The ecological approach to visual perception.* Hillsdale, NJ: Erlbaum. (Original work published in 1979.)

Gigerenzer, G., Todd, P. M., & the ABC Research Group. (1999). *Simple heuristics that make us smart.* New York: Oxford University Press.

Gleick, J. (1992). *Genius: The life and science of Richard Feynman.* New York: Pantheon.

Gluck, K. A., & Pew, R. W. (2005). *Modeling human behavior with integrated cognitive architectures.* Mahwah, NJ: Erlbaum.

Gray, W. D., & Kirschenbaum, S. S. (2000). Analyzing a novel expertise: An unmarked road. In J. M. C. Schraagen, S. F. Chipman, & V. L. Shalin (Eds.), *Cognitive task analysis* (pp. 275–290). Mahwah, NJ: Erlbaum.

Gray, W. D., Schoelles, M. J., & Fu, W. (2000). Modeling a continuous dynamic task. In N. Taatgen & J. Aasman (Eds.), *Proceedings of the Third International Conference on Cognitive Modeling* (pp. 158–168). Veenendal, The Netherlands: Universal Press.

Greeno, J. G. (1987). Situations, mental models, and generative knowledge. In D. Klahr & K. Kotovsky (Eds.), *Complex information processing* (pp. 285–316). Hillsdale, NJ: Erlbaum.

Hammond, K. R., & Stewart, T. R. (Eds.). (2001). *The essential Brunswik.* New York: Oxford University Press.

Hooey, B. L., & Foyle, D. C. (2001). A post-hoc analysis of navigation errors during surface operations: Identification of contributing factors and mitigating strategies. *Proceedings of the 11th Symposium on Aviation Psychology.* Ohio State University, Columbus.

Hutchins, E. (1995). *Cognition in the wild.* Cambridge, MA: MIT Press.

Jagacinski, R. J., & Flach, J. (2003). *Control theory for humans.* Mahwah, NJ: Erlbaum.

Kirlik, A. (1998a). The design of everyday life environments. In W. Bechtel & G. Graham, (Eds.), *A companion to cognitive science* (pp. 702–712). Oxford: Blackwell.

———. (1998b). The ecological expert: Acting to create information to guide action. *Fourth Symposium on Human Interaction with Complex Systems.* Dayton, OH: IEEE Computer Society Press: http://computer.org/proceedings/hics/ 8341/83410015abs.htm.

———. (2006). *Adaptive perspectives on human-technology interaction: Methods and models for cognitive engineering and human-computer interaction.* New York: Oxford University Press.

———, Miller, R. A., & Jagacinski, R. J. (1993). Supervisory control in a dynamic and uncertain environment: A process model of skilled human- environment interaction. *IEEE Transactions on Systems, Man, and Cybernetics,* 23(4), 929–952.

Kirsh, D. (1996). Adapting the environment instead of oneself. *Adaptive Behavior,* 4(3/4), 415–452.

———. (2001). The context of work. *Human-Computer Interaction, 16,* 305–322.

McGill, W. J. (1954). Multivariate information transmission. *Psychmetrika,* 19(2), 97–116.

Monk, A. (1998). Cyclic interaction: A unitary approach to intention, action and the environment. *Cognition, 68,* 95–110.

Neisser, U. (1976). *Cognition and reality.* New York: W. H. Freeman.

Newell, A. (1990). *Unified theories of cognition.* Cambridge, MA: Harvard University Press.

———. (1992). Author's response. *Behavioral and Brain Sciences,* 15(3), 464–492.

Polanyi, M. (1966). *The tacit dimension.* New York: Doubleday.

Raab, M., & Gigerenzer, G. (2004). Intelligence as smart heuristics. In R. J. Sternberg & J. E. Pretz (Eds.),

Cognition and intelligence. New York: Cambridge University Press.

Rothrock, L., & Kirlik, A. (2003). Inferring rule-based strategies in dynamic judgment tasks: Toward a non-compensatory formulation of the lens model. *IEEE Transactions on Systems, Man, and Cybernetics—Part A: Systems and Humans, 33*(1), 58–72.

Rouse, W. B. (1984). *Advances in Man-Machine Systems Research: Vol. 1.* Greenwich, CT: JAI Press.

———. (1985). *Advances in man-machine systems research: Vol. 2.* Greenwich, CT: JAI Press.

Runeson, S. (1977). On the possibility of "smart" perceptual mechanisms. *Scandinavian Journal of Psychology, 18,* 172–179.

Shah, K., Rajyaguru, S., St. Amant, R., & Ritter, F. E. (2003). Connecting a cognitive model to dynamic gaming environments: Architectural and image processing issues. *Proceedings of the Fifth International Conference on Cognitive Modeling (ICCM)* (pp. 189–194).

Sheridan, T. B. (2002). *Humans and automation: System design and research issues.* Santa Monica, CA: Human Factors and Ergonomics Society and Wiley.

———, & Johannsen, G. (1976). *Monitoring behavior and supervisory control.* New York: Plenum Press.

Shiffrin, R. M., & Dumais, S. T. (1981). The development of automatism. In J. R. Anderson, (Ed.), *Cognitive skills and their acquisition* (pp. 111–140). Hillsdale, NJ: Erlbaum.

Simon, H. A. (1956). Rational choice and the structure of environments. *Psychological Review, 63,* 129–138.

———. (1992). What is an "explanation" of behavior? *Psychological Science, 3*(3), 150–161.

Suchman, L. A. (1987). *Plans and situated actions.* New York: Cambridge University Press.

Tolman, E. C., & Brunswik, E. (1935). The organism and the causal texture of the environment. *Psychological Review, 42,* 43–77.

Vygotsky, L. S. (1929). The problem of the cultural development of the child, II. *Journal of Genetic Psychology, 36,* 414–434. (Reprinted as "The instrumental method in psychology." In J. V. Wertsh [Ed.], *The concept of activity in Soviet psychology* [pp. 134–143]. Armonk, NY: M. E. Sharpe, 1981).

Warren, W. H. (1984). Perceiving affordances: Visual guidance of stair climbing. *Journal of Experimental Psychology: Human Perception and Performance, 10,* 683–703.

PART V

INTEGRATING EMOTIONS, MOTIVATION, AROUSAL INTO MODELS OF COGNITIVE SYSTEMS

Vladislav D. Veksler & Michael J. Schoelles

If human cognition is embodied cognition, then surely physiological arousal, motivation, and emotions are part of this embodiment. Current theories argue both that cognition affects emotion (Lazarus, 1991; Gratch & Marsella, chapter 16; Hudlicka, chapter 19) and that emotion affects cognition (Busemeyer, Dimperio, & Jessup, chapter 15; Gunzelman, Gluck, Price, Van Dongen, & Dinges, chapter 17; Ritter, Reifers, Klein, & Schoelles, chapter 18). A major issue in understanding cognition is its integration with emotions, and vice versa. Having opened the Pandora's box of affective states, the world of cognitive science will never be quite the same.

Of course, the same logic that led to 100 years of study of tiny parts of cognitive processes in sterile and unchanging task environments can be used to justify the isolation of the study of cognition from the influence of emotion. However, in the face of mounting evidence that affect is a necessary concomitant of decision making (Damasio, 1995; Mellers, Schwartz, & Ritov, 1999) even the resistance of hard-core experimental

psychology seems to be crumbling. The paradigm has shifted; the issue is not how to avoid affect in accounts of cognition but how to account for behavior as emerging from a cognitive–affective control system.

The chapters in this section provide a glimpse of things to come. Emotions are presented as specific mechanisms that are integral and essential to cognition, as opposed to vague concepts that work in opposition to rational thought.

In chapter 15, Busemeyer et al. expand decision field theory (Busemeyer & Townsend, 1993) to account for the interaction of affective states with attention and decision utilities over time. These modifications allow for the direct implementation of affect within an influential theory of decision making.

In laying out another piece of the puzzle, Gratch and Marsella (chapter 16) discuss how appraisal theory of emotion explains the influence of emotion on cognition. They claim that appraisal theory can provide a unifying conceptual framework for control of disparate cognitive functions. They sharpen this argument by showing how appraisal theory influenced the design of the AUSTIN virtual human architecture.

Gunzelmann et al. (chapter 17) take on the specific task of accounting for the effects of fatigue by manipulating execution-threshold and goal-value parameters within an existing cognitive architecture. The resulting model provides a good fit to both fatigued and non-fatigued human performance.

Ritter et al. (chapter 18) propose a hypothetical set of *overlays* to account for the various effects of stress. In step with Gunzelmann et al., Ritter et al.'s overlays include execution-threshold and goal-value parameter manipulation. They also propose a set of other possible parameter, system, and model manipulations that may be the direct effects of emotional states on cognition.

In describing the MAMID architecture, Hudlicka (chapter 19) agrees with the parameter overlay approach of Ritter et al. and Gunzelmann et al. She additionally focuses on the effects of cognition on emotions, implemented as an affect appraiser module in MAMID.

Throughout the following chapters, emotions are treated as Type 1, Type 2, and Type 3 controls. It may be counter to the intuitions of traditional cognitive scientists to imagine that the influence of affective states can be so pervasive in cognition as not to be able to classify emotions as a single control type. On deeper analysis, however, it is surely possible to think of emotions as Type 3 goals, productions, and declarative knowledge, or as a separate Type 2 emotions module. And while implementing drives like arousal or hunger would be simplest to do using a Type 2 module, the neurophysiological elements associated with emotions (e.g., dopamine) are profuse throughout the brain and partake in all cognitive activity, arguing for emotions as part of a Type 1 systems control. At the very least, the basic pleasure/pain (seeking/preventative) behavior would seem to belong in Type 1 systems — at some level of description, this functionality already exists in many cognitive architectures.

Regardless of whether you subscribe to a Type 1, 2, or 3 implementation of emotions, and regardless of whether you subscribe to the dynamic systems approach (Busemeyer et al., chapter 15), the appraisal theory approach (Gratch et al., chapter 16), the ACT-R approach (Gunzelmann et al., chapter 17; Ritter et al., chapter 18), or the MAMID architecture approach (Hudlicka, chapter 19), analysis of the integration of emotions within the cognitive system does much to progress our understanding of the control of human cognition. Indeed, the chapters in this section may well represent the beginnings of the next-generation *mental* architectures—fully embodied and more capable of modeling a wider range of the human experience.

References

Busemeyer, J. R., & Townsend, J. T. (1993). Decision field theory: A dynamic-cognitive approach to decision making in an uncertain environment. *Psychological Review, 100*, 432–459.

Damasio, A. R. (1995). *Descartes' error: Emotion, reason, and the human brain.* New York: HarperCollins.

Lazarus, R. (1991). *Emotion and adaptation.* New York: Oxford University Press.

Mellers, B., Schwartz, A., & Ritov, I. (1999). Emotion-based choice. *Journal of Experimental Psychology-General, 128*(3), 332–345.

15

Integrating Emotional Processes Into Decision-Making Models

Jerome R. Busemeyer, Eric Dimperio, & Ryan K. Jessup

The role attributed to emotion in behavior has waxed and waned throughout the preceding century. When the recent cognitive revolution hit, theories of mental processes treated the brain as a computer. Models lost sight of the motivations and desires that went into thinking. In this chapter, we review research demonstrating an influential role for motivation and emotion in decision making. Based on these findings, we present a formal model for the selection of goals that integrates emotion and cognition into the decision-making process. This model is a natural extension of decision field theory, which has been successfully used to explain data in traditional decision-making tasks. This model assumes that emotions, motivations, and cognitions interact to produce a decision, as opposed to being processed independently and in parallel. By allowing emotion and cognition to coexist in a single process, we demonstrate a testable model that is consistent with existing findings.

The role attributed to emotion in behavior has waxed and waned throughout the preceding century. When the recent cognitive revolution hit, theories of mental processes treated the brain as a computer. Models lost sight of the motivations and desires that went into thinking. In this chapter, we review research demonstrating an influential role for motivation and emotion in decision making. Based on these findings, we present a formal model for the selection of goals that integrates emotion and cognition into the decision-making process. This model is a natural extension of decision field theory (Busemeyer & Townsend, 1993), which has been successfully used to explain data in traditional decision-making tasks. This model assumes that emotions, motivations, and cognitions interact to produce a decision, as opposed to being processed independently and parallel. By allowing emotion and cognition to coexist in a single process, we demonstrate a testable model that is consistent with existing findings.

During the heyday of neobehaviorism, motivational processes held sway over general system theories of behavior (Hull, 1943; Skinner, 1953; Spence, 1956). Basic drives and learned incentive motives were postulated to guide behavior. Theorizing about unobservable mental processes was shunned (Tolman, 1958, was an exception). Such a stilted understanding of mental processing eventually led to the downfall of these grand and systematic theories.

The rise of the computer-information-processing metaphor in the 1950s paved the way for a cognitive revolution. Cognitive scientists re-aligned their attention on mental processing mechanisms. Short- and long-term memory storage and retrieval were postulated and serial or parallel processes controlled flow of information. A second major attempt to construct general system theories of behavior was initiated (Anderson, Lebiere, Lovett, & Reder, 1998; Meyer & Kieras, 1997; Newell, 1990). However, motivation and emotion was foreign to computer systems, and it was eschewed by information processing theorists. The "goals" of a production rule system had to be hardwired directly by human hand. The cognitive revolution removed the heart from its systems, leaving an artificial intelligence unable to understand the value of its goals. This restricted view of motivation and emotions may eventually lead to the breakdown of the cognitive revolution.

This chapter presents a formal model for integrating emotion and cognition within the decision process that is used to select goals. Emotion enters this decision process by affecting the weights and values that form the basis of decisions. We begin by reviewing some basic facts and concepts from research on emotion. Then we review recent experimental research that examines the influence of emotion and motivation on decisions. Finally, we present a formal model called decision field theory (Busemeyer, Townsend, & Stout, 2002) where motives and emotions dynamically guide the decision process for selecting goals.

What Are We Trying to Integrate?

Let us begin by defining some basic concepts and summarizing some of their characteristics. *Plans* are action–event sequences designed to achieve specific goals, and *problem solving* is a process used to generate potential plans. *Decision-making* processes are used to select one of the plans generated by problem solving for execution. Decisions are based on *judgments*, which evaluate consequences and estimate likelihoods of events. Evaluations of consequences are based on the satisfaction or dissatisfaction of motives.

Motives are persistent biological and cultural needs. These consist of basic drives such hunger, thirst, pain, and sex; but secondary needs are built from these primary needs such as safety and affection; and eventually higher-order needs emerge, such as curiosity and freedom (see Maslow, 1962). *Emotions* are temporary states reflecting changes in motivational levels. For example, joy may be temporarily experienced by a sudden gain of power and wealth; anger may be experienced by a sudden loss of power and wealth; fear may be experienced by threatened loss of power and wealth. *Affect* is an evaluation of an emotional state according to a positive (approach) or negative (avoidance) feeling (movement tendency). For example, joy produces positive affect and anger produces negative affect. Emotions have a dynamic time course, and *moods* reflect lingering affect that can moderate later cognitive processing. For example, joy can produce a lingering positive mood, which can make a person feel optimistic about subsequent events; anger can produce a lingering negative mood, which

can make a person feel pessimistic about subsequent events (Lewis & Haviland-Jones, 2000). The dynamic nature of motives and emotions present a challenge to traditional static theories of decision making, and decision field theory attempts to address these dynamic characteristics formally.

What Are the Bases for Emotions?

Emotional experiences have broad influences across the neural, physiological, and behavioral systems. Emotional experiences produce changes in neural brain activation, increasing activation in some cases (such as fear), and decreasing activation in other cases (such as sadness). Neural transmitters are released, such as GABA inhibitors, or dopamine reward signals. Emotions produce hormonal responses—either adrenaline (epinephrine), causing anxiety and preparation for fleeing; or noradrenaline (norepinephrine), activating aggression and preparation for a fight. Physiological reactions of the autonomic nervous system consist of changes in pupil size, heart rate, respiratory rate, skin temperature, and skin conductance (from perspiration). The behavioral reactions include changes in facial expression and body posture, as well as programmed reactions and coping responses (fight or flight).

The cognitive system has a crucial function in interpreting, appraising, and facilitating these neural, physiological, and behavioral reactions (Schachter & Singer, 1962; Lazarus, 1991; Weiner, 1986). For example, if someone else caused an event to happen, and the person had substantial control over the event, and the event generated a negative effect, then your cognitive system would categorize this emotional experience as anger toward the person who caused this negative result. However, if you personally caused an event to happen, and you had control over the event, and the event generated a negative effect, then your cognitive system would categorize this emotional experience as guilt for your role in causing this negative result (see Roseman, Antonius, & Jose, 1996). Thus, the cognitive system categorizes the emotional experience on the basis of the affect and contextual information about the event.

Single Versus Dual System Views of Emotion

Neurophysiological research on emotions indicates that two neural pathways underlay emotional experiences

(Buck, 1984; Gray, 1994; LeDoux, 1996; Levenson, 1994; Panksepp, 1994; Scherer, 1994; Zajonc, 1980). First there is a subcortical direct route, which is fast, spontaneous, unconscious, physiological, and involuntary reaction. This is mediated through a direct (thalamus $\rightarrow$ amygdala $\rightarrow$ motor cortex) limbic circuit. Second, there is an indirect neocortical route, which has a slower-coping response based on a conscious appraisal of the situation. This is mediated through an indirect (thalamus $\rightarrow$ sensory cortex $\rightarrow$ prefrontal cortex $\rightarrow$ amygdala $\rightarrow$ motor cortex) neocortical circuit. Recently, however, Damasio (1994) has argued for an integration of two systems taking place in the orbital (ventral–medial) prefrontal cortex.

This neurophysiological evidence gives rise to opposing views about how emotions and cognitions interact to influence decision making. Some argue strongly that there are two separate and independent systems for making decisions; while others argue that these two sources are integrated into a single emotional–cognitive decision-making process.

A two-system point of view has been promoted by many theorists (Epstein, 1994; Hammond, 2000; Kahneman & Frederick, 2002; Loewenstein & O'Donoghue, 2005; Metcalfe & Mischel, 1999; Peters & Slovic, 2000; Sloman, 1996; Stanovich & West, 2000). According to this view, the first system is an emotional, intuitive, affective-based system for making decisions. It processes in parallel, is fast, implicit, unconscious, automatic, associative, noncompensatory, highly contextual, and experience based. This system places little demand on working memory. The second system is a rational, analytic, reasoning-based system for making decisions. It is slow, serial, explicit, conscious, controlled, compensatory, comprehensive, and abstractions based. This system places large demands on working memory. The systems operate independently, and only interact by having the second system correct the errors of the first, if needed, and if there is sufficient time and working memory available.

A single integrated system approach has been advocated by a smaller number of theorists (e.g., Damasio, 1994; Gray, 2004; Mellers, Schwarz, Ho, & Ritov, 1997). According to this view, emotions provide dynamic signals that feed into and help guide the cognitive system over time for making decisions. To describe exactly how this temporal integration and interaction occurs is a major challenge for this viewpoint. Decision field theory provides dynamic mechanisms for integrating fast emotional signals with slower cognition information to guide decisions.

Review of Research on Emotions and Decisions

This brief review is organized around a series of questions concerning the relevance of emotions for decision theory. For a more thorough review, see Loewenstein and Lerner (2003).

1. Do we need to change decision theory for emotional consequences?

Early evidence pointing toward a need to include emotion came from studies examining the effects of anticipated regret (Zeelenberg, Beattie, van der Plight, & de Vries, 1996; see also Mellers et al., 1997, for related research). In these experiments, participants were given a series of choices between safe versus risky gambles. On some trials, they were informed that they would receive outcome feedback immediately after the choice, while on other trials they were informed that feedback would not be provided. Standard utility theories predict that the opportunity for outcome feedback should not have any effect on preference; however, the expectation was that regret would be anticipated for not choosing risky options when feedback was presented. In agreement with the latter prediction, preferences tended to reverse and switch toward the riskier gamble when immediate feedback was anticipated.

Another line of evidence petitioning for change came from research on the effects of emotional outcomes on decision weights (Rottenstreich & Hsee, 2001). According to weighted utility theories, the utility of a simple gamble of the form "win x with probability p, otherwise nothing" is determined by the product of the utility of the outcome, x, multiplied by the decision weight associated with the probability p. Both the utility and the decision weight are subjective and depend on an individual's personal beliefs and values. However, a critical assumption is that these two factors are separable, and in particular, the decision weight is a function of p alone and not a function of x. This decision weight function has typically been estimated using monetary gambles, and it is usually found to be an inverse S-shaped function of p (Kahneman & Tversky, 1979). However, Rottenstreich & Hsee (2001) found that the shape of the decision-weight function changed depending on whether the x was a purely monetary outcome versus an outcome with greater affective impact (e.g., avoidance of an electric shock). The decision-weight function was estimated to be flatter in the middle of the probability scale when emotional outcomes were used as compared with monetary outcomes.

Emotions also change the rate of temporal discounting in choices between long-term large rewards over short-term smaller rewards. Gray (1999) found that participants who were shown aversive images (producing a feeling of being threatened) had higher discount rates. Stress focused individuals' attention on immediate returns making them appear more impulsive.

Finally, a third line of evidence comes from research examining the type of decision strategy used to make choices (Luce, Bettman, & Payne, 1997). Compensatory strategies, such as those enlisting a weighted sum of utilities, require making difficult trade-offs and integrating information across all the attributes. Noncompensatory strategies, such as a lexicographic rule, only require rank ordering alternatives on a single attribute and thus avoid difficult trade-offs. Luce, Bettman, & Payne (1997) found that when faced with emotionally difficult decisions, individuals tend to switch from a compensatory to noncompensatory strategies to avoid making difficult negative emotional trade-offs.

2. Can emotions distort or disturb our reasoning processes?

One line of evidence supporting this idea comes from research on emotional carryover effects (Goldberg, Lerner, & Tetlock, 1999; Lerner, Small, & Loewenstein, 2004). For example, in the study by Goldberg et al., participants watched a movie about a disturbing murder. The murderer was brought to trial, and in one condition, the murderer was acquitted on a technicality, but in another condition, the murderer was found guilty. After watching the film, the participants were asked to make penalty judgments for a series of unrelated misdemeanors. Goldberg et al. found that when the murderer was freed on a technicality, anger aroused by watching the murder movie spilled over to produce higher punishments for unrelated crimes, as compared with the condition in which the murderer was convicted.

Shiv and Fedorikhin (1999) examined conflicts between motivation and cognition. Participants were given a choice between a healthy and unhealthy snack under either a high-stimulating condition (real cakes or fruit snacks visibly present) or a low-stimulating condition (symbolic information about cakes and fruits). Also in one condition, they made this decision under a high memory load (they were asked to rehearse items for a later recall test) or under no memory load. Considering the reasons for the choice, participants generally favored the healthy snack. However, when hunger

was stimulated under the vivid condition, and the healthy thoughts were suppressed (by the working memory task), then preferences reversed and the unhealthy snack was chosen most frequently.

A similar line of research was conducted by Markman and Brendl (2000). Habitual smokers were offered the opportunity to purchase raffle tickets for one of two lotteries—one with a cash prize and one with a cigarette prize, to be awarded after a couple of weeks delay. Half of the smokers were approached before smoking a postclass cigarette (and hence they had a strong need to smoke a cigarette). The other half were approached just after smoking their postclass cigarette (and hence the strength of the need to smoke a cigarette was diminished). Those who had not yet smoked purchased more raffle tickets to win cigarettes than did those who had already smoked. In contrast, they purchased fewer raffle tickets for the cash prize than did those who already smoked. Thus the need to smoke exaggerated the value of the cigarette lottery relative to the monetary lottery even though the latter could be used to purchase cigarettes.

3. Does reasoning always improve decision making?

Reasoning does not appear to universally improve decisions. Wilson et al. (1993) asked participants to provide their preference for either posters picturing animals in playful poses or posters of abstract, impressionist paintings. One group of participants was forced to provide reasons for their preference whereas the other group was not. Those who were forced to provide reasons for their preference were more likely to prefer the posters of the cute animals, whereas those who were not compelled to provide reasons preferred the impressionist posters. Participants were given the poster of their choice and a few weeks later were asked how satisfied they were with their selection. Those who were forced to provide reasons for their preference were significantly less satisfied with their selection than those who were not. These results suggest that thinking about reasons led to a focus on information about the domain that was not important to people in the long run.

4. Can we predict the effect of emotions on our decisions?

Research indicates we are not very good at predicting the influence of emotions on our choices. Loewenstein & Lerner (2003) review a number of experiments illustrating what they call *hot–cold empathy gaps*. When in a cold state (not hungry), people underpredict how they will feel in a hot state (hungry) (see Read & van Leeuwen, 1998). When in a hot state (sexually aroused), people cannot accurately predict how they will later feel when in a cold state (morning-after effect) and vice versa. This concludes our brief review illustrating some interesting interactions between motives, emotions, and decision making.

Decision Field Theory

Now we summarize a dynamic theory that describes how to incorporate motivational processes into decision making. First, we introduce a dynamic model of decision making called *decision field theory* (DFT). This theory has been previously used to explain choices between uncertain actions (Busemeyer & Townsend, 1993), multiattribute choices (Diederich, 1997), multi-alternative choices (Roe, Busemeyer, & Townsend, 2001), and the relations between choices and prices (Johnson & Busemeyer, 2005). In this chapter, we build on our previous efforts to extend decision field theory to account for the effects of motivation and emotion on decision making (Busemeyer et al., 2002). DFT advances older static models by providing a dynamic account of the decision process over time. This is important for explaining interactions between emotional and cognitive processes as the product of one integrated system rather than as a two system approach.

Decision Process

It will be helpful to have a concrete decision in mind when presenting the theory. The following example was chosen to highlight the application of the theory to navigational decisions under emergency or crisis conditions producing high time pressure and high emotional stress. A man was on a mission that required riding cross country on his motorcycle. He was cruising around 50 mph down a two-lane state highway when he came up behind a truck full of old car tires. The highway was not in good shape, with many potholes left by snowplows from the previous winter. The truck bumped into of one of these pits, causing a tire to somersault out of the truck and land flat on the road, directly in the motorcyclist's path. Although this example concerns navigating a motorcycle, it contains aspects that are shared in other navigational decisions, such as emergencies that occur during a plane flight.

The motorcyclist assessed the situation and noted that there was no shoulder on the road to serve as an escape route and that there was a line of cars following closely behind him. Thus the man was faced with a difficult problem-solving task, upon which he very quickly generated three potential plans of action: (A) drive straight over the tire, (B) swerve to the side, or (C) slam on the breaks. Each action involved planning a complex sequence of perceptual-motor movements. For example, driving straight across the tire required accelerating a little to push across the tire, hitting the tire dead center with sufficient speed to overcome it, a strong grip on the handlebars, and careful balancing of the bike.[1]

Each course of action could result (for simplicity) in one of four possible consequences: (c_1) a safe maneuver without damage or injury; (c_2) laying the motorcycle down and damaging the motorcycle, but escaping with minor cuts and bruises; (c_3) crashing into another vehicle, damaging the motorcycle, and suffering serious injury; (c_4) flipping the motorcycle over and getting killed.

An abstract representation for this decision problem is shown in Table 15.1, where the rows represent actions, columns represent consequences, and the cells represent the likelihoods that an action produces a consequence. The affective evaluations of the consequences are represented by the values m_j shown in the columns of the table, and the beliefs are represented by decision weights, w_{ij}, shown in the cells of the table. In the motorcyclist's opinion, option A was very risky, with high possibilities for the extreme consequences, c_1 and c_4. Action B was more likely to produce consequence c_2, and action C was more likely to produce consequence c_3.

The basic ideas behind the decision process are illustrated in Figure 15.1. The horizontal axis represents time (in milliseconds), beginning from the onset of the decision (when the tire initially flipped out of the truck) until the action was taken. The vertical axis represents strength of preference for each of the courses of action. Each trajectory represents the evolution of the preference strength for one of the options over time. At each moment in time, the decision maker

TABLE 15.1 Abstract Payoff Matrix for Motorcycle Decision

Actions	Possible Consequences			
	m_1	m_2	m_3	m_4
Act A	w_{A1}	w_{A2}	w_{A3}	w_{A4}
Act B	w_{B1}	w_{B2}	w_{B3}	w_{B4}
Act C	w_{C1}	w_{C2}	w_{C3}	w_{C4}

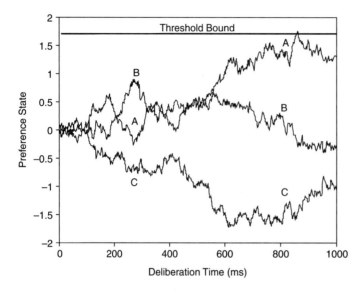

FIGURE 15.1 A simulation of the decision process. Horizontal axis is time, vertical axis is preference strength, and each trajectory represents one course of action. The top bar is the threshold bound. The first option to hit the bound wins the race and is chosen. (See color insert.)

anticipates the possible consequences of an action, and attention switches from one action and consequence to another over time. According to this figure, the man begins (in the region of 100–200 ms) considering advantages favoring option A (e.g., he thinks for a moment that he may be able to safely pass over the tire, and slamming on the breaks may cause the car behind to crash into him, and rapidly swerving to the side could cause the motorcycle to flip over). However, some time later (shortly after 200 ms) his attention switches, and he reconsiders advantages of option B (e.g., he now fears choosing option A may cause the tire to get entangled with the chain of the motorcycle and flip the bike over). These comparisons are accumulated or temporally integrated over time to form a preference state for each course of action. For example, just 250 ms into the deliberation process, the preference state for option B dominates; later (at 600 ms), the preference

state for A overcomes, and after 800 ms, it crosses a threshold bound and wins the race. It is at this point that option A is chosen as the planned course of action (plan to drive straight over the tire). Note that according to this description, emotions and rational beliefs are integrated rapidly and effectively into a single-preference state across time to guide decisions.

The threshold bound for stopping the deliberation process is a criterion that the decision maker can use to control the speed and accuracy of a decision. If the threshold is set to a very high value, then more information is accumulated, but at the cost of longer decision times. If the threshold bound is set to a very low criterion, then less information is accumulated, but with less time. In this example, under severe time pressure, the threshold bound must be set at a relatively low criterion.

This decision process can be formulated as a connectionist model as illustrated in Figure 15.2. Affective

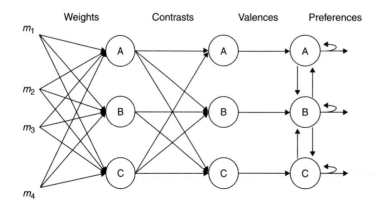

FIGURE 15.2 Diagram illustrating the connectionist interpretation of the decision process. The inputs are the evaluations of consequences, the first layer represents weighted evaluations, the second layer represent valences for each option, and the third layer represents the preference state for each option.

evaluations of the various possible consequences represent the inputs into the decision system. (These are represented by the *m*s shown on the far left.) The input evaluations are filtered by an attention process, which maps the evaluations of consequences into a momentary evaluation for each course of action (represented by the second layer of nodes). Then the momentary evaluations are transformed into valences, one for each course of action, represented by the third layer of nodes. The valence of an action represents the momentary advantage or disadvantage of that action compared to the other actions. Finally, the valences are input to a recursive system at the final layer, which generates the preference states at each moment in time. These preference states are the final outputs, which produce the trajectories shown in Figure 15.1.

More formally, the amount of attention allocated to the *j*th consequence of the *i*th action at time *t* is denoted, $W_{ij}(t)$. This attention weight is assumed to fluctuate from moment to moment, according to a stationary stochastic process. The mean of this process generates the decision weight, $E[W_{ij}(t)] = w_{ij}$. For example, if attention switches in an all or none manner, then $W_{ij}(t) = 1$ or 0, and $w_{ij} = E[W_{ij}(t)]$ is the probability that attention will be focused on a consequence of an action at any moment. Thus, the decision weight is the average amount of time spent thinking about a consequence. It is assumed to be affected by the likelihood of the consequence, but according to this interpretation, other factors that attract attention may also affect these decision weights.

The momentary evaluation of the *i*th action is an attention-weighted average: $U_i(t) = \sum W_{ij}(t) \cdot m_j$, where *j* is an index associated with one of the possible consequences of an action, $W_{ij}(t)$ represents the amount of attention allocated to a particular consequence at any moment, and m_j is the affective evaluation of a consequence. Note that $U_i(t)$ is a random variable (because $W_{ij}[t]$ is a random variable), but its mean is a weighted average $E[U_i(t)] = \sum w_{ij} \cdot m_j = u_i$, which corresponds to a weighted utility commonly used by decision theorists (cf. Luce, 2000).

The valence of an action is defined as the difference $v_i(t) = U_i(t) - U_.(t)$, where $U_.(t)$ is the average evaluation over all actions.[2] The valence represents the momentary advantage/disadvantage for option *i* at time *t* compared with the average of all actions at that moment. The sum across valences always equals zero.

The valences for an action are integrated over time to form a preference state for each action, denoted

P_i for option *i*. This preference state can range from positive (approach) to zero (neutral) to negative (avoidance). Each preference state starts with an initial value, $P_i(0)$, which may be biased by past experience (in Figure 15.1, they start out unbiased). The preference state evolves during the deliberation according to the following linear dynamic stochastic difference equation (where *h* is a small time step):

$$P_i(t + h) = \sum s_{ij} \cdot P_j(t) + v_i(t + h). \qquad (1)$$

The coefficients s_{ij} allow feedback from previous preference states to influence the new state. The self-feedback coefficient, s_{ii}, controls the memory for past valences. The lateral inhibitory links, $s_{ij} = s_{ji}$ for $i \neq j$, produce a competitive system in which strong preferences grow and weak preferences are suppressed. Lateral inhibition is commonly used in artificial neural networks and connectionist models of decision making to form a competitive system in which one option gradually emerges as a winner dominating over the other options (cf. Grossberg, 1988; Rumelhart & McClelland, 1986). The lateral inhibitory coefficients are important for explaining context effects on choice (see Roe et al., 2001).

In summary, a decision is reached by the following deliberation process: As attention switches across consequences over time, different affective values are probabilistically considered, and these values are compared across actions to produce valences, and finally these valences are integrated into preference states for each action. This process continues until the preference for one action exceeds a threshold criterion, at which point the winner is chosen. Note that a single system is postulated to temporally integrate rational beliefs about potential consequences with affective reactions to these consequences over time.[3]

To illustrate the dynamic behavior of the model, consider a decision whether to take a gamble. Suppose action A has an equal chance of winning $250 or losing $100, and action B is just status quo (not gambling, not winning, or lose anything). In this simple case, we set the evaluations to the following values: ($m_1 = 250/250 = 1$, $m_2 = 0$, $m_3 = -100/250 = -.4$). For Action A, we assume a .50 probability of attending to m_1 and .50 probability of attending to m_3; that is, $w_{A1} = E[W_{A1}(t)] = .50$, and $w_{A3} = E[W_{A3}(t)] = .50$. For action B, only one outcome is possible, zero, so that $w_{B2} = E[W_{B2}(t)] = 1$. The time step was set to $h = .01$, self-feedback was set to $s_{ii} = 1 - (.07) \cdot h$, the lateral inhibition was set to $s_{AB} = s_{BA} = 0$, and the initial state was set to $P_A(t) = -1$ (initially biased in favor of not playing).

Under these assumptions, we ran a simulation 5,000 times (see Appendix A) to generate the choice probabilities and the mean deliberation times, for a wide range of threshold parameters (θ ranged from 1 to 5 in steps of .25).[4] Figure 15.3 plots the relation between choice probability and mean decision time for option A, the gamble, as a function of the threshold parameter. Both decision time and choice probability increase monotonically with the threshold magnitude, starting below 50% choice of the gamble (because of the initial bias) and gradually rising above 50% choice for the gamble (because it has a positive expected value). Busemeyer (1985), Diederich (2003), and Diederich and Busemeyer (2006) presents empirical evidence supporting these types of dynamic predictions for choices between gambles.

Affective Evaluation of Consequences

Now we turn to a more detailed analysis of the evaluations, m_j, and how they are affected by emotions. In general, consequences are described and evaluated according to various objectives or attributes that a person is trying to maximize (or, as in this case, minimize). In the motorcycle example, the evaluation of consequences depends on minimizing two attributes: personal injury and motorcycle damage. Note that the motorcyclist may be willing to sacrifice some personal injury

to avoid motorcycle damage. The success of the mission (the cross-country trip) depends on an operational motorcycle, and a few cuts and bruises will heal and can be tolerated.

The effect of an attribute on an evaluation of a consequence depends on two factors: (1) the quality or amount of satisfaction that a consequence can deliver with respect to an attribute and (2) the importance or need for the attribute. For example, suppose a consequence scores high with respect to minimizing personal injury but low with regard to minimizing motorcycle damage. The final evaluation depends on the importance of the motorcycle relative to personal injury. If the mission is very important, and the motorcycle is crucial for completing the mission, then this is evaluated as an unattractive consequence; however, if the mission and the motorcycle are not considered very important, then this is an attractive consequence. Thus attribute importance moderates the effect of attribute quality.

More formally, decision theorists (cf. Keeney & Raiffa, 1976) generally postulate that each consequence can be characterized by a number of attributes, and each attribute has an importance weight, here denoted n_k for the kth attribute. Additionally, each consequence has a quality (amount of satisfaction) that can be gained on an attribute, here denoted as q_{jk} for the value of the jth consequence with respect to the kth attribute. These two factors are combined according to a multiplicative rule, $n_k \cdot q_{jk}$ to produce the net effect

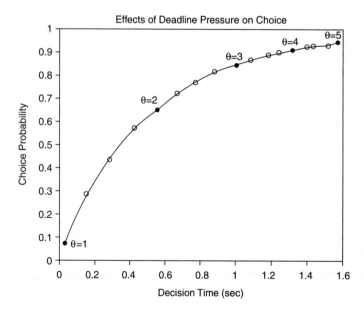

FIGURE 15.3 Multiple simulations demonstrate that the choice probability and the average decision time increase monotonically as the threshold θ is increased.

of an attribute on the evaluation of a consequence. Furthermore, if the attributes are independent, then the effects of each attribute add to form a weighted value of a consequence: $m_j = \sum n_k \cdot q_{jk}$.

So far, this is simply a static representation of an evaluation, which is commonly used by decision theorists. Decision field theory (Busemeyer et al., 2002) departs from this static representation by postulating that importance weights depend on personal needs, which are assumed to vary dynamically across time: $n_k(t)$. DFT also diverges by considering the quality a consequence has on an attribute as the degree of satisfaction a consequence is expected to provide with respect to the attribute: q_{jk}. Consequently, we assume that evaluations are changing across time according to $m_j(t) = \sum n_k(t) \cdot q_{jk}$, and momentary evaluations now involve stochastic attention weights as well as dynamic evaluations: $U_i(t) = \sum W_{ij}(t) \cdot m_j(t)$. The rest of decision field theory (e.g., Equation 1) accommodates this new dynamical feature in a natural way, as it continues to operate in the same manner as previously described for making decisions. This is one of the advantages of using a dynamic model for decision making.

Personal needs, $n_k(t)$, are postulated to change across time. A control feedback loop forms the basis for adjusting these needs over time (Busemeyer et al., 2002; see also Carver & Sheier, 1990; Toates, 1980). We assume that an individual has an ideal point on each attribute, denoted as g_k (for goal state) as well as a current level of achievement or status quo for an attribute, denoted $a_k(t)$. The discrepancy between these two

values, $[g_k - a_k(t)]$, provides a feedback signal for adjusting the need for that attribute, $n_k(t)$. For example, if g_k is the ideal level of hunger, and $a_k(t)$ is the current level of hunger (operationalized as hours without food), then the difference between these two determines the adjustment for the need to eat. Positive discrepancies produce an increase in need, and negative discrepancies produce are decrease in need. Accordingly, the need for an attribute varies across time according to the following difference equation:

$$n_k(t + h) = L_k \cdot n_k(t) + [g_k - a_k(t + h)], \qquad (2)$$

where L_k is a constant that determines the rate of feedback control of needs over time, which may depend on the type of attribute. For example, the consumatory effect of eating when hungry may be slower than the consumatory effect of drinking when thirsty. These differential feedback rates provide a formal means to account for the fast direct versus slow indirect neural pathways for emotion in the brain.

Figure 15.4 provides a depiction of the integrated cognitive–motivational network, illustrating how cognitions and emotions interact over time. Returning to the motorcyclist's decision, we can trace the decision process along the network. We assume that there is an ideal goal state for maintaining the operation of the motorcycle (and completing the mission) g_m as well as a goal for personal safety g_p. Let us focus on changes in the needs for personal safety $n_p(t)$ during the deliberation process. The sudden appearance of the tire in the middle of the road produces an abrupt drop in the

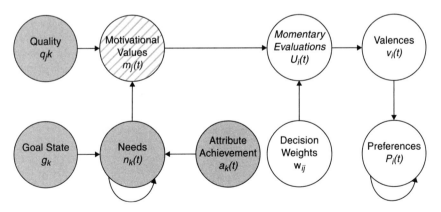

FIGURE 15.4 Cognitive–motivational network. The highlighted regions indicate the parts of the decision process related to emotions. The difference between the *goal state* and the actual *attribute achievement* is used to update the *need* for a given attribute. This need is in turn combined with expected satisfaction (*quality*) to provide a motivational value for all considered consequences.

TABLE 15.2 Decision Weights for Motorcycle Example

Consequence	c_1	c_2	c_3	c_4
Action	No damage; no injury	Damage motorcycle; minimal injury	Damage motorcycle; serious injury	Get killed
Action A (Drive straight)	.55	0	0	.45
Action B (Swerve)	0	.90	.10	0
Action C (Slam brakes)	0	.10	.90	0

current level of personal safety, that is a drop in the variable $a_p(t)$ (emotionally felt as fear). This generates a gap or an error signal, $[g_p - a_p(t)]$, which causes a rapid growth in the need for personal safety, $n_p(t)$. The quality (amount of satisfaction) that a consequence produces for personal safety, q_{ip}, will then be combined with the need for personal safety, $n_p(t)$, to generate a dynamic value $m_j(t)$ of each consequence. These dynamic values are combined with the shifting attention weights, $W_{ij}(t)$, to form momentary evaluations, $U_i(t)$. The momentary evaluation of an action is compared with other actions to produce a valence for each action, $v_i(t)$. Finally, the valences feed into the preference states $P_i(t)$ to determine the selected course of action.

Computation Example Applied to Emergency Decisions

To illustrate an important dynamic property generated by Equation 2, let us return to the motorcyclist's dilemma. Tables 15.2 and 15.3 show the decision weights and the quality values used in this example. According to Table 15.2, action A (driving straight

across the tire) is risky — it is likely to produce either of the two extreme consequences, a c_1 (safe maneuver) or c_4 (getting killed); action B (swerving) is likely to produce an intermediate but safer consequence c_2 (laying down the motorcycle); and action C (slamming the brakes) is likely to produce consequence c_3 (hitting a vehicle). The qualities, q_{ik} (achievement scores), on the personal safety and motorcycle maintenance attributes, are shown in Table 15.3 (higher scores are more desirable). According to Table 15.3, c_1 scores best on both attributes, c_2 score well on the first attribute but very poorly on the second, c_3 score moderately bad on both, and c_4 scores the worst on both. Additionally, the predictions for the mean preferences were computed from Equations 1 and 2 using the following dynamic parameters: we set the gaps equal to $[g_p - a_p(t)]$, = .80 and $[g_m - a_m(t)]$, = .40, indicating a larger gap for personal safety; L_P = .90 and L_M = .70, indicating a larger feedback control parameter for the personal safety attribute; and finally we set s_{ii} = .9 (self-feedback) and s_{ij} = −.05 (lateral inhibition for $i \neq j$) in Equation 1 to control the dynamics of the preference states. (The time step was set to $h = 1$ for simplicity.)

TABLE 15.3 Quality Values for Motorcycle Example

Consequence	Attribute	
	Personal Safety	Motorcycle Maintenance
c_1 (Safe maneuver)	1	1
c_2 (Lay down motorcycle)	.70	0
c_3 (Crash into vehicle)	.20	.40
c_4 (Flip motorcycle)	0	0

Note: In DFT, quality is the degree of satisfaction a consequence is expected to provide with respect to the attribute.

The predictions are shown in Figure 15.5. As can be seen in the top panel of this figure, the need for personal safety grows more slowly and to a much higher asymptote as compared with the need for motorcycle maintenance. This shift in needs produces a reversal in preference over time between actions A and B. As can be seen in the bottom panel, the risky action, A, initially is preferred, but later the safer action B dominates. In other words, as deliberation progresses, the person's preference switches from the risky to the safer action. This shows how the model can explain what is called the chickening out effect (Van Boven, Loewenstein, & Dunning, 2005). In conclusion, DFT allows for preference reversals over time, which cannot occur with a static utility theory.[5]

Applications to Previous Research

Let us briefly outline a DFT account of some of the past findings reviewed earlier. Consider first the emotional carryover effect reported by Goldberg et al. (1999).

In this case, anger aroused in the first part of an experiment carried over to affect punishment decisions in second phase. According to DFT, the anger aroused in the first stage decays exponentially over time as described by Equation 2. This persistence of anger would enhance the need and thus the importance for the retribution attribute of later punishment decisions. The earlier studies did not examine how this effect changes over time, but one testable prediction from the present theory is that the effect should decay exponentially as a function of the time interval between the initial arousal of anger and the subsequent test on irrelevant penalty judgments.

Next consider the experiment by Shiv and Fedorikhim (1999) who examined conflicts between reasons and emotions. According to DFT, hunger stimulation produces an increase in the need to satisfy hunger, increasing the importance of the food taste attribute, and consequently increasing the preference for the unhealthy snack; at the same time, the memory load would decrease the attention weight to the attributes related to health maintenance. A test of this

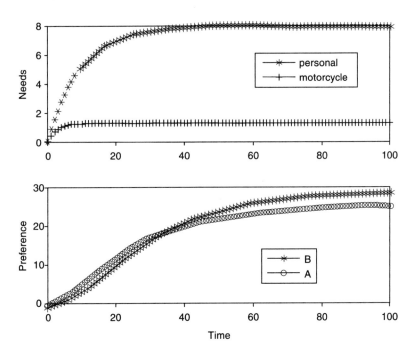

FIGURE 15.5 Results of simulations run under conditions $\lambda_P = .80$ and $\lambda_M = .40$; indicating a larger gap for personal safety; $L_P = .90$ and $L_M = .70$ indicating a larger feedback control parameter for the personal safety attribute; and $s_{ii} = .90$ and $s_{ij} = -.05$ for $i \neq j$. The need for personal safety gradually grows toward an asymptote. This change in need directly causes an increase in preference for the less risky Action B.

theory could be performed by factorially manipulating the taste quality of the unhealthy snack and the degree of hunger. We predict that these two factors should interact according to a multiplicative rule.

In the study by Markman and Brendl (2000), unsatiated smokers were more interested in winning cigarettes than those who had just finished smoking. In the framework of DFT, smoking decreases the need for future smoking and decreases the importance of obtaining more cigarettes. The motivation for winning cigarettes over money is lowered. We predict that the size of the cigarette prize will interact according to a multiplicative rule with the time since smoking a cigarette.

As a final example, consider the study by Rottenstreich and Hsee (2001) who found that emotions cause changes in the probability weighting function. This finding cannot be readily explained in terms of the effects of emotions on needs for attributes. In this case, we may be required to formulate a new mechanism that allows emotions to moderate the amount of attention various consequences receive (i.e., a model for changing the decision weights, w_{ij}, depending on the quality of the emotion produced by an outcome). This has yet to be done within the DFT framework.

Concluding Comments

Emotions and motives are dynamic reactions to environmental challenges. A threat, for example, rapidly generates a fear reaction, promoting actions to seek safety to quell the rising fear. Consequently, dynamic models are required to track their effects on decisions over time. Decision field theory differs prominently from other standard decision theories by providing a dynamic description of decision processes. This characteristic of the theory provides a natural way to incorporate the dynamic effects of emotions and motivations. The goal of this chapter was to present a formal model for integrating cognition and emotion into a single decision process.

At the beginning of this chapter, we presented two opposing views about the way emotions could influence decision making. According to a two system view, there is a Type 1 (emotion-based system) and a Type 2 (reason-based system), and the Type 1 system is corrected by the Type 2 system. Alternatively, according to a single-system view, motives, and emotions control cognition by adjusting the importance weights on the basis of

needs, which vary dynamically over time. This integrated view of emotion and cognition was actually proposed long ago by one of the founders of the cognitive revolution, Herb Simon (1967). Decision field theory provides a formal mechanism for implementing the single system view.

It is worthwhile to step back and try to assess the advantages and disadvantages of each approach. One of the important reasons for advocating a two-system approach is the fact there are at least two separate neural pathways for emotions, the direct versus the indirect path. However, the two-system approach has not been clearly articulated as a detailed neural model and so this remains a fairly rough correspondence at best. Furthermore, it not difficult to incorporate fast and slower emotional signals into a dynamic integrated model of cognition, and so this is not strictly speaking evidence for a two systems approach. In particular, the dynamic model for need (represented by Equation 2) provides differential feedback rates to accommodate fast direct versus slow indirect neural pathways for emotion.

One of the main advantages of the integrated approach presented here is that we have a formal or computational model that is able to derive precise predictions for cognitive and emotional interactions. The separate systems approach fails on this criterion. Although the System 2 part of the theory is precisely worked out (this is just the standard utility theory), there is a lack of formal modeling for the Type 1 (the emotional) component of this approach. Thus one cannot predict a priori whether a decision will be based on the Type 1 versus the Type 2 system, nor is it clear what decision the Type 1 system will make.

Finally, we have tried to emphasize two important points in this chapter: (a) inclusion of motivation and emotional processes are critical and are necessary for the development of a complete computational model of cognition; and (b) it is feasible to formulate computational models that integrate emotion and cognition by having what are commonly thought of as two processes combined in the decision making process. Decision field theory provides one example of how this can be accomplished.

Appendix A

% Simulate DFT predictions for two alternative choice
% Simulation Parameters

N = 1000; % no of reps
% model parameters
h = .01; hh = sqrt(h);   % time step
theta = 5;   % threshold bound
W = [.5; .5]; % Attention weights for two outcome
                                          % gamble
b = .07; c = 0;   % Feedback matrix
S2 = [b c ; c b];
C2 = [1 −1; −1 1];   % Contrast matrix
M2 = [250 − 100; 0 0]/250;   % Value Matrix
P0 = [−1; 1];   % initial preference state
% Model
P2 = []; T2 = [];
U2 = M2*W;
V2 = U2-mean(U2);
for n = 1:N   % Replication loop
    B = 0; t = 0; P = P0;
    while (B < theta)   % Choice Trial
        w = W(1) > rand; w = [w ; (1−w)];
        U2 = M2*W;
        E2 = U2*(w−W) ;
        P = P + (S2*h)*P + V2*h + hh*E2;
        [B,Ind] = max(P); t = t+h;
    end
    P2 = [P2 ; Ind]; T2 = [T2; t];
end
P2 = [sum(P2 = = 1) ; sum(P2 = = 2)]/N; % Choice
                                          % Probability
T2 = mean(T2);   % Mean Choice Time

Appendix B

This appendix presents a detailed derivation from the equations presented earlier. To simplify the analysis, we will examine a special case in which the time step is set to $h = 1$ (this only fixes the time unit and does not change the qualitative conclusions). Furthermore, we commonly assume that the self-feedback coefficients are all equal ($s_{ii} = s$). Usually, we assume that the lateral inhibition coefficient connecting a pair of actions depends on the similarity between the two actions. However, if all the actions are equally dissimilar, which we will assume in this case, then all of the lateral inhibitory coefficients are equal ($s_{ij} = −c$ for $i \neq j$). Finally, we commonly assume the initial preference states sum to zero, $\sum_j P_i(0) = 0$, from which it follows that $\sum_j P_i(t) = 0$ for every t. Under these assumptions, Equation 1 reduces to

$$
\begin{aligned}
P_i(t+1) &= s \cdot P_i(t) - c \cdot \sum_{i \neq j} P_j(t) + v_i(t+1) \\
&= s \cdot P_i(t) - c \cdot [-P_i(t)] + v_i(t+1) \\
&= (s+c) \cdot P_i(t) + v_i(t+1) \\
&= \alpha \cdot P_i(t) + v_i(t+1),
\end{aligned}
\tag{3}
$$

where $\alpha = (s + c)$. The expected value of Equation 3 is equal to

$$
\begin{aligned}
E[P_i(t+1)] &= E[\alpha \cdot P_i(t) + v_i(t+1)] \\
&= \alpha \cdot E[P_i(t)] + E[v_i(t+1)].
\end{aligned}
\tag{4}
$$

Assuming that $0 < \alpha < 1$, then the solution to Equation 4 is

$$
E[P_i(t+1)] = \alpha^{t+1} \cdot P_i(0) + \sum_{\tau=0,t} \alpha^\tau \cdot E[v_i(t-\tau+1)].
\tag{5}
$$

Next consider the expectation of the valence, which is given by

$$
\begin{aligned}
E[v_i(t)] &= E[U_i(t) - U_.(t)] = E[U_i(t)] - E[U_.(t)] \\
&= u_i(t) - u_.(t),
\end{aligned}
$$

where $u_i(t) = E[U_i(t)]$ and $u_.(t) = \sum u_j(t)/N$ for N alternatives. Substituting this into the solution Equation 5 yields

$$
\begin{aligned}
E[P_i(t+1)] &= \alpha^{t+1} \cdot P_i(0) + \sum_{\tau=0,t} \alpha^\tau \cdot [u_i(t-\tau+1) \\
&\quad - u_.(t-\tau+1)] = \alpha^{t+1} \cdot P_i(0) \\
&\quad + \sum_{\tau=0,t} \alpha^\tau \cdot u_i(t-\tau+1) \\
&\quad - \sum_{\tau=0,t} \alpha^\tau \cdot u_.(t-\tau+1).
\end{aligned}
\tag{6}
$$

Choice probabilities are determined by the mean difference between any two preference states. Consider the difference between two actions, i versus i^*. The second sum in Equation 6 cancels out when we compute differences:

$$
\begin{aligned}
E[P_i(t+1)] - E[P_{i^*}(t+1)] &= \alpha^{t+1} \cdot [P_i(0) - P_{i^*}(0)] \\
&+ \sum_{\tau=0,t} \alpha^\tau \cdot [u_i(t-\tau+1) \\
&- u_{i^*}(t-\tau+1)]
\end{aligned}
\tag{7}
$$

Recall that $u_i(t) = E[U_i(t)] = E[\sum W_{ij}(t) \cdot m_j(t)] = \sum w_{ij} \cdot m_j(t)$, so that

$$
\begin{aligned}
u_i(t) - u_{i^*}(t) &= \sum w_{ij} \cdot m_j(t) - \sum w_{i^*j} \cdot m_j(t) \\
&= \sum (w_{ij} - w_{i^*j}) \cdot m_j(t),
\end{aligned}
$$

and inserting this into Equation 7 produces

$$E[P_i(t+1)] - E[P_{i^*}(t+1)]$$

$$= \alpha^{t+1} \cdot [P_i(0) - P_{i^*}(0)]$$

$$+ \sum_{\tau=0,t} \alpha^{\tau} \cdot \left[\sum_j (w_{ij} - w_{i^*j}) \cdot m_j(t - \tau + 1) \right]. \quad (8)$$

At this point, note that if $m_j(t)$ was fixed across time at $m_j = \sum_k n_k \cdot q_{jk}$ (i.e., a static weighted value), then Equation 8 reduces to

$$E[P_i(t+1)] - E[P_{i^*}(t+1)]$$

$$= \alpha^{t+1} \cdot [P_i(0) - P_{i^*}(0)] + \left(\sum_{\tau=0,t} \alpha^{\tau} \right) \cdot \left(\sum_j (w_{ij} - w_{i^*j}) \cdot m_j \right)$$

$$= \alpha^{t+1} \cdot [P_i(0) - P_{i^*}(0)] + \left(\frac{1 - \alpha^{t+1}}{1 - \alpha} \right)$$

$$\cdot \left(\sum_j (w_{ij} - w_{i^*j}) \cdot mj \right)$$

$$= \alpha^{t+1} \cdot [P_i(0) - P_{i^*}(0)] + \left(\frac{1 - \alpha^{t+1}}{1 - \alpha} \right) \cdot (u_i - u_{i^*}). \quad (9)$$

It is informative to compare Equation 9 with a static weighted utility model, where the latter assumes that the preference between actions i and i^* is determined solely by the static difference in weighted utilities $(u_i - u_{i^*}) = \sum_j (w_{ij} - w_{i^*j}) \cdot m_j$. Both theories share a common set of parameters: the decision weights w_{ij} and the values m_j; but DFT adds two new parameters, the initial state $P_i(0)$ and growth-decay rate α. If the initial preference state is zero (neutral), then the first term in Equation 9 drops out, and the mean difference in preference states for DFT is always consistent with the mean difference in weighted utilities. However, if the initial preferences are ordered opposite of the weighted utilities, then preferences will reverse over time (as illustrated in Figure 15.3). To simplify the remaining analyses, we will assume that the initial preference state is zero.

Now let us examine the crucial issue: how are the affective evaluations influenced by the emotional process across time? In this case, the evaluations change dynamically across time according to the needs, $m_j(t) = \sum_k n_k(t) \cdot q_{jk}$, and inserting this into Equation 8 yields the new result (assuming for simplicity hereafter that $P_i[t] = 0$):

$$E[P_i(t+1)] - E[P_{i^*}(t+1)] = \sum_{\tau=0,t} \alpha^{\tau}$$

$$\cdot \sum_j (w_{ij} - w_{i^*j}) \cdot \left(\sum_k n_k(t - \tau + 1) \cdot q_{jk} \right). \quad (10)$$

As can be seen from Equation 10, the dynamics depend on the solution of $n_k(t)$, which is derived from Equation 2. However, the solution for Equation 2 depends on assumptions about changes in the current status on an attribute $a_k(t)$ at each moment in time, which in turn,

depends on past decisions and on the exogenous environmental disturbances that must be specified.

Suppose that before the onset of the decision, the current state matches the goal state so that the need adjustment is zero for each attribute, $[g_k - a_k(t)] = 0$ for $t < 0$, and the need system is at equilibrium. Then suddenly, because of exogenous events, the current status on an attribute $a_k(t)$ drops far below the ideal point g_i at time $t = 0$ (decision onset), so that there is a gap between the current state and the ideal state, symbolized as $\Delta_k = [g_k - a_k(t)] > 0$ for $t > 0$. In this case, we need to solve the simple difference equation

$$n_k(t + 1) = L_k \cdot n_k(t) + \Delta_k$$

and assuming $0 < L_k < 1$, then the solution is given by

$$n_k(t) = \sum_{\tau=0,t-1} L_k^{\tau} \cdot \Delta_k = \Delta_k \cdot \left(\sum_{\tau=0,t-1} L_k^{\tau} \right) = \left(\frac{1 - L_k^t}{1 - L_k} \right) \cdot \Delta_k$$

$$(11)$$

Substituting this solution into the expression for $u_i(t)$ yields

$$u_i(t) = \sum_j w_{ij} \cdot m_j(t) = \sum_j w_{ij} \cdot \left(\sum_k n_k(t) \cdot q_{jk} \right)$$

$$= \sum_j w_{ij} \cdot \left(\sum_k \left(\frac{1 - L_k^t}{1 - L_k} \right) \cdot \Delta_k \cdot q_{jk} \right)$$

Finally, inserting the solution given by Equation 11 into Equation 10 produces the final solution:

$$E[P_i(t+1)] - E[P_{i^*}(t+1)]$$

$$= \sum_{\tau=0,t} \alpha^{\tau} \cdot \sum_j (w_{ij} - w_{i^*j}) \cdot \left(\sum_k \left(\frac{1 - L_k^{t-\tau+1}}{1 - L_k} \right) \cdot \Delta_k \cdot q_{jk} \right).$$

$$(12)$$

It is instructive to compare Equation 12 with the static weighted utility theory, according to which $(u_i - u_{i^*}) = \sum_j (w_{ij} - w_{i^*j}) \cdot (\sum_k n_k \cdot q_{jk})$ completely determines preference. Both theories share a common set of parameters: n_k, q_{jk}, w_{ij}; but DFT adds the following two additional parameters, α and L_k. The critical qualitative property that distinguishes DFT from the static utility model is that DFT allows preferences to reverse across deliberation time, which is impossible with the static theory.

Notes

1. This example is based on a personal experience of the first author, who decided to go straight across the tire, and managed to survive to tell this story.

2. In the past, we defined U as the average of all options other than option i. Here we define it as the average of all options. However, the definition used here produces a valence that is proportional to the previous version: the previously defined valence equals $[N/(N-1)]$ times the currently defined valence, where N is the number of options in the choice set.

3. Formally, this is a Markov process, and matrix formulas have been mathematically derived for computing the choice probabilities and distribution of choice response times (see Busemeyer & Diederich, 2002; Busemeyer & Townsend, 1992; Diederich & Busemeyer, 2003). Alternatively, computer simulation can be used to generate predictions from the model. Normally, we use the matrix computations because they are more precise and faster, but to show how easy it is to simulate this model, we used the simulation program shown in the Appendix A for the analyses presented next.

4. These closely matched the calculations from the Markov chain equations; however, the latter are more accurate and didn't produce the little dip that appears at the end of Figure 15.3. The Markov chain method was also a couple of orders of magnitude faster to compute.

5. Appendix B provides a more formal derivation of this property of the theory.

References

Anderson, J. R., Lebiere, C., Lovett, M. C., & Reder, L. M. (1998). ACT-R: A higher-level account of processing capacity. *Behavioral & Brain Sciences, 21*, 831–832.

Buck, R. (1984). *The communication of emotion.* New York: Guilford Press.

Carver, C. S., & Scheier, M. F. (1990). Origins and functions of positive and negative affect: A control-process view. *Psychological Review, 97*(1), 19–35.

Busemeyer, J. R. (1985). Decision making under uncertainty: A comparison of simple scalability, fixed sample, and sequential sampling models. *Journal of Experimental Psychology, 11*, 538–564.

——, & Diederich, A. (2002). Survey of decision field theory. *Mathematical Social Sciences, 43*, 345–370.

Busemeyer, J. R., & Townsend, J. T. (1992). Fundamental derivations for decision field theory. *Mathematical Social Sciences, 23*, 255–282.

——, & Townsend, J. T. (1993). Decision field theory: A dynamic-cognitive approach to decision making in an uncertain environment. *Psychological Review, 100*, 432–459.

——, Townsend, J. T., & Stout, J. C. (2002). Motivational underpinnings of utility in decision making: Decision field theory analysis of deprivation and satiation. In S. Moore (Ed.), *Emotional cognition* (pp. 197–220). Amsterdam: John Benjamins.

Damasio, A. R. (1994). *Descartes' error: Emotion, reason, and the human brain.* New York: Putnam.

Diederich, A. (1997). Dynamic stochastic models for decision making under time constraints. *Journal of Mathematical Psychology, 41*, 260–274.

——. (2003). MDFT account of decision making under time pressure. *Psychonomic Bulletin and Review, 10*(1), 157–166.

——, & Busemeyer, J. R. (2003). Simple matrix methods for analyzing diffusion models of choice probability, choice response time, and simple response time. *Journal of Mathematical Psychology, 47*(3), 304–322.

——, & Busemeyer, J. R. (2006). Modeling the effects of payoffs on response bias in a perceptual discrimination task: Threshold bound, drift rate change, or two stage processing hypothesis. *Perception and Psychophysics, 97*(1), 51–72.

Epstein, S. (1994). Integration of the cognitive and the psychodynamic unconscious. *American Psychologist, 49*(8), 709–724.

Goldberg, J. H., Lerner, J. S., & Tetlock, P. E. (1999). Rage and reason: The psychology of the intuitive prosecutor. *European Journal of Social Psychology, 29*(5–6), 781–795.

Gray, J. A. (1994). Three fundamental emotion systems. In P. Ekman & R. J. Davidson (Eds.), *The nature of emotion: Fundamental questions* (pp. 243–247). New York: Oxford University Press.

Gray, J. R. (1999). A bias toward short-term thinking in threat-related negative emotional states. *Personality & Social Psychology Bulletin, 25*(1), 65–75.

——. (2004). Integration of emotion and cognitive control. *Current Directions in Psychological Science, 13*(2), 46–48.

Grossberg, S. (1988). *Neural networks and natural intelligence.* Cambridge, MA: MIT Press.

Hammond, K. R. (2000). Coherence and correspondence theories in judgment and decision making. In T. Connolly & H. R. Arkes (Eds.), *Judgment and decision making: An interdisciplinary reader* (2nd ed., pp. 53–65). New York: Cambridge University Press.

Hull, C. L. (1943). *Principles of behavior, an introduction to behavior theory.* New York: D. Appleton-Century.

Johnson, J. G., & Busemeyer, J. R. (2005). A dynamic, computational model of preference reversal phenomena. *Psychological Review, 112*, 841–861.

Kahneman, D., & Frederick, S. (2002). Representativeness revisited: Attribute substitution in intuitive judgment. In T. Gilovich & D. Griffin (Eds.), *Heuristics and biases: The psychology of intuitive judgment* (pp. 49–81). New York: Cambridge University Press.

———, & Tversky, A. (1979). Prospect theory: An analysis of decision under risk. *Econometrica, 47,* 263–291.

Keeney, R. L., & Raiffa, H. (1976). *Decisions with multiple objectives: Preferences and value tradeoffs.* New York: Wiley.

Lazarus, R. S. (1991). *Emotion and adaptation.* London: Oxford University Press.

LeDoux, J. E. (1996). *The emotional brain: The mysterious underpinnings of emotional life.* New York: Simon & Schuster.

Lerner, J. S., Small, D. A., & Loewenstein, G. (2004). Heart strings and purse strings: Carryover effects of emotions on economic decisions. *Psychological Science, 15*(5), 337–341.

Levenson, R. W. (1994). Human emotion: A functional view. In P. Ekman & R. J. Davidson (Eds.), *The nature of emotion: Fundamental questions* (pp. 123–126). New York: Oxford University Press.

Lewis, M., & Haviland-Jones, J. M. (2000). *Handbook of emotions* (2nd ed.). New York: Guilford Press.

Loewenstein, G., & Lerner, J. S. (2003). The role of affect in decision making. In R. J. Davidson, K. R. Scherer, & H. H. Goldsmith (Eds.), *Handbook of affective sciences* (pp. 619–642). New York: Oxford University Press.

———, & O'Donoghue, T. (2005). *Animal spirits: Affective and deliberative processes in economic behavior.* Manuscript in preparation.

Luce, M. F., Bettman, J. R., & Payne, J. W. (1997). Choice processing in emotionally difficult decisions. *Journal of Experimental Psychology: Learning, Memory, & Cognition, 23*(2), 384–405.

Luce, R. D. (2000). *Utility of gains and losses: Measurement-theoretical and experimental approaches.* Mahwah, NJ: Erlbaum.

Markman, A. B., & Brendl, C. M. (2000). The influence of goals on value and choice. In D. L. Medin (Ed.), *The psychology of learning and motivation: Advances in research and theory* (Vol. 39, pp. 97–128). San Diego, CA: Academic Press.

Maslow, A. H. (1962). *Toward a psychology of being.* Oxford: Van Nostrand.

Mellers, B. A., Schwartz, A., Ho, K., & Ritov, I. (1997). Decision affect theory: Emotional reactions to the outcomes of risky options. *Psychological Science, 8*(6), 423–429.

Metcalfe, J., & Mischel, W. (1999). A hot/cool-system analysis of delay of gratification: Dynamics of willpower. *Psychological Review, 106*(1), 3–19.

Meyer, D. E., & Kieras, D. E. (1997). A computational theory of executive cognitive processes in multiple-task performance. Part I: Basic mechanisms. *Psychological Review, 104,* 3–65.

Newell, A. (1990). *Unified theories of cognition.* Cambridge, MA: Harvard University Press.

Panksepp, J. (1994). The basics of basic emotions. In P. Ekman & R. J. Davidson (Eds.), *The nature of emotion: Fundamental questions* (pp. 20–24). New York: Oxford University Press.

Peters, E., & Slovic, P. (2000). The springs of action: Affective and analytical information processing in choice. *Personality & Social Psychology Bulletin, 26*(12), 1465–1475.

Read, D., & van Leeuwen, B. (1998). Predicting hunger: The effects of appetite and delay on choice. *Organizational Behavior & Human Decision Processes, 76*(2), 189–205.

Roe, R. M., Busemeyer, J. R., & Townsend, J. T. (2001). Multi-alternative decision field theory: A dynamic connectionist model of decision-making. *Psychological Review, 108,* 370–392.

Roseman, I. J., Antoniou, A. A., & Jose, P. E. (1996). Appraisal determinants of emotions: Constructing a more accurate and comprehensive theory. *Cognition & Emotion, 10*(3), 241–277.

Rottenstreich, Y., & Hsee, C. K. (2001). Money, kisses, and electric shocks: On the affective psychology of risk. *Psychological Science, 12*(3), 185–190.

Rumelhart, D., & McClelland, J. L. (1986). *Parallel distributed processing: Explorations in the microstructure of cognition* (Vol. 1). Cambridge, MA: MIT Press.

Schachter, S., & Singer, J. (1962). Cognitive, social, and physiological determinants of emotional state. *Psychological Review, 69,* 379–399.

Scherer, K. R. (1994). Toward a concept of modal emotions. In P. Ekman & R. J. Davidson (Eds.), *The nature of emotion: Fundamental questions* (pp. 25–31). New York: Oxford University Press.

Shiv, B. & Fedorikhin, A. (1999) Heart and mind in conflict: The interplay of affect and cognition in consumer decision making. *Journal of Consumer Research, 26,* 278–292.

Simon, H. A. (1967). Motivational and emotional controls of cognition. *Psychological Review, 74*(1), 29–39.

Skinner, B. F. (1953). *Science and human behavior.* New York: Macmillan.

Sloman, S. A. (1996). The empirical case for two systems of reasoning. *Psychological Bulletin, 119*(1), 3–22.

Spence, K. W. (1956). *Behavior theory and conditioning.* New Haven, CT: Yale University Press.

Stanovich, K. E., & West, R. F. (2000). Individual differences in reasoning: Implications for the rationality debate? *Behavioral & Brain Sciences, 23*(5), 645–726.

Toates, F. M. (1980). *Animal behaviour: A systems approach.* Chichester: Wiley.

Tolman, E. C. B. (1958). *Behavior and psychological man; essays in motivation and learning.* Berkeley: University of California Press.

Van Boven, L., Loewenstein, G., & Dunning, D. (2005) The illusion of courage in social predictions: Underestimating the impact of fear of embarrassment on other people. *Organizational Behavior and Human Decision Processes, 96*(2) 130–141.

Weiner, B. (1986). Attribution, emotion, and action. In R. M. Sorrentino & E. T. Higgins (Eds.), *Handbook of motivation and cognition: foundations of social behavior* (pp. 281–312). New York: Guilford Press.

Wilson, T. D., Lisle, D. J., Schooler, J. W., Hodges, S. D., Klaaren, K. J., & LaFleur, S. J. (1993). Introspecting about reasons can reduce post-choice satisfaction. *Personality & Social Psychology Bulletin, 19*(3), 331–339.

Zajonc, R. B. (1980). Feeling and thinking: Preferences need no inferences. *American Psychologist, 35*(2), 151–175.

Zeelenberg, M., Beattie, J., van der Pligt, J., & de Vries, N. K. (1996). Consequences of regret aversion: Effects of expected feedback on risky decision making. *Organizational Behavior & Human Decision Processes, 65*(2), 148–158.

16

The Architectural Role of Emotion
in Cognitive Systems

Jonathan Gratch & Stacy Marsella

In this chapter, we will revive an old argument that theories of human emotion can give insight into the design and control of complex cognitive systems. In particular, we claim that appraisal theories of emotion provide essential insight into the influences of emotion over cognition and can help translate such findings into concrete guidance for the design of cognitive systems. Appraisal theory claims that emotion plays a central and functional role in sensing external events, characterizing them as opportunity, or threats, and recruiting the cognitive, physical, and social resources needed to respond adaptively. Further, because it argues for a close association between emotion and cognition, the theoretical claims of appraisal theory can be recast as a requirement specification for how to build a cognitive system. This specification asserts a set of judgments that must be supported to interpret correctly and to respond to stimuli and provides a unifying framework for integrating these judgments into a coherent physical or social response. This chapter elaborates argument in some detail based on our joint experience in building complex cognitive systems and computational models of emotion.

To survive in a dynamic, semipredictable, and social world, organisms must be able to sense external events, characterize how they relate to their internal needs (e.g., is this an opportunity or a threat?), consider potential responses (e.g., fight, flight or plan), and recruit the cognitive, physical, and social resources needed to adaptively respond. In primitive organisms, this typically involves hardwired or learned stimulus–response patterns. For sophisticated organisms such as humans, this basic cycle is quite complex and can occur at multiple levels and timescales, involve deliberation and negotiation with other social actors, and can use a host of mental functions, including perception, action, belief formation, planning, and linguistic processing. Progress in modeling such complex phenomena depends on a theory of cognitive system design that clearly delineates core cognitive functions, how they interoperate, and how they can be controlled and directed to achieve adaptive ends.

In this chapter, we will revive an old argument that theories of human emotion can give insight into the design and control of complex cognitive systems and argue that one theory of emotion in particular, appraisal theory, helps identify core cognitive functions and how they can be controlled (see also Hudlicka, chapter 19, this volume). Debates about the benefit of emotion span recorded history and were prominent, as well, in the early days of cognitive science. Early cognitive scientists argued that emotional influences that seem irrational on the surface have important social and cognitive functions that would be required by *any* intelligent system. For example, Simon (1967) argued that emotions serve the crucial function of interrupting normal cognition when unattended goals require servicing. Other authors have emphasized how social emotions such as anger and guilt may reflect a mechanism that improves group utility by minimizing social conflicts, and thereby explains people's "irrational" choices to cooperate in social games such as prison's dilemma (Frank, 1988). Similarly, "emotional biases" such as wishful thinking may reflect a rational mechanism that is more accurately accounting for certain social costs, such as the

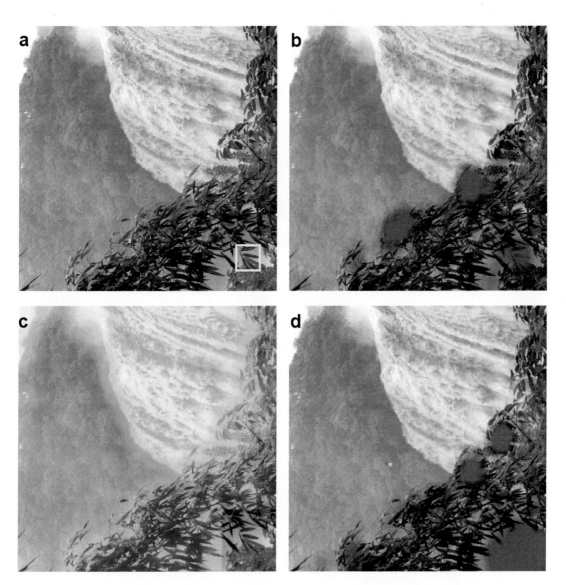

FIGURE 9.3 Study on saccadic selectivity in complex images.

FIGURE 14.1 Experimental participant performing the Scout World task.

FIGURE 14.6 Simulated view of the Chicago O'Hare taxi surface in foggy conditions. Courtesy of NASA Ames Research Center.

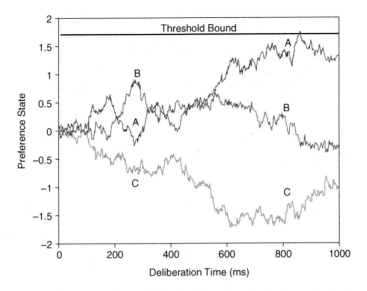

FIGURE 15.1 A simulation of the decision process. Horizontal axis is time, vertical axis is preference strength, and each trajectory represents one course of action. The top bar is the threshold bound. The first option to hit the bound wins the race and is chosen.

FIGURE 16.3 The MRE and SASO-ST systems allow a trainee to interact with intelligent virtual characters through natural language for task-oriented training.

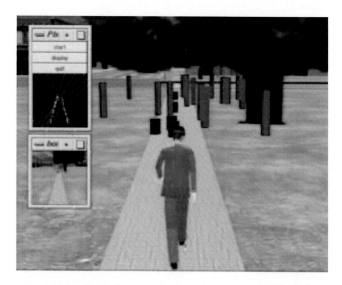

FIGURE 20.1 The Walter simulation. The insets show the use of vision to guide the humanoid through a complex environment. The upper inset shows the particular visual routine that is running at any instant; in this case the lines indicate that the sidewalk is being detected. The lower inset shows the visual field from the perspective of Walter's head-centered frame.

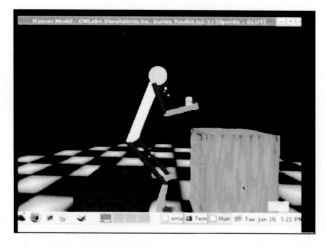

FIGURE 20.2 Motor control of a humanoid figure can be economically commanded with a handful of discrete control points.

Visual Routines

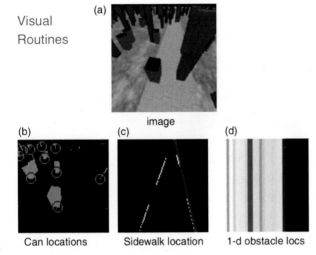

(a)

image

(b) Can locations

(c) Sidewalk location

(d) 1-d obstacle locs

FIGURE 20.4 The visual routines that compute state information. (a) Input image from Walter's viewpoint. (b) Regions that fit the litter color profile. Probable litter locations are marked with circles. (c) Processed image for sidewalk following. Pixels are labeled in white if they border both sidewalk and grass color regions. The red line is the most prominent resulting line. (d) One-dimensional depth map used from obstacle avoidance (not computed directly from the rendered image). Darker stripes are closer.

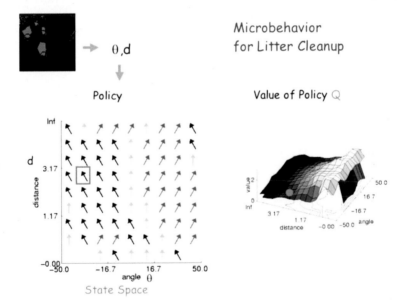

Microbehavior for Litter Cleanup

θ,d

Policy

Value of Policy Q

State Space

FIGURE 20.5 The central portion of the litter cleanup microbehavior after it has been learned. The color image is used to identify the heading to the nearest litter object as a heading angle θ and distance d. Using this state information to index the table allows the recovery of the policy, in this case *heading* = −45 deg, and its associated value. The fact that the model is embodied means that we can assume there is neural circuitry to translate this abstract heading into complex walking movements. This is true for the graphics figure that has a "walk" command that takes a heading parameter.

a)

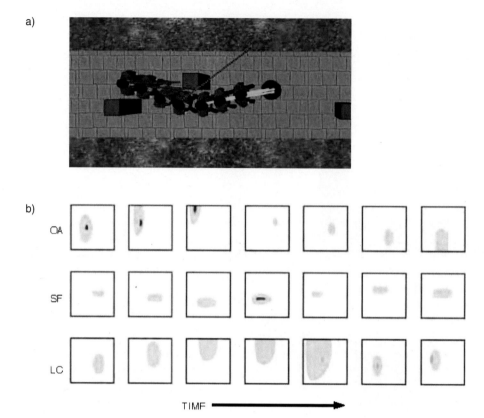

b)

OA

SF

LC

TIME ➡

FIGURE 20.7 (a) An overhead view of the virtual agent during seven time steps of the sidewalk navigation task. The blue cubes are obstacles, and the purple cylinder is litter. The rays projecting from the agent represent eye movements; red corresponds to obstacle avoidance, blue corresponds to sidewalk following, and green corresponds to litter collection. (b) Corresponding state estimates. The top row shows the agent's estimates of the obstacle location. The axes here are the same as those presented in Figure 20.6. The beige regions correspond to the 90% confidence bounds before any perception has taken place. The red regions show the 90% confidence bounds after an eye movement has been made. The second and third rows show the corresponding information for sidewalk following and litter collection.

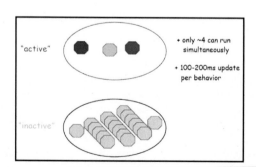

"active"

+ only ~4 can run
simultaneously

+ 100-200ms update
per behavior

"inactive"

FIGURE 20.8 The model assumes that humans have an enormous library of behaviors that can be composed in small sets to meet behavioral demands. When an additional behavior is deemed necessary, it is activated by the "operating system." When a running behavior is no longer necessary, it is deactivated.

Behavior List

■ Follow Sidewalk
■ Avoid Obstacles
■ Pick Up Objects
☐ Look For Corner
☐ Look For Crosswalk
■ Approach Crosswalk
☐ Wait For Light
■ Follow Crosswalk
■ Approach Sidewalk

State Machine Diagram

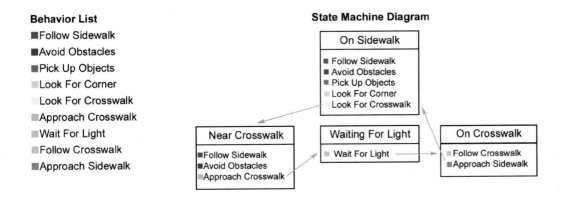

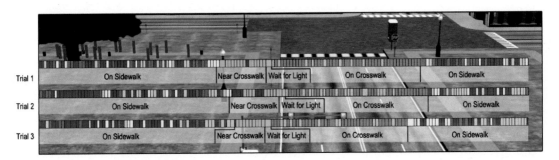

FIGURE 20.9 A list of microbehaviors used in Walter's overall navigation task. (Top right) The diagram for the programmable context switcher showing different states. These states are indicated in the bands underneath the colored bars below. (Bottom) Context switching behavior in the sidewalk navigation simulation for three separate instances of Walter's stroll. The different colored bars denote different microbehaviors that are in control of the gaze at any instant.

FIGURE 22.1 A comprehensive theory of visual search could accurately predict where people are likely to click to find the current postage rates on the U.S. Postal Service Web site (www.usps.com). The theory could also predict other measures such as search time and scanpaths. (© United States Postal Service. Used with permission. All rights reserved.)

cost of betrayal when a parent defends a child despite strong evidence of their guilt in a crime (Mele, 2001). Ironically, after arguing for the centrality of emotion in cognition, Simon and others in the cognitive modeling community went on to develop narrow focused models of individual cognitive functions that assumed away many of the central control problems that emotion is purported to solve.

After some neglect, the question of emotion has again come to the forefront as models have begun to catch up to theory. This has been spurred, in part, by an explosion of interest in integrated computational models that incorporate a variety of cognitive functions (Anderson, 1993; Bates, Loyall, & Reilly, 1991; Rickel et al., 2002). Indeed, until the rise of broad integrative models of mental function, the problems emotion was purported to solve, for example, juggling multiple goals, were largely hypothetical. More recent cognitive systems embody a variety of mental functions and face very real choices on how to allocate resources. A reoccurring theme in emotion research is the role of emotion in addressing such control choices by directing cognitive resources toward problems of adaptive significance for the organism. Indeed, Simon appealed to emotion to explain how his sequential models could handle the multiplicity of motives that underlie most human activity:

> The theory explains how a basically serial information processor endowed with multiple needs behaves adaptively and survives in an environment that presents unpredictable threats and opportunities. The explanation is built on two central mechanisms: 1. A goal-terminating mechanism [goal executor] . . . 2. An interruption mechanism, that is, emotion, allows the processor to respond to urgent needs in real time. (Simon, 1967, p. 39)

Interrupts are part of the story, but contemporary emotion research suggests emotion exact a far more pervasive control over cognitive processes. Emotional state can influence what information is available in working memory (Bower, 1991), the subjective utility of alternative choices (see Busemeyer, Dimperio, & Jessup, chapter 15, this volume), and even the style of processing (Bless, Schwarz, & Kemmelmeier, 1996; Schwarz, Bless, & Bohner, 1991). For example, people who are angry or happy tend to perform more shallow inference and are more influenced by stereotypical beliefs, whereas sad individuals tend to process more deeply and be more sensitive to the true state of the world.

These psychological findings are bolstered by evidence from neuroscience underscoring the close connection between emotion and centers of the brain associated with higher-level cognition. For example, studies performed by Damasio and colleagues suggest that damage to ventromedial prefrontal cortex prevents emotional signals from guiding decision making in an advantageous direction, particularly for social decisions (Bechara, Damasio, Damasio, & Lee, 1999). Other studies have illustrated a close connection between emotion and cognition via the anterior cingulate cortex, a center of the brain often implicated in cognitive control (Allmana, Hakeema, Erwinb, Nimchinskyc, & Hofd, 2001). Collectively, these findings demonstrate that emotion and cognition are closely coupled and suggest emotion has a strong, pervasive and controlling influence over cognition.

We argue appraisal theory (Arnold, 1960; Frijda, 1987; Lazarus, 1991; Ortony, Clore, & Collins, 1988; Scherer, 1984), the most influential contemporary theory of human emotion, can help make sense of the various influences of emotion over cognition and, further, help translate such findings into concrete guidance for the design of cognitive systems. Appraisal theory asserts that emotion plays a central and functional role in sensing external events, characterizing them as opportunity or threats and recruiting the cognitive, physical, and social resources needed to adaptively respond. Further, because it argues for a close association between emotion and cognition, the theoretical claims of appraisal theory can be recast as a requirement specification for how to build a cognitive system—it claims a particular set of judgments must be supported to interpret and respond to stimuli correctly and provides a unifying framework for integrating these judgments into a coherent physical or social response. This chapter elaborates argument in some detail based on our joint experience in building complex cognitive systems and computational models of emotion.

Computational Appraisal Theory

Appraisal theory is the predominant psychological theory of human emotion, and here we argue that it is also the most fruitful theory of emotion for those interested in the design of cognitive systems (Arnold, 1960; Frijda, 1987; Lazarus, 1991; Ortony et al., 1988; Scherer, 1984).[1] The theory emphasizes the connection between emotion and cognition, arguing that emotions are an aspect of the mechanisms by which organisms detect, classify, and adaptively respond to significant changes to their environment. A central tenet is that emotions are associated with patterns of individual judgment that characterize the personal significance of external events (e.g., Was this event expected in terms of my prior beliefs? Is this event congruent with my goals? Do I have the power to alter the consequences of this event?). These judgments involve cognitive processes, including slow deliberative, as well as fast automatic or associative processes.

There are several advantages to adopting an appraisal–theoretic perspective when approaching the problem of cognitive system design. Unlike neuroscience models, appraisal theory is often cast at a conceptual level that meshes well with the level of analysis used in most cognitive systems, as emotions are described in terms of their relationship to goals, plans, and problem solving. In this sense, appraisal theories contrast sharply with categorical theories (Ekman, 1992) that postulate a small set of innate hardwired neuromotor programs that are separate from cognition or dimensional theories that argue that emotions are classified along certain dimensions and make no commitment to underlying mechanism (Russell & Lemay, 2000). Finally, as a paradigm that has seen consistent empirical support and elaboration over the past fifty years, appraisal theory has been applied to a wide range of cognitive and social phenomena and thus provides the most comprehensive single framework for conceptualizing the role of emotion in the control of cognition.

Appraisal and Coping

Appraisal theory argues that emotion arises from the dynamic interaction of two basic processes: appraisal and coping (Smith & Lazarus, 1990). Appraisal is the process by which a person assesses his overall relationship with his environment, including not only current conditions but past events as well as future prospects. Appraisal theory argues that appraisal, although not a deliberative process, is informed by cognitive processes and, in particular, those processes involved in understanding and interacting with the physical and social environment (e.g., planning, explanation, perception, memory, linguistic processes). Appraisal maps characteristics of these disparate mental processes into a common set of terms called *appraisal variable* (e.g., Is this event *desirable*? Who *caused* it? What *power* do I have over its unfolding?). These variables serve as an

intermediate description of the person–environment relationship—a common language of sorts—and are claimed to mediate between stimuli and response (e.g., different responses are organized around how a situation is appraised). Appraisal variables characterize the significance of events from the individual's perspective. Events do not have significance but only by virtue of their interpretation in the context of an individual's beliefs, desires and intention, and past events.

Coping refers to how one responds to the appraised significance of events. People are motivated to respond to events differently depending on how they are appraised (Peacock & Wong, 1990). For example, events appraised as undesirable but controllable motivate people to develop and execute plans to reverse these circumstances. On the other hand, events appraised as uncontrollable lead people toward denial or resignation. Appraisal theories often characterize the wide range of human coping responses into two broad classes: *problem-focused coping strategies* attempt to change the environment; *emotion-focused coping strategies* (Lazarus, 1991) involves inner-directed strategies for dealing with emotions, for example, by discounting a potential threat or abandoning a cherished goal. The ultimate effect of these strategies is a change in the person's interpretation of their relationship with the environment, which can lead to new appraisals (reappraisals). Thus, coping, cognition, and appraisal are tightly coupled, interacting and unfolding over time (Lazarus, 1991): an agent experience fear upon perceiving a potential threat (appraisal), which motivates problem solving (coping), which leads to relief upon deducing an effective countermeasure (reappraisal). A key challenge for any model of this process is to capture these dynamics.

EMA: A Computational Perspective

EMA is a computational model that attempts to concretize the mapping between appraisal theory and cognitive system research (Gratch & Marsella, 2001, 2004, 2005; Marsella & Gratch, 2003).[2] Given appraisal theory's emphasis on a person's evolving *interpretation* of their relationship with the environment, EMA's development has centered on elucidating the mechanisms that inform this interpretation and how emotion informs and controls the subsequent functioning of these mechanisms. At any point in time, the agent's current view of the agent–environment relationship is represented in "working memory," which changes with further observation or inference. EMA treats appraisal as a set of feature detectors that map features of this representation into appraisal variables. For example, an effect that threatens a desired goal is assessed as a potential undesirable event. Coping is cast as a set of control signals that direct the processing of auxiliary reasoning modules (i.e., planning, belief updates) to overturn or maintain those features that yielded the appraisals. For example, coping could resign the agent to the threat by abandoning the desired goal, or alternatively, it could signal the planning system to explore contingencies. Figure 16.1 illustrates this perspective on appraisal theory as a mechanism for the control of cognition.

To a mechanistic account, we have adopted a strategy of using conventional artificial intelligence reasoning techniques as proxies for the cognitive mechanisms that are claimed to underlie appraisal and coping. Appraisal theory posits that events are interpreted in terms of several appraisal variables that collectively can be seen as a requirement specification for the classes of inference a cognitive system must support. This specification is far broader than what is typically supported by conventional artificial intelligence techniques, so to capture this interpretative process within a computational system, we have found it most natural to integrate a variety of reasoning methods. Specifically, we build on the causal representations developed for decision-theoretic planning (Blythe, 1999) and augment them with methods that explicitly model commitments

to beliefs and intentions (Grosz & Kraus, 1996; Pollack, 1990). Plan representations provide a concise representation of the causal relationship between events and states, key for assessing the relevance of events to an agent's goals and for assessing causal attributions. Plan representations also lie at the heart of many autonomous agent reasoning techniques (e.g., planning, explanation, natural language processing). The decision-theoretic concepts of utility and probability are crucial for modeling appraisal variables related to the desirability and likelihood of events. Explicit representations of intentions and beliefs are critical for assessing the extent to which an individual deserves blame or credit for their actions, as such attributions involve judgments of intent, foreknowledge and freedom of choice (Shaver, 1985; Weiner, 1995). As we will see, commitments to beliefs and intentions also play a role in modeling coping strategies.

In EMA, the agent's interpretation of its agent–environment relationship is reified in an explicit representation of beliefs, desires, intentions, plans, and probabilities (see Figure 16.2). Following a blackboard-style model, this representation (corresponding to the agent's working memory) encodes the input, intermediate results, and output of reasoning processes that mediate between the agent's goals and its physical and social environment (e.g., perception, planning, explanation, and natural language processing). We use the term *causal interpretation* to refer to this collection of data structures to emphasize the importance of causal reasoning as well as the interpretative (subjective)

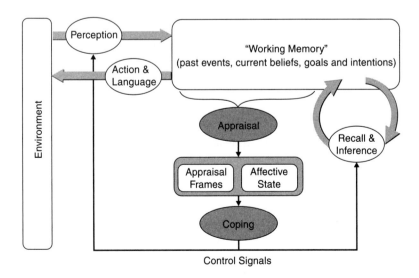

FIGURE 16.1 A view of emotion as "affective control" over cognitive functions.

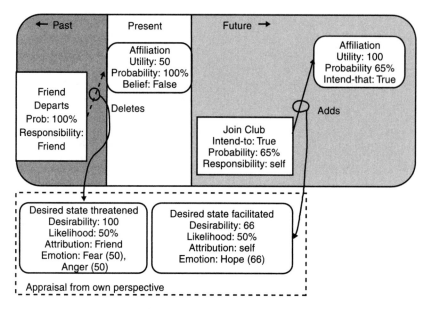

FIGURE 16.2 An instance of a causal interpretation and associated appraisal frames.

character of the appraisal process. Figure 16.2 illustrates an instance of this data structure in which an agent has a single goal (affiliation) that is threatened by the recent departure of a friend (the past "friend departs" action has one effect that deletes the "affiliation" state). This goal might be reachieved if the agent joins a club. Appraisal assesses each case where an act facilitates or inhibits some proposition in the causal interpretation. In the figure, the interpretation encodes two "events," the threat to the currently satisfied goal of affiliation, and the potential reestablishment of affiliation in the future. Associated with each event in the causal interpretation is an *appraisal frame* that summarizes, in terms of appraisal variables, its significance to the agent.

Each event is characterized in terms of appraisal variables by domain-independent functions that examine the syntactic structure of the causal interpretation:

- Perspective: From whose viewpoint is the event judged?
- Desirability: What is the utility of the event if it comes to pass, from the perspective taken (e.g., does it causally advance or inhibit a state of some utility)? The utility of a state may be intrinsic (agent X attributes utility Y to state Z) or derived (state Z is a precondition of a plan that, with

some likelihood, will achieve an end with intrinsic utility).
- Likelihood: How probable is the outcome of the event? This is derived from the decision-theoretic plan.
- Causal attribution: Who deserves credit or blame? This depends on what agent was responsible for executing the action, but also involves epistemic considerations such as intention, foreknowledge and coercion (see Mao & Gratch, 2004).
- Temporal status: Is this past, present, or future?
- Controllability: Can the outcome be altered by actions under control of the agent whose perspective is taken? This is derived by looking for actions in the causal interpretation that could establish or block some effect, and that are under control of the agent who's perspective is being judged (i.e., agent X could execute the action).
- Changeability: Can the outcome be altered by external processes or some other causal agent? This involves consideration of actions believed available to others as well as their intentions.

Each appraised event is mapped into a discrete emotion instance of some type and intensity, following the scheme proposed by Ortony et al. (1988). A simple activation-based focus of attention model computes a current emotional state based on most recently accessed emotion instances.

Coping determines how one responds to the appraised significance of events. Coping strategies are proposed to maintain desirable or overturn undesirable in-focus emotion instances. Coping strategies essentially work in the reverse direction of appraisal, identifying the precursors of emotion in the causal interpretation that should be maintained or altered (e.g., beliefs, desires, intentions, and expectations). Strategies include:

- Action: select an action for execution
- Planning: form an intention to perform some act (the planner uses intentions to drive its plan generation)
- Seek instrumental support: ask someone that is in control of an outcome for help
- Procrastination: wait for an external event to change the current circumstances
- Positive reinterpretation: increase utility of positive side-effect of an act with a negative outcome
- Acceptance: drop a threatened intention
- Denial: lower the probability of a pending undesirable outcome
- Mental disengagement: lower utility of desired state
- Shift blame: shift responsibility for an action toward some other agent
- Seek/suppress information: form a positive or negative intention to monitor some pending or unknown state

Strategies give input to the cognitive processes that actually execute these directives. For example, planful coping will generate an intention to perform the "join club" action, which in turn leads the planning system to generate and to execute a valid plan to accomplish this act. Alternatively, coping strategies might abandon the goal, lower the goal's importance, or reassess who is to blame.

Not every strategy applies to a given stressor (e.g., an agent cannot engage in problem-directed coping if it is unaware of an action that affects the situation); however, multiple strategies can apply. EMA proposes these in parallel but adopts strategies sequentially. EMA adopts a small set of search control rules to resolve ties. In particular, EMA prefers problem-directed strategies if control is appraised as high (take action, plan, seek information), procrastination if changeability is high, and emotion-focus strategies if control and changeability is low.

In developing a computational model of coping, we have moved away from the broad distinctions of problem-focused and emotion-focused strategies. Formally representing coping requires a certain crispness lacking from the problem-focused/emotion-focused distinction. In particular, much of what counts as problem-focused coping in the clinical literature is really inner-directed in an emotion-focused sense. For example, one might form an intention to achieve a desired state—and feel better as a consequence—without ever acting on the intention. Thus, by performing cognitive acts like planning, one can improve ones interpretation of circumstances without actually changing the physical environment.

Appraisal Theory and the Design of Virtual Humans

The question we raise in this chapter is the connection between emotion research and the design of cognitive systems. We have explored this question within our own work in the context of the design of virtual humans. These are software agents that attempt to simulate human cognitive, verbal, and nonverbal behavior in interactive virtual environments. From the perspective of this volume, virtual humans serve to illustrate the complexity of contemporary cognitive systems and the host of integration and control problems they raise. After describing the capabilities of such agents, we show how our understanding of human emotion, and computational appraisal theory in particular, has influenced the design of a general architecture that can detect, classify, and adaptively respond to significant changes to their virtual environment.

Figure 16.3 illustrates two applications of this architecture that support face-to-face multimodal communication between users and virtual characters in the context of interpersonal skills. In the mission rehearsal exercise (MRE), the learner plays the role of a lieutenant in the U.S. Army involved in a peacekeeping operation in Bosnia (Swartout et al., 2001). En route to assisting another unit, one of the lieutenant's vehicles becomes involved in a traffic accident, critically injuring a young boy. The boy's mother is understandably distraught, and a local crowd begins to gather. The learner

FIGURE 16.3 The MRE and SASO-ST systems allow a trainee to interact with intelligent virtual characters through natural language for task-oriented training. (See color insert.)

must resolve the situation by interacting through spoken language with virtual humans in the scene and learn to juggle multiple, interacting goals (i.e., assisting the victim vs. continuing his mission). In the Stability and Support Operations—Simulation and Training (SASO-ST) exercise, the learner plays the role of a captain assisting security and reconstruction efforts in Iraq (Traum, Swartout, Marsella, & Gratch, 2005) and must negotiate with a simulated doctor working with a nongovernmental aid organization and convince him to move his clinic to another location. The learner must resolve the situation by interacting through spoken language and by applying principles of effective negotiation. In both applications, agents must react in real time to user dialogue moves, responding in a way that is appropriate to the agent's goals and subject to the exercise's social and physical constraints.

An Integration Challenge

The AUSTIN virtual human architecture underlying these applications must integrate a diverse array of capabilities.[3] Virtual humans develop plans, act, and react in their simulated environment, requiring the integration of automated reasoning and planning techniques. To hold a conversation, they demand the full gamut of natural language research, from speech recognition and natural language understanding to natural language generation and speech synthesis. To control their graphical bodies, they incorporate real-time graphics and animation techniques. And because human movement conveys meaning, virtual humans draw

heavily on psychology and communication theory to appropriately convey nonverbal behavior. More specifically, AUSTIN integrates:

- A task reasoning module that allows virtual humans to develop, execute, and repair team plans and to reason about how past events, present circumstances, and future possibilities affect individual and team goals. Agents use domain-independent planning techniques incorporating elements of decision-theoretic plan representations with explicit representations of beliefs, intentions, and authority relationships between individuals (Rickel et al., 2002; Traum, Rickel, Gratch, & Marsella, 2003), and must balance multiple goals and multiple alternative plans for achieving them.
- A realistic model of human auditory and visual perception (Kim, Hill, & Traum, 2005) that restricts perceptual updates to information that is observable, given the constraints of the physical environment and character's field of view. Although this has the benefit of reducing perceptual processing and renders the virtual human's behavior more realistic, limited perception introduces the control problem of what features of the environment should be actively attended.
- A speech understanding module that incorporates a finite-state speech recognizer and a semantic parser to produce a semantic representation of utterances (Feng & Hovy, 2005). These interpretations may be underspecified, leading to perceptual ambiguity in the speech processing that raises a host of control decisions (e.g., should the argent

clarify the ambiguity or should it choose the most likely interpretation).

- A dialogue model that explicitly represents aspects of the social context (Matheson, Poesio, & Traum, 2000; Traum, 1994) while supporting multiparty conversations and face-to-face communication (Traum & Rickel, 2002). This module must make a variety of choices in concert with other action selection decisions in the agent: It must choose among many speech acts, including dialogue acts that can influence who has the conversational turn, what topic is under discussion, and whether to clarify or assume.

- A natural language generator that must assemble and choose between alternative utterances to convey the agent's speech act. This can produce nuanced English expressions that vary depending on the virtual human's emotional state as well as the selected content (Fleischman & Hovy, 2002).

- An expressive speech synthesizer capable of choosing between different voice modes, depending on factors such as proximity (speaking vs. shouting) and illocutionary force (command vs. normal speech) (Johnson et al., 2002).

- A gesture planner that that assembles and chooses between alternative nonverbal behaviors (e.g., gestures, head movements, eyebrow lifts) to associate with the speech (Marsella, Gratch, & Rickel, 2003). This module augments the BEAT system (Cassell, Vilhjálmsson, & Bickmore, 2001) to incorporate information about emotional state as well as the syntactic, semantic, and pragmatic structure of the utterance.

- A procedural animation system developed in collaboration with Boston Dynamics, Inc., supports the animation and rendering of the virtual character.

- A control system, based on appraisal and coping, that characterizes the current task and dialogue state in terms of appraisal variables and suggests a strategic response that informs choices made by other modules, including the perception module, task reasoner, dialogue manager, language generator, and gesture planner.

This integration raises serious control and coordination problems are similar to the issues emotion is posited to address. The agent must divide cognitive resources between plan generation, monitoring features of the environment, and attending to a conversation. But because the agent is embodied with a humanlike appearance and communicates through naturalistic methods, this becomes far more than a traditional scheduling problem. For example, if an agent takes several seconds to respond to a simple yes or no questions, users will become annoyed or read too much into the delay (one trainee felt the character was angry with them as a result of a bug that increased dialogue latency). Further, the agent must maintain some sense of consistency across its various behavioral components, including the agent's internal state (e.g., goals, plans, and emotions) and the various channels of outward behavior (e.g., speech and body movements). When real people present multiple behavior channels, observers interpret them for consistency, honesty, and sincerity, and for social roles, relationships, power, and intention. When these channels conflict, the agent might simply look clumsy or awkward, but it could appear insincere, confused, conflicted, emotionally detached, repetitive, or simply fake.

This cognitive architecture builds on prior work in the areas of embodied conversational agents (Cassell, Sullivan, Prevost, & Churchill, 2000) and animated pedagogical agents (Johnson, Rickel, & Lester, 2000) but integrates a broader set of capabilities than such systems. Classic work on virtual humans in the computer graphics community focuses on perception and action in three-dimensional (3D) worlds (Badler, Phillips, & Webber, 1993; Thalmann, 1993) but largely ignores dialogue and emotions. Several systems have carefully modeled the interplay between speech and nonverbal behavior in face-to-face dialogue (Cassell, Bickmore, Campbell, Vilhjálmsson, & Yan, 2000; Pelachaud, Badler, & Steedman, 1996), but these virtual humans do not include emotions and cannot participate in physical tasks in 3D worlds. Some work has begun to explore the integration of conversational capabilities with emotions (Lester, Towns, Callaway, Voerman, & FitzGerald, 2000; Marsella, Johnson, & LaBore, 2000; Poggi & Pelachaud, 2000) but still does not address physical tasks in 3D worlds. Likewise, prior work on STEVE addressed the issues of integrating face-to-face dialogue with collaboration on physical tasks in a 3D virtual world (Rickel & Johnson, 2000), but STEVE did not include emotions and had far less sophisticated dialogue capabilities than our current virtual humans. The tight integration of all these capabilities is one of the most novel aspects of our current work. The AUSTIN cognitive architecture seeks to advance the state of the art in each of these areas but also to explore how best to integrate them into a single agent architecture, incorporating a flexible blackboard architecture to facilitate experiments with the connections between the individual components.

Emotion, Design, and Control

We claim that appraisal theory provides a unifying conceptual framework that can inform the design of complex cognitive systems. We illustrate how it has informed our approach to integration and control of the AUSTIN cognitive architecture. The control and integration issues arising from the AUSTIN are hardly unique to virtual humans. The problem of allocating computational resources across diverse functions, coordinating their activities and integrating their results is common to any complex system. The solutions to such problems, however, have tended to be piecemeal as research has tended to focus on a specific control issue, for example, *exploration versus exploitation* or *planning versus acting*. In contrast, we argue that appraisal theory provides a single coherent perspective for conceptualizing cognitive control.

Adopting an appraisal theoretic perspective translates into several proscriptions for the design of a cognitive system.

Appraisal as a Uniform Control Structure

Appraisal theory suggests a general set of criteria and control strategies that could be uniformly applied to characterize, inform, and coordinate the behavior of heterogeneous cognitive functions. Whether it is processing perceptual input or exploring alternative plans, cognitive processes must make similar determinations: Is the situation/input they are processing desirable and expected. Does the module have the resources to cope with its implications? Such homogenous characterizations are often possible, even if individual components differ markedly. By casting the state of each module in these same general terms, it becomes possible to craft general control strategies that apply across modules.

Further, appraisal theory argues that each appraisal variable provides critical information that informs the most adaptive response. For example, if there is a threat on the horizon that may vanish of its own accord, it is probably not worth cognitive resources to devise a contingency and an organism should procrastinate; if the threat is looming and certain, an organism must act and its response should vary depending on its perceived sense of control: approach (i.e., recruit cognitive or social resources to confront the problem) if control is high or avoid (i.e., retreat from the stressor or abandon a goal) if control is low. From an ecological perspective (see Todd & Schooler, chapter 11, and Kirlik, chapter 14, this volume), these mappings can be viewed as simple control heuristics that suggest appropriate guidance for the situations an organism commonly experiences, and may translate into robust control strategies for cognitive systems.

In AUSTIN, we have explored this principle of control uniformity to the design of two core components, the plan-reasoning module and the dialogue manager. Besides the plan-based appraisal and coping, AUSTIN introduces analogous techniques to characterize the current state of a dialogue in terms of appraisal variables (e.g., What is the desirability of a particular dialogue tactic? How likely it is to succeed and how much control an agent has over this success?) and crafted alternative dialogue strategies that mirror the plan and emotion-focused coping strategies available to the planning system.

Besides simplifying AUSTIN's control architecture, this principle offered insight on how to elegantly model and select amongst alternative dialogue strategies. For example, the SASO-ST system is designed to teach principles of negotiation, including the competitive/cooperative orientation of the parties to the negotiation and the strategies they employ in light of those orientations. Specifically, one oft-made distinction is between integrative and distributive stances toward negotiation (Walton & Mckersie, 1965). A distributive stance occurs when parties interpret a negotiation as zero-sum game, where some fixed resource must be divided, whereas an integrative stance arises when parties view the situation as having mutual benefit. Third, parties may simply believe that there is no possible benefit to the negotiation and simply avoid the negotiation or deny the need for it, what is termed *avoidance* (e.g., Sillars, Coletti, Parry, & Rogers, 1982). Although described with different terminology, there are strong conceptual similarities between this theory of negotiation and appraisal theory: both argue that response strategies are influenced by an appraisal of the current situation. For example, if the outcome of a negotiation seems undesirable but avoidable, the agent adopts a strategy to disengage (e.g., change topics). If these attempts fail, the agent may reappraise the situation as less controllable and thus more threatening, motivating distributive strategies. By adopting an appraisal-theoretic perspective, we are able to recast negotiation stances as alternative strategies for coping with the appraised state of the negotiation, and thereby leverage the existing appraisal/coping machinery.

Appraisal as a Value Computation

Appraisal can be seen as a utility calculation in the sense of decision theory and thus can subsume the role played by decision theory in cognitive systems. For example, it can determine the salience and relative importance of stimuli. The difference is that appraisal can be seen as a multiattribute function that incorporates broader notions than simply probability and utility. In particular, it emphasizes the importance of control—does the agent have the power to affect change over the event—which, according to appraisal theory, is critical for determining response. Thus, appraisal theory can support the value computations presumed by many mental functions, but support subtler distinctions than traditional cognitive systems.

In AUSTIN, appraisal acts as a common currency for communicating the significance of events between the planning, dialogue management, and perceptual modules and facilitates their integration. One example of this is determining linguistic focus. In natural language, people often speak in imprecise ways, and one needs to understand the main subject of discussion to disambiguate meaning correctly. For example, when the trainee encounters the accident scene in the MRE scenario, he might ask the virtual human, "What happened here?" In principle many things have happened: the trainee just arrived, the soldiers assembled at the meeting point, an accident occurred, a crowd formed, and so forth. The virtual human could talk about any one of these and be factually correct, but not necessarily pragmatically appropriate. Rather, people are often focused most strongly on the things that upset them emotionally, which suggests an emotion-based heuristic for determining linguistic focus. Because we model the virtual character's emotions, the dialogue planning modules have access to the fact that he is upset about the accident can use that information to give the most appropriate answer: describing the accident and how it occurred.

Another example is the integration of top-down and bottom-up attention in the control of perception. In AUSTIN, the virtual human must orient its sensors (virtual eyes) to stimuli to perceive certain changes in the environment, which raises the control problem of what to look at next. This decision can be informed by bottom-up processes that detect changes in the environment (e.g., Itti & Koch, 2001) and by top-down processes that calculate the need for certain information. We have been exploring the use of appraisal as a value calculation to inform such top-down processes. Thus, for example, attention should be directed toward stimuli generating intense appraisals.

Appraisal as a Design Specification for Cognition

Appraisal theory presumes that an organism can interpret situations in terms of several criteria (i.e., appraisal variables) and use this characterization to alter subsequent cognitive processing (e.g., approach, avoidance, or procrastination). On the one hand, these assumptions dictate what sort of inferences a cognitive system must support. On the other hand, they argue that inferential mechanisms must support qualitatively different processing strategies, sensitive to the way input is appraised. Traditional cognitive systems consider only a subset of these criteria and strategic responses. In terms of appraisal, for example, cognitive systems do a good job about reasoning about an event's desirability and likelihood but rarely consider the social factors that inform causal attributions. In terms of coping, cognitive systems excel at problem-focused strategies (e.g., planning, acting, seeking instrumental social support) but have traditionally avoided emotion-focused strategies such as goal abandonment and denial.

Adopting this perspective, we identified several missing capabilities in the AUSTIN cognitive architecture, particularly as it relates to human social behavior. In its early incarnation, for example, AUSTIN used physical causality as a proxy for human social inference. In terms of the appraisal variable of causal attribution, this translates into the inference that if a person performed an action with some consequence, they deserve blame for that consequence. However, appraisal theory identifies several critical factors that mediate judgments of blame and responsibility for social activities, including whether the person intended the act, were aware of the consequence, and if their freedom to act was constrained by other social actors. Before making such inferences, AUSTIN would make inappropriate attributions of blame, such as blaming individuals when their actions were clearly coerced by another agent. Subsequent research has illustrated how to incorporate such richer social judgments into the architecture (Mao & Gratch, 2005).

This principle also led to the modeling of emotion-focused coping strategies, important for increasing the cognitive realism of the agent but also of potential value for managing commitments and cognitive focus of attention. Following Pollack (1990), commitments to goals and beliefs can be viewed as control heuristics that prevent the expenditure of cognitive resources on activities inconsistent with these commitments. This notion of commitment is argued to contribute to bounded decision making, to ease the problem of juggling multiple goals and coordinate group problem solving. Appraisal theory suggests a novel solution to the problem of when to abandon commitments that we have incorporated into AUSTIN. The standard solution is to abandon a commitment if it is inconsistent with an agent's beliefs, but coping strategies like denial complicate the picture, at least with respect to modeling humanlike decision making. People can be strongly committed to a belief, even when it contradicts perceptual evidence or their other intentions or social obligations (Mele, 2001). This suggests that there is no simple criterion for abandoning commitments, but rather one must weight the pros and cons of alternative conflicting commitments. Appraisal and coping provide a possible mechanism for providing this evaluation. Appraisal identifies particularly strong conflicts in the causal interpretation, whereas coping assesses alternative strategies for resolving the conflict, dropping one conflicting intention or changing some belief so that the conflict is resolved.

Conclusion

As cognitive systems research moves beyond simple, static, and nonsocial problem solving, researchers must increasingly confront the challenge of how to allocate and focus mental resources in the face of competing goals, disparate and asynchronous mental functions, and events that unfold across a variety of timescales. Human emotion clearly exacts a controlling influence over cognition, and here we have argued that a functional analysis of emotion's impact can profitably inform

the control of integrated cognitive systems. Computational appraisal theory, in particular, can help translate psychological findings about the function of emotion into concrete principles for the design of cognitive systems. Appraisal theory can serve as a blueprint for designing a uniform control mechanism for disparate cognitive functions, suggesting that the processing of these individual components can be uniformly characterized in terms of appraisal variables and controlled through a common mapping between appraisal and action tendency (coping). Appraising the activities of individual components also allows emotion to act as a common currency for assessing the significance of events on an agent's cognitive activities. Finally, as a theory designed to characterize emotional responses to a wide span of human situations, appraisal theory can serve as a requirements specification, suggesting core cognitive functions often overlooked by traditional cognitive systems. These principles have influenced the course of our own work in creating interactive virtual humans and, we contend, can profitably contribute to the design of integrated cognitive systems.

Acknowledgments

We gratefully acknowledge the feedback of Wayne Gray and Mike Schoelles on an earlier draft of this chapter. This work was sponsored by the U.S. Army Research, Development, and Engineering Command. The content does not necessarily reflect the position or the policy of the government, and no official endorsement should be inferred.

Notes

1. *Appraisal theory* is commonly used to refer to a collection of theories of emotion that agree in their basic commitments but vary in detail and process assumptions. Here we emphasize their similarity. See Ellsworth and Scherer for a discussion the similarity and differences between competing strands of the theory (Ellsworth & Scherer, 2003) In our own work, we are most influenced by the conception of appraisal theory advocated by Richard Lazarus.

2. EMA stands for *Emotion & Adaptation*, the title of the book by Richard Lazarus that most influenced the development of the model.

3. AUSTIN is an incremental extension of our earlier STEVE system.

References

Allmana, J., Hakeema, A., Erwinb, J., Nimchinskyc, E., & Hofd, P. (2001). The anterior cingulate cortex: The evolution of an interface between emotion and cognition. *Annals of the New York Academy of Sciences, 935,* 107–117.

Anderson, J. R. (1993). *Rules of the mind.* Hillsdale, NJ: Erlbaum.

Arnold, M. (1960). *Emotion and personality.* New York: Columbia University Press.

Badler, N. I., Phillips, C. B., & Webber, B. L. (1993). *Simulating humans.* New York: Oxford University Press.

Bates, J., Loyall, B., & Reilly, W. S. N. (1991). Broad agents. *Sigart Bulletin, 2*(4), 38–40.

Bechara, A., Damasio, H., Damasio, A. R., & Lee, G. (1999). Different contributions of the human amygdala and ventromedial prefrontal cortex to decision-making. *Journal of Neuroscience, 19*(13), 5473–5481.

Bless, H., Schwarz, N., & Kemmelmeier, M. (1996). Mood and stereotyping: The impact of moods on the use of general knowledge structures. *European Review of Social Psychology, 7,* 63–93.

Blythe, J. (1999). Decision theoretic planning. *AI Magazine, 20*(2), 37–54.

Bower, G. H. (1991). Emotional mood and memory. *American Psychologist, 31,* 129–148.

Cassell, J., Bickmore, T., Campbell, L., Vilhjálmsson, H., & Yan, H. (2000). Human conversation as a system framework: Designing embodied conversational agents. In J. Cassell, J. Sullivan, S. Prevost, & E. Churchill (Eds.), *Embodied conversational agents* (pp. 29–63). Cambridge, MA: MIT Press.

——, Sullivan, J., Prevost, S., & Churchill, E. (Eds.). (2000). *Embodied conversational agents.* Cambridge, MA: MIT Press.

——, Vilhjálmsson, H., & Bickmore, T. (2001). *BEAT: The Behavior Expressive Animation Toolkit.* Paper presented at the SIGGRAPH, Los Angeles.

Ekman, P. (1992). An argument for basic emotions. *Cognition and Emotion, 6,* 169–200.

Ellsworth, P. C., & Scherer, K. R. (2003). Appraisal processes in emotion. In R. J. Davidson, H. H. Goldsmith, & K. R. Scherer (Eds.), *Handbook of the affective sciences* (pp. 572–595). New York: Oxford University Press.

Feng, D., & Hovy, E. H. (2005). *MRE: A study on evolutionary language understanding.* Paper presented at the Proceedings of the Second International Workshop on Natural Language Understanding and Cognitive Science (NLUCS), Miami.

Fleischman, M., & Hovy, E. (2002). *Emotional variation in speech-based natural language generation.* Paper presented at the International Natural Language Generation Conference, Arden House, New York.

Frank, R. (1988). *Passions with reason: The strategic role of the emotions.* New York: W. W. Norton.

Frijda, N. (1987). Emotion, cognitive structure, and action tendency. *Cognition and Emotion, 1,* 115–143.

Gratch, J., & Marsella, S. (2001). *Tears and fears: Modeling emotions and emotional behaviors in synthetic agents.* Paper presented at the Fifth International Conference on Autonomous Agents, Montreal, Canada.

——, & Marsella, S. (2004). A domain independent framework for modeling emotion. *Journal of Cognitive Systems Research, 5*(4), 269–306.

——, & Marsella, S. (2005). Evaluating a computational model of emotion. *Journal of Autonomous Agents and Multiagent Systems, 11*(1), 23–43.

Grosz, B., & Kraus, S. (1996). Collaborative plans for complex group action. *Artificial Intelligence, 86*(2), 269–357.

Itti, L., & Koch, C. (2001). Computational modeling of visual attention. *Nature Reviews Neuroscience, 2*(3), 194–203.

Johnson, W. L., Narayanan, S., Whitney, R., Das, R., Bulut, M., & LaBore, C. (2002). *Limited domain synthesis of expressive military speech for animated characters.* Paper presented at the 7th International Conference on Spoken Language Processing, Denver, CO.

——, Rickel, J., & Lester, J. C. (2000). Animated pedagogical agents: Face-to-face interaction in interactive learning environments. *International Journal of AI in Education, 11,* 47–78.

Kim, Y., Hill, R. W., & Traum, D. R. (2005). *A computational model of dynamic perceptual attention for virtual humans.* Paper presented at the Proceedings of 14th Conference on Behavior Representation in Modeling and Simulation (BRIMS), Universal City, California.

Lazarus, R. (1991). *Emotion & adaptation.* New York: Oxford University Press.

Lester, J. C., Towns, S. G., Callaway, C. B., Voerman, J. L., & FitzGerald, P. J. (2000). Deictic and emotive communication in animated pedagogical agents. In J. Cassell, S. Prevost, J. Sullivan, & E. Churchill (Eds.), *Embodied conversational agents* (pp. 123–154). Cambridge, MA: MIT Press.

Mao, W., & Gratch, J. (2004). *Social judgment in multi-agent interactions.* Paper presented at the Third International Joint Conference on Autonomous Agents and Multiagent Systems, New York, New York.

——, & Gratch, J. (2005). *Social causality and responsibility: Modeling and evaluation.* Paper presented at the International Working Conference on Intelligent Virtual Agents, Kos, Greece.

Marsella, S., & Gratch, J. (2003). *Modeling coping behaviors in virtual humans: Don't worry, be happy.* Paper presented at the Second International Joint Conference on Autonomous Agents and Multi-agent Systems, Melbourne, Australia.

——, Gratch, J., & Rickel, J. (2003). Expressive Behaviors for Virtual Worlds. In H. Prendinger & M. Ishizuka (Eds.), *Life-like characters tools, affective functions and applications* (pp. 317–360). Berlin: Springer.

——, Johnson, W. L., & LaBore, C. (2000). *Interactive pedagogical drama.* Paper presented at the Fourth International Conference on Autonomous Agents, Montreal, Canada.

Matheson, C., Poesio, M., & Traum, D. (2000). *Modeling grounding and discourse obligations using update rules.* Paper presented at the First Conference of the North American Chapter of the Association for Computational Linguistics.

Mele, A. R. (2001). *Self-deception unmasked.* Princeton, NJ: Princeton University Press.

Ortony, A., Clore, G., & Collins, A. (1988). *The cognitive structure of emotions.* Cambridge: Cambridge University Press.

Peacock, E., & Wong, P. (1990). The stress appraisal measure (SAM): A multidimensional approach to cognitive appraisal. *Stress Medicine, 6,* 227–236.

Pelachaud, C., Badler, N. I., & Steedman, M. (1996). Generating facial expressions for speech. *Cognitive Science, 20*(1).

Poggi, I., & Pelachaud, C. (2000). Emotional meaning and expression in performative faces. In A. Paiva (Ed.), *Affective interactions: Towards a new generation of computer interfaces* (pp. 182–195). Berlin: Springer.

Pollack, M. (1990). Plans as complex mental attitudes. In P. Cohen, J. Morgan, & M. Pollack (Eds.), *Intentions in communication* (pp. 77–104). Cambridge, MA: MIT Press.

Rickel, J., & Johnson, W. L. (2000). Task-oriented collaboration with embodied agents in virtual worlds. In J. Cassell, J. Sullivan, S. Prevost, & E. Churchill (Eds.), *Embodied conversational agents* (pp. 95–122). Cambridge, MA: MIT Press.

——, Marsella, S., Gratch, J., Hill, R., Traum, D., & Swartout, W. (2002). Toward a new generation of virtual humans for interactive experiences. *IEEE Intelligent Systems, July/August,* 32–38.

Russell, J. A., & Lemay, G. (2000). Emotion concepts. In M. Lewis & J. Haviland-Jones (Eds.), *Handbook of emotions* (pp. 491–503). New York: Guilford Press.

Scherer, K. (1984). On the nature and function of emotion: A component process approach. In K. R.

Scherer & P. Ekman (Eds.), *Approaches to emotion* (pp. 293–317). Hillsdale, NJ: Erlbaum.

Schwarz, N., Bless, H., & Bohner, G. (1991). Mood and persuasion: Affective states influence the processing of persuasive communications. *Advances in Experimental Social Psychology, 24,* 161–199.

Shaver, K. G. (1985). *The attribution of blame: Causality, responsibility, and blameworthiness.* New York: Springer.

Sillars, A. L., Coletti, S. F., Parry, D., & Rogers, M. A. (1982). Coding verbal conflict tactics: Nonverbal and perceptual correlates of the avoidance-distributive-integrative distinction. *Human Communication Research, 9,* 83–95.

Simon, H. A. (1967). Motivational and emotional controls of cognition. *Psychological Review, 74,* 29–39.

Smith, C. A., & Lazarus, R. (1990). Emotion and adaptation. In L. A. Pervin (Ed.), *Handbook of personality: Theory & Research* (pp. 609–637). New York: Guilford Press.

Swartout, W., Hill, R., Gratch, J., Johnson, W. L., Kyriakakis, C., LaBore, C., et al. (2001). *Toward the Holodeck: Integrating graphics, sound, character and story.* Paper presented at the Fifth International Conference on Autonomous Agents, Montreal, Canada.

Thalmann, D. (1993). Human modeling and animation. In *Eurographics '93 State-of-the-Art Reports.*

Traum, D. (1994). *A computational theory of grounding in natural language conversation.* Unpublished doctoral dissertation, University of Rochester, Rochester, New York.

——, & Rickel, J. (2002). *Embodied agents for multiparty dialogue in immersive virtual worlds.* Paper presented at the First International Conference on Autonomous Agents and Multi-agent Systems, Bologna, Italy.

——, Rickel, J., Gratch, J., & Marsella, S. (2003). *Negotiation over tasks in hybrid human-agent teams for simulation-based training.* Paper presented at the International Conference on Autonomous Agents and Multiagent Systems, Melbourne, Australia.

——, Swartout, W., Marsella, S., & Gratch, J. (2005). *Fight, flight, or negotiate.* Paper presented at the Intelligent Virtual Agents, Kos, Greece.

Walton, R. E., & Mckersie, R. B. (1965). *A behavioral theory of labor negotiations: An analysis of a social interaction system.* New York: McGraw-Hill.

Weiner, B. (1995). *The judgment of responsibility.* New York: Guilford Press.

17

Decreased Arousal as a Result of Sleep Deprivation

The Unraveling of Cognitive Control

Glenn Gunzelmann, Kevin A. Gluck, Scott Price,
Hans P. A. Van Dongen, & David F. Dinges

This chapter discusses recent efforts at developing mechanisms for capturing the effects of fatigue on human performance. We describe a computational cognitive model, developed in ACT-R, that performs a sustained attentional task called the psychomotor vigilance task (PVT). We use neurobehavioral evidence from research on sleep deprivation, in addition to previous research from within the ACT-R community, to select and to evaluate a mechanism for producing fatigue effects in the model. Fatigue is represented by decrementing a parameter associated with arousal in ACT-R, while also reducing a threshold value in the architecture to capture attempts at compensating for the negative effects of decreased arousal. These parameters are associated with the production utility computation in ACT-R, which controls the selection/execution cycle to determine which production (if any) to execute on each cognitive cycle. In ACT-R, this mechanism is linked to the basal ganglia and the thalamus. In turn, portions of the thalamus show heightened activation in attentional tasks under conditions of sleep deprivation. The model we describe closely captures the performance of human participants on the PVT, as observed in a laboratory experiment involving 88 hours of total sleep deprivation.

Until recently, computational cognitive models of human performance were developed with little consideration of how factors such as emotions and alertness influence cognition. However, with increased sophistication in models of cognitive systems, advances in computer technology, and pressure for ever more realistic representations of human performance, cognitive moderators are emerging as an important area of research within the field of computational modeling (e.g., Gratch & Marsella, 2004; Hudlicka, 2003; Ritter, Reifers, Klein, Quigley, & Schoelles, 2004). There is a sense in which this development is both premature and long overdue. Evidence for its prematurity can be found in many of the other chapters in this volume. Cognitive science has yet to unravel many of the intricacies of "normal" human cognition. Therefore, adding additional complexity by including cognitive moderators that influence those thought processes constitutes a substantial challenge. However, cognitive moderators are pervasive in human cognition. It seems essential, therefore, that they be considered in attempts to

understand human cognitive functions. If cognitive architectures are to be viewed as "unified theories of cognition" (Newell, 1990), then they must include mechanisms to represent those factors that have substantial modulatory effects on cognitive performance.

This chapter describes an effort to introduce a theory of degraded cognitive functioning into the adaptive control of thought–rational, or ACT-R, cognitive architecture. In this case, the degradation arises from the combined effect of sleep deprivation and endogenous circadian variation. We describe a computational cognitive model that incorporates mechanisms to represent decreased alertness and describe the impact of those mechanisms on the model's performance on the psychomotor vigilance task (PVT), a sustained attention task that has been extensively validated to be sensitive to variation in sleep homeostatic and circadian dynamics, while being relatively immune to the effects of aptitude and learning (Dorrian, Rogers, & Dinges, 2005). Our modeling effort draws on recent research on partial and total sleep deprivation (e.g., Van Dongen et al., 2003), and leverages recent advances in understanding how sleep deprivation impacts neurobehavioral and brain functioning (e.g., Drummond et al., 1999, 2000; Drummond, Gillin, & Brown 2001; Habeck et al., 2004; Portas et al., 1998).

In the sections that follow, we describe relevant research related to sleep loss. This is followed by a description of the PVT and then the ACT-R model we have developed to perform it. We use the model to demonstrate the effectiveness of our approach for capturing performance decrements as a function of sleep deprivation. In describing the model, we suggest some alternative mechanisms to illustrate how the effects of sleep deprivation can be seen as resulting from impacts to either central control (Type 1 control) or the internal control of functional processes (Type 2 control), which includes processes like memory retrieval or programming motor movements. This distinction constitutes a major theme of this book. Although the mechanistic explanation for the effects of sleep deprivation we have developed is not explicitly defined in terms of Type 1 or Type 2 control, the discussion illustrates how the modeling effort is improved through consideration of this distinction.

Neuropsychological Research on Sleep Deprivation

Unquestionably, sleep deprivation has a negative effect on human performance across a wide array of tasks and situations. Determining the particular impacts of sleep deprivation, both behaviorally and physiologically, has been a significant topic of study in psychological and medical research for quite some time (e.g., Patrick & Gilbert, 1896; von Economo, 1930). Research originally focused on identifying the nature of neurobehavioral incapacitation but shifted to changes in cognitive performance when early studies did not provide conclusive evidence that sleep loss eliminated the ability to perform specific tasks (e.g., Kleitman, 1923; Lee & Kleitman, 1923). Current research directions have been motivated by the desire to uncover the neurophysiologic mechanisms that produce diminished alertness and decrements in cognitive performance, as well as any compensatory mechanisms. Research evaluating behavioral, pharmacological, and technological countermeasures to offset deficits of sleep deprivation has also been a long-standing focus of research (e.g., Bonnet et al., 2005; Caldwell, Caldwell, & Darlington 2003; Caldwell, Caldwell, Smith, & Brown, 2004; Dinges & Broughton, 1989).

At the cortical level, studies have shown inconsistent patterns of regional activation responses to sleep deprivation, depending on the type of cognitive task, its difficulty, and the method used to measure activation (e.g., Chee & Choo, 2004; Drummond et al., 1999, 2001; Habeck et al., 2004). At the subcortical level, a main area that consistently shows sensitivity to sleep deprivation is the thalamus (Chee & Choo, 2004; Habeck et al., 2004; Lin, 2000; Portas et al., 1998). The thalamus typically shows an increase in activation when individuals are asked to perform a task while sleep deprived, relative to performing the task when well rested. For instance, Portas et al. (1998) asked participants to perform a short-duration attention task while activity was measured using fMRI. They found that the thalamus showed increased activation while performing the attention task under conditions of sleep loss, while overall performance (response time) was not significantly different from baseline. From these results, they concluded, "This process may represent a sort of compensatory mechanism. . . . We speculate that the thalamus has to 'work harder' in conditions of low arousal to achieve a performance that is equal to that obtained during normal arousal" (p. 8987). The possibility of such a compensatory mechanism involving the thalamus is discussed further in the section on the computational model later in this chapter.

Biomathematical Models of Sleep Deprivation

In addition to the significant progress that has been made in understanding the neurobehavioral mechanisms of sleep deprivation, researchers studying fatigue have also developed biomathematical models that reflect the influence of sleep history and circadian rhythms on overall cognitive performance, or alertness (Mallis, Mejdal, Nguyen, & Dinges, 2004). Such models provide a means for describing the dynamic interaction of these factors. For instance, Figure 17.1 shows the predictions for one of these models, the circadian neurobehavioral performance and alertness (CNPA) model (Jewett & Kronauer, 1999), for a protocol involving 88 hr of total sleep deprivation. The circadian rhythm component of the model is responsible for the cyclic nature of the predictions and increased sleep loss is responsible for the overall decline across days.

Although there is room for improvement in all current biomathematical models of performance (Van Dongen, 2004), the models have potential value for predicting global changes in alertness over time in a

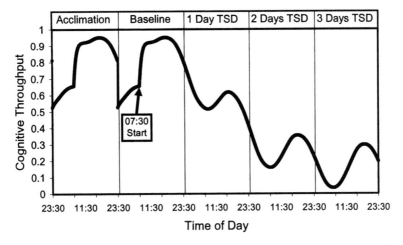

FIGURE 17.1 Predictions of alertness from the circadian neurobehavioral performance and alertness (CNPA) model for a study involving 88 continuous hours awake, beginning at 7:30 a.m. on the baseline day.

variety of circumstances. However, a key limitation is that these models do not make predictions of how changes in alertness will affect performance on particular tasks (e.g., changes in response times or changes in types or frequencies of errors). The fits described in Van Dongen (2004) were produced by scaling the alertness predictions to minimize the deviation from the data. These values had to be computed post hoc. So, while the predictions from the models approximate relative changes in performance, they do not actually provide a priori estimates of how much response times will change in absolute magnitude or how errors will increase over time.

The computational cognitive modeling research described in this chapter will eventually allow us to bridge the gap between biomathematical models and complex cognitive task performance. Computational cognitive models make detailed predictions about human performance, including response times and errors. The goal of the project is to use the predictions from the biomathematical models to drive changes in mechanisms in the ACT-R cognitive architecture. In this way, the predictions of the biomathematical model can be used to produce parameter changes in the cognitive model, which can be used to make specific predictions about how human performance declines as a function of fatigue. Although this latter goal has not yet been reached, this chapter describes the progress we have made toward it, especially the determination of a set of mechanisms in ACT-R to account for changes in alertness. These mechanisms are demonstrated in the context of the PVT, which is described next.

Psychomotor Vigilance Task

The psychomotor vigilance task (PVT; Dinges & Powell, 1985) assesses vigilant/sustained attention and has been used frequently in sleep deprivation research. Its main advantages are that performance is both sensitive to the levels of sleep deprivation and relatively insensitive to either aptitude or learning (Dorrian et al., 2005). During a typical PVT trial, a stimulus appears in a prespecified location on a monitor at random intervals between 2 s and 10 s. The subject's task is to press a response button as fast as possible each time a stimulus appears but not to press the button too soon. When the response button is pressed, the visual stimulus displays reaction time in milliseconds to inform the subject of how well they performed. The duration of a test session is typically 10 min.

The data from a PVT session consist of approximately 90 responses, which can be classified to facilitate understanding how PVT performance changes as fatigue increases (Dorrian et al., 2005). The range for the first category, which we will refer to as "alert" responses, is between 150 ms and 500 ms after stimulus onset (median is typically around 250 ms), indicative of a participant that is responding about as rapidly as neurologically possible to each stimulus. Responses greater than 500 ms but less than 30,000 ms (i.e., 30 s) are considered to be "lapses" of attention (errors of omission; Dinges & Kribbs, 1991; Dorrian et al., 2005). These responses indicate that attention is wavering from the display, but that participants are recovering at some point to detect the stimulus. In some instances, participants fail to respond even after 30 s, which is a dramatic breakdown in performance that is classified as a "sleep attack" (Dorrian et al., 2005). In these cases, the experimenter intervenes to wake the participant. At the opposite end of the response-time continuum are "false starts" (errors of commission), which are responses that occur before the stimulus appears, or within 150 ms of the stimulus onset (i.e., neurologically too fast to be a normal, alert response). These responses represent anticipation of the stimulus's appearance.

As sleep deprivation increases, the proportion of alert responses decreases, and the distribution of reaction times shifts to the right, resulting in increased proportions of lapses and sleep attacks. As participants attempt to compensate based on feedback that they are lapsing (errors of omission) more frequently, the proportion of false starts (errors of commission) increases as well (Doran, Van Dongen, & Dinges, 2001). A sample set of data from the PVT is shown in Figure 17.2 (these data are from Van Dongen, 2004; Van Dongen et al., 2001). In the experiment that provided the data, participants first spent three nights in the laboratory to acclimate to a common sleep cycle of 8 hr for sleep per day. After this, participants were kept awake continuously for 88 hr, until near midnight on the fourth day. This is the same protocol that was used to generate the CNPA predictions in Figure 17.1, which shows alertness predictions for the last day of acclimation and for the 88-hr sleep-deprivation period. The first day of this period, during which no actual sleep loss was yet incurred, was used as a baseline day. Beginning at 7:30 a.m. on the baseline day, participants completed a series of tasks, including the PVT, repeatedly in 2-hr cycles (the set of tasks took approximately 30 min to complete). Note that the PVT data shown in Figure 17.2 are averaged over sessions performed within each day of the protocol, whereas the CNPA data in Figure 17.1 illustrate the dynamic changes in alertness that occur within each of the days (circadian rhythms).

The next section describes the computational cognitive model. The model represents the first step in developing the capability to make detailed a priori predictions about changes in human performance on particular tasks as a function of increased levels of fatigue. The model performs the PVT, and parameter changes in the model impact performance in a manner similar to human performance under conditions of sleep deprivation.

Computational Cognitive Model

The computational model described in this chapter was developed in the ACT-R 5 cognitive architecture (Anderson et al., 2004). Here we will describe only the ACT-R mechanisms that are associated with the

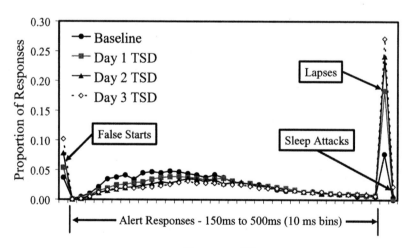

FIGURE 17.2 Human performance on the psychomotor vigilance task. Data are from a study where participants were kept awake for 88 continuous hours, while performing a battery of tests every 2 hr (data from Van Dongen et al., 2001; Van Dongen 2004). Averages across test sessions within each day are shown.

parameters that were manipulated to alter the architecture's level of alertness, which produces the performance decrements exhibited by the model.

We have constrained the selection of appropriate parameters and mechanisms for this effort in several ways. For instance, we have taken into account previous research in the ACT-R community (Belavkin, 2001; Jongman, 1998), and we have used the conclusions from neuropsychological research on the effect of sleep deprivation on the functioning of various brain areas, particularly the thalamus (Chee & Choo; Habeck et al., 2004; Portas et al., 1998). To use the conclusions from this work, we leveraged recent advances in the development of the ACT-R architecture, which have included mapping its components to brain areas (Anderson, chapter 4, this volume). This mapping establishes a "common space," where links between neuropsychological research on fatigue can be putatively linked with aspects of the architecture. The constraints imposed by this research implicate a mechanism in ACT-R that is related to the production selection/execution cycle as a candidate for being impacted by fatigue. This process is associated with the basal ganglia and the thalamus in the current conceptualization of ACT-R (Figure 17.3).

The production/execution cycle involves evaluating alternative productions and then selecting the "best" among them. During the selection process, productions are compared using a value called expected utility (U_i), which is calculated for each candidate production using the equation:

$$U_i = P_i G - C_i + \varepsilon$$

In this equation, P_i is the probability of success if production$_i$ is used and C_i is the anticipated cost. In general, G has been termed *the value of the goal*. However,

Production Execution Cycle in ACT-R
(including hypothesized mapping to brain areas)

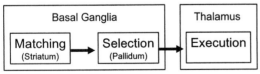

FIGURE 17.3 Production execution cycle in the adaptive control of thought–rational (ACT-R) cognitive architecture, including hypothesized mapping to brain areas. The expected utility equation is associated with the selection component of this process, while execution is controlled by the utility threshold (T_u). Adapted from http://actr.psy.cmu.edu/.

the research cited above uses the G parameter to capture the influence of arousal on performance (Belavkin, 2001; Jongman, 1998). We use this conceptualization of G in our model as well. Noise (ϵ) is added to the calculation to add a stochastic component to the value. The noise is sampled from a Gaussian distribution with a mean of 0 and a variance of about 0.21.[1] A value for U_i is calculated for each production$_i$ that matches the current state on each production cycle. The production$_i$ with the highest value for U_i is selected.

Once a production is selected, the next step is execution. This process is associated with the thalamus in ACT-R (Figure 17.3). Production execution is controlled by a parameter called the *utility threshold*, T_u. The selected production is executed, provided that U_i exceeds T_u. If it does not, no production is executed and the model is "idle" for the duration of that production cycle (approximately 50 ms).[2] The neuropsychological data suggest that fatigue may indirectly affect this process, with individuals trying to offset the adverse effects through an attempt at compensation (Portas et al., 1998). As the behavior of the model illustrates, some compensation may be possible, but it does not completely offset the negative effects associated with sleep loss.

We find it encouraging that research on the neurobehavioral effects of fatigue and research within the ACT-R community both point to a common mechanism for capturing fatigue effects in ACT-R. The convergence of this research on the production selection/execution cycle in ACT-R indicates that one of the impacts of fatigue may be a decreased likelihood of successfully executing an appropriate sequence of productions. This entails both an increased likelihood of having cognitive cycles where the system is idle as well as the execution of inappropriate productions. Next we describe the model we constructed in ACT-R, which is based on this conceptualization of the impact of fatigue.

Model Design

Because the PVT is simple in design, the ACT-R model is relatively straightforward. Before the stimulus appears the model can (1) deliberately wait for the stimulus or (2) errantly make a response (a false start). Once the stimulus has appeared, the model can (1) attend to the stimulus and then respond (this is two productions) or (2) respond without attending the stimulus (a false start that happens to come after the stimulus appears and is therefore counted as an appropriate response). At any point in the task, it is possible for the model to be idle for one or more cognitive cycles.

For nearly all the productions in the model, P_i was set to 1, meaning that the goal would be achieved successfully if that production was fired. The lone exception to this was the production that errantly responds. P_i for this production was 0, on the assumption that it is highly unlikely to result in achieving the goal of successfully responding to the stimulus. The consequence of this is a reduced likelihood of that production firing relative to the other, appropriate productions, since U_i becomes a negative value $(-C_i)$. However, with noise added to the utility computation, this production is occasionally the one with the highest U_i and also rises above T_u.

To produce decrements in performance like those associated with sleep deprivation, we conceptualized increased sleep deprivation as resulting in decreased arousal. Therefore, we implement fatigue in ACT-R by decreasing the G parameter, in line with previous research within the ACT-R community (Belavkin, 2001; Jongman, 1998). Reducing the value of G decreases U_i for all productions, where P_i is greater than 0. This makes it more likely that the expected utilities for those productions will fall below T_u, thereby making them less likely to fire. In turn, this increases the probability of idle cognitive cycles. A key feature of this mechanism is that idle cycles actually become more likely than appropriate actions once U_i (before noise is added) falls below T_u, which happens after the first full day of total sleep deprivation.

In addition to the decreased arousal that is associated with sleep loss, we include a secondary process that can be viewed as an attempt to compensate for the negative effect of fatigue, as suggested by Portas et al. (1998). In ACT-R terms, the increased activity they observed in the thalamus can be seen as an attempt to make it easier for the selected production to fire successfully. The most natural way of representing this in ACT-R is by reducing T_u. With lower values for T_u, productions with increasingly lower U_i values are

able to fire. In cases where there is an idle cognitive cycle, new U_i values are computed at the beginning of the next cycle and the process repeats. The noise added to the calculation of U_i creates the possibility of all matching productions being below threshold on one cycle, followed by a cycle where at least one production rises above threshold.

There is one additional component to the model. A mechanism was implemented to represent the process of falling asleep. The idle cognitive cycles that result when no productions rise above threshold represent situations where arousal is so low that none of the available actions are executed. As an individual falls asleep, the probability of being in such a state should increase. In the model, this is represented by having the value of G decrease when idle cognitive cycles occur during the time when the stimulus is on the screen. On each occasion when this occurs, G is decremented by 0.035. This is limited by the architectural requirement that G remain positive. So, G is decremented by 0.035 on each idle cognitive cycle, unless that decrement results in a negative number, in which case G is set to a minimum value of 0.0001, where it remains stable until the start of the next trial. This progressively reduces the probability of a response with each passing cycle in the model. The value of G is reset to the starting value at the beginning of each trial, reflecting either a successful response or being awakened by the experimenter after a sleep attack. The values used for G and T_u for the four days of the experiment are presented in Table 17.1. As this table illustrates, G falls more rapidly than T_u, meaning that the model's performance deteriorates across the four days of the study. The model's performance is described in more detail in the next section.

Model Performance

Baseline performance in the model reflects the interplay of the knowledge in the system with the various mechanisms described above using the parameter values

TABLE 17.1 Parameter Values for G and T_u in the ACT-R Model for Each Day in the Sleep Deprivation Protocol

Day	G (Arousal)	T_u (Utility Threshold)
Baseline	1.98	1.84
Day 1 of total sleep deprivation	1.80	1.78
Day 2 of total sleep deprivation	1.66	1.70
Day 3 of total sleep deprivation	1.58	1.64

identified above and shown in Table 17.1. All other parameters were kept at their default ACT-R values. Baseline performance represents an alert, well-rested participant. The model's performance in this condition, along with the human data, is illustrated in panel A of Figure 17.4. The model closely captures all of the phenomena of interest. The largest discrepancy between the two data sets is that the human participants' alert responses tend to be somewhat faster than the model's. This may reflect an attentional strategy in use by the participants, but not implemented in the model.[3] The accuracy of the model's predictions at baseline is matched by the predictions it makes for performance after one, two, and three days without sleep, as shown in panels B, C, and D of Figure 17.4, respectively. The model captures the increase in false starts, the shift in alert response times, and the increases in lapses and sleep attacks. For all four days, the performance of the model closely matches the human data. Overall, the correlation between the human data and the model predictions is 0.99 (root mean square deviation = 0.0047).

The impact of decreasing G in the model is to reduce the likelihood that the appropriate productions will have U_i values above T_u. We interpret this situation as reduced arousal (i.e., ACT-R gets sleepy), with the impact being that the appropriate productions are not executed. The reduction in arousal relative to the level of compensation immediately decreases the likelihood that a fast response will be made. Thus, there is a shift in the distribution of reaction times to the right, with progressively fewer fast responses as G gets lower. The decrease in frequency of fast responses has secondary effects. Each time an opportunity to respond is missed (idle cycles), G is decremented. This makes the model less likely to respond at the next opportunity. So the impact of lower arousal accumulates to produce many more lapses, along with more sleep attacks.

Finally, the false starts in the model increase similarly to the human data. This is primarily a side effect of the compensation in the model, reflected by the decreased values for T_u. As T_u decreases, it is more likely that the value for U_i for the production that produces a false start (i.e., *just click*) will have a U that

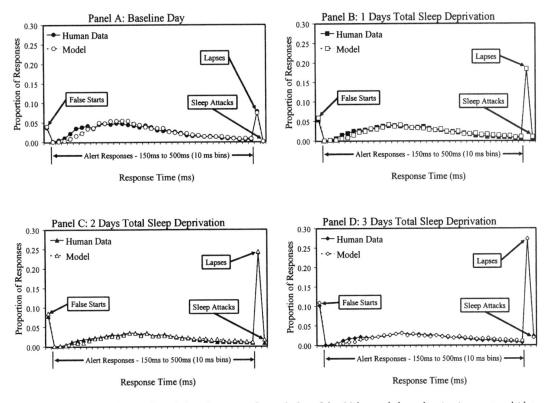

FIGURE 17.4 Human data and model performance for each day of the 88-hr total sleep deprivation protocol (data from Van Dongen et al., 2001; Van Dongen, 2004).

rises above threshold. Thus, the mechanism for compensation produces an undesirable behavioral consequence in the model that matches the human data; namely, an increased likelihood of responding before the stimulus appears. It is also the case that the decrease in G contributes to this effect (though to a lesser extent), since changes in G do not impact U for the *just-click* production (because $P = 0$ for this production). The combination of these two aspects of the model produces an increase in the probability of producing a false start that lines up with the data from the human participants.

General Discussion

The model described accurately captures the impact of sleep deprivation on human performance on the PVT. The mechanisms that were implemented are based on research within the ACT-R community, combined with neurobehavioral evidence concerning the impact of fatigue on brain activity. This illustrates the potential for using multiple constraints from diverse research communities to develop an understanding of cognitive mechanisms that impact performance. The ultimate goal of this work is to develop a general set of mechanisms for understanding how fatigue impacts the human cognitive system. The mechanisms must specify the relationship between fatigue and cognitive control, even if it is not done explicitly. ACT-R has components that correspond to the distinction between Type 1, or centralized, control and Type 2 control, or the internal control of functional processes.

The central production system represents the Type 1 control in ACT-R, while the other modules represent Type 2 control structures. The production cycle drives behavior in ACT-R models and directs the operation of the other modules. Other modules respond to requests from the production system by producing actions (e.g., motor movements or shifts of visual attention) or by performing some operation and making the results available to the production system (e.g., a retrieval from declarative memory).

At one extreme, alertness could be implemented as a separate module in ACT-R that operates as a Type 2 control structure to influence central cognition. At the other extreme, alertness could be viewed as a component of the Type 1 control system.[4] We are not committed to a particular view of how sleep deprivation fits with cognitive control. Our model represents fatigue as a set of mechanisms that operate on the central

production system in ACT-R. Decreased arousal (G) and a lower utility threshold (T_u) both involve the central production system and influence the actions that are taken by the model. The performance decrements in the other modules stem from productions failing to fire to send the appropriate requests. Still, the model has little to say about whether those mechanisms are part of the Type 1 control structure or are a separate, Type 2, control structure. This is because the driving force behind the fatigue mechanisms is not yet integrated into the ACT-R architecture. Specifically, the biomathematical models of alertness, which we intend to use to drive the parameter changes, are currently outside the architecture. How they are represented in the architecture could have important implications for our understanding of how cognitive moderators like fatigue operate.

In the current model, the deficits in performance that develop with diminished alertness represent a breakdown in Type 1 control, where appropriate actions are no longer taken. Increases in the frequency of idle cognitive cycles produces a reduction and shift in alert response times. In addition, when the model is idle arousal continues to fall, making the model progressively less likely to respond. Essentially, the Type 1 control in the model disappears, as the central production system ceases to drive behavior forward. The model has, in essence, gone to sleep. Even though false starts are a consequence of a production becoming more likely to fire, this effect still represents a breakdown in control, since the production has a low probability of success. More generally, with the current mechanism for decreasing arousal and compensation, ACT-R becomes less able to differentiate between successful and unsuccessful productions. The effect is more erratic (i.e., unstable) performance, just as is found in human participants who are deprived of sleep (Doran, et al. 2001).

It is not the case that all of the effects must come through interaction with Type 1 control. We have considered an alternative mechanism for producing false starts that involves processes within the motor module. In ACT-R, motor actions can be preplanned so that they can be executed more rapidly. Type 1 control guides the execution of the planned action. It is possible that one of the effects of fatigue is to increase the likelihood that planned actions are inadvertently executed without direction from the central production system. There is evidence that increased levels of fatigue may reduce cognitive inhibition (e.g., Harrison & Horne, 1998).

Thus, if planned actions are held back using a process like inhibition, then sleep deprivation could result in more frequent false starts through a breakdown in Type 2 control. We have not implemented this alternative, partly because the current ACT-R architecture does not include mechanisms for inhibiting motor movements that could be manipulated to produce this effect. However, it provides an interesting possibility for future research.

This alternative mechanism for false alarms raises an important issue for understanding fatigue. If it is the case that fatigue can directly influence one Type 2 control structure, like a motor module, then presumably other modules may be affected as well. Such a conclusion would suggest that sleep deprivation has effects that transcend the distinction between Type 1 and Type 2 control. That is, it is possible that sleep deprivation influences mechanisms associated with Type 1 control *and* Type 2 control.

Conclusion

We have presented a model that performs the PVT, combined with a set of mechanisms that produces decrements in performance on that task, which mirror the effect of sleep loss in human participants. The mechanisms implemented in the model were based on previous research, both in the ACT-R community and on the neurobehavioral impacts of sleep loss. The data were fit at the level of average performance for each day of the study. We have demonstrated that these mechanisms produce performance changes in the model that closely match the performance changes of human participants who have been denied sleep over a period of nearly four consecutive days. The next step is to fit the model to the experimental data from each 10-minute session, which occurred at 2-hour intervals throughout the study. This effort is already underway and has provided further support for our approach. We are using the results from this process to develop a mapping between alertness predictions generated by CNPA and parameter values in the ACT-R model. Once the appropriate mapping is identified, it will be used to validate our ability to make a priori predictions about human performance in other experimental protocols and for other tasks.

A major challenge for validating our model lies in its ability to account for findings from a variety of tasks. The model's performance on the PVT is almost entirely dependent on procedural components of the architecture; the only declarative knowledge needed is a representation of the goal. It will be interesting to see whether additional mechanisms are required to account for changes in performance associated with tasks that require more declarative knowledge. For instance, we are currently developing a model for a serial addition/subtraction task (SAST), which relies on knowledge of simple (single-digit) addition and subtraction problems. Our initial examinations of the model for this task suggest that the procedural mechanisms that produced the effects in PVT may not be sufficient to account for the changes in response times and accuracy on the SAST. This may indicate that decreased alertness effects many aspects of the cognitive system simultaneously. Certainly, other areas of the brain show changes in activity as a function of sleep loss. As this research progresses and matures, we hope to develop a comprehensive, mechanistic explanation of the changes that occur at the level of the human cognitive architecture as a function of changes in alertness. To do this, we will continue to incorporate as many theoretical and empirical constraints as we can, to guide the identification of mechanisms and to inform our understanding of the processes involved.

Acknowledgments

Cognitive model development was sponsored partly by AFRL's Warfighter Readiness Research Division and partly by grant number 04HE02COR from the Air Force Office of Scientific Research (AFOSR). Human data collection was sponsored by AFOSR grants F49620-95-1-0388 and F49620-00-1-0266, and by NIH grant RR00040. We would like to thank Robert O'Connor for his role in the original experiment described here, and for his more recent assistance in providing the raw data for comparison with the model. We would also like to thank Wai-Tat Fu, Frank Ritter, and Mike Schoelles for their helpful comments on earlier drafts.

Notes

1. In ACT-R, noise is controlled by a parameter, s, whose relation to the variance (σ^2) is defined by the equation: $\sigma^2 = (\pi^{2*}s^2)/3$. In this model, s was set to the default ACT-R value of 0.25.

2. ACT-R does not produce idle cycles by default. This required an enhancement of this component of the architecture, and we are grateful to Dan Bothell for his help implementing this. In addition, cycle times in this model were noisy, which is an available option in ACT-R. The duration of a particular cycle in the model varied between 25 ms and 75 ms, according to a uniform distribution, which is the default function for this noise value in ACT-R.

3. The implemented model does not explicitly maintain attention once it has responded to the stimulus. As a result, when the next stimulus appears, an extra production is required to shift attention to it. With a probabilistic mixture of this process and one that maintains attention on the stimulus location between trials, the distribution of response times can be captured more accurately. However, the better fit comes at the cost of model complexity, which seemed unwarranted for the current modeling effort.

4. It is possible to consider a third option, which is that arousal is outside cognitive control altogether (i.e., it is not under control). However, arousal clearly seem to affect aspects of control in cognition, suggesting that it operates through these structures, if not as one, or part, of them.

References

Anderson, J. R., Bothell, D., Byrne, M. D., Douglass, S., Lebiere, C., & Qin, Y. (2004). An integrated theory of the mind. *Psychological Review, 111*, 1036–1060.

Belavkin, R. V. (2001). The role of emotion in problem solving. In C. Johnson (Ed.), *Proceedings of the AISB '01 Symposium on emotion, cognition, and affective computing* (pp. 49–57). York, UK: Heslington.

Bonnet, M. H., Balkin, T. J., Dinges, D. F., Roehrs, T., Rogers, N. L., & Wesensten, N.J. (2005). The use of stimulants to modify performance during sleep loss: A review by the Sleep Deprivation and Stimulant Task Force of the American Academy of Sleep Medicine. *Sleep, 28*(9), 1144–1168.

Caldwell, J. A., Caldwell, J. L., & Darlington, K. K. (2003). Utility of dextroamphetamine for attenuating the impact of sleep deprivation in pilots. *Aviation, Space and Environmental Medicine, 74*(11), 1125–1134.

———, Caldwell, J. L., Smith, J. K., & Brown, D. L. (2004). Modafinil's effects on simulator performance and mood in pilots during 37 h without sleep. *Aviation, Space and Environmental Medicine, 75*(9), 777–784.

Chee, M. W. L., & Choo, W. C. (2004). Functional imaging of working memory after 24 hr of total sleep deprivation. *Journal of Neuroscience, 24*(19), 4560–4567.

Dinges, D. F., & Broughton, R. J. (Eds.). (1989). *Sleep and alertness: Chronobiological, behavioral and medical aspects of napping.* New York: Raven Press.

———, & Kribbs, N. B. (1991). Performing while sleepy: Effects of experimentally-induced sleepiness. In T. H. Monk (Ed.), *Sleep, sleepiness and performance* (pp. 97–128). Chichester: Wiley.

———, & Powell, J. W. (1985). Microcomputer analyses of performance on a portable, simple visual RT task during sustained operations. *Behavior Research Methods, Instruments, & Computers, 17*(6), 652–655.

Doran, S. M., Van Dongen, H. P., & Dinges, D. F. (2001). Sustained attention performance during sleep deprivation: Evidence of state instability. *Archives of Italian Biology: Neuroscience, 139*(3), 253–267.

Dorrian, J., Rogers, N. L., & Dinges, D. F. (2005). Psychomotor vigilance performance: Neurocognitive assay sensitive to sleep loss. In C. A. Kushida (Ed.), *Sleep deprivation: Clinical issues, pharmacology and sleep loss effects* (pp. 39–70). New York: Marcel Dekker.

Drummond, S. P. A., Brown, G. G., Gillin, J. C., Stricker, J. L., Wong, E. C., & Buxton, R. B. (2000). Altered brain response to verbal learning following sleep deprivation. *Nature, 403*, 655–657.

———, Brown, G. G., Stricker, J. L., Buxton, R. B., Wong, E. C., & Gillin, J. C. (1999). Sleep deprivation-induced reduction in cortical functional response to serial subtraction. *Neuroreport, 10*, 3745–3748.

———, Gillin, J. C., & Brown, G. G. (2001). Increased cerebral response during a divided attention task following sleep deprivation. *Journal of Sleep Research, 10*, 85–92.

Gratch, J., & Marsella, S. (2004). A domain-independent framework for modeling emotion. *Cognitive Systems Research, 5*, 269–306.

Habeck, C., Rakitin, B., Moeller, C., Scarmeas, N., Zarahn, E., Brown, T., et al. (2004). An event-related fMRI study of the neurobehavioral impact of sleep deprivation on performance of a delayed-match-to-sample task. *Cognitive Brain Research, 18*, 306–321.

Harrison, Y., & Horne, J. A. (1998). Sleep loss impairs short and novel language tasks having a prefrontal focus. *Journal of Sleep Research, 7*(2), 95–100.

Hudlicka, E. (2003). Modeling effects of behavior moderators on performance: Evaluation of the MAMID methodology and architecture. In *Proceedings of the 12th Conference on Behavior Representation in Modeling and Simulation* (pp. 207–215). Orlando, FL: Institute for Simulation and Training.

Jewett, M. E., & Kronauer, R. E. (1999). Interactive mathematical models of subjective alertness and alertness in humans. *Journal of Biological Rhythms, 14,* 588–597.

Jongman, L. (1998). How to fatigue ACT-R? In *Proceedings of the Second European Conference on Cognitive Modelling* (pp. 52–57). Nottingham: Nottingham University Press.

Kleitman, N. (1923). The effects of prolonged sleeplessness on man. *American Journal of Physiology, 66,* 67–92.

Lee, M. A. M., & Kleitman, N. (1923). Studies on the physiology of sleep: II. Attempts to demonstrate functional changes in the nervous system during experimental insomnia. *American Journal of Physiology, 67,* 141–152.

Lin, J. S. (2000). Brain structures and mechanisms involved in the control of cortical activation and wakefulness, with emphasis on the posterior hypothalamus and histaminergic neurons. *Sleep Medicine Reviews, 4*(5), 471–503.

Mallis, M., Mejdal, S., Nguyen, T., & Dinges, D. (2004). Summary of the key features of seven biomathematical models of human fatigue and performance. *Aviation, Space, and Environmental Medicine, 75,* A4–A14.

Newell, A. (1990). *Unified theories of cognition.* Cambridge, MA: Harvard University Press.

Patrick, G. T. W., & Gilbert, J. A. (1896). On the effects of loss of sleep. *Psychology Review, 3,* 469–483.

Portas, C. M., Rees, G., Howseman, A. M., Josephs, O., Turner, R., & Frith, C.D. (1998). A specific role for the thalamus in mediating the interaction of attention and arousal in humans. *The Journal of Neuroscience, 18,* 8979–8989.

Ritter, F. E., Reifers, A., Klein, L. C., Quigley, K., & Schoelles, M. (2004). Using cognitive modeling to study behavior moderators: Pre-task appraisal and anxiety. In *Proceedings of the Human Factors and Ergonomics Society* (pp. 2121–2125). Santa Monica, CA: Human Factors and Ergonomics Society.

Van Dongen, H. P. A. (2004). Comparison of mathematical model predictions to experimental data of fatigue and performance. *Aviation, Space, and Environmental Medicine, 75*(3), 15–36.

———, Price N. J., Mullington J. M., Szuba, M. P., Kapoor, S. C., & Dinges, D. F. (2001). Caffeine eliminates sleep inertia: Evidence for the role of adenosine. *Sleep, 24,* 813–819.

von Economo, C. (1930). Sleep as a problem of localization. *Journal of Nervous and Mental Disease, 71,* 249–259.

Wu, J., Gillin, J. C., Buchsbaum, M. S., Hershey, T., Hazlett, E., Sicotte, N., & Bunney, W. E. (1991). The effect of sleep deprivation on cerebral glucose metabolic rate in normal humans assessed with positron emission tomography. *Sleep, 14,* 155–162.

18

Lessons From Defining Theories of Stress for Cognitive Architectures

*Frank E. Ritter, Andrew L. Reifers, Laura Cousino Klein, &
Michael J. Schoelles*

We describe a range of theories of how cognition is influenced by stress. We use a cognitive architecture, ACT-R, to represent these theories formally. The theories make suggestions for developing cognitive architectures, in that nearly all of them require that time-on-task influence performance, and at least one suggests that workload and strategies are monitored to access and cope with stress. By examining the theories as a whole, we can see how the stress theories and the mechanisms that give rise to them can be tested. We can also see that they are incomplete, in that individually and as a group they do not make predictions that are consistent with data. For example, many of them do not predict that repeated serial subtraction (part of the Trier Social Stressor Task) will be affected by stress (and it is).

In this chapter, we look for lessons applicable to cognitive architectures from several popular theories of stress. The goal is to specify mechanisms that can be implemented within an architecture or defined as changes to current mechanisms to simulate the effect of stress on embodied cognition. We examine theories from Wickens, from Hancock and Warm, and from the biophysiology literature. We have chosen to incorporate these theories of stress into the ACT-R architecture because of its modular construction, but the intent is for the ideas presented here to be applicable to a wide range of cognitive architectures.

These theories of stress are typically not cast as additions to the knowledge necessary to perform a task (which would make implementing them be creating a cognitive model) but are described as changes to how people process information under stress. Thus, they make suggestions about process, about how the mechanisms of embodied cognition change across all tasks under stress. Implementing them thus becomes modifying the architecture or changing parameters of existing

mechanisms. Unfortunately, not all the lessons that we have extracted from our survey are specific enough to be directly implemented. In these cases, the lessons serve as specification guidelines for new mechanisms.

Each theory of how stress influences embodied cognition, if specific enough, is presented as an "overlay" to the ACT-R cognitive architecture. These changes are overlaid onto the basic ACT-R theory, modifying how the mechanisms process information, and in one case, giving any ACT-R model a secondary task—to worry. The main idea of an overlay is to change the architecture in such a manner that the behavior of all models developed under that architecture will be affected. In a similar way, the changes to ACT-R to model fatigue by Gunzelmann, Gluck, Price, Van Dongen, and Dinges (chapter 17, this volume) should be applicable to other ACT-R models and thus it is an example overlay.

Including these theories (overlaying these theories) into cognitive architectures offers several advantages. We will be able to construct models that make stress theories more precise. Until now these theories of stress have typically only been verbally stated. We will be able to provide more realistic opponents in competitive environments by making them responsive to stress. We also will extend the scope of cognitive architectures because most architectures have not been developed with these theories of stress in mind. And we will be able to explore more complex concepts, such as how situation awareness is influenced by stress.

Theories of cognition under stress can also benefit. These stress theories are not all complete. Implementing them in a cognitive architecture will force them to interact with other embodied cognitive mechanisms, and this will also make the relationships among these theories can be made more explicit.

Before examining the stress theories, we briefly review cognitive architectures, expand our definition of architectural overlays, and describe a sample task to help explain the application of overlays. After describing the theories of stress, we discuss why the theories of stress require the passage of time as an architectural mechanism. We conclude with a plan for testing the overlays including specifying what currently has been done and what remains to be accomplished.

Cognitive Architectures and Overlays

We start this review by explaining cognitive architectures and developing the idea of an overlay with respect to a cognitive architecture. We also describe an example task that we will use to illustrate and, in later work, to test these overlays.

Cognitive Architectures

Cognitive architectures are an attempt to represent the mechanisms of cognition that do not change across tasks. Typically, cognitive architectures include mechanisms such as long-term procedural and declarative memory, a central processor, and some working memory for that processor, and perceptual and motor components (e.g., see Anderson et al., 2004; Newell, 1990).

Gray (chapter 1, this volume) references three types of control mechanisms in cognitive architectures. Type 1 theories are those that have central control. A central executive processor represents this type of control. Type 2 mechanisms speak to peripheral processors and distributed control. It is likely that the human mind

has both aspects. Current cognitive architectures typically include both types of control. Type 3 control mechanisms are task specific changes. We will also introduce the idea of distributed and physiological implementations, and label these potential changes as Type 4 mechanisms, which reflects the physiological implementation of all mechanisms.

Overlays

An overlay is a technique for including a theory of how a behavioral moderator, such as stress, influences cognition across all models within a cognitive architecture. An overlay, as we propose it, is an adjustment or set of adjustments to the parameters or mechanisms that influence all models implemented in the architecture to reflect changes due to an altered mental state or due to long term changes such as development (e.g., Jones, Ritter, & Wood, 2000). In many architectures, there are a set of mechanisms and a number of global parameters that play a role in the model's functioning; an overlay modifies a combination of parameters and mechanisms to represent situation specific changes to information processing that may last longer than the situation itself. For example, an eyeglasses overlay would allow more inputs to be passed to the vision processor; a caffeine overlay could increase processing speed by 3% and improve vigilance by 30% (and these changes would decrease with time); a fatigue overlay could decrease processing speed by 3% for times on task over an hour.

This overlay approach keeps the architecture consistent across tasks and populations but allows that there may be differences in processing mechanisms and capabilities for individuals or groups in certain contexts. The concept of an overlay and the associated software are useful concepts for developing theories in this area. It provides a way to describe these differences in a way that can be applied to all models in the architecture. With further work, of course, these overlays would migrate into the architecture.

There have been a few previous attempts to generate overlays, although most of their authors did not call them overlays. A brief review notes the scope of possible overlays. The earliest attempts we are aware of are to model fatigue in ACT-R (Jongman, 1998) and Soar (Jones, Neville, & Laird, 1998), and modeling fear in Soar (Chong, 1999). Later work on creating a model of stress and caffeine used simple adjustments of increasing the variability of applying procedural knowledge (Ritter, Avraamides, & Councill, 2002).

Another chapter in this volume (Gunzelmann et al., chapter 17) presents a more developed overlay to model fatigue.

There have also been some complex overlays that add mechanisms rather than modify parameters. Belavkin has worked on several overlays to ACT-R to make models in ACT-R more sensitive to local success and failures, creating a simple theory of how motivation influences problem solving (e.g., Belavkin & Ritter, 2004). Gratch and Marsella (2004) created a rather complex addition to Soar to model task appraisal. It is not clear that their overlay is truly portable, but their work suggests how such an appraisal process can be built in an architecture and provides an example overlay in Soar and increases the understanding of appraisal processes.

Application to ACT-R and Other Architectures

Implementing overlays is easier to do in architectures that are modular, where their source code is available, the implementation language easily allows extension (like Lisp), and the existing mechanisms have parameters. This chapter uses ACT-R as the architecture to implement the theories as overlays because ACT-R (Anderson et al., 2004) has all these features. However, the overlays description could be used fairly easily with other architectures that have enough of these features. For example, COJACK (Norling & Ritter, 2004; Ritter & Norling, 2006) was designed to support the creation of overlays, and EPIC (Kieras, Wood, & Meyer, 1995) should also support including overlays like these.

ACT-R is a hybrid embodied cognitive architecture (Anderson et al., 2004). ACT-R specifies constraints at a symbolic level and a Bayesian-based subsymbolic level. It has perceptual-motor components as well as memory and cognitive control components. All components or modules communicate through buffers. Examples of ACT-R models are available in chapters 2 (Gluck, Ball, & Krusmark), 4 (Anderson), 17 (Gunzelmann, Gluck, Price, Van Dongen, & Dinges), and 24 (Salvucci).

A Sample Task—Serial Subtraction

In this chapter, we will use one of our tasks in the CafeNav suite (Ritter, Ceballos, Reifers, & Klein, 2005), repeated serial subtraction, as a task where performance is modified by stress. We have collected data on this task and have an ACT-R model. It shares components with tasks that use arithmetic knowledge. A large

literature in physiology and biophysiology has used serial subtraction as a task to study stress (e.g., Kirschbaum, Pirke, & Hellhammer, 1993). Other tasks could be used, and, indeed, that is the point of overlays, that these changes would lead to different performance in other tasks as well.

In serial subtraction the subject is asked to repeatedly subtract a small number, like 7 or 13, from a running total that starts as a four-digit number, such as 5,342. If the answer is incorrect, the experimenter informs the subject that the answer is incorrect, and gives the correct starting number.

Subjects can be manipulated to appraise this task as threatening with associated physiological changes to heart rate, blood pressure, and hormonal measures. When this occurs, the number of subtraction attempted drops by about 10% from 61 per 4-minute block to 46, and the number correct only drops from 56 to 42 (Tomaka, Blascovich, Kelsey, & Leitten, 1993, Experiment 2). Thus, threat appraisal appears to impair performance rate, but not performance accuracy.

We have created a model in ACT-R 5 and recently in ACT-R 6 to perform this task. The ACT-R 5 model has been modified with a simple stress and a simple caffeine overlay and the results compared with published data (Ritter, Reifers, Klein, Quigley, & Schoelles, 2004). In both versions of ACT-R, this model uses procedural and declarative knowledge and audio input and output. Overlays that modify these mechanisms will modify the model's performance.

The Theories of Stress and Cognition

We examine six theories of how stress influences performance. We have generated implementations of the first four theories as overlays to ACT-R.[1] We have not generated implementations of the last two theories, but we derive lessons for the development of architectures, in particular, for ACT-R, from the last two theories.

Wickens's Theories

We have taken Wickens's theories of stress from Wickens, Gordon, and Liu's textbook on human factors (1998, chapter 13, pp. 324–349). The theories of how stress influences cognition are not yet very predictive. Wickens et al. note that the amount of stress is difficult to predict for a given situation, as differences can arise due to how the task is performed (e.g., allowing more or less time to appraise the situation), how the task is appraised due to level of expertise (e.g., threatening versus challenging), and whether one perceives that they are in control of the situation. The authors do, however, go on to provide some theoretical statements that can be used to implement theories of stress as overlays. We examine three of them here, perceptual tunneling, cognitive tunneling, and changes to working memory.

Wickens—Perceptual Tunneling

Wickens et al. (1998) and Wickens and Hollands (2000) note perceptual narrowing (or tunneling) as a major effect of stress. We take perceptual narrowing to be where the effective visual perceptual field becomes smaller with stress, such that items in the periphery become less attended. Thus, the item focused on is typically the cause of stress or related to relieving a stressor and other items are less available to cognitive processing.

Perceptual tunneling can be implemented in ACT-R in several ways. One way is to decrease the default distance (from the screen) parameter. This effectively makes objects on the periphery less visible (we do not imply that under stress people lean into the screen, this is purely a way to implement this effect). Another way is to modify the visual attention latency parameter. This makes moving attention to the periphery slower, and some models and humans would use information less from the periphery because it is harder to get, missing more changes there. Another way is to limit what is in the perceptual field by decreasing its width; this is slightly more difficult to implement than the other approaches. Finally, an approach to create perceptual tunneling is to increase the saccade time. ACT-R 5 is based on a theory of moving visual attention and does not include a theory of eye movements. An extension to ACT-R called EMMA (Salvucci, 2001) adds a theory of eye movements to ACT-R. In particular, EMMA computes eye movement times, which can be changed to implement perceptual tunneling.

All of these implementations modify Type 2 mechanisms away from central cognition. These visual mechanisms, as implemented in ACT-R, are not used in the current model of the serial subtraction task, so these changes would have little effect on serial subtraction. For serial subtraction, the inputs, corrections, and outputs are all auditory. Even if tunneling is applied to the auditory information, tunneling should only help performance on this task as tunneling focuses attention

on the primary task and removes attention from a secondary task. The overlay of this popular theory will not be able explain what happens to the performance of serial subtraction under stress, as there is only one task. However, this overlay would be usefully applied to many ACT-R models that use vision to interact with tasks, and predict how performance in more complex environments would change under stress. It may also lead to improved performance, as Wickens and Hollands (2000) review, in tasks where avoiding distractions improves performance.

Wickens—Cognitive Tunneling

Cognitive tunneling occurs where a limited number of options are considered by central cognition. To implement cognitive tunneling, the declarative and procedural retrieval thresholds in ACT-R can be modified by the overlays to represent a greater reliance on well-known and well-practiced knowledge. Alternatively, noise in the procedural rule application process can be decreased (i.e., activation noise and expected gain noise), which would lead to only the most well-practiced materials being retrieved and applied. These overlays modify central cognition, a Type 1 mechanism. We might expect this overlay to hurt performance on a complex task, but on serial subtraction cognitive tunneling should improve performance because it will help focus attention.

Wickens—Working Memory

Wickens and colleagues (1998, p. 385) are among the many that note that working memory capacity appears to decrease under stress. Under stress, working memory appears to be less available for storing and rehearsing information and less useful when performing computations or other attention-demanding tasks. Wickens et al. go on to note that long-term memory appears to be little affected and may even be enhanced. This effect (or lack there of) may be due to focusing on well-learned knowledge. Focusing on well-known knowledge to the exclusion of less well known knowledge would lead to using less of long-term memory.

The implementation of this theory is not completely straightforward. Lovett, Daily, and Reder (2000) showed how working memory can be modeled within the ACT-R architecture and proposed the W (goal activation) parameter as a measure of individual working memory capacity.

A promising approach is to modify the decay rate of working memory objects. This is interesting, as this approach is sensitive to many known effects on working memory, including that it would predict that that the time to report objects in working memory will influence the measurable size of memory. Other ways to decrement working memory are to increase the declarative memory retrieval threshold parameter, which would make fewer memories available, and to decrease the base-level activation of all memory elements when stressed.

All of these overlays modify central cognition, and thus influence a Type 1 mechanism. Different implementations of this theory as an overlay would have different effects on the serial subtraction model. Changes to working memory should decrease the number of calculations per unit of time. The changes might also influence percent correct, but these overlays would need to be implemented to obtain accurate predictions.

Decreased Attention

Hancock (1986) describe the effect of stress on cognition as being decreased attention and provides a theory based on a review of work on stress and attention. Wickens et al. (1998) also note that there is decreased attention with stress.

ACT-R includes several ways that attention can be modified. Working memory capacity is a type of attention that is also a parameter that can be modified in ACT-R. We use it in the Wickens-WM (working memory) overlay, so we exclude it here.

Another way to implement an overlay that decreases attention to the task is to create a secondary task. This secondary task represents worry. This approach to modeling stress is consistent with theories of math anxiety and other studies of anxiety that posit a dual task of worry as the cause of poor math performance in people with math anxiety (e.g., Ashcraft, 2002; Cadinu, Maass, Rosabianca, & Kiesner, 2005; Sarason, Sarason, Keefe, Hayes, & Shearin, 1986). The rules that create this secondary task might be seen as architectural productions, similar to Soar's default rules, but far less useful. The genesis and activation of these rules are likely tied into central cognition and emotion.

This secondary worry task has been implemented as a pair of productions that represent how worrying uses resources and decreases attention. When these rules are chosen and applied, they take time from the primary task, allow the declarative memory of the primary task to decay over the time that they take to apply and can leave the working memory in an incorrect state if they are interrupted themselves.

This overlay will decrease the rate of subtractions on the serial subtraction task. If the secondary worry task was performed quite often and the math facts were poorly known, it could also increase the error rate. If the worry task was performed only occasionally, it is possible that it would not change the error rate. We have used a secondary task as an overlay before to model the effects of worry as a type of stress with our serial subtraction task (Ritter et al., 2004). We have also applied the dual-task worry overlay to a driving model to predict how worry could increase lane deviation for a simple driving task (Ritter, Van Rooy, St. Amant, & Simpson, 2006).

This overlay does not directly modify a mechanism, per se, but its effects are felt primarily in central cognition as the result of including another task. This overlay is classifiable as a Type 3 change (an addition to knowledge). The mechanism of its genesis will be more complicated and may vary across individuals.

The Task as a Stressor

Hancock and Warm (1989) argued that tasks are themselves stressors. They describe the effect of stress on task performance as an individual's interest in maintaining an optimal amount of information flow. They note several negative effects, including that "physiological compensation is initiated at the point at which behavioral response reaches the exhaustive stage." So, they note, as stress increases, the agent modifies its cognitive efforts to maintain performance, then its physiology changes to support these efforts, and then, after a sufficient period of time, there will be a catastrophic collapse due to exhaustion of the physiological level. Hancock and Warm (1989) note that the task itself should be seen a source of stress, with sustained attention as the stress generator. For well-practiced, automatic tasks, which need little attention, there is little effect of stress on performance, and performance of the task does not increase stress.

In their article, they explain why heat stress on a simple task can be better predicted than noise stress. Their approach is to understand the input, the adaptation, and the output. Heat stress has a fairly simple input description. Its effects on cognition, physiology, and adaptation appear to be, they note (p. 532), direct and straightforward. There are few strategies to cope with heat, and these strategies are simple and do not modify cognition.

The empirical results they review show a very nice area where physiology directly affects cognition. Cognition directly suffers when thermoregulatory action can no longer maintain core body temperature.

Hancock and Warm (1989) point out that environmental auditory noise as a stressor is less easy to characterize, because adaptation to noise is more complex. Adaptation to noise varies more across individuals and is more complex cognitively and physiologically than is adaptation to heat stress.

We have not yet created an overlay based on Hancock and Warm's proposal that tasks themselves are stressors. Such an overlay could be quite complex. They describe important aspects of this process, that of input, adaptation, and response. Although we have not modeled the effects of tasks as stressors themselves, we recognize that modeling tasks as stressors is an interesting and important next step in the effort to model the effects of stress. Hancock and Warm's (1989) theory also predicts that performance and resources will degrade faster when performing more complex tasks. We also gain a greater understanding about stress and other moderators: situations where people can use multiple coping strategies will be more difficult to model.

An overlay based on Hancock and Warm's (1989) theory, if created, would need to use central cognition to compute how to modify its behavior. The physiological implementation of the mechanisms and how they would tire from effort is perhaps a new type of mechanism (Type 4). This theory suggests that a detailed model of serial subtraction that includes coping strategies will be challenging to create. At the least, this theory predicts that errors will increase over time on the serial subtraction task because the system will fatigue.

Pretask Appraisal and Stress

Some researchers, particularly biobehavioral health scientists, have been interested in stress that occurs as a result of particular cognitive appraisal processes (e.g., Lazarus & Folkman, 1984). According to appraisal theories, before performing a task, the task is appraised as to how difficult the task will be, and how well the person thinks they will be able to cope with the task's demands (i.e., what coping resources they have). These appraisals are typically classified into two categories, either "threatening" or "challenging." Threatening tasks are those in which the person approaching the task believes that their resources (including task knowledge)

and coping abilities are not great enough for the task, whereas tasks appraised as challenging indicates that their resources and coping abilities are great enough to meet the task demands.

Challenging appraisals give rise to better energy mobilization, and better performance in general than threatening appraisals. In these studies, such results are also found when the appraisal is manipulated and knowledge held constant, so it is not just a knowledge-based effect. This general result has been shown for a several different tasks, but performance on repeated serial subtraction has perhaps most often been used to study this effect especially with regard to the cardiovascular and general physiological changes that occur as a function of these appraisals (e.g., Quigley, Feldman Barrett, & Weinstein, 2002). For example, in these serial subtraction tasks threatening appraisals lead to about a 25% slower performance than "challenging appraisal" conditions, but with the same accuracy (e.g., Tomaka et al., 1993).

This overlay, which would be much larger than the others reported here, has not been implemented in our set. Our previous attempts to create an appraisal overlay for ACT-R have simply modified how accurately rules are applied, providing more accurate rule applications for challenged appraisals and less accurate application for threatened appraisals. This overlay appeared to match the limited data available (Ritter et al., 2004). A complete version of an appraisal overlay might be so extensive that it should not even be called an overlay, but rather a complete extension to the architecture, such as Gratch and Marsella's (2004) work.

The serial subtraction model would be particularly sensitive to this overlay. The difficulty will be translating the physiology results that either co-occur with or are reflective of changes to specific cognitive mechanisms. We note the next steps in this area below in steps towards testing these models.

Discussion

We have presented six interesting theories of how stress influences cognition and created implementations of four of them for applying to cognitive models in ACT-R. Creating this set of overlays provides a rich set of lessons for the further development of ACT-R and other cognitive architectures. Implementing the remaining theories and testing these overlays should provide further lessons about how cognition and performance vary on these tasks. These lessons include lessons on the development of these theories of stress, the important role that time has to play in cognitive architectures, and how to test these theories. After discussing these lessons, we note directions for future work.

Theories of Stress Are Not Complete

Our review suggests the many ways that stress can be conceptualized as changes to cognition. However, most of the theories we examined modify only one aspect of cognition.

In several cases, there were multiple ways to implement these theories. Where this occurred we often included multiple variants. This is useful for the theories, as testing the multiple interpretations will help make these theories more concrete and accurate.

Some of the theories sound different, but become the same when we implement them (e.g., decreased attention and decreased working memory capacity). This may be due to the limitations of us as designers of models or limitations of the architecture we are working with. We hope, however, that it is because these multiple theories are attempting to describe the same phenomena.

Time: An Additional Aspect to Mechanisms

This review suggests that there is an additional aspect to mechanisms that ACT-R needs, namely, how the mechanisms are sensitive to the passage of time. Many stress theories predict that people are sensitive to time. That is, the theories predict that information processing and thus the cognitive mechanisms vary as time passes. Hancock and Warm (1989), for example, explicitly note this effect. Most models in most architectures, including ACT-R, do not account for time's influence on task performance. For an interesting counterexample of a model that is influenced by time see work by Gunzelmann and colleagues, chapter 17, this volume. Their model modifies the value of the goal over time, representing fatigue.

These overlays will also need to incorporate aspects of how time is spent (e.g., sleeping) and how time-on-task for a single, particular task will influence task performance. Thus, we have proposed a Type 4 mechanism to add to Gray's (chapter 1) taxonomy of control mechanisms. Type 4 is related to physiological mechanisms, how they support cognition, and how they become fatigued over time. (It could be called a sub-subsymbolic level.)

Testing These Overlays

These overlays need to be applied to models of a range of tasks and the results compared with subjects' task performance. Fitting and testing the effects of these overlays will refine them. A large battery of tasks will be important because on a single task some of these overlays may make similar predictions. Because this testing process requires a suite of tasks, a range of data, and several models, it will require reuse of models and tasks. Preliminary data (Ritter et al., 2005) on a selected task suite that taxes different cognitive mechanisms shows potential differences in performance that should help differentiate and refine these overlays.

We are in the process of testing the overlays described in this chapter on a number of models in ACT-R of tasks: a working memory task (MODS; Lovett et al., 2000), a signal detection and simple reaction time task (Reifers, Ritter, Klein, & Whetzel, 2005), a serial subtraction task (Ritter et al., 2004), and a complex dynamic classification task (Schoelles & Gray, 2001). Currently, the task environments have been developed and tested. The visual signal detection task (VSDT) model was built from scratch, but the others are revisions to older ACT-R models. All of these models have been compared to normal data, either in their previous version or in their current version. The serial subtraction model and the VSDT model have also generated predictions with a caffeine overlay (Reifers et al., 2005; Ritter et al., 2004).

The models and the overlays for perceptual tunneling, cognitive tunneling, working memory, and decreased attention have been created in ACT-R 6. These overlays are ready to be tested. One of them has been used before as well—we added the decreased attention overlay to a model of a simple driving task and generated predictions (Ritter et al., 2006). The predictions leave the model a considerably poorer driver (as measured by average lane deviation and time between crashes in the task). We do not yet have data from stressed individuals that can be used to test these predictions.

The next phase is to run the models with and without the overlays and then to analyze the model and human data. Data from the CafeNav study ($N = 45$) is now in hand for testing the overlays explained here. (A description of the pilot version study is available; Ritter et al., 2005.) This study measured subjects as they performed these tasks and measured their stress response using a range of measures (including heart rate, blood pressure, and salivary cortisol). The next step is to run the models with the overlays included and to compare the resulting behavior to our fresh data.

Further Work on Overlays

These stress overlays could be extended to model different levels of stress or individual differences such as personality type. They could also be refined. For example, the worry distracter thought overlay could have several different thoughts, some not focused on the task, which would slow down the model and thereby reduce the activation of working memory, and other thoughts that might be more task related, and would slow down task performance but raise the activation of particular memories related to the current task, and thereby lead to different learning or performance.

Several important theories of stress, not examined here in detail, make assumptions about processing that cannot yet be directly or routinely implemented in models or their architectures. Examples of these effects include how under stress, particularly high workload stress, tasks are deliberately shed (e.g., Parasuraman & Hancock, 2001; Wickens, Gordon, & Liu, 1998) and how negative information in particular may be lost or ignored under stress. Moderating behavior in these ways will require more complex models and architectures that represent and use a measure of task load and that include multiple strategies.

These overlays developed here are static overlays, representing the state of processing in a challenged (or neutral) and in a threatened state and are fixed across time. Stress is probably not a binary effect. Stress has a range of values and a more dynamic nature than the theories and these overlays currently provide. Future work will have to grapple with the additional difficulties of creating a model of stress that dynamically adjusts to the effect of the task itself and to the resulting range of stress on task performance.

Summary

We have reviewed the literature on the effects of stress on embodied cognition with the goal of being able to simulate these effects in cognitive models. We described the process of overlaying the ACT-R theory of embodied cognition with these theories of stress to develop more accurate cognitive models. The theories we examined as the basis of our architectural overlays were perceptual and cognitive tunneling, diminished working memory capacity, decreased attention, pretask appraisals, and the task as a stressor.

Creating a set of overlays that start to model the effects of several of the major theories of stress was a first step. The next step is to compare the predictions of the models with overlays to data, including the differences in error rates and error types that these overlays predict. After we test these overlays, there are several lines of research that can be investigated. First, we would hope to create other global overlays modeling other psychological phenomena, demonstrating that the concept of overlays in a cognitive architecture is indeed useful. For example, overlays could be extended to study extreme levels of stressors: of fatigue, of heat, and of dosages of caffeine. Granted, these goals are ambitious, but the possible implications and impact seem to be worth exploring.

Acknowledgments

Susan Chipman has provided several types of support for this work. Roman Belavkin, Wayne Gray, and Karen Quigley have helped our thinking in this area. Comments from Wayne Gray, Glenn Gunzelmann, Karen Quigley, William Stevenson, and Dan Veksler have improved this chapter. This work was supported by the US Office of Naval Research, Award N000140310248.

Note

1. Implementations of the model and these overlays are at acs.ist.psu.edu/cafenav/overlays/

References

Anderson, J. R., Bothell, D., Byrne, M. D., Douglass, S., Lebiere, C., & Qin, Y. (2004). An integrated theory of the mind. *Psychological Review, 111*(4), 1036–1060.

Ashcraft, M. H. (2002). Math anxiety: Personal, educational, and cognitive consequences. *Current Directions, 11*(5), 181–185.

Belavkin, R. V., & Ritter, F. E. (2004). OPTIMIST: A new conflict resolution algorithm for ACT-R. In *Proceedings of the Sixth International Conference on Cognitive Modeling* (pp. 40–45). Mahwah, NJ: Erlbaum.

Cadinu, M., Maass, A., Rosabianca, A., & Kiesner, J. (2005). Why do women underperform under stereotype threat? *Psychological Science, 16*(7), 572–578.

Chong, R. (1999). Towards a model of fear in Soar. In *Proceedings of Soar Workshop 19* (pp. 6–9). University of Michigan Soar Group. Retrieved November 22, 2006, from http://www.eecs.umich.edu/~soar/sitemaker/workshop/19/rchong-slides.pdf

Gratch, J., & Marsella, S. (2004). A domain-independent framework for modeling emotion. *Journal of Cognitive Systems Research, 5*(4), 269–306.

Hancock, P. A. (1986). The effect of skill on performance under an environmental stressor. *Aviation, Space, and Environmental Medicine, 57*(1), 59–64.

———, & Warm, J. S. (1989). A dynamic model of stress and sustained attention. *Human Factors, 31*(5), 519–537.

Jones, G., Ritter, F. E., & Wood, D. J. (2000). Using a cognitive architecture to examine what develops. *Psychological Science, 11*(2), 93–100.

Jones, R. M., Neville, K., & Laird, J. E. (1998). Modeling pilot fatigue with a synthetic behavior model. In *Proceedings of the 7th Conference on Computer Generated Forces and Behavioral Representation* (pp. 349–357). Orlando, FL: Division of Continuing Education, University of Central Florida.

Jongman, G. M. G. (1998). How to fatigue ACT-R? In *Proceedings of the Second European Conference on Cognitive Modelling* (pp. 52–57). Nottingham: Nottingham University Press.

Kieras, D. E., Wood, S. D., & Meyer, D. E. (1995). Predictive engineering models using the EPIC architecture for a high-performance task. In *Proceedings of the CHI '95 Conference on Human Factors in Computing Systems* (pp. 11–18). New York: ACM.

Kirschbaum, C., Pirke, K.-M., & Hellhammer, D. H. (1993). The Trier Social Stress Test—A tool for investigating psychobiological stress responses in a laboratory setting. *Neuropsychobiology, 28*, 76–81.

Lazarus, R. S., & Folkman, S. (1984). *Stress, appraisal and coping.* New York: Springer.

Lovett, M. C., Daily, L. Z., & Reder, L. M. (2000). A source activation theory of working memory: Cross-task prediction of performance in ACT-R. *Journal of Cognitive Systems Research, 1*, 99–118.

Newell, A. (1990). *Unified theories of cognition.* Cambridge, MA: Harvard University Press.

Norling, E., & Ritter, F. E. (2004). A parameter set to support psychologically plausible variability in agent-based human modelling. In *The Third International Joint Conference on Autonomous Agents and Multi Agent Systems (AAMAS04)* (pp. 758–765). New York: ACM.

Parasuraman, B., & Hancock, P. A. (2001). Adaptive control of mental workload. In P. A. Hancock & P. A. Desmond (Eds.), *Stress, workload, and fatigue* (pp. 305–320). Mahwah, NJ: Erlbaum.

Quigley, K. S., Feldman Barrett, L. F., & Weinstein, S. (2002). Cardiovascular patterns associated with threat and challenge appraisals: A within-subjects analysis. *Psychophysiology, 39*, 292–302.

Reifers, A., Ritter, F., Klein, L., & Whetzel, C. (2005). *Modeling the effects of caffeine on visual signal detection (VSD) in a cognitive architecture.* Poster session presented at Attention: From Theory to Practice, Applied Attention Conference, Champaign, IL. (A festschrift for Chris Wickens.)

Ritter, F. E., Avraamides, M., & Councill, I. G. (2002). An approach for accurately modeling the effects of behavior moderators. In *Proceedings of the 11th Computer Generated Forces Conference* (pp. 29–40), 02-CGF-100. Orlando, FL: University of Central Florida.

———, Ceballos, R., Reifers, A. L., & Klein, L. C. (2005). *Measuring the effect of dental work as a stressor on cognition* (Tech. Rep. No. 2005–1). Applied Cognitive Science Lab, School of Information Sciences and Technology, Penn State. Retrieved November 22, 2006, from http://acs.ist.psu.edu/reports/ritterCRK05.pdf

———, & Norling, E. (2006). Including human variability in a cognitive architecture to improve team simulation. In R. Sun (Ed.), *Cognition and multi-agent interaction: From cognitive modeling to social simulation* (pp. 417–427). Cambridge: Cambridge University Press.

———, Reifers, A., Klein, L. C., Quigley, K., & Schoelles, M. (2004). Using cognitive modeling to study behavior moderators: Pre-task appraisal and anxiety. In *Proceedings of the Human Factors and Ergonomics Society* (pp. 2121–2125). Santa Monica, CA: Human Factors and Ergonomics Society.

———, Van Rooy, D., St. Amant, R., & Simpson, K. (2006). Providing user models direct access to interfaces: An exploratory study of a simple interface with implications for HRI and HCI. *IEEE Transactions on System, Man and Cybernetics, Part A: Systems and Humans, 36*(3), 592–601.

Salvucci, D. D. (2001). An integrated model of eye movements and visual encoding. *Cognitive Systems Research, 1*(4), 201–220.

Sarason, I. G., Sarason, B. R., Keefe, D. E., Hayes, B. E., & Shearin, E. N. (1986). Cognitive interference: Situational determinates and traitlike characteristics. *Journal of Personality and social Psychology, 51*(215–226).

Schoelles, M. J., & Gray, W. D. (2001). Argus: A suite of tools for research in complex cognition. *Behavior Research Methods, Instruments, & Computers, 33*(2), 130–140.

Tomaka, J., Blascovich, J., Kelsey, R. M., & Leitten, C. L. (1993). Subjective, physiological, and behavioral effects of threat and challenge appraisal. *Journal of Personality and Social Psychology, 65*(2), 248–260.

Wickens, C. D., Gordon, S. E., & Liu, Y. (1998). *An introduction to human factors engineering.* New York: Addison-Wesley.

———, & Hollands, J. G. (2000). *Engineering psychology and human performance* (3rd ed.). Upper Saddle River, NJ: Prentice-Hall.

19

Reasons for Emotions

Modeling Emotions in Integrated Cognitive Systems

Eva Hudlicka

Research in psychology and neuroscience demonstrates the existence of complex interactions between cognitive and affective processes. Although it is clear that emotions influence cognitive processes, the mechanisms of these interactions have yet to be elucidated. This chapter describes a model of a generic mechanism mediating the affective influences on cognition, implemented within a domain-independent cognitive–affective architecture (MAMID). The proposed mechanism is based on the hypothesis that emotions influence cognition by modifying the speed and capacity of fundamental cognitive processes (e.g., attention, working memory) and by inducing specific content biases (e.g., threat or self-bias). This hypothesis is realized within MAMID by modeling emotion effects in terms of parameters that control these architecture processes. A broad range of interacting emotions and traits can then be represented in terms of distinct configurations of these parameters. While MAMID includes a model of cognitive appraisal, the primary emphasis is on modeling the effects of emotions and affective personality traits on the processes mediating situation assessment and decision making, including the processes mediating cognitive appraisal. Initial evaluation experiments demonstrated MAMID's ability to model distinct patterns of decision making and behavior associated with different emotions and affective personality profiles. The chapter concludes with a discussion of the benefits and shortcomings of the MAMID generic modeling methodology and architecture, briefly outlines some representational requirements for modeling emotions within cognitive architectures, and concludes with reflections on emerging trends in emotion modeling.

The past 2 decades have witnessed an unprecedented growth in emotion research (Forgas, 2001), both in disciplines that have traditionally addressed emotions (e.g., psychology, sociology) and in a variety of interdisciplinary areas, including cognitive science, AI (artificial intelligence), and HCI (human–computer interaction; see also Hudlicka, 2003b; Picard's *Affective Computing*, 1997). In AI and cognitive science, there has been a great deal of interest in modeling emotion within cognitive architectures and exploring the utility of emotions in agent and robot architectures (Canamero, 1998; Fellous & Arbib, 2005; Hudlicka and Canamero, 2004).

The attempts to develop computational models of emotions have necessitated a degree of operationalization that frequently revealed a lack of clear definitions, not to mention gaps in understanding of the mechanisms of emotion generation and emotion–cognition interactions. The attempts to incorporate emotions

within agent and robot architectures have also raised many important questions about the role of emotions within an adaptive architecture and the structures and processes required to implement emotions.

This chapter describes the MAMID[1] cognitive–affective architecture that was developed to evaluate a particular method of modeling emotion effects on decision making. The underlying assumption of the modeling methodology is that a broad range of interacting emotion effects can be represented in terms of distinct configurations of a number of architecture parameters. The parameters, and the functions that map specific emotions onto patterns of parameter values, are defined on the basis of empirical evidence and the associated effects correspond to observed psychological processes. These parameters control processing within individual architecture modules, and induce a range of distinct modes of processing. Different emotions exert characteristic effects on particular processes, influencing the speed, capacity and biasing of attention, situation assessment, or goal selection. The internal *microeffects* are eventually manifested in differences in observable behavior associated with different emotions (e.g., withdraw from a fear-inducing stimulus, approach a joy-inducing stimulus). The view of emotions as distinct modes of processing is consistent with recent evidence from neuroscience (Fellous, 2004).

An initial implementation of this methodology was developed within a peacekeeping scenario, where different types of commanders were modeled by instances of the MAMID architecture. However, both the modeling methodology and the architecture are domain-independent, and MAMID is currently being transitioned to a simulated search-and-rescue team task.

This chapter is organized as follows. Background information from emotion research is provided in the section, "Emotion Research Background," followed by a description of the MAMID cognitive–affective architecture, the modeling methodology, and an evaluation experiment in the section, "The MAMID Cognitive–Affective Architecture and Modeling Methodology." Benefits and shortcomings of the MAMID modeling methodology and architecture are then discussed in the section, "Benefits, Shortcomings, and Future Research," along with plans for future research. The chapter then briefly outlines some representational requirements for emotions and compares MAMID with two more broadly scoped cognitive–affective architectures ("MAMID Architectural Requirements and Relationship With Generic Cognitive–Affective Architectures") and concludes with a few reflections on emerging trends in emotion modeling ("Conclusions").

Emotion Research Background

This section provides an orienting background from emotion research in psychology, focusing on the aspects relevant to the MAMID architecture: terminology, description of cognitive appraisal processes, and effects of emotions on cognition.

Definitions

Many researchers bemoan the lack of terminological clarity in emotion research. This is particularly the case for newcomers to the field and for researchers attempting to build computational models, which require precise definitions of the processes and structures involved. Indeed, we lack the type of unambiguous, architecture-based definitions called for by emotion modelers (Sloman, 2004). Nevertheless, there is sufficient consensus among researchers about many terms, structures, and processes relevant to emotion modeling. Next, I briefly define the emotion terms relevant for the materials discussed in this chapter.

First, we distinguish between *cognitive* and *emotional* mental states. Following Sloman (Sloman, Chrisley, & Scheutz, 2005), cognitive states are defined to refer to beliefs and perceptions that describe the properties of the self or world, without any self-relevant evaluations. Emotional states refer to mental states that reflect an evaluation or judgment of the current state of self or the world as it relates to the agent's goals and desires. Emotional states are considered by many to be the essential components of motivation: "No matter how many beliefs, percepts, expectations, and reasoning skills an [agent] has, they will not cause it to do one thing rather than another . . . unless it also has at least one desire-like state" (Sloman et al., 2005, p. 213).

In emotion research, emotional states are further distinguished based on their degree of *specificity, duration, cognitive complexity, and universality*. At the highest level of generality are *affective states*. These are undifferentiated evaluations of the current stimuli in terms of simple positive evaluations ("good for me") versus negative evaluations ("bad for me") and associated undifferentiated behavioral tendencies: "approach" positive stimuli versus "avoid" negative stimuli. These types of states are often referred to as *primary emotions* (Sloman et al., 2005), or as *proto emotions* by Ortony, Norman, and Revelle (2005).

In contrast to these undifferentiated evaluative states, *emotions proper* reflect more precise evaluations and correspondingly more differentiated behavioral tendencies. Examples of emotions are the familiar joy, sadness, fear, anger, jealousy, contempt, and pride. These states have relatively short durations, lasting on the order of seconds to minutes. Emotions can be further differentiated into *basic* and *complex* on the basis of their cognitive complexity and the universality of triggering stimuli and behavioral manifestations (Ekman & Davidson, 1994).

The set of *basic emotions* typically includes the familiar fundamental emotions such as fear, anger, joy, sadness, disgust, and surprise. These emotions share a number of elicitors (triggering stimuli) and many behavioral manifestations across individuals, cultures and even species (e.g., fear is associated with fleeing or freezing; anger and aggression with fighting; sadness with passivity and withdrawal).

Complex emotions consist of emotions such as guilt, pride, and shame. Complex emotions differ from basic emotions in two important aspects. First, they have a much larger cognitive component, which allows for more idiosyncrasies in both their triggering elicitors and their behavioral manifestations. As a consequence, complex emotions show much greater variability across individuals and cultures than basic emotions. Second, they are associated with an explicit sense of self within its social milieu (Ekman & Davidson, 1994; Lewis & Haviland, 1993).

Moods are similar to emotions in their degree of differentiation and complexity (e.g., angry, happy, sad, jealous) but differ in several important aspects. First, duration of moods is much longer, ranging from hours to days to months. Second, the eliciting stimulus of moods is frequently not apparent. Third, the behavioral tendencies associated with moods may be specific (a pattern of "lashing out" when angry) or very diffuse, more analogous to that of the undifferentiated affective states (Ekman, 1994).

Having described the discrete categories of states above, it is important to realize that emotions rarely exist in isolation. While high intensities of single emotions can exist for short periods of time and allow us to identify an individual as "angry" or "happy" or "sad," typically a number of emotions, moods, and affective states coexist. The affective portrait of a biological agent is thus best thought of as an evolving musical composition, with a number of parallel, evolving melodies and harmonies, as the affective states, emotions, and moods interact and blend in complex, dynamic patterns.

Another important element of understanding emotions is the recognition that emotional states are *multimodal phenomena*, with manifestations across four distinct modalities (e.g., Clore & Ortony, 2000): *physiology* (e.g., increased heart rate, perspiration, enlarged pupils); *behavior* (e.g., smile vs. frown, fight vs. flee); *cognition* (specific patterns of cognition and cognitive biases are associated with different emotions; e.g., positive mood associated with positive thoughts; anxiety associated with threat focus); and *distinct subjective feelings*. Frequently, debates about lack of terminological clarity or confusion regarding the roles of emotions can be resolved when the multimodal nature of emotions is recognized.

When considering affective factors that influence cognition and behavior, it is also important to draw the distinction between traits and states. *Traits* reflect stable, long-lasting, or permanent characteristics, for example, personality traits of extraversion or agreeable-

ness, or basic intellectual capabilities such as working memory speed and capacity (Matthews & Dreary, 1998; Revelle, 1995). *States* reflect transient states of the organism, with durations ranging from seconds to days and months. Both states and traits are associated with characteristic patterns of processing and behavior.

Cognitive Appraisal

Key components of emotional processing are the mechanisms that evaluate the current stimuli (internal and external) in terms of an affective label: an affective state, emotion or mood. This process is referred to as *appraisal* (also *cognitive appraisal*) and has been studied extensively over the past 15 years (Ellsworth & Scherer, 2003; Scherer, 2003). Earlier studies focused on descriptive characterizations of the phenomenon, attempting to identify types of elicitors required for particular emotions (e.g., large, approaching object triggering fear; frustration of goal achievement by an individual triggering anger). Such patterns can be used to develop *black box models* of appraisal (e.g., Scherer, Schoor, & Johnstone, 2001). More recently, attempts have been made to identify the mechanisms mediating these processes, as exemplified by the work of Smith and colleagues (Smith & Kirby, 2001). Such mechanistically oriented models lend themselves to computational implementations of *process models*.

Appraisal processes are typically divided into two stages: automatic and deliberate. *Automatic appraisal* rapidly generates a high-level initial assessment, typically in terms of simple positive or negative affective states. This is followed by a slower, more *deliberate appraisal* process, which generates basic or complex emotions. Because of the increased complexity of the cognition involved in the deliberate appraisal, there is potential for more variability and individual idiosyncrasies in the elicitors-to-emotion mappings. The deliberate appraisal is frequently conceptualized to also include an assessment of the individual's coping potential.

Emotion Effects on Cognition

Emotions and moods (as well as personality traits) influence cognition via a variety of distinct effects on multiple structures and processes, both transient and long term. Effects exist both at the low level: *elementary cognitive task level* and at *complex cognitive processing levels*. Low-level effects influence attentional and memory processes, such as attention orientation and the speed and capacity of working memory. At higher levels of complexity, emotions influence goal selection and planning, situation assessment and expectation generation, and the processes mediating learning and judgment.

As might be expected, *states* (e.g., emotions, moods) tend to produce transient changes that influence the dynamic characteristics of a particular cognitive or perceptual process. In contrast, *traits* tend to exert their influence via more stable structures, such as the content and organization of long-term memory schemas and preferential processing pathways among functional components mediating decision making. Traits also influence the general characteristics of specific affective responsiveness, such as the emotion trigger sensitivity and ramp-up and decay rates. Examples of specific effects are listed in Table 19.1.

The MAMID Cognitive–Affective Architecture and Modeling Methodology

Next, I describe the MAMID cognitive–affective architecture that illustrates one approach for modeling

emotion effects within a cognitive system (Hudlicka, 2002, 2003a). Emotions represent a core component of MAMID: both their generation via a cognitive appraisal process, and the explicit modeling of their effects on distinct stages of decision making, including the processes mediating cognitive appraisal (Hudlicka, 2004).

Emotion effects are modeled via a parameter-based methodology capable of modeling a broad range of interacting individual differences, also including traits (e.g., extraversion) and cognitive characteristics (e.g., fundamental cognitive capabilities such as working memory capacity and speed). The underlying assumption of this methodology is the hypothesis that emotions influence cognitive processes by modifying the nature of the processing itself, that is, by influencing the *speed* and *capacity* of particular processes and by inducing *particular biases* (e.g., bias toward threatening stimuli). This hypothesis has been proposed independently by multiple researchers in psychology (Matthews, 2004; Ortony et al., 2005), neuroscience (Fellous, 2004), and cognitive science (Hudlicka, 1997, 1998; see also Pew & Mavor, 1998). This hypothesis is realized within

TABLE 19.1 Effect of Emotions on Cognition: Examples of Empirical Findings

Anxiety and Attention & WM	*Obsessiveness and Decision Making*
Attentional narrowing	Delayed decision making
Predisposing toward detection of threatening stimuli	Reduced ability to recall recent activities
Reduced working memory capacity	Reduced confidence in ability to distinguish among
(Mineka et al., 2003)	actual and imagined events
	Narrow conceptual categories (Persons & Foa, 1984)
Arousal and Attention	*Mood and Memory*
Faster detection of threatening cues (Mineka et al., 2003)	Mood-congruent memory recall (Blaney, 1986; Bower, 1981)
Positive Affect and Problem Solving	*Negative Affect and Perception, Problem Solving, Decision Making*
Heuristic processing	Depression lowers estimates of degree of control
Increased likelihood of stereotypical thinking, unless held accountable for judgments	Anxiety predisposes toward interpretation of ambiguous stimuli as threatening
Increased estimates of degree of control	Use of simpler decision strategies
Overestimation of likelihood of positive events	Use of heuristics and reliance on standard and
Underestimation of likelihood of negative events	well-practiced procedures
Increased problem solving	Decreased search behaviour for alternatives
Facilitation of information integration	Faster but less discriminate use of information—
Variety seeking	increased choice accuracy on easy tasks, decreased
Less anchoring, more creative problem solving	on difficult tasks
Longer deliberation, use of more information	Simpler decisions, more polarized judgments
Promotes focus on "big picture"	Increased self-monitoring
(Gasper & Clore, 2002; Isen, 1993; Mellers et al., 1998)	Promote focus on details
	(Isen; 1993; Mellers et al., 1999; Williams et al., 1997)

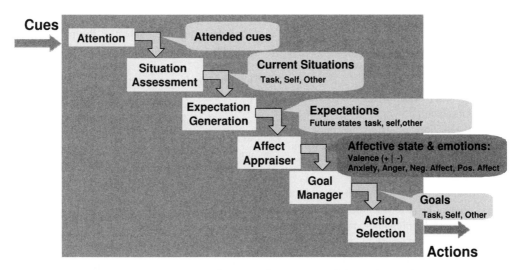

FIGURE 19.1 MAMID cognitive–affective architecture: Modules and mental constructs.

MAMID by modeling emotion effects in terms of parameters that influence particular processes within the architecture. The section, "Generic Methodology for Modeling Effects of Emotions on Cognition," discusses this methodology in more detail.

MAMID Cognitive–Affective Architecture

MAMID is a sequential "see-think-do" architecture, consisting of seven modules (see Figure 19.1):

- Sensory preprocessing: translating the incoming raw data into high-level, task-relevant perceptual cues.
- Attention: filtering the incoming cues and selecting a subset for further processing.
- Situation assessment: integrating individual cues into an overall situation assessment (see Figure 19.2).
- Expectation generation: projecting the current situation onto one or more possible future states.

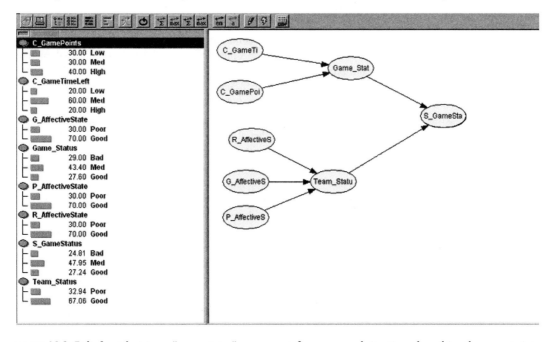

FIGURE 19.2 Belief net deriving a "game status" assessment, from cues and situations describing the *game points, game time left, and the emotional states of the team members.*

- Affect appraiser: deriving the affective state (both valence and four of the basic emotions) from a variety of influencing factors, both *static* (traits) and *dynamic* (current affective state, current situation, goal, expectation).
- Goal manager: selecting the most relevant goal for achievement.
- Action selection: selecting the most suitable action for achieving the current goal within the current context.

These modules map the incoming stimuli (cues) onto the outgoing behavior (actions), via a series of intermediate internal representational structures (situations, expectations, and goals): the *mental constructs*. This mapping is enabled by long-term memories (LTM) associated with each module (with the exception of sensory preprocessing and attention), represented in terms of belief nets or rules. The LTM clusters explicitly represent how the incoming mental constructs are mapped onto the outgoing constructs; cues are aggregated into perceptual schemas (situations), projected onto future states of the world and the self (expectations), and mapped onto goals and eventually actions. Figure 19.2 shows an example of a belief net from an implementation of MAMID modeling individual players within a search-and-rescue game. The belief net derives a situation reflecting the "game status," from cues and situations about the game attributes (points collected so far, game time left), and the emotional states of the team members.

Mental constructs are characterized by a number of attributes. These include the *domain attributes*, which describe the state of the world, self, or other agents, and the *meta-attributes*, which determine the construct's rank and define its processing requirements. Domain attributes represent domain knowledge in terms of a sextuple of attributes describing the state of the world, self, or others. Meta-attributes consist of attributes specifying the construct's processing requirements (time and capacity) and those influencing its rank (threat level, salience and desirability, valence, familiarity, novelty, and confidence). Within each cycle, the most highly ranked constructs within a given module are selected for processing (e.g., cue will be attended, situation derived, goal or action selected). The exact number of constructs selected depends jointly on the processing capacity of the module and the capacity requirements of the constructs. The values of the meta-attributes are a function of, among

other things, the current emotions. As such, these attributes represent a subset of the parameter space available for representing the effects of emotions.

Intermodule Communication

Mental constructs are passed among the modules via a series of buffers. The buffers allow any module access to all constructs generated by the preceding modules, and, in some cases, to constructs generated in a previous processing cycle.

The availability of preceding constructs allows selective implementation of shortcuts among modules, via a set of dedicated parameters. These parameters thus provide additional means of modeling emotion biases. For example, during states of high anxiety, intermediate, more complex processing may be bypassed, and cues can map directly into actions.

The availability of constructs from previous cycles allows for dynamic feedback among constructs and thus departs from a strictly sequential processing sequence. This allows modeling of several observed biases, such as goal-directed priming.

Affect Appraisal Module

The affect appraisal (AA) module is a core component of the MAMID architecture and incorporates elements of several recent appraisal theories: multiple levels and multiple stages (Leventhal & Scherer, 1987; Scherer, 2003; Sloman, 2003; Smith & Kirby, 2001), and aims for consistency with empirical data.

AA integrates external data (cues), internal interpretations (situations and expectation), and priorities (goals), with both the current emotional states and the static individual characteristics (traits) and generates an appraisal of the current stimuli. The appraisal is generated at two levels of resolution. A *low-resolution assessment* is produced via automatic appraisal, using universal emotion elicitors such as novelty and threat level and generates a positive or negative *affective state* (valence). A *higher-resolution categorical assessment* is produced by the expanded appraisal, using more *cognitively complex and idiosyncratic elicitors*, and generates a vector of intensities for four of the basic emotions: anxiety/fear, anger, sadness, happiness (refer to Figure 19.3).

Differences in the triggering elicitors for particular emotions allow for individual idiosyncrasies in emotion triggering. For example, Agent A might react to situation x with anger, Agent B with fear, whereas Agent C might not have an affective reaction at all.

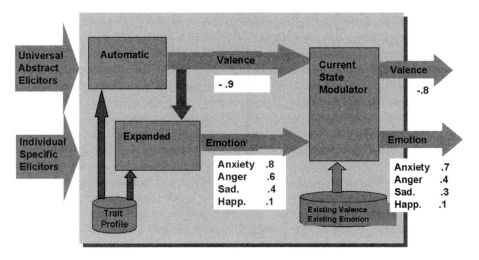

FIGURE 19.3 Structure of the MAMID affect appraisal module.

Figure 19.4 shows examples of simple emotion-generating belief nets for a "social optimist" and "introvert pessimist" players within the search-and-rescue game implementation. These two nets derive the degree of anger in response to being criticized by another player. The belief nets highlight the individual differences in both the situations that contribute to different emotions in these stereotypes, reflected in the net topology and contents (propositions represented by the individual nodes), and the magnitude of emotion intensities generated. The differences in final emotion intensities are due in part due to differences in the prior probabilities associated with the leaf nodes, which reflect default "expectations" regarding the probabilities of certain events and conditions (e.g., "social optimist" player expects the game status and team perception of self to be generally better than the "pessimist" player); and in part to the conditional probabilities linking these nodes to the final emotion node, which determine the degree to which these conditions contribute to the particular emotion.

Generic Methodology for Modeling Effects of Emotions on Cognition

To model the interacting effects emotions on cognitive processing, MAMID uses a generic methodology for modeling a broad range of individual differences within a cognitive architecture (Hudlicka, 1997, 1998, 2002). The core component of this methodology is the ability to represent a broad range of individual differences, including *emotions* and *personality traits*, in terms of a

series of parameters that control the architecture processes and structures. The effects of these factors are then implemented by manipulating the architectural processes and structures via these parameters. The underlying thesis of this approach is that the combined effects of a broad range of factors influencing cognition and behavior can be modeled by varying the values of these parameters. These parameters are defined outside of the architecture proper, influence speed and capacity of its modules (e.g., attention, working memory), and cause biases in recall, interpretation, goal management, and action selection. To the extent possible, the parameter definition and behavior are consistent with existing empirical data. Figure 19.5 provides a high-level view of this methodology, showing a schematic illustration of the general relationship between a representative set of traits and states, the architecture parameters, and the architecture itself. Figure 19.6 provides a more detailed view, showing how a specific configuration of individual differences (high introversion and low emotional stability) contributes to higher intensity of a particular emotion (anxiety), which in turn has specific effects on processing within the architecture modules (threat bias in attention, interpretive threat bias in situation assessment, and threat bias in expectation generation).

Functions implementing the mappings were constructed on the basis of the available empirical data. For example, reduced attentional and working memory capacity, associated with anxiety and fear, are modeled by dynamically reducing the attentional and working memory capacity of the architecture modules,

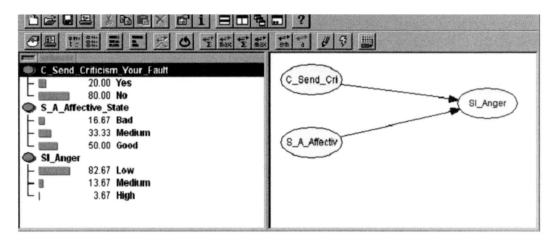

"Social Optimist" Stereotype (considers the critic's own affective state)

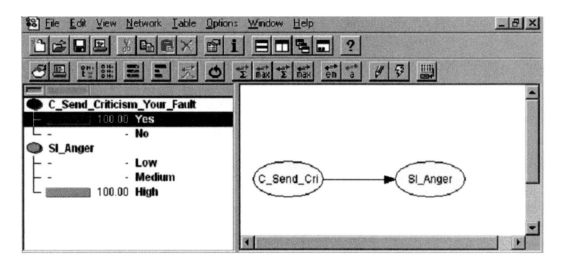

"Introvert Pessimist" Stereotype (considers only own state)

FIGURE 19.4 Examples of simple anger-generating belief nets for "social optimist" and "introvert pessimist" player stereotypes. Note the difference in the factors considered when generating anger (leaf nodes on the left), and the likelihood of deriving low versus high anger intensity. Each player has a number of nets for each type of emotion generated. If multiple nets are triggered during a particular point in the simulation, their values are combined.

which then reduces the number of constructs processed (fewer cues attended, situations derived, expectations generated, etc.). Attentional threat bias is modeled by higher ranking of threatening cues, thus increasing their likelihood of being attended and by higher ranking of threatening situations and expectations, thus increasing the chances of a threatening situation or expectation being derived (see Figure 19.6).

Evaluation Experiment

The first implementation of MAMID was developed and evaluated in the context of a peacekeeping scenario, with separate instances of the architecture controlling the behavior of peacekeepers, reacting to a series of surprise situations (e.g., ambush, hostile crowd; Hudlicka, 2003a). Distinct types of commanders were defined (e.g., anxious, aggressive, normal), which

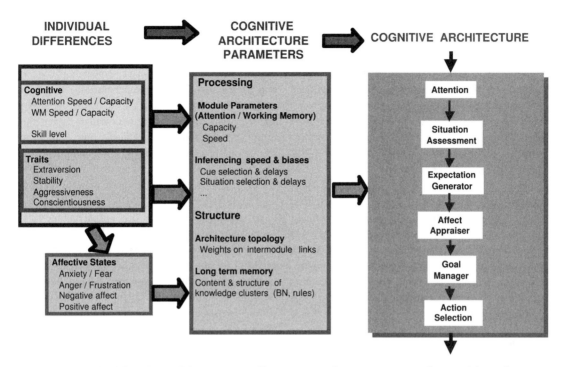

FIGURE 19.5 Methodology for modeling emotion effects in terms of parametric manipulations of the architecture processes and structures.

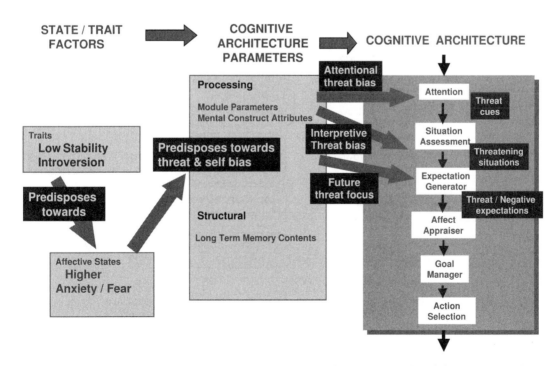

FIGURE 19.6 Modeling biasing effects of traits and states: State and trait anxiety-induced threat processing bias.

then demonstrated distinct processing differences due to the different trait profiles and dynamically generated states, eventually resulting in distinct mission outcomes.

Figure 19.7 illustrates the internal processing of two instances of MAMID architecture, a "normal" and an "anxious" commander, during a "hostile crowd" event. The figure shows the mental constructs produced by each of the architecture modules as the agent processes the incoming cues (presence of hostile crowd), interprets the situation, and eventually selects an action. The "balloons" in the center column indicate, for each module, the types of biases resulting from the trait and state anxiety effects, eventually resulting in distinct decisions and behavioral outcomes for the two commanders: the anxious commander uses excessive force and fires into the crowd, whereas the normal commander uses "nonlethal" crowd control measures. This occurs because the anxious commander exhibits attentional, interpretive, and projective threat-related biases, as well as an interpretive and projective self-bias, which cause a neglect of relevant cues that might lead to a different (less threatening) interpretation (e.g., unit not endangered by crowd), causing a derivation of an overly threatening assessment of the current situation (unit in danger) as well as a focus on

self (own life is in danger; situation assessment). These interpretations then lead to further negative projections about impending danger to the unit, as well as a possible negative career impact (expectation generation). These situations and expectations together then focus on defending the unit and reducing one's own anxiety level (goal selection), which then triggers the use of excessive force (to defend the unit) and excessive communication (means of coping with excessive anxiety).

A key factor contributing to these biases is the anxious commander's level of anxiety, which begins at a higher baseline level than his normal counterpart, occurs more frequently and reaches higher intensities. Figure 19.8 compares the anxiety levels generated by MAMID for the "normal" and the "anxious" commanders, during the course of the scenario. Note the emerging effect of the simulation, where the "illumination" event coincides with the "destroyed bridge" event for the anxious commander, due to his slower speed and slower performance with the bridge repair, thus resulting in two negative events having occurring at the same time. The differences in behavioral outcomes for these two commanders are then shown in Figure 19.9.

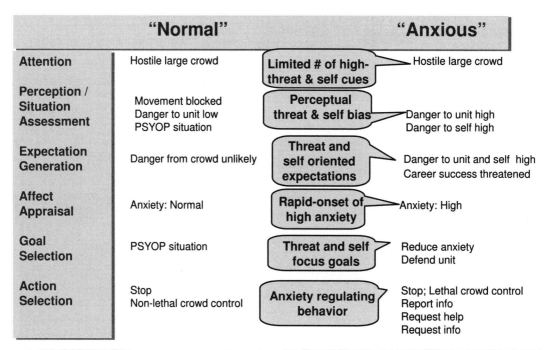

FIGURE 19.7 Contrasting internal processing and behavior for "normal" and "anxious" commanders, in response to encountering a hostile crowd.

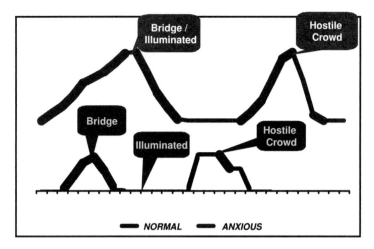

FIGURE 19.8 Anxiety levels over time for the normal and anxious commanders.

These results provide a first step in the evaluation process and demonstrate the ability of the MAMID methodology and architecture to model distinct agent types and generate distinct behavioral outcomes via a series of manipulations of the fundamental cognitive processes comprising the multiple stages of the decision-making process. The affect appraisal module plays a key role, by dynamically generating distinct emotions in response to distinct external and internal contexts.

These emotions, in conjunction with different personality traits, then give rise to different behavior in response to the same set of external circumstances. The MAMID test-bed environment also serves as a research tool for investigating broader issues in computational emotion modeling, by supporting the testing of hypotheses regarding alternative mechanisms of emotion–cognition interactions. The domain-independent MAMID architecture is currently being transitioned

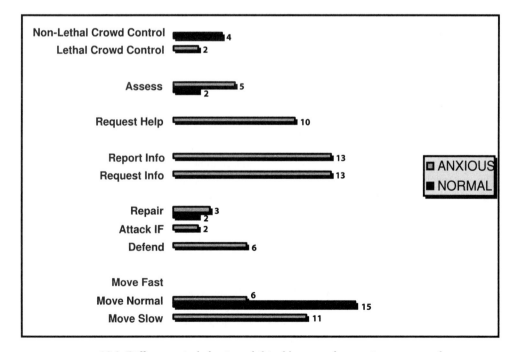

FIGURE 19.9 Differences in behavior exhibited by normal vs. anxious commanders.

into a distributed team task context, where it will be evaluated with respect to human performance data.

Benefits, Shortcomings, and Future Research

Benefits of the MAMID Modeling Methodology and Test-Bed Environment

The evaluation experiment demonstrated the ability of the parameter-based methodology to model a range of interacting emotion effects, as well as the effects of traits. The experiment demonstrated that the parametric manipulation of cognitive processes can produce observable differences in behaviors, which in turn lead to differences in task outcomes. These results have implications for both research and applications.

There are several advantages of modeling the effects of emotions and traits in terms of a uniform parameter space.

- By combining the effects of multiple emotions[2] within a single parameter space, this method provides an efficient means of modeling the complex, interacting effects of multiple emotions, and other individual differences.
- The current parameter space allows the definition of additional emotions (e.g., frustration) and traits (e.g., task- vs. process-focused leadership), as demonstrated in a follow-up transition to the search-and-rescue team game implementation (Hudlicka, 2005b).
- The ability to interactively manipulate the parameters and their calculations allows the exploration of multiple hypotheses regarding the mechanisms of particular emotional biases.
- The methodology is psychologically grounded, with both the emotion effects and the architecture parameters selected on the basis of empirical data and psychological theory. This is in contrast to some recent efforts that have adopted the parameter-based approach, but include parameters such as global system noise to degrade system performance (e.g., Ritter, Avramides, & Councill, 2002). Such parameters can model general degradation of performance and can be useful for some types of modeling applications. However, because they have no direct counterpart

in psychological processes, they cannot be used to generate useful empirical hypotheses that would help elucidate the mechanisms of affective effects on cognition.

- The method provides a computational model which supports the exploration of a recent hypothesis from neuroscience that distinct emotions correspond to distinct *modes* of neuronal functioning (Fellous, 2004).
- The ability to represent a range of biases on multiple cognitive processes offers a means of defining emotions in terms of these distinct biases. Such definitions would provide less ambiguous characterizations of emotions than are currently possible with surface-level behavioral observations.
- Within the more applied realm, the parameter-based method of modeling emotion effects provides an efficient means of rapidly generating a number of distinct agents, each characterized by a unique emotional and personality profile, and associated distinct patterns of decision-making and behavior.

The test-bed environment was effective in facilitating the rapid selection and fine-tuning of alternative individual differences profiles, and observation and analysis of resulting behavior variations and subsequent task outcomes.

The methodology and test-bed environment also provide a computational modeling research tool for exploring alternative mechanisms of specific affective biases. For example, one of the effects of anxiety is an interpretational threat bias: an anxious individual is more likely to assess a given situation as threatening. Two alternative mechanisms may be involved here, acting individually or jointly: (1) available cues are judged to be more threatening or (2) threatening situations are preferentially selected during the interpretation process. MAMID allows the modeling of both alternatives, as well as the associated differences in the magnitude of the bias associated with different degrees of anxiety. Based on these alternative models, MAMID can generate data from simulations of specific tasks, which can be used to formulate targeted experimental hypotheses for further empirical validation. For example, if the mechanism involves assessment of particular cues as more threatening, additional effects of such high-threat cues would be manifested within the system. The consequent ability to characterize the bias mechanisms in terms of the processes and structures involved not only improves our understanding but also has practical applications. For example, knowing that anxiety-linked interpretation bias is mediated by biased selection of situations versus biased selection of cues, helps determine how such biases could be counteracted via appropriate user interface designs in decision-aiding systems in high-stress contexts, such as spacecraft repair or emergency management. There are also implications for the assessment and treatment of a variety of cognitive and affective disorders, such as anxiety disorders.

MAMID Shortcomings and Future Research

The primary shortcoming of MAMID is the limited extent of its evaluation and lack of formal validation studies, where MAMID agent behavior would be compared with human data. We are currently in the process of transitioning MAMID to a different task context, a team search-and-rescue task, where extensive empirical data are available about human performance (Orasanu, Fischer, & Kraft, 2003). This will provide opportunities for additional validation experiments, and for assessing the generality and scalability of both the emotion modeling methodology, and the overall MAMID architecture.

The second shortcoming is that MAMID, like most cognitive architectures, provides a highly *underconstrained method for modeling* cognitive–affective phenomena. In other words, a given set of data could be accounted for by multiple mechanisms. MAMID's utility in elucidating the specific mechanisms that mediate particular cognitive–affective interactions must therefore also be demonstrated. This shortcoming is related to the general problem of *limited empirical data available at the desired level of resolution*; a problem shared by most cognitive architectures. This is a particular problem for architectures attempting to model transient emotional states, and their effects on highly internal cognitive processes that are inherently difficult or impossible to measure (e.g., identify specific situations, expectations, and goals and their generation and use).

Another drawback is the extensive *effort required to construct the long-term memory models*. Finally, MAMID *exhibits brittleness and requires repeated fine-tuning* of both the knowledge-bases (agents' LTM), and the model parameters. All four of these problems represent active areas of research in cognitive modeling.

There are of course many possible functional enhancements to MAMID, including: metacognitive capabilities (Hudlicka, 2005a); enhancements of the appraisal processes; and modeling of additional emotions and traits, and associated biases; and increased parallel processing, to name a few. We would also like to extend the test-bed functionalities to include interactive manipulations of the functions that map the individual differences onto architecture parameters and improved model visualization capabilities.

MAMID Architectural Requirements and Relationship With Generic Cognitive–Affective Architectures

More research is needed to determine the specific structural and functional requirements for representing particular emotions and their interactions with cognition. As stated above, a characterization of emotions in terms of specific architectural components is one of the goals of computational emotion research (Sloman, 2004). Nevertheless, we can begin to make some observations about these structural requirements.

The original rationale for developing MAMID was to demonstrate the feasibility of modeling *emotion effects as patterns of process-modulating parameters*. The structure of the MAMID architecture was thus determined by the objective to demonstrate that a parameter-based methodology is capable of producing the observed phenomena associated with a broad range of emotion effects. Specifically, that the parameter-induced differences in the processes mediating attention, perception, decision-making and action selection will result in differences in behavior.

This objective dictated the specific MAMID *modules, processes,* and *constructs*. The need to dynamically generate emotions required a dedicated *affect appraiser* module, implementing cognitive appraisal, via assessment of the current situation with respect to the agent's goals. This in turn required the explicit representation of *situations*, reflecting the current state of self, others and the world, and *goals*, reflecting the desired states of these entities, required modules generating these constructs: the *situation assessment* and *goal manager* modules. Empirical data provide extensive evidence that expectations influence emotions and vice versa, and can be as powerful in inducing emotions as the actual situations. For example, an expectation of an undesirable event can exert similar effects on cognition as the actual event, and, similarly, the positive emotions generated from an anticipation of some desirable situation exert similar effects as the actual situation. These observations led to the explicit representation of *expectations,* and the associated *expectation generation* module that produces expectations from the current cues and situations. Each of these modules required an associated *long-term memory,* providing the necessary knowledge-base.

MAMID's structure was thus built bottom-up and introduced the mental constructs and modules as necessary, to enable the dynamic generation of emotions and models of their effects on cognition. The rather narrow original objective required components that eventually resulted in an architecture with features in common with more broadly scoped, and more complex, cognitive–affective architectures, such as Sloman's Cog_Aff (Sloman, 2003) and Ortony et al.'s recently proposed generic cognitive–affective architecture design (Ortony et al., 2005). These include most notably multiple-levels of processing complexity within and across modules, and multiple and distributed effects of emotions across architecture modules. (A more thorough comparison of these architectures with respect to their explanatory capabilities, phenomena they can model, and explanations they can generate is beyond the scope of this chapter.)

To the extent that these architectures, as well as MAMID, contain components with psychological analogs, both functional and structural, they can be considered architectures in the Newell sense of a cognitive architecture. It is intriguing to consider the possibility that the systemic effects of emotions across a variety of brain structures, implemented in terms of parametric changes across multiple "cognitive" modules, may begin to provide a framework within which uniform bases for multiple mechanisms of emotions can be explored.

Conclusions

Over the past 15 years important discoveries in neuroscience and psychology have contributed to a growing interest in the scientific study of emotion and have in effect "legitimized" academic emotion research. This has been paralleled by increased interest in emotion modeling in cognitive science. The frequent love–hate relationship that characterized academic emotion research until the 1980s was echoed in the early cognitive science work. Attitude toward emotion was frequently of the all-or-nothing variety. Emotion was either summarily rejected as infeasible, irrelevant, or both, or uncritically embraced as essential for adaptive behavior. As emotion modeling research matures, these attitudes are giving way to more integrated views, and several encouraging trends are emerging.

There are a growing number of interdisciplinary collaborations among cognitive scientists, psychologists, and neuroscientists, frequently involving parallel efforts involving both modeling and empirical studies. Emotion modelers are increasingly asking more fundamental questions about the roles, mechanisms, and architectural requirements of emotions. This leads to more comprehensive models of emotions that go beyond the early models focusing on appraisal, frequently implementing an appraisal model proposed by Ortony et al. (1998) and toward the types of generic architectures discussed above. These architectures do not model emotion in terms of dedicated modules. Rather, emotions represent particular types of information processing within these architectures, closely coupled with "cognitive" processing, and frequently arise as emergent properties, as stated by Sloman et al. (2005): "More human-like robot emotions will emerge, as they do in humans, from the interactions of many mechanisms serving different purposes, not from a particular, dedicated 'emotion mechanism.'" This is consistent with a neuroscience view that there is unlikely to be a clear-cut causal relationship among the cognitive and the emotional processes and that "emotion and cognition are integrated systems implemented by the same brain structures rather than two different sets of interacting brain structures" (Fellous, 2004, p. 44). One may view these perspectives as the beginnings of unified theories of *emotion* and *cognition.*

Notes

1. MAMID = methodology for analysis and modeling of individual differences.

2. As well as additional individual differences, including traits and cognitive factors.

References

Blaney, P. H. (1986). Affect and memory. *Psychological Bulletin,* 99(2), 229–246.

Bower, G. H. (1981). Mood and memory. *American Psychologist, 36*, 129–148.

Canamero, D. (2001). Building emotional artifacts in social worlds: Challenges and perspectives. In *Emotional and intelligent: II. The tangled knot of social cognition* (pp. 976–1009). AAAI Fall Symposium, TR FS-01–02. Menlo Park, CA: AAAI Press.

———. (1998). Issues in the design of emotional agents. In *Emotional and intelligent: II. The tangled knot of cognition* (pp. 45–54). AAAI Fall Symposium, TR FS-98–03. Menlo Park, CA: AAAI Press.

Clore, G. L., & Ortony, A. (2000). Cognition in emotion: Always, sometimes, or never? In R. D. Lane & L. Nadel (Eds.), *Cognitive neuroscience of emotion.* New York: Oxford University Press.

Ekman, P., & Davidson, R. J. (1994). *The nature of emotion.* Oxford: Oxford University Press.

Ellsworth, P. C., & Scherer, K. R. (2003). Appraisal processes in rmotion. In R. J. Davidson, K. R. Scherer, & H. H. Goldsmith (Eds.), *Handbook of affective sciences.* New York: Oxford University Press.

Fellous, J.-M., & Arbib, M. (2005). *Who needs emotions?* Oxford: Oxford University Press.

———. (2004). From human emotions to robot emotions. In *Architectures for modeling emotion.* AAAI Spring Symposium, TR SS-04–02. Menlo Park, CA: AAAI Press.

Forgas, J. P. (2001). Introduction: Affect and social cognition. In J. P. Forgas, (Ed.), *Handbook of affect and social cognition.* Mahwah, NJ: Erlbaum.

Gasper, K., & Clore, G. L. (2002). Attending to the big picture: Mood and global versus local processing of visual information. *Psychological Science, 13* (1), 34.

Hudlicka, E. (1997). Modeling behavior moderators in military HBR models. TR 9716. Lincoln, MA: Psychometrix.

———. (1998). Modeling emotion in symbolic cognitive architectures. In *Emotional and intelligent: II. The tangled knot of social cognition.* AAAI Fall Symposium. TR FS-98–03. Menlo Park, CA: AAAI Press.

———. (2002). This time with feeling: Integrated model of trait and state effects on cognition and behavior. *Applied Artificial Intelligence, 16*, 1–31.

———. (2003a). Modeling effects of behavior moderators on performance: Evaluation of the MAMID methodology and architecture. In *Proceedings of BRIMS-12,* Phoenix, AZ.

———. (2003b). To feel or not to feel: The role of affect in HCI. *International Journal of Human-Computer Studies, 59* (1–2), 1–32.

———. (2004). Two sides of appraisal: Implementing appraisal and its consequences within a cognitive architecture. In *Architectures for Modeling Emotion.* AAAI Spring Symposium, TR SS-04–02. Menlo Park, CA: AAAI Press.

———. (2005a). Modeling interaction between metacognition and emotion in a cognitive architecture. In *Metacognition in Computation.* AAAI Spring Symposium, TR SS-05–04. Menlo Park, CA: AAAI Press.

———. (2005b). *Transitioning MAMID to the DDD task domain.* Report 0507. Blacksburg, VA: Psychometrix.

———, & Canamero, L. (2004). Preface. In *Architectures for Modeling Emotion.* AAAI Spring Symposium 2004, TR SS-04–02. Palo Alto, CA.

Isen, A. M. (1993). Positive affect and decision making. In J. M. Haviland & M. Lewis (Eds.), *Handbook of emotions* (pp. 261–277). New York: Guilford Press.

Leventhal, H., & Scherer, K. R. (1987). The relationship of emotion to cognition. *Cognition and Emotion, 1*, 3–28.

Lewis, M. (1993). Self-conscious emotions. In J. M. Haviland & M. Lewis (Eds.), *Handbook of emotions.* New York: Guilford Press.

———, & Haviland, J. M. (1993). *Handbook of emotions.* New York: Guilford Press.

Marsella, S., & Gratch, J. (2002). A step toward irrationality: Using emotion to change belief. In *Proceedings of the 1st International Joint Conference on Agents and Multiagent Systems.* Bologna.

Matthews, G. (2004). Designing personality: Cognitive architecture and beyond. In *Proceedings of the AAAI Spring Symposium.* Menlo Park, CA: AAAI Press.

———, & Dreary, I. J. (1998). *Personality traits.* Cambridge: Cambridge University Press.

Mellers, B. A., Schwartz, A., & Cooke, A. D. J. (1998). Judgment and decision making. *Annual Review of Psychology, 49*, 447–477.

Mineka, S., Rafaeli, E., & Yovel, I. (2003). Cognitive biases in emotional disorders: Information processing and social-cognitive perspectives. In R. J. Davidson, K. R. Scherer, & H. H. Goldsmith (Eds.), *Handbook of affective sciences.* Oxford: Oxford University Press.

Orasanu, J., Fischer, U., & Kraft, N. (2003). Enhancing team performance for exploration missions. NSBRI Report. Moffett Field, CA: NASA-Ames.

Ortony, A., Clore, G. L., & Collins, A. (1988). *The cognitive structure of emotions.* New York: Cambridge University Press.

———, Norman, D., & Revelle, W. (2005). Affect and proto-affect in effective functioning. In J.-M. Fellous & M. Arbib (Eds.), *Who needs emotions?* New York: Oxford University Press.

Persons, J. B., & Foa, E. B. (1984). Processing of fearful and neutral information by obsessive-compulsives. *Behavior Research Therapy, 22*, 260–265.

Pew, R., Mavor, A. (1998). *Representing human behavior in military simulations.* Washington, DC: National Academy Press.

Picard, R. (1997). *Affective computing*. Cambridge, MA: MIT Press.

Revelle, W. (1995). Personality processes. *Annual Review of Psychology, 46,* 295–328.

Ritter, F., Avramides, M., & Councill, I. (2002). Validating changes to a cognitive architecture to more accurately model the effects of two example behavior moderators. *Proceedings of 11th CGF Conference,* Orlando, FL.

Scherer, K. R. (2003). Introduction: Cognitive components of emotion. In R. J. Davidson, K. R. Scherer, & H. H. Goldsmith (Eds.), *Handbook of affective sciences.* New York: Oxford University.

——, Schorr, A., & Johnstone, T. (2001). *Appraisal processes in emotion: Theory, methods, research.* Oxford: Oxford University Press.

Sloman, A. (2003). How many separately evolved emotional beasties live within us? In R. Trappl, P. Petta, & S. Payr (Eds.), *Emotions in humans and artifacts.* Cambridge, MA: MIT Press.

——. (2004). What are emotion theories about? In *Architectures for modeling emotion.* AAAI Spring Symposium, TR SS-04–02. Menlo Park, CA: AAAI Press.

——, Chrisley, R., & Scheutz, M. (2005). The architectural basis of affective states and processes. In J.-M. Fellous & M. Arbib (Eds.), *Who needs emotions?* New York: Oxford University Press.

Smith, C. A., & Kirby, L. D. (2001). Toward delivering on the promise of appraisal theory. In K. R. Scherer, A. Schorr, & T. Johnstone (Eds.), *Appraisal processes in emotion.* New York: Oxford University Press.

Trappl, R., Petta, P., & Payr, S. (2003). *Emotions in humans and artifacts.* Cambridge, MA: MIT Press.

Williams, J. M. G., Watts, F. N., MacLeod, C., & Mathews, A. (1997). *Cognitive psychology and emotional disorders.* New York: Wiley.

PART VI

MODELING EMBODIMENT IN INTEGRATED COGNITIVE SYSTEMS

Hansjörg Neth & Christopher W. Myers

Beyond the realms of philosophical thought experiments human brains do not normally reside in vats but are embedded in a body that provides an interface between the mind and its environment. Cognition is fundamentally embodied: There simply is no vision without an eye that sees, no sensation without a nervous system that experiences it, and no decision making without muscles that execute or communicate a response.

If the embodied nature of cognition may seem trivial, it is surprising that academic psychology has chosen to disregard it throughout most of its existence. Whereas philosophers have long been exploring the consequences of this mind–matter coupling, the empirical sciences of human behavior have first eschewed all notions of mind altogether and then become too enamored with an ideal of disembodied information processing to exploit the fact that minds experience and act upon the world through molds of matter.

Recently, researchers from disciplines as diverse as biology, ethnography, psychology, and robotics have

begun to embrace an embodied view of cognition. Interestingly, their rediscovery of philosophical insights has been motivated by practical problems, such as designing more usable software interfaces or robots that can navigate their surroundings without elaborate world models. Consequently, the debate about the embodied aspects of human cognition has been reinvigorated and even become fashionable. While embodied cognition comes in many different flavors (see Clark, 1997; Wilson, 2002, for overviews), its advocates share the conviction that the particular shape and sensory–motor capabilities of the human body provide crucial constraints on the nature of cognitive operations.

Viewing the body as an essential interface between cognition and world shifts the research focus from a traditionally functionalist perspective (that determines the input–output functions required for a task, regardless of their concrete implementation) to an analysis of *how* humans achieve a task, given their particular perceptual and motor resources. But merely acknowledging the existence of physical constraints on cognition does not yet provide guidance on when, where, and in which way these constraints are imposed.

Although the three chapters that compose this section address different behavioral phenomena and subscribe to different theoretical frameworks, they have much in common. The first and most obvious unifying element is that their authors adopt an embodied perspective on traditionally disembodied problems (route planning, rapid pointing movements, visual search) and show how this shift in emphasis changes both the problems and their solutions. Second, capitalizing on bodily constraints on the cognitive control of behavior not only provides the authors with theoretical guidance but also allows them to bridge the boundaries between traditionally disparate fields (e.g., visual perception and action, motor movement, and decision making). Finally, the three chapters are united by a firm commitment to formal methodologies (reinforcement learning, mathematical modeling, computational cognitive modeling). This shared commitment to explicit formalisms keeps their theoretical accounts grounded, facilitates the transfer of their methods or findings to other domains, and helps to distinguish theoretical progress from purely verbal speculation.

In "On the Role of Embodiment in Modeling Natural Behaviors" (chapter 20), Ballard and Sprague explore the task of visually guided locomotion by creating a virtual human, named Walter. Walter can negotiate a crowded sidewalk while collecting litter and uses computationally efficient *microbehaviors* to navigate his path, avoid collisions, and recognize and collect litter. Walter's repertoire of behavioral routines is governed by reinforcement-learning algorithms and arbitrated by an uncertainty mechanism that assigns priority to uncertain states. Grounding Walter's microbehaviors in embodiment leads to a computational economy of cognitive mechanisms, but these benefits come at a high cost: harvesting the body's remarkable computational powers is not easy, and implicit somatic processes—by definition—are hard to explicate. The authors demonstrate that this trade-off can work in the researchers' favor, as a solid foundation of embodied routines facilitates the subsequent modeling of cognitive operations.

Maloney, Trommershäuser, and Landy illuminate the intimate relationship between hands and brains

in "Questions Without Words: A Comparison Between Decision Making Under Risk and Movement Planning Under Risk" (chapter 21). By drawing an analogy between decision making under risk and rapid pointing movements, they extend and enrich the notions of planning and decision making to encompass motor actions. Whereas decision making under risk is notoriously prone to various fallacies, participants are surprisingly adept at anticipating the stochastic uncertainty of their own motor system. Thus, the authors do not only point out intriguing parallels between historically distinct fields but propose that the study of motor movements may be a better domain to reveal the true competence of human everyday decision making than traditional laboratory tasks.

Hornof (chapter 22) aspires "Toward an Integrated, Comprehensive Theory of Visual Search." His mixed-density text and hierarchical search models explore how we search a visual layout (e.g., a Web page) in which text appears in various densities or labeled groups. As these are interesting microtheories in themselves, this chapter could have been included in Part III, "Visual Attention and Perception." It owes its inclusion in this section to being embedded within the broader framework of the EPIC cognitive architecture, which endows both models with specific constraints on the movement of eyes and hands.

Very much in the spirit of this volume, Hornof also discusses the prospects and potential pitfalls for the integration of task-specific theories into comprehensive cognitive architectures. As advocates of an embodied perspective typically tout its benefits, it is interesting that Hornof joins Ballard and Sprague in emphasizing both the benefits and burdens associated with an integrated approach.

Together, the three chapters aptly demonstrate how embracing an embodied perspective of cognition can change the ways in which we conduct and conceptualize our science. While the embodied view may entail an integrative view, it does not require a revolutionary departure from previous research. Instead, it is part of an evolutionary process that analyzes the interactions of the human mind with its environment on multiple levels. Similarly, these chapters show that the embodied perspective does not have to abandon the methodological rigor of formal accounts and can often extend and enrich existing cognitive theories.

As embodied cognition is always situated, the embodied perspective is complementary to an environmentally embedded view of cognition (Part IV, "Environmental Constraints on Integrated Cognitive Systems"). Ultimately, the embodied nature of cognition is only a piece in the puzzle of completely integrated cognitive systems. The unified view of mind and its bodily extensions will itself have to be integrated with an even broader view that captures the complex interplay of embodied minds and the surrounding world in which minds with bodies pursue their goals.

References

Clark, A. (1997). *Being there: Putting brain, body, and world together again*. Cambridge, MA: MIT Press.

Wilson, M. (2002). Six views of embodied cognition. *Psychonomic Bulletin & Review, 9*(4), 625–636.

20

On the Role of Embodiment in Modeling Natural Behaviors

Dana Ballard & Nathan Sprague

Early research in artificial intelligence emphasized mechanisms of symbol processing with the goal of isolating an essential core of intelligent behavior. Although huge progress was made in understanding automated reasoning, an unsolved problem has been that of specifying the grounding of the symbols. More recently, it has been appreciated that the human body, largely neglected in symbolic AI, may be the essential source of this grounding. Embracing this view comes with great cost because to properly model the grounding process requires enormous amounts of computation. However, recent technological advances allow progress to be made in this direction. Graphics models that simulate extensive human capabilities can be used as platforms from which to develop synthetic models of visuomotor behavior. Currently, such models can capture only a small portion of a full behavioral repertoire, but for the behaviors that they do model, they can describe complete visuomotor subsystems at a level of detail that can be tested against human performance in realistic environments.

All brain operations are situated in the body (Clark, 1997). Even when the operations are purely mental, they reflect a developmental path through symbols that are grounded in concrete interactions in the world. The genesis of this view is attributed to the philosopher Merleau-Ponty (1962), but more recently it has found articulate advocates in Clark (1999), Nöe (2005), O'Regan and Nöe (2001), and Roy and Pentland (2002). These authors argue that not only is the body a source of mechanisms for grounding the experiences that we describe symbolically in language but, in fact, is a sine qua non. We may have music in sheet form, but we can only experience it with a body to play it. The same goes for symbols.

A great boon occurs when embodiment is taken as a tenet of research programs because once the body is modeled, tremendous subsequent computational economies result. However, embodied models can forgo computation because computation is done implicitly by the body itself. Thus modeling this boon places an enormous demand on the researcher to simulate the body's prodigious computational abilities. To this end, research programs that focus on embodiment have been facilitated by the development of virtual reality (VR) graphics environments. These VR environments can now run in real time on standard computing platforms. The value of VR environments is that they allow the creation of virtual agents that implement complete visuomotor control loops. Visual input can be captured from the rendered virtual scene, and motor commands can be used to direct the graphical representation of the virtual agent's body. Terzoupolous (1999) and Faloutsos, van de Panne, and Terzopoulos (2001) pioneered the use of virtual reality as a platform for the study of visually guided control. Embodied control has been studied for many years in the robotics domain, but virtual agents have enormous advantages over physical robots in the areas of experimental reproducibility, hardware requirements, flexibility, and ease of programming.

Embodied models can now be tested using new instrumentation. Linking mental processing to visually guided body movements at a millisecond timescale would have been impractical just a decade ago, but recently a wealth of high-resolution monitoring equipment has been developed for tracking body movements in the course of everyday behavior, particularly head, hand, and eye movements (e.g., Pelz, Hayhoe, & Loeber, 2001). This allows for research into everyday tasks that typically have relatively elementary cognitive demands but require elaborate and comprehensive physical monitoring. In these tasks, overt body signals provide a direct indication of mental processing.

During the course of normal behavior, humans engage in a variety of tasks, each of which requires certain perceptual and motor resources. Thus, there must be mechanisms that allocate resources to tasks. Understanding this resource allocation requires an understanding of the ongoing demands of behavior, as well as the nature of the resources available to the human sensorimotor system. The interaction of these factors is complex, and that is where the virtual human platform can be of value. It allows us to imbue our artificial human with a particular set of resource constraints. We may then design a control architecture that allocates those resources in response to task demands. The result is a model of human behavior in temporally extended tasks.

We refer to our own virtual human model as Walter. Walter has physical extent and programmable kinematic degrees of freedom that closely mimic those of real humans. His graphical representation and kinematics are provided by the DI-Guy package developed by Boston Dynamics. This is augmented by the Vortex package developed by CM Labs for modeling the physics of collisions. The crux of the model is a control architecture for managing the extraction of information from visual input that is in turn mapped onto a library of motor commands. The model is illustrated on a simple sidewalk navigation task that requires the virtual human to walk down a sidewalk and cross a street while avoiding obstacles and collecting litter. The movie frame in Figure 20.1 shows Walter in the act of negotiating the sidewalk, which is strewn with obstacles (tall objects) and litter (small objects) on the way to crossing a street.

The Role of Embodiment

The essential asset of embodiment is that the human body has been especially designed by evolution to solve data acquisition and manipulation problems so that these do not have to be resolved by the forebrain. Instead the forebrain can access and direct the form of these solutions using very compact, abstract representations. Furthermore, as we will elaborate, the vision and motor system solutions have essential features in common.

Human vision uses fixations that have an average duration of 200 to 300 ms. The fixational system brings home the key role of embodiment in behavior. Although the phenomenological experience of vision may be of a seamless three-dimensional surround, the system that creates that experience is discrete. Furthermore as humans are binocular and make heavy use of manipulation in their behaviors, they spend most of their time fixating objects in the near distance. That is the centers

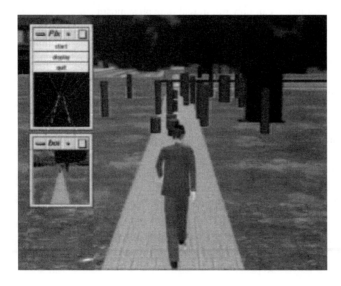

FIGURE 20.1 The Walter simulation. The insets show the use of vision to guide the humanoid through a complex environment. The upper inset shows the particular visual routine that is running at any instant; in this case the lines indicate that the sidewalk is being detected. The lower inset shows the visual field from the perspective of Walter's head-centered frame. (See color insert.)

of gaze of each of the eyes meet at a point in three dimensional space and, to a first approximation, rest on that point for 300 ms.

The human motor system is comprised of an extensive musculoskeletal system that consists of over 200 bones and 600 muscles. One of the most important properties of this system is the passive energy that can be stored in muscles. This springlike system has at least two important properties: (1) it can lead to very economical operation in locomotion, and (2) it can be used in passive compliant collision strategies that save the system from damage. Moreover, it can be driven by a discrete strategy whereby set points for the spring-muscle system are communicated at a low bandwidth. Simulations show that reasonable motions can be obtained with sampling intervals approximating the fixation interval.

Although the vision and motor systems can seem very different, and are different in detailed implementation, at an abstract level they exploit similar kinds of representations.

1. *Routines:* Visual computations have to be efficient since their result has to be computable in 300 ms. Thus for the most part, the computations are task-dependent tests since those can use prior information to simplify the computations. Suppose you are looking for your favorite cup that happens to have an unusual color. Then rather than any elaborate parsing of the visual surround, you can simply look for a place that has that color. Motor routines need to be efficient for the same reason that visual routines are. Since they are also goal directed, they can make extensive use of expectations that can be quickly tested. If you are searching for the same cup on a nearby table in the dark, you can use the remembered shape of the cup to quickly and efficiently interpret haptic data from your fingers and palm.

2. *Dynamic reference frames:* Huge debates have ranged over the coordinate system used in vision—is it head based, eye based, or otherwise? The fixational system shows that it must be dynamic. Depending on the task at hand, it can be any of these. Imagine using a screwdriver to drive a screw into hardwood. The natural reference frame for the task is the screw head where a pure torque is required. Thus, the gaze is needed there, *and* all the muscles in the body are constrained by this purpose: to provide a torque at a site remote to the body. To keep this idea of a remote frame of reference in mind, consider how the visual system codes for disparity. Neurons that have disparity-sensitive receptive fields are code for zero, negative, and positive disparities.

However, zero disparity is at the fixation point, a point not in the body at all. Like visual routines, motor routines use dynamic frames of reference. The screw driving example used for vision also applies for motor control. The multitude of muscles work together to apply a pure torque at the driver end. For upright balance, the motor system must make sure the center of gravity is over the base of the stance. But of course the center of gravity is a dynamic point that moves with postural changes. Nonetheless, the control system must refer its computations to this point.

3. *Simplified computation:* The visual system has six separate systems to stabilize gaze. This of course is an indication of just how important it is to achieve gaze stabilization, but the third consequence is that the visual computations can be simplified as they do not have the burden of dealing with gaze instability. Even though the human may be moving and the object of interest may be independently moving as well, the algorithms used to analyze that object can assume that it remains in a fixed position near the fovea. Many examples could be mentioned to illustrate how the computation done by the motor system makes the job of motor routines easier, but one of the most obvious is the extensive use of passive compliance in grasping. If the motor system was forced to rely on feedback in the way that standard robot systems do, the grasping strategies would have to be far more delicate. Instead, grasp planning can be far easier as the passive conformation of the multifingered hand provides great tolerances in successful grasps (Figure 20.2).

Behavior-Based Control

Any system that must operate in a complex and changing environment must be compositional (Newell, 1990); that is, it has to have elemental pieces that can be composed to create its more complex structures. Figure 20.3 illustrates two broad compositional approaches that have been pursued in theories of cognition, as well as in robotics. The first decomposition works on the assumption that the agent has a central repository of symbolic knowledge. The purpose of perception is to translate sensory information into symbolic form. Actions are selected that result in symbolic transformations that bring the agent closer to goal states. This sense–plan–act approach is typified in the robotics community by early work on Shakey the robot (Nilsson, 1984), and in the cognitive science community by the

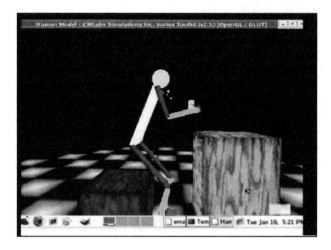

FIGURE 20.2 Motor control of a humanoid figure can be economically commanded with a handful of discrete control points. (See color insert.)

theories of David Marr (1982). In principle, the symbolic planning approach is very attractive, since it suggests that sensation, cognition, and action can be studied independently, but in practice each step of the process turns out to be difficult to characterize in isolation. It is hard to convert sensory information into general-purpose symbolic knowledge, it is hard to use symbolic knowledge to plan sequences of actions, and it is hard to maintain a consistent and up-to-date knowledge base.

The difficulties with the symbolic planning approach have led to alternate proposals. In the robotics community, Brooks (1986) has attempted to overcome these difficulties by suggesting a radically different decomposition, illustrated in Figure 20.3B. Brooks's alternate approach is to attempt to describe whole visuomotor behaviors that have very specific goals. Behavior-based control involves a different approach to composition than

planning-based architectures: simple microbehaviors are sequenced and combined to solve arbitrarily complex problems. The best approach to attaining this sort of behavioral composition is an active area of research. Brooks's own *subsumption* architecture worked by organizing behaviors into fixed hierarchies, where higher-level behaviors influenced lower-level behaviors by overwriting their inputs. Subsumption works spectacularly well for trophic, low-level tasks, but generally fails to scale to handle more complex problems (Hartley & Pipitone, 1991), mainly because of its commitment to hardwired priorities among competing behaviors. For that reason, we have chosen a more flexible control architecture.

Our version of behavior-based control centers around primitives that we term *microbehaviors*. A microbehavior is a complete sensory/motor routine

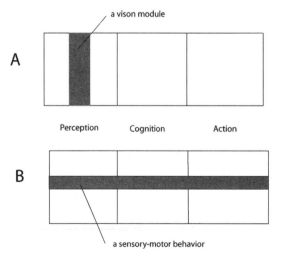

FIGURE 20.3 Two approaches to behavioral research contrasted. (A) In the Marr paradigm individual components of vision are understood as units. (B) In the Brooks paradigm the primitive unit is an entire behavior.

that incorporates mechanisms for measuring the environment and acting on it to achieve specific goals. For example, a collision-avoidance microbehavior would have the goal of steering the agent to avoid collisions with objects in the environment. A microbehavior has the property that it cannot be usefully split into smaller subunits. Walter's microbehavior control architecture follows more recent work on behavior-based control (e.g., Bryson & Stein, 2001; Firby, Kahn, Prokopowicz, & Swain, 1995) that allows the agent to address changing goals and environmental conditions by dynamically activating a small set of appropriate behaviors. Each microbehavior is triggered by a template that has a pattern of internal and environmental conditions. The pattern-directed activation of microbehaviors provides a flexibility not found in the fixed subsumption architecture.

State Estimation Using Visual Routines

Individual microbehaviors are indexed by state information. If they are active and can determine their state, then the corresponding desired motor response is stored as a function of the state. Each microbehavior has an associated visual routine that is especially crafted to compute just the needed state information.

Visual routines can be composed from a small library of special-purpose functions. Although the use of such routines is unremarkable in the field of robotics, the use of such routines as the basis of human vision is more controversial. The arguments for visual routines have be made by Ullman (1985); Roelfsema, Lamme, and Spekreijse (2000); and Ballard, Hayhoe, and Pook

(1997). The main one is that the representations of vision such as color and form are problem neutral in that they do not contain explicitly the data on which control decisions are made and thus an additional processing step must be employed to make decisions. The number of potential decisions that must be made is too large to precode them all. Visual routines address this problem in two ways: (1) routines are composable and (2) routines process visual data in an *as-needed* fashion.

One might think that it would be a good idea to build additional scene descriptions that could be shared by many microbehaviors, but it has proven difficult to describe just what the content of these descriptions should be. And then there is the problem of interrogating these descriptions, which may prove more difficult than interrogating the visual image.

To illustrate the use of visual routines, we describe the ones that create the state information for three of Walter's microbehaviors: collision avoidance, sidewalk navigation, and litter collection. Each of these requires specialized processing. This processing is distinct from that used to obtain the feature images of early vision even though it may use such images as data. The specific processing steps are visualized in Figure 20.4.

1. Litter collection is based on color matching. Litter is signaled in our simulation by purple objects, so that potential litter must be isolated as being the right color and also for being nearby. This requires combining and processing the hue image with depth information. The result of this processing is illustrated in Figure 20.4b.

2. Sidewalk navigation uses color information to label pixels that border both sidewalk and grass regions.

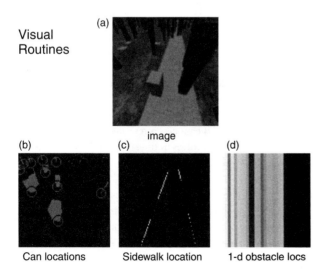

Visual Routines

(a)

image

(b)   (c)   (d)

Can locations   Sidewalk location   1-d obstacle locs

FIGURE 20.4 The visual routines that compute state information. (a) Input image from Walter's viewpoint. (b) Regions that fit the litter color profile. Probable litter locations are marked with circles. (c) Processed image for sidewalk following. Pixels are labeled in white if they border both sidewalk and grass color regions. The red line is the most prominent resulting line. (d) One-dimensional depth map used from obstacle avoidance (not computed directly from the rendered image). Darker stripes are closer. (See color insert.)

A line is fit to the resulting set of pixels, which indicates the estimated edge of the sidewalk. The result of this processing is illustrated in Figure 20.4c.

3. The collision detector uses a depth image. A depth image may be created by any of a number of cues, (stereo, kinetic depth, parallax depth, etc.), but for collisions, it must be processed to isolate potential colliders. The result of this processing is illustrated in Figure 20.4d. A study with human subjects shows that they are very good at this, integrating motion cues with depth to ignore close objects that are not on a collision course (Ballard & Sprague, 2002).

Regardless of the specific methods of individual routines, each one outputs information in the same abstract form: the state needed to guide its encompassing microbehavior.

Each routine exhibits the huge economies obtained by its task dependence and embodiment. In litter collection, only the features of the litter have to be considered in isolating its locations from the background, a vastly easier task than analyzing the entire image. In the same way, obstacles can be easily isolated as nearby depth measurements without analyzing the complete image, and the sidewalk can be recognized by its long linear boundary. Furthermore, this information is in retinal coordinates, but because the retinas are embodied means that the information to turn those coordinates into headings is readily available.

Learning Microbehaviors

Once state information has been computed, the next step is to find an appropriate action. Each microbehavior stores actions in a state/action table. Such tables can be learned by reward maximization algorithms: Walter tries out different actions in the course of behaving and remembers the ones that worked best in the table. The reward-based approach is motivated by studies of human behavior that show that the extent to which humans make such trade-offs is very refined (Trommershäuser, Maloney, & Landy, 2003) as well as studies using monkeys that reveal the use of reinforcement signals in a way that is consistent with reinforcement-learning algorithms (Suri & Schulz, 2001).

Formally, the task of each microbehavior is to map from an estimate of the relevant environmental state, s, to one of a discrete set of actions, $a \in A$, so as to maximize the amount of reward received. For example, the obstacle avoidance behavior maps the distance and heading to the nearest obstacle $s = (d, \theta)$ to one of

three possible turn angles; that is, $A = \{-15°, 0°, 15°\}$. The *policy* is the action so prescribed for each state. The coarse action space simplifies the learning problem.

Our approach to computing the optimal policy for a particular behavior is based on a standard reinforcement learning algorithm, termed *Q-learning* (Watkins & Dayan, 1992). This algorithm learns a value function $Q(s, a)$ for all the state-action combinations in each microbehavior. The Q function denotes the expected discounted return if action a is taken in state s and the optimal policy is followed thereafter. If $Q(s, a)$ is known, then the learning agent can behave optimally by always choosing arg max$_a Q(s, a)$. (See Sprague & Ballard, 2003b, for details.)[1] Figure 20.5 shows the table used by the litter collection microbehavior, as indexed by its state information.

Each of the three microbehaviors has a two-dimensional state space. The litter collection behavior uses the same parameterization as obstacle avoidance: $s = (d, \theta)$, where d is the distance to the nearest litter item, and θ is the angle. For the sidewalk-following behavior, the state space is $s = (\rho, \theta)$. Here θ is the angle of the center line of the sidewalk relative to the agent, and ρ is the signed distance to the center of the sidewalk, where positive values indicate that the agent is to the left of the center, and negative values indicate that the agent is to the right. All microbehaviors use the logarithm of distance to devote more of the state representation to areas near the agent. All these microbehaviors use the same three-heading action space previously described. Table 20.1 shows Walter's reward contingencies. These are used to generate the Q-tables that serve as a basis for encoding a policy. Figure 20.6 shows a representation of the Q-functions and policies for the three microbehaviors.

When running the Walter simulation, the Q-table associated with each behavior is indexed every 300 ms. The action suggested by the current policy is selected and submitted for arbitration. The action chosen by the arbitration process is executed by Walter. This in turn results in a new Q-table index for each microbehavior, and the process is repeated. The path through a Q-table thus evolves in time and can be visualized as a thread of control analogous to the use of the term *thread* in computer science.

Reinforcement learning's state-action tables incorporate embodiment in at least two ways. First, these tables can be simple as they need only direct the underlying process that the body carries out. Sophisticated gaze control and other motoric mechanisms know how to elaborate the state table commands.

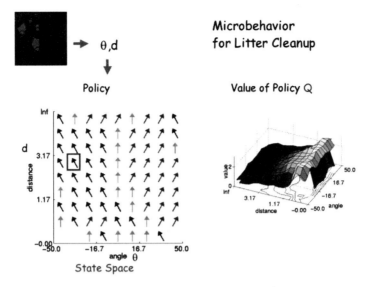

Microbehavior
for Litter Cleanup

FIGURE 20.5 The central portion of the litter cleanup microbehavior after it has been learned. The color image is used to identify the heading to the nearest litter object as a heading angle θ and distance d. Using this state information to index the table allows the recovery of the policy, in this case *heading* = −45 deg, and its associated value. The fact that the model is embodied means that we can assume there is neural circuitry to translate this abstract heading into complex walking movements. This is true for the graphics figure that has a "walk" command that takes a heading parameter. (See color insert.)

The second advantage of the tables occurs because the actions refer to the body itself. This construction tends to avoid nasty side effects that may occur in arbitrary state/action tables. Imagine that the actions are not commensurate as in our case. What can happen in the case of arbitrary tables is that an action in one state table can have indirect and inestimable consequences of another in an unpredictable way. Here again embodiment saves the day by being a final common pathway. Since in our case the different tables are all making heading recommendations, it is easy to know when they may interfere: The different states index different headings.

TABLE 20.1 Walter's Reward Schedule

Outcome	Immediate Reward
Picked up a litter can	2
On sidewalk	1
Collision free	4

Note. The values were chosen to approximate observed human priorities. Different values would result in different behaviors. For example, if "collision free" were zero, Walter would walk through obstacles.

Microbehavior Arbitration

A central complication with the microbehavior approach is that concurrently active microbehaviors may prefer incompatible actions. Therefore, an arbitration mechanism is required to map from the demands of the individual microbehaviors to final action choices. The arbitration problem arises in directing the physical control of the agent, as well as in handling gaze control, and each of these requires a different solution. This is because in Walter's environment, his heading can be a compromise between the demands of different microbehaviors, but his gaze location is not readily shared by them. A benefit of knowing the value function for each behavior is that the Q-values can be used to handle the physical arbitration problem in each of these cases. In this regard, it is the state tables themselves that reflect the future, owing to the way they are constructed. The tables reflect past experiences in reaching goals. Thus the value of a particular action reflects the discounted reward available by taking that action. In this way, the past is prologue. While the traditional symbol processing search used in AI looks for a goal de nuovo, reinforcement learning amortizes this process over all previous

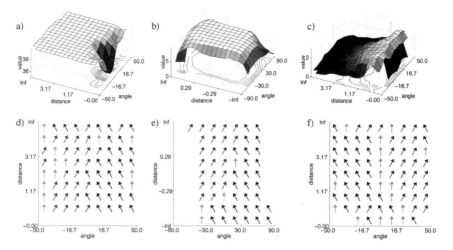

FIGURE 20.6 Q-values and policies for the three microbehaviors. Figures (a)–(c) show $\max_a$ $Q(s, a)$ for the three microbehaviors: (a) obstacle avoidance, (b) sidewalk following, and (c) litter collection. (d)–(f) The corresponding policies for the three microbehaviors. The obstacle avoidance value function shows a penalty for nearby obstacles and a policy of avoiding them. The sidewalk policy shows a benefit for staying in the center of the sidewalk $\theta = 0$, $\rho = 0$. The litter policy shows a benefit for picking up cans that decreases as the cans become more distant. The policy is to head toward them.

trials. The resultant great benefit is the availability of an estimate of the future from the present state.

Heading Arbitration

Because in the walking environment each behavior shares the same action space, Walter's heading arbitration is handled by making the assumption that the Q-function for the composite task is approximately equal to the sum of the Q-functions for the component microbehaviors:

$$Q(s,a) \approx \sum_{i=1}^{n} Q_i(s_i, a), \qquad (1)$$

where $Q_i(s_i, a)$ represents the Q-function for the ith active behavior. Thus the action that is chosen is a compromise that attempts to maximize reward across the set of active microbehaviors. The idea of using Q-values for multiple goal arbitration was independently introduced in Humphrys (1996) and Karlsson (1997).

To simulate that only one area of the visual field may be foveated at a time, only one microbehavior is allowed access to perceptual information during each 300 ms simulation time step. That behavior is allowed to update its state information with a measurement, while the others propagate their estimates and suffer an increase in uncertainty. The mechanics of maintaining state estimates and tracking uncertainty are handled using Kalman filters, one for each microbehavior.[2]

To simulate noise in the estimators, the state estimates are corrupted with zero-mean normally distributed random noise at each time step. The noise has a standard deviation of .2m in both the x and y dimensions. When a behavior's state has just been updated by its visual routine's measurement, the variance of the state distribution will be small, but as we will demonstrate in simulation, in the absence of such a measurement, the variance can grow significantly.

Since Walter may not have perfectly up-to-date state information, he must select the best action given his current estimates of the state. A reasonable way of selecting an action under uncertainty is to select the action with the highest expected return. Building on Equation 1, we have the following: $a_E = n \arg\max_a E[\sum_{i=1}^{n} Q_i(s_i, a)]$, where the expectation is computed over the state variables for the microbehaviors. By distributing the expectation, and making a slight change to the notation, we can write this as:

$$a_E = \arg\max_a \sum_{i=1}^{n} Q_i^E(s_i, a), \qquad (2)$$

where Q_i^E refers to the expected Q-value of the ith behavior. In practice, we estimate these expectations by sampling from the distributions provided by the Kalman filter.

Since the state is transiting through time, it has an internal dynamics that is informed by new state measurements. Kalman filters optimally weight the average of the current state estimate and the new measurement.

Gaze Arbitration

Arbitrating gaze requires a different approach than arbitrating control of the body. Direct reinforcement learning algorithms are best suited for handling actions that have direct consequences for a task. Actions such as eye movements are difficult to put in this framework because they have only indirect consequences: they do not change the physical state of the agent or the environment; they serve only to obtain information.

A much better strategy is to choose to use gaze to update the behavior that has *the most to gain* by being updated. Thus, the approach taken here is to try to estimate the value of that information. Simply put, as time evolves the uncertainty of the state of a behavior grows, introducing the possibility of low rewards. Deploying gaze to measure that state reduces this risk. Estimating the cost of uncertainty is equivalent to estimating the expected cost of incorrect action choices that result from uncertainty. Given that the Q-functions are known, and that the Kalman filters provide the necessary distributions over the state variables, it is straightforward to estimate this factor, $loss_b$, for each behavior b by sampling (Sprague & Ballard, 2003a). The maximum of these values is then used to select which behavior should be given control of gaze.

Figure 20.7 gives an example of seven consecutive steps of the sidewalk navigation task, the associated eye

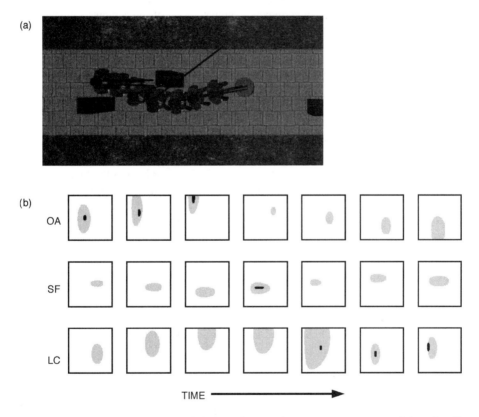

FIGURE 20.7 (a) An overhead view of the virtual agent during seven time steps of the sidewalk navigation task. The blue cubes are obstacles, and the purple cylinder is litter. The rays projecting from the agent represent eye movements; red corresponds to obstacle avoidance, blue corresponds to sidewalk following, and green corresponds to litter collection. (b) Corresponding state estimates. The top row shows the agent's estimates of the obstacle location. The axes here are the same as those presented in Figure 20.6. The beige regions correspond to the 90% confidence bounds before any perception has taken place. The red regions show the 90% confidence bounds after an eye movement has been made. The second and third rows show the corresponding information for sidewalk following and litter collection. (See color insert.)

TABLE 20.2 The Organization of Human Visual Computation From the Perspective of the Microbehavior Model

Abstraction Level	Problem Being Addressed	Role of Vision
Behavior	Need to get state information. The current state needs to be updated to reflect the actions of the body.	Provide state estimation None
Arbitration	Active behaviors may have competing demands for body, legs, eyes; conflicts have to be resolved.	Move gaze to the location that will minimize risk
Context	Current set of behaviors B is inadequate for the task; have to find a new set.	Test for off-agenda exigencies

movements, and the corresponding state estimates. The eye movements are allocated to reduce the uncertainty where it has the greatest potential negative consequences for reward. For example, the agent fixates[3] the obstacle several times as he draws close to it, and shifts perception to the other two microbehaviors when the obstacle has been safely passed. Note that the regions corresponding to state estimates are not ellipsoidal because they are being projected from world space into the agents nonlinear state space.

One possible objection to this model of eye movements is that it ignores the contribution of extrafoveal vision. One might assume that the pertinent question is not which microbehavior should direct the eye, but which location in the visual field should be targeted to best meet the perceptual needs of the whole ensemble of active microbehaviors. There are a number of reasons that we choose to emphasize foveal vision. First, eye-tracking studies in natural tasks show little evidence of "compromise" fixations. That is, nearly all fixations are clearly directed to a particular item that is task relevant. Second, results in Roelfsema, Khayat, and Spekreijse (2003) suggest that simple visual operations such as local search and line tracing require a minimum of 100–150 ms to complete. This timescale roughly corresponds to the time required to make a fixation. This suggests that there is little to be gained by sharing fixations among multiple visual operations.

The Embodied Operating System Model

We think of the control structure as comparable to that used in an operating system as the basic functions are similar in both systems. The behaviors themselves, when they are running, each have distinct jobs to do. Each one interrogates the sensorium with the objective of computing the current state of the process. Once the state of each process is computed then the action recom-

mended by that process is available. Such actions typically involve the use of the body. Thus an intermediate task is the mapping of those action recommendations onto the body's resources. Finally, the behavioral composition of the microbehavior set itself must be chosen. We contend that, similar to multiprocessing limitations on silicon computers, the brain has a multiprocessing constraint that allows only a few microbehaviors to be simultaneously active. This constraint, we believe, is related to that for working memory. Addressing the issues associated with this vantage point leads directly to an abstract computational hierarchy. The issues in modeling vision are different at each level of this hierarchy. Table 20.2 shows the basic elements of our hierarchy highlighting the different roles of vision at each level.

The behavior level of the hierarchy addresses the issues in running a microbehavior. These are each engaged in maintaining relevant state information and generating appropriate control signals. Microbehaviors are represented as state/action tables, so the main issue is that of computing state information needed to index the table. The arbitration level addresses the issue of managing competing behaviors. Since the set of active microbehaviors must share perceptual and motor resources, there must be some mechanism to arbitrate their needs when they make conflicting demands. The context level of the hierarchy maintains an appropriate set of active behaviors from a much larger library of possible behaviors, given the agent's current goals and environmental conditions.

The central tenet of Walter's control architecture is that, although a large library of microbehaviors is available to address the goals of the agent, at any one time, only a small subset of those are actively engaged as shown in Figure 20.8. The composition of this set is evaluated at every simulation interval, which we take to be 300 ms commensurate with the eyes' average fixation time.

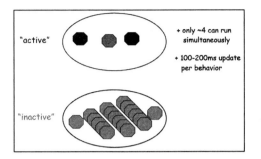

FIGURE 20.8 The model assumes that humans have an enormous library of behaviors that can be composed in small sets to meet behavioral demands. When an additional behavior is deemed necessary, it is activated by the "operating system." When a running behavior is no longer necessary, it is deactivated. (See color insert.)

The issues that arise for vision are very different at different levels of the hierarchy. Moving up the levels:

1. At the level of individual behaviors, vision provides its essential role of computing state information. The issue at this level is understanding how vision can be used to compute state information necessary for meeting behavioral goals. Almost invariably, the visual computation needed in a task context is vastly simpler than that required for general-purpose vision and, as a consequence, can be done very quickly.

2. At the arbitration level, the principal issue for vision is that the center of gaze is not easily shared and instead generally must be allocated sequentially to different locations. Eye-tracking research increasingly shows that all gaze allocations are purposeful and directed toward computing a specific result (Hayhoe, Bensinger, & Ballard, 1998; Johansson, Westling, Backstrom, & Flanagan, 1999; Land, Mennie, & Rusted, 2001). Our own model (Sprague & Ballard, 2003a) shows how gaze allocations may be selected to minimize the risk of losing reward in the set of running behaviors.

3. At the context level, the focus is to maintain an appropriate set of microbehaviors to deal with internally generated goals. One of these goals is that the set of running behaviors be response to rapid environmental changes. Thus the issue for vision at this level is to understand the interplay between agenda-driven and environmentally driven visual-processing demands.

This hierarchy immediately presents us with a related set of questions: How do the microbehaviors get perceptual information? How is contention managed?

How are sets of microbehaviors selected? In subsequent sections, we use the hierarchical structure to address each of these in turn, emphasizing implications for vision.

Microbehavior Selection

The successful progress of Walter is based on having a running set of microbehaviors $B_i, i = 0, \ldots, N$ that are appropriate for the current environmental and task context. The view that visual processing is mediated by a small set of microbehaviors immediately raises two questions: (1) What is the exact nature of the context switching mechanism? (2) What should the limit on N be to realistically model the limitations of human visual processing? Answering the first question requires considering to what extent visual processing is driven in a top-down fashion by internal goals, versus being driven by bottom-up signals originating in the environment. Somewhat optimistically, some researchers have assumed that interrupts from dynamic scene cues can effortlessly and automatically attract the brain's "attentional system" to make the correct context switch (e.g., Itti & Koch, 2000). However, a strategy of predominantly bottom-up interrupts seems unlikely because what constitutes a relevant cue is highly dependent on the current situation. However, there is a strong argument for some bottom-up component: humans are clearly capable of responding appropriately to cues that are off the current agenda.

Our model of the switching mechanism is that it works as a state machine as shown in Figure 20.9. For planned tasks, certain microbehaviors keep track of the progress through the task and trigger new sets of behaviors at predefined junctures. Thus the microbehavior "Look for Crosswalk" triggers the state "Near Crosswalk," which contains three microbehaviors: "Follow Sidewalk," "Avoid Obstacles," and "Approach Crosswalk."

Figure 20.9 (bottom) shows when the different states were triggered on three separate trials. This model reflects our view that vision is predominantly a top-down process. The model is sufficient for handling simple planned tasks, but it does not provide a straightforward way of responding to off-plan contingencies. To be more realistic, the model requires some additions. First, microbehaviors should be designed to error-check their sensory input. In other words, if a microbehavior's inputs do not match expectations, it should be capable

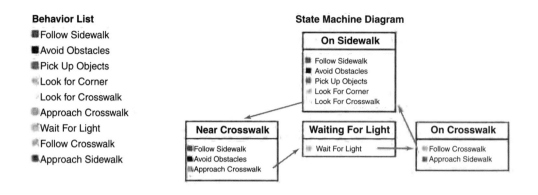

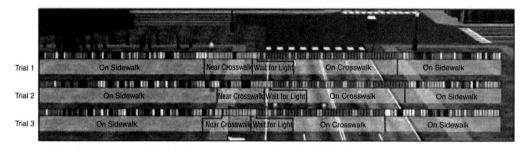

FIGURE 20.9 A list of microbehaviors used in Walter's overall navigation task. (Top right) The diagram for the programmable context switcher showing different states. These states are indicated in the bands underneath the colored bars below. (Bottom) Context switching behavior in the sidewalk navigation simulation for three separate instances of Walter's stroll. The different colored bars denote different microbehaviors that are in control of the gaze at any instant. (See color insert.)

of passing control to a higher-level procedure for resolution. Second, there should be a low latency mechanism for responding to certain unambiguously important signals such as rapid looming.

Regarding the question of the possible number of active microbehaviors, there are at least two reasons to suspect that the maximum number simultaneously running might be modest.

1. A ubiquitous observation is that spatial working memory has capacity limitations. The original capacity estimate by Miller was seven items plus or minus two (Miller, 1956), but current estimates favor the lower bound (Luck & Vogel, 1997). We hypothesize that, with familiar tasks, this limitation is bounded by the number of independently running microbehaviors, which we have termed *threads*. The identification of the referents of spatial working memory has always been problematic because the size of the referent can be arbitrary. This has led to the denotation of the familiar referent as a *chunk*, a jargon word that postpones

dealing with the issue of not being able to quantify the referents. The thread concept is clearer and more specific as it denotes exactly the state necessary to maintain a microbehavior.

2. Large numbers of active microbehaviors may not be possible given that they have to be implemented in a neural substrate. Cortical memory is organized into distinct areas that have a two-dimensional topography. Furthermore, spatial information is usually segregated from feature-based information so that the neurons representing the colors of two objects are typically segregated from the neurons representing their location. As a consequence, there is no simple way of simultaneously associating one object's color with its location together with another object's association of similar properties. (This difficulty is the so-called binding problem [von der Malsburg, 1999].) Some proposals for resolving the binding problem hypothesize that the number of active microbehaviors is limited to one, but this seems very unlikely. However, the demands of a

binding mechanism may limit the number of simultaneous bindings that can be active. Thus, it is possible that such a neural constraint may be the basis for the behavioral observation.

Although the number of active microbehaviors is limited, there is reason to believe that it is greater than one. Consider the task of walking on a crowded sidewalk. Two fast walkers approaching each other close at the rate of 6 m/s. Given that the main source of advanced warning for collisions is visual and that eye fixations typically need 0.3 s and that cortical processing typically needs 0.2–0.4 s, during the time needed to recognize an impending collision, the colliders have traveled about 3 m, or about one-and-a-half body lengths. In a crowded situation, this is insufficient advance warning for successful avoidance. What this means is that for successful evasions, the collision detection calculation has to be ongoing. But that in turn means that it has to share processing with the other tasks that an agent has to do. Remember that by sharing we mean that the microbehavior has to be simultaneously active over a considerable period, perhaps minutes. Several elegant experiments have shown that there can be severe interference when multiple tasks have to be done simultaneously, but these either restrict the input presentation time or the output response time (Pashler, 1998). The crucial issue is what happens to the internal state when it has to be maintained for an extended period.

Conclusions

Our focus has been to review issues and advantages associated with modeling embodied behaviors with our own modeling work on Walter as an illustrative example.

Embodied behaviors introduce a host of constraints that may not be reflected in related pattern-matching systems. First and foremost is the prodigious amount of computation done by the body itself; by "body," we mean the musculoskeletal system and its controlling spinal cord and brain stem circuitry. The extensive reflexes in this circuitry permit stable vision and effortless balance, locomotion and reaching behaviors can operate in the absence of cortical control. The result is that the forebrain structures only have to modulate this circuitry to achieve goals and this modulation takes vastly lower bandwidth than the underlying modulated circuits.

The invariances created by this circuitry have another boon in that they allow for simple expressions of expectations about the world. This means that whether the issue is a particular shape resolved by grasping or the identity of a particular object resolved by foveating, the tests that can settle the issue are simpler than they would have to be otherwise.

As noted, the computation done by the body means that the controlling programs in the forebrain also can be simpler. In our model, they are represented by state action tables where the index to the table is the result of simple tests, such as those just mentioned, and the action is a direction for the body's reflexive structures.

An important open issue in the representation of programs in the forebrain is the extent to which they are multiplexed. That is several programs can be simultaneously running. We argue that various evidence favors the multiplexing result. The main issue raised by multiplexing is the resolution of the different possible demands for the body by different, simultaneously active microbehaviors. Here again embodiment provides a ready answer. Table entries are referred to the body's actions. Thus the conflicts can be directly evaluated. This is in stark contrast to the hypothetical general case where the actions are not commensurate and conflicts might have to be detected by an inference engine, a very expensive proposition indeed.

Finally, the largest open issue associated with the microbehavior hypothesis is how the behaviors are indexed. The claim is that the right small set can be selected for each situation, but how is this done? This unsolved problem is one of indexing. Although it is an open problem, we have a great example of a solution from another domain in Google. By devoting a huge amount of off-line effort, Google is able to index Web pages so that they can be quickly searched. In the same way, the brain may have a way of indexing behaviors over a human's lifetime so that the relevant behaviors for any situation can be quickly accessed.

Notes

1. The term "argmax" means the value of the argument to a function that maximized that function.
2. Because the state is transiting through time, it has an internal dynamics that is informed by new state measurements. Kalman filters optimally weight the average of the current state estimate and the new measurement.
3. Although some visual routines would work in the periphery, we assume that ours *require* fixations.

References

Ballard, D., Hayhoe, M., & Pook, P. (1997). Deictic codes for the embodiment of cognition. *Behavioral and Brain Sciences, 20*, 723–767.

——, & Sprague, N. (2002). Attentional resource allocation in extended natural tasks [Abstract]. *Journal of Vision, 2*, 568a.

Brooks, R. A. (1986). A robust layered control system for a mobile robot. *IEEE Journal of Robotics and Automation, 2*(1), 14–23.

Bryson, J. J., & Stein, L. A. (2001). Modularity and design in reactive intelligence. In B. Nebel (Ed.), *Proceedings of the 17th International Joint Conference on Artificial Intelligence* (pp. 1115–1120). San Francisco: Morgan Kaufmann.

Clark, A. (1997). *Being there: Putting brain, body, and world together again*. Cambridge, MA: MIT Press.

——. (1999). An embodied model of cognitive science? *Trends in Cognitive Sciences, 9*(3) 345–351.

Faloutsos, P., van de Panne, M., & Terzopoulos, D. (2001). The virtual stuntman: Dynamic characters with a repertoire of motor skills. *Computers and Graphics, 25*, 933–953.

Firby, R. J., Kahn, R. E., Prokopowicz, P. N., & Swain, M. J. (1995). *An architecture for vision and action*. International Joint Conference on Artificial Intelligence, Montreal, Canada, pp. 72–79.

Hartley, R., & Pipitone, F. (1991). Experiments with the subsumption architecture. In *Proceedings of the International Conference on Robotics and Automation, 2*, 1652–1658.

Hayhoe, M. M., Bensinger, D., & Ballard, D. H. (1998). Task constraints in visual working memory. *Vision Research, 38*, 125–137.

Humphrys, M. (1996). Action selection methods using reinforcement learning. In P. Maes et al. (Eds.), *Proceedings of the Fourth International Conference on Simulation of Adaptive Behavior*. Cambridge, MA: MIT Press.

Itti, L., & Koch, C. (2000). A saliency-based search mechanism for overt and covert shifts of visual attention. *Vision Research, 40*, 1489–1506.

Johansson, R., Westling, G., Backstrom, A., & Flanagan, J. R. (1999). Eye-hand coordination in object manipulation. *Perception, 28*, 1311–1328.

Karlsson, J. (1997). *Learning to solve multiple goals*. Unpublished doctoral dissertation, University of Rochester.

Land, M. F., Mennie, N., & Rusted, J. (2001). Eye-hand coordination in object manipulation. *Journal of Neuroscience, 21*, 6917–6932.

Luck, S. J., & Vogel, E. K. (1997). The capacity of visual working memory for features and conjunctions. *Nature, 390*, 279–281.

Marr, D. (1982). *Vision*. New York: W. H. Freeman.

Merleau-Ponty, M. (1962). *Phenomenology of perception*. London: Routledge & Kegan Paul.

Miller, G. (1956). The magic number seven plus or minus two: Some limits on your capacity for processing information. *Psychological Review, 63*, 81–96.

Newell, A. (1990). *Unified theories of cognition*. Cambridge, MA: Harvard University Press.

Nilsson, N. (1984). *Shakey the robot* (Tech Rep. No. 223). SRI International.

Nöe, A. (2005). *Action in perception*. Cambridge, MA: MIT Press.

O'Regan, J. K., & Nöe, A. (2001). A sensorimotor approach to vision and visual consciousness. *Behavioral and Brain Sciences, 24*, 939–973.

Pashler, H. (1998). *The psychology of attention*. Cambridge, MA: MIT Press.

Pelz, J., Hayhoe, M., & Loeber, R. (2001). The coordination of eye, head, and hand movements in a natural task. *Experimental Brain Research, 139*, 166–177.

Roelfsema, P. R., Khayat, P. S., & Spekreijse, H. (2003). Subtask sequencing in the primary visual cortex. *Proceedings of the National Academy of Sciences USA, 100*, 5467–5472.

Roelfsema, P., Lamme, V., & Spekreijse, H. (2000). The implementation of visual routines. *Vision Research, 40*, 1385–1411.

Roy, D., & Pentland, A. (2002). Learning words from sights and sounds: A computational model. *Behavioral and Brain Sciences, 26*, 113–146.

Sprague, N., & Ballard, D. (2003a). Eye movements for reward maximization. In *Advances in Neural Information Processing Systems* (Vol. 16), MIT Press Image Sci Vis, 20, 1419–1433.

——, & Ballard, D. (2003b). *Multiple-goal reinforcement learning with modular sarsa(0)* (Tech. Rep. No. 798). Rochester, NY: University of Rochester Computer Science Department.

Suri, R. E., & Schultz, W. (2001). Temporal difference model reproduces anticipatory neural activity. *Neural Computation, 13*, 841–862.

Terzopoulos, D. (1999). Artificial life for computer graphics. *Communications of the ACM, 42*, 32–42.

Trommershäuser, J., Maloney, L. T., & Landy, M. S. (2003). Statistical decision theory and rapid, goal-directed movements. *Journal of the Optical Society A, 20*, 1419–1433.

Ullman, S. (1985). Visual routines. *Cognition, 18*, 97–159.

von der Malsburg, C. (1999). The what and why of binding: The modeler's perspective. *Neuron, 24*, 95–104.

Watkins, C. J. C. H., & Dayan, P. (1992). Q-learning. *Machine Learning Journal 8.*

21

Questions Without Words

A Comparison Between Decision Making Under Risk and Movement Planning Under Risk

Laurence T. Maloney, Julia Trommershäuser, & Michael S. Landy

If you want answers without words, then ask questions without words.
—Augustine of Hippo (translation, Wills, 2001, p. 139)

We describe speeded movement tasks that are formally equivalent to decision making under risk. In these tasks, subjects attempt to touch reward regions on a touch screen and avoid nearby penalty regions, much as a golfer aims to reach the green while avoiding nearby sand traps. The subject is required to complete the movement within a short time and, like the golfer, cannot completely control the outcome of the planned action. In previous experimental work, we compared human performance to normative (optimal) models of decision making and, in marked contrast to the grossly suboptimal performance of human subjects in decision-making experiments, subjects' performance in these experiments was typically indistinguishable from optimal. We conjecture that the key difference between our tasks and ordinary decision making under risk is the source of uncertainty, implicit or explicit. In the movement tasks, the probability of each possible outcome is implicit in the subject's own motor uncertainty. In classical decision making, probabilities of outcomes are chosen by the experimenter and explicitly communicated to the subject. We present an experimental study testing this conjecture in which we introduced explicit probabilities into the movement task. Subjects knew that some regions (coded by color) were stochastic. If they touched a stochastic reward region, they knew that the chances of receiving the reward were 50 percent and similarly for a stochastic penalty region. Consistent with our conjecture, subjects' optimal performance was disrupted when they were confronted with explicit uncertainty about rewards and penalties.

Many everyday decisions are wordless. We slow down in navigating a narrow doorway. We speed up while crossing the street with an eye to oncoming traffic. We swerve to avoid a colleague on the stairs. These decisions likely depend on several factors, notably expectations of pain or embarrassment, but we would be hard pressed to justify our "choices" in words or explain exactly what those factors are. The range of similar "wordless" decisions is endless and much of our day is taken up with them. Yet, as important as these everyday decisions may be, it is not even clear that we are aware that we have made a particular decision or are aware of the factors that led to it. The reader may even hesitate to classify such wordless decisions with the

kind of decision making we engage in while attempting to fatten a stock portfolio or play out a hand of poker.

Moreover, even if we accept that these wordless decisions are decisions, we cannot be sure that we are at all good at making them. If we did not slow down in approaching the doorway, is there any appreciable chance that we would hit either side of it instead of passing through? Or are we being reckless in not slowing even more? The information needed to make this sort of decision well is, on the one hand, the gains and losses that are possible and, on the other, accurate estimates of the possible discrepancy between how we intend to move and how we actually do move. It will turn out to be the latter sort of information that distinguishes ordinary decision making from the wordless decision making considered here.

In this chapter, we discuss results from experiments on human movement planning in risky environments in which explicit monetary rewards are assigned to the possible outcomes of a movement. We will show that these conditions create movement tasks that are formally equivalent to decision making under risk—if subjects can anticipate the stochastic uncertainty inherent in their movements. To anticipate our conclusion, we find that they can do so and that their performance in what we term *movement planning under risk* is remarkably good. These tasks form a promising alternative domain in which to study decision making, and they are distinguished by the fact that the uncertainties surrounding possible outcomes are intrinsic to the motor system and, so far as we can judge, difficult to articulate: These tasks are questions without words where stochastic information is available but not explicitly so.

We first review basic results on decision making under risk and then present experimental results concerning movement planning under risk. We end with a description of an experiment in which we ask subjects to carry out a movement task in an environment where the uncertainties surrounding possible outcomes are a combination of implicit motor uncertainty and explicit uncertainties imposed by the experimenter. As we will see, subjects who accurately compensate for their own implicit motor uncertainty fail to compensate correctly for added explicit uncertainty.

Decision Making Under Risk: A Choice Among Lotteries

Imagine that, when you turn this page, you will find a $1,000 bill waiting for you. It's yours to keep and do with as you like. What alternatives spring to mind? A banquet? Part payment on a Hawaiian vacation? Opera tickets? Each possible outcome of the choice you are about to make is appealing, to some extent, but sadly they are mutually exclusive. You have only one $1,000 bill, and you can only spend it once. Decision making is difficult in part because, by making one choice, we exclude other, desirable possibilities.

There is a further complication, though, that makes decisions even harder. You can choose how to dispose of your windfall, but you can't be completely sure of the outcome. You can plan to have a superb meal in a famous restaurant—and fall ill. You may schedule a vacation—and spend a week in the rain. Not every opera performance is brilliant. What you choose in making a decision is rarely a certain outcome but typically a probability distribution across possible outcomes. If the possible outcomes associated with a particular choice are denoted $O_1, \ldots, O_n$, the effect of any decision is to assign a probability p_i to each possible outcome O_i. The result is called a *lottery* and denoted $(p_1, O_1; p_2, O_2; \ldots; p_n, O_n)$, where

$$\sum_{i=1}^{n} p_i = 1.$$

Decision making is, stripped to its essentials, a choice among lotteries. In choosing a plan of action, the decision maker, in effect, selects a particular lottery. In the previous example, we did not assume that the decision maker knows what the probabilities associated with any plan of action are. Decisions without knowledge of the probabilities associated with each outcome are referred to as *decision making under uncertainty*. When decision makers have access to the probabilities induced by each possible plan of action, they are engaged in *decision making under risk*. Our focus here is on the latter sort of decision. In Table 21.1, we enumerate four possible plans of action that assign probabilities to each of four possible monetary outcomes. If the decision maker selects Lottery 1 (L1), for example, there will be an 80% chance of winning $100 and a 20% chance of losing $100. In contrast, L3 guarantees a gain of $50. The key problem for the decision maker is to select among lotteries such as those presented in Table 21.1.

For the decision maker who prefers more money to less, certain of these lotteries dominate others. L3 evidently dominates L4. Comparison of the probabilities

TABLE 21.1 Four Lotteries

	Possible Outcomes			
	$100	−$100	$50	$0
	Probabilities			
Lottery 1	0.8	0.2	0	0
Lottery 2	0.5	0.5	0	0
Lottery 3	0	0	1	0
Lottery 4	0	0	0	1

Four possible monetary outcomes are listed in the first row. The remaining rows specify a lottery with the given probabilities assigned to the corresponding outcomes above. If the decision maker selects Lottery 4 (L4), for example, $0 will be received with certainty. If the decision maker selects L3, $50 is received with certainty. With L2, it is equally likely that the decision maker will win or lose $100.

associated with each outcome in L1 and L2 shows that L1 guarantees a higher probability of gain and a lower probability of loss in every case. L1 dominates L2. That is, the decision maker intent on winning money should never select L2 if L1 is available or L4 if L3 is available. But the choice between L1 and L3 is not obvious: with L1, there is an evident tension between the high probability of winning $100 with lottery L1 and the 20% chance of losing $100. With L3, in contrast, the maximum possible gain is $50, but it is also the minimum: L3 offers $50 for sure. Which will we pick, given the choice? Which should we pick?

Decision Making Under Risk: Normative Theories

The classical decision-making literature distinguishes between descriptive and normative theories of decision making (Tversky & Kahneman, 1988). A descriptive theory of decision making attempts to predict what choice a decision maker would make if confronted with any collection of lotteries such as Table 21.1. Currently no descriptive theory is widely accepted (Birnbaum, 2004). We will briefly describe previous attempts to develop descriptive theories in the next section.

A normative theory is a rule that orders any set of lotteries from best to worst. We saw that we could order some of the alternatives in Table 21.1 by dominance (L1 > L2, L3 > L4). Any normative rule allows us to complete this ordering. The lottery highest in the resultant ordering is then the "best" lottery according to the normative rule. The two oldest normative rules are maximum expected value (MEV) and maximum expected utility (MEU).

If the outcomes are numerical (e.g., money), then the expected value of a lottery $L = (p_1, O_1; p_2, O_2; \ldots; p_n, O_n)$ is the sum of the values weighted by the corresponding probabilities.

$$EV(L) = \sum_{i=1}^{n} p_i O_i. \qquad (1)$$

The lottery selected by the MEV rule is the lottery with the highest expected value (Arnauld & Nichole, 1662/1992). In Table 21.1, for example, the MEV lottery is L1 with expected value $60.

Some decision makers may not agree with the recommendation of the MEV rule for Table 21.1 and instead prefer the sure win of L3 (with EV $50) to L1 with EV $60. If we were to multiply all of the outcomes by 100 so that

$$LI' = (0.8, \$10,000; 0.2, -\$10,000) \qquad (2)$$

and

$$L3' = (1, \$5,000), \qquad (3)$$

then almost all will choose L3' with EV $5,000 rather than L1' with EV $6,000. Decision makers are often risk averse in this way, especially when confronted with single decisions involving large values.

Daniel Bernoulli (1738/1954) proposed an alternative normative rule based on expected utility, intended to justify risk aversion. If any monetary outcome O_i is assigned a numerical *utility* denoted $U(O_i)$, then we can assign an expected utility (Bernoulli, 1738/1954) to each lottery

$$EU(L) = \sum_{i=1}^{n} p_i U(O_i + W), \qquad (4)$$

where W is the total initial wealth of the decision maker. The decision maker who seeks to maximize expected utility chooses the action whose corresponding lottery has the MEU. When outcomes are numeric and the utility function is a linear transformation with positive slope, MEU includes MEV as a special case.

Bernoulli proposed MEU as both a normative and a descriptive theory, intended to explain risk aversion. If the utility function $U(O)$ is concave (the utility of each successive dollar is less than the preceding dollar), then risk aversion is a consequence of MEU. In addition, the concept of utility has the great advantage that we can potentially describe decision making among outcomes that are nonnumerical by assigning them numerical utilities.

Decision Making Under Risk: Descriptive Theories

Research in human decision making under risk during the past forty years is a catalog of the many, patterned failures of normative theories, notably MEU, to explain the decisions humans actually make (Bell, Raiffa, & Tversky, 1988; Kahneman, Slovic, & Tversky, 1982; Kahneman & Tversky, 2000). These failures include a tendency to frame outcomes in terms of losses and gains with an aversion to losses (Kahneman & Tversky, 1979) and to exaggerate small probabilities (Allais, 1953; Attneave, 1953; Lichtenstein, Slovic, Fischhoff, Layman & Coombs, 1978; Tversky & Kahneman, 1992) as illustrated by the two examples in Figure 21.1.

Figure 21.1A shows data from Attneave (1953). In this study, subjects were asked to estimate the frequency of occurrence of letters in English text. It is evident that

A

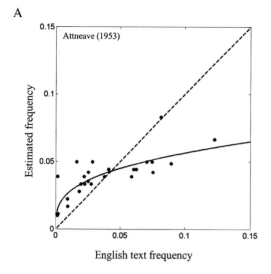

B

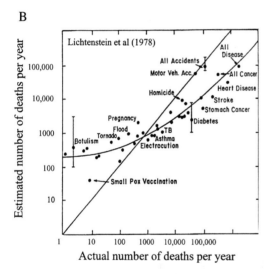

several orders of magnitude and the same pattern of exaggeration of small probabilities emerges. Estimates of the smaller probabilities are occasionally off by factors of 100 or more.

Studies that examine subjects' use of probabilities in decision making under risk draw similar conclusions (Gonzalez & Wu, 1999; Tversky & Kahneman, 1992; Wu & Gonzalez, 1998, 1999): subjects' use of probability or frequency information is markedly distorted and leads to suboptimal decisions, a phenomenon we will return to below.

There are other well-documented deviations from MEU predictions, and the degree and pattern of deviations depends on many factors. How these factors interact and affect decision making is controversial. What is not in dispute is that it takes very little to lead a human decision maker to abandon an MEU rule in decision making tasks and that, in the terminology of Kahneman and Tversky, human decision makers are given to "cognitive illusions."

**Movement Planning Under Risk:
A Different Kind of Decision**

The typical tasks found in the literature on decision making under risk are paper-and-pencil choices with no time limit on responding. Full information about outcomes and probabilities is explicitly specified, and there are usually only two or three possible choices (lotteries). In these tasks, both probabilities and values are selected by the experimenter and are communicated to the subject through numeric representations or by simple graphical devices. It is rare that any justification is given for why a particular probability should be attached to a particular outcome. These sorts of decisions are very far from the everyday, wordless decisions discussed in the introduction.

Here, we introduce a movement planning task that is formally equivalent to decision making under risk, and we describe how subjects perform in these sorts of tasks. As will become evident, one major difference between these tasks and more traditional examples of decision making under risk is that the source of uncertainty that determines the probabilities in each lottery is the subject's own motor variability. No explicit specification of probability or frequency is ever given to the subject. Information about probability or frequency is implicit in the task itself.

Trommershäuser, Maloney, and Landy (2003a, 2003b) asked subjects to make a rapid pointing move-

FIGURE 21.1 Estimated frequencies versus true frequencies. (A) A plot of estimated frequency of occurrence of letters in English text versus actual frequency (redrawn from Attneave, 1953). Subjects overestimated the frequency of letters that occur rarely relative to the frequency of letters that occur frequently. (B) A plot of estimated frequencies of lethal events versus actual frequencies (redrawn from Lichtenstein et al., 1978). Subjects markedly overestimated the frequency of rare events relative to more frequently occuring events.

they overestimated the frequency of letters that occur rarely compared with those that occur more frequently. In Figure 21.1B, we replot data from Lichtenstein et al. (1978). These data are the estimated frequencies of lethal events plotted versus the true frequencies. The data span ment and touch a stimulus configuration on a touch screen with their right index finger. The touch screen was vertical, directly in front of the subject. A typical stimulus configuration from Trommershäuser et al. (2003a) is shown in Figure 21.2A. This stimulus configuration or its mirror image was presented at a random location within a specified target area on the screen. A trial started with a fixation cross. The subject was required to move the index finger of the right hand to the starting position (marked on the space bar of a keyboard). The trial began when the space bar was pressed. The subject was required to stay at this starting position until after the stimulus configuration appeared or the trial was aborted. Next, a blue frame was displayed delimiting the area within which the target could appear and preparing the subject to move. Five hundred milliseconds later the target and penalty circles were displayed. Subjects were required to touch the screen within 700 ms of the display of the circles or they would incur a "timeout" penalty of 700 points. If a subject hit within the green[1] target in time, 100 points were earned. If the subject accidentally hit within an overlapping red circle, points were lost. If the subject hit in the region common to the two circles, both the reward associated with the green and the penalty associated with the red were incurred. If the subject hit the screen within the time limit, but missed both circles, no points were awarded.

The penalty associated with the red circle varied with experimental condition. In one condition, the penalty associated with the red circle was zero (i.e., there were no consequences for hitting within the red circle). At the other extreme, the penalty for hitting within the red circle was 500 points, five times greater than the reward for hitting the green circle. Subjects were always aware of the current penalty associated with the red circle.

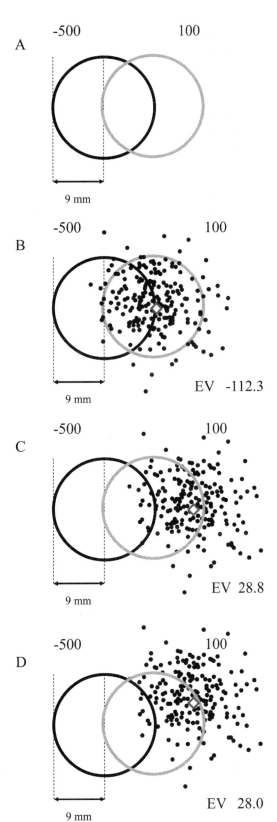

FIGURE 21.2 Possible movement strategies. (A) Stimulus configuration from Trommershäuser et al. (2003a). The reward circle is gray; the penalty circle is black. The reward and penalty associated with hitting within each region are shown. (B) The black dots represent simulated movement end points for a subject with motor variability $\sigma = 4.83$ mm who has adopted a movement strategy with mean end point marked by the diamond, at the center of the reward region. The expected value (see text) is shown and is negative. This maximizes the chances of hitting the reward region. (C) The mean end point is shifted horizontally by 7 mm and the expected value is now positive. (D) A shift upward from the axis of symmetry. The expected value is also positive but less than that of B.

They also knew that the summed points they earned over the course of the experiment would be converted into a proportional monetary bonus. At the end of every trial, the subject saw a summary of that trial (whether the subject had timed out, hit within the green and/or red, and how much was won or lost on that trial). The cumulative total of how much the subject had won or lost so far in the experiment was also displayed.

If subjects could perfectly control their movements, they could simply touch the portion of the green circle that did not overlap the red whenever touching the red region incurred any penalty. However, because of the time limit, any planned movement resulted in a movement endpoint on the screen with substantial scatter from trial to trial (Fitts, 1954; Fitts & Peterson, 1964; Meyer, Abrams, Kornblum, Wright, & Smith, 1988; Murata & Iwase, 2001; Plamondon & Alimi, 1997). In Figure 21.2B, we show a hypothetical distribution of endpoints. These points are distributed around a mean endpoint marked by a diamond which, in Figure 21.2B, is at the center of the green circle. In Figure 21.2C–21.2D, we illustrate endpoint distributions with different mean endpoints. We return to these illustrations below.

The uncertainty in the actual location of the endpoint on each trial is the key problem confronting the subject in these tasks, and if we want to model the subject's behavior, we must have an accurate model of the subject's movement uncertainty. In all of the experiments reported by Trommershäuser et al. (2003a, 2003b), the distributions of endpoints were not discriminable from isotropic Gaussian and the distribution for each subject could therefore be characterized by a single number, σ, the standard deviation of the Gaussian in both the horizontal and vertical directions.

In Figure 21.3A, we plot 1,080 endpoints for one typical naïve subject together with quantile–quantile plots of the deviations in the horizontal and vertical directions. The quantile–quantile plots in Figure 21.3B–21.3C compare the distributions to a Gaussian distribution. Estimated values of σ varied from subject to subject by as much as a factor of 2. Trommershäuser et al. (2003a, 2003b) verified that the value of σ did not vary appreciably across the conditions of each experiment and across the range of locations on the screen where stimuli could be presented. Thus, if the subject displaces the mean endpoint by a small amount, the distribution of endpoints is simply shifted by that amount.

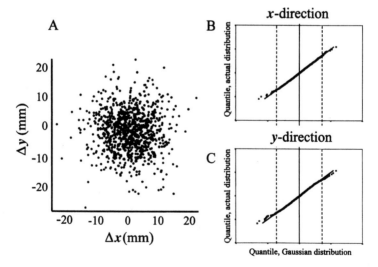

FIGURE 21.3 The distribution of endpoints. (a) Deviations Δx, Δy of 1,080 mean end points from the mean of the corresponding condition for a subject with motor variability $\sigma = 3.62$ mm. The distribution is close to isotropic and Gaussian. (b) Quantile–quantile plot (Gnanadesikan, 1997) of the deviations in the horizontal (x) direction compares the distribution of these deviations to the Gaussian distribution. (c) Quantile–quantile plot of the vertical (y) deviations. The linearity of these plots indicates that the distributions are close to Gaussian.

What should a subject do to maximize his or her winnings in the task just described? The subject can choose a particular movement strategy s, i.e. a plan of movement that is then executed. The relevant outcome of the planned movement in a given trial is the point where the subject touches the screen. Even if the subject executed the same plan over and over, the outcomes would not be the same as illustrated in Figure 21.2. By selecting a movement plan, the subject effectively selects an isotropic bivariate Gaussian density function $\phi_s(x, y; x_c, y_c, \sigma)$ of possible endpoints on the touch screen centered on the point (x_c, y_c) with standard deviation σ,

$$\phi_s(x,y;x_c,y_c,\sigma) = \frac{1}{2\pi\sigma^2}e^{-\frac{(x-x_c)^2+(y-y_c)^2}{2\sigma^2}}. \quad (5)$$

Subjects learn the task by performing more than 300 practice trials so that σ has stabilized before data collection starts. Therefore, the choice of movement strategy is the freedom to choose a mean endpoint (x_c, y_c). In the following, we identify a movement strategy s with its mean movement endpoint (x_c, y_c) and we denote the distribution more compactly as $\phi(x, y; s, \sigma)$.

Under these experimental conditions, the subject's choice among possible movement strategies is precisely equivalent to a choice among lotteries. To see this, first consider the possible outcomes of their movement when there is one penalty and one reward circle present on the screen as in Figure 20.2A, with values of -500 and $+100$ points, respectively. A movement that hits the touch screen within the time limit could land in one of four regions: penalty only (region $R_{R\bar{G}}$, value $V_{R\bar{G}} = -500$), the penalty/reward overlap (region R_{RG}, value $V_{RG} = -400$), reward only (region $R_{\bar{R}G}$, value $V_{\bar{R}G} = 100$), or outside of both circles (Region $R_{\bar{R}\bar{G}}$, value $V_{\bar{R}\bar{G}} = 0$). The probability of each of these outcomes depends on the choice of mean endpoint. For example,

$$p_{RG}(s) = \int_{R_{RG}} \phi(x,y;s,\sigma)dxdy, \quad (6)$$

is the portion of the probability mass of the distribution that falls within the region R_{RG} when the mean endpoint is (x_c, y_c).

The diamonds in Figure 21.2B–21.2D mark possible mean endpoints corresponding to different movement strategies s. For each movement strategy, we can compute the probability of each of the four outcomes, denoted $p_{RG}, \ldots, p_{\bar{R}\bar{G}}$, by Monte Carlo integration. Any

such choice of movement strategy s corresponds to the lottery:

$$L(s) = (p_{R\bar{G}}, V_{R\bar{G}}; p_{RG}, V_{RG}; p_{\bar{R}G}, V_{\bar{R}G}; p_{\bar{R}\bar{G}}, V_{\bar{R}\bar{G}}). \quad (7)$$

The expected values of movement endpoints are given in Figure 21.2B–21.2D. In Figure 21.2B, for example, the mean endpoint is at the center of the green circle. Pursuing this strategy leads to the highest rate of hitting within the reward circle. However, if the penalty associated with the penalty circle is 500 as shown, the expected value of this strategy is markedly negative. In contrast, the strategy illustrated in Figure 21.2C has positive expected gain, higher than that corresponding to Figure 21.2B and 21.2D. However, we cannot be sure that any of the movement strategies illustrated in Figure 21.2B–21.2D maximizes expected value without evaluating the expected value of all possible movement endpoints.

Many other lotteries are available to the subject, each corresponding to a particular movement strategy or aim point and each with an associated lottery and expected value. In choosing among these possible motor strategies and all others, the subject effectively selects among the possible sets of probabilities associated with each outcome and an expected value associated with the associated lottery. This is illustrated in Figure 21.4, which shows the expected value corresponding to each possible mean endpoint as a surface plot (upper row) and as a contour plot (lower row) for three different penalty values (0, 100, and 500). The reward value is always 100. The maximum expected value (MEV) point is marked in each of the contour plots by a diamond. When the penalty is 0, the MEV point is at the center of the reward circle. As the penalty increases, the MEV point is displaced farther and farther from the center of the reward region. For a mean endpoint closer to the center of the reward region, the probability of reward is higher, but the probability of hitting within the penalty region is also higher. This increase in expected penalty more than cancels the increase in reward, i.e. the resulting expected value that corresponds to this mean endpoint is suboptimal. Alternatively, if the MEV point were farther from the center of the reward region, then the probability of hitting the penalty region would decrease but so would the probability of hitting the reward region. The MEV point strikes exactly the correct balance between risk and reward.

The MEV point depends on the geometry, number, and locations of the penalty and reward regions; the magnitudes of rewards and penalties; and the subject's

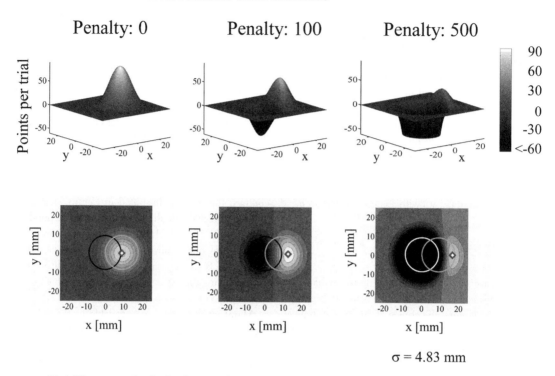

FIGURE 21.4 The expected value landscape. The expected value for each possible mean end point in the *xy*-plane is shown as surface plots (upper row) and corresponding contour plots (lower row) for three different penalty values (0, −100, −500). The maximum expected value (MEV) points in each contour plot are marked by diamonds. The MEV point is the center of the reward circle when the penalty is 0. It shifts away from the penalty circle as the penalty increases. Note that the penalty circle in the rightmost contour plot is coded as white rather than black so as to be visible.

own motor variability σ. These factors combine to create an infinite set of possible lotteries that the subject must choose among. Given the evident complexity of the decision making task implicit in Figure 21.2, it would be surprising if subjects chose strategies that maximized the expected value in these tasks. Moreover, the kinds of failures in decision making tasks that we discussed in the previous section should lead to particularly poor performance in these tasks.

If, for example, subjects interpret the penalty and reward regions in terms of loss and gain on each trial, then the loss aversion documented by Kahneman, Tversky, and others should lead them to move their mean endpoint further from the penalty region than the MEV point. Moreover, when the penalty is large relative to the reward, the probability of hitting the penalty region is very small for mean endpoints near the MEV point. If the subject overestimates the magnitude of this probability (as subjects did in the experiments illustrated in Figure 21.1), the subject will also tend to move too far away from the penalty region to be optimal.

In summary, we have very little reason to expect that subjects will approach optimal performance in movement planning tasks of the kind just described. They are very complex in comparison to ordinary decision tasks. The subject has little time to decide. Known patterns of failure in decision making should lead to poor performance. Given these expectations, the results of the experiments of Trommershäuser et al. (2003a, 2003b), presented next, are remarkable.

Movement Planning Under Risk: Initial Experimental Results

Method and Procedure

In this section we describe the results of Experiment 2 in Trommershäuser et al. (2003b). This experiment included four one-penalty configurations consisting of a reward circle and a single overlapping penalty circle (Figure 21.5A) and four two-penalty configurations

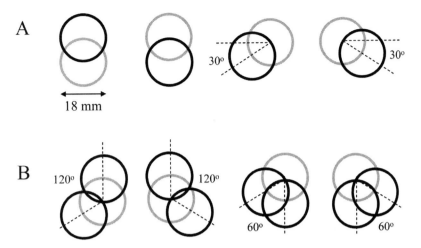

FIGURE 21.5 Stimulus configurations. The stimulus configurations employed in Experiment 2 of Trommershäuser et al. (2003b). The reward region (gray circle) was always assigned a reward of 100 points. The penalty assigned to the penalty regions (black circles) was either 0 (no penalty) or 500. The different spatial configurations were interleaved but the penalty values remained constant within a block of trials. (A) The one-penalty configurations. (B) The two-penalty configurations.

consisting of a reward circle and two overlapping penalty circles (Figure 21.5B). Trials were blocked with 32 trials per block (four repetitions of each of the eight configurations in Figure 21.5). The reward value was always 100 points and the penalty value was either 0 or 500 points (varied between blocks). If subjects hit a region shared by two or more circles, the subject incurred all of the rewards and penalties associated with all circles touched. In particular, if the subject touched within the region shared by two penalty circles, as many as 1000 points were lost, 500 for each penalty circle.

Subjects completed 24 trials in total for each of the 16 conditions of the experiment (two penalties crossed with eight stimulus configurations), a total of 384 experimental trials. Subjects were well practiced, having completed a similar experiment involving only configurations similar to those in Figure 21.2A. Before that first experiment, subjects carried out several training sessions. The purpose of the training trials was to allow subjects to learn to respond within the timeout limit of 700 ms and to give them practice in the motor task. A subject did not begin the main experiment until movement variability σ had stabilized and the timeout rate was acceptably low (see Trommershäuser et al., 2003b, for details).

Results

Figure 21.6A shows the MEV points for one of the one-penalty configurations and a particular subject with $\sigma = 2.99$ mm, the least variable subject in this experiment. These points of maximum expected value are computed by first estimating the expected-value landscape numerically (as in Figure 21.4) for each subject (i.e., each value of σ) and stimulus configuration and picking the mean endpoint that maximizes expected value. When the penalty is 0, the MEV point is the center of the reward circle, marked by the open circles in Figure 21.6. When the penalty is 500, it is displaced from the center of the reward circle, marked by the xs. Since all of the remaining one-penalty configurations are rotations of the first (Figure 21.5A), the MEV points for all four configurations are also rotations of the first. Subjects with higher values of σ had MEV points in the penalty 500 conditions that fell on the same radii but were displaced further from the center of the reward circle. In Figure 21.6B, we show the MEV point for one of the two-penalty configurations (for the same subject).

In Figure 21.7A, we summarize the results of Experiment 2 for the one-penalty conditions in Trommershäuser et al. (2003b), plotting each subject's mean

Predictions

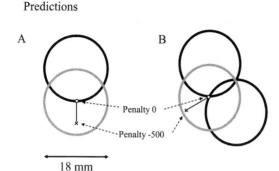

18 mm

FIGURE 21.6 Maximum expected value (MEV) predictions. (A) The predicted MEV point for a one-penalty configuration in Experiment 2 of Trommershäuser et al. (2003b), for subject S5 for whom $\sigma = 2.99$ mm. These predictions are obtained by computing the expected value landscape as in Figure 21.4 for each combination of values, spatial configuration, and subject's motor variability. The mean end point that maximizes expected value is the MEV point. (B) The predicted MEV point for a two-penalty configuration in Experiment 2 of Trommershäuser et al. (2003b), for subject S5 with motor variability $\sigma = 2.99$ mm.

endpoint in each condition of the experiment. The predicted MEV points when the penalty value was 500 are indicated by the *xs*. The mean endpoints of the subjects are shown as solid circles. The subjects' mean endpoints lie close to the predicted values and there is no patterned deviation across subjects. Figure 21.7B contains the corresponding results for the two-penalty configurations.

Trommershäuser et al. (2003b) also computed a measure of the efficiency of each subject's performance. For each subject and each condition, we computed the expected winnings of an ideal subject that used a movement strategy corresponding to the subject's MEV point and had the same movement variability as the actual subject. We computed the ratio between the subject's actual earnings and the expected earnings of the ideal and refer to this ratio (expressed as a percentage) as the subject's *efficiency*. Note, that, if the subject were 100% efficient, then we would expect the measured efficiencies of the subject to be both greater than and less than 100%, since the actual earnings on any one repetition of the experiment could be greater than or less than the expected earning (just as 100 tosses of a fair coin could result in more or less than 50 heads). These calculations of efficiency are specific to the subject. An ideal subject with a larger motor variability than a second ideal subject would be expected to earn

less. Efficiencies are noted in Figure 21.7. They are not significantly different from 100% except for Subject S1 in the two-penalty configurations. Thus, subjects efficiently solve movement planning problems that are equivalent to decision making under risk with an infinite number of four term lotteries (the one-penalty case) or an infinite number of seven- or eight-term lotteries (the two-penalty case). They solve these problems in under 700 ms with estimated efficiencies exceeding 90%.

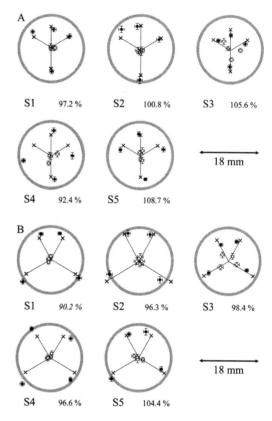

FIGURE 21.7 Results. (A) The predicted maximum expected value (MEV) points and actual mean end points for five subjects for Experiment 2 of Trommershäuser et al. (2003b). The one penalty configurations. (B) The two-penalty configurations. Open circles represent zero-penalty trials for which the MEV point is the center of the circle. Closed circles represent penalty −500 trials for which the MEV locations are indicated by the *xs*. The numbers to the right of each configuration indicate efficiency, computed as actual score divided by the MEV score (see Trommershäuser et al., 2003b, for details). Only one subject had an efficiency that was significantly different from 100% or optimal (italicized).

Trommershäuser et al. (2003b) considered the possibility that subjects only gradually learn the correct mean endpoint as a result of feedback. If this were the case, then we might expect to see trends in the choice of mean endpoint across the earliest trials in a condition. Subjects trained to hit within the center of the green circle during the preexperimental training phase might only gradually adjust their mean endpoints away from the penalty circle when the penalty is 500. Or, alternatively, a subject who is loss averse and given to exaggerating the (small) probability of hitting within the penalty circle might choose a mean endpoint that is initially too far from the penalty circle and only gradually adjust it toward the center of the reward region. In either case, we would expect to see a trend in mean aim point along the axis joining the center of the reward region and the centroid of the penalty region.

Trommershäuser et al. found no significant trends across subjects or conditions and concluded that subjects select movement strategies that maximize expected value almost immediately when confronted with a particular stimulus configuration. There is no evidence for learning in the data. Further, Trommershäuser, Gephstein, Maloney, Landy, and Banks (2005) have demonstrated that subjects can rapidly and successfully adapt to novel levels of movement variability imposed by the experimenter.

Explicit and Implicit Probabilities: An Experimental Comparison

Several factors may have contributed to the near-optimal performance of subjects in Trommershäuser et al. (2003a, 2003b). The movement planner makes a long series of choices and over the course of the experiment winnings increase. Decision makers faced with a series of decisions tend to move closer to MEV (Redelmeier & Tversky, 1992; Thaler, Tversky, Kahneman, & Schwartz, 1997; "the house money effect:" Thaler & Johnson, 1990). Further, the gain or loss associated with each trial is small. Studies of risky choice find that subjects are closer to maximizing expected value for small stakes (Camerer, 1992; Holt & Laury, 2002) and when subjects receive considerable feedback over the course of the experiment (Barron & Erev, 2003).

However, the most evident difference between movement planning under risk and ordinary decision making under risk is that, in the former, the subject is never given explicit information about probability distributions across outcomes associated with each possible movement plan. He or she must in effect "know" the probability of hitting each region of the stimulus configuration in order to plan movements well. The results just presented suggest that human movement planning has access to such implicit probability information. We note that we do not claim that the subject has conscious access, but only that this information seems to be available for planning movement.

In the study reported next, we introduce an element of explicit probability information into the movement task of Trommershäuser et al. (2003a, 2003b). In explicit probability conditions, the outcome of each trial not only depended on where the subject touched the computer screen but also on an element of chance unrelated to motor performance.

Method

As in Trommershäuser et al. (2003a), the stimulus configuration consisted of a reward circle and a penalty circle (Figure 21.8A). The color of the penalty region varied between trials and indicated the penalty value for that trial (white: 0, pink: −200, red: −400 points). The reward circle was always green (drawn here as medium gray) and the reward value was always 100. The target and penalty regions had radii of 8.4mm. The target region appeared in one of four possible positions, horizontally displaced from the penalty region by ±1 or ±2 multiples of the target radius ("near" and "far" in Figure 21.8A). The far configurations were included to keep subjects motivated through easily scored points, but were not included in the analysis. As in previous experiments, the stimulus configuration was displayed at a random location within a specified target region on each trial to prevent subjects from using preplanned strategies.

Explicit–Implicit Manipulation

The key manipulation concerned the certainty or lack of certainty of receiving a reward or penalty when a reward or penalty circle was touched. Either the reward or penalty circle could be stochastic. If a circle was stochastic, then the reward or penalty was obtained only 50% of the time when the circle was touched. The four probability conditions and the terms we use in describing them are shown in Figure 21.8B.[2] The *certainty condition* was similar to the experiment of Trommershäuser et al. (2003a).

The case Both 50% is of special interest. In this case, we scaled both the probability of getting a reward and the probability of incurring a penalty by 50%.

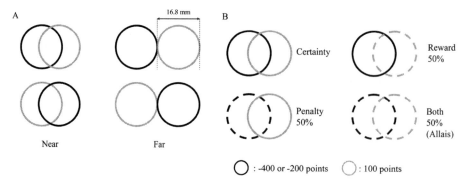

FIGURE 21.8 Experimental conditions. The configurations used in the final experiment. (A) The spatial conditions. (B) The probability conditions. In each condition, the subject received the reward or penalty associated with the region with either probability 1 (solid) or probability 0.5 (dashed). In the actual experiment, the penalty values were coded by color and the different probability conditions were constant across blocks and communicated to the subject at the start of each block.

The net result is to scale the expected value landscape by a factor of one-half. In particular, the location of the MEV point is unaffected. This condition is also of interest since not only should an ideal MEV mover choose the same mean aim point in both conditions, but also the performance of an MEU (utility-maximizing subject) should be invariant as well. To see this, we need only write down the conditions for the MEU subject to prefer lottery $L(s) = (p_1, O_1; \ldots; p_n, O_n)$, over lottery $L(s') = (p'_1, O_1; \ldots; p'_n, O_n)$,

$$\sum_{i=1}^{n} p_i U(O_i + W) > \sum_{i=1}^{n} p'_i U(O_i + W) \qquad (8)$$

and note that multiplying both sides of the inequality by a positive factor to scale all of the probabilities does not affect the inequality. The ideal MEU subject will still prefer s even with all probabilities scaled by a common factor.

There are other connections between the conditions of the experiment that are summarized in Figure 21.9. Consider, for example, the Penalty 50% condition with penalty 400. The MEV subject should choose the same mean aim point as in the certainty condition with penalty 200 since the net effect of the stochastic penalty is to reduce the expected value of the penalty by a factor of 2. We refer to the conditions in which the

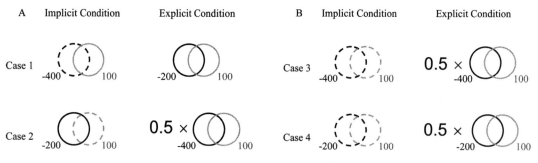

FIGURE 21.9 Equivalent conditions. (A) For the maximum expected value (MEV) movement planner, there are four pairs of probability conditions ("cases") where the expected value for any mean movement end point of the first configuration in the pair is either identical to that of the second or one-half of that in the second. The first condition in each case is explicit (either the reward or penalty or both is stochastic) and the second is implicit. Two cases are shown that are equivalent for an MEV movement planner but that might not be equivalent for an maximum expected utility (MEU) movement planner (with a non-linear utility function). Two cases are shown that would remain equivalent for an MEU movement planner as well as an MEV movement planner. For an MEV planner (A or B) or an MEU planner (B only), the predicted mean movement end points would be the same for the two configurations in each case.

probabilities are all implicit as *implicit conditions*, the conditions with explicit probabilities as *explicit conditions*. Figure 21.9 lists four equivalences between an implicit condition and an explicit condition that should hold for the MEV subject. We will refer to these pairs of equivalent conditions as "cases." We emphasize that the computations involved in carrying out the MEV strategy with added explicit probabilities are very simple. The subject need only replace -400 by -200, -200 by -100, or 100 by 50, depending on the pattern of explicit probabilities and payoff/penalty values in each session, to translate from the explicit condition of each case in Figure 21.9 to the equivalent implicit condition.

Procedure

Each session began with a test to ensure the subject knew the meaning of each color-coded penalty circle. The subject received feedback and had to correctly identify each penalty type twice. After a subsequent calibration procedure, there was a short block of 12 warm-up trials with zero penalty. The score was then reset to zero and data collection began. The time course of each trial was as described for previous experiments and the same feedback was given after every trial except as described below. The experiment comprised five experimental sessions of 372 trials each, 12 warm-up trials (not included in analyses), and 10 blocks of 36 trials. Blocks alternated between blocks containing configurations with penalty values of 0 and 200, and blocks with penalty values of 0 and 400. A penalty 200 block consisted of six repetitions of penalty 200 and three repetitions of penalty zero for each of the four spatial configurations, and the penalty 400 block was organized similarly. Each session corresponded to a stochastic condition in the order: certainty, Penalty 50%, Reward 50%, Both 50%, certainty. The certainty condition was repeated at the end (in the fifth session) to make sure that subjects' reaction to certain rewards and penalties remained stable across the course of the experiment.[3] Trials in which the subject left the start position less than 100 ms after stimulus display or hit the screen after the time limit were excluded from the analysis. Each subject contributed approximately 1,800 data points; that is, 60 repetitions per condition (with data collapsed across spatially symmetric configurations; 120 repetitions in the certainty conditions).

Results

Mean movement endpoints for each condition were compared with optimal movement endpoints as predicted by the optimal movement planning model of Trommershäuser et al. (2003a) based on each subject's estimated motor uncertainty σ. Subjects' *efficiency* was computed as previously described: the ratio between a subject's cumulative score in a condition and the corresponding expected optimal score predicted by the model. We used bootstrap methods to test whether each subject's measured efficiency differed from the optimal performance possible for that subject. In Figure 21.10, we plot actual points won for each subject and for each of two penalties in the certainty conditions. Five out of six subjects' scores were statistically indistinguishable from optimal (their efficiencies were indistinguishable from 100%) replicating previous findings (Trommershäuser et al., 2003a, 2003b). Only one subject (data indicated by the dashed circles)

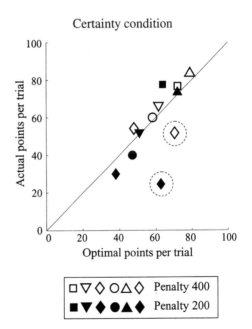

FIGURE 21.10 Results for the certainty conditions. A plot of points won versus the expected maximum expected value MEV points expected for each of the six subjects in the Certainty "near" condition. Different symbols correspond to individual subjects. Open symbols indicate the Penalty 400 condition, and filled symbols the Penalty 200 condition. Data points for which performance was significantly worse than predicted are indicated by the dashed circles.

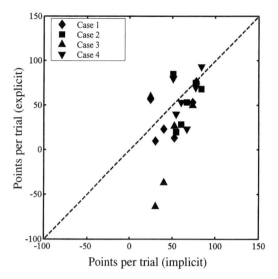

FIGURE 21.11 A plot of the outcomes of equivalent configurations. For each subject the mean number of points earned per trial for explicit probability conditions are plotted versus results for equivalent implicit probability conditions, for each of the cases in Figure 21.9. For the optimal maximum expected value observer, the expected values for the corresponding conditions (where one condition's results are doubled if necessary; see Figure 21.9) are identical. Thus, we would expect the plotted results to be distributed symmetrically around the 45-deg line. Instead, 19 out of 24 plotted points fall below the line. We reject the hypothesis that performance in explicit probability conditions equals that in equivalent implicit probability conditions (binomial test, $p = 0.003$).

differed significantly from optimal performance because the subject did not shift far enough away from the penalty region.

Next, we consider the pairs of conditions that should be equivalent for the MEV movement planner. We emphasize that the introduction of explicit probabilities imposed very little computational burden on subjects. One subject (JM) spontaneously reported after the experiment that she had followed the optimal strategy described above: a penalty of 400 that was incurred 50% of the time in the Penalty 50% condition should be treated exactly as the corresponding certainty condition with penalty 200, and so on. As we will see, however, her results are not consistent with the strategy she claimed to follow.

Figure 21.11 shows the winnings of each subject for each of the equivalent probability conditions shown in Figure 21.9 (doubling the winnings where one

condition is predicted to result in one-half the winnings of the other). In each case, the winnings from the implicit probability condition are plotted on the horizontal axis, the winnings from the equivalent explicit probability condition on the vertical axis. For the MEV movement planner, the plots should be randomly distributed above and below the 45-deg line. Instead, 19 out of 24 fall below, indicating that subjects tended to earn less in stochastic conditions than they did in equivalent certainty conditions. We can reject the hypothesis that subjects do equally well in the explicit as in the equivalent implicit probability conditions (binomial test, $p = .003$). The introduction of explicit probabilities tended to reduce subjects' winnings.

Subjects' performance dropped significantly below optimal when gains or losses were explicitly stochastic. This is also obvious in Fig. 21.12, which shows efficiencies of each subject in each condition. A subject following an MEV strategy would exhibit an expected efficiency of 100, marked by the horizontal line. For the certainty condition, one subject (AL) deviated significantly from MEV in both penalty conditions, as noted above. The estimated efficiencies for the other five subjects are distributed evenly around 100. In the three explicitly stochastic conditions, several other subjects exhibit large drops in efficiency, going as low as −400 (they are losing money at four times the rate they could have won money).

For five out of six subjects in the implicitly stochastic (certainty) conditions, earnings were indistinguishable from those expected from an optimal MEV strategy. Performance dropped significantly below optimal only when subjects were confronted with explicit uncertainty about whether they would incur a reward or penalty on a single trial. The introduction of explicit probabilities disrupted subjects' near-optimal performance even when the added cognitive demands were trivial.

Conclusion

We have introduced a class of movement tasks that are formally equivalent to decision making under risk and that model a very important class of everyday decisions "without words." In these decisions, there is a strong element of uncertainty in the outcome obtained as a consequence of any plan of action we choose, but this uncertainty originates in our own motor system. Our results suggest that, at least in some of these situations, we act as if we had good but "wordless" access to

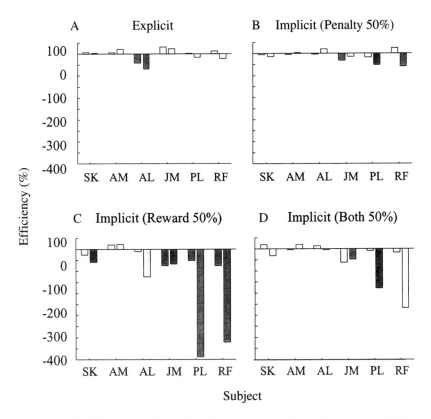

FIGURE 21.12 Efficiency is plotted for all subjects in all penalty and probability conditions. The expected efficiency for a maximum expected value movement planner is 100%. There are two bars for each subject, the left for the penalty 200 condition, the right for the penalty 400 condition. We tested each subject's estimated efficiency against optimal at the .05 level by a bootstrap method (Efron & Tibshirani, 1993). Gray bars indicate significantly suboptimal performance. (A) Certainty. In the certainty condition we could not reject the hypothesis of optimality for five out of six subjects (see also Figure 21.10). (B) Penalty 50%. (C) Reward 50%. (D). Both 50%. In the explicit probability conditions (B, C, D), all but one subject fell short of optimal in one or more conditions. In the five cases where the bar goes below 0, a subject could have won money on average, but instead lost money steadily.

estimates of the probabilities attached to the possible outcomes of any plan of action. Moreover, our use of these implicit probabilities is close to optimal.

Our results are consistent with the findings of Gigerenzer and Goldstein (1996) and Weber, Shafir, and Blais (2004; see also Hertwig, Barron, Weber, & Erev, 2004): decision makers have difficulty reasoning with explicitly stated probabilities. Weber et al. (2004) find that experience-based choices do not suffer from the same suboptimal decisions as pencil and paper tasks involving explicit probabilities. We note however that the absence of learning trends in Trommershäuser et al. (2003b) indicates that experience with a particular

stimulus configuration was not necessary. Subjects seemed to develop enough understanding of their own movement variability to allow them to adapt to novel configurations almost instantly.

The reader may still question whether the study of movement planning, as evinced by the tasks we have presented, has major implications for the classical study of decision making in economic tasks or whether it informs us about human cognitive abilities. Resolution of this issue will only follow considerable experimental work comparing human performance in both kinds of tasks. We will end, though, with two conjectures. First, we conjecture that the *cognitive illusions* that decision makers experience in paper-and-pencil tasks are not representative of performance in the very large number of movement decisions we encounter in any single day. Second, we suggest that the human capacity for decision making bears the same relation to the economic tasks of classical decision making as human language competence bears to solving the Sunday crossword puzzle.

Acknowledgments

This research was funded in part by National Institutes of Health Grant EY08266, Human Frontier Science Program grant RG0109/1999-B, and by the Emmy-Noether-Programme of the German Science Foundation (DFG).

Notes

1. The stimulus configurations in all of the experiments discussed here were coded by colors, typically red ("penalty") and green ("reward"). We replace green by gray and red by black in the illustrations here.

2. The dashed lines used to represent probability condition in Figure 21.8B and following figures are for the convenience of the reader. The subject always saw solid circles with penalty coded by color and probability condition constant across each block of the experiment, communicated to the subject at the beginning of each block.

3. For one subject, mean movement endpoints differed significantly between the two certainty Sessions 1 and 5, indicating that his responses did not remain stable across sessions. His data were excluded from the analysis. Movement endpoints of the remaining six subjects were collapsed for Sessions 1 and 5.

References

Allais, M. (1953). Le comportement de l'homme rationnel devant la risque: critique des postulats et axiomes de l'école Américaine. *Econometrica, 21*, 503–546.

Arnauld, A., & Nicole, P. (1662/1992). *La logique ou l'art de penser.* Paris: Gallimard.

Attneave, F. (1953). Psychological probability as a function of experienced frequency. *Journal of Experimental Psychology, 46*, 81–86.

Barron, G., & Erev, I. (2003). Small feedback-based decisions and their limited correspondence to description based decisions. *Journal of Behavioral Decision Making, 16*, 215–233.

Bell, D. E., Raiffa, H., & Tversky, A. (Eds.). (1988). *Decision making: Descriptive, normative and prescriptive interactions.* Cambridge: Cambridge University Press.

Bernoulli, D. (1738/1954). Exposition of a new theory on the measurement of risk (L. Sommer, Trans.). *Econometrica, 22*, 23–36.

Birnbaum, M. H. (2004). Human research and data collection via the Internet. *Annual Review of Psychology, 55*, 803–832.

Camerer, C. F. (1992). Recent tests of generalizations of expected utility theory. In W. Edwards (Ed.), *Utility: Theories, measurement, and applications* (pp. 207–251). Norwell, MA: Kluwer.

Efron, B., & Tibshirani, R. (1993). *An introduction to the bootstrap.* New York: Chapman-Hall.

Fitts, P. M. (1954). The information capacity of the human motor system in controlling the amplitude of movement. *Journal of Experimental Psychology, 47*, 381–391.

———, & Peterson, J. R. (1964). Information capacity of discrete motor responses. *Journal of Experimental Psychology, 67*, 103–112.

Gigerenzer, G., & Goldstein, D. G. (1996). Reasoning the fast and frugal way: Models of bounded rationality. *Psychological Review, 103*, 650–669.

Gnanadesikan, R. (1997). *Methods for statistical data analysis of multivariate observations* (2nd ed.). New York: Wiley.

Gonzalez, R., & Wu, G. (1999). On the shape of the probability weighting function. *Cognitive Psychology, 38*, 129–166.

Hertwig, R., Barron, G., Weber, E. U., & Erev, I. (2004). Decisions from experience and the effect of rare events in risky choice. *Psychological Science, 15*, 534–539.

Holt, C. A., & Laury, S. K. (2002). Risk aversion and incentive effects in lottery choices. *American Economic Review, 92*, 1644–1645.

Kahneman, D., Slovic, P., & Tversky, A. (Eds.). (1982). *Judgment under uncertainty: Heuristics and biases.* Cambridge: Cambridge University Press.

———, & Tversky, A. (1979). Prospect theory: An analysis of decision under risk. *Econometrica, 47*, 263–291.

———, & Tversky, A. (Eds.). (2000). *Choices, values & frames.* New York: Cambridge University Press.

Lichtenstein, S., Slovic, P., Fischhoff, B., Layman, M., & Coombs, B. (1978). Judged frequency of lethal events. *Journal of Experimental Psychology: Human Learning and Memory, 4*, 551–578.

Meyer, D. E., Abrams, R. A., Kornblum, S., Wright, C. E., & Smith, J. E. K. (1988). Optimality in human motor performance: Ideal control of rapid aimed movements. *Psychological Review, 95*, 340–370.

Murata, A., & Iwase, H. (2001). Extending Fitts' law to a three-dimensional pointing task. *Human Movement Science, 20*, 791–805.

Plamondon, R., & Alimi, A. M. (1997). Speed/accuracy trade-offs in target-directed movements. *Behavioral Brain Sciences, 20*, 279–349.

Redelmeier, D. A., & Tversky, A. (1992). On the framing of multiple prospects. *Psychological Science, 3*, 191–193.

Thaler, R. H., & Johnson, E. J. (1990). Gambling with the house money and trying to break even—the effects of prior outcomes on risky choice. *Management Science, 36*, 643–660.

———, Tversky, A., Kahneman, D., & Schwartz, A. (1997). The effect of myopia and loss aversion on risk taking: An experimental test. *Quarterly Journal of Economics, 112*, 647–661.

Trommershäuser, J., Gepshtein, S., Maloney, L. T., Landy, M. S., & Banks, M. S. (2005). Optimal compensation for changes in task-relevant movement variability. *Journal of Neuroscience, 25*, 7169–7178.

———, Maloney, L. T., & Landy, M. S. (2003a). Statistical decision theory and trade-offs in the control of motor response. *Spatial Vision, 16*, 255–275.

———, Maloney, L. T., & Landy, M. S. (2003b). Statistical decision theory and rapid, goal-directed move-ments. *Journal of the Optical Society A, 20*, 1419–1433.

Tversky, A., & Kahneman, D. (1988). *Risk and rationality: Can normative and descriptive analysis be reconciled?* New York: Russell Sage.

———, & Kahneman, D. (1992). Advances in prospect theory: cumulative representation of uncertainty. *Risk and Uncertainty, 5*, 297–323.

Weber, E. U., Shafir, S., & Blais, A.-R. (2004). Predicting risk-sensitivity in humans and lower animals: Risk as variance or coefficient of variation. *Psychological Review, 111*, 430–445.

Wills, G. (2001). *Saint Augustine's childhood: Confessiones Book One.* New York: Viking.

Wu, G., & Gonzalez, R. (1998). Common consequence effects in decision making under risk. *Journal of Risk and Uncertainty, 16*, 115–139.

———, & Gonzalez, R. (1999). Nonlinear decision weights in choice under uncertainty. *Management Science, 45*, 74–85.

22

Toward an Integrated, Comprehensive Theory of Visual Search

Anthony Hornof

With the goal of moving toward a comprehensive theory of visual search that could predict visual search performance for a wide range of applied visual tasks, this chapter describes how a comprehensive theory of human performance that integrates a Type 1 theory of central cognitive control with function-specific Type 2 theories can be used to generate new task-specific Type 3 theories in the domain of visual search. The chapter briefly describes the particular comprehensive theory that is used, the EPIC (Executive Process–Interactive Control) cognitive architecture and two specific sets of visual search models that were built using the EPIC architecture. The chapter discusses a number of components that may be essential for any comprehensive model of visual search. This includes the integration of existing theories that pertain to visual perception and oculomotor control; the identification and refinement of task strategies; and a means of incorporating or subsuming other "stand-alone" theories of visual search that are not constructed using a comprehensive theory. A major outstanding challenge is to create a comprehensive model of visual search by integrating across numerous individual models that were built and tuned to explain somewhat specific search tasks.

Human–computer interface design would benefit greatly by having a comprehensive computational theory of visual search that could predict how people will interact with visual interfaces before the interfaces are implemented. For example, the theory could be built into a tool that could predict how long it would take people to find the current postage rates on the United States Postal Service Web site, which is shown in Figure 22.1. Web designers could use the tool to help ensure that people can find information quickly and easily. The tool would take as input the Web page, a search task such as "find current postage rates," and a user profile; the tool would produce as output a prediction of what link the user would select, how long it would take them to find it, and scanpaths they would likely take during the search. Predictive design tools (John, Vera, Matessa, Freed, & Remington, 2002; Kieras, Wood, Abotel, & Hornof, 1995) have already been shown to be useful early in the development cycle, before it is practical or feasible to test designs with real people. However, these tools do not yet incorporate a comprehensive

theory of visual search—because such a theory does not yet exist.

Though finding the postage rates would likely involve other cognitive and memory functioning, such as processing the semantic content of the Web pages, much of the task performance—and the usability of the interface—will relate specifically to visual search. Visual search describes how users seeking a target will decide where to look next and how they will process the visual information that is encountered as they move their eyes. It will be useful to develop a comprehensive theory of visual search within a more general theory of human performance (such as a "unified theory of cognition"; Newell, 1990) to explore how various general processing components specifically assist in the activity of visual search.

A comprehensive theory of visual search will need to integrate across numerous theories from various fields of study, including perception, attention, oculomotor control, memory, visual search, and more. Much of the challenge will be to identify which theories need

FIGURE 22.1 A comprehensive theory of visual search could accurately predict where people are likely to click to find the current postage rates on the U.S. Postal Service Web site (www.usps.com). The theory could also predict other measures such as search time and scanpaths. (© United States Postal Service. Used with permission. All rights reserved.) (See color insert.)

to be integrated, which theories need to be developed, which theories are subsumed by others, and which theories are simply not required for predicting basic performance in real-world tasks. For example, a theory of eye movement control needs to be integrated because clearly the eyes move during visual search, but the notion of covert visual attention may not be important to integrate because covert visual attention does not appear to move independently of the eyes in real-world search tasks (Findlay & Gilchrist, 2003).

The development of a comprehensive, integrated theory of visual search will be aided by cognitive architectures. Cognitive architectures are software frameworks that are designed to facilitate the construction of computational models of human information processing and human performance. Cognitive architectures are candidate unified theories of cognition (Newell, 1990) that integrate both Type 1 theories—which pertain to central cognitive control—and Type 2 theories—which explain individual functional processes involved in human behavior, such as theories pertaining to visual perception, memory, or motor activity. Cognitive architectures can be used to explain categories of human activity such as visual search or driving (as in Salvucci, 2006); these explanations will be referred to as task-specific, or Type 3, theories. This chapter discusses the use of a cognitive architecture to develop Type 3 theories of visual search for specific task environments, theories

that will ultimately be integrated into a comprehensive (Type 3) theory of visual search for a wide range of task environments.

Numerous computational models of visual search have been proposed by previous researchers. Some of these models have been constructed with a cognitive architecture, such as (Fleetwood & Byrne, 2006; Hornof, 2004). Others have been constructed independently of cognitive architectures (Pomplun, Reingold, & Shen, 2003; Wolfe, 1994; also see chapters by these authors in this book). Especially for models of visual search that rely on a cognitive architecture, it is useful to consider the distinctions between Type 1, Type 2, and Type 3 theories because the distinction may help to clarify which aspects of the visual search model are inherited and which are generated or synthesized anew. The distinction may also assist in the evaluation of a model and help to identify which components are appropriate for reuse.

This chapter discusses two specific sets of visual search models—mixed density text search models and hierarchical search models. The models are Type 3 theories that have been built using a cognitive architecture. The architecture and models help to identify information processing components (Type 2 theories) that may be essential in a comprehensive, integrated theory of visual search. The chapter also discusses some of the outstanding challenges of integrating across various theories of visual search (both built without and with a cognitive architecture) along the way to developing a comprehensive, unified theory.

The EPIC Cognitive Architecture: An Integration of Type 1 and Type 2 Theories

The EPIC (Executive Process–Interactive Control) cognitive architecture (Kieras & Meyer, 1997) is a framework for building computational simulations of human information processing and human performance. Figure 22.2 shows an overview of the EPIC cognitive architecture. On the left is the simulated task environment. On the right are the various processors associated with the simulated human. Each of the processors is constructed based on function-specific theories pertaining to aspects of human information processing. The theories describe, for example, how visual features are more available near the center of the gaze, and how eye movements are executed to move

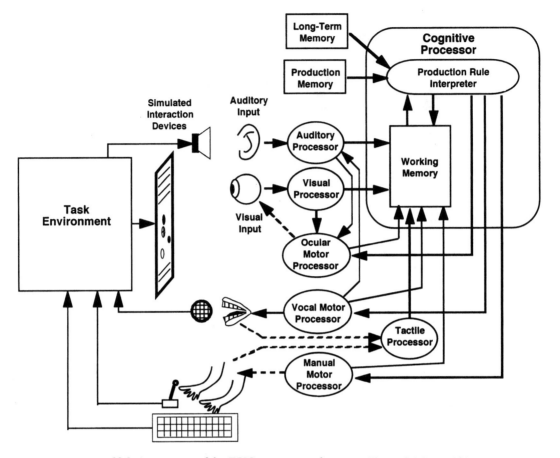

FIGURE 22.2 An overview of the EPIC cognitive architecture (Kieras & Meyer, 1997).

the gaze to new locations. These two function-specific theories, in fact, work together to simulate "active vision" (Findlay & Gilchrist, 2003). Active vision is the primary means in which people visually examine a scene—by moving their eyes.

The EPIC cognitive architecture synthesizes function-specific theories and, in the context of modeling visual search, is used to generate new task-specific theories. When building a model in EPIC, the analyst (the person building the model) must write the computer code associated with (a) the task environment based on the task description and (b) the production rules (production memory) based on task analysis of the human–machine activity to be simulated. Sometimes, the analyst must also add new visual processing features if the task involves visual stimuli with features that are new to the architecture. For example, EPIC's visual–perceptual processor will "see" objects based on characteristics such as an object's distance from the point of gaze but not based on the density of nearby objects.

To explain density effects, the analyst needs to add new processing features.

The architecture constrains the models that can be built based on the human information processing constraints that are built into the architecture (derived from the function-specific theories) and based on how the processes interact within the cognitive architecture. For example, EPIC has only one region of high resolution vision and it changes the location of that region with eye movements. Once the task environment and production rules are added to the architecture, the computer code is executed and the model is run. The model generates a prediction of how information will flow among the processors and how a person will perform the task. The predictions include task execution times as well as the timing and location of eye movements.

Much of the task of building a model in EPIC revolves around the development of the human strategies (in the form of production rules). Production rules are perhaps the simplest and most theoretically noncommittal way to encode procedural or strategic knowledge. They are not entirely theoretically vacuous, however. For example, whether the production rule processor can fire any number of rules at a time (as in EPIC) or is constrained to fire only one rule at a time (as in ACT-R) is a serious theoretical commitment (of the Type 1 theory) that directly affects how an analyst can and will encode the production rules for a task, especially when the task involves multiple subtasks that are executed concurrently.

The cognitive architecture and function-specific (Type 2) theories interact. The function-specific theories that are included in the architecture affect the kind of models that can be built with the cognitive architecture, and the assembled cognitive architecture affects the kind of new task-specific (Type 3) theories that are constructed. Models of visual search built with EPIC show how cognitive modeling can integrate function-specific theories such as those pertaining to visual acuity, oculomotor processing, and cognitive processing, all in the context of a unified cognitive architecture that can then be used to generate new task-specific theories of visual search.

Two specific models of visual search will be discussed to demonstrate this integration, synthesis, and generation of new models. The models are the mixed-density search models and the hierarchical search models. Both sets of models were constructed using the EPIC cognitive architecture. This cognitive-architecture-based approach to modeling visual search will then be compared and contrasted with other computational models of visual search (Type 3 theories) that were constructed without the benefit and burden of a comprehensive cognitive architecture.

Mixed-Density Text Search Models

The mixed-density text search models are task-specific (Type 3) theory built using the EPIC cognitive architecture based on established principles for applying the architecture and generating a model.

The mixed-density text search task is designed to explore the effects of searching a visual layout in which text appears in various densities, such as in the USPS.gov home page in Figure 22.1. Participants are presented with a visual layout that is divided into six visual groups of words. Each visual group is either sparse (five words in a nine-point font) or dense (10 words in

an 18-point font). There are three layout conditions: sparse (six sparse groups), dense (six dense groups), or mixed (three spare groups and three dense groups). Each layout type is presented in a different block of trials. The target position and words in the layout are randomly chosen for each trial. The participant is precued with the exact target. More details on the task are available in Halverson and Hornof (2004a, 2004c).

When modeling, there is sometimes an implicit assumption that the function-specific theories assembled into a cognitive architecture are adequate, or close enough, or correct enough, if the task-specific theory (Type 3 theory) accurately predicts the data. This is not ideal because the overall correct behavior could be generated either by correct subcomponents or by incorrect subcomponents that correctly compensate for one another's incorrectness. For example, the visual–perceptual subcomponent could (incorrectly) move an entire display of text into working memory all at the same time and could account for search times with a scan of the text in working memory. One of the challenges in developing these models is to find appropriate data points to evaluate intermediary assumptions made by the function-specific theories in the model. In models of visual search, for example, eye movement data can be used. Models that directly account for eye movements can potentially be evaluated based on human eye movement data. There are many challenges in using eye movement data to evaluate models, such as aggregating or summarizing the data in a useful manner and prioritizing which eye movement measures are the most important for the model to predict. Predicting the mean number of fixations and the mean fixation duration, for example, might be more important than predicting the scanpaths—the locations visited and the order they are visited—simply because the number of fixations and the fixation duration are much more straightforward to measure and compare. Given that each scanpath unfolds uniquely over time and space, there is no established means of calculating the "average" scanpath.

Figure 22.3 shows the data collected from 24 human participants who completed the mixed-density search task. Note that the first graph shows the standard measure of reaction time, but the other two measures pertain to eye movements—fixations per trial and mean fixation duration. Including these last two measures enables us to evaluate some aspects of the function-specific theories that are integrated within EPIC.

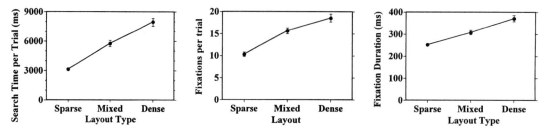

FIGURE 22.3 The observed human performance for the mixed-density text search task, including search time, fixations-per-trial, and fixation duration, all as a function of layout type.

The general trend in the data in Figure 22.3 is that all three measures increase as layouts become more dense. This is not surprising for search time and fixations per trial because the number of items in the layout increases from each condition to the next, so more time and more fixations are required to find the target. It is slightly surprising, though, to see an increase in fixation duration across the three conditions.

A series of models were constructed using the EPIC architecture. They are discussed in detail in Halverson and Hornof (2004a) and more briefly here. The models were developed iteratively. The initial iterations of the model did not explain the data and were clearly wrong, but they did help to identify aspects of a Type 3 theory of visual search, as well as subtle details of the Type 2 theories pertaining to visual perception that are incorporated into EPIC, that need to be further developed to explain the data. When building a cognitive model, especially when explaining existing data, it is common for the initial model to not explain all of the data. Though the long-term goal is a priori predictive modeling, intermediary post hoc explanatory models must be developed to refine the architecture and the cognitive strategies. The point is to learn from the process of developing and fine-tuning the models in a principled manner.

The models were developed primarily by (a) identifying visual perceptual parameters appropriate for the two densities of text and (b) developing a set of production rules to represent and capture aspects of the cognitive strategy used in the visual search task.

All of the strategies randomly choose the next gaze location. Keeping this aspect of the strategy constant allowed us to focus on other important strategic issues besides search order. A random search order was used in previous modeling efforts, such as Hornof and Kieras (1997) and Hornof (2004), both of which showed that a random search strategy with two or three items exam-

ined per fixation is a good first approximation for predicting the average search time per display. Random search also has benefits for a priori engineering models because it requires relatively few visual features to be encoded from the visual display, primarily just the absolute location of objects. Features such as relative position or visual groupings, which are more difficult to automatically extract, are not needed. A random search is not perfect, though. A more refined strategy is required to predict search time per position, scanpaths, and other trends such as how people move their eyes directly to a known target location.

Figure 22.4 shows a flowchart that represents a production rule strategy developed for the final mixed-density text search model presented in Halverson and Hornof (2004a). The strategy works as follows: The model looks at and then clicks on the precue. Then, in parallel, the model (a) selects the next gaze location and prepares an eye movement to that location and

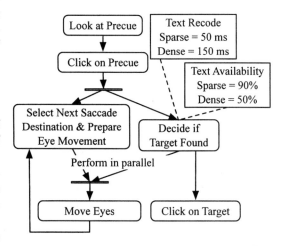

FIGURE 22.4 Flowchart summarizing the final mixed-density search strategy.

(b) decides if the target has been perceived with the current fixation. In Figure 22.4, the dashed lines connect visual–perceptual parameters to the rules that are affected by those parameters. The first parameter is text recoding time. All sparse text that lands in EPIC's fovea (1° in radius) is recoded in 50 ms, whereas dense text requires 150 ms. These are adjustments to EPIC's standard 100 ms text recoding time. The second parameter is the availability of text in the fovea. On any given fixation, almost all (90%) of sparse text that appears in the fovea is processed, where each dense word that appears in the fovea has only a 50% chance of being processed. The settings were inspired in part by Bertera and Rayner (2000) who found that the effective field of view (at the point of gaze) does not change with density, though based on the data we were observing, some sort of perceptual constraint created a need for more fixations. The flowchart shows how the time required to visually process sparse versus dense text directly affects the time required to decide whether the target has been found. The strategy assumes that people will not move their eyes to a new location until both (a) the next saccade is prepared and (b) it is decided that the target is not yet found. All of these assumptions in the model work together to explain why the three measures shown in Figure 22.3 increase as layout density increases.

The flowchart shows two refinements made to the Type 2 theories in the EPIC architecture that pertain to the perceivability of text. The parsimonious assumption of the EPIC architecture is that all text within 1° of visual angle of the point of gaze gets processed with 100 ms of recoding time. This theoretical assumption has worked very well for a number of EPIC models that did not specifically deal with text of varying densities. Now that the architecture is being used to develop Type 3 theories of visual search of mixed density layouts, the associated Type 2 theoretic assumptions must be

refined to account for the fact that sparse text can evidently be perceived more quickly than dense text. Furthermore, the perceptual processing parameters are modified to accommodate a theoretical assumption (asserted in this modeling effort) that only two or three items, and not everything within 1° of visual angle, is perceived with each fixation. These modifications to the Type 2 theory were deemed to be the most straightforward and parsimonious way to explain the data and are refinements to the Type 2 theory embedded within EPIC that we would anticipate using for future a priori modeling.

Figure 22.5 shows the predictions made by the final mixed-density search model, after the previously discussed adjustments were made to the function-specific theories associated with visual perception that are built into the EPIC architecture. The model correctly explains the data for all three layouts. The three data measures include fixation duration, search time, and number of fixations per trial. The model also provides a number of fine-tunings to the function-specific theories associated with the EPIC architecture. These enhancements will be used in future visual search models for layouts with mixed-density text.

Eye movement data helped to guide the construction of the mixed-density text search models. Working with eye movement data is not as straightforward as working with reaction time data because there are many different measures that can be made, but eye movement data do provide an excellent opportunity to probe the accuracy of the model at a smaller grain size, close to the level of detail of the component function-specific theories.

Hierarchical Search Models

The second set of visual search models examines how people search for text in a visual layout when

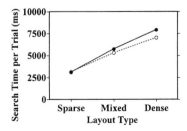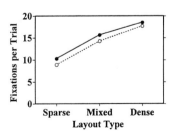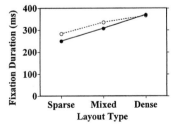

FIGURE 22.5 The performance that was observed in humans (solid lines) and predicted by the final mixed-density search model (dotted lines).

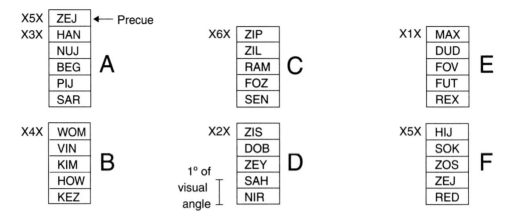

FIGURE 22.6 The visual layout used in the hierarchical search task, annotated with the letters A–F to identify the groups.

information is arranged in a visual hierarchy. That is, information is in groups, and the groups have useful descriptive headings. Many Web pages use visual hierarchies, such as on www.nytimes.com, where headlines are arranged under headings such as Business, National, International, and Sports. Visitors to the Web site do not need to search every individual headline to find an article on their favorite sports team, but can just scan the group headings until they find Sports. A comprehensive, predictive model of visual search will need to account for the effects of such a visual hierarchy.

Figure 22.6 shows the basic screen layout used in the hierarchical search task. Participants are shown a precue on the top left. The precue disappears and the layout appears. In some layouts, the groups have headings (such as X5X), and the precue includes the heading of the target group. Layouts vary in size, having one, two, four, or six groups. A two-group layout, for example, only includes groups A and B in Figure 22.6. Each unique combination of group headings (present or absent) and layout size is presented in a different block of trials. Sixteen participants completed the task. More details on the experiment are available in Hornof (2001).

Participants took significantly longer to find the target in layouts without the group headings. Clearly, when group headings were present, participants used them to speed their search. With headings, participants effectively had two search targets, first the precued group heading and then the target itself. Participants took longer to find a target in the same group location (as in Group B) in layouts that had more groups. This

effect can be seen in the space between the curves in Figure 22.7, and was disproportionately larger for layouts without group headings. As discussed in Hornof (2001, 2004), the effect suggests that people follow a less systematic search when presented with more items to examine and when presented with a less usefully structured layout.

Models were constructed in EPIC to explain the search time data. As suggested by the data, a fundamentally different search strategy was implemented for the layouts with group headings. A number of different search strategies were also implemented to explore how people might try to be systematic but inadvertently include randomness in their search, with increased randomness for layouts without group headings. The progression of model development is discussed in detail in Hornof (2004). Figure 22.7 shows the predictions of two best-fitting models.

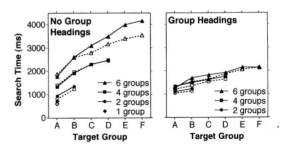

FIGURE 22.7 The observed (solid lines) and predicted (dashed lines) search times for the hierarchical search task, for layouts without (left frame) and with (right frame) useful group headings.

The models demonstrate that visual search strategies will be a key component of a comprehensive theory of visual search. A relatively small modification to the physical layout resulted in participants selecting a fundamentally different search strategy. Whereas search time typically increases with the number of distractors, in this case adding the headings reduced the search time because participants clearly selected a different search strategy. A key component of a comprehensive theory of visual search will also be the way in which a physical layout gets translated into a set of objects and features that are then input into the model. These translations, combined with an encoding of the simulated users' expectations, will affect the strategy that the model selects and executes. Toward this end, the hierarchical search models identify an appropriate set of perceptual features for a hierarchical search made by an experienced user.

The hierarchical search models help to establish appropriate motor processing parameters for generating accurate eye movement timings. The models were initially built without the benefit of eye tracking data, and later evaluated with eye tracking data. Figure 22.8 shows typical scanpaths predicted by the models and observed with people. The models accurately predicted many aspects of the observed eye movements, including the use of group labels, the number of fixations per

trial, fixation duration, anticipatory fixations, response to the layout onset, the effects of white space and text shape, and the number of items examined with each fixation. The models did not accurately predict the variety of scanpaths that were observed, target overshoot, the number of fixations per group, or the number of groups revisited per trial.

A very detailed comparison between the observed eye movement data and the predictions of the hierarchical search models is available in Hornof and Halverson (2003). The observed and predicted eye movements are compared on eleven different measures. Most but not all aspects of the eye movement data were predicted accurately. The overall predictions suggest that the cognitive architecture holds promise for making accurate a priori predictions of visual search behavior.

There are several outstanding challenges in the construction of predictive models of visual search, including that people will demonstrate great variance and sometimes clearly different strategies for tasks. It may be that different people use different strategies, the same people use different strategies at different times, or both. Regardless, it is clear that strategies will be an essential component of theories of visual search. The notion that people use strategies is not usually articulated separately as a theory, though it is by Newell

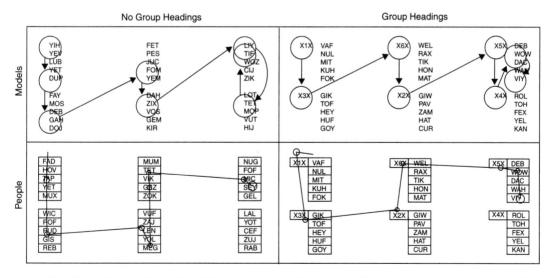

FIGURE 22.8 Typical scanpaths predicted by the hierarchical search models (top two frames) and observed in people (bottom two frames). In the models, the diameter of the circles represents the area in which text can be perceived near the point of fixation. In the observed data, the diameter represents fixation duration.

and Simon (1972), who state that the application of methods (task strategies) is central to human task execution and problem solving.

Current Challenge: Integrating Models From Different Tasks

Other visual search models are written in EPIC for other visual search tasks. One model simulates positionally constant search, in which people know the target location and thus move their eyes directly to the target (Hornof & Kieras, 1999). The model accounts for many details such as how there is more error in eye movements to more distant target locations (Abrams, Meyer, & Kornblum, 1989). Another model, which has not been built yet but appears to be relatively straightforward, conducts a color-based search by ignoring non-target-colored items to explain data such as that collected by Halverson and Hornof (2004b). There have also been a number of visual search models built using other cognitive architectures, such as Fleetwood and Byrne (2006).

Though these visual search models integrate across various theories of perception, attention, motor behavior, and so forth, to a large extent each model explains a single isolated phenomenon such as the effect of mixed-density text, a hierarchical layout, a positionally constant layout, or icon complexity. Real-world visual layouts (see Figure 22.1) clearly incorporate all these visual features at the same time. To develop a comprehensive model to predict search behavior on such layouts, existing models need to be integrated. Despite the progress we are making toward integrating function-specific theories into a cognitive architecture and creating new integrated task-specific theories, there is still much progress to be made to integrate the models we have thus far generated to explain a wider range of visual search tasks.

Integrating various visual search models requires the integration of task strategies. EPIC as a theory emphasizes production rule strategies as an important component of a model. Russo (1978) emphasizes the importance of (and yet the lack of research investigating) visual search strategies. In some task domains, such as a simulated air traffic control task, it will be possible to generate the production rules directly from the task instructions (Anderson et al., 2004). However, for many visual search tasks, the task instructions are simply as

"Find X," and the production rules will need to be selected based on the physical details of the layout and a user profile including their familiarity with the layout.

Previous models explore how different characteristics of a visual layout or task will affect how a person does the task. These models will be integrated into a comprehensive model of visual search for different levels of user experience and for different visual features in real-world tasks, such as finding postage rates on USPS.gov. Users who know the exact location would likely conduct a positionally constant "search" by just moving their eyes to the known location. Users who do not know the exact location might examine various physical features in some sort of order, such as first the moving things, then the pictures, then the colored items, then the big text, then the smaller text. Perhaps, if users quickly perceive a clear visual hierarchy to the page, they would look at the group headings to decide which group, if any, to examine in more detail.

It is not entirely clear how we will integrate across strategies from various models. Sometimes the nature of the task can help to determine which strategic component should take precedence. It is not clear whether and how such precedences will be able to be generated automatically by a model. Previous dual-task model integration (Kieras & Meyer, 1997) may provide some guidance, though there are now more than just two strategies to integrate, and task constraints and individual differences will weigh heavily. Visual search, even of complex layouts, is not generally considered a dual task. Features, expectations, and strategies do compete, but it is all the same task of visual search. Perhaps an "executive visual search process" is needed to coordinate among other strategies. The author and his students are actively pursuing the challenges of model integration.

Modeling Visual Search With and Without a Cognitive Architecture

The models presented above use an entire cognitive architecture (which integrate Type 1 and Type 2 theories) to build task-specific (Type 3) theories of visual search. Some models of visual search are built without using a cognitive architecture. This includes two models that are discussed in this volume—the guided search model (discussed by Wolfe in chapter 8) and the area activation model (discussed by Pomplun in chapter 9). Each model presents an extensive theory of feature-based visual search. The models incorporate a number of stages of processing. They create feature-based activation maps, and use the maps to select either items (in the guided search model) or screen regions (in the area activation model) for further processing. But the models are limited in terms of the low-level features that are currently supported, do not readily support the "top-down" influences of a search strategy, and are limited in their predictions of eye movements and search time.

The guided search and area activation models are task-specific models in that they incorporate multiple perceptual, decision, and perhaps motor processes, but they might be considered "stand-alone" models in that they are not built using a cognitive architecture. The models do not enjoy the benefits or suffer from the burdens of using a cognitive architecture.

There are many benefits to building a visual search model using a comprehensive cognitive architecture: The model gets a lot of components, such as perhaps an eye movement processor, "for free" from the architecture. This helps to ensure that the major processing components are accounted for. To the extent that these components have been built based on function-specific theories that are believed to be correct, the overall model is perhaps more likely to be correct. Ultimately, it will be necessary to integrate across multiple theories. Multiple models built using a common theoretical framework should be easier to integrate than multiple stand-alone models. Models built using a cognitive architecture should also help to expand and refine the theory to explain a broader range of phenomena. If a community of researchers is using an architecture, function-specific theories that are integrated into the architecture for one task will be useful in models built for other tasks. For example, adding a computational semantic system as another Type 2 process to an architecture would be useful to a wide range of models that interact with semantic content.

There are also many burdens to using a comprehensive cognitive architecture to build a model of visual search: The components provided for free do not always work as needed. For example, these architectures do not at present specifically provide perceptual feature activation maps that would lend themselves for use by the guided search or area activation models (though perhaps the architectures should). Sometimes, the components in the architecture are inadequately

specified, sometimes simply because the function-specific theories that are encoded are not complete. Nonetheless, an analyst using an architecture is somewhat obligated to keep it intact. That is, the analyst cannot arbitrarily change all sorts of modules and settings. This is both in keeping with the guiding principles of using such an architecture and with developing the architecture as a candidate unified theory of cognition. For example, this keeps the architecture intact so that it can still be used to accurately simulate and predict performance in other nonvisual tasks that are performed concurrently with the visual tasks. Though the architecture constrains the exploration of models, it generally does so in a helpful way, ensuring that all necessary processes are accounted for in the model.

It remains to be seen how task-specific theories of visual search such as the guided search and area activation models could be integrated a cognitive architecture. Ultimately, a unified theory of visual search should account for or incorporate all of these models, or at least account for all of the phenomena explained by the models.

Conclusion

This chapter discussed two specific sets of models of visual search, a framework for building models of visual search and a plan for applying that framework to develop a comprehensive theory of visual search that could account for and predict many aspects of human search behavior for a wide range of visual search tasks. A comprehensive model would be useful to interface designers because it would enable the designers to predict, early in the design process, how people will interact with a visual layout. Two essential components of a comprehensive theory of visual search will be (a) basic perceptual parameters that relate to how much information can be perceived with a single gaze fixation and (b) cognitive strategies that move the eyes based on the users' expectations and the physical structure of the visual layout. The chapter discussed some work that has been done to evaluate visual search models with eye tracking. This comparison is only possible with models that predict eye movements. The chapter suggests that a comprehensive model of visual search should account for eye movements because moving the eyes is a integral—and observable—activity in visual search.

The chapter discusses integrating models of cognitive systems to build a comprehensive visual search model. Theories of central cognitive control (Type 1 theories) have already been integrated with a limited set of function-specific (Type 2) theories to create cognitive architectures, which are candidate unified theories of cognition; that is, comprehensive frameworks for building a wide range of human performance models and task-specific (Type 3) theories of performance for tasks such as driving and visual search. Ongoing challenges for integrating cognitive systems include determining the necessary and sufficient function-specific theories to include in a cognitive architecture, reconciling differences among Type 1, 2, and 3 theories across different architectures and, most relevant to the current chapter, across existing models of visual search that were built both with and without the benefits and burdens of a cognitive architecture.

References

Abrams, R. A., Meyer, D. E., & Kornblum, S. (1989). Speed and accuracy of saccadic eye movements: Characteristics of impulse variability in the oculomotor system. *Journal of Experimental Psychology: Human Perception and Performance, 15*(3), 529–543.

Anderson, J. R., Bothell, D., Byrne, M. D., Douglass, S., Lebiere, C., & Qin, Y. (2004). An integrated theory of the mind. *Psychological Review, 111*(4), 1036–1060.

Bertera, J. H., & Rayner, K. (2000). Eye movements and the span of effective stimulus in visual search. *Perception & Psychophysics, 62*(3), 576–585.

Findlay, J. M., & Gilchrist, I. D. (2003). *Active vision: The psychology of looking and seeing.* New York: Oxford University Press.

Fleetwood, M. D., & Byrne, M. D. (2006). Modeling the visual search of displays: A revised ACT-R/PM model of icon search based on eye tracking data. *Human-Computer Interaction, 21*(2), 153–197.

Halverson, T., & Hornof, A. J. (2004a). Explaining eye movements in the visual search of varying density layouts. *Proceedings of the Sixth International Conference on Cognitive Modeling,* Carnegie Mellon University, University of Pittsburgh, Pittsburgh, PA, July 30–August 1 (pp. 124–129).

———, & Hornof, A. J. (2004b). Link colors guide a search. *Extended abstracts of ACM CHI 2004: Conference on Human Factors in Computing Systems* (pp. 1367–1370). New York: ACM.

———, & Hornof, A. J. (2004c). Local density guides visual search: Sparse groups are first and faster. *Proceedings of the 48th Annual Meeting of the*

Human Factors and Ergonomics Society, New Orleano, LA, September 20–24 (pp. 1860–1864).

Hornof, A. (2004). Cognitive strategies for the visual search of hierarchical computer displays. *Human-Computer Interaction, 19*(3), 183–223.

Hornof, A. J. (2001). Visual search and mouse pointing in labeled versus unlabeled two-dimensional visual hierarchies. *ACM Transactions on Computer-Human Interaction, 8*(3), 171–197.

———, & Halverson, T. (2003). Cognitive strategies and eye movements for searching hierarchical computer displays. *Proceedings of ACM CHI 2003: Conference on Human Factors in Computing Systems* (pp. 249–256). New York: ACM.

———, & Kieras, D. E. (1997). Cognitive modeling reveals menu search is both random and systematic. *Proceedings of ACM CHI 97: Conference on Human Factors in Computing Systems* (pp. 107–114). New York: ACM.

———, & Kieras, D. E. (1999). Cognitive modeling demonstrates how people use anticipated location knowledge of menu items. *Proceedings of ACM CHI 99: Conference on Human Factors in Computing Systems* (pp. 410–417). New York: ACM.

John, B., Vera, A., Matessa, M., Freed, M., & Remington, R. (2002). Automating CPM-GOMS. *Proceedings of ACM CHI 2002: Conference on Human Factors in Computing Systems* (pp. 137–154). New York: ACM.

Kieras, D. E., & Meyer, D. E. (1997). An overview of the EPIC architecture for cognition and performance with application to human-computer interaction. *Human-Computer Interaction, 12*(4), 391–438.

———, Wood, S. D., Abotel, K., & Hornof, A. (1995). GLEAN: A computer-based tool for rapid GOMS model usability evaluation of user interface designs. *Proceedings of the ACM Symposium on User Interface Software and Technology, UIST '95* (pp. 91–100). New York: ACM.

Newell, A. (1990). *Unified theories of cognition.* Cambridge, MA: Harvard University Press.

———, & Simon, H. A. (1972). *Human problem solving.* Englewood Cliffs, NJ: Prentice-Hall.

Pomplun, M., Reingold, E. M., & Shen, J. (2003). Area activation: A computational model of saccadic selectivity in visual search. *Cognitive Science, 24,* 299–312.

Russo, J. E. (1978). Adaptation of cognitive processes to the eye movement system. In J. W. Senders, D. F. Fisher, & R. A. Monty (Eds.), *Eye movements and the higher psychological functions* (pp. 89–109). Hillsdale, NJ: Erlbaum.

Salvucci, D. D. (2006). Modeling driver behavior in a cognitive architecture. *Human Factors, 48*(2), 362–380.

Wolfe, J. M. (1994). Guided Search 2.0: A revised model of visual search. *Psychonomic Bulletin & Review, 1*(2), 202–238.

PART VII

COORDINATING TASKS THROUGH GOALS AND INTENTIONS

Michael J. Schoelles

Humans are constantly performing tasks or activities, and the continuous sequence in which these tasks are executed is determined by goals and intentions. The chapters in this section are organized in the following manner. The first three chapters are focused on coordinating tasks with goals from a cognitive architecture perspective. The next chapter argues against a symbol manipulation approach by considering goals as control signals. The final chapter sketches a theoretical view of intentions as real-time control constructs.

The key issue to tackle is what is a goal and what is an intention and how are they related. Carlson (chapter 27) and Altmann (chapter 26) both point out that the concept of *goal* has many senses. It can be used in a motivational sense; it is something we want to achieve in the future, something we are aspiring to. It can also be an endpoint for a problem-solving experience, or as Altmann states "simply declarative control information that guides the organism's immediate interactions with the environment." It is this last sense that Carlson wants to call an *intention* to distinguish it from

the other senses of the term *goal*. Kieras (chapter 23), Salvucci (chapter 24), and Taatgen (chapter 25) explain goals from a cognitive architecture perspective, and all three refer to goal in the sense of Carlson's intention. Thus a unifying idea among these authors is that goals or intentions are declarative objects that are acted on in real time to direct behavior continuously.

Since goals are declarative, another issue relevant to goals is the nature of the executive processes that act on these declarative structures. Here Salvucci and Kieras are at odds. Salvucci proposes to make executive control a part of the architecture. Kieras contends that control of executive processes can be learned and influenced by changes in task priorities or instruction and therefore should not be hardwired into the architecture but can be implemented as procedural knowledge. Altmann has a third take on this issue in that he claims that control such as task switching can emerge from memory system constraints.

Another issue that arises in considering the process of controlling task sequences is how much knowledge is explicitly required to carry out this function. Carlson and Taagen unite on this issue and each propose a minimum control principle, which states that efficient control is achieved by minimizing the amount of explicit knowledge. Taatgen turns to the environment to achieve this minimization, whereas Carlson proposes deitic schema.

Read Kieras (chapter 23) to get a view of control from a cognitive architecture perspective. See if you agree with his claim that the operating system metaphor is more relevant now than ever. Kieras starts the chapter off with a proposal for how cognitive control might work by describing how cognitive control is implemented in the executive process interact and control (EPIC) architecture. The key idea in EPIC is that the architecture applies few constraints at the executive level, allowing the constraints imposed by the perceptual-motor system to emerge to circumscribe human behavior. He describes EPIC's perceptual and motor systems. He then separates EPIC's control mechanisms into basic control over these input and output systems versus higher-level or executive control.

Salvucci (chapter 24) uses a task that the majority of us are very familiar with—driving—to illustrate how and why he has integrated a general executive into the ACT-R architecure. He begins with his integration of a two-point steering model into his cognitive driving model. He then describes initial attempts at coordinating the driving tasks with other tasks such as dialing a cell phone. The initial solution was domain specific and therefore limited. This shortcoming led to the generalization of the goal-switching mechanisms that he has incorporated into ACT-R.

Cognitive model developers will be interested in Taagen's chapter 25. He claims that cognitive modelers are incorporating too much state information into their goal structures and this leads to more productions than are necessary. By following his minimal control principle, top-down control can be minimized to obtain optimal flexibility.

Altmann (chapter 26) takes on the metaphor that task switching is like "changing gears," which comes from the standard "reconfiguration" model of cognitive control. Altmann wants to provide less of a mechanistic approach and more of "an analogue account," which represents the variability one might expect to find in a biological system. He presents a signal detection model to explain five empirical findings using a simple task switching paradigm. He shows that it is possible to switch tasks without explicitly determining whether the current task cue is the same as or different from the prior task cue.

The chapter ends with a theoretical view of intentions. Carlson's (chapter 27) definition of intentions is compatible with the term *goal* as used in cognitive architectures. Carlson uses his concept of intention to link "computational views of cognitive control with analyses of the subjective experience of control." He uses the process of metacognitive monitoring to bring to light some properties of intentions.

The authors of these chapters are successful in showing that research on goals and intentions is on the critical path to understanding control of cognition. However, the issues contained in this chapter are by no means settled and this area will be to be a productive area of research in cognitive science.

23

Control of Cognition

David Kieras

Cognitive architectures are the current form of the traditional computer metaphor in which human cognition is analyzed in information-processing terms. This theoretical chapter uses a radically updated version of the computer metaphor to present how cognition is controlled. Input and output control are briefly presented, but the emphasis is on the control of sequential and parallel execution in cognition and the mechanisms of executive processes, which are cognitive processes that control the system as a whole, including other cognitive processes. The EPIC (executive process-interactive control) architecture provides specific examples of alternative mechanisms. Unlike EPIC, many extant architectures are based on concepts that are less powerful than current computer technology. Instead, the modern computer metaphor should be setting the theoretical *lower bound* on the sophistication of human cognitive mechanisms.

Where We Were

Inspired by the computer metaphor, a few decades ago psychologists were describing the mind with "box models" like those illustrated in Figure 23.1 (e.g., see Bower & Hilgard, 1981, chap. 13; Norman, 1970). Such models were supposed to illustrate the flow of information between processing stages or storage mechanisms. Sensory and motor organs were not in the picture, or if they were, they were not represented in any detail. Since most of the experiments addressed by the models involved memory paradigms, the storage systems were emphasized, but the contents of them were collections of items loosely specified in terms of *features*. The processes by which information moved between the storage systems was not elaborated in any detail, and the "executive" was somehow responsible for overseeing all of the processing and changing it to suit the needs of the task. To the extent there was rigor in these models, it was supplied by simple probability theory models for the individual boxes, such as the probability that an

item would be lost as a function of time, or information was successfully transferred from one box to another.

Where We Are

At present, many psychologists are describing the mind with models built within a cognitive architecture whose top-level structure is illustrated also with a box diagram such as Figure 23.2, which illustrates the EPIC architecture that will be described in somewhat more detail later in this chapter. But there are many critical differences between what is implied by these current diagrams and the earlier box models. The sensory–motor peripherals are in the diagram, reflecting how their properties are an explicit part of the model. Both memory systems and processing systems are shown, corresponding to the concept that a psychological theory must include both how knowledge is represented and how it is processed. There is no mysterious "executive" because all control of activity is done by the cognitive

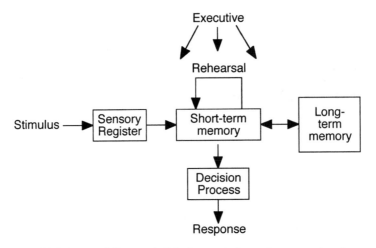

FIGURE 23.1 A typical "box model" in human information-processing theory.

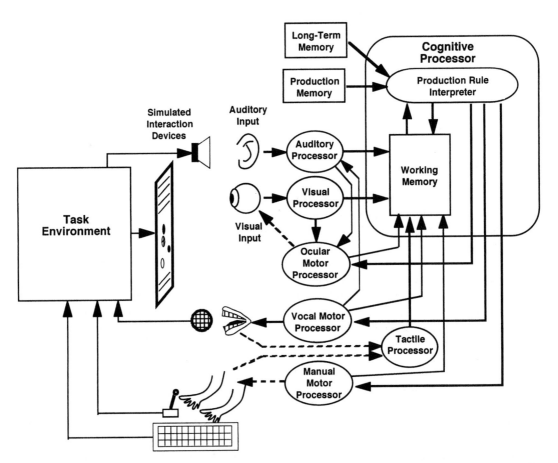

FIGURE 23.2 The EPIC architecture in simplified form. The simulated environment, or device, is on the left; the simulated human on the right.

processor, which in addition to performing cognition, also controls the system as a whole, including itself.

Figure 23.2's content is an example of a cognitive architecture in that the diagram represents an hypothetical collection of fixed mechanisms, analogous to a computer's general-purpose hardware, that are "programmed" by information in memory with the knowledge of how to perform a specific task, analogous to a particular piece of software that might be running on a computer. A complete model for performance in a particular task thus consists of the hypothetical task-independent fixed architecture together with the task-specific knowledge in memory for how to perform the task. This attempt to partition psychological models into task-independent and task-dependent portions is the hallmark of current cognitive theory.

Overview of the Chapter

The subject of this chapter is the cognitive processor and how it controls the architecture as a whole and itself as well. The basic control functions are to control input into the system, output from the system, and cognition in terms of basic sequential processing, parallel processing, interrupt handling, and executive processes that monitor and control cognitive activity. The presentation is strictly theoretical (or metatheoretical); no data are presented or compared to any models. The EPIC architecture will be used as the basic framework for discussion because it illustrates many of the issues well. Examples will be drawn from actual working EPIC models.

The discussion will be based on the computer metaphor, which historically underlies the development of information-processing psychology, but it is actually more useful than ever. That is, the computer system was an obvious metaphor for human information-processing activity and thus helped usher in a postbehavioristic "mentalism" that made it possible to discuss mental activity in a rigorous scientific fashion. However, the original computer metaphor was based on early computer technology circa 1960, and it was not updated as computers evolved. Current computer systems (see Tucker, 2004) are much more sophisticated and have been so for a long time: they perform simultaneous input/out and computation, parallel and concurrent processing of multiple tasks, and are controlled by a general-purpose "executive"—an operating system (OS). Thus the modern computer system is a valuable

metaphor for cognitive control processes in humans because the corresponding psychological theory of control and executive processing is poorly developed and vague; in contrast, these are very well developed topics in computer technology.

The obvious objection to the computer metaphor is that the brain is not a computer! The "hardware" of the brain is fundamentally different and is vastly more complex and sophisticated. However, the computer system is the only information processing system on the planet that has fully understood control and executive mechanisms. The concepts involved provide both a working concrete example of such mechanisms and set a lower bound on the complexity and sophistication of mechanisms that should be considered by psychological theory for similar functions. Thus applying the modern computer metaphor should be useful even if we have profound reservations about its applicability in detail. The relevant computer-metaphor concepts will be introduced as needed in the discussion.

First, the EPIC architecture will be summarized. Then how the cognitive processor controls input and output will then be presented, followed by the bulk of the chapter, on the control of cognition. This is divided into two sections: The first covers the basic control of cognition, which concerns sequential, hierarchical, parallel, and interrupt flow of control. The second discusses executive control of cognition. Executive processes are illustrated with an example of two different executive regimes for a dual-task situation.

Description of the EPIC Architecture

Basic Structure

Figure 23.2 shows the overall structure of the EPIC architecture. The cognitive processor consists of a production rule interpreter that uses the contents of production memory, long-term memory, and the current contents of a *production-system working memory* to choose production rules to fire; the actions of these rules modify the contents of working memory, or instruct motor processors to carry out movements. Auditory, visual, and tactile processors deposit information about the current perceptual situation into working memory; the motor processors also deposit information about their current states into working memory. The motor processors control the hands, speech mechanisms, and eye movements. All of the processors run in parallel with each other. Note that the *task environment* (also called the *simulated device*, or simply the *device*) is a separate module that runs in parallel as well. The pervasive parallelism across perception, cognition, and action motivated the design of EPIC and is reflected in the acronym: Executive Processes Interact with and Control the rest of the system by monitoring their states and activity. Because of this fundamental focus, EPIC is a good choice for a discussion of cognitive control mechanisms.

Figure 23.2 is a simplified view; each of the processors is fairly complex, as illustrated by Figure 23.3, which shows additional detail related to the visual system. The environment controls the current contents of the *physical store*; changes are sent to the *eye processor*, which represents the retinal system and how the visual properties of objects in the physical store are dif-

ferentially available, depending on their physical properties such as color, size, and eccentricity—the distance from the center of the fovea. The resulting "filtered" information is sent to the *sensory store*, where it can persist for a fairly short time, and comprises the input to the *perceptual processor*, which performs the processes of recognition and encoding. The output of the perceptual processor is stored in the *perceptual store*, which is the *visual working memory* and whose contents make up the visual modality-specific partition of the production-system working memory. Thus production rules can test visual information only in terms of the current contents of the perceptual store. Production rules can send commands to *involuntary* and *voluntary ocular processors*, which control the position of the eyes. The voluntary ocular processor is directly controlled by the *cognitive processor*, which

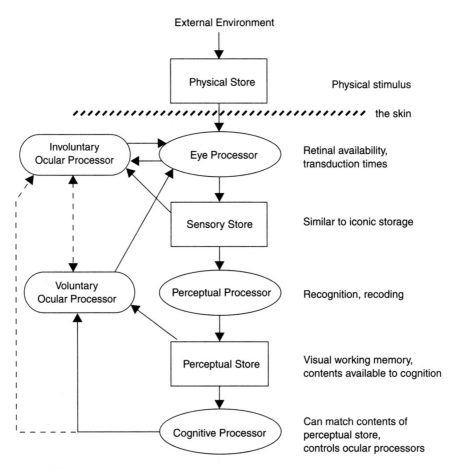

FIGURE 23.3 The internal structure of EPIC's visual processor system. This is an expansion of the visual processor module shown in Figure 23.2.

can command an eye movement to a designated object whose representation is in the perceptual store. The involuntary ocular processor generates "automatic" eye movements, such as reflex saccades to a new or moving object or smooth movements to keep the eye foveated on a slowly moving object, using the location information present in the sensory store. The cognitive processor can only enable or disable these automatic activities and so only indirectly controls the involuntary processor. More detail can be found in Meyer and Kieras (1997a), Kieras and Meyer (1997), and Kieras (2004b).

Production Rules

The cognitive processor uses a very simple production system interpreter called the *parsimonious production system* (PPS). This interpreter has simple syntax and semantics but is implemented with a full rete-match algorithm so that it scales well in terms in of performance to large sets of production rules. The items in the production system working memory are simply lists of symbols—the interpreter does not assume any particular structure of memory; rather these are defined in terms of what inputs to working memory are provided by the perceptual and motor processors and programming conventions for the production rules themselves.

Like most production systems, PPS operates in cycles, which have a 50-ms period. During a period, the perceptual and motor processors continually update the production system working memory. At the beginning of each cycle, the additions and deletions of working memory items specified by the last rule firing are carried out, and then all production rules whose conditions are currently met are "fired"—their actions are executed. Note that *all* rules currently satisfied are fired on each cycle. Also, if there is more than one specific instantiation of a rule's condition present in working memory, the rule will be fired for each of them in the same cycle. There is no preset limit on how many rules can fire, or how many instantiations of them can fire, on a single cycle.

This full parallelism is often mistaken for a claim of unlimited processing power, but it is not. The peripheral mechanisms limit how much information can be simultaneously present from perceptual and motor sources, and some limits are normally observed for the amount of information in working memory created and maintained by the production rules, but this is not enforced by the architecture. Note that as argued by

Kieras, Meyer, Mueller, and Seymour (1999) the true limits of working memory are not at all well understood empirically—for example, compare Baddeley and Logie (1999) with Ericsson and Kintsch (1995) and Ericsson and Delaney (1999). Working memory is either the traditional small-capacity modality-specific "scratchpad" memory, or it is a huge-capacity fast-write usage of long-term memory enabled by well-learned, task-specific retrieval strategies. These two concepts are probably compatible, but at this date, there is not a theoretical consensus on what the "real" working memory is like. So the fact that EPIC does not impose working memory limits is a tactical decision on our part not to "hardwire" our collective ignorance into the architecture but rather to let experience with modeling various tasks determine what kinds and amounts of information are required for plausible modeling of important phenomena and then to use this knowledge to specify and to explain memory limitations and their sources. For example, the work in Kieras et al. (1999) suggests that verbal working memory limitations have their basis not in a direct limit on how many items can be kept active in memory but rather is due to how long phonological representations persist in auditory memory relative to the speed with which they can be refreshed through rehearsal (see also Mueller, Seymour, Kieras, & Meyer, 2003) and how well the recall strategies can recover information from the resulting mélange of partially decayed representations (Mueller, 2002).

The partitions of the production-system working memory are one for each perceptual modality (e.g., auditory and visual), which are synonymous with the perceptual stores for the modality and another partition for the states of the motor processors. A *Control Store* partition contains *Goal*, *Step*, and other similar items that are used to specify the flow of control through production rules. The *Tag Store* partition contains *Tag* items, which simply associate an item in a modality working memory with a symbol designating a role in the production rules, analogous to a variable and its binding in a traditional programming language. That is, one set of production rules might be responsible for identifying which visual object is "the stimulus" for a task, and add a Tag accordingly. Other production rules could then refer to the current stimulus using the Tag instead of having to re-identify the stimulus repeatedly.

Figure 23.4 shows two sample production rules from working models to illustrate how PPS rules look.

```
(Top-see-fixation-point
If ((Goal Do Visual_search) (Step Waitfor Fixation-present)
    (Visual ?object Shape Cross_Hairs) (Visual ?object Color Red))
Then ((Add (Tag ?object fixation-point))
    (Delete (Step Waitfor Fixation-present)) (Add (Step Waitfor probe-present))))


(Top-make-response
If ((Goal Do Visual_search) (Step Make Response)
    (Tag ?target target) (Tag ?cursor cursor)
    (Motor Manual Modality Free))
Then ((Send_to_motor Manual Perform Ply ?cursor ?target Right)
    (Delete (Step Make Response)) (Add (Step Make Response2))))
```

FIGURE 23.4 Example parsimonious production system (PPS) production rules. See text for explanation.

Various aspects of these rules will be explained later; however, for now it suffices to provide a simple overview of the basic syntax. The first rule identifies the visual object for a fixation point consisting of a red crosshair. The second rule moves the cursor to a visual object. Each rule begins with a rule name, then a set of conditions after the *if*. The conditions are implicitly a conjunction; all of them must be satisfied for the rule to be fired. The actions listed after the *then* are executed if the rule is fired. The *add* and *delete* actions modify the contents of the production system working memory. *Send_to_motor* is a command to a motor processor. Terms starting with a *?* character are *variables*. In a condition, the variables are assigned whatever values necessary to make the condition item match an item in working memory. Variables appearing in actions take on the values assigned to them in the condition when the action is executed. Note that the variables are scoped only within the rule; once the rule is fired and executed, the variable values are discarded; there is no memory of them from one rule to the next, or one cycle to the next.

Control of Input

In computers, the process of input is under the ultimate control of the *central processing unit* (CPU). That is,

input from peripheral devices is normally only brought into the main memory if the CPU has instructed it. In early machines, instructions executed by the CPU performed the input/output processing, but in modern machines (since about 1970), a hardware subsystem does all of the detail work. Basically, the CPU executes an instruction that commands an input subsystem to deliver specified information to an area of memory. Thus input involves a bidirectional flow of information: Control and specification information goes from the CPU to the input device, and the input device responds by transferring information from the external environment to the CPU's memory.

In a cognitive architecture, the input processing is similarly done by perceptual subsystems, but the control of input depends on some basic assumptions stemming from the apparent fact that only a fraction of the current environmental input should be responded to. That is, cognition must be protected from an overwhelming flood of external input, so the amount of information that cognition takes as input must be limited in some way. In EPIC, the input to cognition is the information in the perceptual store, so the question is whether and how the cognitive processor must deal with unneeded information in the perceptual store.

The traditional assumption in cognitive psychology is that a selection mechanism of some sort, usually named *selective attention*, controls what information is allowed to influence cognitive processing. In the box model era, the mechanism was usually presented as some sort of filter between perception and cognition (see Norman, 1976). In current cognitive architectures, the attention mechanism typically uses activation levels to determine which condition items can trigger production rules. The idea of a selective attention mechanism has become so pervasive in cognitive theory that it seems odd not to simply adopt it, but it should be clear that the idea is not without its puzzles and problems, as consulting any textbook on the voluminous literature on classic topics, such as early versus late selection and overt versus covert attention, should make clear. However, the actual phenomena of attention may be a result of a much wider range of mechanisms than an internal filter or activation. In particular, Findlay and Gilchrist (2003) point out how the concept of covert attention as been so heavily overapplied to visual perception that it has thoroughly obscured the more fundamental and powerful role played by the nonhomogeneity of the retina and voluntary eye movements that orient the high-resolution area of the retina on objects of interest. They argue that the role and effects traditionally assigned to covert visual attention are either subsumed by overt attention in the form of eye movements, or are very small compared to the effects of retinal nonhomogeneity and eye movements. Thus, a proper theory of attentional effects in vision would start with these fundamentals, and consider covert mechanisms only if there are important effects remaining.

As an alternative, EPIC assumes that when the characteristics of the peripheral systems, and how they can be cognitively controlled and accessed, are taken into account, the need for a separate information-limiting precognitive selection mechanism disappears. In the case of vision, EPIC assumes that the perceptual store has a very large capacity, but the nonhomogeneity of the retina "automatically" results in the visual system having the most information about the object being fixated and substantially less about other objects. Thus the total amount of visual information is limited at the peripheral level, and the objects that have the most information are determined by the current eye location. The selective function is implemented not by some kind of special attentional mechanism, but simply by production rules that decide where more detail is needed and command a corresponding eye movement, using relevant portions of the available visual information to guide the choice. Thus selection is handled simply in terms of what visual object the production rules single out for further processing, and what visual properties are used in this processing.

The first example rule in Figure 23.5, from a choice reaction task, illustrates this concept. The rule waits for a choice stimulus to appear, signaled by the detection of a visual *onset* event, which has been characterized as

```
(Waitfor_ChoiceStimulus
If ((Goal Do ChoiceTask) (Step Waitfor ChoiceStimulus)
    (Visual ?stimulus Detection Onset)
    (Not (Tag ?stimulus Choice_stimulus))))
Then ((Add (Tag ?stimulus Choice_stimulus))
    (Delete (Step Waitfor ChoiceStimulus)) (Add (Step Lookat ChoiceStimulus))))


(Top_Find_Hostile_blip
If ((Goal Monitor Situation) (Step Find Hostile_blip)
    (Visual ?blip Color Red) (Visual ?blip Shape Triangle)
    (Randomly_choose_one))
Then ((Add (Tag ?blip Current_track))))
```

FIGURE 23.5 Example of input selection in production rules. See text for explanation.

"capturing attention" (e.g., Yantis & Jonides, 1990). Visual onset events are assumed to be available over almost all of the visual field, so they will be detected even if the eyes are on some other object. This production rule adds a tag that the onset object is the choice stimulus, and then sets up the next production rules to fire. These will match on the tag for the *Choice_stimulus* to command the ocular processor to move the eyes to that stimulus object, and subsequent rules will use the tag to identify the object whose appearance will determine the response to be made. The second example rule in Figure 23.5 shows that the concept of selecting an object is not limited to events like onsets that "capture attention." This rule is from a radar-operator task model in which red triangular objects are "blips" that represent hostile aircraft and so should be given priority. If the visual perceptual store contains an object that matches these perceptual properties, it is tagged as the current "track" for subsequent processing. In case there is more than one, one is chosen at random. In this case, other rules, following a different control structure, set up the next rules to fire, but these later rules will simply use the tag item to identify the selected object. Still other rules would deal with the possibility that a red triangular object might be present but not visible given the current eye location.

The point of the examples is that there is no need for a distinct attention mechanism that prefilters objects before cognition has access to them. Because the onset event, or the available perceptual properties, resulted in tagging a particular object as the stimulus, the subsequent response production rules can effectively ignore all other objects; the cognitive processor has been protected from excess and irrelevant information by a combination of peripheral limitations and a simple designation in working memory of which object is relevant for further processing.

Tagging a particular object as the one of interest could be described as a form of covert attention, and moving the eyes to an object could be called a case of overt attention. Regardless of these labeling issues, there is no actual "attention" mechanism in EPIC; there are only objects tagged for further processing and information that is available or not, depending on the eye position. In line with previous warnings in the history of psychological theory, the concept of *attention* as used in current cognitive psychology may be a case of *reification:* just because there are a set of phenomena labeled "attention" does not mean that there is a distinct internal mechanism that *does* attention. Rather, attention could simply be our label for the processing performed by the human system when it responds selectively to events in the environment. EPIC shows how attention can be implemented without a dedicated internal mechanism that selects what information will be supplied to the cognitive processor.

Control of Output

In computers, output requires a bidirectional flow of information: The CPU must know when an output device is ready to accept commands before it issues them. Again in modern machines, dedicated hardware subprocessors handle all the detail work, but the basic concept is that of *handshaking*—the CPU checks for whether the output subsystem is free, sends the next output command if so, and then waits for the output subsystem to signal that the operation is complete before sending the next output command. This process became more advanced over the years, so that, for example, the output subprocessor needs only to be told where the data are in memory, and upon command, it begins to autonomously output each item and signals the CPU when it is finished outputting the entire set of data.

Similarly, in a cognitive architecture, the output is done by motor subsystems, but the cognitive processor must check the states of the motor processors and coordinate the usage of different output modalities to meet the task demands. In EPIC, as originally proposed (Kieras & Meyer, 1997; Meyer & Kieras, 1997a), the motor processor for each modality required a relatively large amount of time to prepare each movement by programming its movement features, and then the prepared movement could be separately initiated and executed.[1] Preparation of the next movement could commence while the previous movement was being executed, allowing EPIC to make a series of movements at high speed. However, a slow and deliberate style would wait for each movement to be completed before commanding the next movement. Thus EPIC's motor processors have two especially useful states: The *Modality Free* state means that the output modality (e.g., manual) is currently idle—a movement is neither being prepared nor executed. The *Processor Free* state means that any previously commanded movement has already started execution, so a new movement command will be accepted and prepared and will execute as soon as the current movement (if any) is complete. Since the motor processor can prepare only one movement at a time, an error condition (*jamming*) will arise if it is commanded to prepare a new movement while it is still preparing one. By convention, a production rule that commands a movement always includes a condition that checks for whether the motor processor is in an appropriate state to prevent jamming.

Figures 23.6 and 23.7 illustrate the extremes in motor sequence programming. Note how by deleting and adding Step items, each rule disables itself and enables the next rule in the sequence. In the slow movement sequence, shown in Figure 23.6, each rule waits for the previously commanded movement to be complete before commanding the next movement, and the last rule waits for the last movement to be complete before proceeding. In the fast movement sequence shown in Figure 23.7, each movement is commanded as soon as the processor is willing to accept it, and the last rule does not wait for the movement to be complete. As one might suspect, skilled performance in a task appears to involve the fast movement regime, while the usual assumptions of simple forms of GOMS models (see below) appear to be consistent with a slow movement regime (see Kieras & Meyer, 2000; Kieras, Wood, & Meyer, 1997).

The point of these examples is that the cognitive strategy for how movements are marshaled and controlled can make a large difference in how rapidly a task can be performed, showing that some of the motoric aspects of a task are also a cognitive matter. Incorporating cognitive aspects of motor control is essential in modeling human performance, especially in highly interactive tasks such as computer operation.

Basic Control of Cognition

This section presents concept on the basic aspects of the flow of control in cognition, covering sequential, parallel, and interrupt-driven flow of control. The next section will consider complex control at the level of executive processes.

Sequential Flow of Cognition

In a computer, certain aspects of the sequential control of execution are primitive hardware functions. In particular, the machine instructions that make up a

```
(Slow_wait_to_start
If ((Goal Make Responses) (Step make first)
    (Motor Manual Modality Free))
Then ((Send_to_motor Manual Perform Punch J Right Index)
    (Delete (Step make first))(Add (Step make second))))


(Slow_wait_for_first_done
If ((Goal Make Responses) (Step make second)
    (Motor Manual Modality Free))
Then ((Send_to_motor Manual Perform Punch K Right Middle)
    (Delete (Step make second))(Add (Step waiton second))))


(Slow_wait_for_second_done
If ((Goal Make Responses) (Step waiton second)
    (Motor Manual Modality Free))
Then ((Delete (Step waiton second))(Add (Step continue  working))))
```

FIGURE 23.6 Example of rules that produce a slow sequence of movements by waiting for each movement to be complete before performing the next.

```
(Fast_start_first
If ((Goal Make Responses) (Step make first)
    (Motor Manual Processor Free))
Then ((Send_to_motor Manual Perform Punch J Right Index)
    (Delete (Step make first))(Add (Step make second))))


(Fast_start_second_and_continue
If ((Goal Make Responses) (Step make second)
    (Motor Manual Processor Free))
Then ((Send_to_motor Manual Perform Punch K Right Middle)
    (Delete (Step make second))(Add (Step continue working))))
```

FIGURE 23.7 Example of rules that produce a fast sequence of movements by commanding each movement as soon as the motor processor is free.

program are normally laid out in a series of consecutive memory locations. The hardware responsible for fetching and then executing each instruction assumes that the next instruction is normally in the next memory location; the hardware *program counter* or *instruction address register* is thus automatically incremented to point to the next instruction in memory. A program *branch* or *go to* instruction specifies an address for the next instruction that is not simply the next location; dedicated hardware will then set the program counter to that next address directly. However, there have been computers whose memory was organized differently so that sequential memory locations could not be assumed. An example is early machines that used a rotating magnetic drum as the main memory, where the delay waiting on the drum to rotate to any one position was substantial. Each instruction included the address of the next instruction, allowing them to be located on the drum in positions that minimized the delay in fetching the next instruction.

Production rule architectures are more like the early drum storage machines than modern machines, in that the results of the actions in each production rule normally control which rule will fire next. In EPIC, in fact, the flow of control must be fully explicit in the rules, because no flow of control is implicit in the PPS interpreter. That is, since any number of rules can fire on any cycle, the only way to control which rules will fire is to write the rule conditions and actions

so that the contents of working memory cause the right rules to fire on the right occasions in the right sequence.

To bring some organization to PPS rules, Kieras and Polson (1985) and Bovair, Kieras, and Polson (1990) worked out how to organize production rules in terms of GOMS models (Card, Moran, & Newell, 1983; John & Kieras, 1996; Kieras, 1988, 1997, 2004a), which have been extremely useful in representing human procedural knowledge in models of human–computer interaction. The basic idea can be explained by explaining the acronym: procedural knowledge is well represented in terms of *Goals* that can be accomplished by executing a series of *Operators* organized in terms of *Methods*, which are sequences of operators. *Selection rules* choose between alternate methods for the same goal. Kieras and Polson realized that it is useful to distinguish goals and operator sequence; that is, a method for accomplishing a goal specifies a series of operators, conveniently grouped into steps. Thus basic sequential flow of control for EPIC models is provided using the scheme shown in Figure 23.8 (and anticipated by some of the earlier examples). If working memory starts with the contents (Goal Top) and (Step T1), the first rule, named Top1, will fire, and its actions will delete (Step T1) and add (Step T2). At this point, Top1 can no longer fire, since one of its condition items has been deleted, but the rule Top2 can now fire. Similarly, it disables itself and enables the next rule,

```
(Top1

If ((Goal Top)(Step T1))

Then ((Delete (Step T1))(Add (Step T2))))


(Top2

If ((Goal Top)(Step T2))

Then ((Delete (Step T2))(Add (Step T3))))


(Top3

If ((Goal Top)(Step T3))

Then ((Delete (Step T3))(Add (Step T4))))
```

FIGURE 23.8 Schematic illustration of basic sequential flow of control.

Top3, and so forth. The entire series of rules is under the control of the goal item in memory (Goal Top); other rules can manipulate the goal item to control this method. For example, suppose the first rule fires at the same time as the goal item is removed by some other rule; the (Step T2) item will be present, but rules will be disabled and the next rule Top2 will not fire. If the goal item is replaced, execution will resume where it left off; Top2 will fire, followed by Top3. EPIC's theory of executive processes is based on this concept of controlling methods by removing or replacing their controlling goals.

The other basic form of flow of control is subroutine call/return, whose role in reusing and organizing program code is fundamental to computer programming methodology. On a computer, when a call to a subroutine is made, the address of the next instruction is saved, and a branch is made to the address of the first instruction of the subroutine. When the subroutine completes its work, it returns by branching to the saved address of the next instruction in the calling code. In addition to this basic flow of control scheme, arrangements are made to give the subroutine temporary memory space and pass input and output values between the calling routine and the subroutine. While subroutine call and return can be done with simple instructions, most computer hardware has specialized instructions to allow this very common process to be executed quickly and with only a few instructions.

Again in the GOMS model, accomplishing goals typically involves accomplishing subgoals, and the resulting hierarchy of methods describe the natural units or modules of procedural knowledge. Thus GOMS models make heavy use of an analogue of subroutine call and return, and EPIC models are programmed correspondingly using a set of conventions illustrated in Figure 23.9. The basic scheme can be termed *hierarchical-sequential*. The methods are arranged in a strict calling hierarchy, and the execution is strictly sequential. Each rule that is part of a method or submethod includes a condition that names the goal for the method. Thus the rules for a method can only fire if the corresponding goal is being accomplished.

Rule Top1 calls a submethod by adding the goal for the submethod (Goal Sub) in this example. It disables itself and enables the next rule in the calling sequence. This next rule includes a negated condition (Not (Goal Sub)), which means that the rule can fire only if the (Goal Sub) item is *not* present in working memory. Thus the next rule will wait until the subgoal item has been removed. By convention, the step items for a submethod include the subgoal name, so when the (Goal Sub) item is placed in working memory by the Top1 rule, the rule Sub_startup fires because there are no step items belonging to this method. The startup rule enables the next rule in the method, which enables the next rule, as usual. Thanks to this scheme, the calling method does not need to know which step item governs the first step of the submethod, which makes it easier to write reusable methods that represent sharable pieces of procedural knowledge. Note how the submethod rules are all governed by the same (Goal Sub) item, so when the final submethod rule Sub2_terminate fires, the (Goal Sub) item and the last step are deleted; this indicates that the submethod has accomplished its goal, and the submethod rules are now all disabled. The disappearance of the subgoal triggers the next rule in the calling sequence, Top2, which then goes on to execute its next step.

Like most production-system cognitive architectures, EPIC assumes that working memory is shared across all of the production rules. Thus, the concepts in conventional programming languages of local variables, parameters, and return values do not apply—it is as if all the information is in global variables, accessible to all parts of the program. This presents some technical problems in writing reusable sets of production rules,

```
(Top1_call_submethod
If ((Goal Top)(Step T1))
Then ((Add (Goal Sub))(Delete (Step T1))(Add (Step T2))))


(Top2_wait_for_return
If ((Goal Top)(Step T2)(Not(Goal Sub))
Then ((Delete (Step T2))(Add (Step T3))))

    ... ... ... ... ... ...


(Sub_startup
If ((Goal Sub) Not(Step ???))
Then ((Add (Step S1))))


(Sub1
If ((Goal Sub) (Step S1))
Then ((Delete (Step S1))(Add (Step S2))))


(Sub2_terminate
If ((Goal Sub) (Step S2))
Then ((Delete (Step S2))(Delete (Goal Sub))))
```

FIGURE 23.9 Schematic illustration of hierarchical-sequential flow of control used in subroutine calls.

but it is not necessary to discuss them for the purposes of this chapter.

Parallel Flow

Parallelism Is Routine in Computers

Perhaps the most distorted part of the computer metaphor is the notion that computers do only one thing at a time. While the capacity for parallel operation in computers has been steadily increasing, it is hardly a new capability. In fact, even the classic ENIAC was actually a parallel multiprocessor machine in which the separate computational modules were operating simultaneously.

Current computers are usually classified as being either *uniprocessor* machines, containing only a single CPU, or *multiprocessor* machines, with multiple CPUs. However, even uniprocessor computers have considerable parallelism. Computers have long had subprocessors for input/output that run in parallel with the CPU. The internals of the CPU typically have also involved parallel data movement, but in modern CPUs, the subparts of the CPU also operate in parallel. For example, in *pipelining* and *superscalar architectures*, the CPU is executing more than one instruction simultaneously, and not necessarily even in the original order!

Multiprocessor machines either have a shared main memory, or communicate data over a network, and display true parallel program execution with their multiple CPUs—they can literally execute two machine instructions simultaneously and thus execute two programs truly simultaneously. Such machines are not new (commercial versions appeared in the 1960s), but they have become quite inexpensive, as shown by the common dual-CPU desktop computer.

Parallel Flow of Control in EPIC

Despite its routine presence in computers for decades, and the obviously massively parallel operation of the

```
(Top1_start_submethod

If ((Goal Top)(Step T1))

Then ((Add (Goal Sub))

    (Delete (Step T1))(Add (Step T2))))


(Top2_keep_processing

If ((Goal Top)(Step T2))

Then ((Delete (Step T2))(Add (Step T3))))


    ... ... ... ... ... ...


(Top8_wait_for_submethod_complete

If ((Goal Top)(Step T8)

    (Not(Goal Sub))

Then ((Delete (Step T8))(Add (Step T9))))
```

FIGURE 23.10 Schematic illustration of overlapping hierarchical parallel flow of control.

brain, parallel processing in cognitive theory has traditionally been treated as an exotic capability, with sequential processing assumed to be the normal mode of cognitive operation. There might be a variety of reasons for this bias, but most directly relevant to this chapter is the issue pointed out by Meyer and Kieras (1997a, 1997b, 1999): There has been a long-standing tendency to mistake optional strategies adopted by subjects for "hardwired" architectural limitations that impose sequential processing. For example, in many dual-task paradigms, the task instructions and demands actually encourage unnecessarily sequential processing; in fact, different instructions and payoffs can encourage parallel processing (Meyer, Kieras, Lauber, et al., 1995; Meyer, Kieras, Schumacher, Fencsik, & Glass, 2001; Schumacher, Lauber, et al., 1999; Schumacher, Seymour, et al., 2001). Thus rather than a fundamental architectural feature, the presence of cognitive sequentiality is a result of a lack of strategic sophistication in the production rules in effect for a task. Against this background, it is useful to explore a different theoretical paradigm in which cognitive parallelism is *normal* rather than exotic and unusual.

EPIC directly supports full parallel processing since multiple production rules can fire on a cycle

simultaneously and independently, and since goals and steps are simply working memory items, multiple ones can be present simultaneously—there is no *stack* that requires only one goal to be active at a time. Thus different *threads of execution* can be spawned and terminated at will. Figure 23.9, which illustrates hierarchical–sequential flow of control can be compared with Figure 23.10, which shows simultaneous execution of a top-level method and a submethod. The first rule in Figure 23.10 activates a submethod by adding its goal to working memory, but execution simultaneously continues with the next step in the calling method, until the rule for Step 8, which tests for the submethod having accomplished and removed its goal. Thus both the top-level goal and the subgoal are simultaneously pursued, each in its own thread of execution (see p. 342 for the corresponding *thread* concept in computing), but these are brought back into synchrony before Step T9 at the top level.

A few more schematic examples will illustrate the flexibility of parallel control in EPIC. Figure 23.11 shows how two subgoals could be simultaneously executed in parallel with the top level, which in this example waits for both of them to complete before proceeding; of course, this is optional—the submethods could simply be allowed to complete while other processes continue. Figure 23.12 shows single-step multiple threads within a method. The first rule enables all three different rules for the next step; one always fires to continue to the next step, but either, both, or neither of other two rules might fire as well. Figure 23.13 shows multiple threads within a method. The first rule spawns two threads in the form of different step items,

```
(Top1_start_two_submethods

If ((Goal Top) (Step T1))

Then ((Add (Goal Sub1))(Add (Goal Sub2))

    (Delete (Step T1))(Add (Step T2))))


(Top2_wait_for_both_complete

If ((Goal Top)(Step T2)

    (Not(Goal Sub1))(Not(Goal Sub2)))

Then ((Delete (Step T2))(Add (Step T3))))
```

FIGURE 23.11 Schematic illustration of parallel submethods.

```
(Top1
If ((Goal Top)(Step  T1))
Then ((Delete (Step  T1))(Add (Step  T2))))


(Top2_branch_1
If ((Goal Top)(Step  T2) /* some condition */)
Then (/* some action */))


(Top2_branch_2
If ((Goal Top)(Step  T2) /* some condition */)
Then (/* some action */))


(Top2_continue
If ((Goal Top)(Step  T2))
Then ((Delete (Step  T2))(Add (Step T3))))
```

FIGURE 23.12 Schematic illustration of multiple threads within a single step.

each of which starts a chain of rules governed by the same goal item. If desired, these chains can simply be left to terminate on their own.

One often-cited advantage of a production rule representation for procedural knowledge is *modularity*—the independence of each production rule from other production rules. By allowing more than one rule to fire at once, it is possible to take more advantage of this modularity to provide a variety of flexible control structures.

```
(Top1_spawn_threads
If ((Goal Top)(Step  T1))
Then ((Delete (Step  T1))(Add (Step T2))
      (Add (Step ThrA1))(Add (Step ThrB1))))


(TopThreadA
If ((Goal Top)(Step ThrA1))
Then ((Delete (Step ThrA1))(Add (Step ThrA2))))


(TopThreadB
If ((Goal Top)(Step ThrB1))
Then ((Delete (Step ThrB1))(Add (Step ThrB2))))
```

FIGURE 23.13 Schematic illustration of multiple threads within a method.

Interrupt Control of Cognition

Computers have long had special *interrupt* hardware to allow a rapid response to external events that happen *asynchronously,* that is, at any time not necessarily in synchrony with activities of the computer. An interrupt signal will force a branch to a special address, where interrupt-handling code determines the cause of the interrupt, takes appropriate action, and then branches back to the original execution address. The hardware can be very elaborate, for example, automatically giving precedence to the highest-priority signal. Interrupts can be routine, or extraordinary. For example, control of input/output devices is routinely done with interrupts that signal that the device has completed an operation and is ready for the next datum or instructions. In contrast, some error conditions, such as an attempt to execute an invalid instruction, might cause an interrupt that results in a process being terminated, or even the OS entering an error state and waiting for a restart. Thus handling an interrupt spans a wide range of possible activities.

A production-system cognitive architecture can support interrupt handling but only if the architecture has the appropriate properties. For example, it has to be possible for a rule to fire regardless of whatever else is going on, and the actions of the rule must be able to modify the state of the current processing, such as suspending other activities and starting new ones, essentially a form of executive process, which will be discussed later. Not all architectures can support interrupt rules in a straightforward way. EPIC's cognitive parallelism and representation of control items in working memory makes interrupt rules very simple.

The schematic first rule in Figure 23.14 illustrates a simple form of interrupt handling. This rule will fire unconditionally when a red object is present in the visual working memory, regardless of what other activity is under way. The second rule is governed by the goal of processing a blip in a radar-console task. If a visual object tagged as the current blip begins to disappear from the screen, the rule will delete the goal of processing the blip and start a method that aborts the processing. A rule like this is needed in tasks where a visual object that is the subject of extended processing might disappear at any time; if the object disappears, the processing of it needs to be terminated and some "cleanup" work needs to be done. Such a rule is a simple way to handle what would otherwise be extremely awkward to implement in a strictly sequential architecture;

```
(Watchfor_red
If ((Visual ?obj Color Red))
Then (/* some action */))


(Watchfor_disappearing_blip
If ((Goal Process Blip)
    (Tag ?blip Current_blip)(Visual ?blip Status Disappearing))
Then ((Delete (Goal Process Blip))(Add (Goal Abort Processing)))))
```

FIGURE 23.14 Sample interrupt rules.

it works because it can fire at any time its conditions are met, and its actions can control the state of other processing.

Executive Control of Cognition

What Is Executive Control?

In psychology, the term *executive* is vague and wide ranging, having been applied in ways as suspect as a synonym for the homunculus, to being a specialized activity such as working memory management (see Monsell & Driver, 2000). In the sense used here, the executive is the process that controls what the human system as a whole is doing, such as what tasks are being performed, what resources are being used for them, which tasks take priority over others, and how simultaneous tasks should be coordinated. In computers, this function is most similar to OSs, so it is worthwhile to present a quick survey of the key concepts. See Tucker (2004) or any modern OS textbook, such as Stallings (1998), for more detail.

Survey of Operating System Concepts

OSs originated at a time when computer hardware, especially the CPU, was vastly more expensive than it is now, so there were huge economic incentives to wring every last bit of performance from the investment by clever programming. Today, of course, the hardware is cheaper, but the demands are higher, so OS technology has continued to develop. The key is that because the CPU is tremendously faster than peripheral input/out-

put devices, it is possible for the CPU to execute instructions for one task while waiting for an input/output operation for a different task to complete. Thus, two data processing tasks could progress simultaneously by overlapping CPU execution of one with input/output waits for the other.

This overlapping could be accomplished by custom building a single complex program for each possible combination of computing jobs, but for almost all purposes, such detailed, difficult, and tedious programming work would be prohibitively expense. After various partial solutions (e.g., spooling), a grand generalization emerged around 1970, termed *multiprogramming*, or *multiprocessing*, that made it possible to write software for simultaneous, or *concurrent*, task execution quite easily.

The basic idea is that a control program would always be present in memory and would keep track of which other programs were present in memory, what their states were, and which resources were currently associated with them. This control program was at first called a *resident monitor*, later the OS. With the help of interrupts, the OS code can switch the CPU from one program to another by controlling where execution resumes after handling an interrupt.

The OS manages programs in terms of *processes*, which are a set of resources, code in memory, data in memory, and an execution state, or pathway. The resources include not only peripheral devices such as disk drives but also space in memory and CPU time. When a program needs access to a resource, it makes a request to the OS and waits until the request is granted. Because it is thus informed of resource needs and has all the required information available about all the processes, the OS can manage multiple processes in such a way that they can be simultaneously active as long as their resource requirements do not conflict and that, when conflicts do occur, they are resolved according to the specified priorities.

Two processes are *noncooperating* if they do not attempt to communicate with each other or share resources. In this case, as long as the protocol for accessing resources through the OS is followed, the program for each process can be written independently, as if it was going to be the only program executed; the OS will automatically interleave its resource usage with that of other independent processes, and will ensure that they will not interfere with each other, and that the maximum throughput consistent with process priorities will be obtained. This is the major contribution of OS technology—enabling concurrent operation with simple programming techniques. Anybody who has written and run even simple programs on a modern computer has in fact used these facilities.

Cooperating processes are used in specialized applications and communicate with each other or access a shared resource such as a database. For such situations, the OS provides additional services to allow processes to communicate reliably with each other and to ensure mutual exclusion on critical resources— such as preventing two processes from attempting to modify the same area of memory simultaneously. Using these interprocess communication and mutual exclusion facilities properly in specialized applications is the hard part of multiprogramming technology.

One last distinction concerns the relationship between execution pathways and resources. Switching the CPU between processes can be very time consuming since all of their resources might have to be switched as well. For example, if a low-priority process gets suspended to free up required memory space for a high-priority process, the OS will need to copy the program code, data, and execution state information for the low-priority process out to disk storage and then read it back into memory when it is time to resume the process (virtual memory hardware helps automate this process). In contrast, a *thread* is a *lightweight process*— it is an execution pathway that shares resources with other threads in the same process. The OS can switch execution between threads very quickly because some critical resources such as memory space are shared. Multithreaded programming is becoming steadily more common as a way to implement high-performance applications, but because the threads are usually coop-

erating, OS facilities for interthread communication and mutual exclusion are usually involved.

A key insight from OS methodology is that the key algorithms work for both uniprocessor and multiprocessor machines. While not explicitly acknowledged in OS textbooks, multiprogramming on a uniprocessor machine is essentially simulated parallel processing, so OS operations are similar for both true versus simulated parallel processing. Closely related is that all of an OS's capabilities can be implemented with very little specialized hardware support, although such hardware can improve efficiency and simplify the programming.

How Should Executive Processes Be Represented in a Cognitive Architecture?

Traditionally, executive processes are represented in a cognitive architecture as specialized built-in mechanisms. A well-known example is the *Supervisory Attentional System* in the Norman and Shallice (1986) proposal, which is a specialized brain system whose job it is to pick the most activated schema and cause it to take control of cognition. A more contemporary and computational example is Salvucci, Kushleyeva, and Lee's (2004) proposals to add OS-like facilities into the core functionality of the ACT-R architecture. Proposals of this sort, which propose special-purpose "hardware" support in the architecture for executive processes, overlook the lesson from computing that minimal hardware support is necessary for OSs to work well. More seriously, they also overlook a wealth of literature on how executive control is a strategic process that is a result of learning and can be modified on the fly by changes in task priorities or instructions (for reviews, see Kieras, Meyer, Ballas, & Lauber, 2000; and Meyer & Kieras, 1997a, 1997b, 1999), which suggests that executive processes are similar to other forms of cognitive skill, and so should be represented simply as additional procedural knowledge rather than specialized architectural components.

EPIC's solution is that simply by making task goals another kind of working memory item and allowing more than one production rule to fire at a time, it is easy to write production rules that control other sets of production rules by modifying which goals are currently present in memory. Thus, no special-purpose architectural mechanisms are required for executive control.

The parallelism is critical because it ensures that the executive rules and the individual sets of task rules can all execute simultaneously, meaning that the executive process rules can intervene whenever they need to, and the individual task rules do not need to constantly check for the states of other processes to allow for changes in which processes have control.

Thus, the cognitive modeler can work like the modern computer programmer in that as long as the production rules for noncooperating processes observe a protocol of asking the executive process for access to resources, they can be simply written as if they were the only production rules being executed at the time. Again analogously to modern computer programming, this allows sets of production rules for different tasks to be written as modules that implement principles of abstraction, encapsulation, and reusable code; the fact that production rules can thus be organized into independent modular sets and flexibly combined has implications not just for the practicalities of modeling but also for how humans might learn multitasking skills (see Kieras et al., 2000).

Executive Processes in EPIC

In EPIC, an executive process is simply a set of production rules that control other sets of production rules by manipulating goal information in working memory, along with possibly other information as needed to control the execution of other production rules. This section discusses some distinctions relevant to the representation of executive processes.

Specialized Versus General Executives

Kieras et al. (2000) made a distinctive between *specialized executive* processes that contain task-specific rules for coordinating two or more specific tasks. For example, the executive rules could preposition the eyes to the locations where input for the tasks will appear, improving performance. A specialized executive could monitor the state of one task and modulate the other task to improve performance by changing specific control items in memory. The executive processes describing in the Meyer and Kieras (1997a, 1997b) models for the psychological refractory period used this mechanism to account for the fine quantitative details in a high-precision data set. Making use of task-specific information means that the executive process production rules can be tuned to permit the maximum possible performance, but the details of the rules would only apply to that specific pair of tasks and their exact conditions. In addition, the production rules for each task will have to be written to take into account how they will be controlled by the executive process, typically in very specific ways.

In contrast, a *general executive* process can coordinate any set of subtasks because the executive process rules do not have to be changed to suit particular subtasks. The production rules for each task will have to follow the general executive's protocol to allow them to be controlled by the general executive but do not otherwise have to take into account which other tasks might be running at the same time. Usually, a general executive will be less efficient than a specialized executive because it will not take advantage of any special characteristics of the situation to enable higher performance; however, the general executive production rules are fully general and do not have to be acquired and tuned for each new combination of tasks.

Sequential Versus Concurrent Executives

The overall pattern of executive control of two or more tasks can be ranged from most sequential to most concurrent. A *single-task executive* ensures good performance on the highest-priority task simply by ignoring the lower-priority tasks; the result is best-possible performance on the most important task. A *sequential executive* runs one task, then the other, and then repeats. This is a simple scheme for dealing with resource conflicts—only one task runs at a time, so conflicts are not possible. However, both of these simple solutions will underuse the system resources since they will take no advantage of the possibilities of interleaving tasks.

An interesting case is a *time-slicing executive*: on each cycle, an extremely simple executive removes the goal for one task and inserts the other so that each task is enabled on alternate cycles. To prevent resource conflicts, the production rules for each task must follow the protocol of not attempting to use a resource (such as a motor processor) that is already busy but do not otherwise have to interact with the executive process or each other. Each task is thus allowed to proceed as long as it uses resources that are free. This is a purely sequential executive that interleaves resource utilization between the tasks on an as-needed first-come/first-served basis. This scheme requires that each task can be suspended at any point without conflicts or loss of information and the concurrent tasks be of equal priority. In some sample dual-task models, this scheme works surprisingly well because most of the time for the tasks consists of waiting on motor processors to complete their activity; the one-cycle delay in executing the production rules for each task impairs individual-task performance only very slightly, while the overall dual-task performance benefits by the automatic and very efficient interleaving of motor processor usage.

A fully *parallel executive* runs both tasks simultaneously and intervenes only when necessary to prevent resource conflicts and enforce task priorities. Such an executive requires that each task interact with the executive before attempting to use a resource; only in this way can the executive allocate resources based on the relative priorities of the tasks.

Executive Protocols

To run in an environment in which an executive process is in control, the task rules have to be written according to a protocol of some sort. Under EPIC, this requires that the tasks rules be governed by a single goal that an executive can manipulate to suspend or resume the task as a whole. In addition, some protocol governing access to resources must be followed. Under EPIC, a standard protocol is that a motor processor must not be commanded unless it is in a free state, as described previously. These basic assumptions lead to three kinds of executive protocols, described in terms of what kind of executive can successfully manage the task rules.

Single-task compatibility means that the task rules are written as if all resources were available for use by this set of rules at any time. The single-task, sequential, and time-slicing executives can successfully coordinate tasks that follow only this level of compatibility, but parallel executives cannot.

Specialized-executive compatibility means that the task rules assume the presence of a specific specialized executive that, for example, places specific control items in memory or do some of the task-specific processing. Such task rule sets are completely tied to a specialized executive, which in turn is bound to a specific combination of task rules.

General-executive compatibility means that the task rules follow a general executive protocol for resource access; before attempting to use a resource, they always request allocation of that resource from the general executive, wait for the resource to be allocated, and then release the resource as soon as possible, or

accept preemption. If the task rule sets follow this protocol, then a general executive can reliably coordinate resource usage for any pair of tasks.

Examples of Executive Processes

This section provides concrete examples of the executive concepts and distinctions just presented. A model for a dual task situation will be presented in each of two executive control regimes: a specialized executive and a general executive. First the task will be described, then the models presented.

The "Wickens" Dual-Task Experiment

The Martin-Emerson and Wickens (1992) task was first used for EPIC modeling as reported in Kieras and Meyer (1995), and since then has been reused by other modelers. The task is a simple dual task, involving a two-choice reaction task and a basic compensatory tracking task. Kieras and Meyer originally chose it for modeling because it involved a manipulation of the visual relationship between the two tasks, with eye movements sometimes being required.

Figure 23.15 shows the view of the display in the EPIC simulation package. The small square at the top

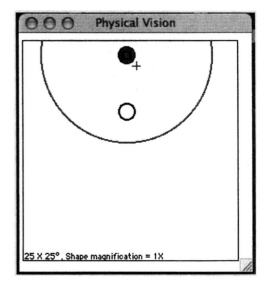

FIGURE 23.15 The "Wickens" task during tracking, as shown in EPIC's display of the physical visual field. The small gray circle represents the position of the fovea; the larger concentric circle is a 10-deg calibration circle.

of the display is the tracking target; the small cross is the tracking cursor. The subject's task is to keep the cross on the target by moving a joystick, despite a forcing function that tries to randomly displace the cursor. The dark circle imposed on the target represents the 1-deg foveal radius, showing that the eye is currently fixated on the target. The large concentric circle is a 10-deg calibration circle. The 1-deg circle near the middle of the display marks where the choice stimulus will appear, which it does at random intervals.

In Figure 23.16, the choice stimulus has appeared; it is either a right- or left-pointing arrow, and the subject is required to press the corresponding one of two buttons under fingers on the left hand. However, whether the arrow orientation can be discriminated visually depends on the distance between it and the current eye position; Martin-Emerson and Wickens manipulated this over a wide range. For reasonable retinal availability characteristics, the eye must be moved for eccentricities like those illustrated. Figure 23.17 shows the position of eye on the choice target. Once the choice stimulus is identified, the response button can be pressed and the tracking task resumed, returning to the situation shown in Figure 23.15. Dependent variables were the choice response latencies and tracking error accumulated during a 2-s period starting when the choice stimulus appeared.

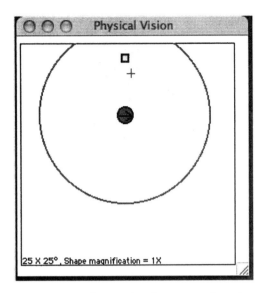

FIGURE 23.17 The eye has been moved to the choice stimulus.

Resource Allocation in the Wickens Task Both tasks require the use of the ocular motor processor for acquiring the stimulus (which implies use of fovea for each stimulus) and the manual motor processor for making the response. The motor processors are thus the critical resources, since under EPIC's assumptions, a motor processor cannot simultaneously perform two different actions. Thus, the two tasks must share them without conflict, and the efficiency of this sharing determines the overall dual-task performance. If each task is implemented with its own set of production rules, then there must be an executive process that coordinates the execution of each set of task rules and their access to the ocular and manual motor processors.

Before going further, it is worth discussing an alternative to an executive process: Suppose the human simply learns a single rule set for the Wickens task as a single complex combined task. That is, instead of trying to concurrently perform two distinct tasks, corresponding to two goals (Goal Do Tracking) and (Goal Do Choice), the human has only a single goal and its corresponding production rules: (Goal Do Combined_Task), and these production rules emit the appropriate sequence of motor movements for joystick manipulations, button pushing, and eye movements as required to respond to cursor movements and arrow appearances. As Kieras et al. (2000) argue, such a combined production rule set might well be a final end state of learning after

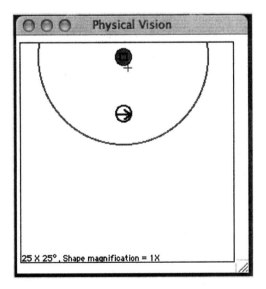

FIGURE 23.16 The choice stimulus has appeared in the marker circle.

considerable practice, but the dual-task representation seems more likely to be the state of learning in typical dual-task experiments, where only a moderate amount of practice has been done. At a purely theoretical level, if there are not separate production rules for each task, it means that we do not have a dual-task situation, but rather a complex single task that does not require the services of an executive. Thus, at least for the sake of theoretical exploration of executive processes, we need to continue with the assumption of separate task rule sets for each task.

Model Components   To construct an EPIC model for the Wickens dual task, we need four components:

1. A set of startup rules that run at the beginning of the experiment to identify and tag the different screen objects for use in the other rule sets.
2. A set of rules for tracking, governed by (Goal Do Tracking). In this example, these can be written in two threads: one keeps the eye on the tracking target, while another moves the joystick whenever the cursor is too far from the target. Since this task is inherently continuous, these rules should be normally enabled by the more-or-less continuous presence of their governing goal in working memory.
3. A set of rules for the choice reaction task, governed by (Goal Do Choice). When the choice stimulus appears, the eye must be moved to the choice stimulus location, and then these rules select and execute the response, depending on the shape property of the stimulus. This task is intermittent, but according to the experiment instructions, it is higher priority than the tracking task.
4. A set of executive rules for task coordination. In general form, they contain no information about the specific tasks, but in specialized form, they will need to use the choice stimulus onset as a trigger to determine when the choice task needs to run. For brevity, the following will focus on the executive rules, and show rules for the two subtasks only where important to explain the executive process.

Sequential Specialized Executive

Figure 23.18 shows an example of how a sequential specialized executive can coordinate the two tasks.

For clarity of exposition, key items in the rules are emphasized with boldface. The first rule, Top_startup, invokes the Identify Objects submethod, whose rules are not shown for brevity. This adds tags to memory identifying the visual objects corresponding to the tracking target and the cursor, which the tracking task rules refer to. The next rule, Top_start_tasks, adds the tracking task goal, which "turns on" the tracking task rules and then enables the next rule, Top_waitfor_ choicestimulus. This rule fires when there is a visual onset event, and tags the visual object as the choice stimulus, and suspends the tracking task by removing its goal. The next rule in sequence starts the choice task. When the choice task is complete, the final executive rule, Top_resume_tracking, replaces the tracking task goal and returns to waiting for the choice stimulus.

Figures 23.19 and 23.20 show the tracking and choice task rules written assuming that a specialized executive is present but as if no other task was present. The tracking task rules are very simple but actually run as two separate one-rule threads. The first, Watch_target, simply moves the eye to the tracking target whenever the target is outside the fovea. The second requires that the target be in the fovea, and moves the joystick with a Ply movement to place the cursor on the target whenever the distance between the target and the cursor (supplied by the perceptual processor) is greater than a specified amount. Note that the two rules can fire independently and repeatedly.

The choice task rules are more sequential in nature, being written for presentation clarity in the slow-response style described previously. The first is a standard startup rule. The second moves the eye to the choice stimulus—which the specialized executive had detected and tagged as part of its activity. Again for simplicity, these rules always move the eye to the choice stimulus, even when it might be close enough to the tracking target to be discriminated in parafoveal vision. For brevity, of the next rules to fire, only one of the response selection rules is shown, the one that maps a left-pointing arrow to the corresponding response button press. The final rule waits for the key-press response to be complete and then terminates the method. Note again how these rules do not depend on any content of the other task of tracking but do depend on the executive process to determine when these rules should run and identify their stimulus in the process.

The sequential specialized executive thus solves the resource allocation problem by forcing the tasks to execute sequentially—the higher-priority choice task

```
(Top_startup
If ((Goal Do Dual_task)(Not (Step ??? ???)))
Then ((Add (Goal Identify Objects))(Add (Step Start Tasks))))


(Top_start_tasks
If ((Goal Do Dual_task)(Step Start Tasks)
     (Not (Goal Identify Objects)))
Then ((Add (Goal Do Tracking))
     (Delete (Step Start Tasks))(Add (Step Waitfor ChoiceStimulus))))


(Top_waitfor_choicestimulus
If ((Goal Do Dual_task)(Step Waitfor ChoiceStimulus)
     (Visual ?stimulus Detection Onset))
Then ((Delete (Goal Do Tracking))
     (Add (Tag ?stimulus Choice_stimulus))
     (Delete (Step Waitfor ChoiceStimulus))(Add (Step Start ChoiceTask))))


(Top_start_choice_task
If ((Goal Do Dual_task)(Step Start ChoiceTask))
Then ((Add (Goal Do ChoiceTask))
     (Delete (Step Start ChoiceTask))(Add (Step Resume Tracking))))


(Top_resume_tracking
If ((Goal Do Dual_task)(Step Resume Tracking)
     (Not (Goal Do ChoiceTask)))
Then ((Add (Goal Do Tracking))
     (Delete (Step Resume Tracking))(Add (Step Waitfor ChoiceStimulus))))
```

FIGURE 23.18 Specialized sequential executive rules for the Wickens task.

runs to completion while the tracking task is suspended. Each subtask rule set can then follow a simple protocol for accessing the motor processors, relying on the specialized executive to ensure that they are not in conflict. The disadvantage is that the tracking task is suspended for the entire time required for the choice task, which in the example presented here, is substantial. This executive is necessarily specialized because it must "know" something about the choice task—namely, what stimulus is the trigger for starting execution of

the choice task. There is little additional harm done by having the specialized executive provide the stimulus specification to the choice task.

Parallel General Executive

This form of executive is most like a modern computer OS in that it performs general resource management during parallel task execution. Both of the tasks are running constantly—both goals are always present

```
(Watch_target

If ((Goal Do Tracking)(Step Watch Target)

      (Tag ?target Target)(Not (Visual ?target Eccentricity Fovea))

      (Motor Ocular Modality Free))

Then ((Send_to_motor Ocular Perform Move ?target)))


(Ply_cursor

If ((Goal Do Tracking)(Step Track Target)

      (Tag ?cursor Cursor)(Tag ?target Target)

      (Visual ?target Eccentricity Fovea)

      (Visual ?cursor Distance ?error)(Greater_than ?error .1)

      (Motor Manual Modality Free))

Then ((Send_to_motor Manual Perform Ply ?cursor ?target Right)))
```

FIGURE 23.19 Tracking task rules that assume a specialized sequential executive.

in memory. One task goal is marked as higher priority than the other. Each task asks for permission to use a motor processor, whereupon the executive grants the request if the resource is free or withholds permission if not. Each task releases a resource as soon as it is finished with it, and the executive reallocates the released resource to a task that is waiting for it. In case of conflict, the executive causes a higher-priority task to preempt a lower-priority task. The advantage to this general executive is that the task rules are independent of each other, and do not rely on the executive to handle any of the task specifics as in the previous example. The disadvantage is that the ask/wait/release processing each task must perform prior to using a resource constitutes some overhead. However, at least in the case of two tasks, this overhead can be minimized with a broad-but-shallow set of executive rules.

Logically, it is impossible to do away with specialized executive knowledge completely, in that there must always be some executive rules that specify which specific tasks should be performed in a given situation and what their priority relationship is. These are shown in Figure 23.21. As before, the first step (not shown here) is identifying the visual objects. The next rule, Top_start_tasks, then adds the goals for both the tracking task and the choice task and marks the tracking task as being lower priority than the choice task.

At this point, the general executive and the two sets of task rules can handle all of the task coordination, so no further top-level rules are shown.

Figure 23.22 shows the rules for the tracking task. A startup rule requests use of the ocular and manual motor processors by adding a *Status* item to memory. The Watch_target rule is the same as before, except its condition includes a Status item requiring that the tracking task has been allocated the ocular processor. Likewise, the Ply_cursor rule checks for permission to use the manual processor. Since the tracking task runs continuously but at a low priority, it never releases either processor—it is content to be preempted as necessary.

Figure 23.23 shows an excerpt of the rules for the choice task. The Waitfor_ChoiceStimulus rule is triggered by the stimulus onset and requests use of the ocular processor. The executive process will note that the request is coming from the higher-priority process and will deallocate the ocular processor from the tracking task (which disables the Watch_target rule) and assign it to the choice task. The rule Lookat_choice_stimulus waits until permission is granted then moves the eye to the choice stimulus and releases the ocular processor. At this point, the executive process reallocates the ocular processor back to the tracking task, reenabling the Watch_target rule. The choice task rule, Waitfor_shape, fires when the shape of the choice stimulus becomes available and requests the manual processor. In a similar manner, the executive withdraws the use of the manual processor from the tracking task and assigns it to the choice task, which then releases it in the same rule that commands the manual response.

To complete the example, Figure 23.24 shows a fragment of the general executive for coordinating any two tasks. The executive consists of many rules like those illustrated, one of which fires for each possible request or release situation, giving single-cycle response by the executive for resource requests and releases.

```
(Choice_task_startup
If ((Goal Do ChoiceTask)(Not (Step ??? ???)))
Then ((Add (Step Lookat ChoiceStimulus))))


(Lookat_choice_stimulus
If ((Goal Do ChoiceTask)(Step Lookat ChoiceStimulus)
    (Tag ?stimulus Choice_stimulus)
    (Motor Ocular Modality Free))
Then ((Send_to_motor Ocular Perform Move ?stimulus)
    (Delete (Step Lookat ChoiceStimulus))(Add (Step Waitfor EyeDone))))


. . .


(Select_response_left
If ((Goal Do ChoiceTask)(Step Select Response)
    (Tag ?stimulus Choice_stimulus)
    (Visual ?stimulus Shape Left_Arrow)
    (Motor Manual Modality Free))
Then ((Send_to_motor Manual Perform Punch B1 Left Middle)
    (Delete (Step Select Response))(Add (Step Finish Task))))


. . .


(Finish_choice_task
If ((Goal Do ChoiceTask)(Step Finish Task)
    (Tag ?stimulus Choice_stimulus)
    (Motor Manual Modality Free))
Then ((Delete (Tag ?stimulus Choice_stimulus))
    (Delete (Step Finish Task))
    (Delete (Goal Do ChoiceTask))))
```

FIGURE 23.20 Choice task rules that assume a specialized sequential executive.

```
(Top_start_tasks
If ((Goal Do Dual_task)(Step Start Tasks)
    (Not (Goal Identify Objects)))
Then (
    (Add (Goal Do Tracking))
    (Add (Status Trackingtask Priority Low))
    (Add (Goal Do ChoiceTask))
    (Add (Status Choicetask Priority High))
    (Delete (Step Start Tasks))
    (Add (Step Continue Running))))
```

FIGURE 23.21 Specialized task startup rule using a general executive.

```
(Tracking_task_startup
If ((Goal Do Tracking)(Not (Step ??? ???)))
Then (
    (Add (Status Trackingtask Request Ocular))
    (Add (Status Trackingtask Request Manual))
    (Add (Step Watch Target))(Add (Step Track Target))))


(Watch_target
If ((Goal Do Tracking)(Step Watch Target)(Tag ?target Target)
    (Not (Visual ?target Eccentricity Fovea))
    (Status Trackingtask Has Ocular)
    (Motor Ocular Modality Free))
Then ((Send_to_motor Ocular Perform Move ?target)))


(Ply_cursor
If ((Goal Do Tracking)(Step Track Target)
    (Tag ?cursor Cursor)(Tag ?target Target)(Visual ?target Eccentricity Fovea)
    (Visual ?cursor Distance ?error)(Greater_than ?error .1)
    (Status Trackingtask Has Manual)
    (Motor Manual Modality Free))
Then ((Send_to_motor Manual Perform Ply ?cursor ?target Right)))
```

FIGURE 23.22 Tracking task rules using a general executive.

. . .

(Waitfor_ChoiceStimulus

If ((Goal Do ChoiceTask)(Step Waitfor ChoiceStimulus)

    (Visual ?stimulus Detection Onset)(Not (Tag ?stimulus Choice_stimulus)))

Then ((Add (Tag ?stimulus Choice_stimulus))(Add (Status Choicetask Request Ocular))

    (Delete (Step Waitfor ChoiceStimulus))

    (Add (Step Lookat ChoiceStimulus))(Add (Step Waitfor Shape))))

(Lookat_choice_stimulus

If ((Goal Do ChoiceTask)(Step Lookat ChoiceStimulus)(Tag ?stimulus Choice_stimulus)

    (Status Choicetask Has Ocular)(Motor Ocular Processor Free))

Then ((Send_to_motor Ocular Perform Move ?stimulus)

    (Add (Status Choicetask Release Ocular))(Delete (Step Lookat ChoiceStimulus))))

(Waitfor_shape

If ((Goal Do ChoiceTask)(Step Waitfor Shape)

    (Tag ?stimulus Choice_stimulus)(Visual ?stimulus Shape ???))

Then ((**Add (Status Choicetask Request Manual)**)

    (Delete (Step Waitfor Shape))(Add (Step Select Response))))

(Select_response_left

If ((Goal Do ChoiceTask)(Step Select Response)

    (Tag ?stimulus Choice_stimulus)(Visual ?stimulus Shape Left_Arrow)

    (Status Choicetask Has Manual)(Motor Manual Processor Free))

Then ((Send_to_motor Manual Perform Punch B1 Left Middle)

    (Add (Status Choicetask Release Manual))

    (Delete (Step Select Response))(Add (Step Finish Task))))

. . .

(Finish_choice_task . . .

Then ((Add (Step Waitfor ChoiceStimulus))))

FIGURE 23.23 Excerpt of choice task rules using a general executive.

For example, the first rule fires when the high-priority task requests a resource currently in use by the low-priority task, and its actions reassign the resource to the high-priority task and note that the low-priority task use of that resource has been suspended. The second rule fires when a task releases a resource when another task using that resource has been suspended and reallocates the resource back to the suspended task.

Thus the use of the ocular and manual processors by the two sets of task rules are "automatically" interleaved by the general executive rules on an as-needed basis, and the resulting performance can be very high,

```
(GenEx_HP_Request_LP_Has_Preempt
If
((Status ?high_task Request ?resource)(Status ?high_task Priority High)
(Not (Status ?high_task Has ?resource))(Not (Status ?high_task Suspended ?resource))
(Not (Status ?high_task Waiting ?resource))(Not (Status ?high_task Release ?resource))
(Status ?low_task Has ?resource)(Status ?low_task Priority Low)
(Not (Status ?low_task Request ?resource))(Not (Status ?low_task Release ?resource)))
Then
((Add (Status ?high_task Has ?resource))
(Delete (Status ?high_task Request ?resource))
(Delete (Status ?low_task Has ?resource))
(Add (Status ?low_task Suspended ?resource))))


(GenEx_Reallocate_Resource_to_Suspended
If
((Status ?user Release ?resource)(Status ?user Has ?resource)
(Status ?low_task Suspended ?resource)
(Not (Status ??? Request ?resource))(Not (Status ??? Waiting ?resource)))
Then
((Delete (Status ?user Has ?resource))
(Delete (Status ?user Release ?resource))
(Add (Status ?low_task Has ?resource))
(Delete (Status ?low_task Suspended ?resource))))
```

FIGURE 23.24 Excerpt from the rule set for a parallel general executive.

even though there is some overhead imposed by the general executive. This overhead can be minimal in the case of two tasks because the rules to adjudicate resource conflicts between two tasks are relatively simple and limited in number, and so can be written in the illustrated manner to involve no more than a single cycle of overhead. To implement the same single-cycle process for three or more tasks leads to a combinatorial explosion in rules and rule complexity, and the multiple-cycle equivalent would be very complicated and slow executing.

Following up on the discussion of multitasking skill learning in Kieras et al. (2000), the complexity of all but the simplest general executive suggests that people have difficulty in multiple-task performance of novel task combinations primarily because they have not

had occasion to tackle the difficult task of learning an efficient high-capacity general executive, leaving them to flounder in a complex multiple-task situation. Perhaps true general multitask executives are so difficult to learn that they rarely appear except in very simple situations, meaning that complex multiple tasks might always require learning a specialized executive before good performance can be produced.

Which Executive Process Model Should Be Used?

The previous sections present several basic aspects of executive processes and illustrate in detail how the different models of executive processes can be implemented as production rules. One obvious question is the choice of executive process representation: Which executive process should be used in a model? As warned at the beginning, this is not an empirical chapter, and this question is basically an empirical question.

But there are several reasons why it may be difficult or impractical to identify exactly what executive process is at work in a particular multiple-task data set. As alluded to previously, Kieras et al. (2000) point out how some executive strategies seem to be compatible with different levels of skill development, especially at the extremes. However, the executive process that should make it easiest to combine novel tasks would be the general parallel executive, but which in turn seems to require considerable learning. Thus the relationship of executive process type to state of practice may not be straightforward. In addition, it should be kept in mind that in a parallel cognitive architecture, the overt performance of the system will be more determined by degree of optimization of the perceptual-motor processes, and only slightly or not at all by the details of the control regime that implements that optimization. In other words, the differences between of the executive processes may not be on the critical path for determining task execution time. For example, the general parallel executive in the above example imposes only an occasional one-cycle overhead, and depending on the exact timings of the perceptual and motor activities, this overhead might not be visible in the manifested performance at all, and perhaps only in an extreme stable and precise data set. Finally, as argued above, much of the details in task performance are influenced by the subjects' strategies adopted under the influence of the task demands, which can be hard to control or not controlled at all in many experimental paradigms.

Thus the different concepts of executive control are not offered necessarily as a better way to account for data, but rather as a better way to construct models. For example, the general parallel executive makes it very simple to construct a multitask model that uses resources efficiently, and as argued in this chapter, there is no good reason to limit the cognitive processor in ways that would prevent its use. If it is difficult to demonstrate that any other executive model fits the data better, why not use the one that is the most theoretically productive and easy to use?

Another application of the executive process models would be in applying the *bracketing heuristic* proposed by Kieras and Meyer (2000) to models of multiple-task performance. This heuristic addresses the problem of how to predict performance in a task that depends heavily on individual task strategies, especially under conditions where no behavioral data is available to fit a model to, as in predicting human performance for purposes of system design. Alternatively, the heuristic provides a relatively easy way to explain effects in observed data where a well-fitting model would be hard to construct. The basic idea of the bracketing heuristic is to construct two a priori models of the task: one is the *fastest-possible* model that uses a task strategy that produces the fastest possible performance allowed by the architecture. The other is the *slowest-reasonable* model uses a straightforward task strategy with no performance-optimizing "tricks." The predicted performance from these two models should bracket the actual performance. In a design analysis context, these predictions can be used to determine whether the system would perform acceptably despite the uncertainty about what level of performance the actual users will produce. The alternative use of the bracketing models is as an aid to interpreting data. Since the two bracketing models are well understood, the data can be compared to them to arrive at a useful qualitative understanding of the effects in the data without the difficulties of iterating a single model to fit the data quantitatively, as illustrated in Kieras, Meyer, and Ballas (2001) who analyzed the trade-offs manifested in two different interfaces for the same task.

The different executive process models can help specify the bracketing models: the fastest-possible model could either be some form of specialized parallel executive, but using a general parallel executive would probably produce very similar performance and be much easier to construct. The slowest-reasonable model could be represented as a simple sequential executive, reflecting a deliberate do-one-thing-at-a-time approach. This approach would make the bracketing heuristic easier and more uniform to apply to the theoretical analysis of multiple-task performance.

Conclusion

Application of the modern computer metaphor to basic functions of cognitive control both clarifies some of the theoretical issues and opens up new areas for theoretical exploration. For example, the relation between the perceptual system and the cognitive processor can be characterized rather differently than the traditional attentional selection approach. Also, the heavy cognitive involvement in motor activity could be much further developed. But the main contribution of the modern computer metaphor is to open up the possibilities for how cognition itself is organized and controlled, both in terms of basic flow-of-control issues such as parallelism and interrupts, and complex executive control of activity and resource allocation.

For many years, cognitive theory has discussed cognitive control, executive processes, and resource allocation as fundamental concepts in human multiple-task performance. Now we can directly model them, so that theories about these topics can develop more concretely, rigorously, and quantifiably. EPIC's flexible flow-of-control mechanisms make these concepts easy to implement, meaning that we can propose a variety of executive strategies and explore their consequences, such as what is actually required to coordinate multiple arbitrary tasks. The currently popular application of an oversimplified computer metaphor grossly underestimates the sophistication of both computers and humans. In contrast, the ease of theorizing about cognitive control with EPIC suggests that a good strategy for the cognitive architecture community would be to keep the control characteristics of architectures open as more is learned about complex task performance. In fact, the concepts used in modern computer systems should place a *lower* bound on the mechanisms we consider for human cognitive control processes.

Acknowledgments

Work on this chapter was supported by the Cognitive Science Program of the Office of Naval Research, under Grant N00014-03-1-0009 to David Kieras and David Meyer. David Meyer's contribution to these ideas has been substantial and essential, as indicated by the many coauthored articles cited in this chapter.

Note

1. A reconsideration of the literature on motor feature programming currently under way suggests that aimed movements to objects visually identified do not require a substantial amount of time for motor feature preparation, but the situation for other kinds of movements is not yet clear.

References

Baddeley, A. D., & Logie, R. H. (1999). Working memory: The multiple-component model. In A. Miyake & P. Shah (Eds.), *Models of working memory: Mechanisms of active maintenance and executive control* (pp. 28–61). New York: Cambridge University Press.

Bovair, S., Kieras, D. E., & Polson, P. G. (1990). The acquisition and performance of text editing skill: A cognitive complexity analysis. *Human-Computer Interaction, 5,* 1–48.

Bower, G. H., & Hilgard, E. R. (1981). *Theories of learning* (5th ed.). Englewood Cliffs, NJ: Prentice-Hall.

Card, S. K., Moran, T. P., & Newell, A. (1983). *The psychology of human-computer interaction*. Hillsdale, NJ: Erlbaum.

Ericsson, K. A., & Delaney, P. F. (1999). Long-term working memory as an alternative to capacity models of working memory in everyday skilled performance. In A. Miyake & P. Shah (Eds.), *Models of working memory: Mechanisms of active maintenance and executive control* (pp. 257–297). New York: Cambridge Univerity Press.

———, & Kintsch, W. (1995). Long-term working memory. *Psychological Review, 102*(2), 211–245.

Findlay, J. M., & Gilchrist, I. D. (2003). *Active vision*. Oxford: Oxford University Press.

John, B. E., & Kieras, D. E. (1996). The GOMS family of user interface analysis techniques: Comparison and contrast. *ACM Transactions on Computer-Human Interaction, 3,* 320–351.

Kieras, D. E. (1988). Towards a practical GOMS model methodology for user interface design. In M. Helander (Ed.), *Handbook of human–computer interaction* (pp. 135–158). Amsterdam: North-Holland Elsevier.

———. (1997). A Guide to GOMS model usability evaluation using NGOMSL. In M. Helander, T. Landauer, & P. Prabhu (Eds.), *Handbook of human–computer interaction* (2nd ed., pp. 733–766). Amsterdam: North-Holland.

———. (2004a). GOMS models and task analysis. In D. Diaper & N. A. Stanton (Eds.), *The handbook of task analysis for human-computer interaction* (pp. 83–116). Mahwah, New Jersey: Erlbaum.

———. (2004b). *EPIC architecture principles of operation*. Document available via anonymous ftp at ftp:// www.eecs.umich.edu/people/kieras/EPICtutorial/ EPICPrinOp.pdf.

———, & Meyer, D. E. (1995). Predicting performance in dual-task tracking and decision making with EPIC computational models. In *Proceedings of the First International Symposium on Command and Control Research and Technology* (pp. 314–325). National Defense University, Washington, DC, June 19–22.

Kieras, D., & Meyer, D. E. (1997). An overview of the EPIC architecture for cognition and performance with application to human-computer interaction. *Human-Computer Interaction, 12,* 391–438.

Kieras, D. E., & Meyer, D. E. (2000). The role of cognitive task analysis in the application of predictive models of human performance. In J. M. C. Schraagen, S. E. Chipman, & V. L. Shalin (Eds.), *Cognitive task analysis* (pp. 237–260). Mahwah, NJ: Erlbaum.

Kieras, D., Meyer, D., & Ballas, J. (2001). Towards demystification of direct manipulation: Cognitive modeling charts the gulf of execution. In *Proceedings of the CHI 2001 Conference on Human Factors in Computing Systems* (pp. 128–135). New York: ACM.

Kieras, D. E., Meyer, D. E., Ballas, J. A., & Lauber, E. J. (2000). Modern computational perspectives on executive mental control: Where to from here? In S. Monsell & J. Driver (Eds.), *Control of cognitive processes: Attention and performance XVIII* (pp. 681–712). Cambridge, MA: MIT Press.

———, Meyer, D. E., Mueller, S., & Seymour, T. (1999). Insights into working memory from the perspective of the EPIC architecture for modeling skilled perceptual-motor and cognitive human performance. In A. Miyake and P. Shah (Eds.), *Models of working memory: Mechanisms of active maintenance and executive control* (pp. 183–223). New York: Cambridge University Press.

———, & Polson, P. G. (1985). An approach to the formal analysis of user complexity. *International Journal of Man-Machine Studies, 22*, 365–394.

———, Wood, S. D., & Meyer, D. E. (1997). Predictive engineering models based on the EPIC architecture for a multimodal high-performance human-computer interaction task. *ACM Transactions on Computer-Human Interaction, 4*, 230–275.

Martin-Emerson, R., & Wickens, C. D. (1992). The vertical visual field and implications for the head-up display. In *Proceedings of the 36th Annual Meeting of the Human Factors and Ergonomics Society*, 1408–1412. Santa Monica, CA: Human Factors and Ergonomics Society.

Meyer, D. E., & Kieras, D. E. (1997a). A computational theory of executive cognitive processes and multiple-task performance: Part 1. Basic mechanisms. *Psychological Review, 104*, 3–65.

———, & Kieras, D. E. (1997b). A computational theory of executive control processes and human multiple-task performance: Part 2. Accounts of psychological refractory-period phenomena. *Psychological Review, 104*, 749–791.

———, & Kieras, D. E. (1999). Precis to a practical unified theory of cognition and action: Some lessons from computational modeling of human multiple-task performance. In D. Gopher & A. Koriat (Eds.), *Attention and performance XVII* (pp. 15–88). Cambridge, MA: MIT Press.

———, Kieras, D. E., Lauber, E., Schumacher, E., Glass, J., & Zurbriggen, E., et al.. (1995). Adaptive executive control: Flexible multiple-task performance without pervasive immutable response-selection bottlenecks. *Acta Psychologica, 90*, 163–190.

———, Kieras, D. E., Schumacher, E. H., Fencsik, D., & Glass, J. M. B. (2001). *Prerequisites for virtually perfect time sharing in dual-task performance*. Paper presented at the meeting of the Psychonomic Society, Orlando, FL.

Monsell, S., & Driver J. (2000). Banishing the control homunculus. In S. Monsell & J. Driver (Eds.), *Control of cognitive processes: Attention and performance XVIII* (pp. 3–32). Cambridge, MA: MIT Press.

———. (2002). *The roles of cognitive architecture and recall strategies in performance of the immediate serial recall task*. Unpublished doctoral dissertation. University of Michigan, Ann Arbor.

Mueller, S. T., Seymour, T. L., Kieras, D. E., & Meyer, D. E. (2003). Theoretical implications of articulatory duration, phonological similarity, and phonological complexity effects in verbal working memory. *Journal of Experimental Psychology: Learning, Memory, and Cognition, 29*, 1353–1380.

Norman, D. A. (1970). *Models of human memory*. New York: Academic Press.

———. (1976). *Memory and attention* (2nd ed.). New York: Wiley.

———, & Shallice, T. (1986). Attention to action: Willed and automatic control of behavior. In R. J. Davidson, G. E. Schwartz, & D. Shapiro (Eds.), *Consciousness and self-regulation* (Vol. 4). New York: Plenum Press.

Salvucci, D., Kushleyeva, Y., & Lee, F. (2004). *Toward an ACT-R general executive for human multitasking*. Proceeding of the 2004 International Conference on Cognitive Modeling, Pittsburgh, PA, July 30–August 1.

Schumacher, E. H., Lauber, E. J., Glass, J. M. B., Zurbriggen, E. L., Gmeindl, L., & Kieras, D. E., et al. (1999). Concurrent response-selection processes in dual-task performance: Evidence for adaptive executive control of task scheduling. *Journal of Experimental Psychology: Human Perception and Performance, 25*, 791–814.

———, Seymour, T. L., Glass, J. M., Fencsik, D., Lauber, E. J., & Kieras, D. E., et al. (2001). Virtually perfect time-sharing in dual-task performance: Uncorking the central cognitive bottleneck. *Psychological Science, 12*, 101–108.

Stallings, W. (1998). *Operating systems: Internals and design principles* (3rd ed.). Upper Saddle River, NJ: Prentice-Hall.

Tucker, A. (Ed.). (2004). *The computer science and engineering handbook* (2nd ed., pp. 46-1–46-25). Boca Raton, FL: CRC Press.

Yantis, S. & Jonides, J. (1990). Abrupt visual onsets and selective attention: Voluntary vs. automatic allocation. *Journal of Experimental Psychology: Human Perception and Performance, 16*, 121–134.

24

Integrated Models of Driver Behavior

Dario D. Salvucci

Our work on modeling driver behavior in a cognitive architecture has benefited greatly from two types of integration: composition of independently developed theories and models into the framework of a cognitive architecture, and generalization of common elements of theories and models into higher-level constructs within the architecture. This chapter highlights three ways in which integration by composition and generalization have arisen in the modeling of highway driving, driver distraction, and executive control within driving. Such integration has played a critical role in the incremental development of new theories of driver behavior and the implications of these theories for other domains. At the same time, this integration has facilitated the development of practical systems that use these theories in real-world applications, such as predicting the distraction potential of novel in-vehicle devices.

As cognitive architectures continue to move forward toward more truly "unified theories of cognition" (Anderson, 1983; Newell, 1990), integration has played and will continue to play a key role in their development. At least two distinct types of integration, which I shall call *integration by composition* and *integration by generalization*, have become evident in recent work on cognitive architectures. Integration by composition is the incorporation of independently developed theories for specific domains or phenomena into a broader cognitive architecture. For example, the EMMA eye-movement model (Salvucci, 2001a) for the ACT-R architecture (adaptive control of thought–rational; Anderson et al., 2004) was largely derived from the E-Z Reader model of eye-movement control (Reichle, Pollatsek, Fisher, & Rayner, 1998), which in turn was developed specifically for the domain of reading. The idea behind the development of EMMA was that, rather than "re-inventing the wheel" of eye-movement theories, an existing rigorous theory could be incorporated into the ACT-R architecture. Such an incorporation is

nontrivial in that it requires adaptation of the theory to fit within the broader architecture—for instance, the initiation of eye movements had to be tied to an existing module that directs visual attention (Byrne, 2001), which in turn is tied to production-rule firings in ACT-R. This type of integration has great benefits for the cognitive architecture: It extends the range of domains or phenomena potentially addressed by the architecture (in EMMA's case, separating observable eye movements from unobservable movements of attention), while perhaps inspiring new ways to think about the existing architecture and how it fits with other psychological theories.

Integration by generalization, in some ways related to but distinct from integration by composition, is the unification of separate models or theories within the architecture into a single, more general model or theory. For example, a recent treatment of list memory (Anderson, Bothell, Lebiere, & Matessa, 1998) proposed a general model that would serve as a basis for any cognitive model involving declarative representations of lists and procedural representations that operated on them. As another example, the path-mapping theory of analogy (Salvucci & Anderson, 2001) provides a common representation for declarative structures in analogical reasoning as well as the procedures that map the "paths" of one structure to another, thus inferring associated objects and relations. In both cases, the integration arises across models developed in the architecture: rather than models each relying on their own domain-specific representations, models can share a single common representation that has been independently validated on a cross section of tasks. Such efforts are even more critical as researchers use cognitive architectures to model increasingly complex tasks, where it is sometimes difficult to tease out the effects of lower-level phenomena (like studying list memory in the context of air traffic control) and thus it is greatly beneficial to have well-tested, lower-level models to allow the modeler to focus on the higher-level aspects of the task.

My colleagues and I have been working for several years on integrated models of driver behavior in the ACT-R cognitive architecture. Driving has proven a fascinating domain for the application of a cognitive architecture like ACT-R: the complex, dynamic nature of driving has pushed the architecture well past simple psychological experiments to more realistic everyday tasks, while the architecture has benefited the driving community by providing a rigorous framework for computational modeling. Integration, both by composition and generalization, has played an extremely important role in the long-term development of the models—perhaps an indication that, as for any complex domain, model development is necessarily a step-by-step process of integrating and building on previous work. In this chapter, I highlight three examples of integration within this work on driver behavior: (a) integration by composition of a lower-level control model into a production-system model for highway driving, (b) integration by composition of the driver model with models of in-vehicle secondary tasks to predict driver distraction, and (c) integration by generalization of the multitasking aspects of the previous models into a general executive for handling multitask performance.

Modeling Highway Driving

The first example of integration in our work on driving involves the composition of a "lower-level" control model

into a production-system model of highway driving. A number of control models of steering have evolved over the past several decades (e.g., Donges, 1978; Godthelp, 1986; Hildreth, Beusmans, Boer, & Royden, 2000). We have developed our own control model (Salvucci & Gray, 2004), described next, that derives from this previous work and formulates basic control using near and far road information. For purposes of developing a model of highway driving, we call these lower-level models in the sense that they focus on a particular aspect of the driving task, namely, steering down a single lane of travel (as opposed to lane changing, turning, etc.). However, such models require a significant conceptual leap for modeling highway driving in at least two significant ways: (1) specifying how the vision system acquires information and how the motor system outputs response and (2) specifying how basic steering can be incorporated into a model that must perform other tasks such as environmental monitoring and higher-level decision making. The ACT-R driver model (Salvucci, 2006) is an effort to do exactly this, specifying a fuller model of highway driving within the ACT-R cognitive architecture (Anderson et al., 2004).

The "two-point" model of steering (Salvucci & Gray, 2004) uses the perceived visual direction of two visual points: a *near point* in the near region of the roadway, used primarily to maintain a central position within the lane; and a *far point* in the far region of the roadway, used primarily to guide steering with respect to upcoming road curvatures. The model derived to some extent from parsimonious accounts of visual guidance in locomotion and steering (e.g., Rushton, Harris, Lloyd, & Wann, 1998; Wilkie & Wann, 2003). The critical distinction between our model and most previous models is that our model explicitly uses near and far information and uses only perceived visual direction to these points to guide steering. The model's two-point nature was inspired by a two-level model by Donges (1978), though Donges's model is much more complex and requires estimation of road curvature, which has been shown to be difficult for human observers to estimate accurately (e.g., Fildes & Triggs, 1985). The model was also inspired by empirical studies showing the two-level nature of visual attention during steering, most notably that of Land and Horwood (1995).

The two-point model is specified as follows. The near point in the model is defined as the center of the roadway at a convenient nearby distance ahead; this distance was set to 7 deg down from the horizon or roughly 6 m ahead of vehicle center, reported by Land and Horwood (1995) as the optimum for acquiring nearby lane-position information. The far point is defined as one of three possible points depending on the current scenario, shown in Figure 24.1: (a) the vanishing point of an approaching straight road segment, (b) the tangent point of an approaching curved segment, or (c) the center of a lead vehicle when one is present. Defining the visual angles θ_n and θ_f as the

visual angles to the near and far points, respectively, we can specify a continuous control law steering angle φ as:

$$\dot{\varphi} = k_f \dot{\theta}_f + k_n \dot{\theta}_n + k_l \theta_n. \tag{1}$$

Alternatively, we can formulate an analogous discrete control law updated at intervals Δt:

$$\Delta \varphi = k_f \Delta \theta_f + k_n \Delta \theta_n + k_l \theta_n \Delta t. \tag{2}$$

From these equations, we can see that the control law attempts to maintain three criteria: a stable far point such that $\dot{\theta}_f \approx 0$, a stable near point such that $\dot{\theta}_n \approx 0$, and a near point centered on the roadway such that $\theta_n \approx 0$.

As a control model of steering behavior, the model nicely fits various aspects of human steering behavior found in recent empirical studies (see Salvucci & Gray, 2004, for details). For example, Land and Horwood (1995) found that when viewing only a far region of the road, drivers exhibited smooth but inaccurate lane keeping (i.e., far from road center); in contrast, when viewing only a near region of the road, drivers exhibited more accurate but "jerky" control. The model reproduces this feature through its separation of near and far road points, the near region helping to guide accurate steering whereas the far region helping to smooth out driving based on the upcoming roadway. As another example, Hildreth et al. (2000) examined driver behavior in cases where the vehicle is veering off-center and the driver must make a quick corrective maneuver to guide the vehicle back to center. Figure 24.2a shows steering profiles for two individual drivers performing a corrective steering maneuver at different vehicle heading angles with respect to road heading; as is evident in the graph, larger heading angles resulted in larger steering magnitudes but roughly the same overall maneuver time. The model's behavior, shown in Figure 24.2b, reproduces this trend because the larger heading angle led to larger visual angles to the near and far points, and thus larger resulting steering angles. In addition, the model is able to capture basic individual differences by setting the scaling constants k in the control law to different parameter values; each model "driver" in the figure incorporates different parameter values that in this case are estimated to best fit the individual drivers in Figure 24.2a.

(a)

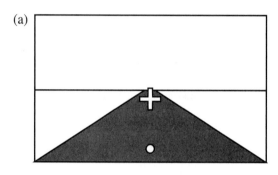

(b)

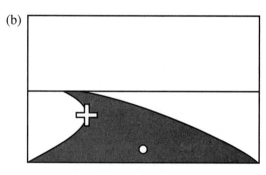

(c)

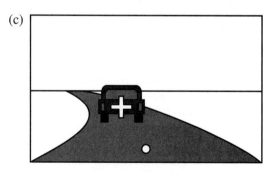

FIGURE 24.1   Near and far points for (a) straight road segment with vanishing point, (b) curved road segment with tangent point, and (c) any road with lead car (from Salvucci & Gray, 2004, with permission from Pion Limited, London).

While the two-point control model can form the backbone of an integrated driver model, the modeling of real behavior in a complex environment such as highway driving clearly requires much more than the

(a)

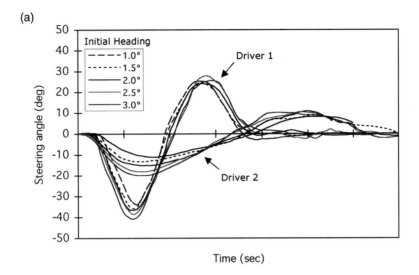

Time (sec)

(b)

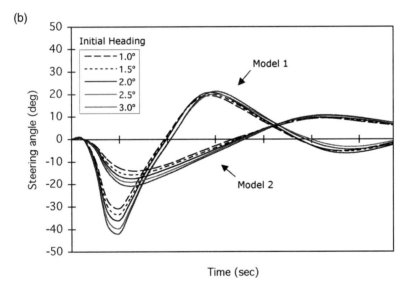

Time (sec)

FIGURE 24.2 Corrective steering profiles given the indicated initial vehicle heading for (a) human drivers in Hildreth et al. (2000), and (b) model simulations (adapted from Salvucci & Gray, 2004). Driver 1 and Driver 2 are two human drivers in the empirical study; Model 1 and Model 2 represent the two-point model with estimated parameter settings to best fit the data from Drivers 1 and 2, respectively.

control model; the control model says nothing about, for example, how the driver acquires visual information and produces motor responses, how the driver monitors her environment and makes higher-level decisions, or how the driver divides her cognitive "attention" to these various tasks efficiently and safely. To this end, the ACT-R integrated driver model (Salvucci, 2006) proposes a fuller account of driver behavior in the context

of multilane highway navigation. The driver model follows integration by composition by embedding the two-point control law into a tight control loop implemented as ACT-R production rules; these control rules iterate a process of (1) acquiring visual information through ACT-R's visual processor (Byrne, 2001); (2) computing an updated steering angle using the discrete form of the two-point control law, while also computing an updated accelerator/brake depression based on a similar control law; and (3) sending these updates to the motor system through the ACT-R motor system, modified for steering- and pedal-specific motor movements. At the same time, the model also integrates additional rule sets for monitoring, specifically looking out at the roadway and noting the current position of other vehicles, and decision making, specifically deciding whether and when to change lanes given time headway to a lead vehicle and distances to adjacent vehicles.

When validating a model of driving (or any complex task) with human empirical data, no one measure will provide a complete picture of the quality of the model; instead, we must validate the model across a number of relevant measures. The ACT-R driver model has been validated for several measures of driver behavior in a highway environment, such as steering and vehicle-position profiles during curve negotiation, analogous profiles for lane changing, and steering and braking performance during distracted driving (described in the next section). To highlight one measure here, Figure 24.3 shows the distribution of gaze time to various components of the visual scene for both human drivers and the integrated driver model. This measure is an example of the additional information derived from the integrated model above and beyond that in the control law: while the control law says nothing about how visual attention should be allocated, the integrated model accurately accounts for significantly

more eye movements (distinct from visual attention) to the far point versus the near point (since near information is acquired peripherally); the integrated model also accounts for gaze time to vehicles in the other lane and the mirror because of its incorporation of the monitoring and decision-making processes. Thus, integration by composition of the two-point control law into the fuller ACT-R model allows us to capture a significantly larger array of measures of real-world human behavior.

Modeling Driver Distraction

A second aspect of integration in our work involves the composition of the ACT-R driver model with models of secondary tasks to account for driver distraction. Driver distraction—inattention to the driving task typically related to secondary in-vehicle tasks such as cell-phone dialing—has received a great deal of media attention because of its dangers on today's roadways. Given that the ACT-R driver model already multitasks among three basic component processes (control, monitoring, decision making), my first attempt to model driver distraction (Salvucci, 2001b) asked a very straightforward question: if we integrate this model with yet another task, such as secondary in-vehicle task, can we account for effects of driver distraction with this model? The ACT-R cognitive architecture is critical to this endeavor in that it specifies the constraints of human

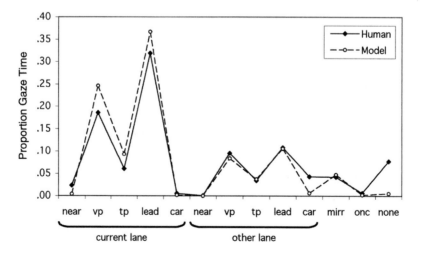

FIGURE 24.3 Proportion gaze time for human data and model simulations. Key: *near* = lane near point, *vp* = vanishing point, *tp* = tangent point, *lead* = lead vehicle, *car* = other vehicle, *mirr* = rear-view mirror, *onc* = oncoming vehicle, *none* = none of the above (adapted from Salvucci, 2006).

behavior, most importantly the constraints on the cognitive processor: the architecture posits a serial, single-threaded cognitive processor that can only "think about" one task at one time. Thus, the integration by composition here generates predictions of distraction that fall directly from both the cognitive architecture and the instantiation of the driver model in the architecture (as described in the previous section).

We have used this "integrated-model approach" (Salvucci, 2001b) to perform several studies of driver distraction. The first such study Salvucci, 2001b) explored how an integrated model could account for differences in distraction arising from cell-phone dialing using different input modalities, namely, manual versus voice dialing. Before performing the empirical study, the driver model (an older version of the current model: Salvucci, Boer, & Liu, 2001) was integrated with an ACT-R model of phone dialing in four conditions, each condition representing a combination of two factors: *full* versus *speed*, indicating whether the driver inputs the entire seven-digit number or a single speed number/code; and *manual* or *voice*, indicating whether the driver types digits manually or speaks the digits out loud (for processing by a speech-recognition system). After the model predictions were generated, the human-driver experiment was run with drivers performing in the same dialing conditions both during driving and as a single task (baseline). One result for the human drivers showed that the total time needed to dial the phone in all four conditions was slightly (1–2 s)

higher while driving; the model also needed more time because of the interleaved driving but also only slightly more time because the interleaving was done rather efficiently. An even more significant measure is that of driver performance in each condition, illustrated in Figure 24.4 as the average lateral deviation from lane center. Again, the model captures the most important qualitative effect in the human data: the manual-dialing conditions produce a significant effect on performance while the voice-dialing conditions produce no significant effect—especially surprising given that the *full-voice* condition incurred the most total time for all conditions for both human drivers and model. The model's a priori predictions in this case were somewhat off target quantitatively, although small changes to the control-law parameter values (as performed in Figure 24.2b) significantly improve the quantitative fit.

Another study of phone dialing while driving highlighted the interaction of driver distraction and age (Salvucci, Chavez, & Lee, 2004). This study used a recent result from Meyer, Glass, Mueller, Seymour, and Kieras (2001) that some aspects of modeling older populations can be accounted for with a 13% slowdown of the cognitive processor; Meyer et al. found this result for simpler laboratory tasks, but the effects of this slowdown when generalized to complex tasks such as driving are often not clear. In our study, we modeled phone dialing and driving as before, but, for the "older" drivers (roughly 60–70 years of age), we incorporated a 13% slowdown in ACT-R's cognitive processor cycle

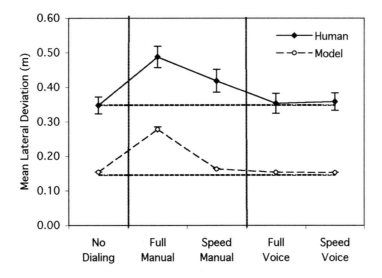

FIGURE 24.4 Distraction from phone dialing in different modalities as measured by lateral deviation for human drivers and model predictions (adapted from Salvucci, 2001).

time (raising it from the default 50 ms to 56.5 ms). Figure 24.5 shows the results as compared with results from human drivers measured by Reed and Green (1999); the graph plots side-to-side lateral velocity as a measure (like lateral deviation) of steering performance while driving. For both model and human drivers, the age-related slowdown has no effect while driving without a task—the control loop runs frequently enough that a slightly longer delay does not have observable effects when filtered through the complex dynamics of the vehicle. However, again for both model and human drivers, the slowdown has a significant effect in the presence of a task: both younger and older drivers are negatively affected by the dialing task, but the older drivers exhibit a significantly larger adverse effect on performance. In this study, integration by composition not only manifests itself in the composition of the driver and dialing models but also in the composition of the age-related slowdown theory into the cognitive architecture to produce immediate predictions from the integrated theory.

While these two studies emphasize distraction from the primarily perceptual-motor task of phone dialing, another study (Salvucci, 2002) highlights how the model can account for *cognitive distraction*—distraction from a primarily cognitive task. In this study, drivers performed a *sentence-span task* involving sentence processing and word recall: drivers listened to five sentences of the form "X does Y" (e.g., "The boy brushed

his teeth."); judged whether the sentence made sense; and after five sentences, recalled and stated the final word of each sentence. The ACT-R model of this intense cognitive task was largely derived from an ACT-R model previously developed by Lovett, Daily, and Reder (2000) for a similar task. As before, the integration of this sentence-span model with the ACT-R driver model immediately made predictions about potential cognitive distraction resulting from performing both tasks at once. The model's predictions were compared with empirical results from a driving study by Alm and Nilsson (1995) and again the model performed well in accounting for effects of distraction, this time for both lateral measures (lateral deviation) and longitudinal measures (brake reaction time to an external stimulus) of driver performance. Again, integration by composition is central to this study, particularly in that most of the secondary task model's declarative and procedural representations were re-used from an existing, independently validated model.

Modeling Executive Control

A third aspect of integration in our driving work, specifically integration by generalization, has come in the development of a "general executive" for human multitasking. All our previous models of driving, like almost all models of other complex tasks reported in

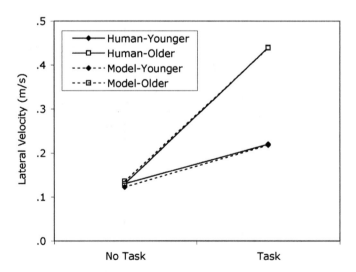

FIGURE 24.5 Distraction from phone dialing for older and younger drivers as measured by lateral velocity for human drivers and model predictions (adapted from Salvucci, Chavez, & Lee, 2004).

the literature, have used "customized executives" (Kieras, Meyer, Ballas, & Lauber, 2000) with a specialized, domain-dependent executive process for switching among tasks. A customized executive has two significant drawbacks: the individual task models must be modified to provide awareness of and switching capabilities to the other tasks and the executive process cannot easily generalize to other domains or even different situations in the same domain. We set out to develop a domain-independent general executive that could take two or more well-learned task skills and automatically perform both tasks together. Like the previous work, the general executive would be situated in the ACT-R cognitive architecture, which (as seen in the distraction work) has a single-threaded cognitive process; to perform multitasking, the general executive must interleave small portions of execution for each task to create a balanced distribution of processing. Thus, the integration by generalization here extracts the common elements of multitasking and interleaving from previous driving work and generalizes it into a domain-independent theory and mechanism.

The recently developed ACT-R general executive (Salvucci, 2005) arose from three guiding principles. First, the general executive (GE) acts as an architectural mechanism: rather than being implemented in ACT-R production rules as a learned cognitive skill, the GE is embedded in the architecture "hardware" as a core domain-independent process. The rationale behind the architectural mechanism comes from observations that implementing an interrupting, scheduling GE is extremely difficult within a production-rule process and that rule-based procedural knowledge and control processes seem to be centered in different regions of the brain (basal ganglia vs. dorsolateral prefrontal cortex, respectively: Anderson et al., 2004; Fincham, Carter, van Veen, Stenger, & Anderson, 2002). Second, the GE must be dependent on time (see, e.g., Kushleyeva, Salvucci, & Lee, 2005), since people are clearly aware of how much time they spend on one task and how soon they should switch to another—for example, checking a flight instrument gauge with a frequency appropriate to the expected frequency of fluctuations in the gauge's measurements. Third, the GE must be dependent on goal representations: rather than switching among tasks at arbitrary points, people switch at reasonable, or logical points, as dictated by the task and/or mental representations of task goals (see, e.g., Gray & Schoelles, 2003). While these three principles are not meant to be exhaustive in describing a general executive, they

do, as described next, cover enough properties of a GE to be useful for modeling a range of complex dynamic tasks.

The ACT-R general executive can be summarized in terms of four core points (see Salvucci, 2005, for a detailed exposition):

- Rules can create multiple goals, all of which are placed in a "goal queue" and remain active until completion (unlike standard ACT-R, which maintains only a single goal).
- Rules can specify a goal's desired start time if desired; by default, goals are set to start at the current time, but rules may also defer goals to a later time.
- Goals run uninterrupted until completion; this assumes a fairly small grain size for goal representations, and iterating processes (such as updating a car's steering control) are treated as iteratively generated new goals to allow other goals to interleave.
- Upon completion of the current goal, the most due (or overdue) goal—that is, the active goal with the earliest desired start time—is selected as the next goal.

In essence, if all goals have a default (immediate) desired start time, the GE reduces to a first-in, first-out queue. However, the GE allows for later start times and a special "now" start time (which starts the goal regardless of the goal-queue state) and also incorporates temporal noise that produces variability in task interleaving.

We have used this general executive in a dual attempt to unify the executive mechanisms in the driving work and, at the same time, propose a mechanism general enough to extend to other complex domains. Given that the earlier modeling efforts used customized executives to integrate control with monitoring (for highway driving) or secondary tasks (for driver distraction), the most recent effort (Salvucci, 2005) aimed to use the same general executive for all such integration. In particular, the same control model described earlier was integrated with another model in three separate studies: one focused on control and monitoring in highway driving, one on control while tuning a radio, and one on control while dialing a phone. The proposed GE nicely captured a number of interesting aspects in the data, including both aggregate measures of task-switching performance (e.g., distributions of gaze to different regions of the road, replicating the results of Figure 24.3 with the general executive) and specific measures of when drivers switch tasks (e.g., time spent on a particular task before switching to the other).

To illustrate one important result from this effort, the empirical study described in Salvucci (2005) further elucidates when people switch between driving and phone dialing. While replicating some of the modeling results of the earlier distraction model (Salvucci, 2001b) with the general executive, this work also examines the step-by-step task switching as observed in the individual key presses during phone dialing: as each

digit of the 10-digit phone number is pressed, the *key delay* records the time elapsed before the key is pressed. Figure 24.6a shows the key delays for human drivers both in the driving and baseline (nondriving) conditions: significantly more time is spent in the driving condition at the digit-block boundaries of the 10-digit number (i.e., at the first position of each block in the three-block form *xxx-xxx-xxxx*), indicating that subjects are interleaving some driving-related processing at these positions; at the same time, the delays at the non-boundary positions are not significantly different. The model's results in Figure 24.6b show a similar

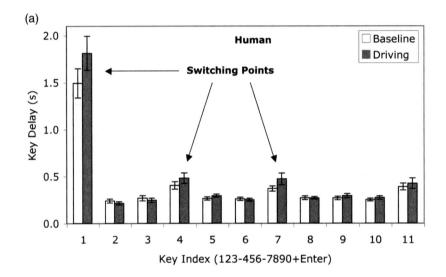

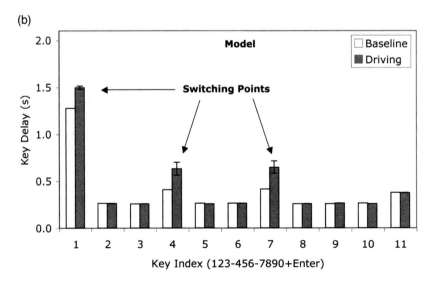

FIGURE 24.6 Task-switching points as illustrated by key delay times for (a) human drivers and (b) model simulations (from Salvucci, 2005).

effect: because of the declarative representation of the phone numbers as blocks of three, three, and four digits and the goal representation of dialing one block at a time, the model switches to driving at the block boundaries and exhibits slightly longer key delays at these positions. This result, combined with various other aggregate validation measures detailed in Salvucci (2005), demonstrates that the executive nicely captures multitasking performance across different driving and secondary tasks. More broadly, we see here that the integration by generalization has succeeded in generalizing a theory of multitasking for various aspects of the driving domain, and our current and future work aims to further validate the theory across other complex dynamic domains.

Theoretical and Practical Implications

The integration of cognitive theories and models by composition and generalization has significant implications for theory development and practical applications. Integration by composition is an extremely important tool for moving toward more unified cognitive theories, allowing for reuse of existing theories and models to help maintain theoretical parsimony across a unified account of cognition. Composition contributes especially to the (necessarily) incremental validation of unified theories: as more basic elements of a theory are validated with smaller-scale laboratory experiments, the broader integrated theories can rely on the earlier lower-level validations and broaden the scope to higher-level validations. For example, simply by using the ACT-R cognitive architecture, our driving work takes advantages of basic theories of memory, perceptual-motor processes, and so on, freeing us from validating such phenomena in such a complex task (where isolating these phenomena would be difficult) and allowing us to focus on broader measures that result from their integration. (Similar work on driving has been and is still under development for other cognitive architectures; e.g., Aasman, 1995; Tsimhoni & Liu, 2003.) Composition thus bootstraps the theory-development process and integrates prior work to facilitate the development of more comprehensive theories.

Integration by generalization sometimes occurs along with composition, in that as an existing theory/model is incorporated into a broader theory/model, certain aspects have to be reformulated and generalized to accommodate the new components (e.g., the generalization of EMMA from the reading domain to other domains; Salvucci, 2001a). At other times, generalization comes later from a realization that different theories/models share common, similar components that might be unified into a more general concept. The latter scenario better characterizes the origins of the general executive model presented here: Only after developing a number of customized executives for different domains did we recognize the potential benefits of unification into a general theory. In fact, this work also involved integration by composition, incorporating ideas from work in other cognitive architectures, including EPIC (see Kieras et al., 2000) and queuing network modeling (Liu, 1996)—again, with generalization and composition working hand-in-hand. Regardless of the origins, integration by generalization helps to ensure that all parts of the unified theory fit together in a parsimonious way.

Whether by composition or generalization, integration also has important implications for practical applications in real-world system design and development. By composing validated models (e.g., the ACT-R driver model) and validated general mechanisms (e.g., the ACT-R general executive), integration provides immediate a priori predictions about task behavior and performance. In the best case, these a priori predictions are accurate in quantitative and qualitative ways; however, even when they miss the mark quantitatively, many times they closely predict the qualitative effects that are often just as, if not more, important than an exact quantitative match. In the driving work, we have attempted to make a priori predictions (i.e., zero-parameter predictions with no data fitting) in several studies, and we often find that the models nicely predict the main effects and interactions. A recent study of the Distract-R system (Salvucci, Zuber, Beregovaia, & Markley, 2005) is an excellent example: With no parameter fitting, we were able to predict the effects of driver distraction from different input modalities and at different ages. After acquiring these results, we then adjusted one parameter to scale the model and achieve the best quantitative fit, but even in its raw a priori form the model performed very well in capturing the qualitative effects. Thus, for practical applications such as predicting the distraction potential of a set of in-vehicle interfaces, these qualitative results are extremely important in that they give us a rank order of interfaces with respect to distraction potential. By incorporating the power of integrated architectures, theories, and models, such tools have great potential for opening up the fundamental theoretical work to a much broader audience of users, designers, and nonmodelers in general.

Acknowledgments

This work is supported by Office of Naval Research Grant N00014-03-1-0036 and National Science Foundation Grants IIS-0133083 and IIS-0426674.

References

Aasman, J. (1995). *Modelling driver behaviour in Soar.* Leidschendam, The Netherlands: KPN Research.

Alm, H., & Nilsson, L. (1995). The effects of a mobile telephone task on driver behaviour in a car following situation. *Accident Analysis & Prevention, 27,* 707–715.

Anderson, J. R. (1983). *The architecture of cognition.* Cambridge, MA: Harvard University Press.

———, Bothell, D., Byrne, M. D., Douglass, S., Lebiere, C., & Qin, Y. (2004). An integrated theory of the mind. *Psychological Review, 111,* 1036–1060.

———, Bothell, D., Lebiere, C., & Matessa, M. (1998). An integrated theory of list memory. *Journal of Memory and Language, 38,* 341–380.

Byrne, M. D. (2001). ACT-R/PM and menu selection: Applying a cognitive architecture to HCI. *International Journal of Human-Computer Studies, 55,* 41–84.

Donges, E. (1978). A two-level model of driver steering behavior. *Human Factors, 20,* 691–707.

Fildes, B. N., & Triggs, T. J. (1985). The effect of changes in curve geometry on magnitude estimates of road-like perspective curvature. *Perception & Psychophysics, 37,* 218–224.

Fincham, J. M., Carter, C. S., van Veen, V., Stenger, V. A., & Anderson, J. R. (2002). Neural mechanisms of planning: A computational analysis using event-related fMRI. *Proceedings of the National Academy of Sciences, 99,* 3346–3351.

Godthelp, H. (1986). Vehicle control during curve driving. *Human Factors, 28,* 211–221.

Gray, W. D., & Schoelles, M. J. (2003). The nature and timing of interruptions in a complex cognitive task: Empirical data and computational cognitive models. In *Proceedings of the 25th Annual Meeting of the Cognitive Science Society* (p. 37). Mahwah, NJ: Erlbaum.

Hildreth, E. C., Beusmans, J. M. H., Boer, E. R., & Royden, C. S. (2000). From vision to action: Experiments and models of steering control during driving. *Journal of Experimental Psychology: Human Perception and Performance, 26,* 1106–1132.

Kieras, D. E., Meyer, D. E., Ballas, J. A., & Lauber, E. J. (2000). Modern computational perspectives on executive mental processes and cognitive control: Where to from here? In S. Monsell & J. Driver (Eds.), *Control of cognitive processes: Attention and performance XVIII* (pp. 681–712). Cambridge, MA: MIT Press.

Kushleyeva, Y., Salvucci, D. D., & Lee, F. J. (2005). Deciding when to switch tasks in time-critical multitasking. *Cognitive Systems Research, 6,* 41–49.

Land, M., & Horwood, J. (1995). Which parts of the road guide steering? *Nature, 377,* 339–340.

Liu, Y. (1996). Queueing network modeling of elementary mental processes. *Psychological Review, 103,* 116–136.

Lovett, M. C., Daily, L. Z., & Reder, L. M. (2000). A source activation theory of working memory: Cross-task prediction of performance in ACT-R. *Cognitive Systems Research, 1,* 99–118.

Meyer, D. E., Glass, J. M., Mueller, S. T., Seymour, T. L., & Kieras, D. E. (2001). Executive-process interactive control: A unified computational theory for answering twenty questions (and more) about cognitive ageing. *European Journal of Cognitive Psychology, 13,* 123–164.

Newell, A. (1990). *Unified theories of cognition.* Cambridge, MA: Harvard University Press.

Reed, M. P., & Green, P. A. (1999). Comparison of driving performance on-road and in a low-cost simulator using a concurrent telephone dialing task. *Ergonomics, 42,* 1015–1037.

Reichle, E. D., Pollatsek, A., Fisher, D. L., & Rayner, K. (1998). Toward a model of eye movement control in reading. *Psychological Review, 105,* 125–157.

Rushton, S. K., Harris, J. M., Lloyd, M. R., & Wann, J. P. (1998). Guidance of locomotion on foot uses perceived target location rather than optic flow. *Current Biology, 8,* 1191–1194.

Salvucci, D. D. (2001a). An integrated model of eye movements and visual encoding. *Cognitive Systems Research, 1,* 201–220.

———. (2001b). Predicting the effects of in-car interface use on driver performance: An integrated model approach. *International Journal of Human-Computer Studies, 55,* 85–107.

———. (2002). Modeling driver distraction from cognitive tasks. In *Proceedings of the 24th Annual Conference of the Cognitive Science Society* (pp. 792–797). Hillsdale, NJ: Erlbaum.

———. (2005). A multitasking general executive for compound continuous tasks. *Cognitive Science, 29,* 457–492.

Salvucci, D. D. (2006). Modeling driver behavior in a cognitive architecture. *Human Factors, 48,* 362–380.

———, & Anderson, J. R. (2001). Integrating analogical mapping and general problem solving: The path-mapping theory. *Cognitive Science, 25,* 67–110.

———, Boer, E. R., & Liu, A. (2001). Toward an integrated model of driver behavior in a cognitive architecture. *Transportation Research Record, 1779,* 9–16.

———, Chavez, A. K., & Lee, F. J. (2004). Modeling effects of age in complex tasks: A case study in driving. In *Proceedings of the 26th Annual Conference of the Cognitive Science Society* (pp. 1197–1202). Mahwah, NJ: Erlbaum.

———, & Gray, R. (2004). A two-point visual control model of steering. *Perception, 33,* 1233–1248.

———, Zuber, M., Beregovaia, E., & Markley, D. (2005). Distract-R: Rapid prototyping and evaluation of in-vehicle interfaces. In *Human Factors in Computing Systems: CHI 2005 Conference Proceedings* (pp. 581–589). New York: ACM Press.

Tsimhoni, O., & Liu, Y. (2003). Modeling steering using the queuing network—model human processor (QN-MHP). In *Proceedings of the Human Factors and Ergonomics Society 47th Annual Meeting* (pp. 1875–1879). Santa Monica, CA: Human Factors and Ergonomics Society.

Wilkie, R. M., & Wann, J. P. (2003). Controlling steering and judging heading: Retinal flow, visual direction and extra-retinal information. *Journal of Experimental Psychology: Human Perception and Performance, 29,* 363–378.

25

The Minimal Control Principle

Niels Taatgen

Control in cognitive models is usually fully internal and tied to a goal representation. To explain human flexibility and robustness in task performance, however, control should be shared between the goal (top-down control) and perceptual input (bottom-up control). According to the *minimal control principle*, top-down control should be minimized to obtain optimal flexibility with the smallest set of task knowledge. In cognitive models based on productions, the amount of control can be quantified by the number of control states needed. Support for the principle consists of an analysis that shows that the number of productions needed in a model increases linearly with the number of control states and by examining a number of examples of small and complex tasks in which minimal control leads to better models. Finally, the interaction between learning and control and the consequences for the representation of instructions will be analyzed.

Cognitive models based on production rules have often been criticized (e.g., Clark, 1997; Dreyfus, 1979) for disregarding the need to interact with the outside world and solving problems completely "in the head." In these models, interaction with the outside world only consists of perceiving the initial state and producing a sequence of actions to reach the goal. Many studies (e.g., Gray & Fu, 2004; Hutchins, 1995; Larkin & Simon, 1987) have stressed the importance of external representations in cognition and the need for cognitive models to be able to interact with such representations. In parallel, developments in robotics and situated cognition (Clark, 1997) have made it clear that interaction with the world is an essential element of cognition that cannot be dismissed even in models of higher cognition. In recent years, we have seen a development in which many cognitive architectures have been outfitted with modules to interact realistically with the outside world (e.g., ACT-R/PM: Byrne & Anderson, 2001; EPIC-Soar: Chong & Laird, 1997) or architectures whose sole focus is on this interaction

(e.g., EPIC: Meyer & Kieras, 1997). These new architectures enable models that accurately predict human performance and learning in many situations in which humans have to interact with the outside world, usually a computer screen, keyboard, and mouse. Despite their success, most models only partially achieve the embodied interaction goal because the *control* of action is still a strictly internal affair. This is not necessarily a limitation of the underlying architecture, but maybe more due to current practice in cognitive modeling. The goal of this chapter is to promote a different way constructing cognitive models. I hope to convince the reader that humans not only offload as much as possible of the problem representation to the environment but also as much control as possible, and that our models should do the same to be accurate. This idea can be formulated as the *minimal control principle* (Taatgen, 2005):

People organize their behavior for a task such that the cognitive model that corresponds to this behavior is the model with the minimal number of control states.

The minimal control principle connects an informal notion about behavior, the amount of control, to a formal notion in a cognitive model, the number of control states. When it holds, it can help to add constraints to cognitive architectures, because it states that when faced with a potential large set of models for a task, the model(s) with the minimal number of control states should be chosen.

The Example of Making Coffee

Before discussing evidence for the minimal control principle, I would like to go through an example by Larkin (1989) to clarify the issues. Larkin describes the task of making coffee, which on the surface seems to be an easy and effortless task but proves to be complex and detailed. Figure 25.1 shows some of the subtasks of making coffee with their respective dependencies. Given the complexity of the graph, it is amazing that the process itself, assuming it is carried out by an experienced coffee maker, is almost completely effortless. Larkin states that the following properties require an explanation:

1. The process is easy.
2. It is largely error free.

3. It is not degraded by interruption: The task can be picked up at any stage without extensive problem solving, and it is even possible to complete the coffee-making process when someone else has started it.
4. The steps are performed in a variety of orders. The constraints on the task only partially order the subtasks, allowing several orders of executing the steps and in theory even parallel execution of steps.
5. The process is easily modified. For example, if there is still water in the reservoir, the plan can effortlessly be modified to skip filling the reservoir but otherwise carry out all the other steps.
6. Performing the task smoothly and easily requires learning. This means that the reasoning process doesn't come "for free." However, experts can adapt the plan in novel ways, so it is not necessarily a case of caching old solutions.

The important aspect of Larkin's model DiBS (display-based solver) is that it does not retain information that can be observed in the environment in its working memory. It is therefore very flexible in its planning and can handle interruptions very well. Larkin also observes that errors made in making coffee typically involve properties that cannot be perceived. For example, if the reservoir of the coffee maker is transparent, people never try

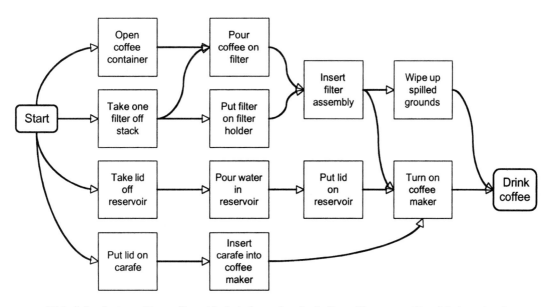

FIGURE 25.1 Subtasks in making coffee with their dependencies indicated by arrows. Simplified version from an example by Larkin (1989).

to fill it twice, but when it is opaque, a frequent error is to try to fill a reservoir that is already full.

Larkin's example of making coffee illustrates a case in which both the representation of the task and part of the control over the task are offloaded to the world. Our actions are not solely determined by our internal goals but partly by the environment as well. The example also shows that some external control on our actions is highly beneficial.

Another example in which external control is useful is in multitasking situations. People generally are quite proficient in executing multiple well-practiced tasks in parallel as long as they do not use the same perceptual and motor resources at the same time, even if those tasks have been learned separately. One explanation is that people swap mental goals at the appropriate moment (Salvucci, 2005). However, finding the appropriate moment for swapping goals is nontrivial and requires a sophisticated planning algorithm. An alternative is to mentally structure the task in such a way that goal changes are unnecessary. If the latter is possible, the minimal control principle urges us to prefer that solution.

Control and the Representation of Instructions

Let us analyze the slightly more simple example of brewing tea. An elementary cookbook might give the following recipe:

1. Put water in kettle
2. Put kettle on stove
3. Put leaves in teapot
4. Wait until water boils
5. Pour water in teapot

If we knew nothing about the function of all these steps (which can be the case in some more involved recipes), the only thing that we could do is carry out the instructions in their listed order, while keeping the last step we have performed in memory, which then serves as a cue for determining the next step. Such a problem representation requires six control states—internally represented symbols or markers of where we are in the process—for the start and end state, and the four states in-between the five steps. This representation leads to inflexible behavior and easily breaks down. For example, if we would lose track of the state due to some interruption, it would be hard to pick up the task again.

Also, this representation hides the fact that Step 3 can be done in parallel with any of Steps 1, 2, and 4, allowing many different orderings that are potentially more efficient.

A better representation is to include in each step the conditions under which that step can be taken, producing the following representation for the tea example:

1. [empty kettle] put water in kettle
2. [kettle with cold water] put kettle on stove
3. [empty teapot] put leaves in teapot
4. [water boils and leaves are in teapot] pour water in teapot

The advantage of this representation is that control is shifted to the environment, so it is no longer necessary to keep track of where you are: The environment will cue the next step. It also allows for multiple orderings of the steps and saves out one instruction because it is no longer necessary to specify the "wait" step. Instead of six control states, we only need a single control state.

From this example, and from Larkin's analysis of making coffee, it does indeed seem that minimizing control states is a good idea for humans and models. To support this idea, I will first present a formal argument that states that the amount of knowledge we need to perform a task flexibly increases linearly with the number of control states. Second, I will discuss some models in which minimal control works particularly well. The first model will entail a simple choice-reaction task that can be easily extended to a dual choice-reaction task paradigm. The advantage will be a good fit of the data that is hard to obtain with a model with more top-down control. The second model will be of a complex dynamic task, in which minimal control improves the flexibility of the model. This improved flexibility helps account for certain aspects of the data and would be hard to explain by a model with more control, except when the additional flexibility would be explicitly encoded.

The Relationship Between Control States and the Amount of Knowledge

The following analysis assumes that we use some sort of production-based system to represent knowledge and that productions are responsible for control. Production rules match a combination of internal states (and knowledge) and information through sensory inputs, and prescribe a set of actions. More formally, the set of production rules that we specify produces the following mapping:

P: {internal states} × {sensory inputs} → {actions}.

It is useful to subdivide the internal state of the system into a *control state* and a *problem state*. The problem state stores specific information about the problem (i.e., what are the numbers to be added or what is the flavor of the tea we are making), and the control state keeps track of where we are in the problem-solving process. The expanded mapping is

P: {control states} × {problem states}
× {sensory inputs} → {actions}.

One way to judge the robustness of a production system is to see whether the rules in the system have appropriate actions for any conceivable combination of control state, problem state, and sensory input, preferably with as few individual productions as possible. This means that number of control states, problem states, and possible sensory inputs all influence the eventual complexity and robustness of the production system. Sensory input is more or less out of our control, although an appropriate attentional mechanism might be able to filter out irrelevant input. The effect of the number of possible problem states can potentially be limited because a single production rule can handle many problem states by using variables. For example, one addition production rule can handle any combination of addends in the problem state. That leaves us with the number of control states. The simplest way of representing the control state, which is used frequently, is to associate each possible state with a specific symbol. In such a representation, production rules have to be either specific to a specific control state or generally applicable to all control states. From the assumption that productions applicable to all control states are uncommon, it can be concluded that the number of production rules needed to provide a full coverage of all possible states and inputs increases linearly with the number of control states. Effectively, this means that we have to create rules for exceptions that may or may not occur in reality, and rules for picking up the task after any type of interruption. For example, in the case where we are in the start state of making tea, but there is still enough water in the kettle, we would need a separate rule to cover this case if we have six control states but not if we have only one control state.

The corollary of this analysis is that the model with the minimal number of control states is also the model with the minimal number of production rules. Philosophers of science have made the general argument that simple theories are to be preferred to complex theories when both are equivalent in explaining the phenomena (Sober, 1975). More specifically, Chater (1999) argues that the simplicity principle also applies to the domain of psychological theory, not only for general philosophy of science reasons but also because of its role in *rational analysis* (Anderson, 1990). According to rational analysis, a theorist in cognitive science should derive the optimal behavioral function given the goals, environment, and computational limitations of the cognitive system, precisely because the cognitive system itself applies such an optimization. The minimal control principle can be seen as a specific instantiation of this more general principle.

Although minimal control seems to be a sound scientific practice, it is nevertheless at odds with present modeling practice. Many methods for describing the structure of tasks organize their descriptions in hierarchies of steps. GOMS (Card, Moran, & Newell, 1983), for example uses methods to organize the structure of the task in which each method consists of a sequence of operators. Each operator can be either a primitive step or a call to a new method. When a GOMS-like analysis serves as the basis for a cognitive model, the result is a model in which each operator in each method needs its own control state. Card et al. (1983, p. 147) note the problem themselves:

> For a general treatment of errors and interruptions of the user, the hierarchical control structure of a GOMS model is inadequate; a more general control structure is required. The use of the stack-discipline GOMS model... should be taken as an approximation here because of its greater simplicity.

Minimal Control in Models of Simple Tasks

The minimal control principle is independent of the particulars of the modeling paradigm or architecture that is used. Our assumption for the models that we will discuss is that production rules have the following form:

{goal state} × {problem state}
× {declarative memory} × {perceptual input}
→ {state modifications} × {perceptual/motor actions}
× {declarative memory access}

The models that will be discussed have been implemented in the ACT-R architecture (Anderson et al., 2004), but details about the architecture will only be discussed when relevant for the model in question. The aspect most specific to ACT-R in the rule pattern is the presence of declarative memory, where actions make requests to declarative memory that can then be matched in the condition of a rule that fires later.

Suppose we have a simple choice reaction task paradigm, in which a letter appears on the screen, and the subject has to push D if the letter is an A, and K when the letter is a B. Table 25.1 lists two production systems to do this task; the first row of the table is a system with three control states, and the second row is a production system with one state. The condition of each of the rules (before the arrow) consists of several matches. For example, the condition part of the first *perceive rule* have two conditions. The first condition checks the control state, and the second checks whether a new visual element has appeared on the screen. Assuming these conditions hold and the rule fires, it will have two actions: direct attention to the new visual element and change the control state. Once the stimulus has been attended and the letter has been perceived this information will match one of the conditions of the *retrieve-response rule*. This rule matches the perceived letter and then tries to retrieve the letter from declarative memory (note that variables start with a "?"). We assume that the letter mappings A to D and B to K are stored as facts in declarative memory. Once declarative memory has found the appropriate fact and has made

it available, the press-key rule can match it and issue the appropriate key press. The two production systems are virtually identical, except that the first system updates the control state after each rule. The second system has no control at all, but relies on events, either perceptual or produced by declarative memory, to structure its sequence.

Figure 25.2 shows the activity of the different components and the rule system in the execution of this model for the case where an "A" is displayed. ACT-R's timing assumptions (largely overlapping with EPIC's) are used: a production rule firing takes 50 ms, the time to attend a visual stimulus 85 ms. The time to retrieve a fact from memory depends on its activation (see Anderson et al., 2004, for details), and the manual time depends on the current position of the hand and the number of features that have to be prepared. In this example, execution is fully serial, which can be attributed to the task because no step can be done before the previous step has finished.

Despite the small differences between the two models, the single state model has a huge advantage: It can be interrupted because there is no control state to reconstruct after the interruption. This property also makes it possible to perform a secondary task in parallel to this task without too much performance loss. Figure 25.2 shows that most of the modules have time to spare during the task, making it in principle possible to do more tasks at the same time. Suppose we have a second task, which involves responding to a tone by saying a word, where different tones map onto different words.

TABLE 25.1 Two Production Systems for a Choice Reaction Task

Perceive-rule If the control state is start And a new visual stimulus has appeared → Move attention to that stimulus And set the control-state to attend	Retrieve-response-rule If the control-state is attend And the attended stimulus is the letter ?letter → Ask declarative memory for the response key associated with ?letter And set the control-state to retrieving	Press-key-rule If the control-state is retrieving And declarative memory has retrieved that ?key is associated with ?letter → Send out a manual command to press key ?key And clear the current goal
Perceive-rule If a new visual stimulus has appeared → Move attention to that stimulus	Retrieve-response-rule If the attended stimulus is the letter ?letter → Ask declarative memory for the response key associated with ?letter	Press-key-rule If declarative memory has retrieved that ?key is associated with ?letter → Send out a manual command to press key ?key And clear the current goal

Note. The productions on the first row implement a three-state model, and the productions on the second row implement a one-state model.

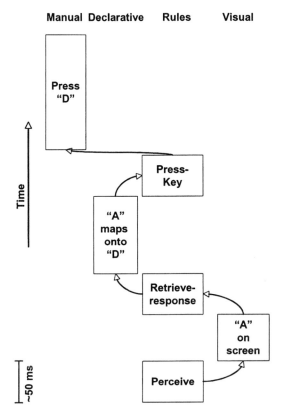

Manual Declarative   Rules   Visual

FIGURE 25.2 Time diagram of reading "A" on the screen and pressing the "D" key. The height of a block in the diagram represents the time that a module requires to complete the particular step, and arrows represent dependencies among the steps.

This paradigm, where a visual-manual task and an aural-vocal task have to be done in parallel, has been used by Schumacher et al. (2001) and by Hazeltine, Teague, and Ivry (2002) to show that with sufficient practice subjects can perfectly parallelize two tasks as long as there is no overlap in perceptual/motor resources. To handle these two tasks, our one control-state model needs three additional rules to handle an audio input and produce a vocal output, which are otherwise identical to the three rules for a visual input and manual output. Augmenting the three-state model would be much harder. One solution is to increase the number of control states to nine for every possible combination of the three states of the individual tasks. For example, we would need a control state that represents the fact that we are retrieving a visual-manual mapping and are perceiving the tone. Two production rules for each of these states would then be needed, one

for the case in which the retrieval of the mapping finishes first, and one for the case in which the perception of the tone finishes first. The single state model has no such disadvantages and can respond in a natural way to any order that steps finish or stimuli are presented. This flexibility is needed to account for all the variations of the tasks that Hazeltine et al. have done, the details of which have been reported by Anderson, Taatgen, and Byrne (2005). These variations consisted of varying the stimulus onset times of the two stimuli (with the tone 50 ms earlier or later than the visual stimulus), making the visual stimulus fuzzy to make it harder to perceive, and varying the mapping between the visual stimulus and the keys that have to be pressed. All the manipulations hardly affected the subjects' ability to do both tasks perfectly in parallel. Figure 25.3 illustrates

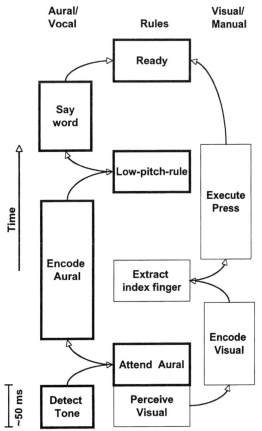

FIGURE 25.3 Time diagram of doing an aural-vocal (thick lines) and a visual-manual (thin lines) task at the same time for a killed performer. In this example the two tasks can be interleaved perfectly. From Taatgen (2005).

the most frequent order of execution. This diagram is similar to Figure 25.2, except that it depicts two simultaneous choice reaction tasks, an aural-vocal (thick lines) and a visual-manual (thin lines). This model also assumes that declarative retrieval is no longer necessary because specific response rules have been learned for each of the stimuli.

Control States in More Complex Tasks

The simple task model in the previous section has no need for control states because the environment and internal events contain enough information to determine what to do next. In more complex cases, this is often not true, making it necessary to have a number (albeit minimal) of control states. An extra control state is necessary if two situations that are identical with respect to perceptual input and problem state require different actions, for example, in the case of an opaque coffee reservoir, where it cannot be determined whether the reservoir is full or empty. Also, even if the next action can in principle be derived from the environment, it might take effort to collect this information, in which case an extra control state is more efficient (see Gray & Fu, 2004, for the case of using internal versus external representations). A control state may also be needed

if we do not want a certain process to be interrupted by external events. One way to achieve this is to structure knowledge in a *weak hierarchy* (Taatgen, 2005). In this representation, instructions are organized in *rule sets*, where each rule set has one control state that is shared by all rules that are in that set. An example of this representation can be found in a model of a radar-screen operator, who has to choose radar tracks on a screen, identify them, and enter the classification into the system (the CMU-ASP system; see Anderson et al., 2004). For each of these stages, the model uses a separate rule set, but within such a set, a single control state is used (Figure 25.4), and each of the rules is triggered by an event, which can be either external (e.g., perceptual) or internal (e.g., a successful retrieval).

The interesting aspect of this model is that it can exhibit new strategies that were not encoded in the original instructions. For example, part of the radar task involves choosing the track on the screen to be classified. The rule set that specifies this part of the task involves two steps: visually attending an unclassified track on the screen and clicking it with the mouse. To accomplish each of the two steps, three rules are needed (Table 25.2). Attending an unclassified track involves finding a track in peripheral vision, shifting attention to that track, and deciding to classify it by storing it in the problem state. Clicking a track with

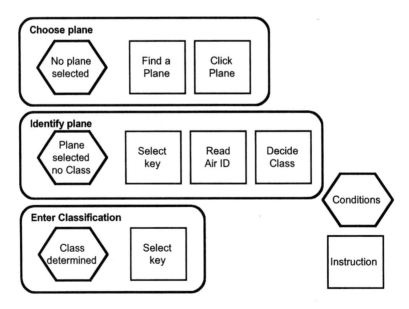

FIGURE 25.4 Three of the rule sets of the CMU-ASP task. The hexagon shows the condition under which the rule set is applicable. Instructions within a rule set can be carried out multiple times if necessary.

TABLE 25.2   Rules to Implement the Choose-Track Subtask in the CMU-ASP Task

Find-track-location	Find-track-attend	Find-track-store
If control-state is choose-track	If control-state is choose-track	If control-state is choose-track
Then find a location with an unclassified track	And an unclassified track has been found	And an unclassified track is attended
	Then attend the unclassified track	And no track is stored in the problem-state
		Then store the track in the problem-state

Click-track-hand	Click-track-mouse	Click-track-click
If control-state is choose-track	If control-state is choose-track	If control-state is choose-track
And the hand is not on the mouse	And the hand is on the mouse	And a track has been stored
Then move the hand to the mouse	And a track has been stored	And the mouse is on the track
	And the mouse is not on the track	Then click the mouse
	Then move the mouse to the track	And set the control-state to top-goal

the mouse involves moving the hand to the mouse, moving the mouse to the track, and clicking the mouse. Carrying out the six rules in the subtask in this order—three perceptual rules followed by three motor rules—is inefficient: Moving the hand to the mouse can be done right away because it does not depend on the choice of the track. A model with a single control state for the whole choose track subtask can interleave the mouse action efficiently with the perceptual actions, while a model with separate control states would have to wait for perception to be done, change the control state, and proceed with the motor actions.

The single control state representation of the choose track subtask has another advantage. Subjects in the CMU-ASP task are instructed to not just classify tracks in any order but to start with more important tracks (defined by distance to home ship, speed, and direction). The initial model previously described can be easily expanded to incorporate this. One property of the rules in Table 25.2 is that even when a track has been chosen, the perceptual rules keep attending new tracks as long as the model remains in the present control state. The only addition needed specifies that a new track that is better than the one that already has been stored in the problem state should replace the stored track:

Find-track-store-better

If control-state is choose-track

And an unclassified track is attended

And that track is better than the track stored in the problem-state

Then Replace the track in the problem-state with the attended track

With the addition of this rule, the model behaves as follows: It finds an unclassified track on the screen while moving the hand to the mouse. As soon as an unclassified track is found, it is stored in the goal, and a mouse movement is initiated to move the hand to the track (assuming the hand is on the mouse by now). During this movement, the model tries to find another unclassified track, and when found compares it to the one that is already in the goal. Assuming the new track is better than the old one, the track in the problem state is replaced, and a new mouse-movement is initiated to move to the new track, while the visual system again tries to find even better candidates. This process continues until the rule that initiates the mouse click fires, because that rule changes the control state and finalizes the subtask.

Searching for a better track in a fully controlled model is much harder to implement, because some stopping criteria would be needed to cease the search for a better track at some point. In the present model, the stopping criterion is implicit. In addition, a more explicitly controlled search process is hard to reconcile with the data, which show that track selection takes less than 2 s. Another aspect of the data is that the quality of the selection is initially consistent with a random track selection and gradually improves later on. Apparently, novices do not compare tracks at all but start doing it after some practice. In ACT-R, this can be explained by production rule learning, which I will discuss in the next section.

The minimal control model exhibits various other qualitative improvements in behavior shared by participants, such as adopting a strategy of using the hands and keyboard and skipping attending parts of the display that are always the same. These improvements can also be attributed to the loose control structure that allows perceptual input to guide behavior.

In the CMU-ASP model the structure of the control states is already determined in the model, largely mirroring the instructions. An experiment in which the control structure first has to be inferred by the subjects is programming a VCR by Gray (2000) and Gray and Fu (2004). The idea behind the minimal control principle is that when people approach a new task they try to find a task representation that requires as few control states as possible. This fits in well with what Gray calls the least-effort principle of cognitive engineering. Programming a VCR involves entering four pieces of information: the channel number, the day of the recording, the start time, and the end time. This VCR has a mode switch with three positions: start, end, and clock-set. The channel, day, and start time have to be entered while the switch is in the start position. The end time has to be entered while the switch is in the end position using the same keys and display as are used for entering the start time. When all the information has been entered, the switch has to be set to *clock-set*, after which the *prog rec* button has to be pushed. The consequence is that this design enforces a three control state representation. In the Gray experiment, subjects first have to use the interface without instruction, and in that phase, they often do not use the mode switch at all, even though they use all the other task-based controls. Errors that are made, even after instruction, are often control state related. For example, once the mode switch is set to end time, it is no longer possible to change the channel, but subjects still sometimes try to do this.

Structuring a task in terms of control states shows a close resemblance to the definition of a unit task. Although in many practical applications, unit tasks are defined by modelers and designers, Card, Moran, and Newell (1983, p.140) originally defined the unit task as a construct from the perspective of a user:

> Although it is often possible to predict the user's actual segmentation of the task into subtasks from the way the instructions are expressed…, it is worth emphasizing that the definition of the subtasks is a decision of the user. We use the term *unit task* to denote these user-defined subtasks.

Card, Moran, and Newell give three reasons for structuring tasks into subtasks: working memory capacity, information horizons, and error control. Working memory capacity, which is related to maintaining the problem state in ACT-R (Lovett, Daily, & Reder, 2000), can be a reason to divide a task into subtasks because the task as a whole needs too many temporary data elements. A control state can help to partition these elements. For example, in the VCR task, it is in principle still possible to use one control state, if attending to the setting of the mode switch is part of every step, but it is more efficient to incorporate it in a control state. Information horizon refers to the problem that the number and length of steps to be taken have to be within reasonable limits; otherwise, the number of potential steps that can be taken becomes to large and searching for the right step takes too long. Having too many steps in a single subtask means the search for the right step can grow out of control. In the examples discussed, search for the right next step was never an issue, but it might be in other problems. Delimiting unit tasks finally makes it easier to localize errors, assuming the unit tasks have been chosen well. This seems to suggest that credit assignment should take place at unit tasks boundaries, that is, during changes in control state.

Each of these reasons are pressures to increase the number of control states, while the minimal control principle is a pressure to keep the number of control states as low as possible. The VCR study by Gray suggests that people approach new tasks by starting with one control state and then gradually add more if the need arises.

How Skilled Performance Is Learned

In the examples discussed, we have looked at skilled behavior in which the task knowledge is represented by production rules. Although the models discussed are implemented in ACT-R, the general idea of minimizing control states should be applicable to any architecture based on production rules. We will now move to the discussion of how skilled behavior is learned, where is becomes more ACT-R specific.

Two properties of novice behavior are that it is largely serial, and that it cannot be assumed that knowledge is already represented in production rules. An assumption in ACT-R is that initial task knowledge is instead represented in declarative memory (Anderson et al., 2004; Taatgen & Lee, 2003). Because the knowledge has to be retrieved from declarative memory to be interpreted, declarative memory becomes the serial bottleneck in novice behavior.

In the dual-task situation, the two tasks compete for two central resources: the production system and declarative memory. Contrary to the perceptual and motor resources, whose execution times are more or less fixed by the physical limitations of the system, ACT-R has learning mechanisms to reduce the load on central resources. Indeed, the production systems in Table 25.1 only models a certain stage in the learning, where it is assumed that the general task is represented as production rules, but the specifics of the stimulus-action mappings are stored in declarative memory. It is important to incorporate learning in our discussion of the minimal control principle because of Larkin's observation that smooth performance requires learning, which is, of course, a well-known fact of skill acquisition. The explanation for this fact is that the smooth behavior produced by the production rules in Tables 25.1 and 25.2 is only the end stage of a learning process. The parallel matching process ensures the rule that that matches the right combination of control state and external input is selected. This is however not possible when task knowledge is still in declarative memory.

Table 25.3 shows a declarative representation of the choice reaction task. For explanation purposes, this representation has been kept very simple, consisting of just the task, the condition, and the action. Although this representation of instructions is similar to our earlier work (Anderson et al., 2004; Taatgen & Lee, 2003), it diverges in one aspect: the instructions are unordered. A difference between interpreted declarative instructions and production rules is that the instruction first has to be retrieved before its condition can be checked. The consequence is that when a model is interpreting instructions it may have to search for the right instruction first. Another possibility, which I will discuss later, is that there is no applicable instruction because it has been forgotten or was never given. In that case a proper next step has to be inferred or discovered. Summarizing, initial behavior is neither smooth nor fast, because declarative retrieval is slow, the next step sometimes has to be searched for, or worse, reconstructed, inferred or discovered.

TABLE 25.3 A Declarative Representation of the Choice Reaction Task

Step1	Step2	Step3
isa instruction	isa instruction	isa instruction
task crt	task crt	task crt
condition stimulus-present	condition visual	condition retrieved-map
action attend-visual	action retrieve-map-visual	action press-key-map

The learning mechanism that can turn the slow and brittle novice behavior into skilled behavior is called *production compilation* (Taatgen & Anderson, 2002). Production compilation combines pairs of rules that fire in sequence into new rules that combine the conditions and actions of both rules. If the first rule involves a declarative retrieval, which is then used in the condition of the second rule, this retrieval is substituted into the new rule. As a consequence, the learned rules are more specialized than the original rules, and declarative retrievals are eliminated from the process. Apart from the advantage of saving both the retrieval time of the instruction the execution time of a production rule, this rule is applicable whenever its conditions are satisfied. This shift in representation can be quite significant in situations in which the order of actions is not fixed. In the case of the dual-task paradigm, a declarative representation does not necessarily produce the optimal order of steps, but the procedural representation does, because each event is handled as soon as possible. Taatgen (2005) and Anderson, Taatgen, and Byrne (2005) describe models that start out with declarative instructions and learn to perfectly interleave a visual-manual and an aural-vocal task in such a way that dual-task performance equals single-task performance.

Procedural knowledge for the CMU-ASP task is learned a similar manner: a rule retrieves an instruction from declarative memory, and interpreting rules carry out these instructions. As a consequence, the flexibility that the model exhibits in interleaving the selection and comparison of the tracks with motor actions only gradually emerges with experience (see Taatgen, 2005, for detailed model/data comparisons).

Inferring Steps in Incomplete Task Representations

The discussion up to this point assumes that all the steps that have to be done to accomplish the goal are either represented as production rules or present in declarative memory. This assumption is too optimistic: In many cases, steps have to be inferred. Sometimes not all the steps are part of an instruction, sometimes parts of the instruction are forgotten or not even read, and sometimes certain steps are part of an error-recovery procedure not part of the original task representation. In such a situation, missing steps have to be inferred or guessed. To be able to infer or guess steps successfully, we have to again augment our representation of steps to include what the purpose and consequences of the step are. The tea example should be augmented to:

1. [empty kettle] put water in kettle [kettle with cold water]
2. [kettle with cold water] put kettle on stove [water boils]
3. [empty teapot] put leaves in teapot [leaves in teapot]
4. [water boils and leaves are in teapot] pour water in teapot [have tea]

Given this representation it is possible to deduce from the goal of having tea which steps have to be taken. Although this is not needed in the case of a complete problem representation where forward reasoning suffices, goal-directed reasoning becomes necessary when some step is missing. For example, if we for some reason have forgotten that putting leaves into the teapot is part of the recipe, we can infer that there is a missing gap between having an empty teapot and a teapot with leaves, in which case it is easy to infer the missing step. In many HCI settings, inferring a step is not always possible but just trying out controls might help you find the right step as long as you know what the step for which you are searching is supposed to do (e.g., wanting to enter the channel on a VCR but not knowing which control to use).

This representation is very similar to STRIPS operators (Fikes & Nilsson, 1971). However, the instruction representation is not used to plan the entire path to the goal, only the upcoming step. The projected outcome of a step is used to build an expectation or as a help to fill in missing steps in the case of insufficient knowledge. If the expectation does not match the real world, this may be an indication that the knowledge is incomplete or overly general in which case an additional control state may be needed.

Conclusions

In this chapter, I put forward arguments to support the notion of the minimal control principle. On the one hand, I argued that it leads to the smallest model and that smaller models are to be preferred over larger models on the basis of the more general principle of simplicity. On the other hand, I have demonstrated in a number of examples how reducing the number of control states leads to better models.

The most important message of the chapter is a modeling recommendation to rethink the way we construct our models, and attempt to decrease the reliance on control states. Although I have used ACT-R as an example, this is a guideline that can and should be used in other architectures. It will not always be easier than the usual practice of many control states, and some architectural changes may be necessary to make control state lean models easier to construct. Something that is slightly harder to do with a more loose control scheme is to conclude that the goal has been achieved. In a pure top-down model, the goal is achieved whenever the last step has been done. In a mixed top-down/bottom-up scheme, this is harder to do.

Although I have discussed how the flexible behavior can gradually emerge through learning, these models already assume an initial control structure, an unrealistic assumption when there is more than one control state. That means there is an exciting area of modeling to be explored, where the challenge is to construct models that learn unit tasks the way Card, Moran, and Newell (1983) intended it: by discovering the unit task boundaries on the basis of knowledge, experience, and trial and error. Part of this discovery process is finding the steps themselves because realistic instructions are almost never complete and almost never fully remembered.

Acknowledgments

This research was supported by Office of Naval Research Grant N00014-04-1-0173 and NASA Grant NRA2-38169. I would like to thank Wayne Gray, Chris Sims, Mike Schoelles, and Stefani Nellen for their comments on the first draft of the manuscript.

References

Anderson, J. R. (1990). *The adaptive character of thought.* Hillsdale, NJ: Erlbaum.

———, Bothell, D., Byrne, M., Douglass, D., Lebiere, C., & Qin, Y. (2004). An integrated theory of mind. *Psychological Review, 111*(4), 1036–1060.

———, Taatgen, N. A., & Byrne, M. D. (2005). Learning to achieve perfect time sharing: Architectural implications of Hazeltine, Teague, & Ivry (2002). *Journal of Experimental Psychology: Human Perception and Performance, 31*(4), 749–761.

Byrne, M. D., & Anderson, J. R. (2001a). Serial modules in parallel: The psychological refractory period and perfect time-sharing. *Psychological Review, 108,* 847–869.

Card, K. C., Moran, T. P., & Newell, A. (1983). *The psychology of human-computer interaction.* Hillsdale, NJ: Erlbaum.

Chater, N. (1999). The search for simplicity: A fundamental cognitive principle? *Quarterly Journal of Experimental Psychology, 52A,* 273–302.

Chong, R. S., & Laird, J. E. (1997). Identifying dual-task executive process knowledge using EPIC-Soar. In *Proceedings of the nineteenth annual conference of the cognitive science society* (pp. 107–112). Hillsdale, NJ: Erlbaum.

Clark, A. (1997). *Being there: Putting brain, body and world together again.* Cambridge, MA: MIT Press.

Dreyfus, H. (1979). *What computers can't so: A critique of artificial reason.* New York: Harper & Row.

Fikes, R. E., & Nilsson, N. J. (1971). STRIPS: A new approach to the application of theorem proving to problem solving. *Artificial Intelligence, 2,* 189–208.

Gray, W. D. (2000). The nature and processing of errors in interactive behavior. *Cognitive Science, 24*(2), 205–248.

———, & Fu, W. T. (2004). Soft constraints in interactive behavior: The case of ignoring perfect knowledge in-the-world for imperfect knowledge in-the-head. *Cognitive Science, 28*(3), 359–382.

Hazeltine, E., Teague, D., & Ivry, R. B. (2002). Simultaneous dual-task performance reveals parallel response selection after practice. *Journal of Experimental Psychology: Human Perception & Performance* 28(3), 527–545.

———. (1995). How a cockpit remembers its speed. *Cognitive Science, 19,* 265–288.

Larkin, J. H. (1989). Display-based problem solving. In D. Klahr & K. Kotovsky (Eds.), *Complex information processing: The impact of Herbert A. Simon* (pp. 319–341). Hillsdale, NJ: Erlbaum.

———, & Simon, H. A. (1987). Why a diagram is (sometimes) worth ten thousand words. *Cognitive Science, 11,* 65–99.

Lovett, M. C., Daily, L. Z., & Reder, L. M. (2000). A source activation theory of working memory: Cross-task prediction of performance in ACT-R. *Cognitive Systems, 1,* 99–118.

Meyer, D. E., & Kieras, D. E. (1997). A computational theory of executive cognitive processes and multiple-task performance. Part 1: Basic mechanisms. *Psychological Review, 104,* 2–65.

Salvucci, D. D. (2005). A multitasking general executive for compound continuous tasks. *Cognitive Science, 29*(3), 457–492.

Schumacher, E. H., Seymour, T. L., Glass, J. M., Fencsik, D. E., Lauber, E. J., Kieras, D. E., et al. (2001). Virtually perfect time sharing in dual-task performance: Uncorking the central cognitive bottleneck. *Psychological Science, 12*(2), 101–108.

Sober, E. (1974). *Simplicity.* Oxford: Clarendon Press.

Taatgen, N. A. (2005). Modeling parallelization and flexibility improvements in skill acquisition: From dual tasks to complex dynamic skills. *Cognitive Science, 29,* 421–455.

———, & Anderson, J. R. (2002). Why do children learn to say "broke"? A model of learning the past tense without feedback. *Cognition, 86*(2), 123–155.

———, & Lee, F. J. (2003). Production compilation: A simple mechanism to model complex skill acquisition. *Human Factors, 45*(1), 61–76.

26

Control Signals and Goal-Directed Behavior

Erik M. Altmann

The psychological notion of a "goal" can take various forms, but at bottom all goal-directed activity seems to require that the correct control signal be detectable by the cognitive system against a background of old or alternative signals. A simple signal-detection model based on this premise explains a variety of empirical phenomena from the domain of task switching that might otherwise seem unrelated and that have no obvious explanation in terms of standard, but somewhat naïve, "reconfiguration" accounts of cognitive control. The model can be used to frame discussion of a variety of memory- and attention-related processes, including encoding, retrieval, priming, and inhibition.

The psychological notion of a goal is more diverse than one might expect from sifting any one literature that examines it. One sense of a goal is motivational, for example, to eat when one is hungry; here, the goal organizes a suite of behaviors ranging from initial high-level choices about where to head for food (restaurant, vending machine) down to motor actions (opening a menu, putting coins in a slot). A second, more conventionally cognitive sense of a goal arises in problem solving, for example, in a puzzle task like the Tower of Hanoi, in which a goal to shift a tower of disks from one peg might spawn a sequence of subgoals that pull the tower apart disk by disk and ultimately reconstruct it in its target destination. The distinction between these two senses of goal is less clear than it might seem, as there is implicit in the latter, problem-solving sense a motivational aspect as well, in that people focused on solving a problem are necessarily not focused on other, alternative behaviors. Moreover, even within the domain of problem solving, goals can exist at multiple levels; the goal to move a disk to a peg in the Tower of

Hanoi is a fine-grained goal compared with the overall goal of altering the initial state of the puzzle (composed of multiple goals and pegs) to match the final state specified for that trial. However, a common thread to all these senses of *goal* is that they involve control information of some kind that guides the organism's immediate interactions with the environment in one direction rather than another when multiple paths are available.

This sense of goal as short-term control information touches much of what one might consider goal-directed behavior in everyday life. Examples include searching for one's keys; recalling where one parked the car, or what the point of an errand was to the supermarket, or even to the garage or some other room in the house; maintaining a legal driving speed when the speed limit keeps fluctuating; staying on point during a conversation; or, more dramatically, applying the most recent rules of engagement to decide whether to shoot at an approaching person or vehicle. In all these cases, coherence of perception and action depends on guidance from some kind of control information represented in memory that is relevant now but will likely be eclipsed by new, perhaps only slightly different control information in the near future. Even much more high-level goals, at the motivational or regulatory level, ultimately boil down to sequences of low-level interactions between agent and environment that reflect such goal-related contingencies.

Given that short-term goals, whatever their precise nature and provenance, sometimes fail to guide our behavior, it seems useful to ask what dynamics might make such failures more or less common. This chapter surveys some results from a line of work that treats this as a signal-detection problem: The current goal is the signal that has to be detected against a background of other signals that might represent old goals that are now moot, for example, or alternative goals to the current, correct one that would lead to inappropriate actions.

Task Switching

An experimental paradigm relevant to studying low-level control signals is *task switching*. The moniker is slightly wishful because what are called tasks in this paradigm are simpler than what one might consider tasks in everyday life and are arguably so similar to one another that there is very little switching involved. Nonetheless, these weaknesses are also a strength, in that with two very simple and very similar tasks, and frequent shifts between them, task-switching paradigms amplify the problem of clutter or interference among goal-related mental signals enough for performance measures like response time to reveal subtle effects of whatever cognitive processes keep the currently correct signal detectable.

In the specific variant of task switching of interest here, the experimental participant performs a large number of trials in sequence. A trial starts with the onset of a simple stimulus, such as a digit (0 to 9, typically, excluding 5). The participant determines either whether the digit is even or odd (one task) or whether the digit is higher or lower than five (the other task), depending on which task is currently appropriate. After a run of a half-dozen trials or so, a task cue is presented to indicate what task to perform for the next run of trials and so it goes with runs of trials interspersed with task cues. The task indicated by the cue (even-odd vs. high-low) is randomized, such that some cues happen to require the participant to switch tasks relative to the

previous run, whereas others require the participant to repeat the task he or she was already doing for another run of trials. The task cue is perceptually available only for a brief period (perhaps half a second) between two trials, the digit stimulus presented for each trial affords either task, and the same response keys are used for both tasks, so accurate performance depends on encoding the cued task as some kind of mental signal, and on detecting that signal during trial performance to interpret the stimulus correctly and select a response. The challenge in performing correctly in this task environment is *roughly* that of correctly completing a simple errand every few seconds for an hour with the target of a given errand always being one of the same two things and with no external cues to serve as a reminder of which thing is relevant *now*.

Theoretical developments in task switching have largely been driven by the kind of "boxes in the head" metaphor popular earlier in the cognitive revolution. Thus we have a "little signal person in the head" (Rogers & Monsell, 1995, p. 217) charged with "a sort of mental 'gear-changing'" (Monsell, 2003, p. 135) when the cued task is different from the task it replaces. Support for such metaphors comes from the empirical finding of *switch cost*, which is the difference in performance on switch trials (those following a switch cue) as compared with repeat trials (those following a repeat cue). Switch cost is pervasive in task-switching studies, supporting the reconfiguration metaphor on the logic that a pervasive effect must measure some kind of functional process. Thus, "switch cost might seem to offer an index of the control processes involved in reconnecting and reconfiguring the various modules in our brains, so as to perform one task rather than another" (Monsell & Driver, 2000, p. 16). In a similar vein, Logan (2003) writes of executive processes "programming" (p. 48) the component processes responsible for performing the substantive work of a task. All current task-switching models that are process oriented enough to run as computational models of some kind explain switch cost in terms of a functional reconfiguration mechanism, though they differ somewhat in the details (De Jong, 2000; Gilbert & Shallice, 2002; Kieras, Meyer, Ballas, & Lauber, 2000; Meiran, 2000; Sohn & Anderson, 2001).

The approach described here leans away from this perhaps somewhat naive approach in which discrete mental structures have to be reconfigured to perform one task rather than another and toward an analogue account that represents the variability one might expect to find in a biological system. This alternative view pares away the central assumption of the reconfiguration view—that it matters whether the current and previous tasks are different.

A Signal Detection Model

To formalize the notion of a goal as a cognitive control signal, it is useful to speak in terms of the strength, or *activation*, of signals and to assume that the strongest or most active control signal at any given moment is the one that guides behavior. Performance in task switching requires the ability both to focus on one task and to switch to a new one, which in terms of activation means keeping the correct task the most active in memory for

as long as necessary but no longer. Figure 26.1 depicts these assumptions using a graphical formalism adapted from signal detection theory (Wickens, 2002; for applications to task switching, see Altmann, 2002, 2004; Altmann & Gray, 2002). Within each panel, the abscissa shows activation and the ordinate shows the probability of an item having a given activation level. Thus, each probability density function shown in the figure characterizes the activation of one control signal. The variance associated with each density function reflects fluctuation in activation levels from one moment to the next, presumably due to variations in the state of the biological substrate that implements the signal. Thus, at any given moment, a control signal will most likely be at its mean activation because that is the activation level that is most probable (i.e., has the highest value

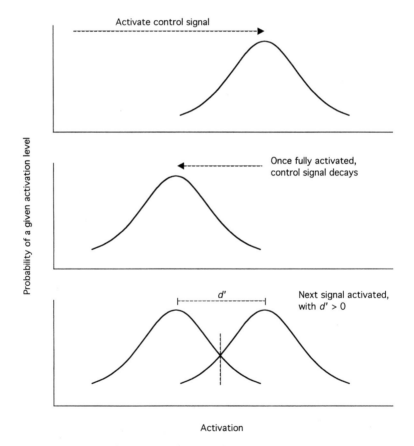

FIGURE 26.1 A model of control signal detection. Each density function represents the activation of one control signal. (Top panel) The system responds to a task cue by activating a new signal. (Middle panel) As that signal guides behavior on trials, it also decays. (Bottom panel) This decay has ensured that when next task cue is activated, it can be distinguished from the decayed one ($d' > 0$).

on the ordinate). However, a signal's activation level could also be above or below its mean, with smaller probability the farther that level is from the mean.

The bottom panel of Figure 26.1 shows a situation in which two control signals are represented in the system, each corresponding to one task cue. The density function on the left represents the old cue, which is no longer relevant because the system has since encountered a new cue. The density function on the right represents the new cue that has replaced the old cue. The new signal is more active than the old signal (the signal-detection quantity d' is positive), which is the desired situation because the new signal represents the most recent task cue. Note that if d' were zero, this would represent a situation of catastrophic interference in which old and new control signals cannot be distinguished. Also note that, in the general case, there will be not one old signal but many, one corresponding to each instance (Logan, 2002) of exposure to a task cue (and each represented by its own density function, in a series that would trail off to the left of the figure). However, there is no loss of generality to reason in terms of two cues represented in memory, one relevant (on the right) and one not (on the left), because the irrelevant one can be taken to represent the most active of a set of irrelevant ones (a simple application of extreme value statistics; see Logan, 2002).

At the intersection of the two densities is an *activation threshold*, which corresponds to the response criterion in signal detection theory but, in this context, serves as a filter to block out weak signals. This filtering function allows the system to trade between control *failures* and control *errors*, where a failure (or miss, in signal detection terms) occurs when the system tries to detect a control signal but none is above threshold, and an error (or false alarm, in signal detection terms) occurs when the system tries to detect a control signal but the wrong signal is at that moment the most active. Thus, in this model, the activation threshold is a system parameter that can be adjusted leftward or rightward in response to task demands that might alter the relative costs of failures and errors. The activation threshold also has a natural interpretation in terms of response latencies, with time to consult a control signal taking longer when the threshold is farther to the right. This interpretation rests on the assumption that trying to detect a control signal is a time-consuming operation that can be repeated if it fails, in that the moment-to-moment variability in activation levels can cause one attempt to detect a signal to fail but the next to succeed. Thus, the

further to the right the threshold, the more likely it is that a given attempt to detect a control signal will fail and lead to another attempt that extends the overall time it takes for detection to succeed.

The top and middle panels of Figure 26.1 show key supporting processes through which the system can maintain the functional situation in the bottom panel across any number of task cues presented by the environment. The top panel represents a control signal being activated in response to presentation of a task cue; the density function simply shifts rightward along the activation axis from some initial default level (the tail of the arrow) to a level at which detection will be accurate and fast enough to meet performance requirements. The middle panel of Figure 26.1 shows decay, which, by this logic, must ensue once the signal has been fully activated. Decay, here, refers to a gradual loss of activation, and is necessary to maintain the functional situation represented in the bottom panel ($d' > 0$). Without decay, each new signal would have to be made more active than its predecessor to make d' positive, making new shifts of control increasingly difficult and ultimately impossible (assuming some sort of biological or other upper bound on activation of signals). Alternatively, without decay, interference from old control signals would build up monotonically, again ultimately making shifts of control impossible. Thus, signals rising and falling (decaying) in tandem ensure that cognitive control can be sustained indefinitely—or at least that the limiting factor is not how often the control signal changes.

Phenomena

This section briefly describes five empirical phenomena from the task-switching literature and explains them in terms of the signal detection model. The phenomena fall into two classes. In the first class are two effects that can be explained in terms of the activating process in the top panel of Figure 26.1. The *preparation effect*, of which an example appears in Figure 26.2, is the decrease in response latency on the first trial of a run with increases in the interval between task cue onset and trial stimulus onset (the cue-stimulus interval, or CSI). In the signal detection model, the preparation effect can be interpreted as the time needed to activate a control signal, with more of this work being done ahead of time the longer the CSI. *Conventional switch cost* is the increase in first-trial latency when the first

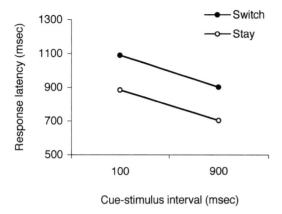

FIGURE 26.2 Response latencies for the first trial of a run, separated by cue-stimulus interval and by whether or not the cue preceding the first trial switched the task relative to the previous run. Data are from Altmann (2004) Experiment 4.

trial follows a switch cue compared with when it follows a repeat cue (the difference between the curves in Figure 26.2). This effect and its moderators are by far the most prominent issues discussed in the task-switching literature (for reviews, see Logan, 2003; Monsell, 2003), reflecting the influence of the reconfiguration metaphor on design and discussion. Applied to the signal detection model as a constraint, conventional switch cost simply suggests that a new control signal takes less time to activate when it duplicates the previous control signal, an effect that one could attribute to repetition priming of some kind. Conventional switch cost would thus be a side effect of a pervasive process in human memory that has little or no specific relevance to cognitive control.

The second class of empirical phenomena includes three within-run effects related to the decay of the current control signal, as depicted in the middle panel of Figure 26.1. *Within-run slowing* is a gradual but regular increase in latencies across successive trials, starting with the second trial position, illustrated in Figure 26.3. This is attributable directly to signal decay; as the density shifts leftward along the activation axis (Figure 26.1), less of it will be above threshold, extending the average time to sample the signal on a given trial. *Within-run error increase* is a corresponding trend in error rates, also illustrated in Figure 26.3, and also attributable to signal decay, assuming a fixed speed-accuracy trade-off function. As the density shifts leftward and average time to sample the signal increases, all else being equal,

the number of occasions on which the system "times out" and moves on will increase, causing an increase in guessing as manifested in performance error. This effect is often noisier than within-run slowing, perhaps because the mechanism is somewhat indirect, but errors need only be flat as a function of position to be relevant to the interpretation of within-run slowing, because a flat or increasing trend in errors rules out a speed-accuracy trade-off account of latencies and errors together. Finally, *full-run switch cost*, as distinct from conventional switch cost, manifests in terms of errors, again illustrated in Figure 26.3, and spans the full run of trials instead of only the first. The effect is caused by the old control signal being transiently more active then the current control signal when the system samples. Such interference is more likely to cause errors on switch runs than on repeat runs because on

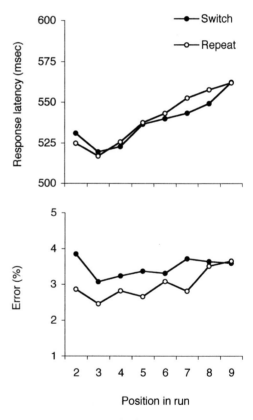

FIGURE 26.3 Response latencies and error percentages for the second through ninth trials of a run, separated by whether or not the cue preceding the first trial switched the task relative to the previous run. Data are from Altmann (2004) Experiment 4.

switch runs the old signal codes the wrong task. Such interference may or may not affect response latencies, depending on the precise mapping from the model in its abstract form in Figure 26.1 to a set of computational mechanisms that actually do the task.

Discussion

The phenomena summarized above are a diverse lot, which on the surface might seem to have little if anything to do with one another and which are difficult to relate to one another functionally in terms of the standard reconfiguration view of cognitive control. A "little signal person in the head" charged with "a sort of mental gear-changing" in response to a shifted task certainly accounts for conventional switch cost, but for the other four effects, it seems one would have to improvise. The signal detection model, in contrast, accounts for all five phenomena largely through implication from a single underlying set of constraints.

In contrast to the reconfiguration view of cognitive control, this signal detection approach requires no decision on the part of the cognitive system about whether the task indicated by the current cue is the same as or different from the previous task. Such a decision is necessary from the reconfiguration perspective because the need for system reconfiguration is by definition limited to situations in which there is a change to the current task. The signal-detection approach posits instead that the system responds rather blindly to any task cue or perhaps to any goal more generally, by activating a corresponding control signal. This seems to simplify the nuts and bolts of cognitive control in a significant way because assessing the difference or similarity of two representations—of tasks in this case, but more generally of any two semantic representations—is a separate challenge that would require its own powerful process to compute. It seems a lesser burden to require only that control processes recognize a task cue when they encounter one and raise a control signal in response, without also having to make comparative judgments about activities the system was previously engaged in.

The language throughout this chapter has been cast in terms of "signals," inspired in part by earlier similar framing of attentional processes (Posner & Boies, 1971), but in fact the dynamics of the signal detection model sketched here are also the dynamics of memory activation in ACT-R (Anderson et al., 2004). One reason to prefer reference to signals rather than memory is that the general notion of memory comes with its own set of baggage, implying weightier notions of encoding, retrieval, and storage than are implied here; it may seem odd, for example, to speak of "retrieving" a memory for a task cue every 500 ms or so (once per trial), when "retrievals" are also what people undertake when they try to reconstruct the list of 50 states of the United States, say, with each individual retrieval taking perhaps seconds. However, framing the current model in terms of memory constructs would allow a cleaner theoretical integration with priming effects, which seem to be the most likely source of conventional switch cost. As we move toward implementation of this model as a cognitive simulation (in ACT-R), the memory formulation will likely be the one that dominates.

Two other well-known psychological constructs are relevant to examine in light of a signal detection model of control. The more straightforwardly integrated of the two is the mechanism of *priming*. Priming takes diverse forms, including associative, perceptual, and repetition, to name just three. However, all such positive instances, in which priming facilitates access to the primed information, can be interpreted in simply terms of the activation of a target being increased. In Figure 26.1, then, priming can be interpreted as shifting the density function of a target, on the right in the bottom panel, further to the right along the activation axis, increasing system sensitivity. Thus, for example, a reminding of some kind—"Yes: *That's* what I was looking for!"—can be viewed as a sudden dose of activation delivered to some target that had either fallen below the activation threshold or been "masked" by interfering signals with high activation. Indeed, there is reasonably strong evidence (Altmann & Trafton, 2002; Hodgetts & Jones, 2006) that priming is critically necessary to allow retrieval of suspended subgoals in problem solving domains, the logic being that once a signal has decayed, there is no mechanism *other* than priming that can make it the most active again.

A second relevant construct to consider is *inhibition*, broadly defined. A theoretical debate has raged about whether inhibitory mechanisms are theoretically distinct from activation mechanisms (e.g., MacLeod, Dodd, Sheard, Wilson, & Bibi, 2003), but at some level, it seems unrealistic to dispute the importance of inhibitory *function* at some higher level, as, for example, in the problem of trying not to think of something (Wegner, 1994), and in work on inhibition in social cognition (e.g., von Hippel & Gonsalkorale, 2005). A related construct is negative priming, in which a stimulus that was irrelevant on the previous trial and had to be suppressed in some way is now relevant on the current trial but seems to retain some of the effect of its earlier suppression (examples from the task switching domain include Mayr, 2002; Mayr & Keele, 2000; Waszak, Hommel, & Allport, 2003). By analogy to the interpretation of positive priming, one might view negative priming in terms of activation spreading to a *distractor* signal and thus shifting rightward the *left-hand* density function in the bottom panel of Figure 26.1. Inhibitory functions more generally might then be considered in terms of a variety of compensatory and perhaps strategic responses triggered by a distractor becoming strongly activated. Thus, the system may shift the activation threshold to the right to filter distracting signals more stringently or try to bring new targets to activation levels above that of the distractor to mask it effectively. Indeed, this characterization of inhibition as a masking process is one account of how people respond when they are asked not to think of a specific thing: They try to comply by thinking about other things instead (Wegner, 1994). However, this sketchy account of inhibitory functioning raises more questions than it answers, concerning, for example, how the system might detect an increase in activation of a distractor and how quickly the activation threshold can be adjusted in one direction or the other. Nonetheless, there is theoretical substance to framing inhibitory functioning in this way, in that it seems congruent with the position that inhibitory function is effortful or controlled (e.g., Engle, Conway, Tuholski, & Shisler, 1995; Lavie, 2004) more than it is a cognitive "off" switch (Anderson & Green, 2001).

In conclusion, the suggestion here is that interference among control signals is the basic constraint on goal-directed behavior and that decay of such signals is an architectural housecleaning process that prevents this interference from becoming catastrophic. This analysis provides a reasonably direct account of within-run effects and first-trial effects in task switching; in contrast, it is quite unclear what mechanisms would have to be incorporated in traditional reconfiguration models of cognitive control, which focus on switch cost as reflecting the functional control processes, to account for within-run effects in particular. Perhaps one of the important general questions for future research involving this model is to ask how it might be applied to gain analytical traction in other of the many domains in which people exhibit goal-directed behavior.

Acknowledgments

This research was supported by Grant N00014-03-1-0063 from the Office of Naval Research. I thank Wayne Gray, Hansjorg Neth, and Chris Simms for their very useful suggestions for improving this chapter.

References

Altmann, E. M. (2002). Functional decay of memory for tasks. *Psychological Research, 66,* 287–297.

——. (2004). Advance preparation in task switching: What work is being done? *Psychological Science, 15,* 616–622.

——, & Gray, W. D. (2002). Forgetting to remember: The functional relationship of decay and interference. *Psychological Science, 13,* 27–33.

——, & Trafton, J. G. (2002). Memory for goals: An activation-based model. *Cognitive Science, 26,* 39–83.

Anderson, J. R., Bothell, D., Byrne, M. D., Douglass, S., Lebiere, C., & Qin, Y. (2004). An integrated theory of the mind. *Psychological Review, 111,* 1036–1060.

Anderson, M. C., & Green, C. (2001). Suppressing unwanted memories by executive control. *Nature, 410,* 366–369.

De Jong, R. (2000). An intention-activation account of residual switch costs. In S. Monsell & J. Driver (Eds.), *Attention and performance XVIII: Control of cognitive processes* (pp. 357–376). Cambridge, MA: MIT Press.

Engle, R. W., Conway, A. R. A., Tuholski, S. W., & Shisler, R. J. (1995). A resource account of inhibition. *Psychological Science, 6,* 122–125.

Gilbert, S. J., & Shallice, T. (2002). Task switching: A PDP model. *Cognitive Psychology, 44,* 297–337.

Hodgetts, H. M., & Jones, D. M. (2006). Interruption of the Tower of London task: Support for a goal activation approach. *Journal of Experimental Psychology: General, 135,* 103–115.

Kieras, D. E., Meyer, D. E., Ballas, J. A., & Lauber, E. J. (2000). Modern computational perspectives on executive mental processes and cognitive control: Where to from here? In S. Monsell & J. Driver (Eds.), *Control of cognitive processes: Attention and performance XVIII* (pp. 681–712). Cambridge, MA: MIT Press.

Lavie, N. (2004). Load theory of selective attention and cognitive control. *Journal of Experimental Psychology: General, 133,* 339–354.

Logan, G. D. (2002). An instance theory of attention and memory. *Psychological Review, 109,* 376–400.

Logan, G. D. (2003). Executive control of thought and action: In search of the wild homunculus. *Current Directions in Psychological Science, 12,* 45–48.

MacLeod, C. M., Dodd, M. D., Sheard, E. D., Wilson, D. E., & Bibi, U. (2003). In opposition to inhibition. In B. H. Ross (Ed.), *The psychology of learning and motivation* (Vol. 43, pp. 163–214). San Diego, CA: Academic Press.

Mayr, U. (2002). Inhibition of action rules. *Psychonomic Bulletin & Review, 9,* 93–99.

———, & Keele, S. W. (2000). Changing internal constraints on action: The role of backward inhibition. *Journal of Experimental Psychology: General, 129,* 4–26.

Meiran, N. (2000). Modeling cognitive control in task-switching. *Psychological Research, 63,* 234–249.

Monsell, S. (2003). Task switching. *Trends in Cognitive Sciences, 7,* 134–140.

———, & Driver, J. (2000). Banishing the control homunculus. In S. Monsell & J. Driver (Eds.), *Control of cognitive processes: Attention and performance XVIII* (pp. 3–32). Cambridge, MA: MIT Press.

Posner, M. I., & Boies, S. J. (1971). Components of attention. *Psychological Review, 78,* 391–408.

Rogers, R. D., & Monsell, S. (1995). Costs of a predictable switch between simple cognitive tasks. *Journal of Experimental Psychology: General, 124,* 207–231.

Sohn, M.-H., & Anderson, J. R. (2001). Task preparation and task repetition: Two-component model of task switching. *Journal of Experimental Psychology: General, 130,* 764–778.

von Hippel, W., & Gonsalkorale, K. (2005). "That is bloody revolting!" Inhibitory control of thoughts better left unsaid. *Psychological Science, 16,* 497–500.

Waszak, F., Hommel, B., & Allport, A. (2003). Task-switching and long-term priming: Role of episodic stimulus-task bindings in task-shift costs. *Cognitive Psychology, 46,* 361–413.

Wegner, D. M. (1994). Ironic processes of mental control. *Psychological Review, 101,* 34–52.

Wickens, T. D. (2002). *Elementary signal detection theory.* New York: Oxford University Press.

27

Intentions, Errors, and Experience

Richard A. Carlson

Intentions are schematic states of working memory that instantiate goals by specifying desired outcomes, actions, and the objects of those actions, thus serving a role in real-time control of behavior. Fluent performance may be achieved in part by concise representation of intentions, which may entail representing some elements *deictically*—specifying only the time and place at which information is available—rather than semantically. This chapter reviews theoretical considerations and empirical results that support this deictic specification hypothesis, demonstrating that deictic specification has both benefits for performance and costs in terms of limits on metacognition. When regularities in the performance environment, such as consistency in the time and place at which information is available, allow deictic specification, performance is superior but is characterized by predictable error patterns and poor error monitoring. Studies of event counting illustrate these phenomena. This analysis of intentions provides a basis for integrating theoretical understanding of cognitive control and of metacognition.

My goal in this chapter is to sketch a theoretical view of intentions. By *intention*, I mean a mental state that specifies a goal or desired outcome and means for achieving that outcome and that serves a role in real-time control. This concept can serve as an integrating concept by linking computational views of cognitive control with analyses of the subjective experience of control, by clarifying theoretical distinctions between automatic and control processes, and by linking approaches from cognitive psychology with insights from philosophy and linguistics. I will focus on properties of intentions related to mechanisms underlying metacognitive monitoring and some common errors in routine skills.

Why Intentions?

Why use the term *intention?* To some, this term carries unwanted baggage, being associated too closely with folk psychology, subjective experience, and prescientific

explanations of behavior (Malle, Moses, & Baldwin, 2001). Isn't the term *goal*, which is common in computational models—for example, *goal* has been a central concept in the ACT family of theories from the beginning (Anderson, 1983)—sufficient? There are several reasons for focusing on *intention*. First, some ambiguities are associated with the term *goal*. "Goal" refers sometimes to a desired outcome alone (Austin & Vancouver, 1996) and sometimes also to means for achieving it (Mandler, 1984), sometimes to a currently active representation and sometimes to an element of long-term memory, sometimes to an active representation controlling current behavior, and sometimes to one that is part of a plan for future behavior, and so on. The term *intention* refers to active representations instantiating goals currently controlling behavior, distinguishing them from goals not currently doing so.

A second reason for focusing on intentions is to emphasize the theoretical link between the conscious experience of control and the cognitive function of control. This link is important both in its own right, for developing a theory of conscious experience, and for understanding the role of metacognitive experience in the control of cognition. I argue that understanding the representation of intentions can help in understanding the nature of and limits on metacognition.

A third reason for focusing on intention is to clarify the distinction between automatic and controlled processes (e.g., Schneider & Shiffrin, 1977), a distinction that is cast either explicitly or implicitly in terms of conscious experience and apparent control by intentions (Shallice, 1994). Automatic processes are sometimes described as occurring without intention, but a careful examination of the literature suggests that automaticity typically depends on current intentions (Bargh & Chartrand, 1999), even for relatively low-level information processing (Folk, Leber, & Egeth, 2002), in that external stimuli often have automatic effects only when they are relevant to currently active goals. Hommel (2000) suggested that automaticity be understood in terms of information processes set up or prepared by intentions, consistent with the approach described here.

Finally, focusing on intentions provides a basis for establishing links between theoretical work in cognitive psychology and cognitive modeling and other approaches to understanding control, such as the well-developed analyses of intention in philosophy, developmental psychology, and linguistics. The theoretical approach described here is based in part on insights into the nature and function of intentions derived from philosophical analyses by Dennett (1991, 2003), Metzinger (2003), and Searle (1983).

Representing Intentions

Intentions are states of working memory with schematic structures that specify their elements in varying ways and have characteristic temporal dynamics. By describing intentions as states of working memory, I emphasize that they are active information structures subject to the capacity limits, brief duration if not refreshed, and other properties commonly attributed to working memory (Miyake & Shah, 1999). This view is compatible with a variety of conceptions of working memory but perhaps fits best with a view of working memory as a combination of limited capacity for activation together with a repertoire of strategies (e.g., verbal rehearsal) for refreshing that activation.

Intentions serve a control function by specifying mental or physical actions and the mental or physical objects on which those actions operate. Actions and objects may be specified at varying levels of description, depending on skill and circumstances (Carlson, 1997; Vallacher & Wegner, 1987). Instantiating a goal as an intention also specifies, at least implicitly, a particular place ("here") and time ("now"), providing a basis for the temporal and spatial frames of reference that organize behavior (Carlson, 2002). The experience of control results from the points of view embodied in intentions, and from memory for intentions and outcomes. I focus here on single intentions, but how these intentions are situated in larger goal structures and informational contexts is also crucial to their roles in control (Carlson, 2002).

Intentions as Schemas

The idea that intentions are best described as schematic is a familiar one. Norman (1981) classified action slips by identifying sources of error in an activation–trigger–schema (ATS) model that described control as activation and triggering in hierarchies of schematic intentions. Cooper and Shallice (2000) described a computational model of control descended from Norman's analysis (e.g., by way of Norman & Shallice, 1986), though their theory dissociates the representations of goals and of actions. My analysis is based in part on Searle's (1983) philosophical analysis of intentionality,[1] and is more similar to the "atomic" approach of ACT-R (e.g., Anderson & Lebiere, 1998) than to the Cooper and Shallice approach in focusing on the linked representation of component goals and actions rather than on broader hierarchical networks of goals with separate action representations.

The schematic structure of intentions is depicted in the top panel of Figure 27.1. This structure has several important aspects: an implicitly specified *agent* ("I"), a *mode* ("intend"), and a content. The basis for this general structure is discussed in more detail elsewhere (Carlson, 1997, 2002). Most important for the present purpose is the schematic structure of the content, which specifies three kinds of elements: an action, or *operation*; objects, or *operands*, of that operation; and an *outcome*. Describing this content as schematic refers both to its structuring in terms of relations among operation, operands, and outcome and to the idea that these elements may be specified abstractly. This analysis is similar to that described by Anderson (1993) in his account of the relations between goal chunks in declarative working memory and in productions. Anderson

---

General schematic structure of intentions

{ I } intend that

by (my) performing *operation x* on

objects y₁, y₂, ...

outcome z be accomplished

Schematic structure of intentions for each step of fluent event counting

{ I } intend that

by (my) *assigning*

the *next number* to the *next event*

the *count be incremented to obtain a new total*

FIGURE 27.1 The schematic structure of intentions. The general schematic structure of intentions is shown in the top panel, and the specific schematic structure hypothesized for event counting is shown in the bottom panel.

notes that generality in productions is achieved by variabilizing the chunks involved. For example, a production for adding digits may specify that *some* digit is available in the knowledge under focus, rather than specifying a particular digit (Anderson, 1993, pp. 36–43).

This content is represented over time, as with other kinds of working memory content. In particular, specifying an operator provides a procedural frame to which operands can be assimilated to generate an outcome. It is therefore more efficient to construct or to retrieve representations such that operators are considered before operands. For example, individuals solve simple arithmetic problems more quickly when operator symbols (e.g., "+") appear in advance of numbers (Carlson & Sohn, 2000). Control of complex processes can be understood as repetition of an intention–outcome cycle, similar to other proposals that recognize the cyclic nature of cognition (e.g., Neisser, 1976; Norman, 1988), including the production cycle in computational models such as ACT-R (adaptive control of thought–rational). In each cycle,

intentional elements are assimilated to a schematic intention.

Implications and Empirical Phenomena

It is well established that skill acquisition changes the nature of control, reducing the need for deliberate control. Yet even highly skilled, routine performances are intentional—done on purpose—although expertise may entail representing intentions at higher levels of abstraction than is possible for novices. I have suggested four principles for understanding intentional control in skilled activity (Carlson, 1997, p. 270):

1. The *goal instantiation* principle: Cognitive control depends on current, instantiated goals whose contents specify outcomes to be achieved by one's own action.
2. The *juxtaposition* principle: Cognitive activity results from the juxtaposition (or synchronous activation) of mental state contents and available information.

3. The *minimal deliberation* principle: Cognitive control by instantiated goals involves minimal deliberation or planning, at least for routine activities.

4. The *minimal control* principle: Fluency is achieved by minimizing the amount of explicit information involved in the cognitive control of activity.

The first principle distinguishes goals as represented in intentions from goals represented in long-term memory or as parts of plans to be achieved in the (nonimmediate) future, and the second principle suggests that intentions cause behavior by the standard mechanism of activation as realized in many computational models. The third and fourth principles point to important ways in which skill acquisition results in increased fluency: In skilled activity, intentions may be generated associatively rather than by reasoning or problem solving (Bargh & Chartrand, 1999), and those intentions constitute concise, efficient representations. This is similar to Taatgen's (chapter 25, this volume) independently developed minimal control principle in that both views emphasize the need to minimize the explicit representation of control parameters.

The increased fluency that results from practice depends on regularities or consistencies in the task environment. Most research has focused on consistency in stimulus–response mappings (e.g., Schneider & Shiffrin, 1977), or on sequential consistencies in perceptual-motor tasks (e.g., Willingham, Nissen, & Bullemer, 1989), but there is also evidence that various kinds of consistency in more complex cognitive tasks allow greater improvements with practice (e.g., Blessing & Anderson, 1996; Carlson & Lundy, 1992; Woltz, Gardiner, & Bell, 2000). Most theoretical accounts of these benefits suggest that regularities in task environments allow the development of more concise representations for task control (e.g., Blessing & Anderson, 1996; Haider & Frensch, 1999). For example, individuals may restructure tasks such that fewer steps—and thus fewer intentions—are required (Blessing & Anderson, 1996; Carlson, 1997). Representations of individual intentions may also become more concise with practice, increasing fluency by reducing the time and capacity required to represent those intentions in working memory. These concisely represented intentions may specify the means of achieving their outcomes only at relatively high levels of abstraction (Carlson, 1997; Vallacher & Wegner, 1987).

Deictic Specification of Intentional Elements

One way to make an intention more concise is to rely on *deictic* specification of intentional elements. *Deixis* is a concept from the study of linguistic pragmatics, meaning something like "reference by pointing" (Fillmore, 1997). The idea is that when we communicate by speaking, we often refer to objects, people, places, or times by figuratively pointing rather than by conceptual or semantic description. For example, I might say "could you hand me *that?*" or *"now* we're ready to go." "That" and "now" refer to an object and a time available to speaker and audience on the basis of shared context, and perhaps literal pointing, but their referents cannot be recovered from a transcript of the utterances. Ballard, Hayhoe, Pook, and Rao (1997) suggested that much perceptual-motor activity is accomplished by deictic binding of perceptual information to motor schemes. They described such routines as "'do-it-where-I'm-looking' strategies" (p. 725). Deictic specification of intentional elements can be contrasted with *semantic specification*, in which intentions include representations that describe or conceptually symbolize intentional elements.

An important implication of this *deictic specification hypothesis* is that intentional control depends on temporal synchrony to coordinate the intention–outcome cycle with the availability of information that instantiates the elements of each intention because semantic bases for associating inputs with schematic slots are not available. Because semantic matching is not available as a basis for binding information to the slots of intention schemas, information must be selected for such binding on the basis of the spatial and temporal specification characteristic of deixis. Temporal coordination has been studied extensively in the domain of motor control (e.g., Rosenbaum & Collyer, 1998) but only rarely in cognitive domains (but see Carlson & Stevenson, 2002). Ballard et al. (1997) noted that sensory– motor primitives such as eye fixations could serve to establish the origins of the spatial frames of reference needed to organize actions such as grasping. Similarly, the times at which intentional elements are instantiated can establish the origins of temporal frames of reference that organize mental activities.

Ballard et al. (1997) described in substantial detail the adaptive computational roles that could be played by what they called "deictic codes." Here, I focus on some consequences of deictic specification that are not always adaptive, resulting in both errors and limits on error monitoring.

Norman's Analysis of Action Slips

Norman (1981) classified action slips on the basis of their hypothesized causes within his ATS model. He suggested three major categories: faulty intention formation, faulty activation, and faulty triggering. Norman focuses on the second and third of these categories, describing a large number of errors that result from inappropriate activation, loss of activation, false (inappropriate) triggering, or failures of triggering. Here, I focus on one subcategory of faulty intention formation, which Norman called *description errors*. These errors are said to be due to insufficient specificity in the formation of intentions. For example, Norman suggests that "throwing a soiled shirt in the toilet rather than the laundry basket" (p. 7) results from insufficient specificity in representing the "container" element of this intention.

The deictic specification hypothesis sketched in the previous section—that the intentions guiding fluent performance specify at least some of their elements deictically—suggests that such errors result from a feature of skilled intentional control that is usually adaptive. When all goes well, the environmental regularities that support deictic specification—for example, the presence of the laundry basket when one's intention is to deposit a soiled shirt—are sufficient for successful performance on the basis of a minimal representation of one's intention. When those regularities fail, and an element of an intention is instantiated by the wrong object—for example, the toilet rather than the laundry basket—error results. Thus, given an analysis of the regularities in a task environment, it should be possible to predict the nature of errors that may result from deictic specification.

Error Monitoring

We are essentially always monitoring our behavior in relation to our intentions, and some authors have suggested that such monitoring is the central function of consciousness (Mandler, 1984; Reason, 1990). An error in the execution of an intention is a mismatch of intended and actual outcomes. The best-developed theories concerning error monitoring and detection are in the domain of speech production (Postma, 2000). In these theories, speech monitoring is accomplished by explicit comparison of intentions with outcomes on the basis of a semantic match—does the actual speech produced match my communicative intent? The present analysis suggests, however, that explicit error monitoring on the basis of semantic matching will often be difficult or impossible because intentions often specify their elements deictically rather than semantically. In such cases, error monitoring will instead be *implicit*, based not on explicit, semantic comparisons of intentions and outcomes but on monitoring processing fluency. Errors will be detected only when something disrupts the smooth flow of processing, for example, when no information is available to instantiate an element of an intention. Implicit monitoring appears to be the basis for comprehension monitoring in reading (Forlizzi, 1988), explaining the common experience of suddenly realizing that one has read several paragraphs without comprehension (Schooler, 2002). As long as the process proceeds fluently, and information is available at the appropriate point in time to instantiate the deictically specified intentional elements, individuals will fail to notice errors. However, when fluency is disrupted, for example, by a scrambled sentence (Forlizzi, 1988) or an external probe (Schooler, 2002) while reading, errors will be detected. The failure to notice errors is, of course, a commonly noted phenomenon with important practical implications (e.g., Reason, 1990).

An Illustration: Intentional Control of Event Counting

Event counting—enumerating events that occur one at a time—provides a useful paradigm for studying cognitive control. The knowledge involved is trivial for most adults (although there are interesting issues concerning the nature and acquisition of that knowledge; e.g., Gallistel & Gelman, 1992), but the task itself is surprisingly difficult. Counting any substantial number of events is cognitively demanding and error prone, suggesting that attention and control are significant issues in determining performance.

In a series of studies, my students and I have explored several aspects of event counting (Carlson & Cassenti, 2004; Carlson & Smith, 2005; Cassenti, 2004; Cassenti & Carlson, in press). In our experimental paradigm, participants count events—such as briefly presented asterisks or other characters—displayed on a computer screen. At the end of an event sequence, the participant reports his or her count and confidence in that count. Participants in some experiments also have the option of pressing a "panic button" to indicate that they have lost count during the sequence, which stops the sequence and requires immediate entry of an answer. This paradigm allows us to control the timing of events—their pacing—and whether the interval between events is constant (rhythmic) or varied. It provides data about the nature of participants' counting errors, whether they report the correct count and how far and in what direction erroneous counts deviate from the correct count, and about their success in error monitoring, whether they detect errors, as indicated by reporting lower confidence for errors than for correct counts or by pressing the panic button.

Carlson and Cassenti (2004) explored the implications of the deictic specification hypothesis in a series of experiments using this paradigm. Consider the regularities or consistencies that may be available in the task environment: events to be counted may have consistent identities and may appear in consistent locations at predictable times. Carlson and Cassenti manipulated the availability of temporal regularity by displaying events at consistent, rhythmic intervals or at varied intervals (equating either average or minimum time per event). Second, consider a regularity embodied in the skill of counting itself: the ability to reliably generate the number sequence. This aspect of counting is learned at an early age and is practiced extensively by most members of our culture. Individuals appear to generate the number sequence by rapid retrieval of numbers from memory, together with well-practiced rules for updating the tens (or higher) digits (Healy & Nairne, 1985).

These regularities allow fluent counting to be controlled by the schematic intention represented in the bottom panel of Figure 27.1. This figure illustrates the deictic specification hypothesis for counting rhythmic events—both the next number to be used and the event to which the number is to be assigned are specified deictically. Control will be successful to the extent that temporal synchrony allows appropriate assignment of numbers to events (see Carlson & Cassenti, 2004, for further detail).

The deictic specification hypothesis makes several predictions about event counting performance. First, it should be much easier to count events that occur rhythmically, because the temporal regularity makes it easier to rely on deictic specification—as long as external events (the appearance of items to be counted) and internal events (the generation of numbers and their

consolidation as counts) occur at predictable intervals, counting can proceed smoothly with minimal semantic representation of intentional elements. When events do not occur rhythmically, a deictic specification of an element (e.g., *next event*) may need to be semantically elaborated, for example, with features that distinguish perceived from remembered events. Second, deictic specification when counting relatively rapid rhythmic events should result in errors that arise from *intention-outcome confusions*—cases in which an outcome, a number just assigned to an event, is mistaken for an element of an intention, the number to be used next. Such confusions will result in undercounts because the same number will be assigned to more than one event. Third, error monitoring should be poor when counting rhythmic events because deictic specification of intentional elements does not allow for explicit comparisons of intentions and outcomes. As long as the elements are available in sufficiently close temporal synchrony to allow the counting process to proceed fluently, with each schematic slot filled at the appropriate moment, there will be no mismatch of intention and outcome and thus no informational basis for detecting errors. Fourth, when errors *are* detected because of disruptions in fluent performance, individuals are likely to adopt repair strategies that lead to different patterns of errors. Buchner, Steffens, Immen, and Wender (1998) argued that common repair strategies in event counting will typically produce overcount errors. The idea is that individuals will attempt to correct their count by adding to the most-available number in working memory, often assuming incorrectly that this number is several steps old.

The results of these experiments supported these predictions. Counting was substantially more accurate when events appeared at rhythmic rather than varied intervals if average interval was equated, and achieving equivalent accuracy required average intervals approximately 40% longer for varied than for rhythmic intervals. As expected, most errors in rhythmic counting were undercounts, whereas most errors with varied intervals were overcounts. For example, when intervals were adjusted for approximately equal overall accuracy with rhythmic and varied intervals (Carlson & Cassenti, 2004, Experiment 3), 71% of errors were undercounts on rhythmic-interval trials, compared to 41% on varied-interval trials.

Perhaps most interesting, error monitoring was much poorer when counting rhythmic events that allowed deictic specification. Figure 27.2 shows the probability

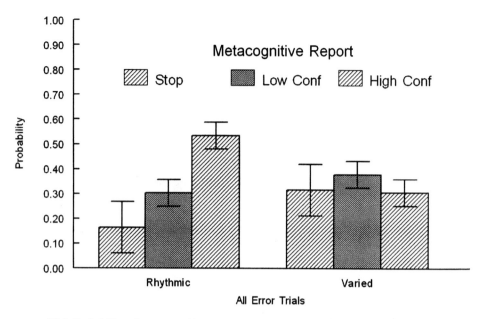

FIGURE 27.2 Probability of metacognitive report types on error trials from Carlson and Cassenti (2004) Experiment 3. With rhythmic intervals, the probability of stopping is low and the probability of reporting high confidence is large.

of pressing the panic button to indicate a lost count (*stop*) or of expressing high versus low confidence, given that an error was made. Again, these data are from an experiment in which intervals were adjusted for approximately equal overall accuracy with rhythmic and varied intervals. Participants were less likely to press the panic button, and more likely to express high confidence, when errors were made while counting rhythmic events. On varied-interval trials, error monitoring was much better, presumably because the lack of temporal regularity prevented a smooth flow of processing. Confidence on correct trials did not differ for rhythmic and varied interval conditions.

These findings are illustrative of our research program on counting and other repetitive sequential tasks. They demonstrate that when the conditions for deictic specification are met, performance is better—all else being equal—but that deictic specification leads to characteristic error types and poor error monitoring.

Other Examples

The hypothesis that the intentions controlling skilled activity may specify their elements deictically allows us to reinterpret a number of other results. Consider the possibility that under some circumstances intentions

may specify operators rather than (or as well as) operands deictically. Wenger and Carlson (1996) examined the effects of practice on performance of relatively complex arithmetic routines with hierarchical goal structures that required participants to sometimes hold results for several steps before using them. Figure 27.3 illustrates one variation of the task used in their study. Participants performed a series of cascaded arithmetic steps, pressing the space bar on a computer keyboard to display the information needed for each step. Each step required one of four operations—MIN, MAX, SUM, or DIF (absolute difference)—be performed on two numbers. Each number was either a constant displayed on the screen or a variable name (indicated by letters A–M) referring to a prior result held in working memory. Two aspects of the task could be consistent or varied, depending on the experimental condition: (a) the assignment of operators to positions in the sequence and (b) the abstract goal structure, defined as the relations among steps specified by the use of variable names (A, B, C, etc.). The middle and rightmost columns of the figure illustrate the abstract goal structure for this example, indicating the information used from working memory to complete each step and the results to be held in working memory at the completion of the step. In this paradigm, the operator to be used at each step was available at a predictable time

INFORMATION DISPLAYED AT STEP	INFORMATION USED FROM WM	WM CONTENT AFTER STEP
A = X1	None	A
B = f (X2, X3)	None	A, B
C = f (B, X4)	B	A, C
D = f (A, C)	A, C	D
E = f (D, X5)	D	E
F = f (E, X6)	E	F
G = f (X7, X8)	None	F, G
H = f (F, X9)	F	G, H
I = f (G, H)	G, H	I
J = f (X10, X11)	None	I, J
K = f (I, X12)	I	J, K
L = f (J, K)	J, K	L
M = f (L, X13)	L	M

FIGURE 27.3 Task structure for multiple-step arithmetic problems used in Wenger and Carlson (1996). At each step, the operator function (f) was one of MIN, MAX, SUM, or DIF, and X1–X13 were integer constants in the range of 2–19.

and place, but operands had to be selected at each step from the screen or from values held in memory.

Three aspects of the results of these studies stand out. First, there was a very large benefit (as measured by problem-solving time) of practicing with consistent abstract goal structures, compared with practicing with problems with abstract goal structures that varied from problem to problem. Second, there was little or no benefit of practicing with consistent sequences of operators, compared with practicing with varied sequences of operators. Third, participants showed little if any ability to remember the sequences of operators in the problems they practiced, even when those operator sequences were consistent throughout practice. Wenger and Carlson (1996) pointed out that these results demonstrated that participants' improved performance did not depend primarily on learning the binding of operators to steps but on learning the regularities in working memory demands. However, as in many studies of skill acquisition, regularities in the presentation of the task—for example, when and where operator information appeared—were treated as defining the task environment, not as consistencies crucial to learning. In hindsight, ignoring the theoretical import of these regularities limited our insight into our findings.

The deictic specification hypothesis developed here helps to make sense of these results. The symbols indicating operators appeared in predictable places and at predictable times (relative to participants' key-press requests). Given limited working memory capacity for representing intentions, it is presumably adaptive to devote that capacity to representing the most difficult part of this task—managing intermediate results in working memory—and to specify only where and when to pick up operator information.

The deictic specification hypothesis may also help to explain the benefits of external task cues. Consider a result from the task-switching literature—even when participants have reliable foreknowledge of which task to perform, explicit task cues result in superior performance and reduced switch costs. For example, Koch (2003) demonstrated that even given predictable task switches, the availability of explicit external cues reduced switch costs by approximately 35% when participants had time to process those cues. Koch cites other evidence that procedural learning of task sequences does not reduce switch costs in the absence of external task cues.

This result might be understood in terms of loading or reconstructing an intention in working memory.

Assuming that the time to do so depends on the complexity of the representation, the more concise representation possible when an intention specifies its elements deictically might reduce task switching time. Of course the condition for deictic specification is the reliable availability of the information in the environment, a condition met by Koch's experimental paradigm.

These examples illustrate a general point relevant to the deictic specification hypothesis: cognitive performance, especially skilled performance, must be understood in part by analyzing the structure of the task environment (e.g., Kirlik, chapter 14, this volume). The event counting, mental arithmetic, and task-switching paradigms all have sequential structures that require rapid changes of intention and other working memory content. Those sequential structures also have regularities, sometimes treated as background rather than objects of analysis that support deictic specification. Individuals may rely on the environment rather than memory whenever they can (Ballard et al., 1997), but that reliance depends on spatial and temporal consistency in the availability of environmental information (Carlson & Stevenson, 2002; Cary & Carlson, 2001). In the words of Ballard et al. (1997, p. 740), the deictic specification hypothesis suggests that we "shift the focus to the ongoing *process* of cycling information through working memory."

Experience of Control

It has become clear to many researchers that integrated theorizing about cognitive systems, especially in the context of complex task performance, will require understanding metacognition and its functions in guiding performance, strategy selection, and learning. Focusing on intentions and how they specify their elements suggests some bases for addressing these issues.

Levels of Experience

The term *metacognition* refers to a range of experiences and functions (Metcalfe & Shimamura, 1994; Nelson, 1996; Schooler, 2002). For example, researchers often distinguish between "on-line" metacognition or monitoring and longer-term metacognitive knowledge. Long-term metacognitive knowledge is clearly a result of development and learning; for example, the knowledge that one's ability to remember new information can be enhanced by rehearsal is typically not acquired until the early school years (Flavell & Wellman, 1977),

and many college students have not learned that spaced practice is superior to massed practice.

The real-time experience of an individual fully engaged with a task—that is devoting all available attention or whatever flavor of "cognitive resources" you prefer—is always *implicitly* metacognitive. Because purposeful behavior is controlled by intentions with the structure outlined above, an experiencing, acting agent is continuously if implicitly specified. This is the foundational condition for higher-order metacognition, and this base-level experience is truly in parallel, rather than interleaved, with task performance. However, this view also implies that higher-order, more reflective forms of metacognition will compete with task performance and thus be interleaved with it, as suggested by Anderson (chapter 4, this volume). In particular, higher-order forms of metacognition—such as explicit error monitoring by comparison of an intention with its outcome—must rely on memory for intentions, even if on occasion that memory is very recent.

Error Monitoring as Metacognition

Most discussions of error monitoring imply that it is a special function evoked (or not) as needed (Reason, 1990). In fact, however, implicit error monitoring is likely ubiquitous, fully integrated with the intentional control of routine activities (Mandler, 1984). An interesting possibility in the context of this volume is that momentary affect, the fleeting emotionality of experience, is the signature of implicit monitoring. As the intention–outcome cycle unfolds, positive affect may reflect a match between current circumstances and the conditions for performing an intended operation (Kuhl & Kazén, 1999) or progress toward a desired outcome (Carver & Scheier, 1998). The role of affect as agent-specific information involved in the control of activity at a fine-grained level (Carlson, 1997, 2003) is currently under investigation in my laboratory (Carlson & Smith, 2005).

Limits on Experience

The view that intentions are states of working memory that may specify their elements deictically implies limits on the subjective experience of control. When performance is rapid and fluent, working memory capacity will be occupied by intentions and fleeting representations of their elements. Thus higher-order reflection must compete with on-line control (Anderson, chapter 4, this volume), suggesting one kind of limit on the experience of control. A second kind of limit comes from well-known limits on working memory and its relation to more durable, longer-term memory—not everything that is briefly represented in working memory is later retrievable from long-term memory.

A third kind of limit on the experience of control is associated with deictic specification. The results reported by Carlson and Cassenti (2004) demonstrate poor error monitoring in fluent counting that results from deictic specification of operands, and those reported by Wenger and Carlson (1996) demonstrate poor metacognitive memory for a crucial aspect of the sequences their participants practiced, the sequence of operators. The deictic specification hypothesis thus provides an alternative to the view that knowledge underlying skill is represented unconsciously. Awareness is typically tested by requiring reports in a context that does not include the regularities supporting deictic specification. Implicit knowledge inferred from a lack of correspondence between such reports and observed performance may simply reflect the development of concise representations whose successful use depends on the regularities available in the performance context.

Experience and Control

We experience ourselves as in control of our own behavior, but the relation of that experience to actual cognitive control has been controversial in cognitive science. Few if any cognitive scientists would endorse a *full correspondence* view—that the informational content of experience is coextensive with the computational description of control processes. Like naïve realism in the domain of perception, such "naïve intentionalism" is seriously flawed as a theoretical perspective. A number of authors have recently argued for a *no correspondence* view—that the informational content of experience has no functional overlap with the computational description of control processes. This view is associated with Libet's (1985) efforts to demonstrate that awareness of intention lags behind the neural events that are truly causal in voluntary action, and has recently been expounded at length by Wegner (2002, 2004).[2]

The analysis of intentions as states of working memory suggests instead a *partial correspondence* view—the subjective experience of control is constituted by a systematic subset of the computational description of control processes (Metzinger, 2003). This view, I have argued, provides a basis for linking theoretical descriptions of conscious experience and of cognitive control (Carlson, 2002).

Conclusion

Thinking about control from a representational perspective, in which intentions are conceived as states of working memory, has some advantages over more common focus on capacity or architectural distinctions. I have argued here that some phenomena of control—in particular, error mechanisms and on-line error monitoring—can be understood in terms of how intentions are represented and how that representation changes as a function of skill acquisition that specializes in those representations for regularities in the task environment and in the function of associative memory. Fluent performance is guided by, and experienced via, concisely represented intentions that specify at least some of their elements deictically, relying on environmental regularities and spatial–temporal coordination for control. This normally adaptive feature of skilled activity will result in occasional errors that can be attributed to what Norman (1981) described as underspecification. Under these circumstances, error monitoring will be implicit, and individuals will not detect errors that fail to disrupt the smooth flow of processing. One example of such errors is intention–outcome confusions. Such confusions are likely responsible for some errors of omission that may have drastic consequences in high-performance environments (Reason, 1990).

An important question in the context of the work reported in this volume is how the theoretical ideas and phenomena discussed in this chapter—intentions as schematic structures in working memory, deictic specification of intentional elements, implicit monitoring, and limits on metacognition—might be realized in a computational model. For example, is a separate "intentional module" as suggested by Taatgen (chapter 25, this volume) necessary? Some aspects of these ideas are, of course, already represented in computational models. For example, I have already noted the similarity of the deictic specification hypothesis to the variabilization of goal chunks in ACT-R, and Ballard et al. (1997) described ways in which sensory-motor primitives can provide bases for spatial frames of reference. In other cases, previously implemented ideas might be adapted to fit the hypotheses sketched here. For example, the idea that individuals select microstrategies to reduce response times by milliseconds (Gray & Boehm-Davis, 2000) might be extended to examine how such microstrategies are used to achieve precise temporal synchrony between cognitive processes and the pickup of information from the environment.

Other ideas—that implicit error monitoring is based on detecting failures in fluency while explicit error monitoring involves comparing intentions with outcomes—might require additional representational resources. It is unclear, however, that dramatic changes in architectural resources would be required to implement any of the ideas developed here.

In setting the stage for the workshop on which this volume is based, Gray (chapter 1, this volume) described several senses of *integrated*—"integration of cognition, perception, and action to yield embodied cognition," "integration of implicit with explicit knowledge," "integration of mind with environment," and "integration of emotion with cognition." The theoretical perspective developed in this chapter suggests ways to think about each of these senses of integration and suggests that thinking about *intentions* can point the way to seeing metacognition as integrated with the control of behavior.

Notes

1. In Searle's work, and that of other philosophers, *intentionality* refers to a broader concept, the idea that mental states are distinguished from physical states by the property of being directed at or about other states, their objects. Searle provides an extensive analysis of this concept, noting the similarities and differences between intentions, perceptual states, beliefs, desires, and so on. In his analysis, all mental states are about or directed at their *conditions of satisfaction*, conditions that must hold if the states are to be satisfied. For example, the conditions of satisfaction for an intention include that a desired outcome occur as a result of one's intended action.

2. Both of these views have been critiqued; for example, see Dennett and Kinsbourne (1992) on the Libet studies and responses to Wegner (2004).

References

Anderson, J. R. (1983). *The architecture of cognition.* Cambridge, MA: Harvard University Press.

——. (1993). *Rules of the mind.* Hillsdale, NJ: Erlbaum.

——, & Lebiere, C. (1998). *The atomic components of thought.* Mahwah, NJ: Erlbaum.

Austin, J. T., & Vancouver, J. B. (1996). Goal constructs in psychology: Structure, process, and content. *Psychological Bulletin, 120*(3), 338–375.

Ballard, D. H., Hayhoe, M. M., Pook, P. K., & Rao, R. P. N. (1997). Deictic codes for the embodiment of cognition. *Behavioral and Brain Sciences, 20*(4), 723–767.

Bargh, J. A., & Chartrand, T. L. (1999). The unbearable automaticity of being. *American Psychologist, 54*(7), 462–479.

Blessing, S. B., & Anderson, J. R. (1996). How people learn to skip steps. *Journal of Experimental Psychology: Learning, Memory, and Cognition, 22*(3), 576–598.

Buchner, A., Steffens, M. C., Irmen, L., & Wender, K. F. (1998). Irrelevant auditory material affects counting. *Journal of Experimental Psychology: Learning, Memory, and Cognition, 24*, 48–67.

Carlson, R. A. (2002). Conscious intentions in the control of skilled mental activity. In B. Ross (Ed.), *The psychology of learning and motivation* (Vol. 41, pp. 191–228). San Diego, CA: Academic Press.

——. (1997). *Experienced cognition.* Mahwah, NJ: Erlbaum.

——, & Cassenti, D. N. (2004). Intentional control of event counting. *Journal of Experimental Psychology: Learning, Memory, and Cognition, 30*, 1235–1251.

——, & Lundy, D. H. (1992). Consistency and restructuring in learning cognitive procedural sequences. *Journal of Experimental Psychology: Learning, Memory, and Cognition, 18*, 127–141.

——, & Sohn, M.-H. (2000). Cognitive control of multiple-step routines: Information processing and conscious intentions. In S. Monsell & J. Driver (Eds.), *Control of cognitive processes: Attention and performance XVIII* (pp. 443–464). Cambridge, MA: MIT Press.

——, & Smith, R. E. (2005). *Counting on emotion.* Paper presented at the 46th annual meeting of the Psychonomic Society, Toronto, Ontario, Canada, November 11, 2005.

——, & Stevenson, L. M. (2002). Temporal tuning in the acquisition of cognitive skill. *Psychonomic Bulletin & Review, 9*(4), 759–765.

Carver, C. S., & Scheier, M. F. (1998). *On the self-regulation of behavior.* Cambridge: Cambridge University Press.

Cary, M., & Carlson, R. A. (2001). Distributing working memory resources in problem solving. *Journal of Experimental Psychology: Learning, Memory, and Cognition, 27*, 836–848.

Cassenti, D. N. (2004). *When more time hurts performance: A temporal analysis of errors in event counting.* Unpublished doctoral dissertation, Pennsylvania State University.

——, & Carlson, R. A. (in press). The role of pacing and working memory loads on error type patterns in routine skill. *American Journal of Psychology.*

Cooper, R., & Shallice, T. (2000). Contention scheduling and the control of routine activities. *Cognitive Neuropsychology, 17*, 297–338.

Dennett, D. C. (1991). *Consciousness explained.* Boston: Little, Brown.

——. (2003). *Freedom evolves.* New York: Viking.

———, & Kinsbourne, M. (1992). Time and the observer: The where and when of consciousness in the brain. *Behavioral and Brain Sciences, 15*(2), 183–247.

Fillmore, C. J. (1997). *Lectures on deixis.* Stanford, CA: Center for the Study of Language and Information.

Flavell, J. H., & Wellman, H. M. (1977). Metamemory. In R. V. Kail & J. W. Hagen (Eds.), *Perspectives on the development of memory and cognition* (pp. 3–33). Hillsdale, NJ: Erlbaum.

Folk, C. L., Leber, A. B., & Egeth, H. E. (2002). Made you blink! Contingent attentional capture produces a spatial blink. *Perception and Psychophysics, 64,* 741–753.

Forlizzi, L. (1988). *Relationships among use, predicted use, and awareness of use of comprehension-repair strategies: Converging evidence from different methodologies.* Unpublished doctoral dissertation, Pennsylvania State University, University Park.

Gallistel, C. R., & Gelman, R. (1992). Preverbal and verbal counting and computation. *Cognition, 44,* 43–74.

Gray, W. D., & Boehm-Davis, D. A. (2000). Milliseconds matter: An introduction to microstrategies and to their use in describing and predicting interactive behavior. *Journal of Experiment Psychology: Applied, 6,* 322–335.

Haider, H., & Frensch, P. A. (1999). Information reduction during skill acquisition: The influence of task instruction. *Journal of Experimental Psychology: Applied, 5*(2), 129–151.

Healy, A. F., & Nairne, J. S. (1985). Short-term memory processes in counting. *Cognitive Psychology, 17,* 417–444.

Hommel, B. (2000). The prepared reflex: Automaticity and control in stimulus-response translation. In S. Monsell & J. Driver (Eds.), *Control of cognitive processes: Attention and performance XVIII* (pp. 247–273). Cambridge, MA: MIT Press.

Koch, I. (2003). The role of external cues for endogenous advance reconfiguration in task switching. *Psychonomic Bulletin and Review, 10,* 488–492.

Kuhl, J., & Kazén, M. (1999). Volitional facilitation of difficult intentions: Joint activation of intention memory and positive affect removes Stroop interference. *Journal of Experimental Psychology: General, 128,* 382–389.

Libet, B. (1985). Unconscious cerebral initiative and the role of conscious will in voluntary action. *Behavioral and Brain Sciences, 8,* 529–566.

Malle, B. F., Moses, L. J., & Baldwin, D. A. (2001). *Intentions and intentionality: Foundations of social cognition.* Cambridge, MA: MIT Press.

Mandler, G. (1984). *Mind and body.* New York: W.W. Norton.

Metcalfe, J., & Shimamura, A. P. (1994). *Metacognition: Knowing about knowing.* Cambridge, MA: MIT Press.

Metzinger, T. (2003). *Being no one.* Cambridge, MA: MIT Press.

Miyake, A., & Shah, P. (1999). *Models of working memory: Mechanisms of active maintenance and executive control.* Cambridge: Cambridge University Press.

Neisser, U. (1976). *Cognition and reality.* San Francisco: Freeman.

Nelson, T. O. (1996). Consciousness and metacognition. *American Psychologist, 51*(2), 102–116.

Norman, D. A. (1981). Categorization of action slips. *Psychological Review, 88,* 1–15.

———. (1988). *The psychology of everyday things.* New York: Basic Books.

———, & Shallice, T. (1986). Attention to action: Willed and automatic control of behavior. In R. J. Davidson, G. E. Schwartz, & D. Shapiro (Eds.), *Consciousness and self-regulation: Advances in research and theory* (Vol. 4, pp. 1–18). New York: Plenum Press.

Postma, A. (2000). Detection of errors during speech production: a review of speech monitoring models. *Cognition, 77,* 97–131.

Reason, J. T. (1990). *Human error.* Cambridge: Cambridge University Press.

Rosenbaum, D. A., & Collyer, C. E. (1998). *Timing of behavior: Neural, psychological, and computational perspectives.* Cambridge, MA: MIT Press.

Schneider, W., & Shiffrin, R. M. (1997). Controlled and automatic human information processing: I. Detection, search, and attention. *Psychological Review, 84,* 1–66.

Schooler, J. W. (2002). Re-representing consciousness: Dissociations between experience and meta-consciousness. *Trends in Cognitive Sciences, 6,* 339–344.

Searle, J. R. (1983). *Intentionality: An essay in the philosophy of mind.* Cambridge: Cambridge University Press.

Shallice, T. (1994). Multiple levels of control processes. *Attention and Performance, 15,* 395–420.

Vallacher, R. R., & Wegner, D. M. (1987). What do people think they're doing: Action identification and human behavior. *Psychological Review, 94,* 3–15.

Wegner, D. M. (2002). *The illusion of conscious will.* Cambridge, MA: MIT Press.

———. (2004). Precis of the Illusion of conscious will. *Behavioral and Brain Sciences, 27,* 649–692.

Wenger, J. L., & Carlson, R. A. (1996). Cognitive sequence knowledge: What is learned? *Journal of Experimental Psychology: Learning, Memory, and Cognition, 22*(3), 599–619.

Willingham, D. B., Nissen, M. J., & Bullemer, P. (1989). On the development of procedural knowledge. *Journal of Experimental Psychology: Learning, Memory, and Cognition, 15,* 1047–1060.

Woltz, D. J., Gardner, M. K., & Bell, B. G. (2000). Negative transfer errors in sequential cognitive skills: Strong-but-wrong sequence application. *Journal of Experimental Psychology: Learning, Memory, and Cognition, 26,* 601–625.

PART VIII

TOOLS FOR ADVANCING INTEGRATED MODELS OF COGNITIVE SYSTEMS

Wayne D. Gray

There may be no greater heroes in the history of science than those researchers who turn a new idea or technology into a tool for the advancement of science. In the hands of Galileo and Anthony Leeuwenhoek, the newly invented compound lens led to major advances in astronomy and biology. In a similar manner, the growth of cognitive science is the result of tools that have allowed us to model cognitive processes. In the 1950s, these tools took the form of information processing languages such as IPS and Lisp (Simon). In the 1960s, the field discovered stronger formalisms such as production systems (Newell, 1973), which provided a control structure that was congenial to the models being developed at that time. The invention of the personal computer and the rise of tools for end-user programming greatly helped the widespread dissemination and adoption of connectionist modeling (Bechtel & Abrahamsen, 1991). Indeed, a milestone in symbolic modeling occurred in 1993 when Anderson's book *Rules of the Mind* (Anderson, 1993) was published that included a floppy disk with a version of ACT-R that could be run on any Macintosh computer. With both connectionist and symbolic modeling available to anyone who owned a personal computer, the enterprise of cognitive modeling became independent of the few large research universities in which it had been nurtured and was set for worldwide dissemination.

The two chapters in this section continue the spirit of building better tools to enable us to build better models with which to study cognitive processes. In chapter 28, Howes, Lewis, and Vera discuss an approach for computing the optimal method or strategy given the constraints of the task, the task environment, the user's knowledge, and the postulated cognitive architecture. The predicted optimal human strategy can be compared with those strategies used by human subjects. The match or mismatch between optimal performance predicted by the theory and asymptotic human performance becomes an important test of cognitive theory and an important spur to additional research and theory development.

Cooper's COGENT (chapter 29) is a tool of a different sort. Most of the tools created for developing computational cognitive models are specific to one type of model (such as connectionist models) or one architectural approach (such as is the case for Soar, ACT-R, and EPIC). The COGENT approach is different. COGENT seeks to provide modelers with a common interface to a range of tools so that different modeling paradigms may be explored without the requirement to learn different program-specific interfaces or languages. Furthermore, COGENT seeks to open the world of computational cognitive modeling to nonprogrammers by providing its users with a graphical "box and arrow" programming language to build working cognitive models.

Our colleagues in engineering and in the natural sciences have a range of tools that help them model the phenomena of interest to their field. For cognitive modeling to become as widespread and common as modeling in these other disciplines, we need to have a variety of similar tools available for both novice and experienced modelers. The constraint modeling approach of Howes and his colleagues and Cooper's COGENT are important developments that should increase the spread of computational cognitive modeling while providing new tools to develop cognitive theory.

References

Anderson, J. R. (1993). *Rules of the mind*. Hillsdale, NJ: Erlbaum.

Bechtel, W., & Abrahamsen, A. (1991). *Connectionism and the mind: An introduction to parallel processing in networks*. Cambridge, MA: Blackwell.

Newell, A. (1973). Production systems: Models of control structures. In W. G. Chase (Ed.), *Visual information processing* (pp. 463–526). New York: Academic Press.

Simon, H. A. (n.d.). "Allen Newell: Biographical Memoirs," 140–172. Retrieved from http://stills.nap.edu/readingroom/books/biomems/anewell.html

28

Bounding Rational Analysis

Constraints on Asymptotic Performance

Andrew Howes, Richard L. Lewis, & Alonso Vera

Critical of mechanistic accounts of cognition, Anderson showed that a demonstration that cognition is optimally adapted to its purpose and environment can offer an explanation for its structure. Simon, in contrast, emphasized that the study of an adaptive system is not a "logical study of optimization" but an empirical study of the conditions that limit the approach to the optimum. In response, we sketch the requirements for an approach to explaining behavior that emphasizes explanations in terms of the optimal behavior given not only descriptions of the objective and environment but also descriptions of the human cognitive architecture and knowledge. A central assumption of the proposal is that a theory explains behavior if the optimal behavior predicted by the theory shows substantial correspondence to asymptotic human performance.

How can we explain human behavior? Our first purpose in this chapter is to articulate an approach that emphasizes explanations in terms of the optimal behavior given not only constraints imposed by the objective and environment but also constraints imposed by knowledge and the human cognitive architecture. A second purpose is to describe techniques that realize this approach by supporting the formal reasoning about such constraints. We take as a starting point Simon's (1992) observation that,

> behaviour cannot be predicted from optimality criteria without information about the strategies and knowledge agents possess or acquire. The study of behaviour of an adaptive system is not a logical study of optimization but an empirical study of the side conditions that place limits on the approach to the optimum. (p. 160)

Before articulating our proposal we first review two existing classes of explanation of behavior: (1) rational explanations of the functions of cognition (rational analysis; Anderson, 1990) and (2) cognitive architecture-based simulations of the mechanisms by which the functions are achieved. These two lines of work provide much of the intellectual framework within which to motivate and understand the present proposal, in particular, understanding both how it is continuous with prior work, and how it departs in significant ways.

Rational Analysis

Anderson (1990) emphasised the value of explanations in terms of environment (or at least its experience) and the goals of the cognitive system. He stated a general principle of rationality: "The cognitive system operates at all times to optimize the adaptation of the behaviour of the organism" (p. 28). Anderson started with the assumption that evolution has to some extent optimized cognition to its environment. He argued that within the limits set by what evolution can achieve, a species is at some stable point in time at a local maximum.

Anderson proposed that, if the principle of rationality were applied to the development of a theory of cognition, then substantial benefits would accrue. In particular, the rational approach (1) offers a way to avoid the identifiability problem because the theory depends on the structure of an observable world and not on the "unobservable structure in the head"; (2) offers an explanation for why people behave the way they do rather than just how they behave (i.e., because they gain benefit from optimization); and (3) offers guidance to the construction of a theory of the mechanism.

One important implication of these benefits is that rational analysis allows the theorist to avoid the pitfalls of assuming that the human mind is a random collection of mechanisms that are poorly adapted to many of the tasks that people want to achieve. Indeed, explicit in articles advocating rational analysis (including Anderson's) is a critique of mechanistic accounts of cognition on just these grounds. For example, according to Chater and Oaksford (1999):

> From the perspective of traditional cognitive science, the cognitive system can appear to be a rather arbitrary assortment of mechanisms with equally arbitrary limitations. In contrast, rational analysis views cognition as intricately adapted to its environment and to the problems it faces. (p. 57)

A central and distinguishing feature of rational analysis is that it demands a thorough analysis of the task environments to which cognition is adapted. For example, in light of rational analyses such as Oaksford and Chater (1996), normative accounts of the Wason four-card task can be seen to fail precisely because they do not reflect the structure of the general environment experienced by people, but rather are tuned to the simple, unrepresentative tasks studied in the laboratory. At the same time, mechanistic accounts fail when they implicitly adopt the normative analysis and attribute departures from normative behavior to arbitrary limitations of the underlying cognitive mechanisms. Such considerations led Anderson to critique of his own mechanistic theory of cognition, ACT* (Anderson, 1983; though see Young & Lewis, 1998, for a description of an alternative functional approach to cognitive limitations pursued in the Soar architecture).

Bounding Rational Analysis

Understanding the relationship between rational analysis and Simon's seminal work on bounded rationality is instructive. Although rational analysis may at first appear at odds with bounded rationality, Anderson (1990) argued that there was no incompatibility between rational analysis and satisficing: It might be rational and optimal to find a normatively satisfactory solution when rational is defined relative to time and processing constraints. Despite this possible in-principle compatibility, Simon (1991, 1992) was critical of rational analysis. His critique focused on the fact that Anderson placed emphasis on the analysis of the environment and backgrounded the role of what Simon called *side conditions*. By side conditions Simon meant the constraints that were placed on cognition by knowledge, by strategies, and by the human cognitive architecture—the very constraints that rational analysis was intended to abstract away from. Simon (1992) stated:

> There is no way to determine a priori, without empirical study of behaviour, what side conditions govern behaviour in different circumstances. Hence, the study of the behaviour of an adaptive system like the human mind is not a logical study of optimization but an empirical study of the side conditions that place limits on the approach to the optimum. Here is where we must look for the invariants of an adaptive system like the mind. (p. 157)

Accounting for the Side Conditions: Simulations of Behavior Based on Cognitive Architectures

The most sophisticated current techniques for representing and reasoning about such side conditions are based on computational cognitive architectures (Anderson & Lebiere, 1998; Meyer & Kieras, 1997). We focus here on ACT-R (adaptive control of thought–rational) because it uniquely represents the confluence of rational analysis and mechanistic approaches to cognition. Anderson (1990) was clear that there are benefits to both and that the two approaches are complementary. Accordingly, Anderson modified ACT* to reflect the insights gained from rational analyses of memory and choice (Anderson & Milson, 1989; Lovett & Anderson, 1996). The resulting theory, ACT-R, combines a model of the decay of activation in declarative memory, derived from the rational analysis of Anderson and Milson (1989), with a model of production rule conflict-resolution derived from a rational analysis of the selection of action on the basis of history of success (Lovett and Anderson, 1996).

One of the strengths of a theory of the cognitive architecture is that it shows how the mechanisms of cognition, perception, and action work together as a single integrated system to produce behavior. ACT-R has been a spectacular success in just this way. But despite the grounding of ACT-R in rational analysis, models of specific tasks situations constructed in ACT-R are still subject to the rational analysis critique of mechanistic explanations, for two reasons. First, ACT-R integrates a range of mechanisms that are necessary to complete a comprehensive model of human cognition but that are not directly motivated by rational analysis. Examples include the model of perceptual/motor processing (ACT-R/PM) that was motivated by efforts to build architectures capable of interaction (Byrne & Anderson, 2001; Meyer & Kieras, 1997) and the limits on source activation, which is used to model working memory constraints (Anderson, Reder, & Lebiere, 1996). The problem here is not simply that some components are motivated by rational analyses and others are not. Perhaps the more fundamental problem is that motivations of individual components fail to take into account that it is the system as a whole that is adapting to the environment, not components. As cognitive architectures make exceptionally clear, components only have behavioral consequences in conjunction with the set of other components required to produce behavior.

Second, ACT-R must be provided with specific strategies in the form of production rules to perform specific tasks. This is a theoretically necessary feature of architectures (Newell, 1990), but in practice this variable content provides theoretical degrees of freedom to the modeler that may obscure the explanatory role of the architecture in accounting for psychological phenomena. Although Newell's (1990) timescale analysis was in principle correct—that the architecture shows through at the level of immediate behavior—recent work is making it increasingly clear that considerable strategic variability is evident even at the level of tasks operating in the subsecond range. A prime example is elementary dual-tasking situations (Meyer & Kieras, 1997), which we turn to next to illustrate our new approach.

Local and Global Adaptation and the Role of Mechanism and Strategy

The current state of affairs can be summarized as follows and clearly points to new directions for modeling research:

1. A strength of rational analysis is that it provides deeper explanations for both components of the

cognitive architecture and behaviors in particular task situations. These explanations take the form of demonstrations that the components and behaviors are rational adaptations to the structure of the environment viewed from a sufficiently *global* perspective, rather than departures from local optima that point to arbitrary limitations in the underlying cognitive mechanisms.

2. A weakness of rational analysis is that it does not provide a way to systematically and incrementally take into account the side conditions that bound the approach to optimality in any given *local* task situation. Although it may in fact provide an explanation for some of the side conditions themselves (to the extent that rational analysis explanations of architectural components are successful), there is no way to systematically draw out the detailed implications of these side conditions for understanding what behavior is adaptive in a particular task environment.

3. A related weakness of rational analysis is that it does not provide a way to explore the adaptation of the system as a *whole*—and the interaction of all its parts—as opposed to the individual components, or classes, of behaviors.

4. A strength of computational cognitive architectures is that they provide a way to explore the interactions of what Simon called the *side conditions* on the approach to optimality, including knowledge and basic mechanisms of cognition, perception, and action. They can thereby be applied in detail to a wider range of specific task situations than can rational analysis and, as ACT-R demonstrates, offer one way to explore architectural mechanisms that may themselves be motivated by rational analysis.

5. A weakness of cognitive architectures is that they do not yield the deep explanations of rational analysis (even if partially grounded in rational analysis–motivated components) because behavior arises as a function of both architecture and posited strategies, and the latter represents a major source of theoretical degrees of freedom, where strategies may be posited, not because they are maximally adaptive but because they match the empirical results.

In short, we believe that cognitive architectures have made only partial progress in addressing the critique of rational analysis, and rational analysis has made only partial progress in addressing Simon's critique that it has backgrounded the role of mechanistic and knowledge constraints. What we seek is a framework that will permit us to reason about what behaviors are adaptive in a specific local task situation, given a posited set of constraints (architectural and knowledge constraints) on the approach to optimal behavior and an explicit payoff function.

The framework we propose is an initial attempt to achieve this goal. The framework is consistent with rational analysis but differs from cognitive architectures in that it values calculation of what is optimal. It differs from rational analysis and is consistent with cognitive architectures in that it directly takes into account the complex interaction of architectural mechanisms and how they give rise to the details of behavior. It differs from both approaches in that it seeks explanations of specific behaviors as optimal adaptations to both external task constraints and internal system constraints. More precisely, the proposed framework demands that (a) optimality is defined relative to the entire set of constraints acting on the behaving system (internal as well as external constraints); (b) there is an explicit payoff function, known as an *objective function*; and (c) the optimal performance predicted by the theory corresponds to the empirically asymptotic level of adaptation.

In the discussion section, we briefly consider other related approaches in psychophysics and cognitive modeling. We will also make recommendations for the types of data collection and reporting that are needed for the sorts of analysis we are proposing. We turn first to a description of the approach and its application to modeling a specific task.

How to Explain Behavior

In our recent work, we have developed an approach designed to complement rational analysis and architectural simulation (Howes, Vera, Lewis, & McCurdy, 2004; Vera, Howes, McCurdy, & Lewis, 2004). The three commitments are (1) a commitment to exploring the implications of constraints for the asymptotic bounds on adaptation; (2) a framework for representing theories as sets of constraints; and (3) a computational mechanism for calculating the implications of constraints.

Exploring the Bounds on Adaptation

When people acquire a skill, they are able to adapt behavior so as to improve incrementally the value of

some payoff or objective function. With practice, the scope for improvement attenuates and performance asymptotes. It may asymptote at a level that is consistent with constraints imposed by the environment or perhaps at a level determined by the knowledge that is brought to the task. The bounds may instead be imposed by the human cognitive architecture, including its resource limits (Norman & Bobrow, 1975, 1976). More plausibly, the asymptote may be determined by a combination of constraints, including the stochastic and temporal profiles of the particular task environment and the human cognitive, perceptual, and motor systems.

We assume that under such circumstances, people seek to iteratively improve payoff. Effort is oriented toward increasing the value of an objective function that specifies the perceived costs and benefits of action. In seeking to improve a payoff, people adapt performance by adopting specific strategies. Improvement eventually asymptotes, and if we ignore for the moment the possibility that people are trapped by local maxima, performance should asymptote at a level where it generates the optimal payoff, given the constraints (including those imposed by architecture and knowledge). For a theory of the human cognitive architecture to explain an empirically observed asymptotic bound, substantial correspondence is required between the asymptote and the optimal performance, given the theory. A theory that predicts better performance than the observed asymptote is underconstrained; a theory that predicts worse performance is over constrained.

Importantly, explanations of the causal role in behavior of, for example, memory, must account for the fact that such component systems have behavioral consequences only when working together with the entire cognitive system. In contrast to rational analysis, the idea is to explore the implications for asymptotic performance of theories of an integrated set of mechanisms applied to a local task environment. The approach can therefore be thought of as a *bounded* rational analysis.

Following Card, Moran, and Newell (1983), we can think in terms of behavior as being determined by the objective function plus three sets of constraints:

Objective + Task Environment + Knowledge
+ Architecture → Behavior.

Each set of constraints generally *underspecifies* behavior in the absence of an explicit objective function. Thus, the task environment alone affords a large space of possible behaviors; this space is further constrained by architecture constraints. For our present purposes, this space of possible behaviors represents the space of possible *strategic variations* and may be constrained yet further by knowledge constraints. The objective function then selects a single surface in this subspace that represents the optimal set of possible behaviors satisfying the joint set of constraints.

Obviously, an explanation in terms of objective and environment (i.e., a rational analysis) is to be preferred to a bounded rational analysis on the grounds of parsimony. Such explanations are possible when the objective and the environment constraints alone yield a subspace of behaviors that corresponds with observed behavior. However, in many circumstances such explanations are not possible.

A Framework for Representing Theories as Sets of Constraints

The second requirement is for a theoretical ontology that provides a language for expressing constraints on information processing mechanisms. By definition, information processes receive, transform, and transmit information. A process receives and transmits information from and to other processes. We assume that for two processes to exchange information they must overlap in time or each must overlap in time with a common mediating, or buffering process, which stores the information for, perhaps, a short period of time. McClelland (1979) introduced the hypothesis that information processes were cascaded, that is, they overlapped in time, and the quality of information passed from one process to another increased with time.

Our version of cascade theory commits to the following assumptions: (1) processes must overlap in time if they are to transfer information; (2) a process is executed by a processor (also known as a resource); (3) some function relates the accuracy of information produced to the duration since the process started (Howes et al., 2004). (It follows from 3 that a process has a minimum duration, before which no transmission occurs, and a maximum duration, after which no transmission occurs.)

In addition to a framework for representing the constraints on information flow we need some commitment to the process and processing capabilities of the human cognitive architecture. What processes characterize human cognition? What kinds of processors are they executed by? Here, we are interested in an account in which information processing is conceived

of in terms of an interacting set of processes each with defined resource requirements, temporal duration, and input/output characteristics. As a starting point, we take Card, Moran, and Newell's (1983) model–human processor (MHP). The representation of processes abstracts over representation and algorithm. For the purposes of explaining behavior, it is not always necessary to define the precise mapping between input and output representations of individual processes. It is sufficient, for example, to state that a stimulus is perceived and that a response is retrieved in some mean time with some standard deviation.

Calculating the Implications of Constraints on Behavior

The third requirement is for a language in which constraints on human information processing (see the section, "A Framework for Representing Theories as Sets of Constraints") can be specified in a computable form and the implications for the asymptotic bound on performance calculated.

Human performance depends on a multiplicity of complex interacting constraints derived from the environment, from the human cognitive architecture, and from knowledge, which makes calculating the implications of constraints difficult. Skilled performance of a routine task usually involves the execution of a number of parallel but interdependent streams of activity. For example, one hand may move to a mouse, while the other finishes typing a word. The eyes begin to fixate on a menu, while the required menu label is retrieved from memory. Each of these processes takes a few hundred milliseconds, but together they form behaviors that take many seconds. Importantly, details of how processes are scheduled have significant consequences for the overall time and resource requirements.

Vera et al. (2004) proposed that one response is to represent theories as sets of constraints using predicate calculus constraint logic. A constraint is simply a logical relation between variables. Constraint satisfaction has the potential to provide a formal framework for the specification of theories of interactive cognition and thereby for the construction of mathematically rigorous tools for supporting the prediction of the bounds that the constraints imply for adaptation. Of central importance is that constraints are declarative and additive. They are declarative in that relationships between variables can be stated in the absence of a mechanism for computing the relationship. They are additive in the sense that the order in which constraints are specified does not matter. These properties should allow theoretical assumptions to be expressed in a computable form that is relatively independent of the arbitrary constraints that are sometimes imposed by the machine, or software algorithms, with which computation is conducted.

Constraints allow the specification of what is to be computed without specification of how the computation is carried out (the algorithm), which means that considerable flexibility is enabled in the desired properties of the schedule. It does not matter which algorithm is used to derive the optimal solution (as long as it works!). It happens that our previous work has made use of a branch-and-bound algorithm (Howes et al., 2004; Vera et al., 2004), but tools based on dynamic programming or whatever other algorithm would do just as well. Similarly, Monte Carlo simulation can be used to generate an estimate of the optimal adaptation as long as care is taken to search the space of possible strategies within which the optimal solution is located.

Our point is not to argue for the value of a particular optimization algorithm but to argue for the scientific utility of considering the relationship between the optimum under constraints and the empirically observed asymptote. From the perspective of the scientific aims, the set of possible optimal solutions is defined precisely by the payoff function (the objective) and by the set of (declarative) constraints: There is no need to specify the optimization algorithm to specify a theory; the optimization algorithm is simply the means by which one derives the implications of the theory.

Example: Constraints on Dual-Task Performance

Consider how we might predict performance on simple psychological refractory period (PRP) tasks. For example, in Schumacher et al.'s (1999) Experiment 3, participants were required to respond to a tone and a visual pattern (simple or complex) with key presses that depended on whether the tone was high or low and whether the pattern contained a particular feature. The tone and the pattern were presented with a small gap of between 50 and 1,000 ms (stimulus onset asynchrony, or SOA). Participants were asked to prioritize the tone task (Task 1) over the pattern task (Task 2). The tone task response times were, on average, unaffected by SOA. In contrast, the mean pattern task response time, at a short SOA (50 ms), was less than

the sum of the tone task and pattern response times at long SOAs (>500 ms). This finding has been taken as evidence that some elements of tone and pattern task were performed in parallel at short SOAs. Byrne and Anderson were interested in modeling Schumacher's data using ACT-R/PM to demonstrate that cognitive parallelism is not required to explain these results. They argued that the results can be modeled with either the EPIC or ACT-R/PM assumptions and that Schumacher's data provides evidence for strategic deferment of the pattern task response.

Optimizing Over the Statistics of Interaction

Our previous work, Howes et al. (2004), demonstrated the potential analytic role of optimization (as described in "How to Explain Behavior," this chapter) in exploring the space of possible adaptations. What it did not do was articulate how constraint analyses can be used to explore how people adapt to the statistics of interaction with an uncertain environment and with uncertainty in the duration of internal processes. One such adaptation is required in the PRP task where participants need to ensure the ordering of Task 1 and Task 2 responses despite fluctuations in the durations of each response. Here, we develop a constraint model of a generic PRP task and describe its predictions.

Asked to respond as quickly as possible to a single stimulus an individual will produce a range of approximately normally distributed responses. In a dual-task situation, such as the psychological refractory period task, each response has its own distribution. If in the dual-task situation there is some benefit, a gain, from responding rapidly and some cost to making a response reversal, responding to Task 2 (the pattern task) before Task 1 (the tone task), then participants will weigh the costs and benefits of fast and slow responses. Parameterized with the response means for individual responses, their standard deviations, and estimates of the costs and benefits of the space of possible behaviors, an adequate theory of the human cognitive architecture must predict the asymptotic mean and standard deviations of the response times in a dual-task scenario.

Imagine an architecture A′ in which there are no mechanisms by which the processing of Task 1 can influence the processing of Task 2. That is, there are no necessary task interactions and no shared cognitive or perceptual/motor resources. How do we test whether A′ can predict and explain performance on a PRP task? According to the assumptions of our framework, we need to determine whether the best possible performance predicted by the architecture corresponds to the asymptotic human performance. If the processing for each task is entirely independent, then the extent to which the response distributions overlap will determine the frequency with which response reversals occur. If participants intend to avoid response reversals, then a strategy is required. In the case of the very simple architecture A′, the only strategy available that mitigates against response reversals is to delay the response to Task 2. By delay, we mean to wait a fixed, trial independent amount of time that is added to what would otherwise be required to make the Task 2 response. This simple strategy will temporally separate the Task 1 and Task 2 response distributions.

If the best available strategy is to delay the Task 2 response, then the next question is by how much? If we know what the payoff function is for participants (i.e., how much they gain for a correct response and how much they lose for a response reversal), then we can derive exactly the optimal amount of time to delay Response 2. According to the theory, a participant should select a value to delay Task 2 that is consistent with the constraints and which maximizes the value of the payoff.

We can specify the constraints and the objective function (the payoff) for different values of the delay as a constraint model. However, rather than fixed durations, here we sample duration from a normal probability distribution.

> Constraints
> SOA in {SOAmin . . . SOAmax}
> $RT1_i$ = normal(M1, SD1)
> $RT2_i$ = normal(M2, SD2) + DELAY.

In addition to the constraints, the objective captures a speed/accuracy trade-off between going fast and avoiding response reversals. The payoff for a trial is the gain minus the total time cost and minus the cost of reversal. The time cost is defined as a weight times the duration of the latest of the two responses. The cost of reversal is defined as some weight times 1 if a response reversal occurred and 0 otherwise. Higher values of the strategically set delay variable will tend to decrease the proportion of response reversals but at the cost of increasing the total time required to perform the task.

> Payoff: For trials 1 to N,
> Average Payoff
> = (ΣI = 1 to N : GAIN − C_t − C_r) /N   (1)

> Time cost $C_t = W_n \times \max(RT1_i, (SOA + RT2_i)))$
> Cost of reversal $C_r = W_m \times f(SOA + RT2_i − RT1_i))$
> $f(X > 0) = 0$
> $f(X < 0) = 1$

Despite the simplicity of this model, to our knowledge none of the reported PRP data sets are suitable for testing its validity. While some experiments, such as those reported by Schumacher et al. (1999), were controlled for the cost/benefit trade-off between a fast response and a response reversal, neither the payoff achieved by participants, nor details of standard deviations and reversals rates are reported.

The constraint model makes predictions of dual-task performance given parameters determined from single task performance. What needs to be calculated is a prediction of the asymptotic performance time and error rate at short SOAs given (a) mean and standard deviations of performance time at long SOAs and (b) an experimental paradigm in which participants are subject to a payoff regime enforcing a trade-off between RT and rate of response reversal. Importantly, to test the model, each trial would take a fixed amount of time independently of the response time. This assumption eliminates the additional benefit of a rapid response, beyond the reward that determines the weightings in the objective function, though other payoff regimes are possible.

Calculation of the optimal value of the Task 2 delay requires Monte Carlo simulations for each potential duration. The payoff achieved on each trial can then be aggregated to give a total payoff for each possible strategy (value of the delay). For A′ there will be an N-shaped relationship between duration of delay and payoff. The optimal performance (maximal payoff) implied by A′, and therefore the predicted asymptote on human performance, given task environment, constraints, and payoff function will correspond to the peak of this curve.

Note that unlike in model fitting methodologies (e.g., as used in Meyer & Kieras, 1997, to determine the length of the defer process) we are not proposing to choose a value of the delay parameter so that the model fits the data. Rather we are claiming that the model predicts that participants will delay Task 2 by a particular duration (the optimal duration of the delay process), subject to an estimable confidence interval, and given only parameters set from single task performance. If at asymptote participants delay by more or less than the predicted duration then the A′ theory is wrong or at least incomplete. To the extent that the theory cannot be successfully modified by adding or removing constraints, we have learned something interesting about the limits on the mechanisms of adaptation.

Discussion

The example that we have developed demonstrates a method for calculating the asymptotic bound predicted by a theory. However, the constraints are based on overly simplistic assumptions about the internal processing mechanisms. Calculating the implications of more elaborate theories such as Byrne and Anderson's (2001) or Meyer and Kieras's (1997) over the statistics of interactive behavior requires constraint models such as those developed in Howes et al. (2004).

Comparison of the predicted asymptote to the observed asymptote provides a test of the adequacy of the theory. If people do not perform as well as the predicted asymptote, then the implication is that the theory is underconstrained. If people perform better than the predicted asymptote then the implication is that the theory is overconstrained. It follows that there is no role for the notion of suboptimality within the explanatory framework that we have described. If global maxima are discoverable and if optimality is defined relative to the entire set of constraints and the objective function rather than relative to the task and environment, then the extent to which the optimal performance corresponds to the asymptotic human behavior is a measure of the goodness of the theory, not of the suboptimality of the human behavior.

General Discussion

We have sketched the requirements for an approach to explaining behavior that emphasises the importance of explanations in terms of the optimal behavior given not only descriptions of the objective and environment *but also* descriptions of the human cognitive architecture and knowledge. We illustrated the approach with a model of strategic processing in PRP tasks. The model made limited assumptions about response variance, and we described how a prediction of dual-task response separation given an objective function that traded time taken against response reversal could be derived. We claimed that a theory could be said to explain the data if it could be established that there was substantial

correspondence between the optimal performance implied by the theory and the asymptotic performance observed in human behavior. In the remainder of the general discussion we (a) describe related work, (b) describe how to design experiments that provide data amenable to constraint-based explanations, and (c) reflect further on how a bounded rational analysis complements rational analysis (Anderson, 1990).

Related Work

The approach to explaining cognition was motivated in part by Roberts and Pashler (2000) and also by Kieras and Meyer (2000). Both have observed the potential problems with failing to explore the contribution of strategies and architectural constraints to the range of possible models of human performance. Kieras and Meyer (2000) responded by proposing the use of a *bracketing heuristic*. A bracket was defined by the speed of the fastest-possible strategy for the task and the slowest-reasonable strategy. Kieras and Meyer predicted that observed performance should fall somewhere between the performance of these two strategies. They also articulated the importance of exploring the space of strategies to explaining the phenomena being modeled. While there are similarities, there are two differences to our approach. First Kieras and Meyer (2000) focused on bracketing the *speed* of strategies rather than their payoff. Second, they saw bracketing as a means of coping with the problem that optimizations cannot be forecast. Kieras et al. state that bracketing was a way to construct "truly predictive models in complex task domains where the optional strategy optimizations users would devise cannot be forecast" (p. 131).

Others have also used analyses of optimal performance to bracket predictions. There is a long and active tradition in analyses of optimal performance in psychophysics (Geisler, 2003; Swets, Tanner, & Birdsall, 1961; Trommershäuser, Maloney, & Landy, 2003). More recently authors such as Kieras and Meyer (2000) and Neth, Sims, Veksler, and Gray (2004) have contrasted human performance to optimal performance on more complex cognitive tasks. Neth et al. (2004) used the analysis of the best possible performance, given a particular strategy, to predict a bracket for behavior on a decision-making task. For example, Fu and Gray (2004) and O'Hara and Payne (1998) have exposed apparent suboptimalities in behavior and sometimes offered explanations that allow those behaviors to be interpreted as rational adaptations given additional constraints.

Exploring optimality criteria has been particularly fruitful in psychophysics. An ideal observer theory is a computational theory of how to perform a perceptual/cognitive task optimally, given properties of the environment and the costs/benefits associated with different outcomes (Geisler & Diehl, 2003). However, Geisler and Diehl (2003) state: "While ideal observer theory provides an appropriate benchmark for evaluating perceptual and cognitive systems, it will not, in general, accurately predict the design and performance of real systems, which are limited by a number of factors. . . ." Also, Geisler (2003) stated: "Organisms generally do not perform optimally, and hence one should not think of an ideal observer as a potentially realistic model of the actual performance of the organism." Geisler and Diehl (2003) particularly focus on the fact that the real observer may correspond to a local maximum in the space of possible solutions, whereas the ideal observer corresponds to the global maximum. The ideal observer corresponds to Marr's computational theory and is a theory of what the organism should compute given the task and stimuli (Geisler & Diehl, 2003), not what it is rational to compute given the entire set of constraints.

Kieras and Meyer (2000) and Geisler and Diehl (2003) may be right, in general, to be pessimistic about the prevalence of task domains in which it is possible to forecast people's strategy optimizations. However, if task domains where it is possible can be identified and if it is accepted that architectures show through at the limit, particularly when resources are limited (Norman & Bobrow, 1975, 1976), then these task domains may be particularly useful for evaluating theories of the human cognitive architecture: The optimal solution given the theory can be taken as a forecast of the asymptote. In addition, Kieras and Meyer's view may have been influenced by the lack of available techniques for calculating the optimal solution given a complex and heterogeneous set of constraints on information processing models. Our previous work (Howes et al., 2004; Vera et al., 2004) has articulated general purpose analytic techniques for predicting strategy optimizations.

Experimental Methodology

The predictions made by the analyses could not be tested against results from experiments that failed to control for the trade-off between speed and accuracy. Unfortunately, despite the work of Meyer and Kieras (1997), the absence of controls on speed/accuracy trade-offs is widespread in experimental cognitive psychology. With a few exceptions, error rates tend to be dismissed as small or are excluded from analysis presumably motivated by the view that they are an aberration, just noise that distracts from the main picture. The reality is that human adaptation to the objective function within the limits set by the constraints is pervasive. People adapt enthusiastically and continuously (Charman & Howes, 2003). They adapt tasks that take hundreds of seconds to complete, and they adapt tasks that take hundreds of milliseconds to complete. To understand this adaptation, it is critical to understand fully the objective to which participants are adapting.

Knowing what the participant was instructed is probably not sufficient (Kieras & Meyer, 2000). Participants do not merely do what they are told; rather they interpret instructions to generate objective functions that are consistent with longer-term traits. A challenge is to find experimental paradigms for resource-limited tasks that expose participants' objective functions and thereby support the rigorous calculation of the predicted asymptote. While the work of Trommershäuser, Maloney, and Landy (2003) provides an example of what can be done for pointing tasks, more needs to be done for tasks that involve sequential ordering.

Bounding Rational Analysis

The role of optimality criteria in cognitive science has been controversial. Indeed, one objection to our approach might be people do not optimize, they satisfice. This assessment would seem consistent with Simon's critique of the assumption that people make optimal economic decisions (Simon, 1957). But this would be to miss the fundamental distinction between the idea that Simon rejected (i.e., the hypothesis that people are optimally adapted to the environment) and the idea proposed in this chapter: That given the adaptive nature of human cognition, an explanation of behavior must explain why people do not do better than they do. It must explain the approach to the asymptote in terms of the implications of psychological bounds. Despite a shared value in determining optimal adaptations, the approach that we have described is not rational analysis. Where Anderson emphasized optimality in terms of the task and environment, the approach that we have articulated equally emphasizes constraints on architecture and knowledge. Our approach is more closely aligned with Simon (1992) who emphasized the need to investigate the side conditions that place limits on the approach to the optimum.

In fact, to the extent that predictions made through optimization are constrained by hypotheses concerning internal processing limits, the predictions are not optimal relative to the goal and environment. Our approach is therefore consistent with Simon's reminder that explaining behavior requires reference to internal processing limits and capabilities, and it is consistent with the idea that people satisfice. The challenge that we are addressing could be characterized, perhaps, as how precisely to articulate what it means to satisfice.

We also expect that there are many task environments where incremental improvement is unlikely to lead to an optimal solution. In these cases, suboptimal performance may result from too many local maxima. Here incremental improvement may lead to an asymptote but not the asymptote that corresponds to the optimal solution given the theory.

In general, task environments where the optimal solution is within the grasp of incremental improvement may be more suitable for evaluating the consequences of theories of psychological resources for the asymptotic bound on the adaptation of behavior. For these environments, the absence of a correspondence between the optimum implied by a theory and the behavioral asymptote is evidence for the inadequacy of the theory. In contrast, in task environments where the optimal solution is unlikely to be generated incrementally, the absence of correspondence could be due to either the shape (availability of local maxima) of the task environment or to the inadequacy of the theory. Behavior in these task environments is unlikely to offer a good basis for empirical tests of a theory of what bounds adaptation.

Analyses similar to that which we have described in this chapter could assist the development of rigorous answers to questions of optimality and therefore rationality in interactive cognitive skill. Questions have been raised by a number of authors about the extent to which people make optimal adaptations (Fu & Gray, 2004; Gray & Boehm-Davis, 2000; Taatgen, 2005). First, to explain cognition, optimality must take into account constraints on architecture and knowledge, not only constraints on the environment. Second, a claim that behavior is suboptimal or biased, or that it is

not rational, does not explain behavior. A claim of suboptimality carries little content in the absence of an explicit theory of what *is* optimal and a means for calculating the implications of that theory. An apparent suboptimality raises the question about what modification is required to the theory to align the predicted performance bound with the empirically observed asymptotic bound.

Conclusion

To conclude, we have argued that neither rational analysis nor computational simulation are sufficient approaches to explaining cognition. Another promising approach may be to test for correspondence between theories of optimal performance given both environmental and psychological constraints and empirically observed asymptotic bounds in particular task environments. Such an approach requires not only theories of the constraints imposed by the task environment but also theories of the constraints imposed by the cognitive architecture and by knowledge, it not only requires exploration of the trajectories through the space of possible adaptations but also systematic analysis of the bounds on that space.

References

Anderson, J. (1983). *The architecture of cognition.* Cambridge, MA: Harvard University Press.

Anderson, J. R. (1990). *Rational analysis.* Mahwah, NJ: Erlbaum.

———, & Lebiere, C. (1998). *The atomic components of thought.* Mahwah, NJ: Erlbaum.

———, & Milson, R. (1989). Human memory: An adaptive perspective. *Psychological Review, 96*(4), 703–719.

———, Reder, L. M., & Lebiere, C. (1996). Working memory: Activation limitations on retrieval. *Cognitive Psychology, 30*, 221–256.

Byrne, M. D., & Anderson, J. R. (2001). Serial modules in parallel: The psychological refractory period and perfect time sharing. *Psychological Review, 108*(4), 847–869.

Card, S. K., Moran, T. P., Newell, A. (1983). *The psychology of human computer interaction.* Hillsdale, NJ: Erlbaum.

Charman, S. C., & Howes, A. (2003). The adaptive user: An investigation into the cognitive and task constraints on the generation of new methods. *Journal of Experimental Psychology: Applied, 9*, 236–248.

Chater, N., & Oaksford, M. (1999). Ten years of the rational analysis of cognition. *Trends in Cognitive Science, 3*(2), 57–65.

Fu, W.-T., & Gray, W. D. (2004). Resolving the paradox of the active user: Stable suboptimal performance in interactive tasks. *Cognitive Science, 28*, 901–935.

Geisler, W. S. (2003). Ideal observer analysis. In L. Chalupa & J. Werner (Eds.), *The visual neurosciences* (pp. 825–837). Boston: MIT Press.

——, & Diehl, R. L. (2003). A Bayesian approach to the evolution of perceptual and cognitive systems. *Cognitive Science, 27*(3), 379–402.

Gray, W. D., & Boehm-Davis, D. A. (2000). Milliseconds matter: An introduction to microstrategies and to their use in describing and predicting interactive behavior. *Journal of Experiment Psychology: Applied, 6*(4), 322–335.

Howes, A., Vera, A. H., Lewis, R. L., & McCurdy, M. (2004). Cognitive constraint modelling: A formal approach to supporting reasoning about behavior. In K. D. Forbus, D. Gentner, & T. Regier (Eds.), *Proceedings of the Twenty-sixth Annual Meeting of the Cognitive Science Society, CogSci2004* (pp. 595–600). Hillsdale, NJ: Erlbaum.

Kieras, D. E., & Meyer, D. E. (2000). The role of cognitive task analysis in the application of predictive models of human performance. In J. M. Schraagen, S. F. Chipman, & V. L. Shalin (Eds.), *Cognitive task analysis* (pp. 237–260). Mahwah, NJ: Erlbaum.

Lovett, M., & Anderson, J. R. (1996). History of success and current context in problem solving: Combined influences on operator selection. *Cognitive Psychology, 31*, 168–217.

McClelland, J. L. (1979). On the time relations of mental processes: An examination of systems of processes in cascade. *Psychological Review, 86*, 287–330.

Meyer, D. E., & Kieras, D. E. (1997). A computational theory of executive cognitive processes and multiple-task performance: Part 1. Basic mechanisms. *Psychological Review, 104*, 3–65.

Neth, H., Sims, C. R., Veksler, V., & Gray, W. D. (2004). You can't play straight TRACS and win: Memory updates in a dynamic task environment. In K. D. Forbus, D. Gentner, & T. Regier (Eds.). *Proceedings of the Twenty-sixth Annual Meeting of the Cognitive Science Society* (pp. 1017–1022). Hillsdale, NJ: Erlbaum.

Newell, A. (1990). *Unified theories of cognition.* Cambridge, MA: Harvard University Press.

Norman, D. A., & Bobrow, D. G. (1975). On data-limited and resource-limited processes. *Cognitive Psychology, 7*, 44–64.

——, & Bobrow, D. G. (1976). On the analysis of performance operating characteristics. *Psychological Review, 83*(6), 508–510.

Oaksford, M., & Chater, N. (1996). Rational explanation of the selection task. *Psychological Review, 103*, 381–391.

O'Hara, K. P., & Payne, S. J. (1998). The effects of operator implementation cost on planfulness of problem solving and learning. *Cognitive Psychology, 35*, 34–70.

Roberts, S., & Pashler, H. (2000). How persuasive is a good fit? A comment on theory testing. *Psychological Review, 107*(2), 358–367.

Schumacher, E. H., Lauber, E. J., Glass, J. M., Zurbriggen, E. L., Gmeindl, L., Kieras, D. E., et al. (1999). Concurrent response-selection processes in dual-task performance: Evidence for adaptive executive control of task scheduling. *Journal of Experimental Psychology: Human Perception and Performance, 25*, 791–814.

Simon, H. A. (1957). *Models of man.* New York: Wiley.

——. (1991). Cognitive architectures and rational analysis: Comment. In K. Van Lehn (Ed.), *Architectures for Intelligence: The 22nd Carnegie Mellon Symposium on Cognition.* Hillsdale, NJ: Erlbaum.

——. (1992). What is an "explanation" of behavior? *Psychological Science, 3*, 150–161.

Swets, J. A., Tanner, W. P., Jr., & Birdsall, T. G. (1961). Decision processes in perception. *Psychological Review, 68*, 301–340.

Taatgen, N. A. (2005). Modeling parallelization and speed improvement in skill acquisition: From dual tasks to complex dynamic skills. *Cognitive Science, 29*, 421–455.

Trommershäuser, J., Maloney, L. T., & Landy, M. S. (2003). Statistical decision theory and tradeoffs in motor response. *Spatial Vision, 16*, 255–275.

Vera, A., Howes, A., McCurdy, M., & Lewis, R. L. (2004). A constraint satisfaction approach to predicting skilled interactive cognition. In *CHI '04: Proceedings of the SIGCHI conference on Human factors in computing systems* (pp. 121–128). New York: ACM Press.

Young, R. M., & Lewis, R. L. (1998). The Soar cognitive architecture and human working memory. In A. Miyake & P. Shah (Eds.), *Models of working memory: Mechanisms of active maintenance and executive control* (pp. 224–256). New York: Cambridge University Press.

29

Integrating Cognitive Systems

The COGENT Approach

Richard P. Cooper

This chapter describes COGENT, a graphical object-oriented system for the development and exploration of computational models. COGENT builds on the box and arrow notation popular within information processing psychology and provides a graphical editor, standardized notation, and operational semantics for box and arrow diagrams. Control within COGENT models is distributed and is manifest via local interactions between connected parameterized boxes. Some consequences of this approach to control within COGENT models are discussed. COGENT's generality is then illustrated through a partial implementation of Norman and Shallice's contention scheduling/supervisory system theory of controlled and automatic processing. The implementation combines multiple control mechanisms within a single cognitive model and demonstrates COGENT's ability to model neurological damage through modification of box parameters.

As many of the chapters in this volume demonstrate, cognitive architectures serve dual roles within cognitive modeling: as theories of the global organization of cognitive processes and as frameworks within which models of specific tasks or cognitive faculties might be implemented. The second role is a consequence of the first, and there is much to be gained from it. For example, it is frequently argued that cognitive architectures constrain models developed within them (Newell, 1990; but see also Cooper & Shallice, 1995). However, there are also limitations to the use of cognitive architectures for general modeling. How should one develop a model if one does not wish to subscribe to any specific architecture, for example? And once a model has been developed within an architecture, how can one determine whether critical aspects of the model's behavior are a result of architectural assumptions, model assumptions, or some interaction between the two? For these and similar reasons, it is sometimes appropriate to develop architecture-neutral models, but the development of such models presents its own

issues. If one is to develop such a model, must one have expertise in both cognitive psychology and computer programming? And assuming one has the necessary computational expertise, how might the model be developed so as to be readily communicable to other cognitive scientists? COGENT (cognitive objects in a graphical environment; Cooper, 2002; Cooper & Fox, 1998) is a general simulation environment for cognitive modeling that has been designed specifically to address these issues.

As a general simulation environment, COGENT supports the implementation and evaluation of both theories of specific cognitive domains (e.g., sentence processing or syllogistic reasoning; see Cooper, 2002) and complete cognitive architectures (like Soar, ACT-R, 4-CAPS, EPIC, etc.). COGENT's basic design philosophy derives from the tradition of box and arrow diagrams within cognitive psychology. Such diagrams were introduced with the cognitive revolution of the 1950s (see, e.g., the seminal work of Broadbent, 1958) and remain popular (for recent examples, see Shallice, 2002, and Anderson et al., 2004). They have substantial intuitive appeal for specifying processes or functions in terms of interacting subprocesses, but in their basic form, box and arrow diagrams are inadequate for modern cognitive theorizing because they lack well-defined rules of interpretation—a single diagram may be interpreted in any number of ways by different theorists. COGENT thus formalizes the use of box and arrow diagrams by providing them with an operational semantics. This allows users with relatively little programming experience to develop box and arrow diagrams into computationally complete cognitive models and to then execute the models to test or generate predictions.

This chapter describes some of the key features of COGENT, concentrating on issues of integration and control within COGENT models. The next section describes the box and arrow notation used by COGENT and its operational semantics as defined by COGENT's box and arrow model interpreter. This is followed by a discussion of some key computational issues in the control and integration of modular cognitive systems as they relate to models developed within COGENT. The final section uses a partial COGENT implementation of Norman and Shallice's model of controlled and automatic behavior to demonstrate how two distinct control mechanisms may be combined within a single integrated COGENT model.

The COGENT Cognitive Modeling Environment

The COGENT software package consists of four principal components: (1) a tool for creating, editing, and fleshing out cognitive models in the form of box and arrow diagrams; (2) an interpreter for running fleshed out box and arrow diagrams; (3) a system for designing and running computational experiments on cognitive models; and (4) a project management tool for maintaining a series of related cognitive models as the series is developed. This chapter is primarily concerned with just the first two of these components. For further details of all aspects of COGENT, see Cooper and Fox (1998) and Cooper (2002).

The Box and Arrow Language

Within COGENT's box and arrow language a model is a set of boxes of varying types with arrows joining those boxes. Different types of box correspond to different types of information processing or storage device (e.g., rule-based processes, buffers, and connectionist networks). Arrows represent communication between boxes, with different types of arrow representing different communicative functions (e.g., reading information or sending information). For both boxes and arrows, the type is represented diagrammatically by its shape. Thus, a rule-based process is represented by a hexagonal box, while a send arrow is represented by an arrow with a standard triangular arrowhead. The box and arrow language is supported by a graphical box and arrow diagram editor (Figure 29.1), which provides the user with a palette of box and arrow types. The user may construct a model by selecting from this palette and placing elements on the main canvas, or editing elements already there.

The information processing function of a box (as defined by the box's type) specifies broad characteristics of the box's processing, but it does not fully determine the box's computational behavior. The specific computational behavior of each instance of each box is determined by a set of properties (which are specific to each box type) and the box's initial contents. The current version of COGENT provides six primary types of box:

- *Buffers* store information in the form of discrete representations. Their properties specify the capacity limit (if any) on the buffer, whether access to the buffer's elements is determined by recency or primacy (or neither), the decay function (if any) that operates on the buffer's contents, and so on. Thus, in a model of the control and interaction of memory processes, buffers with properties set to different values might be used to implement both a limited capacity short-term store and an unlimited capacity decay-prone long-term store. Buffers may also be initialized with an ordered list of elements.
- *Processes* transform or manipulate information according to sets of symbolic rules. These rules specify how the process produces outputs on the basis of either inputs sent to the process or the contents of buffers read by the process. The process's properties specify gross aspects of the interpretation of these rules (e.g., whether they should be applied on each and every occasion they are applicable or only on a random subset of such occasions).

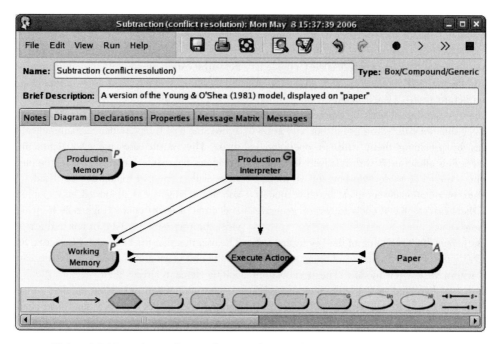

FIGURE 29.1 A COGENT box and arrow diagram, showing the functional structure of a production system model.

- *Networks* consist of arrays of nodes with weighted connections between them. They work with vectors of feature-based representations and may function as pattern associators (*feed-forward networks*), pattern completers (*associative networks*), or constraint satisfaction devices (*interactive activation networks*). Network properties specify the network's size (i.e., number of nodes) and activation propagation parameters (e.g., learning rate, decay rate, noise).
- *Compounds* encapsulate the processing of a set of related boxes. Thus, in a model of a complex system that comprises multiple interacting subsystems, each subsystem might be implemented as a compound, with the functioning of each subsystem specified within each of those compounds. This approach clarifies the gross structure of the model and allows the interactions between subsystems to be examined more clearly.
- *Sockets* allow models to interact with external devices through the TCP Internet Protocol (TCP/IP). A *client socket* box may issue TCP/IP requests to any network address. It will then wait for a response. The network address may correspond to an external device running as a server (i.e., an external TCP/IP device that waits for

and responds to requests), another process on the same computer, or another COGENT model. In a complementary manner, a *server socket* box waits for and responds to TCP/IP requests.
- *Data boxes* support simple input and output. Thus *data sources* feed a preset stream of input data into a model at a prespecified rate, while *data sinks* collect a model's output and present it in text, table, or graphical form.

Most types of box also have properties that determine features of the box's on-screen representation. Thus *graphical buffers* (which are buffers that display their contents in graphical form) have properties to specify the scales and labels of axes, while *tabular data sinks* (which are data sinks that represent their contents in tabular form) have properties that specify the width, height and order of rows and columns.

The Box and Arrow Interpreter

COGENT's box and arrow interpreter provides box and arrow diagrams produced within COGENT with an operational semantics. The interpreter implements a cyclic execution model in which each box has a state, a state transition function, and an input/output function.

On each cycle each box may produce output (in accordance with its input/output function) and its state may change (in accordance with its state transition function). Output on each cycle consists of messages—packets of information that are written to a global blackboard. These messages may serve as input on the next cycle provided the box that produced the message is appropriately connected to a box that may process it as input. The blackboard is wiped between cycles. Full details of the interpreter's processing cycle are given in Cooper (1995).

Individual types of box extend this basic model in a variety of ways. For example, the state transition function of a propositional buffer specifies how the buffer's contents change when elements are added to or deleted from the buffer, taking into account the buffer's properties (e.g., its capacity and its behavior when its capacity is exceeded). Buffers do not produce output, so their input/output function is always null. Processes, on the other hand, have a constant state (defined as the set of symbolic rules that determine their behavior), so their state transition function is the identity function. Their input/output function is determined by the rules they contain. These rules may be dependent on both inputs received (via the blackboard) and the contents of buffers that the process reads from (as determined by arrows enabling such operations between the process and buffers). Networks differ from both buffers and processes in that they have both a changing state (a weight matrix, which may be adjusted in response to training exemplars) and an output function (which is based on applying the input to the current weight matrix).

Strengths of COGENT's Box and Arrow Approach

The computational foundation of COGENT's formal interpretation of box and arrow diagrams is rooted in the object-oriented methodology of computer science (see, e.g., Graham, 1994). This methodology is highly appropriate for implementing modular and semimodular systems, such as those commonly postulated within the contemporary cognitive literature, for several reasons. Most critically, many current theorists (particularly those who employ a box and arrow approach) adopt the view that central cognition is at least partially modular and hence decomposable (see Shallice, 1988) and that a key task of cognitive science is to establish a functional decomposition of central cognition that is consistent with experimental and neuropsychological findings.

Computationally, this partial modularity may be modeled with systems that employ information encapsulation with restricted information sharing via message passing—basic premises of the object-oriented approach.

The use of configurable components by COGENT also has advantages with respect to psychological theory. First, the components and their properties are deliberately inspired by the kinds of entities postulated in existing box and arrow style theories. They are therefore at a level commensurate with psychological theory (i.e., buffers and processes) rather than at a level derived from computer science (e.g., the level of input/output streams or the level of window widgets, as in other object-oriented languages and systems). Second, many properties of COGENT boxes fit well with the distinction between competence and performance (Chomsky, 1965), with competence being an idealization of performance. Thus, buffer capacity limitations and decay may be viewed as performance factors that (a) impact upon an idealized competence system to yield actual behavior and (b) may vary across individuals. If one subscribes to the competence/performance distinction, one may explore the effect of performance factors within a COGENT model by varying the values of relevant properties from the idealized values (e.g., no decay and no capacity limitations). Variation of parameter values from an hypothesized norm may also be used to explore individual differences or pathological models. Thus, the effects of specific neural damage may be modeled by, for example, increasing buffer decay within a model of normal performance.

A further advantage of the approach, particularly when modeling behavior on complex tasks, stems from the use of box and arrows diagrams to specify the functional components arising from a high-level task analysis. The internal functioning of the boxes may then be specified in different ways to implement (and then compare) different theoretical accounts of the behavior within the common task analysis. In this way, COGENT supports comparative modeling. (See Cooper, Yule, & Fox, 2003, for a concrete example in the domain of category learning.)

Finally, COGENT's approach to model execution also fits well with the requirements of cognitive modeling. Model execution simulates parallel cyclic processing, with all boxes within a model effectively running in parallel. On each cycle, all boxes apply their output function to generate output messages from their properties, their current state, and any input messages, and all boxes apply their state transition function to generate a new state based on their properties, their previous state, and any input messages. Only once all state changes and outputs have been calculated are box states and message lists updated. True parallelism is therefore ensured, and for this reason COGENT may simulate symbolic and connectionist processes, as well as hybrid processes that merge both forms of processing. However, and despite the cyclic nature of processing, boxes may still function asynchronously in that they need not produce output or change state on every cycle: a process may be configured so as to only produce output in response to a particular event (e.g., when the process is triggered by an appropriate message or when a related buffer's contents change).

Specific Issues

Control: Triggered Versus Autonomous Rules

All boxes within a COGENT model function in parallel. On each processing cycle, each box reads all messages sent to it, adjusts its state in response to those messages and the passing of time, and generates a set of further messages for other boxes to process on the next cycle. There is, therefore, no central controller is beyond a clock. However, global patterns of control may be determined locally by the way in which individual boxes process their inputs.

The main arbiters of local control are *rule-based process* boxes, whose processing may be contingent on the structure of rules within the box and the contents of

other boxes read by those processes. Two types of rule are supported by rule-based processes, triggered rules and autonomous rules. Triggered rules respond to messages sent to a process, while autonomous rules monitor other boxes and fire when specified patterns are detected. Figure 29.2a shows a typical triggered rule, while Figure 29.2b shows a typical autonomous rule.

The triggered rule (Figure 29.2a) applies only on cycles where the box containing it receives a message of the form *recall*. If this occurs, the rule tests a further condition *recall(Word)*, a user-defined condition whose definition is not given here. COGENT adopts Prolog's conventions for variable naming. The initial capital W therefore indicates that *Word* is a variable that may be instantiated with a specific value if the condition succeeds. If on any cycle the trigger occurs and the condition succeeds, the rule fires and sends a message of the form *recalled(Word)* (with *Word* bound to the value obtained from successful satisfaction of the initial condition) to the box named *Experimenter: Administer Task* and a message of the form *recall* to the box named *Input/Output*.

Autonomous rules differ from triggered rules only in that they do not have a trigger and hence may in principle fire on any cycle. Thus, the autonomous rule in Figure 29.2b tests its two conditions on each processing cycle, and the rule fires whenever those conditions are met. Note that, in this example, the first condition tests that a buffer called *Possible Operators* does not contain an element of the form *operator (AnyOperator, AnyState)*. While the second condition tests whether any operators have satisfied preconditions

(a)

> **Rule 1 (refracted; once):** *The recall rule (serial version)*
> TRIGGER: recall
> IF: recall(Word)
> THEN: send recalled(Word) to Experimenter: Administer Task
>   send recall to Input/Output

(b)

> **Rule 2 (unrefracted):** *Propose all operators whose preconditions are satisfied*
> IF: not   operator(AnyOperator, AnyState) is in Possible Operators
>   preconditions_satisfied(Operator)
> THEN: add operator(Operator, possible) to Possible Operators

FIGURE 29.2 (a) A triggered rule. See text for details. (b) An autonomous rule. See text for details.

by calling a user-defined condition. If both conditions are satisfied, *Operator* will be instantiated and a term of the form *operator(Operator, possible)* will be added to the buffer named *Possible Operators*.

Rules of either type may fire multiple times on a single processing cycle if there are multiple ways of instantiating their conditions. Independently, they may also be refracted so as to only apply once for any particular instantiation.

The different types of rule correspond to different types of control, but as rules of both types may be mixed freely within a single rule-based process, and as a model may contain any number of rule-based processes, COGENT effectively supports a continuum of control modes, with sequential triggering of processing at one end and parallel firing of autonomous processes at the other. In the former case, the COGENT box and arrow diagram will represent a flowchart with arrows indicating flow of control, while in the latter, it will represent a fully distributed parallel system with arrows indicating flow of data.

Integration: Recombining Processing Results

A significant issue in any system comprising multiple parallel subprocesses concerns how the results of those subprocesses may be combined to yield a single output or response. This is common in dual-route models in which a response might be produced by either of two routes and some mechanism is required for selecting or merging the results of each route. It also arises in the ACT-R architecture (adaptive control of thought–rational), where in the current version processing is coordinated through temporary storage of critical data in a series of buffers (the goal, retrieval, manual, and visual buffers; Anderson et al., 2004).

Different types of boxes within COGENT provide different possible solutions to the problem of recombining results. At a purely programmatic level, rules within a rule-based process may monitor multiple buffers or may be contingent on multiple elements within a buffer, and this can allow a rule to combine the results of multiple processes into a single response. More interestingly, from the perspective of contemporary cognitive psychology, both input selectivity (where one of several inputs is selected for further processing) and output selectivity (where one of several generated outputs is selected for further processing) may be implemented and combined within COGENT models.

Input selectivity is appropriate when a process may receive multiple, possibly conflicting inputs, and must select one of those inputs for further processing. That selection might be based on the current processing mode of the process or some characteristic of one of the inputs. Within COGENT, both cases require that a process works in concert with a buffer (or an interactive activation network as described in the following paragraph). In the former case, the buffer will contain a single element representing the processing mode, the process will contain one triggered rule for each possible processing mode, each rule will include in its conditions a test of the buffer for the processing mode, and the tests will be mutually exclusive. In the latter case, the inputs may all be temporarily stored in a buffer to allow deliberate selection of one input on the next processing cycle. Inputs in this case, might include a numerical component representing strength, and deliberate selection might involve selecting the input with greatest strength.

Output selectivity enables the results of multiple processes to be combined into a single response. Interactive activation network boxes support this function. Nodes within such boxes have activation values that vary subject to excitation and inhibition according to standard principles of interactive activation (McClelland, 1992). Thus, an interactive activation network may implement a competitive filter, whereby multiple processes provide excitation to different nodes, with the nodes corresponding to discrete potential outputs and lateral inhibitory interactions between nodes ensuring that only one node is highly active at any moment. An output process may then select the most active node (or any nodes above a threshold) and act accordingly.

External Interaction

Since COGENT is not a cognitive architecture, it imposes no specific constraints on input or output processes (i.e., it does not enforce the use of specific perceptual or motor modules). Instead, COGENT supports external interaction in a variety of ways. The simplest approach is via *data boxes*. A COGENT model may include *data sources* and *data sinks* to specify and collate model input and output, respectively. These boxes may be thought of as proxies for perceptual and motor modules.

A second approach to external interaction is provided through COGENT's *client socket* and *server socket* box types. Thus, a COGENT model might initiate interaction with a stand-alone environment simulation (written in the programming language of the developer's choice) by issuing TCP/IP requests to the environment simulation through a client socket box (with the environment taking the role of server) or respond to externally generated TCP/IP requests from the environment via a server socket box.

The third approach to external interaction draws on the power and expressiveness of COGENT's rule-based modeling language. Essentially, one may use the same techniques to simulate both the cognitive processes that make up a model and the environment in which the model is embedded. Compound boxes (i.e., boxes whose contents are themselves box and arrow diagrams) may be used to encapsulate the simulation of the environment and the simulation of the cognitive processes, with arrows between the two licensing communication between them. Thus, many of our own COGENT simulations share the top-level structure shown in Figure 29.3. The figure shows two compound boxes (one labeled *Experimenter* and the other *Subject*) with send arrows indicating that each box may send information to the other box.[1]

In a simulation that uses the top-level structure of Figure 29.3, the *Experimenter* box will typically perform a number of functions. Minimally, these are generation and presentation of stimuli at the required rate to the *Subject* box, and collection and logging of *Subject* responses. Stimulus generation may be a complex process that involves randomization of items drawn from a pool. It may also be sensitive to

Subject responses. Indeed, an experiment may require that some stimuli are generated in response to requests from the *Subject* box. All of these situations may be programmed within COGENT's box and arrow language with appropriate use of processes and buffers. Response logging may be equally complex, with COGENT being used for postprocessing and the generation of graphical output and statistical analyses (including common univariate and multivariate statistical tests, which are provided as primitive functions within generic compound boxes).

This third approach to external interaction, combined with the presence of client and server socket boxes in the inventory of COGENT box types, raises a further possibility for external interaction, namely, the use of COGENT to implement an environment that interacts via TCP/IP with models developed in other systems. Thus, an ACT-R or EPIC model might interact with a COGENT environment, making use of COGENT's data tabulation and analysis facilities. Indeed, once a COGENT environment model has been developed it might be used with a range of alternative cognitive models developed within various systems or architectures to facilitate comparison of the models.

Putting It All Together: A COGENT Model of Routine and Deliberative Behavior

To demonstrate integration and control issues within COGENT, this section provides some details of a

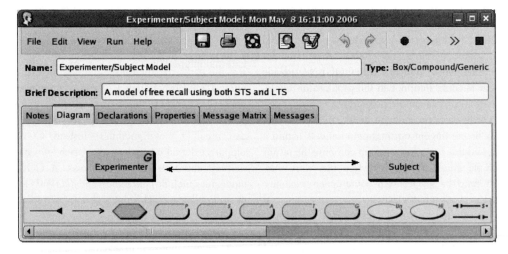

FIGURE 29.3 A common top-level structure appropriate for many COGENT simulations.

partial COGENT implementation of Norman and Shallice's dual-system theory of the control of routine and deliberative behavior. The theory was chosen as an illustrative example for two reasons. First, it is a theory of the human cognitive architecture. Although it is less developed at the computational level than other theories of the human cognitive architecture represented in this volume, it is of comparable breadth. Second, the theory requires the combination of two control modes (routine and deliberative), and hence its implementation demonstrates how multiple control modes may be integrated within a single COGENT model.

The CS/SS Theory: A Verbal Description

A number of theorists have argued for a qualitative difference between controlled or deliberative processing/behavior and routine or automatic processing/behavior (e.g., Norman & Shallice, 1986; Schneider & Shiffrin, 1977). Norman and Shallice (1986) argued on the basis of the qualitative difference between the two that behavior is the result of the interaction of two systems. The routine system, *contention scheduling* (CS), is held to consist of a network of abstract behavioral routines (schemas) that compete for control of behavior through a process of interactive activation. In contrast, the nonroutine system, the *supervisory system* (SS), operates only indirectly on behavior by selectively biasing the activations of schemas within the contention scheduling system. The nonroutine system is held to correlate phenomenologically with will or intention, and is called into play when existing behavioral routines are insufficient or inappropriate for controlling behavior (e.g., in novel situations or situations requiring particular care or attention).

Within contention scheduling schemas have associated activation values and when a schema's activation is sufficiently high, the schema is selected and behavior is controlled by that schema. Lateral inhibition between schemas with conflicting requirements for special purpose cognitive and effective subsystems (e.g., language processing subsystems or the hands) ensures that only one such schema may be highly active at a time, while self-excitation of schemas prevents activations decaying to zero. At the same time, a schema's node may receive an excitatory influence from the representation of the environment if the environment is in a state in which the schema is commonly performed. This "triggering" excitation biases the routine system towards behaviors that are situationally

appropriate, and if not regulated through input from the supervisory system will result in situationally appropriate behaviors being selected and performed.

The internal functioning of the supervisory system was not the focus of Norman and Shallice's original work, but recently Shallice and colleagues (e.g., Shallice, 2002, 2004; Shallice & Burgess, 1996) have begun to specify some of the supervisory system's subprocesses, and Glasspool (2005; see also Glasspool & Cooper, 2002) has provided a mapping between those processes and Fox and Das's (2000) "domino" architecture from the intelligent agents literature. For now, it is sufficient to conceive of the supervisory system as a general-purpose problem-solving and planning system that is able to represent problems, generate goals, evaluate candidate strategies for achieving those goals, and enact a strategy by biasing appropriate schemas within contention scheduling. The critical feature of the supervisory system in comparison with other theories of higher-cognitive functioning concerns the evidence base: While Soar was founded on problem solving and ACT-R was founded on associative memory, the functional processes of the supervisory system are derived from patterns of cognitive breakdown following neural damage (cf. Shallice, 2002, 2004).

The Task and the Global Structure of the Model

The abstract CS/SS model is developed here as a model of behavior on the Wisconsin card sorting task (WCST). In this task, participants are presented with a sequence of cards where each card shows several colored shapes. Cards vary according to three features: the color, shapes, and number of shapes shown. Thus, one card might show four red squares while another might show two green triangles. Participants are required to sort the cards according to a criterion feature that is chosen by the experimenter, with the only information available for determining the sorting criterion provided by positive or negative feedback from the experimenter, and with the experimenter changing the criterion feature once the participant has correctly sorted six cards in succession. Patients with frontal brain damage tend to perform poorly on this task, often failing to switch away from a sorting criterion when it changes. Within the CS/SS model supervisory processes are held to be localized within the prefrontal cortex, and frontal damage is argued to lead to impairments in such processes (see Shallice, 1988, for further details).

One may therefore hypothesize that frontal behavior may be simulated by selectively impairing the SS component of the CS/SS model.

To separate clearly the model from the environment in which it may be tested, the model as described here assumes an approach to external interaction similar to that shown in Figure 29.3. Specifically, the top level of the COGENT model assumes three boxes: two compound boxes with internal structure—*Experimenter* and *Subject*—and a propositional buffer called *Card Table*. *Experimenter* and *Subject* can both read and write to *Card Table*, which mediates all interaction between the two. The role of *Experimenter* is to place cards on the table, wait for a response from *Subject*, provide feedback to *Subject* on that response, and score the responses. Details of its implementation are available from the author.

The COGENT Model of Contention Scheduling

Figure 29.4 shows the functional components of the COGENT implementation of contention scheduling.[2] The principal component of the model is *Schema Hierarchy*, an interactive activation network containing nodes for each schema. For the WCST, it is assumed that there is one schema (and hence one node) for each sorting criterion. Nodes within *Schema Hierarchy* are triggered or excited by features of things in the environment. *Triggering Functions* implements this mechanism. For the WCST, we assume that *Triggering Functions* ensures that all three schema nodes receive the same level of excitation whenever a card is presented to the system (an excitation of 0.10 units proves appropriate), as any one of the three could be used to control behavior. *Selection Process* operates in parallel with the other processes and monitors *Schema Hierarchy*. It ensures that the buffer *Selected Schemas* is up to date by recording the names of any nodes whose activations are above the selection threshold and removing the names of any nodes that were previously selected but whose activation has since fallen. The final process, *Act*, monitors the environment (i.e., *Card Table*) and *Selected Schemas*. Within the WCST, when a card is present and a schema is selected, *Act* applies the selected schema thus sorting the card. The rules for all three processes of the contention scheduling model are shown in Figure 29.5.

All that is required to complete the basic CS model is specification of appropriate activation parameter

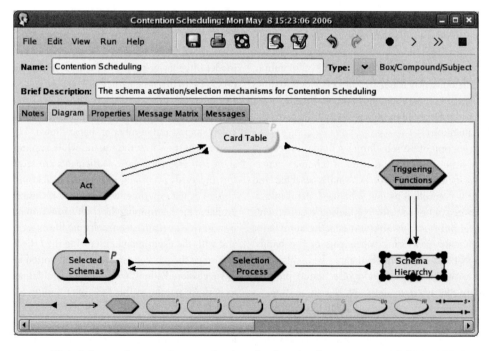

FIGURE 29.4 A box and arrow diagram showing the functional components of the contention scheduling model and their connectivity.

(a)

> **Rule 1 (unrefracted):** *Excite each card sorting schema*
> IF: request(sort(Card)) is in Card Table
>   node(Criterion, _) is in Schema Hierarchy
> THEN: send excite(Criterion, 0.100) to Schema Hierarchy

(b)

> **Rule 1 (unrefracted):** *Select the most active schema*
> IF: not   selected(_) is in Selected Schemas
>   node(S, A) is in Schema Hierarchy
>   the value of the "Selection Threshold" property is T
>   A is not less than T
>   not   node(_, B) is in Schema Hierarchy
>   B is greater than A
> THEN: add selected(S) to Selected Schemas
>
> **Rule 2 (unrefracted):** *Deselect a schema if it is no longer the winning schema*
> IF: selected(S) is in Selected Schemas
>   node(S, A) is in Schema Hierarchy
>   exists   node(_, B) is in Schema Hierarchy
>   B is greater than A
> THEN: delete selected(S) from Selected Schemas

(c)

> **Rule 1 (unrefracted):** *If a schema is selected, then act on it!*
> IF: selected(Criterion) is in Selected Schemas
>   request(sort(Card) is in Card Table
>   not   action(_) is in Card Table
> THEN: add action(match(Card, Criterion)) to Card Table

FIGURE 29.5   (a) The rule for *Triggering Functions*. (b) The rules for *Selection Process*. (c) The rule for *Act*.

settings in *Schema Hierarchy.* With self-excitation and lateral inhibition both set to 0.50, and noise set to 0.005, the CS model is able to respond to cards by adopting a sorting rule at random (using interactive activation with lateral inhibition to select from the three rules available) and then use that rule for the duration of the task. Thus, the CS model can perform the task without any additional machinery, but clearly it is a poor model of human performance—it ignores feedback and will always sort the entire pack of cards with the first rule it chooses. However, the model does demonstrate the basic operation of contention scheduling in the absence of supervisory functions.

The COGENT Model of the Supervisory System

Given that the CS model is able to perform the WCST, but only by adopting a single sorting rule, the role of the supervisory system is to respond appropriately to negative feedback by biasing schemas within CS to switch and explore alternative sorting rules. Figure 29.6 shows the functional components of a generic model of the supervisory system. *Perception* and *Monitoring* are best conceived of as fully autonomous processes that may function in parallel with other subcomponents of the system. *Perception* copies

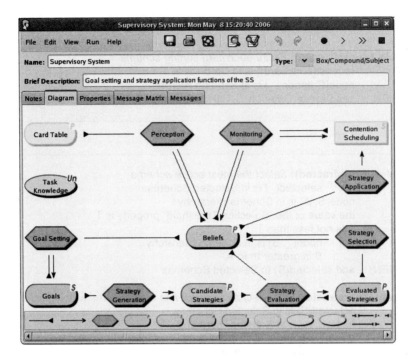

FIGURE 29.6 A box and arrow diagram showing the functional components of the supervisory attention model and their connectivity.

information presented to *Subject* via *Card Table* directly into *Beliefs*, while *Monitoring* monitors *Beliefs* for negative feedback. When such feedback is detected *Monitoring* inhibits the currently selected schema (within *Contention Scheduling*) until it is deselected.

The interaction between the *Perception* and *Monitoring* processes and the *Beliefs* buffer is sufficient to ensure that the model disengages from a sorting strategy when the strategy proves to be incorrect. The remainder of the model implements a processing loop that ensures that the model also selects an alternative strategy. *Goal Setting* raises a goal to sort the next card. *Strategy Generation* generates the set of strategies (i.e., sorting criteria) that might be employed to achieve the goal. *Strategy Evaluation* attaches numeric evaluations to the possible strategies, referring to *Beliefs* about good and bad strategies. When a single best strategy emerges (either because all other strategies have received negative feedback or through random selection from equal alternatives) *Strategy Application* implements the strategy by selectively exciting the appropriate schema node within *Contention Scheduling*.

Because of space limitations, the rules that implement each of the processes are not provided here, but

they are not significantly more complex than those employed by the CS model, and in each case they are generic (rather than task specific). *Task Knowledge*, a globally accessible knowledge base, provides the detailed task specific knowledge that fleshes out the generic rules for the WCST. Thus, *Task Knowledge* specifies the possible strategies to use when sorting the cards and the conditions under which the goal of sorting a card can be considered to have been achieved.

While the principal processes of the supervisory system model are motivated by neuropsychological disorders, the system as implemented here is derived from the Fox and Das (2000) "domino" agent architecture (see also Glasspool, 2005; Glasspool & Cooper, 2002). That architecture implements a cyclic decision-making agent, but processing in the supervisory system model also shares similarities with processing in the Soar cognitive architecture (Newell, 1990). Thus, *Strategy Generation, Evaluation,* and *Selection* mirror the Soar processes of operator proposal, evaluation, and selection.

The complete, fully functioning model is able to perform the WCST with few errors and, crucially, with no perseverative errors. Some errors are inevitable in the task because (at least in the version of the task modeled here) the subject cannot know when the experimenter will alter the target sorting criterion, and when the criterion does change, the subject must select at random between the two other criteria. Following negative feedback, the fully functioning model always switches to a different criterion, and so errors are never perseverative in nature. However, perseverative errors do arise when the functioning of the supervisory system is impaired. Such impairments may be explored within COGENT by altering key properties of boxes within the supervisory system model. Thus, introducing decay to *Beliefs*, or limiting the firing of rules within *Monitoring*, both lead to perseverative errors. This is consistent with neuropsychological evidence which suggests that while frontal patients tend to produce perseverative errors on the WCST, the occurrence of such errors is not predictive of damage to specific regions of the prefrontal cortex. That is, damage to different regions of the prefrontal cortex can lead to a similar propensity toward perseverative errors (but see Stuss et al., 2000, for evidence of dissociations between frontal patient groups when finer grained measures and variants of the WCST are employed).

Key Features of the Complete Model

In the context of the current volume, the key features of the complete model concern the way in which reactive and deliberative control are combined. The CS model implements simple reactive control of behavior, responding to stimuli by applying the most active schema. The SS model is able to modulate or bias this behavior when appropriate, by inhibiting one schema and exciting an alternative. The functioning of SS is, however, deliberative rather than reactive: SS actively generates alternatives and evaluates those alternatives before selecting one and then triggering the corresponding schema within CS.

At the implementation level, the combination of reactive and deliberative control is achieved through a combination of interactive activation and rule-based processing. Interactive activation provides the basis for reactive control within CS, while rule-based processing with message passing between processes and buffers provides the basis for deliberative control within SS. The interface between the two is provided by the mapping between the strategies considered by SS and the schemas available within CS. Since the representations within each system are symbolic and the mapping is one-to-one, SS may bias CS toward a desired

behavior by exciting the corresponding node and away from an undesired behavior by inhibiting the corresponding node. This simple approach provides a seamless interface between the distinct computational methods employed within each system.

If the representations within one system were not symbolic (e.g., if CS employed distributed feature-based representations), then a more complex arrangement involving the transduction of information between the two systems would be required. Nevertheless, the basic interaction could till be implemented provided that the mapping between schemas and strategies is one-to-one.

Conclusion

We have seen that COGENT provides a successful operationalization of the box and arrow notation through an object-oriented approach to computation. With respect to integration and control, COGENT's provision of triggered and autonomous rules combined with boxes such as buffers and networks supports a variety of control regimes. Distinct control regimes may even be mixed within a single model, and COGENT has been used for the development of hybrid symbolic/connectionist cognitive models. Equally, COGENT provides a rich basis for cognitive modeling across a range of domains. This is illustrated by the partial implementation of the CS/SS cognitive architecture presented here, but COGENT has also been successfully applied in specific cognitive domains such as memory, problem solving, reasoning, and language processing (see Cooper, 2002, for details).

The COGENT implementation of the CS/SS model is intended only as an illustration of COGENT's use. Many challenges for the model remain. First, it remains to demonstrate its generality. As a theory of the human cognitive architecture, the CS/SS model should be capable, with appropriate task knowledge, of accounting for behavior across a wide range of tasks. To this end, work has begun on applying the model to other standard neuropsychological tasks (e.g., the Tower of London and sentence completion tasks). While it is anticipated that many modifications will be required, this work will seek to maintain the core assumptions of the CS/SS distinction and the principal processes of each system. Second, it is necessary to move toward simulating quantitative effects such as error rates and response times. The demonstration of perseverative errors following breakdown of SS components as described provides only weak support for the model. Finally, it is necessary to provide a mapping of the functional components of the model onto neural regions to tie the simulation results in with the neuropsychological findings underlying Shallice's fractionation of the supervisory system. One would anticipate, given the origins of the CS/SS theory, that this should not be a difficult task. It is, however, essential for validating the model.

Acknowledgments

I am grateful to John Fox, David Glasspool, Tim Shallice, and Peter Yule for many discussions over a number of years that have strongly influenced both the COGENT modeling environment and the specific modeling work presented here. I am also grateful to Wayne Gray and Chris Sims, for their valuable feedback on an initial draft of this chapter, and attendees of the Integrated Models of Cognitive Systems workshop for feedback on the material when presented at that workshop.

Notes

1. The G and S superscripts in the boxes indicate that the *Experimenter* box is a *generic compound* (which may contain any code, including output boxes that support graphical representations and processes that call statistical functions), while the *Subject* box is a *subject compound* (which can contain only a subset of COGENT functions, namely, those that might reasonably be employed as cognitive modeling primitives).

2. The model as presented here has been deliberately simplified and includes little beyond what is required for the WCST. For a more complete model of contention scheduling, see Cooper and Shallice (2000) or Cooper, Schwartz, Yule, and Shallice (2005).

References

Anderson, J. R., Bothell, D., Byrne, M., Douglass, S., Lebiere, C., & Qin, Y. (2004). An integrated theory of mind. *Psychological Review, 111,* 1036–1060.

Broadbent, D. E. (1958). *Perception and communication.* New York: Pergamon Press.

Chomsky, N. (1965). *Aspects of the theory of syntax.* Cambridge, MA: MIT Press.

Cooper, R. P. (1995). Towards an object-oriented language for cognitive modelling. In J. D. Moore & J. F. Lehman (Eds.), *Proceedings of the 17th Annual Conference of the Cognitive Science Society* (pp. 556–561), Pittsburgh, PA.

———. (2002). *Modelling high-level cognitive processes.* Mahwah, NJ: Erlbaum.

———, & Fox, J. (1998). COGENT: A visual design environment for cognitive modelling. *Behavior Research Methods, Instruments, & Computers, 30,* 553–564.

———, Schwartz, M. F., Yule, P., & Shallice, T. (2005). The simulation of action disorganisation in complex activities of daily living. *Cognitive Neuropsychology, 22,* 959–1004.

———, & Shallice, T. (1995). Soar and the case for unified theories of cognition. *Cognition, 55,* 115–145.

———, & Shallice, T. (2000). Contention scheduling and the control of routine activities. *Cognitive Neuropsychology, 17,* 297–338.

———, Yule, P., & Fox, J. (2003). Cue selection and category learning: A systematic comparison of three theories. *Cognitive Science Quarterly, 3,* 143–182.

Fox, J., & Das, S. (2000). *Safe and sound: Artificial intelligence in hazardous applications.* Cambridge, MA: MIT Press.

Glasspool, D. W. (2005). The integration and control of behaviour: Insights from neuroscience and AI. In D. N. Davis (Ed.), *Visions of mind: Architectures for cognition and affect* (pp. 208–234). Hershey, PA: IDEA Group.

———, & Cooper, R. P. (2002). Executive processes. In R. P. Cooper (Ed.), *Modelling high-level cognitive processes* (pp. 313–362). Mahwah, NJ: Erlbaum.

Graham, I. (1994). *Object oriented methods.* Wokingham, UK: Addison-Wesley.

McClelland, J. L. (1992). Toward a theory of information processing in graded, random, interactive networks. In D. E. Meyer & S. Kornblum (Eds.), *Attention and performance XIV: Synergies in experimental psychology, artificial intelligence, and cognitive neuroscience* (pp. 655–688). Cambridge, MA: MIT Press.

Newell, A. (1990). *Unified theories of cognition.* Cambridge, MA: Harvard University Press.

Norman, D. A., & Shallice, T. (1986). Attention to action: Willed and automatic control of behavior. In R. Davidson, G. Schwartz, & D. Shapiro (Eds.), *Consciousness and self regulation: Advances in research and theory* (Vol. 4, pp. 1–18). New York: Plenum Press.

Schneider, W., & Shiffrin, R. M. (1977). Controlled and automatic processing: I. Detection, search and attention. *Psychological Review, 84,* 1–66.

Shallice, T. (1988). *From neuropsychology to mental structure.* Cambridge: Cambridge University Press.

———. (2002). Fractionation of the supervisory system. In D. T. Stuss & R. Knight (Eds.), *Principles of frontal lobe function* (pp. 261–277). Oxford: Oxford University Press.

———. (2004). The fractionation of supervisory control. In M. Gazzaniga (Ed.), *The cognitive neurosciences III* (pp. 943–956). Cambridge, MA: MIT Press.

———, & Burgess, P. W. (1996). The domain of supervisory processes and temporal organisation of behaviour. *Philosophical Transactions of the Royal Society of London, B351,* 1405–1412.

Stuss, D. T., Levine, B., Alexander, M. P., Hong, J., Palumbo, C., Hamer, L., et al. (2000). Wisconsin Card Sorting Test performance in patients with focal frontal and posterior brain damage: Effects of lesion location and test structure on separable cognitive processes. *Neuropsychologia, 38,* 388–402.

PART IX

AFTERWORD

Wayne D. Gray

Someone has to go last and in this case the honor falls to Mike Byrne. I did not commission Mike to write "a last chapter" but that is what he did. In a way his chapter 30 and my chapter 1 can be viewed as bookends for the 28 chapters that lie between. I began the book with a short essay on three types of control of cognition that need to be considered when building integrated models of cognitive systems. Not wanting to overfit the data, I did not try to sharply define what these distinctions were. Instead my goal was to sketch some broad distinctions that might make conversations between research camps easier and, indeed, might give those who build integrated models a new vocabulary to discuss their models and to communicate their needs to those who build single-focus models of cognitive functions or, as Mike puts it, local theories.

Before writing chapter 1, my discussion of Type 1, 2, and 3 control was limited to several paragraphs on the now defunct workshop Web site. I am pleased that, in addition to Mike's chapter, three other chapters have found a good use for this vocabulary (Gluck, Ball, and Krusmark in chapter 2; Gunzelman, Gluck, Price, Van Dongen, and Dinges in chapter 17; and Ritter, Reifers, Klein, and Schoelles in chapter 18). I am especially pleased that Mike seems to have used the Type 1 versus Type 2 distinction as a call-to-arms for the cognitive community to come together and start building the types of models needed to create integrated models of cognitive systems. His perspective is that of a Type 1 modeler in search of Type 2 components that can be plugged and played in his models. His appeal is very direct, very scholarly, and very personal.

30

Local Theories Versus Comprehensive Architectures

The Cognitive Science Jigsaw Puzzle

Michael D. Byrne

Reasons for considering integrated models of cognitive systems as a desideratum are myriad, ranging from deeply theoretical interest in the structure of cognition for its own sake to the desire to gain leverage on difficult applied problems to pedagogical interests in blending cognitive psychology with related disciplines like artificial intelligence. I would like to concentrate on the more applied reasons, many of which are also articulated elsewhere (see Byrne, 2003; Byrne & Gray, 2003). Consider a modern jetliner pilot. The task faced by the pilot is complex, safety critical, and takes many years of intensive training to master. The environment in which this task is done is staggeringly complex, visually rich, places extreme time demands on the flight crew, and is partially managed by a fairly opaque piece of automation. This is not only a challenging domain for the pilots, but for human factors engineers who have the task of trying to make the pilot's job easier and safer.

As much of this volume might suggest, it is the integration of all these capabilities which is particularly pertinent here. Scientists and engineers who want to understand pilots, or emergency room physicians, or even just regular people doing fairly routine office work (e.g., surfing the Web) do not have the luxury of isolating one aspect of human performance. Real people in real settings—to borrow Ed Hutchins's (1995) terrific turn of phrase, those exercising "cognition in the wild"—bring to bear an integrated set of capabilities. Within the space of a minute, a pilot or an office worker might retrieve something from memory, plan, conduct a visual search, execute a routine procedure, make a linguistic inference, produce a judgment, and solve a problem. These activities are not circumscribed but rather highly dependent on one another. A key inference in resolving a language comprehension ambiguity might be driven by the outcome of a visual search and the new information comprehended lead to a new plan of action. People who study pilots, doctors, and office workers cannot specialize in one small area of psychology but rather must confront human cognition, perception, and action as an integrated unit. Furthermore, this serves as a reminder: the human mind, the thing we are trying to understand and model, actually functions as an integrated whole.

Oddly, experimental psychology has tended not to view the problem this way. The dominant approach has been "divide and conquer." This may serve well in some contexts, but there are potentially serious hazards. Even though it is true that human visual attention can operate in impoverished visual contexts for trivially simple tasks, it is not necessarily the case that visual attention operates the same way in those circumstances as it does when the person is simultaneously trying to monitor multiple flight instruments and make a decision about whether to abort a landing. Nonetheless, experimental participants in psychology laboratories have done literally millions of trials where the entire experience consists of making a present/absent judgment on red or green horizontal or vertical bars on uniform black or white backgrounds. I would be unsurprised to find out that the number of paired associates memorized in psychology laboratories numbered in the billions. I am not trying to suggest that such research has no value; I think we have learned a great deal about visual attention from search experiments and have learned useful things about human memory from list memorization experiments. No, the problem is that accounts based on experiments like these tend to stop at the boundaries of those experiments. They do not fit in with other theories and accounts; there is no coherent picture being formed.

This is precisely the problem Allen Newell (1973) warned us about more than 30 years ago in his famous "twenty questions" paper. This paper should be required reading for anyone studying human cognition because of prophetic comments like this one:

> Suppose that in the next thirty years we continued as we are now going. Another hundred phenomena, give or take a few dozen, will have been discovered and explored. Another forty oppositions will have been posited and their resolutions initiated. Will psychology then have come of age? Will it provide the kind of encompassing of its subject matter—the behavior of man—that we all posit as a characteristic of a mature science? And if so, how will this transformation be accomplished by this succession of phenomena and oppositions? (pp. 287–288)

I submit that, to a first order of approximation, the majority of cognitive psychology has indeed continued on for the past 30-odd years more or less as it was in 1973 (though certainly with a variety of improvements,

particularly methodological) and that it has largely not made it to the level of mature science. Theories in cognitive psychology are generally too informal and too piecemeal to be considered on the same playing field as physics, chemistry, and even modern biology. This is because the divide-and-conquer approach is still the prevalent mode of operation. Perhaps surprisingly, this is often understandable. The human cognitive/perceptual/motor system is obviously incredibly complex, perhaps the most complex system we humans have ever tried to understand. The problem of understanding it all is so vast; it seems intractable taken all at once. Specialization happens in all sciences by necessity, and the science of cognition is no exception. It is legitimately impossible to be an expert in every aspect of human performance. As I said at the workshop, let the vision folks figure out the gory details of texture segmentation and let the memory folks figure out the gory details of modality-specific short-term memory decay. Divide and conquer has legitimate appeal; I certainly do not want to have to become an expert in coarticulation effects or a myriad of other topics in cognitive psychology.

However, there can be serious drawbacks to the divide-and-conquer approach as an overall research strategy. Divide and conquer is only appropriate to the extent that susbsystems are truly independent. There is obviously some validity to this assumption, as there are clearly susbsystems that are indeed somewhat independent. For example, it would seem a very poor way to build a motor control system if the machinery that computes the trajectory my arm takes when I reach for the mouse is dependent on whether the sentence I just comprehended was heard versus seen. The issue, then, is how to recover an integrated system from a collection of subdivided parts and to show the value of bothering with the integration.

One integrated approach that is intended to provide insight into numerous interactive domains (and commercial jetliners are most certainly interactive, at least from the perspective of the pilot if not the passengers) is a framework known as the interactive behavior triad, or IBT. The interactive behavior triad is depicted in Figure 30.1 and can be considered a broader variant of a perspective first sketched elsewhere (Byrne, 2001; Gray, 2000; Gray & Altmann, 1999).

The idea behind the IBT is that fully understanding human interactive behavior requires understanding of three crucial sets of constraints. The *task* is taken to be the goal, or frequently set of goals, the human (or, to adopt human factors jargon, the "operator")

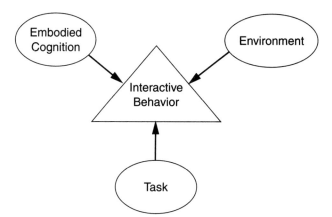

FIGURE 30.1 The interactive behavior triad (IBT).

operator is trying to accomplish and the knowledge required to fulfill those goals. This is not unlike the kind of decomposition found in Card, Moran, and Newell (1983) though perhaps broader in some senses. Misunderstanding the task faced by people is a common source of failure for technological systems. Engineers and computer scientists often invent technologies that, while interesting in their own right, fail to address appropriately any real task actually done by people "in the wild" (more on this shortly). The *environment* is taken to be the set of constraints and affordances available to the operator from the task environment. This includes not only the various artifacts in the environment but also the properties of any system being controlled directly or indirectly by the operator (e.g., the aircraft flight dynamics, which may set severe limits on how the pilot can attempt to accomplish his/her goals), as well as the broader environment in which the interaction is situated (e.g., general properties of airports; see also chapter 14 by Kirlik and chapter 11 by Todd and Schooler, this volume).

Why are these important to a theory of cognition? A wonderful illustration of a series of fundamental failures to understand both the tasks and environments occupied by real office workers is presented in Sellen and Harper's (2001) *The Myth of the Paperless Office*. They describe how efforts to achieve the goal of a paperless office environment (explicitly stated by multiple information technology companies over the years) are seriously misguided because such efforts are based on a shallow and inaccurate understanding of both the tasks done by office workers and the important physical and cultural characteristics of offices and the paper in them.

This brings us at last to *embodied cognition*. This is taken to be the capabilities and limitations of the

integrated human perceptual–cognitive–motor system. It is impossible to construct a device (such as a cockpit) and expect smooth interaction with humans without some understanding of how humans themselves work. What is remarkable is how much progress has been made by human factors researchers and practitioners in domains like commercial aviation despite the profound lack of a good integrated theory of perception, cognition, and motor control. In fact there is a tendency for people in such domains to specialize in the task domain rather than in what kind of cognition is done, hence journals and organizations such as "aviation psychology." Now, it is obviously possible to specialize somewhat within such domains—there are certainly people who focus on specific problems in the visual tasks faced by pilots or radiologists or whatever—but, realistically, complete solutions to problems of interactivity require an integrated view. For reasons laid out in some detail in Byrne (2003) and Byrne and Gray (2003), my strong preference is that this integrated view also meets the standard of being an executable simulation model that therefore makes quantitative predictions.

While one of the aims of the interactive behavior triad is to provide leverage on applied problems, since such problems often hinge on critical interactions between two or more of the IBT components, its use as a guiding framework would also address Newell's warning. The interactive behavior triad is intended to foster both cumulation and integration because it encourages a "big picture" approach. Consideration of the effects of different tasks and environments discourages the relentless pursuit of repeated minor variations of the same task in the same context. Furthermore, the view of cognition as fully embodied encourages pursuit across long-standing subdiscipline boundaries that are surprisingly high in experimental psychology; for example, the divide between perception and cognition is amazingly vast. The IBT is intended to serve as a reminder that all of the pieces of human capability must all fit together.

The notion of everything fitting together raises what I have come to call "the jigsaw puzzle problem." The analogy here is that the whole cognitive science endeavor has as one of its goals the construction of a complete and detailed picture of human cognition and performance (which would nicely support the interactive behavior triad). There are multiple ways one might go about trying to generate such a picture. The divide-and-conquer approach characteristic of the field has subdivided the picture into some large number of little pieces. Most individual researchers in the cognitive sciences, or more specifically, most theorists of Type 2 (Gray, chapter 1, this volume), have taken responsibility for different pieces. That is, they have taken on the task of figuring out what their part of the overall picture, and only that part of the picture, looks like.

Of course, if we want to understand the complex behaviors we see in the real world (e.g., pilots), what we need is a comprehensive theory that encompasses, subsumes, or includes many of these Type 2 theories, or what has been termed a Type 1 theory. The job of the Type 1 theorist, then, is to take all the little jigsaw puzzle pieces and assemble them into a coherent picture. There are certainly multiple approaches one could take to this, and several have indeed been tried by different Type 1 theorists. One approach is to start with the idea that a central controller will be necessary and then gradually incorporate more and more of the Type 2 pieces into the monolithic controller. Type 2 theories will thus be subsumed into a larger system. This appears to be the Soar approach (Newell, 1990). A vaguely similar approach was taken with ACT* (Anderson, 1983) and early versions of ACT-R (Anderson, 1993), which treated declarative memory as a separate component but put all other functionality into the central controller. This produced a fairly monolithic architecture looking something like the one depicted in Figure 30.2.

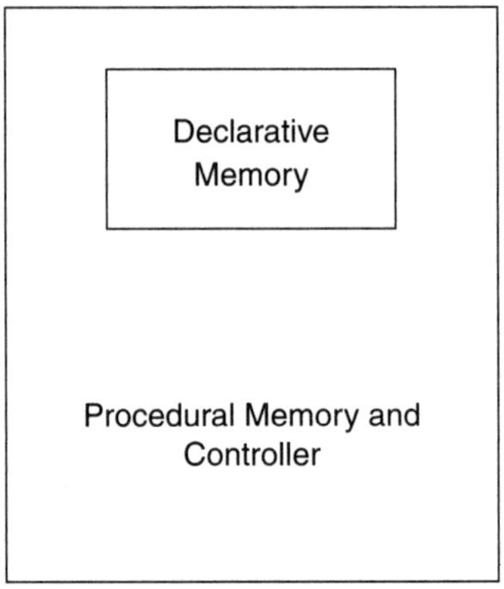

This figure already looks dated by its simplicity. Indeed, the early 1980s architectures were almost entirely concerned with modeling central cognition and not perceptual–motor activities and thus such capabilities were not included in these architectures (though they perhaps could have been). I suspect that in the long run this monolithic kind of approach will

FIGURE 30.2 Generalized monolithic cognitive architecture in which the complexity lives almost entirely in the controller.

not scale up to cover the entire picture very well and has the added drawback of not mapping easily on the more modular conceptualization of the brain offered by modern cognitive neuroscience.

An alternative approach, one represented by EPIC (Kieras & Meyer, 1997) and more recent versions of ACT-R (Anderson et al., 2004) is to embrace the notion of modularity and simply coordinate multiple modules with a central controller, rather than trying to do everything within the context of a central controller. This yields an architecture organized more like what is depicted in Figure 30.3. Given my research history, I unsurprisingly see this as a framework with a good chance of ultimately supporting the full picture. Not every architecture will include the same number of subsystems (neither ACT-R or EPIC currently include a module for emotion; different architectures are likely to disagree about the number of visual subsystems to represent, etc.), but this general organizational strategy still applies. The resemblance of this style of organization to the model human processor of Card, Moran, and Newell (1983) is not accidental, and this bears a meaningful similarity to Baddeley's (1986) well-known conceptualization of working memory, with the controller in the "central executive" role. Thus, despite the relatively recent emergence of EPIC and the current organization of ACT-R, this is neither a new or especially radical conceptualization.

Another salient point is that several modules (e.g., vision) in this scheme have a more or less direct

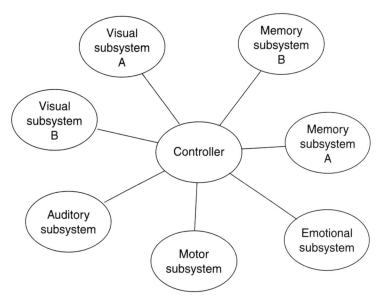

FIGURE 30.3 Modular architecture where all subsystems are coordinated by a single central controller.

connection to the external environment, though that is not depicted in Figure 30.3. I chose to omit it from this diagram simply to note that some architectures may give much higher priority to understanding the internal components than to the role of interaction with the external world. Of course, the interactive behavior triad suggests that this would not be an optimal approach, but reasonable theorists could disagree on the centrality of such a connection.

Alternately, there are camps within cognitive science that eschew the notion of a central controller and see global control as an emergent property of a modular organization (e.g., the ICS system of Barnard, 1999). Figure 30.4 is a general representation of how such an architecture might be organized. Such a decentralized architecture is probably more representative of connectionist approaches such as Atallah, Frank, and O'Reilly (2004), though there are some interesting symbolic/connectionist fusion approaches which have something of this feel (e.g., Just, Carpenter, & Varma, 1999). Approaches of this kind are interesting and definitely have merit, but I have yet to see any such system scaled up to a task as intricate and complex as piloting. Whether it is ultimately possible is still an open question.

Again, the inclusion of connections to the external world is not intrinsically a property of this kind of organization; I am simply illustrating how one might

construct such a system. I do generally agree with the notion that the environment plays a key role in assuring good overall control flow in the human cognitive system, but I do not intend to claim that systems without central controllers are better suited to addressing such issues.

I am not suggesting that other organizations are impossible or unlikely. My goal here is not to produce a catalog of all styles of cognitive architecture, but to point out that there are multiple approaches for integrating various Type 2 theories into a larger Type 1 architecture. Unfortunately, this is happening on only a very small scale; cognitive scientists who are engaged in building Type 1 theories are few and far between. On the basis of this analogy, the task here should be somewhat like solving a jigsaw puzzle; simply fit together all the relevant pieces, and then the picture should be complete. If it were that easy, I suspect more researchers would be engaged in this task.

Of course, it is in fact much more complex than this. Why? Why are the theorists and developers behind Soar or ACT-R or EPIC unable to simply assemble a collection of pieces that have been generated by Type 2 theorists? As someone who has tried to do this, I think there are multiple reasons. The first problem is that Type 1 theories are generally designed to support modeling and are thus executable simulation systems. Why it is that Type 1 theories tend to be computational

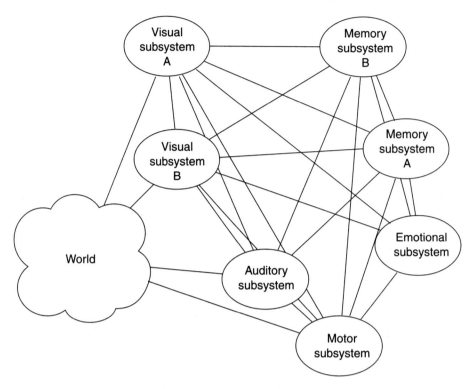

FIGURE 30.4 Modular architecture with no central controller (distributed control).

is itself an interesting issue, but it is beyond the scope of the current discussion; take it as a given that most Type 1 theories are computational or at least computationally oriented. To produce such a system, all of the pieces of the system must be specified in enough detail that an implementation is possible.

Lack of complete specification is, in fact, not unusual in cognitive science. Of course, there have always been exceptions and some things cleanly quantified; in fact, some quantitative and implementable formalisms describing certain aspects of behavior have even been elevated to "law" status, such as Fitts' law for aimed motor movements (Fitts, 1954) or the Hick-Hyman law for choice reaction (Hick, 1952; Hyman, 1953). Unfortunately, these truly have been the exception rather than the rule. Returning to the jigsaw puzzle analogy, assembling a collection of informal and underspecified theories into a coherent whole is essentially impossible; it would be like trying to piece together a collection of amorphous, shifting amoebae. The good news is that this is changing, as several of the chapters in this volume clearly demonstrate. But it is not clear that there are yet enough well-specified Type 2

theories to assemble a collection of them into a meaningful Type 1 theory. I suspect—or at least hope—that this situation will change substantially in the next few decades.

Unfortunately, even if the cognitive science landscape were populated with well-specified and empirically convincing Type 2 theories, those of us in the business of constructing Type 1 systems would still face an enormous uphill battle in trying to assemble them. Consider the perspective of a Type 1 theory builder. A general organization is chosen, and there is some idea about what the overall picture might look like. The landscape of Type 2 theories is surveyed and the builder then goes out in search of puzzle pieces that can be fit together to form an overall picture. The unfortunate reality of the situation is that right now, the chance that the right pieces will all be available and will actually fit together is quite small.

Why? Because unlike real jigsaw puzzles where a picture is cut up one time in such a way that all pieces fit with something else, in the cognitive sciences there has been no central oversight of how the pieces are cut. It is as if everyone is given access to an extremely low-resolution version of the total picture, allowed to cut out one's own piece, and then asked to work on improving the resolution of just that piece.

This leads to three closely related and challenging problems. The first is what I call the *piece identification problem*, the second the *piece fit problem*, and finally the *piece distribution problem*. With the piece identification problem, researchers not only get to pick out the part of the picture to call their "piece," but they also get to choose the label for that piece. In building a Type 1 theory, in an attempt to have broad coverage, one would ideally want one piece from each area of the picture. But it is hard to know what part of the picture a piece represents because there is such little uniformity in terminology or input/output relations. Terminology is probably the easiest to illustrate. Considering all the literature that contains the term, it seems apparent that any Type 1 theory is going to have to address phenomena associated with "attention." But what is attention? According to various Type 2 researchers in vision, we have selective attention (Broadbent, 1954), divided attention (Treisman & Gelade, 1980), object-based versus location-based attention (Duncan, 1984), and attentional sets (Folk, Remington, & Johnston, 1992). In central or cognitive work on attention, there is the limited attention in dual tasking (Pashler, 1994), attention given to various features in categorization (Kruschke, 1992), the attentional blink (Broadbent & Broadbent, 1987), and the attentional capacity that is freed when tasks are automatized (Shiffrin & Schneider, 1977). Not only are there many different senses of attention, but there is little contact and even less unification between various senses. So even identifying which pieces to try to fit together is a challenge for the Type 1 theorist. This is the most obvious of the three problems, so I will not elaborate further.

The piece fit problem has to do with input/output relations. Because Type 2 work has become so compartmentalized, there is often little or no discussion or specification in Type 2 theories of where the inputs to a particular mechanism come from and what form they take, nor where the outputs go and what form they take. Take, for example, models of visual search, of which there are many (see chapters by Wolfe [chapter 8] and Pomplun [chapter 9] in this volume for examples of solid Type 2 visual search theories). Most (but not all) models of visual search stipulate that somewhere in the connection to the visual search subsystem (or *visual selective attention*) there exists some information about the target being searched for, for example, that it

is red. But what these models leave unsaid is what can be specified or how it is specified. In the typical visual search experiment, objects in the visual field appear entirely at random (or random from among a fixed set of locations). Experimental participants typically know either nothing in advance about the location of the target or they have specific information about the target location (e.g., from a cue), which may or may not be accurate. Assuming the visual search system is at least sometimes driven in a top-down fashion, what can the "top" module specify? Can approximate location be specified? People often have vague but meaningful expectations about where things will be located. For example, when searching a Web page for a link, people can reasonably expect that the target should be "somewhere on the left." Can that kind of expectation be passed to the visual selective attention system, and if so, how?

The situation is not much better on the output side. Some more coarse models predict only overall response times for present/absent judgments. Other models generate simulated movements of visual attention and/or point-of-gaze unfolding over time. The piece fit problem, then, is that there is no simple way to gather Type 2 theories or models and match them up with other system components.

I suspect that the extent of this problem is grossly underestimated by most cognitive scientists, so I want to provide a somewhat detailed illustration of the problem. To do so, I want to draw on my experience in constructing the perceptual–motor subsystems for ACT-R. At the time I began working with ACT-R, it was a monolithic single-channel architecture focused primarily on central cognition (see Figure 30.2). The good news was that some useful work on giving ACT-R elementary visual and motor capabilities had already been done, under the ACT-R visual interface described in Anderson, Matessa, and Lebiere (1997). However, that system was entirely single channel; that is, if the motor system was busy moving the hands, then everything else in the system stopped. This may be a reasonable approximation in some contexts, particularly tasks where the bulk of the task is pure cogitation, but that approach is inadequate for multiple-task situations or highly interactive domains where cognition, perception, and action are all interleaved. The power of this kind of interleaving has been apparent for some time, as laid out in the model human processor and as clearly demonstrated by the model human processor–based models of telephone operators presented in Gray, John, and Atwood (1993).

At the time I began my work on ACT-R, the EPIC architecture, which has a modular organization and thus allowed for parallel cognitive, perceptual, and motor activity, was just beginning to attract attention. Our first thought on how to modularize ACT-R was to simply replace the "cognitive processor" in EPIC, also a production system, with ACT-R's production system. We knew that a similar effort was under way using the Soar architecture as a cognition alternative (e.g., Chong & Laird, 1997). After considerable analysis of EPIC, I decided that this strategy would more or less work for some modules (particularly motor and vocal modules) but that it could not work for others, particularly vision. What I determined was that the input/output properties of the cognitive systems were different enough that the interface between vision and cognition had to be fundamentally different for ACT-R and for EPIC's cognitive processor. Why? EPIC's philosophy of minimal commitment to assumptions, which is a powerful research strategy, allowed a great deal of flexibility in how vision and cognition could interact in EPIC. ACT-R, however, had already made a great many long-standing commitments on numerous fronts, especially with respect to memory. This meant that the communication between vision and cognition had to be much more constrained.

A more concrete description of one of the issues may help clarify. In EPIC, the visual system deposits representations of the visual scene (e.g., working memory elements representing visible objects) directly into the production system's working memory. When anything changes (for example, a new object appears, or something as mundane as an eye movement rendering new objects visible and old ones not), new representations are added and, if necessary, old ones simply deleted. This seems straightforward and obvious. However, it also happens not to work for ACT-R. Working memory elements in ACT-R have more regulated structure than the arbitrary lists used by EPIC, but more importantly, they have interactions with other parts of the system. They have activations that decay, strengths of association with other items, and counts of times they have been accessed to compute base-level activation. Simply deleting such elements is not straightforward in ACT-R.

In fact, what we ended up with in ACT-R is a system where the vision module does have an EPIC-like representation where things are deleted and added as necessary; this store is visual iconic memory. But unlike in EPIC, contents of this store cannot simply be

copied into declarative memory and deleted when no longer current; in ACT-R, gatekeeping between the iconic store and declarative memory is provided by an attentional mechanism. No such mechanism is present in EPIC, which has no need for a "covert attention" construct. The EPIC design is somewhat cleaner and definitely more parsimonious, but it does not provide for things like a decaying memory for what was seen or associative learning of object locations over time.

The real point here is not to argue about which architecture has the better visual system; the real point here is that *your theory of vision is constrained by the properties of your theory of cognition*, and vice versa. But since most Type 2 vision theorists do not worry much about cognition, and often use tasks in their research that are carefully designed (either implicitly or explicitly) to minimize the role of cognition, their theories will not reflect such constraints. The same applies to many Type 2 cognition theorists as well, but of course visual constraints are being neglected. Thus, even when it is possible to find two puzzle pieces that cover adjoining sections of the picture, they are unlikely to smoothly fit together.

In contrast to the piece identification and piece fit problems, which occur because of *how* Type 2 research is done, the piece distribution problem occurs because of *where* Type 2 research is done. In trying to collect as many pieces to assemble as possible, a Type 1 theory builder would find that certain parts of the picture have many pieces, many of which overlap, thus creating a small pile of pieces. However, there would also be portions of the picture with no pieces whatsoever. There are large areas that are simply not covered. Some of these are understandable, while some of them are quite surprising. To take an example from my conversations with Dave Kieras, there is the issue of *retinal availability functions*. We know, for instance, that color perception is "best" in the fovea and "worst" at higher eccentricities. What we do not know is the shape of the function mapping eccentricity to availability for numerous visual features such as size and orientation. It is striking that this kind of basic information about the perceptual system is not readily available in the research literature.

To again take an example from my own research, consider the two displays depicted in Figure 30.5. In my laboratory participants are trained to execute relatively simple routine procedures on each of these simulated devices. Critically, each of these procedures is isomorphic in terms of the goal–subgoal–action

(a)

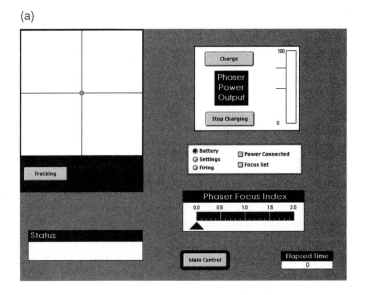

(b)

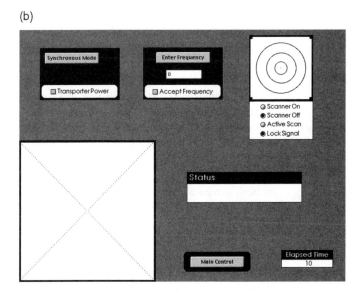

FIGURE 30.5 (a) Display for the "phaser" task. (b) Display for the "transporter" task.

hierarchy. That is, both tasks contain the same number of subgoals and steps. For example, in the "phaser" task, the first subgoal is to charge the phaser. To do this, subjects click the *power connected* button, then the *charge* button, then they wait for the charge indicator (the bar on the right) to fill, then click *stop charging*, and finally click *power connected*. The "transporter" task starts with a subgoal with the same structure but with a different name and using different buttons. Both tasks have the same number of subgoals, and each subgoal has the same number of steps. Subjects are required to practice the tasks to an

accuracy criterion so these tasks are essentially routine procedures.

An approach for analyzing and predicting behavior in such tasks is GOMS analysis (goals, operators, methods, and selection rules; Card, Moran, & Newell, 1983). GOMS models have an extensive history and strong empirical track record of successfully capturing human behavior for routine procedural tasks (see John & Kieras, 1996, for a review). Thus, these tasks should already be covered by an extant account, because according to a GOMS analysis, these tasks should be equivalent (other than some minor differences in mouse pointing time based on Fitts's law). Because of this, in my lab we refer to these tasks as "GOMS-isomorphic."

In fact, human performance on these tasks differs *substantially* in terms of time to execute individual steps of the procedures and error rates at each step. For example, the error rate on the first step differs by more than a factor of three between the two isomorphs. (For more details on this, see Byrne, Maurier, Fick, & Chung, 2004). What is obviously different between the two displays is the layout of the controls, clearly something in the visual domain but which may affect how the controls are functionally characterized by the cognitive system. However, there is essentially nothing in either the visual attention literature or the literature on control of sequential behavior that predicts how these two tasks will be different or explains why. This is the kind of gap alluded to earlier. While this particular gap is on the boundary of vision and cognition, I strongly suspect that there are a preponderance of similar gaps along other boundaries between other subareas.

This problem is not simply getting the input/output relations between two puzzle pieces to match up; there are important parts of the overall picture for which no pieces exist. Additionally, there are other parts of the picture, such as how people perform analogical mapping or supervised category learning, for which there are multiple available pieces. Obviously, this makes assembling the pieces constructed by Type 2 theorists a difficult task, as anyone who has ever tried to solve a difficult jigsaw puzzle with a substantial number of missing pieces and many nearly identical ones can attest.

Between these three problems, one might be inclined to conclude that building comprehensive Type 1 theories is impossible and that realizing an even more complete view of human performance like the interactive behavior triad is simply a pipe dream. While I think the challenges are substantial, I do believe the last decade has seen enormous progress on this front. Theories such as ACT-R and EPIC continue to push boundaries, and this volume positively indicates the serious interest in further progress, not only from cognitive scientists but also from funding agencies. I am encouraged, but not surprised, by interest from funding agencies. Integrated models of human performance within a broad context such as the IBT ultimately will be necessary to address many classes of real-world problems of interest to such agencies.

So, in the spirit of further progress, what can be done to help address this problem? Many things could be done by both Type 1 and Type 2 theorists to facilitate progress on assembling the big picture. I do not expect a sudden burst of productivity as a result of these suggestions, but rather I hope both groups will occasionally take these considerations to heart. The first issue is accessibility of the Type 1 architectures. Accurate or not, these architectures are widely perceived as monolithic, complex, and unapproachable. It is unlikely that this notion can be easily dismissed in the context of journal articles. To address this, Type 1 theorists need to do a better job of outreach. Material describing the architecture and how to think about it needs to be clear and approachable. Having worked with several Type 1 architectures, it is fair to claim that none have scored very well on usability for novices, which potentially could be helped by more GUI-based software tools to not only develop models but also to test and understand a model's behavior as it runs. Good documentation is also crucial but often lacking.

Another recommendation would be if Type 1 theorists were explicit about what pieces their systems lack and why extant Type 2 models in those areas have not been incorporated. Admitting that one's architecture is incomplete can be difficult; however, it may prompt Type 2 researchers to realize that their piece is needed to fill an existing gap or to consider how their model might fit in with the larger picture.

Finally, the software architecture of most Type 1 systems do not lend themselves to inclusiveness. Most of these systems are relatively closed, and the code base is indeed monolithic. Obviously, modular architectures are likely to have an advantage here, since Type 2 theories can more easily be incorporated into a modular system. However, this is not enough. There should be well-documented rules for intermodule interaction—even an application program interface (API)—so that it is clear how a new module could be incorporated. The existence of such a thing might encourage Type 2 researchers and might lend credence to the idea that the Type 1 theorists are actually interested in covering areas outside of their traditional specialty.

Alternatively, what could Type 2 theorists do? This is a more difficult question because it raises another problematic issue. While it is relatively obvious why Type 1 system builders (especially those interested in comprehensive frameworks like the interactive behavior triad) would want assistance from Type 2 theorists, what is the incentive for Type 2 researchers? Unless journal editors start mandating that experimental tasks require more integrated performance or that theories and models have clear connections to a larger cognitive system, change here is likely to be slow. The psychology laboratory, by its nature, tends not to be a place that engenders complex tasks that require diverse capabilities because getting experimental participants up to speed on tasks of high complexity is expensive.

However, most (or at least many) Type 2 researchers are honest with themselves; they realize that their domain is but a part of a larger picture and are at least open to trying to understand the whole of cognition and performance. While most Type 2 theorists might not be interested in pursuing such work themselves, many can at least see that such endeavors have scientific value. So, for those Type 2 researchers who are interested in supporting such activities, several things could be done to help Type 1 theorists synthesize Type 2 models into a more comprehensive architecture. (Note that these are intended as suggestions, no matter how much this may sound like a list of demands.)

Type 1 researchers would be helped if Type 2 theorists elucidated exactly what parts of the overall picture are and are not covered by each Type 2 model. The idea is to try to locate the puzzle piece within the larger picture and, just as important, to try to provide information about its shape so it is clear how it might be fit together with other pieces. Saying that "this is a signal detection model of selective visual attention" or "this is a diffusion model of decision-making" is not enough. Explicit context, delineation, and specification are needed. For example:

- What is and what is not included in this account? For those things that might have been included but were not, why were they not included? That is, was it theoretical, technical, pragmatic? Type 2 theorists often know a lot about where *not* to go in a domain, and Type 1 researchers would benefit from such wisdom in not repeating mistakes.
- What cognitive–perceptual–motor function does the model address? How would that be used by a larger system? What kinds of larger tasks would contain that function as a subtask?
- What does the model take as input and in what form does it take it? Does the model take inputs strictly from perception or from other cognitive subsystems? If the model takes perceptual input, what aspects of that input have to be preprocessed and how? What is the representation over which the model works?
- What does the model output? Products of the process (e.g., decisions), or timing, or both?

Does the model produce intermediate products (e.g., saccades) or a final time for the trial (or problem or whatever is being modeled)? Are there other subsystems (particularly motor) which are supposed to receive output form the model? If so, what information is passed on and in what form?

Unfortunately, many of these things obviously do not easily fit into journal articles. However, provision of such information (perhaps on a Web site) would make it substantially easier for Type 1 system architects to incorporate the relevant ideas, which has the potential of providing wide distribution of the ideas and thus increasing research impact. And who doesn't want to increase their research impact?

Increasing impact is a goal that serves all members of the community. When Type 2 theories can be integrated into Type 1 systems and those systems improved by the result, everybody wins. The Type 2 theorist wins because their work will be seen by an audience beyond their particular specialization. Furthermore, the Type 2 theory will be not merely disseminated within the Type 1 community, but also it will have a chance to affect new tasks and new environments. The Type 1 theorist wins because the theory becomes more complete and, it is hoped, more accurate with a wider range of applicability. Broader and higher fidelity Type 1 human performance theories support stronger IBT-based analysis and modeling and thus should have significant potential for application to highly interactive real-world domains like aviation and medicine.

Thus, the effort involved in uniting Type 1 and Type 2 research is worth making, as the long-run payoffs should be substantial. The work in this volume represents an important step in the direction of this payoff, so I hope further workshops and volumes that reflect these concerns will become increasingly common. While cognitive science has not entirely avoided falling into the trap forecast by Newell some 30 years ago, if enough researchers are willing to commit to the endeavor, we have the potential to successfully build integrated theories.

Acknowledgments

I would like to thank the Office of Naval Research for its support under Grant N00014-03-1-0094 and the National Aeronautics and Space Administration for its support under Grant NDD2-1321. The views and conclusions contained herein are those of the author and should not be interpreted as representing the official policies or endorsements, either expressed or implied, of NASA, ONR, the U.S. government, or any other organization.

References

Anderson, J. R. (1983). *The architecture of cognition.* Mahwah, NJ: Erlbaum.

———. (1993). *Rules of the mind.* Hillsdale, NJ: Erlbaum.

———, Bothell, D., Byrne, M. D., Douglass, S., Lebiere, C., & Quin, Y. (2004). An integrated theory of the mind. *Psychological Review, 111,* 1036–1060.

———, Matessa, M., & Lebiere, C. (1997). ACT-R: A theory of higher level cognition and its relation to visual attention. *Human-Computer Interaction, 12,* 439–462.

Atallah, H. E., Frank, M. J., & O'Reilly, R. C. (2004). Hippocampus, cortex, and basil ganglia: Insights from computational models of complementary learning systems. *Neurobiology of Learning and Memory, 82,* 253–267.

Baddeley, A. D. (1986). *Working memory.* Oxford: Oxford University Press.

Barnard, P. J. (1999). Interacting cognitive subsystems: Modeling working memory phenomena within a multi-processor architecture. In A. Miyake & P. Shah (Eds.), *Models of working memory: Mechanisms of active maintenance and executive control* (pp. 298–329). New York: Cambridge University Press.

Broadbent, D. E. (1954). The role of auditory localization in attention and memory span. *Journal of Experimental Psychology, 47,* 191–196.

———, & Broadbent, M. (1987). From detection to identification: Response to multiple targets in rapid serial visual presentation. *Perception & Psychophysics, 42,* 105–113.

Byrne, M. D. (2001). ACT-R/PM and menu selection: Applying a cognitive architecture to HCI. *International Journal of Human-Computer Studies, 55*(1), 41–84.

———. (2003). Cognitive architecture. In J. A. Jacko & A. Sears (Eds.), *The human-computer interaction handbook: Fundamentals, evolving technologies and emerging applications* (pp. 97–117). Mahwah, NJ: Erlbaum.

———, & Gray, W. D. (2003). Returning human factors to an engineering discipline: Expanding the science base through a new generation of quantitative methods—preface to the special section. *Human Factors, 45,* 1–4.

———, Maurier, D., Fick, C. S., & Chung, P. H. (2004). Routine procedural isomorphs and cognitive control structures. In C. D. Schunn, M. C. Lovett, C. Lebiere, & P. Munro (Eds.), *Proceedings of the Sixth International Conference on Cognitive Modeling* (pp. 52–57). Mahwah, NJ: Erlbaum.

Card, S. K., Moran, T. P., & Newell, A. (1983). *The psychology of human-computer interaction.* Hillsdale, NJ: Erlbaum.

Chong, R. S., & Laird, J. E. (1997). Identifying dual-task executive process knowledge using EPIC-Soar. In M. Shafto & P. Langley (Eds.), *Proceedings of the Nineteenth Annual Conference of the Cognitive Science Society* (pp. 107–112). Hillsdale, NJ: Erlbaum.

Duncan, J. (1984). Selective attention and the organization of visual information. *Journal of Experimental Psychology: General, 113,* 501–517.

Fitts, P. M. (1954). The information capacity of the human motor system in controlling the amplitude of movement. *Journal of Experimental Psychology, 47,* 381–391.

Folk, C. L., Remington, R. W., & Johnston, J. C. (1992). Involuntary covert orienting is contingent on attentional control settings. *Journal of Experimental Psychology: Human Perception & Performance, 18*(4), 1030–1044.

Gray, W. D. (2000). The nature and processing of errors in interactive behavior. *Cognitive Science, 24*(2), 205–248.

———, & Altman, E. M. (1999). Cognitive modeling and human-computer interaction. In W. Karwowski (Ed.), *International encyclopedia of ergonomics and human factors.* New York: Taylor & Francis.

———, John, B. E., & Atwood, M. E. (1993). Project Ernestine: A validation of GOMS for prediction and explanation of real-world task performance. *Human-Computer Interaction, 8,* 237–309.

Hick, W. E. (1952). On the rate of gain of information. *Quarterly Journal of Experimental Psychology, 4,* 11–26.

Hutchins, E. (1995). *Cognition in the wild.* Cambridge, MA: MIT Press.

Hyman, R. (1953). Stimulus information as a determinant of reaction time. *Journal of Experimental Psychology, 45,* 188–196.

John, B. E., & Kieras, D. E. (1996). The GOMS family of user interface analysis techniques: Comparison and contrast. *ACM Transactions on Computer-Human Interaction, 3,* 320–351.

Just, M. A., Carpenter, P. A., & Varma, S. (1999). Computational modeling of high-level cognition and brain function. *Human Brain Mapping, 8,* 128–136.

Kieras, D. E., & Meyer, D. E. (1997). An overview of the EPIC architecture for cognition and performance with application to human-computer interaction. *Human-Computer Interaction, 12,* 391–438.

Kruschke, J. K. (1992). ALCOVE: An exemplar-based connectionist model of category learning. *Psychological Review, 99,* 22–44.

Newell, A. (1973). You can't play 20 questions with nature and win: Projective comments on the papers of this symposium. In W. G. Chase (Ed.), *Visual information processing* (pp. 283–308). New York: Academic Press.

Newell, A. (1990). *Unified theories of cognition.* Cambridge, MA: Harvard University Press.

Pashler, H. (1994). Dual-task interference in simple tasks: Data and theory. *Psychological Bulletin, 116*(2), 220–244.

Sellen, A. J., & Harper, R. H. R. (2001). *The myth of the paperless office.* Cambridge, MA: MIT Press.

Shiffrin, R. M., & Schneider, W. (1977). Controlled and automatic human information processing: II. Perceptual learning, automatic attending and a general theory. *Psychological Review, 84,* 127–190.

Treisman, A. M., & Gelade, G. (1980). A feature-integration theory of attention. *Cognitive Psychology, 12,* 97–136.

Author Index

Aasman, J., 365
Abotel, K., 314
Abrahamsen, A., 401
Abrams, R. A., 7, 302, 322
Abrett, G. A., 40
Adams, D., 87
Adams, M. J., 40
Ahissar, M., 97, 101
Alimi, A. M., 301
Allais, M., 299
Allender, L., 35
Allmana, J., 231
Allport, A., 386
Alm, H., 362
Altman, E. M., 432
Altman, J. W., 35
Altmann, E. M., 51, 159, 382, 384, 385
Alvarez, G., 108
Anderson, E., 106
Anderson, G., 21
Anderson, J. R., v, 5, 7–8, 10n.2, 15, 17, 38–39, 49,
    51, 53, 54, 64, 65, 67, 77, 86, 88, 148, 151, 152,
    158, 166, 168, 173, 199, 207, 213, 231, 246, 255,
    256, 322, 356, 357, 363, 368, 371–73, 376–77,
    381, 385, 388–91, 401, 403–5, 410, 434, 437
Anderson, L. K., 100
Anderson, M. C., 386
Andersson, P., 154
Antoniou, A. A., 214
Arani, T., 109
Arbib, M., 263
Archer, S., 35
Ariely, D., 101, 137
Armstrong, K. M., 107
Arnauld, A., 299
Arnell, K. M., 9, 108

Arnold, M., 231
Ashcraft, M. H., 258
Atallah, H. E., 435
Attneave, F., 299, 300
Atwood, M. E., 437
Austin, J., 67
Austin, J. T., 388
Avraamides, M., 255, 274

Baars, B. J., 80
Backstrom, A., 293
Bacon, W. F., 105
Baddeley, A. D., 50, 74n.1, 331, 434
Badler, N. I., 237
Baillargeo, N., 77
Baldassi, S., 106
Baldwin, D. A., 388
Ball, J. T., 17, 20, 22
Ballard, D. H., 8, 122, 145, 287–88, 293, 391, 396,
    397
Ballas, J., 342, 353, 363, 381
Banks, M. S., 300, 307
Bar, M., 181
Bargh, J. A., 389, 391
Barnard, P. J., 435
Barnes, G. H., 33
Baron, S., 30, 34, 41, 196
Barron, G., 307, 311
Barto, A. G., 8, 71, 166
Bartroff, L., 108
Bates, J., 65, 231
Bauer, B., 102
Beattie, J., 215
Bechara, A., 231
Bechtel, W., 401
Bekesy, G. von, 38

Belavkin, R. V., 247, 248, 256
Belky, E. J., 106, 121
Bell, B. G., 391
Bell, D. E., 299
Bellenkes, A. H., 14
Benaglio, I., 107
Bennett, S. C., 102, 104, 108
Bensinger, D., 293
Beregovaia, E., 365
Berglan, L. R., 108
Bernoulli, D., 299
Berretty, P. M., 156, 157
Bertera, J. H., 121, 123, 319
Bettman, J. R., 153, 157, 216
Beusmans, J. M. H., 357
Beutter, B., 108
Bibi, U., 385
Bichot, N. P., 102, 106, 125, 137
Bickmore, T., 237
Biederman, I., 181
Biele, G., 154
Billari, F. C., 157
Birdsall, T. G., 410
Birmingham, H. P., 31
Birnbaum, M. H., 299
Birnkrant, R. S., 100
Bisantz, A., 205
Bjork, E. L., 158
Bjork, R. A., 158
Blais, A.-R., 311
Blaney, P. H., 266
Blascovich, J., 256
Blaser, E., 106
Bless, H., 231
Blessing, S. B., 53, 391
Blythe, J., 233
Blythe, P. W., 156
Bobrow, D. G., 406, 411
Bock, K., 181
Bodner, G. E., 181
Boehm-Davs, D. A., 4, 9, 397, 412
Boer, E. R., 15, 17, 357
Bohner, G., 231
Boies, S. J., 385
Bonnet, M. H., 244
Boot, W. R., 139
Booth, K. S., 145
Borges, B., 154
Bothell, D., 158, 356
Botvinick, M. M., 187
Bovair, S., 336
Bower, G. H., 39, 231, 266, 327
Boyd, M., 155
Boyer, M., 63, 67

Boyton, G. M., 52
Braver, T. S., 182
Brawn, P., 107
Brendl, C. M., 216, 224
Bridgeman, B., 138
Broadbent, D. E., 38, 39, 414, 437
Broadbent, M., 437
Bröder, A., 161, 162
Brogan, D., 121
Brooks, R. A., vii, 4, 200, 206, 286
Broughton, R. J., 244
Brown, D. L., 244
Brown, G. G., 244
Brown, P., 188
Bruner, J., 67
Brunswik, E., 165, 205
Bryson, J. J., 287
Buchner, A., 393
Buck, R., 215
Buckner, R. L., 52
Bucy, R. S., 34
Bugajska, M., 79
Bullemer, P., 391
Bullock, S., 152
Burgess, P. W., 421
Burr, B. J., 7
Burr, D. C., 106
Busemeyer, J. R., 67, 211, 213, 214, 217, 220, 221, 227n.3
Buss, R., 69
Butcher, S. J., 114
Byrne, M. D., 38, 51, 194, 207, 208, 315, 322, 356, 359, 368, 373, 377, 405, 410, 431–33, 440

Cadinu, M., 258
Caldwell, J. A., 244
Caldwell, J. L., 244
Callaway, C. B., 237
Camerer, C. F., 307
Cameron, E. L., 106
Campbell, C. S., 3
Campbell, D. T., 3
Campbell, L., 237
Canamero, D., 263
Canamero, L., 263
Cappa, S., 107
Card, S. K., 7, 17, 38–40, 139, 169, 172, 174, 336, 371, 376, 378, 406, 407, 433, 434, 439
Carlson, R. A., 389–97
Carpenter, P. A., 53, 435
Carpenter, R. H. S., 147n.6
Carr, K., 121
Carrasco, M., 106, 108

Carter, C. S., 51, 53, 363
Carver, C. S., 221, 396
Cary, M., 396
Cassell, J., 237
Cassenti, D. N., 392–94, 397
Cassimatis, N. L., 78, 79, 83, 84
Castelhano, M. S., 101
Cavanagh, P., 134
Cave, K. R., 100, 102, 120, 134, 137
Ceballos, R., 256
Chaiken, S., 63
Change, I., 108
Charman, S. C., 411
Chartrand, T. L., 389, 391
Chase, W. G., 77
Chater, N., 166, 371, 404
Chavez, A. K., 361, 362
Chee, M. W. L., 244, 247
Chi, E. H., 176
Chipman, S. F., 17, 40
Cho, R. Y., 182
Chomsky, N., 417
Chong, R. S., 255, 368, 438
Chong, S. C., 101
Choo, W. C., 244, 247
Chrisley, R., 264
Chubb, G. P., 37
Chun, M. M., 100–101, 108, 112, 138, 181
Chung, P. H., 440
Churchill, E., 237
Clancey, W. J., 145
Clark, A., 67, 279, 283, 368
Clark, A. J., 145, 146, 147n.5
Clark, J. J., 107, 135
Cleermans, A., 63, 67
Clore, G. L., 265, 266
Cohen, J. D., 182, 187
Cohen, M. M., 3
Cohen, M. S., 52
Colagrosso, M. D., 182
Coletti, S. F., 238
Collins, A., 73
Collyer, C. E., 391
Colombo, L., 188
Coltheart, M., 180, 188
Confer, H. A., 14, 18
Conway, A. R. A., 386
Cook, T. D., 3
Cooley, J. W., 32
Coombs, B., 299
Cooper, R. P., 389, 414, 415, 417, 421, 424, 425,
    426n.2
Councill, I. G., 255, 274
Cowan, N., 109

Cowan, W. B., 102
Cramer, N. L., 40, 41
Csikszentmihalyi, M., 200
Czerlinski, J., 155

Daily, L. Z., 257, 362, 376
Dale, A. M., 52
Dallas, M., 159
Damasio, A. R., 211, 215, 231
Damasio, H., 231
Daneman, M., 138
Darlington, K. K., 244
Das, S., 421, 424
Davidson, B. J., 135
Davidson, R. J., 264, 265
Davis, J. N., 152
Dayan, P., 288
De Jong, R., 381
Delaney, P. F., 331
Denhiére, G., 7
Dennett, D. C., 389, 398n.2
Derr, M., 7
Desimone, R., 102
Destrebecqz, A., 63, 67
Deubel, H., 107
Deutsch, S. E., 40
de Vries, N. K., 215
Dewey, J., 198, 203, 209n.2
Diederich, A., 217, 220, 227n.3
Diehl, R. L., 411
DiLollo, V., 101, 134, 135
DiMase, J. S., 100, 104, 106
Dinges, D. F., 243–46
Doane, S. M., 27n.1, 87, 89–91, 94
Dodd, M. D., 385
Domangue, T., 70
Donald, M., 198
Donges, E., 357
Doran, S. M., 246, 250
Dorrian, J., 243, 245, 246
Dosher, B. A., 103, 106
Douglass, S., 51
Doyle, J., 77
Dreary, I. J., 265
Dreyfus, H., 368
Driver, J., 107, 341, 381
Drummond, S. P. A., 244
Drury, C. G., 109
Dudey, T., 157
Duket, S., 37
Dumais, S. T., 203
Duncan, J., 101, 108, 437
Dunning, D., 223
Durso, F. T., 91

Eckstein, M. P., 106, 108
Edelman, G., 65
Edman, J., 154
Efron, B., 311
Egeth, H. E., 100, 102, 105, 108, 136, 137, 389
Egeth, H. W., 99
Ekman, M., 154
Ekman, P., 232, 264–65
Elkind, J. I., 30, 32, 40
Ellsworth, P. C., 240, 265
Engel, S. A., 52
Engle, R. W., 386
English, W. K., 7
Enns, J. T., 101, 104, 134, 137, 146n.2
Epstein, S., 215
Erev, I., 161, 307, 311
Ericsson, K. A., 21, 331
Erkelens, C. J., 121
Ernst, A., 71
Erwin, B. J., 231
Evert, D. L., 108

Fajen, B. R., 196
Fallah, M., 107
Faloutsos, P., 283
Farah, M., 187
Farrell, M., 107
Farrell, P. S. E., 35, 38
Fedorikhin, A., 216, 223
Feehrer, C. E., 40
Feigenbaum, E. A., 38
Feldman Barrett, L. F., 259
Fellous, J.-M., 263, 264, 266, 274, 276
Fellows, L., 187
Fencsik, D., 339
Feng, D., 236
Ferguson, T. S., 157
Fick, C. S., 440
Fiddick, L., 153
Fiedler, K., 165
Fikes, R. E., 378
Fildes, B. N., 357
Fillmore, C. J., 391
Fincham, J. M., 363
Findlay, J. M., 107, 121, 123, 315, 316, 333
Firby, R. J., 287
Fischer, U., 275
Fischhoff, B., 299
Fisher, B. D., 145
Fisher, D. L., 356
Fitts, P. M., 7, 31, 39, 302, 436
FitzGerald, P. J., 237
Flach, J. M., 30, 34, 35, 203
Flanagan, J. R., 293

Flavell, J. H., 72, 396
Fleetwood, M. D., 315, 322
Fleischman, M., 237
Foa, E. B., 266
Folk, C. L., 105, 389
Folkman, S., 259
Forgas, J. P., 263
Forlizzi, L., 392
Foster, D. H., 104
Fox, J., 414, 415, 417, 421, 424
Foyle, D. C., 194, 207
Franconeri, S. L., 97
Frank, M. J., 435
Frank, R., 230
Franzel, S. L., 100, 102, 120, 134, 137
Frazier, L., 77
Frederick, S., 215
Freed, M., 314
Frensch, P. A., 391
Friedman-Hill, S. R., 102
Frijda, N., 231
Fu, W.-T., 4, 166, 168, 169, 171–74, 178n.4, 194, 368, 374, 376, 410, 412

Gale, A., 121
Gallistel, C. R., 392
Gancarz, G., 100
Garbart, H., 100, 102
Gardner, M. K., 391
Gasper, K., 266
Geisler, W. S., 107, 410, 411
Gelade, G., 80, 99, 107, 437
Gelman, R., 392
Genter, D., 73
Gepshtein, S., 300, 307
Gibbs, B., 99
Gibson, J. J., 199, 200
Gigerenzer, G., 152–55, 158, 161, 197, 207, 311
Gigley, H. M., 40
Gilbert, J. A., 244
Gilbert, S. J., 381
Gilchrist, I. D., 107, 315, 316, 333
Gillin, J. C., 244
Glass, A. L., 181
Glass, J. M. B., 339, 361
Glasspool, D. W., 421, 424
Gleick, J., 198
Glenberg, A. M., 40
Glenn, F., 38
Glover, G. H., 52, 59, 60
Gluck, K. A., 17, 20–22, 194
Gnanadesikan, R., 302
Godijn, R., 106, 108
Godthelp, H., 357

Goldberg, J. H., 216, 223
Goldstein, D. G., 154–55, 158, 311
Gomez, P., 108
Gonsalkorale, K., 385
Gonzalez, R., 300
Goodale, M. A., 122, 138, 145
Goode, A., 51
Goodie, A. S., 152
Goodnow, J., 67
Gordon, S. E., 256, 261
Gormican, S., 99, 135
Graham, D., 33
Graham, I., 417
Gratch, J., 232, 236, 237, 240, 243, 256, 259
Gray, J. A., 215
Gray, R., 8, 15, 34, 357–59
Gray, W. D., 3–4, 8–10, 15, 17, 159, 169,
        171–73, 178n.4, 194, 195, 215, 216, 260,
        363, 368, 374, 376, 382, 397, 410, 412,
        431–33, 437
Green, B. F., 100
Green, C., 386
Green, D. M., 34
Green, P., 41
Green, P. A., 362
Greeno, J. G., 204
Griffin, Z. M., 181
Gronlund, S. D., 91
Grossberg, S., 219
Grosz, B., 233
Guidry, C., 70
Guttmann, H. E., 36

Habeck, C., 244, 247
Haider, H., 391
Hakeema, A., 231
Halverson, T., 317, 318, 321, 322
Hammond, K. R., 205, 215
Han, S., 103
Hancock, P. A., 258–61
Hansen, M. B., 77
Harnad, S., 82
Harper, R. H. R., 433
Harris, J. M., 357
Harrison, Y., 250
Hartley, R., 286
Haviland, J. M., 265
Haviland-Jones, J. M., 214
Hayes, B. E., 258
Hayhoe, M. M., 8, 122, 145, 284, 287, 293, 391
Hazeltine, E., 373
Headley, D., 35
Healy, A. F., 393
Heeger, D. J., 52

Hellhammer, D. H., 256
Henderson, J. M., 101, 137
Hendry, D., 138
Hendy, D. B., 35, 38
Hertwig, R., 153, 157–60, 311
Hick, W. E., 436
Hildreth, E. C., 357, 359
Hilgard, E. R., 327
Hill, R. W., 236
Hillman, P., 100
Hillstrom, A. P., 114
Ho, K., 215
Hochberg, J. E., 40, 138
Hochstein, S., 97, 100–101
Hockley, W. E., 108
Hodgetts, H. M., 385
Hofd, P., 231
Hoffrage, U., 156–58
Holcombe, A. O., 114
Hollands, J. G., 257
Holt, C. A., 307
Holte, R. C., 157
Holyoak, K. J., 87
Hommel, B., 386, 389
Hooey, B. L., 207
Hooey, R., 194, 207
Hooge, I. T., 121
Horne, J. A., 250
Hornof, A. J., 314, 315, 317–18, 320–22
Horowitz, T. S., 100, 101, 104, 108–10, 112
Horwood, J., 357, 358
Houtmans, M. J. M., 106
Hovy, E. H., 236, 237
Howes, A., 406, 408, 411
Hsee, C. K., 215, 224
Huang, L., 114
Huber, D. E., 182
Hudlicka, E., 243, 263, 266, 269, 270, 274, 275
Huettel, S., 59
Huey, B. M., 40
Hull, C. L., 38, 64, 72, 213
Hummel, J. E., 107
Humphreys, G. W., 100, 101
Humphrys, M., 290
Hutchins, E., 198, 368, 431
Hyle, M., 114
Hyman, R., 436

Irmen, L., 393
Irwin, D. E., 109
Isen, A. M., 266
Itti, L., 105, 135, 136, 239, 293
Ivry, R. B., 373
Iwase, H., 301

Jacoby, L. L., 159
Jagacinski, R. J., 30, 34, 35, 199, 203
Jentzsch, I., 181
Jewett, J. E., 244
Jiang, Y., 100, 108, 138, 181
Jodlowski, M., 94
Joffe, K., 121
Johannsen, G., 194
Johansson, R., 293
John, B. E., 314, 336, 437, 439
Johns, E. E., 181
Johnson, E. J., 153, 157, 307
Johnson, J. G., 217
Johnson, W. L., 237
Johnstone, T., 265
Jolicoeur, P., 102
Jones, A. D., 181, 182, 187
Jones, D. M., 385
Jones, G., 255
Jones, R. M., 16, 255
Jongman, G. M. G., 255
Jongman, L., 247, 248
Jonides, J., 334
Joordens, S., 138
Jose, P. E., 214
Julesz, B., 105, 133
Juslin, P., 165
Just, M. A., 53, 435

Kahn, R. E., 287
Kahneman, D., 99, 152, 215, 299, 307
Kalman, R. E., 34
Karlsson, J., 290
Karmiloff-Smith, A., 64, 67, 69
Karwan, M. H., 109
Kastrup, A., 59
Katz, S. M., 108
Kazén, M., 396
Keefe, D. E., 258
Keele, S. W., 386
Keeney, R. L., 220
Kello, C. T., 188
Kelsey, R. M., 256
Kemmelmeier, M., 231
Kenner, N. M., 100
Kestenbaum, R., 77
Khayat, P. S., 292
Kieras, D. E., 6, 38, 39, 213, 256, 314–16, 318,
    322, 331, 334–36, 339, 342–44, 352–53,
    361, 363, 365, 368, 381, 404, 405, 409–11,
    434, 439
Kiesner, J., 258
Kiger, J. I., 181
Kim, W., 3

Kim, Y., 236
Kinoshita, S., 180, 188, 191
Kinsbourne, M., 398n.2
Kintsch, W., 86–88
Kirby, L. D., 265, 268
Kirlik, A., 194, 199, 202–8
Kirschbaum, C., 256
Kirschenbaum, S. S., 17, 195
Kirsh, D., 204
Kitajima, M., 87
Klein, L. C., v, 199, 243, 256, 260
Klein, R. M., 107, 109
Kleinman, D. L., 30, 34
Kleitman, N., 244
Koch, C., 105, 239, 293
Koch, I., 395
Kornblum, S., 7, 302, 322
Kotler-Cope, S., 69
Kowler, E., 106
Kraft, N., 275
Kramer, A. F., 14, 105, 109, 181, 188
Kraus, S., 233
Krauss, S., 153
Krebs, J. R., 169
Krendel, E. S., 30, 33
Kribbs, N. B., 245
Kristjansson, A., 109
Kronauer, R. E., 244
Krüger, G., 59
Kruschke, J. K., 437
Krusmark, M. A., 20, 22
Kuhl, J., 396
Kunar, M. A., 100
Kushleyeva, Y., 342, 363

LaBore, C., 237
Laham, D., 7
Laird, J. E., 77, 86, 94, 255, 368, 438
Lamme, V., 287
Lamy, D., 105
Lancraft, R., 41
Land, M., 357, 358
Land, M. F., 97, 293
Landauer, T. K., 7
Landy, M. S., 288, 300, 307, 410, 411
Lane, N., 38
Larkin, J. H., 368, 369
Lauber, E. J., 339, 342, 363, 381
Laughery, R., 37
Laury, S. K., 307
Lavie, N., 386
Layman, M., 299
Lazarus, R. S., 211, 214, 231–32, 259
Leber, A. B., 389

Lebiere, C., 10n.2, 38–40, 49, 64, 65, 67, 77, 151, 152, 158, 173, 199, 207, 213, 356, 389, 404, 405, 437
LeDoux, J. E., 215
Lee, C., 114
Lee, F. J., 15, 17, 342, 361–63, 376, 377
Lee, G., 231
Lee, M. A. N., 244
Leitten, C. L., 256
Lemaire, B., 7
Lemay, G., 232
Lerner, J. S., 215–16
Lester, J. C., 237
Levelt, W. J. M., 188
Levenson, R. W., 215
Leventhal, H., 268
Levin, D. T., 107, 141
Levison, W. H., 41
Levy, G. W., 37
Lewis, M., 214, 265
Lewis, R. L., 404, 406
Li, Z., 105
Libet, B., 397
Lichtenstein, S., 299, 300
Liddell, B. J., 145
Lin, J. S., 244
Liu, A., 15, 17
Liu, Y., 41, 256, 261, 365
Lloyd, M. R., 357
Lockhead, G. R., 181
Loeber, R., 284
Loewenstein, G., 215–16, 223
Logan, G. D., 381, 383, 384
Logie, R. H., 331
Lovett, M. C., 213, 257, 260, 362, 376, 404
Loyall, A., 65, 231
Lu, Z. L., 103
Luce, M. F., 216
Luce, R. D., 111
Luck, S. J., 100, 102, 107, 108, 294
Lundy, D. H., 391
Lupker, S. J., 180, 181, 188, 190
Lyon, D. R., 14, 18

Maass, A., 258
Mack, A., 138
Mackinlay, J. D., 139
MacLeod, C. M., 80, 385
Macuga, K. L., 34
Maioli, C., 107
Maljkovic, V., 114, 181
Malle, B. F., 388
Mallis, M., 244
Maloney, L. T., 181, 288, 300, 307, 410, 411

Mandler, G., 388, 392, 396
Mannes, S. M., 86–87
Manning, C. D., 174
Mao, W., 240
Markley, D., 365
Markman, A. B., 216, 224
Markman, E. M., 77
Marr, D., 133, 134, 286
Marsella, S., 232, 236–37, 243, 256, 259
Martignon, L., 156–57
Martin, E. L., 14, 18
Martin-Emerson, R., 344
Maslow, A. H., 65, 72, 214
Massaro, D. W., 3
Masson, M. E., 181
Matessa, M., 158, 314, 356, 437
Matheson, C., 237
Mathews, R., 69, 70
Matthews, G., 265, 266
Maurier, D., 440
Mavor, A. S., v, 38, 266
Mayr, U., 386
Mazzoni, G., 72
McCarl, R., 86
McCarley, J. S., 109
McCarthy, G., 59
McClelland, J. L., 66, 183, 219, 407, 419
McCurdy, M., 406
McElree, B., 5, 9, 106
McGill, W. J., 206
Mckersie, R. B., 238
McKoon, G., 108, 181, 183
McLean, J., 103, 106
McLeod, P., 97
McMains, S. A., 102, 137
McNamara, D., 87
McRuer, D. T., 30, 33, 35
Meiran, N., 381
Mejdal, S., 244
Melcher, D., 102
Mele, A. R., 231, 240
Mellers, B. A., 211, 215, 266
Mennie, N., 293
Merikle, P. M., 138
Merleau-Ponty, M., 283
Metcalfe, J., 73, 215, 396
Metzinger, T., 389, 397
Mewhort, D. J. K., 181
Meyer, A. S., 188
Meyer, D. E., 5, 7, 38, 39, 213, 256, 302, 315, 316, 322, 331, 334, 335, 339, 342, 343, 353, 361, 363, 368, 381, 404, 405, 409–11, 434
Michod, K. O., 110, 112
Miller, D. P., 35

Miller, G., 157, 294
Miller, R. A., 199
Milner, A. D., 122, 138, 145
Milson, R., 166, 404
Mineka, S., 266
Minsky, M., 5
Mischel, W., 215
Mitsudo, H., 100
Miyake, A., 389
Monk, A., 198
Monsell, S., 181, 341, 381, 384
Moore, C. M., 103, 108
Moore, T., 107
Moraglia, G., 104
Moran, T. P., 17, 38, 39, 336, 371, 376, 378, 406, 407, 433, 434, 439
Moray, N., 107
Morton, A., 161
Moseley, M. E., 59
Moses, L. J., 388
Motter, B. C., 106, 121
Mozer, M. C., 182, 184, 186, 188, 191
Mueller, S. T., 331, 361
Muralidharan, R., 41
Murata, A., 301
Murdock, B. B., Jr., 108
Muter, P., 108
Myung, I. J., 3, 67

Nairne, J. S., 393
Najemnik, J., 107
Nakayam, K., 114, 181
Nakayama, M., 81, 100
Neisser, U., 38, 39, 99, 197, 390
Nellen, S., 161
Nelson, T., 72
Nelson, T. O., 396
Nerb, J., 71, 72
Neth, H., 410
Neumann, O., 135
Neumann-Haefelin, T., 59
Nevarez, G., 139
Neville, K., 255
Newell, A., v, v, vii, 4, 5, 9, 17, 38, 39, 77, 86, 87, 94, 151, 158, 167, 177, 199, 200, 213, 243, 255, 285, 314, 315, 321, 336, 356, 371, 376, 378, 401, 405–7, 414, 424, 432–34, 439
Nguyen, T., 244
Nicole, P., 299
Nilsson, L., 362
Nilsson, N., 285
Nilsson, N. J., 378
Nimchinskyc, E., 231
Nissen, M. J., 391

Nöe, A., 283
Norling, E., 256
Norman, D., 264
Norman, D. A., 264, 327, 333, 342, 389, 390, 392, 397, 406, 411, 421
Norretranders, T., 138, 143, 145
Nothdruft, H. C., 104
Noy, I., 41
Nystrom, L. E., 182

Oaksford, M., 166, 404
O'Connell, K. M., 102
O'Donoghue, T., 215
O'Hara, K. P., 410
Oliva, A., 101, 122, 137
Olzak, L. A., 104
O'Neill, P. E., 108
Orasanu, J., 275
O'Regan, J. K., 107, 135, 283
O'Reilly, R. C., 435
Ortmann, A., 152, 154
Ortony, A., 235, 264–66, 276
Otto, P. E., 161, 162

Pachur, T., 154
Paelke, G., 41
Palmer, E. M., 112
Palmer, J., 101, 103, 106
Palmer, S. E., 133–35, 147n.6
Panksepp, J., 215
Papathomas, T. V., 102
Parasuraman, B., 261
Parry, D., 238
Pashler, H. E., 3, 50, 100, 114, 295, 410, 437
Patrick, G. T. W., 244
Pavel, M., 106
Payne, D., 35
Payne, J. W., 153, 156, 157, 161, 216
Payne, S. J., 410
Peacock, E., 232
Pearl, J., 78
Pelachaud, C., 237
Pelisson, D., 138
Pelz, J. B., 8, 284
Pentland, A., 283
Persons, J. B., 266
Peters, E., 215
Peterson, J. R., 302
Peterson, M. S., 109
Pew, R. W., v, 38, 40, 194, 266
Phillips, C. B., 237
Piattelli-Palmarini, M., 152
Picard, R., 263
Pipitone, F., 286

Pirke, K.-M., 256
Pirolli, P., 169, 172, 174
Pitt, M. A., 3
Plamondon, R., 301
Plaut, D. C., 188
Po, B. A., 145
Poesio, M., 237
Poggi, I., 237
Polanyi, M., 199, 203
Pollack, M., 233, 240
Pollatsek, A., 356
Polson, P. G., 87, 336
Pomplun, M., 106, 121–23, 125, 126, 130, 315
Pook, P. K., 122, 145, 287, 391
Portas, C. M., 244, 247, 248
Posner, M. I., 135, 181, 385
Postma, A., 392
Potter, M. C., 101
Poulton, E. C., 30
Powell, J. W., 245
Prablanc, C., 138
Pratt, J., 108
Prevost, S., 237
Pritsker, A. A. B., 37
Prokopowicz, P. N., 287
Purtee, M. D., 20
Puterman, M. L., 167
Pylyshyn, Z., 135–37

Qin, Y., 51, 53, 60n.2
Quigley, K. S., v, 199, 243, 256, 259
Quinlan, P. T., 100

Raab, M., 207
Raiffa, H., 220, 299
Rajyaguru, S., 194
Rao, R. P. N., 122, 130, 145, 391
Rapoport, A., 157, 168
Rastle, K., 188
Ratcliff, R., 103, 108, 111, 181, 183, 192n.4
Rauschenberger, R., 105
Raymond, J. E., 9, 108
Rayner, K., 77, 121, 123, 319, 356
Read, D., 217
Reason, J. T., 392, 396, 397
Reber, A., 63, 66, 67
Redelmeier, D. A., 307
Reder, L. M., 213, 257, 362, 376, 405
Reed, M. P., 362
Rees, G., 135
Reichle, E. D., 356
Reifers, A. L., v, 199, 243, 256, 260
Reilly, W. S. N., 65, 231
Reinecke, A., 107

Reingold, E. M., 106, 121, 122, 315
Reisener, W. Jr., 33
Remington, R., 314
Rensink, R. A., 101, 104, 107, 133–42, 144, 145, 146n.2, 147n.3, 147n.4
Revelle, W., 264, 265
Rickel, J., 231, 236–37
Rieskamp, J., 156, 161, 162
Ritov, I., 211, 215
Ritter, F. E., v, 194, 199, 243, 255–56, 258–60, 274
Ritter, H., 123
Roberts, S., 3, 410
Rock, I., 138
Rodgers, J. L., 3
Rodgers, S. M., 20, 22
Rodriguez, T., 3
Roe, R. M., 217, 219
Roelfsema, P. R., 287, 292
Roelofs, A., 188
Rogers, M. A., 238
Rogers, N. L., 243
Rogers, R. D., 181, 381
Rosabianca, A., 258
Roseman, I. J., 214
Rosenbaum, D. A., 391
Rosenbloom, P. A., 39, 86, 94
Rosenbloom, P. S., 77
Rosenholtz, R., 101, 102
Rossi, A. F., 102
Roth, A., 161
Rothrock, L., 205
Rottenstreich, Y., 215, 224
Rouse, W. B., 194
Roussel, L., 70
Rowe, D. C., 3
Roy, D., 283
Royden, C. S., 357
Rumelhart, D., 66, 219
Runeson, S., 197
Rushton, S. K., 357
Russell, J. A., 232
Russell, L., 30, 32
Russo, J. E., 322
Rusted, J., 293
Ryder, J., 38

Salvucci, D. D., 8, 15, 17, 34, 41, 257, 315, 342, 356–65, 370
Salzman, M. C., 3
Sanders, A. F., 106
Santarelli, T., 38
Sarason, B. R., 258
Sarason, I. G., 258
Sato, S., 100

Sato, T. R., 125
Schachter, S., 214
Schall, J. D., 106, 107
Scheier, M. F., 221, 396
Scherer, K. R., 215, 231, 240, 265, 268
Scheutz, M., 264
Schmalhofer, F., 87
Schmidt, H., 107
Schneider, W., 389, 391, 421, 437
Schneider, W. X., 107
Schoelles, M. J., v, 4, 15, 17, 194, 199, 243, 256, 260, 363
Schooler, J. W., 392, 396
Schooler, L. J., 7, 8, 148, 158–60, 166
Schorr, A., 265
Schraagen, J. M. C., 17
Schreiber, B. T., 14, 18
Schuetze, H., 174
Schultz, A., 79
Schultz, W., 288
Schumacher, E. H., 339, 373, 408, 409
Schwartz, A., 211, 215, 307
Schwartz, M. F., 426n.2
Schwarz, N., 231
Schyns, P. G., 122
Scialfa, C. T., 121
Scolaro, D., 38
Seale, D. A., 157, 168
Searle, J. R., 389
Seashore, R. H., 30
Seger, C., 63, 66, 67
Seifert, D. J., 37
Sellen, A. J., 433
Selten, R., 153
Serafin, C., 41
Seum, C. S., 37
Seymour, T. L., 331, 339, 361
Shafir, S., 311
Shah, K., 194
Shah, P., 389
Shalin, V. L., 17
Shallice, T., 342, 381, 389, 414, 417, 421, 426n.2
Shapiro, K. L., 9, 101, 102, 108
Sharpe, S. H., 144
Shaver, K. G., 233
Shaw, J. C., v
Sheard, E. D., 385
Shearin, E. N., 258
Sheinberg, D., 107
Shen, J., 106, 121, 122, 125, 315
Sheridan, T. B., 194, 199
Shettel, M., 191
Shiffrin, R. M., 203, 389, 391, 421, 437
Shimamura, A. P., 396

Shimizu, Y., 81
Shisler, R. J., 386
Shiv, B., 216, 223
Shneiderman, B., 139
Shore, D. I., 109
Siegel, A. I., 36, 37
Silk, E., 53
Sillars, A. L., 238
Silverman, G. H., 100
Simão, J., 157
Simon, H. A., v, 9, 21, 38, 64, 65, 71, 72, 77, 152, 157, 166, 167, 169, 177, 200, 204, 224, 230, 231, 322, 368, 401, 403, 404, 411
Simons, D., 77
Simons, D. J., 97, 107, 139, 147n.4
Simpson, K., 258
Sims, C. R., 4, 410
Singer, J., 214
Siri, S., 107
Skinner, B. F., 213
Sloman, A., 64, 65, 72, 264, 268, 276
Sloman, S. A., 215
Slovic, P., 152, 215, 299
Slusarz, P., 63, 69
Small, D. A., 216
Smith, C. A., 232, 265, 268
Smith, J. E. K., 7, 302
Smith, J. K., 244
Smith, R. E., 392, 396
Smith, S. L., 100
Snyder, C. R., 135
Sober, E., 371
Sohn, M.-H., 51, 381, 390
Sohn, Y. W., 27n.1, 87, 89–91, 94
Somers, D. C., 102, 137
Sommer, W., 181
Sosta, K., 107
Souther, J., 102
Spada, H., 71
Spekreijse, H., 287, 292
Spelke, E. S., 77
Spence, K. W., 213
Sprague, N., 288, 293
St. Amant, R., 194, 258
Stallings, W., 341
Stanikiewicz, B. J. V. C., 107
Stanley, W., 69
Stanovich, K. E., 215
Stark, L., 138
Steedman, M., 237
Steffens, M. C., 393
Stein, L. A., 287
Stenger, V. A., 51, 53, 363
Stephens, D. W., 169

Sternberg, S., 99
Stevenson, L. M., 391, 396
Stewart, M. I., 102
Stewart, N., 181
Stewart, T. R., 205
Stigler, G. J., 169
Stokes, J., 38
Stone, L., 108
Stout, J. C., 214
Strayer, D. L., 181, 188
Strieb, M., 38
Stroop, J. R., 80
Stuss, D. T., 425
Styles, E. A., 100
Suchman, L. A., 203
Sullivan, J., 237
Sun, R., 63–70, 72–74
Suri, R. E., 288
Sutton, R. S., 8, 71, 166
Swain, A. D., 35, 36
Swain, M. J., 287
Swartout, W., 235, 236
Swensson, R. G., 123
Swets, J. A., 34, 410

Taatgen, N. A., 51, 54, 368, 373, 376–77, 412
Tai, J. C., 106
Takahashi, K., 81
Tanner, W. P., Jr., 410
Tatler, B. W., 138
Taylor, F. V., 31
Taylor, T. E., 180, 181, 188, 190
Teague, D., 373
Terry, C., 63
Terzopoulos, D., 283
Tetlock, P. E., 216
Thagard, P., 87
Thaler, R. H., 307
Thalmann, D., 237
Theeuwes, J., 105, 106, 108
Thomas, J. P., 104
Thompson, K. G., 106, 107, 125
Thornton, T., 103
Tibshirani, R., 311
Toates, F. M., 64, 65, 70, 221
Todd, P. M., 152–53, 155–58, 161, 197
Todd, S., 105
Tolman, E. C. B., 205, 213
Tomaka, J., 256, 258
Tomasello, M., 65
Torralba, A., 101
Towns, S. G., 237
Townsend, J. T., 103, 211, 213, 214, 217, 227n.3
Trafton, J. G., 51, 79, 385

Traum, D. R., 236–37
Treisman, A. M., 80, 99–102, 107, 134–35, 437
Triggs, T. J., 357
Trommershäuser, J., 288, 300–302, 304–7, 309, 311, 410, 411
Trope, Y., 63
Tschaitschian, B., 87
Tsimhoni, O., 365
Tsotsos, J. K., 135
Tucker, A., 329, 341
Tuholski, S. W., 386
Tukey, J. W., 32
Turney, P., 7
Turvey, M. T., 196
Tustin, A., 30, 31
Tversky, A., 152, 157, 215, 299, 307
Tyrell, T., 71

Ullman, S., 105, 287
Usher, M., 183
Uttal, W. R., 51

Vallacher, R. R., 391
Van Boven, L., 223
Vancouver, J. B., 388
van de Panne, M., 283
van der Pligt, J., 215
Van Dongen, H. P. A., 243, 249
van Leeuwen, B., 217
VanLehn, K., 89, 94
Van Rooy, D., 258
van Veen, V., 363
Van Zandt, T., 3, 111
Varma, S., 53, 435
Vecera, S. P., 100, 107, 181, 191
Veksler, V. D., 410
Velichkovsky, B. M., 123
Vera, A. H., 314, 406–8, 411
Verghese, P., 106
Vidnyánszky, Z., 102
Vilhjálmsson, H., 237
Virzi, R. A., 100, 102
Viviani, P., 123
Voerman, J. L., 237
Vogel, E. K., 102, 294
von der Malsburg, C., 99, 294
von Economo, C., 244
von Hippel, W., 385
Vygotsky, L. S., 198

Walton, R. E., 238
Wang, D. L., 107
Wang, R. F., 109
Wann, J. P., 357

Ward, P. A., 104
Ward, R., 108
Ware, C., 139
Warm, J. S., 258–60
Warren, W. H., 200
Wasow, J. L., 77
Waszak, F., 386
Watkins, C., 67, 68, 168
Watkins, C. J. C. H., 288
Webber, B. L., 237
Weber, E. U., 311
Wegner, D. M., 385, 386, 391, 397, 398n.2
Wein, D., 77
Weiner, B., 65, 198, 214, 233
Weinstein, S., 259
Wellman, H. M., 396
Wen, C., 41
Wender, K. F., 393
Wenger, J. L., 394, 395, 397
Wenger, M. J., 103
Werner, G. M., 152
West, R. F., 215
Westling, G., 293
Wetzel, P. A., 21
Whalen, P. J., 145
Wherry, R., 35, 37, 38
Whetzel, C., 260
Wickens, C. D., 13, 14, 19, 41, 256–57, 261,
     344
Wickens, T. D., 382
Wilkie, R. M., 357
Williams, D. E., 121, 122

Williams, J. M. G., 266
Willingham, D. B., 391
Wills, G., 297
Wilson, D. E., 385
Wilson, T. D., 216
Wolf, J. J., 36, 37
Wolfe, J. M., 100–110, 112, 114, 120, 125, 134, 135,
     137, 140, 181, 315
Woltz, D. J., 391
Wong, P., 232
Wood, D. J., 255
Wood, S. D., 256, 314, 335
Wortman, D. R., 37
Wright, C. E., 7, 302
Wu, G., 300

Yan, H., 237
Yantis, S., 100, 105, 136, 137, 334
Yeshurun, Y., 108
Yeung, N., 187
Young, R. M., 404
Yule, P., 417, 426n.2

Zacharias, G., 41
Zachary, W., 35, 38
Zajonc, R. B., 215
Zeelenberg, M., 215
Zelinsky, G. J., 107, 121–23
Zhang, S. B., 3
Zhang, X., 69
Zohary, E., 100
Zuber, M., 365

Subject Index

action-centered subsystem (ACS), 66–70
action rule store (ARS), 68
actions, contiguity of, 71
action selection, 83, 268
action slips, Norman's analysis of, 392
action units, 36
activation, 382. *See also* area activation; rational
    activation theory
  proportional, 71
activation map, 105–6, 121
activation threshold, 383
activation–trigger–schema (ATS) model, 388, 392
ACT-R (adaptive control of thought–rational), vi, viii,
    ix, 401, 419, 434, 437–38, 440
  Anderson's work on, v
  architecture of, vi, 49–51, 194, 342, 372, 402, 414
    computational cognitive model developed in,
      243–51
    interconnections among modules in, 49–50
    modular organization of cognition and, 46, 434
    rationality underlying, 197
    Type 1 and 2 components of, 24
  "atomic" approach of, 389
  attention and, 258
  Bayesian networks and, 78, 84n.1
  BOLD response and, 55
  brain imaging and, 49–53, 61
    brain imaging data, 55, 57, 59
    model fitting, 57, 59–60
    use of brain imaging to provide converging data,
      51–52
  central controller of, 8
  CLARION and, 64
  COGENT and, 420
  cognitive control in pilots and, 15–18
  cognitive–metacognitive interaction and, 65

construction-integration and, 86–88, 94
construction-integration architecture and, 87
declarative and procedural knowledge in, 86, 87,
    372, 376, 434
EMMA and, 356
emotions and, 212, 434
EPIC and, 372
historical perspective on, 38, 39, 434
idle cycles in, 252n.2
Imprint and, 40
map-navigation task and, 173
memory activation in, 378
model fatigue in, 255
modeling the environment and, 199–200
modeling the recognition heuristic within, 151,
    152, 158–61
noise and, 251–52nn.1–2
perceptual-motor subsystems for, 437
problem state in, 376
production execution cycle in, 247, 390
production rules and, 5, 317, 375
rational activation theory and, 10n.2
rational analysis and, 404–5
reinforcement learning model and, 8
skill learning and, 376–77
SNIF-ACT and, 175
softmax method and, 168
stress and, 254, 256, 257, 259–61
task instructions given to, 54
task switching and, 8
time's influence on task, 260
timing assumptions of, 372
Type 1 control and, 5, 8
Type 2 theories and, 435
variabilization of goal chunks in, 397
visual attention module of, 8, 437

ACT-R 5, 246, 257
ACT-R 6, 260
ACT-R driver model, 8, 326, 357–62, 365
ACT-R general executive, 363–64
ACT-R Monte Carlo analysis, 207
ACT-R motor system, 360
ACT-R/PM, 368, 405, 408
ACT-R production system, 77
adaptation
    bounds on, 406
    local and global, 405–6
adaptive control of thought–rational. *See* ACT-R
adaptive toolbox, 153
    keeping it under control, 160–62
    pulling apart the, 153–57
    putting back together the, 157–60
adaptivity, 64
ADAPT model(s), viii, 86, 87, 94, 194
    and constructing individual knowledge bases, 89
    examples of knowledge in, 88
    execution of, 91
        plan selection, 89–91
    goal of, 89
    knowledge representation in, 88–89
    memory constraints in, 91
    simulation procedures in, 89–90
    testing and training individual, 91
    validation of, 92–94
affect, 214. *See also* emotion(s)
affect appraisal, 268–70, 275
affective evaluation, 220–22
affective states, 264, 268. *See also* emotional states
affordance, 199–204
aftereffect, 140
Air Force Manual 11-217, 19, 20, 24
Air Force Office of Scientific Research (AFOSR),
        vii, 15
air traffic control (ATC), 207–8
algebra, adult learning of artificial, 53–54
algebra model, Anderson (2005), 53
alternation of responses, 187
anterior cingulate cortex (ACC), 187
anti-air warfare coordinator (AAWC) system, 53
anxiety, 269–73
appraisal, 265
    automatic, 265
    black box models of, 265
    and coping, 232–33, 235
    deliberate, 265
    as design specification for cognition, 239–40
    process models of, 265
    threatening *vs.* challenging, 259
    as uniform control structure, 238–39
    as value computation, 239

appraisal frames, 234
appraisal theory, 238, 239, 240n.1. *See also* EMA
    computational, 231–32
    and design of virtual humans, 235–36. *See also*
        virtual humans
appraisal variables, 232
architectures of cognition approach, 5. *See also*
        cognitive architecture(s)
area activation, 126, 130
area activation model, 122, 129–30, 322–23
    original, 122–25
    toward a general model of eye movements in visual
        search, 122, 125–29
arithmetic problems, multistep
    task structure for, 394–95
arousal and attention, 266
artificial intelligence (AI), 166
aspiration level, 169
associative memory, 39, 69–70
associative memory networks (AMNs), 70
associative rules, 69
asynchronous diffusion model, 103, 108–9
attended items, buildup of, 140
attention, 266, 267. *See also* coherence theory;
        psychomotor vigilance task
    "capturing," 333–34
    eye movements and, 137
    meanings of, 100, 102
    objects of, 114
    stress and, 258
attentional blink (AB), 101–2, 108–9
attentional capture, 104
attentional dwell time, 107
attentional processing, 143
attentional selection, implicit, 102. *See also* guidance
attentional system, 142
attention selection hypothesis, 80–81
attribute achievement, 221
AUSTIN, 236–40
automatic processes, 389
avoidance, 238

back-propagation network, 66–67
basic activities, 4
basic emotions, 264–65
Bayesian-based subsymbolic level, 256
Bayesian belief revision, 182
Bayesian design, 197
Bayesian inference, 81, 82, 187
Bayesian learning mechanism, 175, 177
Bayesian networks, 78, 79, 84n.1, 183
Bayesian satisficing model (BSM), 150, 165, 167,
        169, 171–72
    local decision rule in, 171

structure of, 171
    tested against human data, 172–76
Bayes's rule, 184
BBN, 40
behavior, explaining. *See* rational analysis
behavior-based control, 285–87
biases, 152, 230–31
    anxiety-induced threat processing, 269–71
biasing effects of traits and states, modeling, 269–71
binding, 99, 107
biomathematical models of sleep deprivation,
        244–45
Birmingham, Henry, 31
blink, attentional, 101–2, 108–9
blocking effect, 188
blocks world task, 8
blood oxygen level dependent (BOLD)
    functions, 52, 57, 60
    response, predicting, 52–53, 55, 57–61
body posture, 214. *See also* embodiment
BOLD. *See* blood oxygen level dependent (BOLD)
Bosnia, 235
bottleneck, 100–103, 106–7, 114
bottom-up learning, 64, 68–70
bottom-up processing, 104–5, 128
bounded rationality, 152, 169
box models (information-processing theory), 327,
        328
bracketing heuristic, 353, 410
brain imaging, 187. *See also under* ACT-R
Brooks paradigm, 4
buffers, 415, 416

CafeNav, 256, 260
caffeine, 255, 256, 260
capacity, 91, 109, 137
categorical processing, 102
categorization by elimination, 156–57
categorization mechanisms that use more than one
        cue, simple, 156–57
causal interpretation, 233–34
causal texture, 205
central controller, 5–8, 434–36
central processing unit (CPU), 332, 334, 338, 341
challenging appraisals, 259
change blindness, 107, 135–36, 141, 144
Chicago O'Hare airport, 206–8
CHI Systems, 38
choice reaction task, 372, 377
choice task rules, 348–49, 351
chunks, 50, 68, 70
cingulate cortex, anterior, 187
circadian neurobehavioral performance and alertness
        (CNPA) model, 244–46

CLARION, viii, 46, 63–67, 74
    action-centered subsystem (ACS) in, 66–70
    architecture of, 63–67
    metacognitive subsystem (MCS) in, 63, 66, 70,
            72–74
        structure of, 72, 73
    motivational subsystem (MS) in, 66, 67, 70–72
        structure of, 71
    non-action-centered subsystem (NACS) in, 66, 67,
            69–70
    simulations conducted with, 73–74
CMU-ASP task, 374–77
    rule sets in, 374
    rules to implement choose-track subtask in, 374–75
coding time, 92
coercive graphics, 144–45
coffee, making, 369–70
    subtasks in, and their dependencies, 369
COGENT (cognitive objects in a graphical
        environment), x, 194, 401–2, 414–15,
        425–26
    cognitive modeling environment in, 415
        approach to model execution, 417–18
        box and arrow interpreter, 416–17
        box and arrow language, 415–16
        control: triggered *vs.* autonomous rules, 418–19
        external interaction, 419–20
        integration: recombining processing results, 419
        strengths of box and arrow approach, 417–18
    model of contention scheduling in, 422–23
    model of routine and deliberative behavior in,
            420–25
        key features, 425
        the task and global structure of, 421–22
    model of supervisory system in, 423–25
COGENT/iGEN, 38
cognition. *See also specific topics*
    implicit *vs.* explicit, 63
    unified theories of, 366
cognition–motivation–environment interaction, 64.
        See also CLARION
cognitive–affective architectures, 38. *See also*
        MAMID cognitive–affective architecture
cognitive appraisal. *See* appraisal
cognitive architecture(s), 5–6, 29, 255, 256
    generalized monolithic, 434
    integrated theory of, 83–84
        architectural hypotheses, 79–80
        computational principles, 79
        higher-order cognition as attention selection
                hypothesis, 80–81
        integrative cognitive focus of attention
                hypothesis, 80
    local theories *vs.* comprehensive, 431–41

cognitive architecture(s) (*continued*)
    modular, 434–35
    rise of, v–vi
    roles of, within cognitive modeling, 414
    simulations of behavior based on, 404–5
cognitive architecture thread, 38
cognitive bestiary, 5, 10
cognitive complexity, 264, 265, 268
cognitive control, 182. *See also* control, of cognition;
        EPIC
cognitive map, 76–78, 80, 82–84
cognitive models, v
cognitive moderators, 243, 250
cognitive–motivational framework, 221–22
cognitive objects in a graphical environment. *See*
        COGENT
cognitive problems, 81
cognitive process models, 38–40
cognitive processor, 330–31
cognitive psychology, 432
cognitive science, 5, 45
    dichotomies in, 63–64
cognitive shortcuts, 166
cognitive tunneling, 257
coherence field, 139, 140
coherence theory (of attention), 136, 139, 140
    basics of, 139–40
    implications of, 140
COJACK, 256
common function principle, 79
common functions, 79
common functions hypothesis, 80
complete processing models, 4
complex emotions, 264–65
compounds, 416
comprehensive architectures. *See under* cognitive
        architecture(s)
computer–human interaction. *See* human–computer
        interaction
computer metaphor, 327, 329. *See also* control, of
        cognition
connectionist approaches, 435
connectionist integrator model, 183–84
connectionist models, 91, 401, 402, 418
    of decision making process, 218, 219
connectionist networks, 415
connectionist processes, 418
connectionist theory, 9
constraints, 10
    on behavior, implications of, 407–8
    on dual-task performance, 408
    environmental, 149–50, 196–97
    framework for representing theories as sets of,
        407

construction–integration (C-I) architecture, 46,
        86–88, 90, 94. *See also* ADAPT
    testing and training individual models with, 91
construction–integration (C-I) cycle, 89–91
construction–integration (C-I) theory, 86–87
contention scheduling (CS), 421–25
contiguity of actions, 71
contrast guidance, 128, 129
control. *See also* EPIC
    of action, 368
    of cognition, 182, 340–41, 353–54
        basic, 335–41
        executive, 341–42. *See also* executive processes
    emotional, 145
    experience and, 397
    experience of, 396–97
    flow of, 335–38
    of input, 332–34
    of nonattentional visual processes, 138
    of output, 334–35
    and representation of instructions, 370–71
    types of, 5–10, 136–37
control and performance concept, 19–20
control and performance model (Model CP), 22
control errors, 383
control failures, 383
control focus and performance model (Model CFP),
        19–22
control module, 49–51, 61
control process (controlled processes), 184–85
control signal, guiding, 100–101
control signal detection, model of, 382–83
control states. *See also* minimal control
    amount of knowledge and, 370–71
    in complex tasks, 374–76
    and minimal control in models of simple tasks,
        371–74
Control Store, 331
control theory
    classical, 29, 31–33
    modern, 29, 34. *See also* optimal control model of
        manual control
cooks, use of tools and action to shape work
        environment, 204–6
cooperating *vs.* noncooperating processes, 342
coping
    appraisal and, 232–33, 235
    computational model of, 235
coping strategies, problem- *vs.* emotion-focused, 232
cross-check unit task, 20
crossover model, 33–34
CS/SS theory, 421–25
current accuracy trace (CAT), 188–89
customized executives, 363

data boxes, 416, 419
data sinks, 416, 419
data sources, 416, 419
Dawes' rule, 155
decision field theory (DFT), 217, 224–25
   affective evaluation of consequences and,
      220–22
   applications of, to previous research, 223–24
   decision process and, 217–20
decision making, 152–53, 266. *See also* adaptive
      toolbox
   elimination by aspects, 157
   reasoning and, 216–17
   under risk, 298–300
   sequential, 168
   under uncertainty, 298
decision-making models, integrating emotional
      processes into, 213–14
decision-making processes, 214
decision mechanisms. *See also* adaptive toolbox
   one-reason, 155–56
decisions, emergency
   computation example applied to, 222–23
decision theory for emotional consequences, need for
      change in, 215–16
declarative memory, rational activation theory of,
      7–8
degraded cognitive function, theory of, 243
deictic codes, 391
deictic representation, 145
deictic specification hypothesis, 391–93, 395–97
deictic specification of intentional elements, 391–92,
      397
deliberate appraisal, 265
deliberative behavior. *See under* COGENT
dependency, sequential. *See* sequential dependencies
describing function, quasi-linear, 31
description errors, 392
DiBS (display-based solver), 369
DI-Guy, 284
direct support of action, 145
display-based solver (DiBS), 369
distractor heterogeneity (visual search), 101
distractor-ratio effect, 125, 126, 128
distractor signal, 386
distractor threshold, 108
distributed perception, 141
distributed representation, 66–68, 70
divide-and-conquer approach, 13, 432
domain attributes, 268
domain-general attention selection hypothesis, 81
D-OMAR, 40
dowry problem, 157
driver behavior. *See also* driving

integrated models of, 356–62
   modeling executive control with, 362–65
   theoretical and practical implications or,
      365–66
driver distraction, 360
   modeling, 360–62
driver model(s)
   ACT-R, 357–62
   integrated, 41
drives, 71–72
driving. *See also* driver behavior
   modeling highway, 357–60
   dual-task experiment, "Wickens," 344–45
   model components in, 346
   resource allocation in, 345–46

ecological approach, 166. *See also*
      exploration/exploitation trade-off
ecological rationality, 152, 153
efficiency, 306, 309
elimination
   by aspects, 157
   categorization by, 156–57
El Mar's Vision 2000 Eye-Tracking System, 21
EMA, 232–36
embedded cognition, 208–9
   modeling interactive behavior and, 199
      modeling environment with dynamic affordance
         distributions, 199–204
      modeling origins of airport taxi errors, 206–8
      using tools and action to shape work
         environment, 204–6
   theoretical issues in modeling, 195–96
      knowing as much or more than the performer,
         197–98
      mind and world function in concert, 198–99
      modeling sensitivity to environmental
         constraints and opportunities, 196–97
embodied cognition, 433
embodied operating system model, 292–93
embodiment, 279–81, 283–84. *See also*
      microbehavior(s)
   role of, 284–85
EMMA eye movement model, 257, 356
"emotional biases," 230–31
emotional–cognitive decision-making process,
      215
emotional control, 145. *See also* control
emotional processes, integrated into decision-making
      models, 213–14, 224. *See also* decision field
      theory
emotional states, 266, 268
   vs. cognitive states, 264
   as multimodal phenomena, 265

emotion effects
  on cognition, 265–66
    generic methodology for modeling, 269–71
  as patterns of process-modulating parameters, 275
emotion-focused coping strategies, 232
emotion-generating belief nets, 269, 270
emotion research
  background of, 264–66
  definitions in, 264–65
emotion(s). *See also* appraisal theory; EMA; MAMID
  as "affective control" over cognitive functions, 233
  bases for, 214
  basic *vs.* complex, 264–65
  design, control, and, 238–40
  primary, 264
  research on decisions and, 215–17
  single *vs.* dual system views of, 214–15
emotions proper, 264
emotion- *vs.* problem-focused coping strategies, 232
endogenous control, 136–37
environment, functional model of
  for perception and action, 205
environmental constraints and opportunities,
    modeling sensitivity to, 196–97
environmental constraints on integrated cognitive
    systems, 149–50
environmental modeling, 199–206
EPIC (executive process–interactive control), 38, 39,
    194, 280, 320, 322, 333, 408, 434, 440
  ACT-R and, 372
  attention and, 334
  central controller of, and Type 3 control, 5
  COGENT and, 420
  cognitive architecture of, ix, 314–17, 326, 327–29,
    354, 365, 368, 402, 414, 438
    basic structure, 329–31
    internal structure of visual processor system, 330
    production rules, 331–32
  COJACK and, 256
  control of output and, 334
  ease of theorizing about cognitive control with, 354
  executive processes in, 343–44
  mixed-density text search models and, 317–19
  motor processors in, 334
  parallel flow of control in, 338–39
  sequential flow of cognition and, 336, 337
  Type 2 theories and, 435
  Type 3 theories embedded in, 10
  Wickens dual-task experiment and, 344–46
EPIC-Soar, 368
error monitoring, 392, 393, 396
establish-control unit task, 20
evaluation function, 34
executable knowledge, 88–89

executive control of cognition, 341–42
executive processes
  examples of, 344–52
  representation of, in cognitive architecture,
    342–53
  sequential *vs.* concurrent, 343–44
  specialized *vs.* general, 343
executive process-interactive control. *See* EPIC
executive process model, choice of, 352–53
executive protocols, 344
exogenous control, 136
expectation generation, 267
expectations, 276
experimental psychology, 432
explicit cognition, 63
explicit conditions, 309
exploration/exploitation trade-off, 165–67, 177, 178.
    See also Bayesian satisficing model
  optimal exploration in diminishing-return
    environment and, 169
  sequential decision making and, 168
  when search cost matters (rational analysis),
    168–69
eye-movement model for visual search tasks, 122,
    125–29. *See also* area activation model
eye movements, 106–7, 137. *See also* EMMA eye
    movement model; gaze arbitration
  general model of, in visual search, 122, 125–29
eye processor, 330
eye tracking, 21, 126, 292, 293, 321

"fast and frugal" heuristic strategies, 207, 208
feature guidance, 128, 129
feature integration theory (FIT), 99, 107, 135
feature-ratio effect, 128
features (visual search), 101
feedback, external, 65
feed-forward neural networks, 79
Feynman, Richard, 198
Fitts' law, 7
Fitts' module, 7
fixation field, 123
fixations. *See* saccadic endpoints
flanking/linear separability, 102
flicker paradigm, 135, 136
fluency, 391
fluency heuristic, 159–60
fMRI. *See* functional magnetic resonance imaging
focus, 64. *See also* control focus and performance
    model; single-focus models of cognitive
    functions
forward inference, 79, 83
F-16s, 16
full-run switch cost, 384

functional magnetic resonance imaging (fMRI) data, 51, 61, 187
functional transactions, 198

Gaussian function, 123, 126–27, 247, 302–3
gaze arbitration, 291–92
gaze tracking. *See* eye tracking
general-executive compatibility, 344
general executive (GE), 362–64
general knowledge, 88
general knowledge store (GKS), 70
general problem solver (GPS), 38
gist of a scene, 137, 142
goal instantiation principle, 390
goal manager, 268, 275
goal (module), 49–51, 61
goals, 368, 380–81. *See also* GOMS; intentions
Goal Setting, 424
goal state, 21
goal system, 51
GOMS (goals, operators, methods, and selection rules), 38, 39, 335–37, 371, 439–40
graphical buffers, 416
greedy heuristic, 123
guidance (visual search), 102. *See also* guided search; visual search
   bottom-up, 104–5, 128
   constraining parameter values, 109–11
   modeling, 103–4
   modeling bottleneck in, 107–13
   reasons for proposing bottleneck in, 106–7
   signal detection theory and, 106
   target-absent trials and errors in, 111–13
   top-down, 101, 105
Guided Search (GS), 97–99, 123, 130. *See also* guidance; visual search
   activation map in, 105–6
   asynchronous diffusion model of, 103
   mechanisms of search in, 99, 103, 109
   reaction time (RT) in, 99, 100, 108–14
Guided Search 1 (GS1), 100
Guided Search 2 (GS2), 100
Guided Search 3 (GS3), 100
Guided Search 4.0 (GS4), 97, 104–14, 129
   modeling guidance, 103–4
   signal detection theory model and, 106–7
   state of, 113–14
   structure or, 100–103
   what is explained by, 100
guided search theory, 120–22, 129
guiding representation, 100

hazard functions, 111, 112
heading arbitration, 290–91

heads-up display (HUD), 14, 15
heat stress, 258–59
heuristic mechanisms. *See* adaptive toolbox; decision mechanisms
heuristic(s), 157. *See also* recognition heuristic
   bracketing, 353, 410
   greedy, 123
   lexicographic, 156, 162
   Minimalist, 155, 156, 161
   one-reason decision, 155–56
   simple, 152–56, 158, 161, 166
heuristics-and-biases program, 152, 153
heuristic strategies, "fast and frugal," 207, 208
hidden Markov model (HMM), 183, 192n.3
hierarchical search models, 319–22
hierarchical-sequential flow of control, 337, 338
hierarchy, weak, 374
hierarchy model, reverse, 101
higher-order cognition
   as attention selection hypothesis, 80–81
   integrating mechanisms of, 78
   integrating mechanisms of lower-order and, 78
   through common functions hypothesis, 80
high-level procedure language, 37
historical accuracy trace (HAT), 188–90
HOPROC, 37
HOS (human operator simulator), 37–38
human associative memory (HAM), 39, 69–70
human–computer interaction (HCI), 39, 87, 263, 336
human–machine systems, 29, 37, 133, 144, 146, 194, 197. *See also* task networks; visual perception, component systems of
human movement performance, early studies of, 29–30
human operator simulator (HOS), 35, 37–38
human performance modeling, 29, 41. *See also* specific topics
human reliability models, 35–36
hybrid models (human performance modeling), 40–41

ideal controller, 34
ideal observer theory, 411
identity matching, 79
ignorance-based reasoning, 154
illusions, cognitive, 300
illusory conjunctions, 107
imaginal module, 49–51, 61
imaging research, 51
implicit attentional selection, 102. *See also* guidance
implicit cognition, 63
implicit conditions, 309
implicit decision networks (IDNs), 68
implicit perception, 138

IMPRINT, 35, 37
inattention, 121
inattentional blindness, 138
indexical representation, 145
individual differences, methodology for analysis and
      modeling of. *See* MAMID
information forging theory (IFT), 174
information-processing pathways, 182
information-processing theory, 327, 328
information scent, 174
inhibition, 385
inhibitory function, 385
"inner loop" control, 35
inner zombie, 145, 146
instruction address register, 336
instructions, control and representation of, 370–71
integrated driver model, 41
integrated-model approach, 361
integrating models of cognitive systems, 4–5. *See also*
      specific topics
integration
   by composition, 356
   by generalization, 356–57, 365
integrative cognitive focus of attention hypothesis, 80
intentional control
   of event counting, 392–94
   in skilled activity, principles for understanding,
      390–91
intentional elements, deictic specification of, 391–92
intentionality, 398n.1
intention–outcome confusions, 393
intentions, 397–98
   control function served by, 388
   coordinating tasks through goals and, 325–26
   defined, 388
   reasons for focusing on, 388–89
   representing, 389–90
      implications and emotional phenomena,
         390–91
   as schemas, 389–90
   schematic structure of, 389–90
interactive behavior, 10, 194–96, 198, 208–9, 410
interactive behavior triad (IBT), 432–34
interactive routines, 4
interface, 17, 139, 146, 279, 314, 402
interference, 80
Internet. *See* World Wide Web
interrupt control of cognition, 340–41
interruption, 71
"introvert pessimist," 269, 270

JACK, 40
judgments, 214
juxtaposition principle, 390

Kalman estimator and predictor, 34
Kanfer–Ackerman air traffic control task, 15
knowledge
   classes of, 88–89
   implicit *vs.* explicit, 67
knowledge bases, constructing individual, 89

layout, memory for, 137–38, 142
learning. *See also specific topics*
   bottom-up *vs.* top-down, 64, 68–70
left-hand density function, 386
LEX, 156
lexicographic heuristics, 156, 162
LEXSEMI, 156
Licklider, J. C. R., 32
"linear law," 31
local decision rule, 172
localist representation, 67, 68, 70
local-minimum environment, 172
local theories, 8, 429
locomoting affordances, 201–2
long-term memory (LTM), 268, 276
lotteries, 298–99
lower-order cognition, 78, 83

magical displays, 144
MAMID cognitive–affective architecture, ix, 212,
      263, 264, 267–68
   architecture requirements of, and relation to
      generic cognitive–affective architectures,
      275–76
   and modeling methodology, 266–67
      affect appraisal (AA) module and, 268–70
      benefits of, and test-bed environment, 274–75
      evolution experiment, 270, 272–74
      intermodule communication and, 268
      modules, processes, controls, and, 275
      shortcomings and future research and, 275
Man-Machine Integrated Design and Analysis
      Software (MIDAS), 38–40
manual control models, 29, 30
   current status of, 34–35
map-navigation task, 172–74
Markov chain method, 227n.4
Markov decision process (MDP), 167, 178, 227n.3
Markov model. *See* hidden Markov model
Marr paradigm, 4, 286
maximum expected utility (MEU), 299, 300, 308
maximum expected value (MEV), 299, 303–6,
      308–11
McRuer, Duane (Mac), 32–33
mean accuracy trace (MAT), 188–89
memory, 79. *See also* associative memory; working
      memory

external, 147n.5
long-term, 268, 276
"medium-term," 137–38
mood and, 266
in search, 108
mental constructs, 268
meta-attributes, 268
metacognition, 394, 396. *See also* motivational and
    metacognitive control
metacognitive subsystem (MCS), 63, 66, 70, 72–74
methodology for analysis and modeling of individual
    differences. *See* MAMID
Micro Analysis and Design (MAAD), 37, 38
microbehavior model, 292
microbehavior(s), 292, 294
    arbitration of, 289–92
    defined, 286–87
    learning, 288–89
    maximum number of simultaneously running,
        294–95
    selection of, 293–95
micromodules, 37
Micro Saint, 35, 37, 38
MIDAS (Man-Machine Integrated Design and
    Analysis), 38–40
mindsight, 138–39
minimal control in models of simple tasks, 371–74
minimal control principle, 368–71, 375–76, 378, 391
minimal deliberation principle, 391
Minimalist heuristic, 155, 156, 161
minimum expected value (MEV), 299, 303–10
mission rehearsal exercise (MRE), 235–36
mixed-density text search models, 317–19
Model CFP. *See* control focus and performance
    model
model evaluation, 3
model human processor (MHP), 39, 407, 434, 437
modularity, 340
Monitoring, 423–25
Monte Carlo integration, 303
Monte Carlo simulations, 36, 207, 408, 409
moods, 265, 266. *See also* emotional states
motivational and metacognitive control, 64
    reasons for modeling, 64–65
motivational processes, 213, 217
motivational subsystem (MS), 66, 67, 70–72
motivation *vs.* cognition, 216
motives, 214
motorcycle, 217–18, 220–23
motor noise, 34
motor system, 34, 145, 280, 284–85, 298, 326, 360
motor variability, 34
movement performance, early studies of human,
    29–30

movement planning under risk, 298, 300–307
movement time (MT), 7
MRI (magnetic resonance imaging), functional, 51,
    61, 187
multiple implementation principle, 79, 80
multiprocessor *vs.* uniprocessor machines, 338
multiprogramming/multiprocessing, 341

navigation errors, 207, 208
need hierarchy, 72
network model thread, 35
network/reliability models, 35–38
networks, 416
neural network models, 187
neural transmitters, 214
New World Vistas, 15
NGOMSL, 39
noise, auditory, 259
noise terms, 34
non-action-centered subsystem (NACS), 66, 67,
    69–70
noncompensatory environments, 156, 215, 216

object file, 135
objective function, 406
objects of attention, 114
obsessiveness, 266
ocular processes, voluntary and involuntary, 330–31
ocular processors, 330–31, 334, 348
off-line stream (visual system), 138
OMAR (operator model architecture), 40
one-reason decision mechanisms, 155–56
one-reason heuristic, 155, 156
on-line stream (visual system), 138
opacity, 100
open-ended cognitive processes, 76–77
operating systems (OSs) concepts, 341–42
opportunism, 71
optimal controller, 34
optimal control model of manual control, 29, 34, 35
optimality, 9, 10, 197, 403, 405, 406, 410–12
optimal performance, 150
    vs. suboptimal performance, 166, 167, 169,
        172–74, 177, 178
optimal stopping rule, 168
overlays, 255–57. *See also* stress and cognition

parallel executive processes, 344
parallel flow, 338–40
parallel general executive, 347–52
parallelism, as routine in computers, 338
parallel process, 100, 103, 107, 213, 275, 329, 339,
    342
parallel search, 99, 103

parsimonious production system (PPS) production rules, 331–32, 336
pattern recognition, 77
perceive rule, 372
perception, 79, 423–24
  types of, 138
perceptual/motor processing, 405
perceptual pathway, 182
perceptual processor, 316, 330
perceptual tunneling, 257
performance only model (Model P), 22–23, 26
performance shaping factors (PSFs), 36
persistence, 71
personality traits and cognition, 269. *See also* traits *vs.* states; *specific traits*
pessimists. *See* "introvert pessimist"
phone use, 41, 360–65
piece distribution problem, 437
piece fit problem, 437
piece identification problem, 437
pilots, cognitive control in, 13–14, 26–27. *See also* control focus and performance model
  naive model of, 16–19
    conceptual flow of cognitive control in, 17, 18
    limiting factors, 16–17
    model validity, 18–19
    unit task representation, 17–18
  Predator synthetic task environment (STE) and, 14–16
  variations in knowledge and strategy in, 22
    sensitivity analysis, 24–25
    three Type 3 theories, 22–24
plan element knowledge, 88–89
plan elements, human pilot, 92
plans, 214
pointwise mutual information (PMI), 174
Polyscheme, viii, 46, 83, 84
pop-out, 104, 134, 135
Postal Service, 314, 315
preattentive object files, 104
preattentive processes, 147n.2
precognitive model, 33
Predator synthetic task environment (STE). *See under* pilots
preferences, combination of, 71
presence/absence (visual search), 101
pretask appraisal, 259
primal sketch, 134
primary drives, 71–72
primary line (visual processing), 133
primary processing (vision), 133
priming, 181, 184, 385
  positive *vs.* negative, 385–86
primitives, visual, 133–34

Pritsker, Alan, 37
probabilistically textured environment, 165
probabilistic information transmission (PIT) model, 182–88
probabilities, explicit and implicit, 307–10
probability matching, 69
problem-focused coping strategies, 232
problem solving, 214. *See also* reasoning and problem solving
  affect and, 266
"problem space," 167
problem state module, 49–51
problem states, 371
procedural knowledge, 88–89
procedural memory, 40, 87, 434
PROCRU, 41
production compilation, 377
production execution cycle in ACT-R, 247
production rule firing, 79
production system, 50, 77, 250, 371, 372, 376–77, 438
production-system cognitive architecture, 6, 9, 337, 340, 357
production-system model, 357
  functional structure of, 416
production system working memory, 329–32
program counter, 336
"programmed" behavior, 33
proportional activation, 71
proto emotions, 264
proto-objects, 104, 134
psychological refractory period (PRP) tasks, 408, 409
Psychology of Human-Computer Interaction, The (Card et al.), 39
psychomotor vigilance task (PVT), 243–46
  computational cognitive model that performs, 246–47, 250–51
    model design, 247–48
    model performance, 248–50
  human performance on, 246
purposefulness, 64

Q-function, 288, 290, 291
Q-learning algorithm, 68, 288
quasi-linear control model, 34, 35
quasi-linear describing function, 31
quasi-linear human operator model, 33
quasi-linear transfer function, 33
QuickEst heuristic, 157

rapid serial visual presentation paradigm, 9
rapid vision, 133, 137, 146n.2. *See also* vision, low level
rational activation theory, 7–8, 10n.2
  of declarative memory, 8

rational analysis, 371, 403–4, 412
  bounded, 404, 406, 411–12
  and explaining behavior, 403, 406–10
  local and global adaptation and role of mechanism
    and strategy in, 405–6
  strengths and weaknesses of, 405–6
rational–ecological approach, 167, 168, 177. *See also*
    exploration/exploitation trade-off
rationality, 152. *See also* ACT-R
  principle of, 403–4
reaction time (RT), 187, 188, 409. *See also* response
    time
  in guided search, 99, 100, 108–14
reasoning, 76–77, 83–84. *See also* cognitive
    architecture(s)
  as attention selection and cognitive self-regulation,
    83
  control and, 77
  as control problem and solution, 76–78
  and decision making, 216–17
  defined, 76–77
  emotions and, 216
  explaining integration, control, and, 81–83
  integrated with other cognitive processes, 82–83
  strategy choice in, 78
reasoning and problem solving, 80, 83,
reasoning strategies, 78
  deciding which to deploy, 82–83
  integrating diverse, 82
recognition heuristic, 153–55
  modeled within ACT-R, 158–61
recognition values, using continuous, 159–60
reentrant processing, 101. *See also* guidance,
    top-down
reference frames, 135
  dynamic, 285
reinforcement learning, 166, 288–90
reliability/network models. *See* network/reliability
    models
remnants, 147n.5
resident monitor, 341
resource allocation, 8, 284, 345–46, 354
response accuracy, 9, 188–90
  estimate of, 185
response pathway, 182
response repetition. *See under* sequential effects
response time (RT), 244–46, 249–51. *See also*
    reaction time
response utility, 184
retinal availability functions, 438
retrieve-response rule, 372
reverse hierarchy model, 101
risky environments. *See* decision making, under risk;
    movement planning under risk

robotics, 279, 283, 285–87
root mean squared deviations (RMSDs), 15, 23
routines, 285. *See also under* COGENT
  interactive, 4
  visual, 284, 285, 287–88
rule-based process, 418
rule-based reasoning, 70
rule–extraction–refinement (RER) algorithm,
    69
rules. *See* action rule store
run-time control of functional modules, 7–8
Russell, Lindsay, 32
  measurement system, 41–42

saccades, 147n.6
saccadic endpoints, 121, 130. *See also* fixation field;
    saccadic selectivity
saccadic selectivity, 121, 123–29
SAINT (systems analysis of integrated networks of
    tasks), 35, 37
salience, 136
Sandia National Laboratories, 35
satisficing, 157, 169. *See also* Bayesian satisficing
    model
"scaling up" cognitive modeling, 195
scanpath, 314, 315, 317, 318, 321
scene
  dynamic *vs.* static aspects of a, 147n.4
  gist of a, 137, 142
Schema Hierarchy, 422, 423
schemas, 422
  intentions as, 389–90
Scout World, 199–204
search. *See also* guided search
  in a problem space, 177
search asymmetry, 102
searching affordances, 201–2
secondary processing (vision), 133
secretary problem, 157
Selected Schemas, 422
selection, visual. *See* visual selection
Selection Process, 422, 423
selection rules, 336
selective attention, 332–33
selective hold/change detection, 135–36. *See also*
    change blindness
selective integration, 135
self-regulation, cognitive, 81
semantic specification, 391
sense–plan–act approach, 285–86
sensory preprocessing, 267
sensory store, 330, 331
sequential decision making (SDM), 168
sequential dependencies, 180–82

sequential effects
  involving response repetition, 182, 185–87
    control processes and speed-accuracy trade-off,
      184–85
    model details, 184
    probabilistic information transmission (PIT)
      model, 182–88
  involving task difficulty, 187–90
    explaining task blocking effect, 188–89
    simulation details, 190
    testing model predictions, 190–91
sequential executive processes, 343
sequential flow of cognition, 335–38
sequential search heuristics, 157
sequential specialized executive, 346–47
serial addition/subtraction task (SAST), 251
serial process, 99, 103
serial search, 99, 103, 109
serial subtraction, 251, 256, 259
servomechanisms, 29, 31. *See also* control theory,
    classical
set size (visual search), 101
setting system, nonattentional, 142
short-order cooking, 10, 150, 204, 209
side conditions, 404, 405
  accounting for, 404–5
Siegel and Wolf network model, 36–37
  computational steps, 37
signal detection model, 382–83
signal detection theory (SDT), 106–7, 260
similarity, 125–26
similarity-based reasoning, 70
simple heuristics, 152–56, 158, 161, 166
simulated device, 330
simulating alternate worlds, 79
single-focus modelers, 3–4
single-focus models of cognitive functions, 4
  off-line control of functional modules in, 8–9
  run-time control of functional modules in, 7–8
single-task compatibility, 344
situation assessment, 267, 275
skill acquisition, 30
skilled performance, learning, 376–77
sleep attacks, 246
sleep deprivation, 250–51. *See also* psychomotor
    vigilance task
  biomathematical models of, 244–45
  neuropsychological research on, 244–45
"smart heuristics," 207
SNIF-ACT model, 174–76
SOAP (Symbolic Operating Assembly Program), 36
Soar
  appraisal and, ix, 256
  chunking, problem solving, and, 86

cognitive architecture, v, ix, 64, 194, 199, 402,
    404, 414, 424, 438
cognitive–metacognitive interaction and, 65
construction-integration compared with, 87,
    94
default rules of, 258
EPIC-Soar, 368
episodic knowledge and, 94
historical perspective on, 38, 39, 42n.2
mechanisms to handle selection and firing of
    production rules, 5
modeling fatigue and fear, 255
models of reasoning in, 77, 81
Tac-Air Soar system, 16
Type 2 theories and, 434, 435
was founded on problem solving, 421
"social optimist," 269, 270
society of mind approach, 5
sockets, 416
soft alerts, 145
soft constraints hypothesis, 10
softmax model, 168
spatial arrangement. *See* layout
specialist-common function implementation
    hypothesis, 80
specialists (processors), 80
specialized-executive compatibility, 344
speed-accuracy tradeoff paradigm, 9
speeded discrimination paradigm, 182
spotlight of attention, 135
Stability and Support Operations—Simulation and
    Training (SASO-ST), 236
standard times, 39
states. *See also* emotional states
  vs. traits, 265, 266, 269–71
state-specification action, 68
statistical summaries, 137
steering. *See also* driver behavior
  two-point model of, 357–60
steering angle, 8
stimulus onset asynchrony (SOA), 408, 409
stimulus selection asynchrony (SSA), 108
Stochastic simulation, 79
stores, perceptual, 331
strategic variations, 407
Strategy Application, 424
Strategy Generation, 424
Strategy Selection, 424
stress and cognition, theories of, 254–61
  are not complete, 260
  decreased attention and, 258
  further work on stress overlays, 261
  pretask appraisal and stress and, 259
  and the task as a stressor, 258–59

testing stress overlays and, 260
  Wickens's, 256–58
STRIPS operators, 378
Stroop effect, 80
subgoaling, 79
subject matter experts (SMEs), 17–24
subliminal perception, 138
subsumption architecture, 286
subsymbolic level, 256, 260
subsymbolic processing, 63, 66
Supervisory Attention System (Norman & Shallice),
  342
supervisory system (SS), 421–25
sustainability, 64
switch cost, 381
Symbolic Operating Assembly Program (SOAP), 36
symbolic planning approach, 286
symbolic processing, 63, 66, 82
symbolic representation, 67, 68
synthetic task environment (STE), 14–16
systems analysis of integrated networks of tasks
  (SAINT), 35, 37

tabular data sinks, 416
Tac-Air Soar system, 16
tacit knowledge, 203
Tag Store, 331
Take The Best, 155, 156, 161
target-distractor similarity, 101
targets, easy to identify but hard to find, 106
target threshold, 108
task environment, 330
  synthetic, 14–16
Task Knowledge, 424
task networks, 29, 35–38
task representations, incomplete
  inferring steps in, 377–78
tasks, structured into subtasks, 376
task switching, 181, 364, 381–83, 386, 395–96
  phenomena related to, 383–85
taxiing, 207
technique for human error rate prediction
  (THERP), 36
telephone use, 41, 360–65
texons, 134
thalamus, 215, 244, 247, 248
thread(s), 288, 342
  of execution, 339
threatening tasks, 259. *See also* decision making,
  under risk; movement planning under risk
three-dimensional (3D) worlds, perception and
  action in, 237
time-slicing executive processes, 343–44
toolbox. *See* adaptive toolbox

top-down control (vision), 135
top-down learning, 64, 70
top-down processing, 101, 105
Tower of Hanoi, 69, 73, 87, 380
tracking, defined, 30–31
tracking paradigm, 30–31
tracking task rules, 348–50
traits *vs.* states, 265, 266, 269–71
transduction, 133
transfer function, 31–33
triadic architecture, 141–43
trial-and-error learning, 64
triggered rule, 418
Triggering Functions, 422, 423
tunneling, 257
twenty questions, 432
2½D sketch, 134
two-point model of steering, 357–60
two-systems theory (vision), 138
Type 1 (central) control, 5–6, 8
Type 1 theories, 315–17, 434–38, 440–41
Type 2 (functional processes) input, output, and
    control
  functional modules and, 6–9
    issues for Type 2 control, 9
    noncognitive modules, 7
Type 2 theories, 315–17, 434–38, 440–41
Type 3 control, 9–10
Type 3 theories, 13, 22–24, 315, 317–19, 322, 324

unbounded rationality, 152
uninhabited air vehicle (UAV) operator. *See* pilots,
    cognitive control in
unit task, 376
utility of information, 166
utility threshold, 247

VCR, programming, 376
verbal reports, 21
vigilance, 145. *See also* psychomotor vigilance task
virtual humans, 235–40
virtual reality (VR), 283
virtual representation, 140, 143
  basics of, 140–41
  implications of, 141
    dependence on knowledge and task, 141
  requirements for successful operation of, 141
vision
  early, 107, 142, 146n.2
  low level, 133, 146n.2. *See also* rapid vision
visual attention, 97–98, 135. *See also under* visual
    perception, component systems of
visual computation, organization of, 292
visual elements, commonality of, 134–35

visual intelligence, 134
visual perception, 97–98
visual perception, component systems of, 132, 133,
    138, 139
  attentional processes in, 135. *See also* visual
    attention
    basis of selection, 137
    capacity, 137
    control, 136–37
    relation between attention and eye movements,
      137
    selective access, 135
    selective hold/change detection, 135–36
    selective integration, 135
  integration of, 139, 143–44
    coherence theory, 139–40
    coordination *vs.* construction, 139, 143, 145
    triadic architecture, 141–43
    virtual representation, 140–41
  integration of external systems and, 144–46
  nonattentional processes in, 137, 139, 143
    commonality in processes, 138
    control, 138
    implicit perception, 138
    "medium-term" memory, 137–38
    mindsight, 138–39
    rapid vision, 137
    visuomotor guidance, 138
  preattentive processes in, 133
    commonality of visual elements, 134–35
    complex properties, 134
    control, 134
    influence of top-down control, 135
    reference frame, 135
    simple properties, 133–34

visual primitives, tests for, 133–34
visual processing, schematic of early, 133
visual routines, 284, 285
  state estimation using, 287–88
visual search, 120. *See also* guided search
  comprehensive theory of, 314–15, 323–24
  general model of eye movements in, 122,
    125–29
visual search models. *See also* EPIC; Guided
    Search 4.0
  from different tasks, integrating, 322
  hierarchical, 319–22
  mixed-density text, 317–19
  phenomena that should be accounted for by,
    101–2
visual selection, basis of, 137
visual selective attention, 437
visual signal detection theory (VSDT), 260
visuomotor guidance, 138
Vortex, 284

Walter (virtual human), 280, 284, 288, 290,
  293
weak hierarchy, 374
Wickens task. *See* dual-task experiment
Wickens-WM (working memory) overlay, 258
Wisconsin card sorting task (WCST), 421–25
within-run error increase, 384
within-run slowing, 384
working memory (WM), 74n.1, 233, 257, 260, 266.
    See also EPIC, cognitive architecture
  production system in, 329–32
  visual, 329
world knowledge, 88
World Wide Web (WWW), searching on, 172–77